Operations Research Proceedi

GOR (Gesellschaft für Operations Research e.V.)

More information about this series at http://www.springer.com/series/722

Natalia Kliewer · Jan Fabian Ehmke
Ralf Borndörfer
Editors

Operations Research Proceedings 2017

Selected Papers of the Annual International Conference of the German Operations Research Society (GOR), Freie Universiät Berlin, Germany, September 6–8, 2017

Editors
Natalia Kliewer
Department of Information Systems
Freie Universität Berlin
Berlin
Germany

Ralf Borndörfer
Department of Mathematics
Freie Universität Berlin
Berlin
Germany

Jan Fabian Ehmke
Management Science Group
Otto-von-Guericke Universität Magdeburg
Magdeburg
Germany

ISSN 0721-5924 ISSN 2197-9294 (electronic)
Operations Research Proceedings
ISBN 978-3-319-89919-0 ISBN 978-3-319-89920-6 (eBook)
https://doi.org/10.1007/978-3-319-89920-6

Library of Congress Control Number: 2018938366

© Springer International Publishing AG, part of Springer Nature 2018
This work is subject to copyright. All rights are reserved by the Publisher, whether the whole or part of the material is concerned, specifically the rights of translation, reprinting, reuse of illustrations, recitation, broadcasting, reproduction on microfilms or in any other physical way, and transmission or information storage and retrieval, electronic adaptation, computer software, or by similar or dissimilar methodology now known or hereafter developed.
The use of general descriptive names, registered names, trademarks, service marks, etc. in this publication does not imply, even in the absence of a specific statement, that such names are exempt from the relevant protective laws and regulations and therefore free for general use.
The publisher, the authors and the editors are safe to assume that the advice and information in this book are believed to be true and accurate at the date of publication. Neither the publisher nor the authors or the editors give a warranty, express or implied, with respect to the material contained herein or for any errors or omissions that may have been made. The publisher remains neutral with regard to jurisdictional claims in published maps and institutional affiliations.

Printed on acid-free paper

This Springer imprint is published by the registered company Springer International Publishing AG part of Springer Nature
The registered company address is: Gewerbestrasse 11, 6330 Cham, Switzerland

Preface

This book contains a selection of refereed short papers presented at the Annual International Conference of the German Operations Research Society (OR2017), which took place at the Freie Universität Berlin, Germany, September 6–September 8, 2017. Over 900 participants attended the conference—practitioners and academics from mathematics, computer science, business administration and economics, and related fields. The scientific program included about 600 presentations. The conference theme, Decision Analytics for the Digital Economy, placed emphasis on the process of researching complex decision problems and devising effective solution methods toward better decisions. This includes mathematical optimization, statistics, and simulation techniques. Yet, such approaches are complemented by methods from computer science for the processing of data and the design of information systems. Recent advances in information technology enable the treatment of big data volumes and real-time predictive and prescriptive business analytics to drive decision and actions. Problems are modeled and treated under consideration of uncertainty, behavioral issues, and strategic decision situations.

Altogether, 100 submissions have been accepted for this volume (acceptance rate 63%), including papers from the GOR doctoral dissertation and master's thesis prize winners. The submissions have been evaluated by the stream chairs for their suitability for publication with the help of selected referees. Final decisions have been made by the editors of this volume.

We would like to thank the many people who made the conference a tremendous success, in particular the members of the organizing and the program committees, the stream chairs, the 14 invited plenary and semi-plenary speakers, our exhibitors and sponsors, our host Freie Universität Berlin, the many people organizing the conference behind the scenes, and, last but not least, the participants from about 46 countries. We hope that you enjoyed the conference as much as we did.

Berlin, Germany Natalia Kliewer
Magdeburg, Germany Jan Fabian Ehmke
Berlin, Germany Ralf Borndörfer
January 2018

Part XX Supply Chain Management

Joint Optimization of Reorder Points in n-Level Distribution Networks Using (R, Q)-Order Policies 657
Christopher Grob, Andreas Bley and Konrad Schade

An Integrated Loss-Based Optimization Model for Apple Supply Chain 663
P. Paam, R. Berretta and M. Heydar

Simulating Fresh Food Supply Chains by Integrating Product Quality 671
Magdalena Leithner and Christian Fikar

Window Fill Rate with Compound Arrival and Assembly Time 677
Michael Dreyfuss and Yahel Giat

Part XXI Traffic, Mobility and Passenger Transportation

Demand-Driven Line Planning with Selfish Routing 687
Malte Renken, Amin Ahmadi, Ralf Borndörfer, Güvenç Şahin and Thomas Schlechte

Scheduling of Electric Vehicles in the Police Fleet 693
Kerstin Schmidt, Felix Saucke and Thomas S. Spengler

Location Planning of Charging Stations for Electric City Buses Considering Battery Ageing Effects 701
Brita Rohrbeck, Kilian Berthold and Felix Hettich

On the Benefit of Preprocessing and Heuristics for Periodic Timetabling 709
Christian Liebchen

Structure-Based Decomposition for Pattern-Detection for Railway Timetables 715
Stanley Schade, Thomas Schlechte and Jakob Witzig

Timetable Sparsification by Rolling Stock Rotation Optimization 723
Ralf Borndörfer, Matthias Breuer, Boris Grimm, Markus Reuther, Stanley Schade and Thomas Schlechte

Traffic Management Heuristics for Bidirectional Segments on Double-Track Railway Lines 729
Norman Weik, Stephan Zieger and Nils Nießen

Traffic Speed Prediction with Neural Networks 737
Umut Can Çakmak, Mehmet Serkan Apaydın and Bülent Çatay

Part XXII Business Track

Delivering on Delivery: Optimisation and the Future of Vehicle Routing 747
Christina Burt, Paul Hart, Desislava Petrova and Adam West

Improving on Time Performance at Deutsche Bahn 753
Christoph Klingenberg

Part I
Awards

Solving the Time-Dependent Shortest Path Problem Using Super-Optimal Wind

Adam Schienle

1 Introduction

With air travel steadily on the rise and the increased fuel burn associated to it, it is ever more important that aircraft fly efficient routes. Planning such routes is a fundamental process of flying: commonly, a route is planned a few hours before the flight, focussing on key factors such as overfly costs and fuel burn. According to the Air Transport Action Group [1], around 1.5 billion barrels of fuel are burnt every year, corresponding to 93.75 billion USD [6]. A decrease of just 0.25% would add up to 234.375 million USD. There is also a visible impact for airlines: Lufthansa's total fuel consumption in 2016 amounted to 9 055 550 tons [7]. Decreasing this by 0.25% leads to 22 639 tons less fuel being burnt, or savings of almost 11.67 million USD per year. In terms of CO_2, this is equivalent to a reduction of more than 70 tons per year [7].

The need for efficient routes gives rise to the Flight Planning Problem (FPP), which is the problem of finding a minimum cost trajectory between two airports on the Airway Network, a directed graph. In general, the objective function consists of several summands, such as fuel costs, overfly costs and crew costs. In this paper, however, we shall concentrate on minimising the fuel costs. We further assume that aircraft fly levelly on a given altitude and neither climb nor descend. In this setting, fuel consumption is equivalent to flight time, which reduces FPP to the *Horizontal Flight Planning Problem* (HFPP). Since winds have a strong impact on flight time and because of the time-dependency of the weather, we can model HFPP as a Time-Dependent Shortest Path Problem (TDSPP).

TDSPP has been extensively studied in the literature, with particular emphasis on road networks. Dijkstra's algorithm yields an optimal solution in polynomial time; however, for large networks, several speedup techniques have been developed,

---

A. Schienle (✉)
Zuse Institute Berlin, Takustraße 7, 14195 Berlin, Germany
e-mail: schienle@zib.de

allowing to curb runtimes by several orders of magnitude with respect to Dijkstra's algorithm [2]. Most of them rely on a preprocessing phase, in which either some shortest paths or other auxiliary data is precomputed and stored to speed up the query. For a comprehensive survey, see [2].

Throughout this paper, a *weighted graph* will always refer to a pair (G, T), consisting of the actual (directed) graph G and a possibly time-dependent weight function $T: A \times [0, \infty) \to [0, \infty)$, mapping an arc $a \in A$ and a time $\tau \in [0, \infty)$ to the travel time $T(a, \tau)$ on a.

The ground distance $d_G(a)$ of an arc $a \in A$ on the Earth's surface is constant, and we assume that aircraft fly with constant air speed[1] v_A. In contrast, the *ground speed* $v_G(a, \tau)$ of an aircraft is dependent on the prevailing wind conditions on the arc and given by the formula

$$v_G(a, \tau) = \sqrt{v_A^2 - w_C(a, \tau)^2} + w_T(a, \tau) \quad \forall a \in A, \tau \in [t_0, t_r], \qquad (1)$$

where $w_C(a, \cdot)$ and $w_T(a, \cdot)$ are the crosswind and trackwind components of the wind vector, i.e., the components perpendicular and parallel to the current flight direction. Ground speed and ground distance are linked via the relation

$$T(a, \tau) = \frac{d_G(a)}{v_G(a, \tau)}. \qquad (2)$$

2 Super-Optimal Wind

We are looking to solve the TDSPP model of HFPP to optimality by using an appropriate shortest path algorithm. A natural choice would be Dijkstra's algorithm; in practice, however, the time to plan a flight is limited and for the most part, this process takes place shortly before the aircraft departs. In particular, this means that query times should be as short as possible. In this paper, we restrict ourselves to the discussion of the A* algorithm, introduced in [5]. For an overview of other algorithms and their applicability to HFPP, see [3].

The intricacy with A* is to find a suitable potential function $\pi_t: V \to [0, \infty)$, which for every $v \in V$ underestimates the cost of a shortest v-t-path in (G, T). We define the reduced cost of an arc $(u, v) \in A$ at time τ as

$$T'((u, v), \tau) = T((u, v), \tau) - \pi_t(u) + \pi_t(v), \qquad (3)$$

and call π_t *feasible on* (G, T) if for every arc $(u, v) \in A$ and for every $\tau \geq 0$, we have $T'((u, v), \tau) \geq 0$. If π_t is feasible, running A* is equivalent to running Dijkstra's algorithm on G using the reduced costs.

---

[1] Speed relative to the surrounding air mass.

To obtain a feasible potential function, we have to find a lower bound for the travel time on the arcs. To this end, we introduce the concept of *Super-Optimal Wind* to underestimate the travel time. While it is possible to minimise the travel time function directly, this takes too long for practical purposes. Furthermore, it requires knowledge of the airspeed in advance, as opposed to constructing the Super-Optimal Wind vector.

We assume that weather is given for a finite set of times $\{t_0, t_1, \ldots, t_r\}$, and between the t_i, the weather data is interpolated to obtain the wind vector $w(a, \tau)$. Let $t_0 = \tau_0 < \tau_1 < \cdots < \tau_n = t_r$ be a discretisation of $[t_0, t_r]$ such that $\tau_i - \tau_{i-1} = \Delta$ for some $\Delta > 0$ and for all $i = 1, \ldots, n$. To ensure that for every $i \in \{0, \ldots, n-1\}$ we always find a $j \in \{0, \ldots, r-1\}$ such that $[\tau_i, \tau_{i+1}] \subset [t_j, t_{j+1}]$, we require that $r \mid n$. We then define for $i = 1, \ldots, n$

$$\underline{w}_C^{(i)}(a) = \min_{\tau \in [\tau_{i-1}, \tau_i]} |w_C(a, \tau)| \quad \text{and} \quad \overline{w}_T^{(i)}(a) = \max_{\tau \in [\tau_{i-1}, \tau_i]} w_T(a, \tau),$$

which are the minimum crosswind and maximum trackwind on each discretisation step. The vector defined through its cross- and trackwind components

$$w_{\text{s-opt}}^{(i)}(a) = (\underline{w}_C^{(i)}(a), \overline{w}_T^{(i)}(a))$$

is called *Super-Optimal Wind* vector, and is used to overestimate the ground speed (note that by (2), this is equivalent to underestimating the travel time). We define

$$\overline{v}_G^{(i)}(a) = \sqrt{v_A^2 - \underline{w}_C^{(i)}(a)^2} + \overline{w}_T^{(i)}(a),$$

and let $\overline{v}_G(a) := \max_{i \in \{1, \ldots, n\}} \overline{v}_G^{(i)}(a)$. It is easy to prove the following lemma:

Lemma 1 *The inequality $v_G(a, \tau) \leq \overline{v}_G(a)$ holds for all $\tau \in [t_0, t_r]$.*

Note that in particular, if $v_G^*(a) = \max_{\tau \in [t_0, t_r]} v_G(a, \tau)$ denotes the maximum ground speed in $[t_0, t_r]$, we also have

$$\overline{v}_G(a) \geq v_G^*(a). \tag{4}$$

Define $r_a^* := \max_{\tau \in [t_0, t_r]} \sqrt{w_C(a, \tau)^2 + w_T(a, \tau)^2}$, the maximum overall wind speed on $a \in A$. Under the condition that $v_A \geq 2r_a^*$, which in practice is always the case, we obtain

Theorem 1 *Suppose $v_A \geq 2r_a^*$. Then there exists a constant $C > 0$ such that*

$$0 \leq \overline{v}_G(a) - v_G^*(a) \leq C\Delta.$$

The first inequality follows directly from (4), and the proof for the second inequality can be found in [3]. Analogous to the ground speed, we define

$$\underline{T}(a) = \min_{i \in \{1,\ldots,n\}} \underline{T}^{(i)}(a) := \min_{i \in \{1,\ldots,n\}} \frac{d_G(a)}{\overline{v}_G^{(i)}}.$$

Letting $T_a^* = \min_{\tau \in [t_0, t_r]} T(a, \tau)$ and following (2), one readily obtains

Corollary 1 *Suppose $v_A \geq 2r_a^*$. Then there exists a constant $C' > 0$ such that for any arc $a \in A$, we have*

$$0 \leq T_a^* - \underline{T}(a) \leq C' \Delta.$$

In particular, $\underline{T}(a)$ underestimates the travel time needed to traverse an arc, and the error is bounded linearly in the discretisation step.

2.1 The Super-Optimal Wind Potential Function

For the A* algorithm, we seek to find a good and feasible potential function. For HFPP, we can exploit the fact that in our application, there is a small number of possible target nodes (corresponding to airports). Since our objective in HFPP is to minimise travel time, we construct the weighted graph $(G, \underline{T})$, where $\underline{T}\colon A \to [0, \infty)$ maps an arc $a \in A$ to the underestimated travel time $\underline{T}(a)$ obtained through the Super-Optimal Wind computation, i.e., $\underline{T}(a) \leq T(a, \tau)$ for all $\tau \in [t_0, t_r]$ and all arcs $a \in A$. Note that $(G, \underline{T})$ is a weighted graph with static arc weights, and we can without effort compute an all-to-one shortest path tree for each target node t. We then define a potential function for HFPP as

$$\pi_t(v) = \min \left\{ \sum_{a \in P} \underline{T}(a) \colon P \text{ is a } (v, t) - \text{path} \right\}.$$

Note that this is equivalent to running the ALT-Algorithm [4] with the target node as the only landmark.

Theorem 2 *The following two statements hold:*

(i) $\pi_t(\cdot)$ *is feasible in* $(G, \underline{T})$.
(ii) $\pi_t(\cdot)$ *is feasible in* (G, T).

For details on the proof, see [8]. In particular, Theorem 2 yields that running the A* algorithm on (G, T) is equivalent to running Dijkstra's algorithm on the reduced cost graph (G, T') obtained from (3), and A* visits at most as many nodes as Dijkstra's algorithm.

2.2 Validation of Super-Optimal Wind

Theorem 1 and Corollary 1 state that the absolute error of the overestimated ground speed with respect to the optimum ground speed is bounded linearly in the dis-

Solving the Time-Dependent Shortest Path Problem Using Super-Optimal Wind

Table 1 Errors and runtimes of Super-Optimal Wind computation

Altitude (ft)	Segments (#)	Av. error (%)	Max. error (%)	Computation time (s)
37000	344936	0.041	5.263	2.50
34000	344920	0.045	5.882	1.59
31000	338567	0.045	8.333	2.52

cretisation step. To assess the quality of the travel time underestimation with Super-Optimal Wind computationally, we ran it on several real-world instances (cf. [3]), each instance using 28 threads.

As our weather prognoses are given at times t_i all spaced three hours apart, a natural choice for the discretisation step is $t_{i+1} - t_i = \tau_{i+1} - \tau_i = \Delta = 3h$. We found this choice to already yield excellent results, as shown in Table 1, which contains the average and maximum values of the relative error $\rho(a) = \frac{T(a) - T_a^*}{T_a^*} \forall a \in A$. The results show that the Super-Optimal Wind is an excellent underestimator in practice, and can be computed fast.

3 A Case Study

In the following, we investigate the effect of wind on a route. In particular, we consider a flight between Taipei-Taoyuan (TPE) and New York-John F. Kennedy (JFK). We use weather data from the 25th April 2017, starting the route on the same day at 0300 UTC. We assume an aircraft flying at 37 000ft ($\approx 11\,277$ m).

Often, routes lie close to the geodesic, but if aircraft can take advantage of strong tailwinds, they commonly divert to areas with more favourable winds. In Fig. 1, we observe that the search space for A* is doughnut-shaped, which is due to the fact that on that day, there was an unusually strong jetstream on the Northern Pacific, rendering the Pacific route shown in green more efficient than the polar route (red), which would seem a more natural choice. When one compares the ground distances of the northerly route to the Pacific route, one finds the red route to be almost 1880 km shorter than the green route – but considering wind, the green route is 131 s faster than the red route, or roughly 0.26% of the total travel time. As this translates directly to fuel burn, it makes sense to favour the seemingly longer Pacific route over the polar route.

In Fig. 1, we also observe that A* visits significantly fewer arcs than Dijkstra's algorithm. This also impacts the runtime: between TPE and JFK, A* yields a speedup factor of 11 over Dijkstra's algorithm. For a more detailed discussion on the speedup of A* over many instances, we refer the reader to [8].

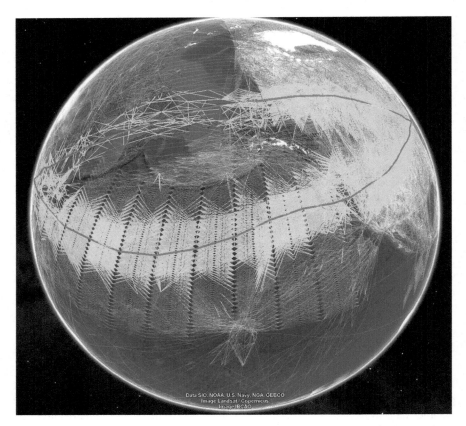

Fig. 1 Search spaces for Dijkstra's algorithm (white) and A* (yellow) between TPE and JFK. The route closest to the geodesic is marked red, the shortest route shown in green (Map data: Google, Landsat/Copernicus/IBCAO)

References

1. Air Transport Action Group (ATAG). (2017). Facts and figures. http://www.atag.org/facts-and-figures.html. Retrived 07 Aug 2017.
2. Bast, H., Delling, D., Goldberg, A., Müller-Hannemann, M., Pajor, T., Sanders, P., Wagner, D., & Werneck, R. F. (2015). Route planning in transportation networks. Technical report, Microsoft Research. Updated version of the technical report MSR-TR-2014-4.
3. Blanco, M., Borndörfer, R., Hoang, N. D., Kaier, A., Schienle, A., Schlechte, T., & Sclobach, S. (2016). Solving time dependent shortest path problems on airway networks using super-optimal wind. *16th Workshop on Algorithmic Approaches for Transportation Modelling, Optimization, and Systems (ATMOS 2016), 54*. (epub ahead of print).
4. Goldberg, A. V., & Harrelson, C. (2004). Computing the shortest path: A* search meets graph theory. Technical Report MSR-TR-2004-24, Microsoft Research, Vancouver, Canada, July 2004.
5. Hart, P. E., Nilsson, N. J., & Raphael, B. (1968). A formal basis for the heuristic determination of minimum cost paths. *IEEE Transactions on Systems Science and Cybernetics, 4*, 100–107.

6. International Air Transport Association. IATA Price Analysis. (2017). https://www.iata.org/publications/economics/fuel-monitor/Pages/price-analysis.aspx. Retrived 07 Aug 2017.
7. Lufthansa Group. (2017). Balance – Nachhaltigkeitsbericht der Lufthansa Group 2016. https://www.lufthansagroup.com/fileadmin/downloads/de/verantwortung/LH-Nachhaltigkeitsbericht-2017.pdf. Retrived 07 Aug 2017.
8. Schienle, A. (2016). Shortest paths on airway networks. Master's thesis, Freie Universitt Berlin.

Anticipation in Dynamic Vehicle Routing

Marlin W. Ulmer

1 Motivation

Decision making in real-world routing applications is often conducted under incomplete information. Vehicle dispatchers deal with uncertainty in travel times, service times, customer demands, and customer requests. This information is only revealed successively during the execution of the route. Technological advances allow dispatchers to adapt their decisions to new information [13]. These developments pave "the way for models of a dynamic nature" [1]. Nevertheless, current decisions influence later outcomes. Anticipation, that is, "incorporating information about the uncertainty of future events" [8] is necessary to avoid myopic decision making. These advances and challenges lead to the field of stochastic and dynamic vehicle routing problems (SDVRPs), a field gaining growing attention in the research community. This attention is reflected in the increasing amount of research on SDVRPs [5]. As [7] state, addressing these new developments and therefore SDVRPs "may necessitate new views, paradigms, and models for decision support." In essence, the field of SDVRPs poses many challenges for the research community in both models and algorithms and has not been studied comprehensively yet.

The canonical model for SDVRPs is a Markov decision process (MDP, [6]). MDPs model subsequent decision states connected by decisions and stochastic realizations of information. Solving the MDP for SDVRPs is challenging due to the "Curses of Dimensionality" [4]. Generally, state space, decision space, and transition space are vast. Methods of approximate dynamic programming (ADP) address these challenges. Still, these methods are not yet established in the field of SDVRP due to the high complexity of the routing problems [10]. In the following, we recall the func-

M. W. Ulmer (✉)
Technische Universität Braunschweig, Braunschweig, Germany
e-mail: m.ulmer@tu-braunschweig.de

tionality and notation of the MDP. We then define and tailor methods of ADP to the specific needs of SDVRPs. We show how ADP enables substantial improvements compared to state-of-the-art benchmark policies.

2 Markov Decision Process

Within the (finite) MDP, a number of decision points $\mathcal{K} = \{0, \ldots, K-1\}$ occurs subsequently. Here, K can be a random variable. For each decision point $k \in \mathcal{K}$, a set of states $\mathcal{S}_k$ is given, combined in the finite set of states $\mathcal{S}$. State $S_0 \in \mathcal{S}$ denotes the initial state and state $S_K \in \mathcal{S}$ denotes the termination state. For each decision point $k \in \mathcal{K}$ and for each state $S_k \in \mathcal{S}$, a subset of decisions $\mathcal{X}(S_k) \subseteq \mathcal{X}$ of the overall decision space $\mathcal{X}$ is given. The combination of a state S_k and a decision $x \in \mathcal{X}(S_k)$ leads to a (deterministic) post-decision state (PDS) $S_k^x \in \mathcal{P}$ with $\mathcal{P}$ the overall set of post-decision states. It further leads to an immediate reward (or costs) $R(S_k, x)$ with $R : \mathcal{S} \times \mathcal{X} \to \mathbb{R}$. Given PDS S_k^x, a stochastic transition $\omega_k \in \Omega$ leads to the next state $(S_k^x, \omega_k) = S_{k+1} \in \mathcal{S}$.

A solution for the MDP is a *decision policy* $\pi \in \Pi$. Decision policies determine the decision to be selected given a specific state. A decision policy $\pi \in \Pi$ is a sequence of decision rules $(X_0^\pi, X_1^\pi, \ldots, X_{K-1}^\pi)$ for every decision point $k \in \mathcal{K}$. Each decision rule $X_k^\pi(S_k)$ specifies the decision to be selected in state S_k. Optimal decision policies $\pi^* \in \Pi$ select decisions leading to the highest expected rewards and therefore maximize the sum of expected rewards. In a specific state S_k, the optimal decision $X_k^{\pi^*}(S_k)$ can be derived by maximizing the sum of immediate and expected future rewards as shown in the Bellman Equation (1):

$$X_k^{\pi^*}(S_k) = \arg\max_{x \in \mathcal{X}(S_k)} \left\{ R(S_k, x) + \mathbb{E}\left[\sum_{j=k+1}^{K} R(X_j^{\pi^*}(S_j)) | S_k \right] \right\}. \quad (1)$$

The expected future rewards are also known as the value $V(S_k^x)$ of PDS S_k^x.

3 Approximate Dynamic Programming

For small MDPs, the values can be calculated recursively to eventually obtain an optimal policy. Still, for SDVRPs, this is usually hardly possible due to the "Curses of Dimensionality" [4]. The state, decision, and transition spaces are generally vast. Thus, solution methods aim on approximating the values by means of simulations and approximate dynamic programming (ADP). In the following, we present two ADP-methods, the dynamic lookup table (DLT) and the offline–online rollout algorithm, capturing the complexity of SDVRPs.

3.1 The Dynamic Lookup Table

One way of approximating the values is value function approximation (VFA).[1] The VFA procedure starts with initial values $\widehat{V}^0$. These values define an initial policy π^0 with respect to the Bellman Equation (1). The VFA then frequently simulates MDP-realizations. In every simulation run i, the VFA uses the current policy π^{i-1} for decision making within the simulation. After the simulation run, the values $\widehat{V}^{i-1}$ are updated with respect to the observed values. The new values $\widehat{V}^i$ then define a new policy π^i. This procedure is continued until a stopping criterion is reached. Subsequently, the VFA approximates the real values and the optimal policy.

The advantage of VFAs is that the simulations are conducted only once *offline* before the actual implementation of the policy. Thus, these methods allow immediate responses to new information, for example, customer requests. Still, to apply VFA, the value for every PDS needs to be stored. For SDVRPs, the number of PDSs is vast and an aggregation is necessary. Thus, PDSs are reduced to a vector of state features (like point of time). This vector space is then partitioned to a lookup table (LT). Conventional LT-partitionings are static. The partitioning is defined a-priori. This leads to disadvantages in the approximation process because "important" LT-areas are represented in insufficient detail while other areas are not sufficiently observed. To alleviate these shortcomings, we propose a dynamic lookup table. This table starts with an initial partitioning and subsequently refines the partitioning in "important" areas. Areas are important if a sufficient number of observations allows and a high variance in the observed values demands a refinement. Thus, the DLT is able to adapt to the approximation process. For a detailed definition and algorithmic procedure of the DLT, we refer to [14].

3.2 Offline–Online Rollout Algorithm

One shortcoming of VFA in general is that not all but only a few state features can be considered in the evaluation. For SDVRPs, VFAs are usually not able to capture spatial information such as customer and vehicle locations [14]. To integrate these details in the evaluation of a PDS, the simulation needs to be conducted *online* in the actual decision state. One prominent online simulation method is the post-decision rollout algorithm (RA). Originating from a particular PDS, an RA simulates a number of trajectories into the future. To determine decisions within the simulations, a base policy is used. The PDS is evaluated with respect to the observed rewards in the simulation. This evaluation is then used in the Bellman Equation to determine the actual decision.

Because the simulations are conducted online, RAs have the disadvantage that the time for simulations is highly limited. Usually, the base policy is a runtime-

---

[1] Notably, in the following, we present *non-parametric* VFA because parametric VFAs are often not able to capture the complex value function structure of SDVRPs [11].

efficient rule of thumb. This inferior decision making within the simulations leads to a discrepancy between simulated and realized outcome. Thus, the evaluation of the PDS may be distorted. To alleviate this disadvantage, we integrate the DLT-policy as base policy in the RA. This leads to an offline–online RA. Within the simulation, decision making is conducted by the offline DLT. The simulation's outcome is then used to determine the actual decision. Thus, the simulation is reinforced and provides better approximation and/or less simulation runs. We further improve the RA's performance by integrating the well-known *Fully Sequential Procedure for Indifference Zone Selection* (IZS) by [3]. For a detailed definition of IZS and the offline–online RA, we refer to [15].

4 Case Study

In the following, we apply DLT and offline–online RA to the dynamic vehicle routing problem with stochastic requests (VRPSR) by [9].

4.1 Problem Definition and Markov Decision Process

In the VRPSR, a vehicle serves customers in a service area within a shift. The vehicle starts and ends its tour at a depot. The customers request service during the shift and are unknown beforehand. Decisions are made about the acceptance or rejection of the new requests and the according routing update. The objective is to maximize the expected number of accepted requests. In the according MDP, a state occurs when the vehicle served a customer. A state S_k consists of the point of time t_k, the currently planned tour θ_k, and the set of new requests C_k^r. Tour θ_k starts at the vehicle's current location, traverses the customers still to serve, and ends at the depot. A decision x determines the subset of requests to accept C_k^a and the according routing update θ_k^x. The reward is the number of accepted requests: $R(S_k, x) = |C_k^a|$. The PDS contains the point of time t_k, and the new routing θ_k^x. The stochastic transition ω_k updates the origin of θ_k^x and provides a set of new requests.

4.2 Computational Experiments

In the following, we describe how we tune DLT and offline–online RA to the needs of the VRPSR. We present the benchmark policies and the results. For the DLT, we select the features point of time t_k and free time budget b_k^x. The free time budget reflects the amount of time available to serve additional requests. It is defined as the difference between remaining time in the shift and the tour duration of θ_k^x. The DLT is therefore two-dimensional. We run 1 million approximation runs and update the

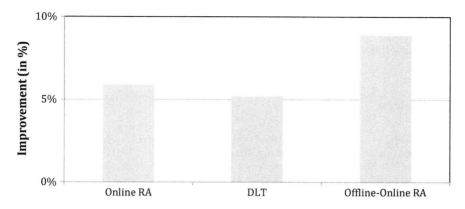

Fig. 1 Average improvement of the policies compared to AI

values with a running average. The partitioning starts with equidistant intervals of 16 min and refines entries of the DLT by splitting them equally into 2×2 new entries. The smallest entry-size is 1 min. For the offline–online RA, we run 16 simulations runs in every state. As benchmark policies, we draw on the current state-of-the-art policy, anticipatory insertion (AI) by [2]. We also compare our methods with the online RA by [12]. The online RA draws on a myopic base policy within the simulations. We compare our methods for a variety of instance settings differing in the number of dynamic requests and the spatial distribution of the requests. The average improvement of the three policies DLT, offline–online RA, and online RA compared to AI are depicted in Fig. 1. We observe that all three ADP-methods outperform the benchmark policy by more than 5%. Notably, the offline DLT is able to achieve similar results as the online RA while not requiring any online runtime. The offline–online RA combines the advantages of the DLT's extensive offline simulations with the online RA's detailed state consideration and achieves improvements of about 9%.

5 Conclusion

Stochastic dynamic vehicle routing problems gain significant interest in the research community. To solve the according MDPs, we have proposed two novel methods of ADP. For the dynamic vehicle routing problem with stochastic requests, we have shown how these methods significantly outperform conventional policies. Future research in stochastic dynamic vehicle routing should focus on applications and methodology. Promising research areas with high dynamism are the growing fields of same-day delivery and shared mobility. The proposed ADP-methodology can be generalized. Instead of a LT-structure, the state space may be dynamically partitioned by means of clustering algorithms. Further, the combination of online and offline ADP may be determined based on a state's value variance and number of observations.

References

1. Gendreau, M., Jabali, O., & Rei, W. (2016). 50th anniversary invited article-future research directions in stochastic vehicle routing. *Transportation Science, 50*(4), 1163–1173.
2. Ghiani, G., Manni, E., & Thomas, B. W. (2012). A comparison of anticipatory algorithms for the dynamic and stochastic traveling salesman problem. *Transportation Science, 46*(3), 374–387.
3. Kim, S.-H., & Nelson, B. L. (2001). A fully sequential procedure for indifference-zone selection in simulation. *ACM Transactions on Modeling and Computer Simulation (TOMACS), 11*(3), 251–273.
4. Powell, W. B. (2011). *Approximate dynamic programming: solving the curses of dimensionality* (2nd ed.)., Wiley series in probability and statistics New York: Wiley.
5. Psaraftis, H. N., Wen, M., & Kontovas, C. A. (2016). Dynamic vehicle routing problems: Three decades and counting. *Networks, 67*(1), 3–31.
6. Puterman, M. L. (2014). *Markov decision processes: discrete stochastic dynamic programming.* New York: Wiley.
7. Savelsbergh, M., & Van Woensel, T. (2016). 50th anniversary invited article-city logistics: Challenges and opportunities. *Transportation Science, 50*(2), 579–590.
8. Speranza, M. G. (2016). Trends in transportation and logistics. *European Journal of Operational Research.*
9. Thomas, B. W. (2007). Waiting strategies for anticipating service requests from known customer locations. *Transportation Science, 41*(3), 319–331.
10. Ulmer, M. W. (2017). *Approximate dynamic programming for dynamic vehicle routing* (Vol. 61)., Operations research/computer science interfaces series Berlin: Springer.
11. Ulmer, M. W. & Thomas, B. W. (2017). Meso-parametric value function approximation for dynamic customer acceptances in delivery routing (submitted).
12. Ulmer, M. W., Mattfeld, D. C., Hennig, M., & Goodson, J. C. (2015). A rollout algorithm for vehicle routing with stochastic customer requests. *Logistics management* (pp. 217–227)., Lecture notes in logistics Berlin: Springer.
13. Ulmer, M. W., Heilig, L., & Voß, S. (2017). On the value and challenge of real-time information in dynamic dispatching of service vehicles. Business & Information. *Systems Engineering, 59*(3), 161–171.
14. Ulmer, M. W., Mattfeld, D. C., & Köster, F. (2017). Budgeting time for dynamic vehicle routing with stochastic customer requests. *Transportation Science.*
15. Ulmer, M. W., Goodson, J. C., Mattfeld, D. C., & Hennig, M. Offline–online approximate dynamic programming for dynamic vehicle routing with stochastic requests. *Transportation Science* (to appear).

Shapley Value Based Allocation for Multi-objective Cooperative Problems

Igor Kozeletskyi

1 Multi-objective Cooperative Games in the Literature

We start this paper with an overview on contributions to the literature that deal with multi-objective cooperative games and distinguish between concepts for transferable and non-transferable utilities for multiple objectives. However, what we can state for both is that there are only very few contributions that consider multi-objective allocations from a game-theoretical perspective and they all are strictly focused solely on a one type of the utility. For transferable utilities Fernandez et al. in [3] introduced set-valued TU games on an example of a multi-criteria minimum cost spanning tree problem. Fernandez et al. in [4] present for set-valued TU games two different core solution concepts, explore the differences among them and also consider multi-objective linear production game as another application of set-valued TU games. Nishizaki and Sakawa in [8] also study a multi-objective representation of linear production games with only common objectives, defined in terms of transferable utilities.

In terms of non-transferable utilities Christinsen et al. in [2] study a class of NTU games resulting from multi-objective linear programs with one objective per player. They consider an extension of the nucleolus as a solution concept and provide an algorithm for its computation. Andersen and Lind in [1] solve the allocation problem for this class of games with the Shapley NTU value and present a simplex-based computational algorithm for a two players case.

The bi-allocation game has been introduced by Kimms, Kozeletskyi and Meca in [6] for a general multi-objective optimization problem with one objective per player and a common objective function for all players. This game was illustrated using a multi-objective extension of linear production games and for allocation problem a new solution concept, inspired by the Shapley value for NTU games (see

---

I. Kozeletskyi (✉)
Hermes Germany, Essener Str. 89, 22419 Hamburg, Germany
e-mail: i.kozeletskyi@gmail.com

[9]), has been proposed. In this paper this solutions concept is described using the multi-objective cooperative TSP. Kimms and Kozeletskyi in [5] define a core-based allocation for the bi-allocation game and propose a computational algorithm for it.

2 A Multi-objective Cooperative Traveling Salesman Problem

In this section a cooperative planning situation in terms of the traveling salesman problem where players besides the traveling costs consider their individual objectives is formulated. Individual objectives are associated with utilities that players can gain from the fulfillment of available orders. Let u_i^k be a utility for player k from order i, if this order is fulfilled by k, meaning that player k visits node i. We assume that this utility can not be negative. This utility can be seen as a measure of importance of order i to player k. For instance, the higher the value of u_i^k the higher are chances that player k will get a new order from the same customer that placed order i in the future. Therefore in our perception every player k seeks to minimize his traveling costs as well as maximize the total utility from the assigned orders. In the case of a given coalition this representation leads to a multi-objective problem with one cost function and utility functions for every player. The cost function represents a common objective for all players.

For the formal definition of the problem additional notation is required. Let $N = \{1, \ldots, n\}$ be a set of salesmen (also here referred to as players) with depots $\{0_1, \ldots, 0_n\}$, where 0_k corresponds to the depot of salesman k. For every non-empty coalition $S \subseteq N$ we take $V(S)$ as the set of orders, $D(S) := \cup_{k \in S} 0_k$ as the set of depots and denote $s := |S|$. Orders set are taken as disjoint, i.e. $\forall k, l \in N$, $k \neq l : V(\{k\}) \cap V(\{l\}) = \emptyset$. And for two coalitions $S, T \subseteq N$ we take that $V(S \cup T) = V(S) \cup V(T)$. The problem can be represented as a directed graph $G(S) = (V(S) \cup D(S), A(S), C(S), T(S))$ with the set of arcs $A(S)$, representing connections between the nodes, matrix of traveling costs $C(S) = (c_{ij})$ and matrix of traveling times $T(S) = (t_{ij})$. Furthermore t_{max} defines the maximal length of a tour. Regarding the decision variables we use the binary variable x_{ijk}, that indicates whether the arc (i, j) is used by salesman k ($x_{ijk} = 1$), or not, and the real-valued decision variable δ_i for the arrival time in node i. This variable is also in combination with a large number M a part of subtour elimination constraints.

Using the introduced notation the objective functions described above can be stated for a non-trivial coalition $S \subseteq N$ as follows:

$$f_0^S = \sum_{k \in S} \sum_{i \in V(S) \cup \{0_k\}} \sum_{j \in V(S) \cup \{0_k\}} c_{ij} x_{ijk} \quad (1)$$

$$f_k^S = \sum_{j \in V(S)} u_j^k \sum_{i \in V(S) \cup \{0_k\}} x_{ijl} \quad k \in S \quad (2)$$

where f_0^S defines the cost function and $\{f_k^S \mid k \in S\}$ are the utility functions. Finally the optimization problem looks as follows:

$$\text{optimize } \{f_0^S, f_{i_1}^S, f_{i_2}^S, \ldots, f_{i_s}^S\} \tag{3}$$

s.t.

$$\sum_{k \in S} \sum_{j \in V(S) \cup \{0_k\}, j \neq i} x_{ijk} = 1 \quad i \in V(S), \tag{4}$$

$$\sum_{k \in S} \sum_{i \in V(S) \cup \{0_k\}, i \neq j} x_{ijk} = 1 \quad j \in V(S), \tag{5}$$

$$\sum_{j \in V(S) \cup \{0_k\}} x_{ijk} + \sum_{j \in V(S) \cup \{0_l\}} x_{jil} \leq 1 \quad i \in V(S),\ k, l \in S, k \neq l, \tag{6}$$

$$\delta_j^k + M \geq \delta_i^k + t_{ij} + M x_{ijk} \quad k \in S,\ i \in V(S),\ j \in V(S) \cup \{0_k\},\ j \neq i, \tag{7}$$

$$\delta_{0_k}^k \leq t_{max} \quad k \in S, \tag{8}$$

$$\delta_i^k \geq d_i \quad i \in V(S),\ k \in S, \tag{9}$$

$$\sum_{j \in V(S)} x_{ijk} = 1 \quad k \in S,\ i \in T^k, \tag{10}$$

$$x_{ijk} \in \{0, 1\} \quad k \in S, i, j \in V(S) \cup \{0_k\}, \tag{11}$$

$$\delta_i^k \geq 0 \quad i \in V(S) \cup \{0_k\},\ k \in S. \tag{12}$$

For a given coalition S and $l \in [1, \ldots, |S|]$, i_l denotes a player on the position l in the coalition S. More precisely this notation should be $i_k(S)$, but as it is clear from the formulation which coalition S is considered, we skip S for notational brevity. The problem (3)–(12) is a multi-objective optimization problem and its solutions are considered under the notion of Pareto optimality. In the remainder of this paper for $S \subseteq N$ we refer to the set of feasible solutions, defined through (4)–(12) as $X(S)$ and to the set of Pareto optimal solutions as $P(S)$. The optimization problem (3)–(12) will be referred to as the multi-objective cooperative TSP.

3 A Game-Theoretic Represantation of the Problem

The above described cooperative scenario includes, besides the cost function, additional objectives that express a metric interpretation of individual preferences of players towards customer's orders. As these preferences vary for different players, values of individual objective functions have non-identical meaning to the players and therefore will be treated as a non-transferable utility in the allocation problem. For this reason the proposed cooperative scenario of a multi-objective cooperaive

TSP represents a combination of transferable and non-transferable utilities, which cannot be treated by a common cooperative TU game. To handle allocation problems from a game-theoretical perspective in such cooperative scenarios a new class of cooperative games, called bi-allocation games, is applied. Kimms, Kozeletskyi and Meca in [6] introduced the bi-allocation game resulting from multi-objective optimization problems with all objectives to be maximized. In this section the definition of the bi-allocation game will be adapted to the case of the multi-objective cooperative TSP, with the cost function as a common objective.

For the multi-objective cooperative TSP the tuple

$$< N, \{f_0^S, (f_k^S)_{k \in S}, X(S)\}_{\substack{S \subseteq N \\ S \neq \emptyset}} >$$

denotes all feasible outcomes for all possible subcoalitions of N. Based on feasible outcomes the corresponding bi-allocation game for the multi-objective cooperative TSP can be defined as a pair $(N, \mathcal{V})$, with the characteristic set $\mathcal{V}(S)$ for $S \subseteq N$, $S \neq \emptyset$:

$$\mathcal{V}(S) = \{\chi \in \mathbb{R}^{2s} \mid \exists x \in X(S), (\exists \pi \in \mathbb{R}^s : \sum_{k \in S} \pi_{i(k)} = f_0^S(x)), \text{ and}$$
$$(\chi_1, \ldots, \chi_s, \chi_{s+1}, \ldots, \chi_{2s}) \leq (f_{i_1}^S(x), \ldots, f_{i_s}^S(x), -\pi_1, \ldots, -\pi_s)\} \quad (13)$$

and $\mathcal{V}(\emptyset) = \emptyset$. Every element $\chi \in \mathcal{V}(S)$ represents a feasible payoff attainable for coalition S that consists of individual utilities χ_k (that have a non-transferable nature) and an allocation of total traveling costs $-\chi_{n+k}$. As it can be seen, this definition is related to the one of an non-transferable ultity (NTU) game, and the characteristic set $\mathcal{V}(S)$ is also a set-valued mapping. However in the bi-allocation game, every characteristic set $\mathcal{V}(S)$ is a subset of a $2s$-dimensional space, as two allocations per player are considered. This formally distinguishes bi-allocation games from NTU games.

Besides the definition of the characteristic set the notion of its boundary is important for the computation of allocation and formal definitions of allocation concepts. The boundary consists of non-dominated allocations from a set $\mathcal{V}(S)$. Using the notion of Pareto optimality and the introduced notation for $P(S)$ we can define the boundary as:

$$\partial \mathcal{V}(S) = \{\chi \in \mathbb{R}^{2s} \mid \exists x \in P(S), (\exists \pi \in \mathbb{R}^s : \sum_{k \in S} \pi_{i(k)} = f_0^S(x)) \text{ and}$$
$$(\chi_1, \ldots, \chi_s, \chi_{s+1}, \ldots, \chi_{2s}) = (f_{i_1}^S(x), \ldots, f_{i_s}^S(x), -\pi_1, \ldots, -\pi_s)\}. \quad (14)$$

For further discussion on bi-allocation games and especially on their properties the reader is referred to [6, 7]. In the next section an allocation concept called the Bi-allocation Shapley value we'll be presented.

4 The Bi-allocation Shapley Value

For the definition of the Bi-allocation Shapley value a transferable correspondence to the bi-allocation game $(N, \mathcal{V})$ is required. This correspondence represents a scalar value associated with every coalition $S \subseteq N$ and hence with every characteristic set $\mathcal{V}(S)$ of the bi-allocation game. The transferable correspondence will be defined through weighting of allocation vectors components. We consider the set of weights

$$\Lambda = \{\lambda \in (0,1)^{2n} \mid \sum_{i \in 2N} \lambda_i = 1 \wedge \forall k \in N : \lambda_{n+i} = \lambda_0, \lambda_0 \in (0,1)\}. \tag{15}$$

A weighting vector $\lambda \in \Lambda$ defines weights associated with allocations from the characteristic set $\mathcal{V}(N)$. In every allocation vector $\chi \in \mathcal{V}(N)$ components $(\chi_{n+1}, \ldots, \chi_{2n})$ correspond to an allocation of the transferable utility f_0^N and are measured on the same scale. Therefore it is reasonable to consider same weights for the transferable part of the allocation and we take λ, such that $\lambda_{n+k} = \lambda_0$ for all $k \in N$ and for some scalar $\lambda_0 \in (0,1)$. Let for every weighting vector $\lambda \in (0,1)^{2n}$ λ^S be a subvector of components of λ corresponding to allocations associated with coalition S, i.e. $\lambda^S = (\lambda_{i_1}, \ldots, \lambda_{i_s}, \underbrace{\lambda_0, \ldots, \lambda_0}_{s})$.

Then for every $S \subseteq N$ and $\lambda \in \Lambda$ the transferable correspondence for the characteristic function $\mathcal{V}(S)$ is

$$\nu_\lambda(S) = \max_{\chi \in \mathcal{V}(S)} \lambda^S \cdot \chi \tag{16}$$

with $\nu_\lambda(\emptyset) = 0$. From the definition of the characteristic set (13) and its boundary (14) it follows that the maximum in (16) is always reached in the boundary $\partial \mathcal{V}(S)$ and $\nu_\lambda(S) = \lambda^S \chi$, for some $\chi \in \partial \mathcal{V}(S)$. This means that $\nu_\lambda(S)$ corresponds to a Pareto optimal solution of the underlying optimization problem, which in our case is the multi-objective cooperative TSP. Furthermore, as all elements of $\partial \mathcal{V}(S)$ are finite-valued, we have that $\nu_\lambda(S) < \infty$ for all $S \subseteq N$ and all $\lambda \in \Lambda$.

From (16) follows that for a given $\lambda \in \Lambda$ ν_λ is a mapping $\nu_\lambda : 2^N \to \mathbb{R}$, with $\nu_\lambda(\emptyset) = 0$ and therefore it represents a well-defined characteristic function of a cooperative TU game. Thus we can define for every $\lambda \in \Lambda$ a cooperative game (N, ν_λ), which can be interpreted as a transferable correspondence of the bi-allocation game $(N, \mathcal{V})$ for a given value λ. For the cooperative game (N, ν_λ) we can compute the Shapley value and define it as $\phi(\nu_\lambda) = (\phi_k(\nu_\lambda))_{k \in N}$, where for every $k \in N$ $\phi_k(\nu_\lambda)$ represents a payoff allocated to player k.

Then following the idea of Shapley for NTU games (see [9]), the allocation vector $\chi \in \mathcal{V}(N)$ is called a Bi-allocation Shapley value of the game $(N, \mathcal{V})$, if there exists $\lambda \in \Lambda$, such that

$$\lambda_k \chi_k + \lambda_0 \chi_{n+k} = \phi_k(\nu_\lambda) \text{ for all } k \in N. \tag{17}$$

From the definition follows that, in contrary to the Shapley value for games with transferable utilities, the Bi-allocation Shapley value is not unique and there can be multiple λ's and hence χ's that satisfy condition (17). We denote by $\Phi(N, \mathcal{V})$ the set of all Bi-allocation Shapley values of the game $(N, \mathcal{V})$. As the Shapley value ϕ of the game (N, ν_λ) is efficient we have that

$$\sum_{k \in N} (\lambda_k \chi_k + \lambda_0 \chi_{n+k}) = \sum_{k \in N} \phi_k(\nu_\lambda) = \nu_\lambda(N).$$

Therefore the Bi-allocation Shapley value χ is an allocation vector that satisfies the definition of $\nu_\lambda(N)$ and hence belongs to the boundary $\partial \mathcal{V}(N)$. So that, the Bi-allocation Shapley value corresponds to a Pareto optimal solution of the multi-objective cooperative TSP with an efficient allocation of the cost function f_0^N. From this follows, that the set of Bi-allocation Shapley values $\Phi(N, \mathcal{V})$ contains only Pareto optimal allocations.

For further discussion of the Bi-allocation Shapley value including its existence, we refer to [6] and regarding a computational approach with an application to the multi-objective cooperative TSP see [7].

5 Conclusions

The introduced cooperative game as well as the Bi-allocation Shapley value represent a general methodology for cooperative scenarios with a common and individual objectives of players. The described concept of weights in the definition of the Bi-allocation Shapley value is in its nature a weighting vector for the objective functions regarding the multi-objective optimization problem. Therefore this allocation concept addresses the multi-objective problem from an a priori perspective and weights can be interpreted as preferences of players towards the objective functions.

References

1. Andersen, K. A., & Lind, M. (1999). Computing the NTU-Shapley value of NTU games defined by multiple objective linear programs. *International Journal of Game Theory*, 28, 585–597.
2. Christinsen, F., Lind, M., & Tind, J. (1996). On the Nucleolus of NTU-games defined by multiple objective linear programs. *Mathematical Methods of Operations Research*, 43, 337–352.
3. Fernandez, F. R., Hinojosa, M. A., & Puerto, J. (2003). Multi-criteria minimum cost spanning tree games. *European Journal of Operational Research*, 158, 399–408.
4. Fernandez, F. R., Hinojosa, M. A., & Puerto, J. (2004). Set-valued TU games. *European Journal of Operational Research*, 159, 181–195.
5. Kimms, A., & Kozeletskyi, I. (2017). Consideration of multiple objectives in horizontal cooperation with an application to transportation planning. *IEEE Transactions*(to appear).
6. Kimms, A., Kozeletskyi I., & Meca, A. (2014). Bi-allocation games: A new class of cooperative games with transferable and non-transferable utilities. Working Paper.
7. Kozeletskyi, I. (2016). *Game-theoretic approaches to allocation Problems in cooperative routing*. Books on Demand.

8. Nishizaki, I., & Sakawa, M. (2001). On computational methods for solutions of multiobjective linear production games. *European Journal of Operational Research, 129*, 386–413.
9. Shapley, L. S. (1988). Utility comparison and the theory of games. In A. E. Roth (Ed.), *The Shapley value: Essays in honor of Lloyd S. Shapley* (pp. 307–319). Cambridge: Cambridge University Press.

Handling Critical Jobs Online: Deadline Scheduling and Convex-Body Chasing

Kevin Schewior

1 Introduction

In online optimization, the input is only revealed incrementally and parts of the output already have to be specified during this process, without knowledge of the future. This addresses a shortcoming of classical optimization techniques, for which one usually assumes full knowledge of the input before the optimization process is started. Indeed, the outcome of decisions made in real-world applications often depends on future events, and certain information about the future is impossible to obtain.

There are important such applications in which possibly unforeseen tasks arrive that are critical in the sense that they must be finished either immediately or until a certain deadline: For instance, in a hospital, patients arrive and require to be treated within a certain time frame. Another example is an on-board computer of an aeroplane or car that needs to perform safety-relevant tasks such as checking for obstacles. The two fundamental mathematical problems considered in this paper, online deadline scheduling and convex-body chasing, are abstractions of such applications. While both problems were not understood very well before, we answered different long-standing open questions and thus makes considerable progress towards understanding them in the thesis of the same title as this paper [16]. In this paper, we review these results.

K. Schewior (✉)
Universidad de Chile, Santiago de Chile, Chile
e-mail: kschewior@gmail.com

K. Schewior
Max-Planck-Institut für Informatik, Saarbrücken, Germany

2 Online Deadline Scheduling with Machine Augmenation

In this problem, an online scheduler receives jobs at their specific release dates and needs to schedule them preemptively on a machine for a specific processing time until their specific deadline. The goal is to minimize the resource usage, which is in this case the number of used identical parallel machines. To evaluate the performance of online algorithms, we compare with offline solutions: In the following, let $m \geq 2$ be the minimum number of machines on which the instance has a feasible schedule. It is known that no online algorithm can guarantee to produce feasible schedules on m machines, even if m is known [10]. Henceforth, we call an online algorithm an m'-machines algorithm if it is guaranteed to produce feasible schedules on m' machines.

We are considering two settings regarding the resumption of preempted jobs: In the migratory setting, they can be resumed on any machine; in the non-migratory setting, they must be resumed on the same machine they were running on before. Indeed, migration may cause a significant overhead in real-world applications. Since the two settings of the problem were first considered [7, 15], it was an open question whether there is an m'-machines algorithm for any constant m', even if $m = 2$ is fixed. Simple greedy algorithms have been shown to fail [15].

We answer this question in the affirmative for the migratory setting: We present an $\mathcal{O}(m \log m)$-machines algorithm [8]. Towards this, we first observe that we can assume m is known at the cost of a constant-factor loss in the required number of machines, which is achieved by a doubling approach. We also prove that jobs whose processing time is less than a constant factor of their respective lifespan can be scheduled via a simple greedy policy on $\mathcal{O}(m)$ machines.

To handle the other class of jobs, the tight jobs, we consider the laxity of each job, that is, the time that the job can be delayed after its release date before it must be started to meet its deadline. We interpret this quantity as the scarce budget for delaying the job. Since a greedy spending of this budget may lead to jobs with too little leftover budget and thus fail, we develop a more sophisticated balancing scheme: We divide the budget of each job into $O(m \log m)$ sub-budgets, each reserved to be only used in a specific situation depending on the other jobs that require to be run.

To analyze this algorithm, we assume that the algorithm fails to schedule some instance feasible on m machines and identify a critical set of jobs responsible for the failure. By applying a novel lower bound on the optimal number of machines needed to schedule sets of tight jobs, we derive a contradiction to the fact that the critical job set was feasible on m machines.

Using similar techniques, we show that our algorithm is an $\mathcal{O}(m)$-machines algorithm for laminar and agreeable instances, two important types of instances that have a special interval structure: In laminar instances, any two job intervals that intersect are contained in one another. Somewhat complementarily, in agreeable instances job intervals can only be contained in one another if they share one endpoint.

For the non-migratory setting of the problem, we show that there exists no $f(m)$-machines algorithm for any function f [9]. This is achieved by a recursive construction forcing any non-migratory algorithm to spread jobs over machines in such a way that releasing another job forces the algorithm to open yet another machine. Not only does this negative result contrast the positive result for the migratory setting, it also contrasts the fact that migration is only of limited power in the offline setting: Any feasible migratory schedule on m machines can be transformed into a non-migratory one on $\mathcal{O}(m)$ machines [12]. On the positive side, we design non-migratory $f(m)$-machines algorithms for agreeable and laminar instances as well as instances that do not contain tight jobs.

Although settled up to a logarithmic factor in this work, it remains an important open question whether $\mathcal{O}(m)$-machines algorithms exist in the migratory setting.

3 Online Deadline Scheduling with Speed Augmenation

Alternatively to increasing the number of machines from m to m' in the online setting, it has also been proposed to increase the speed on each of the machines from 1 to s [15]. An algorithm is then called a speed-s algorithm if it produces feasible schedules on speed-s machines, and the goal is to find such an algorithm with minimal s. It was known that simply prioritizing by smaller deadline or smaller laxity is a speed-$(2 - 1/m)$ algorithm [15] and that there is no speed-s algorithm for $s < (1 + \sqrt{2})/2 \approx 1.207$ [14]. No algorithm with an asymptotical guarantee smaller than 2 is known.

When using speed augmentation, one can either define laxity with respect to a unit-speed or a speed-s machine. Independent of this choice, we give the first formal definition of an algorithm prioritizing by laxity in our continuous-time model. At each time, our algorithm checks if there are more than m unfinished jobs available. If not, each of them is run at speed s; otherwise, we consider the jobs in decreasing order of their laxities. We run all jobs with a strictly smaller laxity as the mth job in this order at speed s, and all jobs with the same laxity as the mth job share the remaining machines. We also show that our definition is essentially the unique one that satisfies certain natural properties.

While it was known that the bound of $2 - 1/m$ is best-possible when prioritizing by deadline, we show the same lower bound for prioritizing by laxity. This is independent of the speed that the laxity depends on. The bound is achieved by a construction that consists of many short jobs with early deadlines that delay a long job with a comparably large laxity, so that at some point the long job is left with almost zero laxity and thus blocks an entire machine. Nesting such instances eventually blocks a number of machines sufficient for the algorithm to fail.

We then refute a claim [1] that the algorithm defined in the same work requires a speed of no more than $e/(e-1) \approx 1.58$. We propose a new algorithm, a combination of ideas from the latter algorithm and prioritizing by laxity, and we conjecture that it is a speed-$(e/(e-1))$ algorithm.

It remains a very interesting research goal to obtain a $(2 - \epsilon)$ speed algorithm for some $\epsilon > 0$ independent of m.

4 Convex-Body Chasing

We consider the following online problem in d-dimensional Euclidean space. The server initially located in the origin receives an online sequence of convex bodies. In response to each body, the task is to immediately move to a point within that body so as to minimize the total moved distance. We evaluate the performance of online algorithms by competitive analysis. This is a special case of the very general online problem of metrical task systems [6], and a constant-competitive algorithm for $d = 2$ is known [11].

We first develop a simple constant-competitive algorithm for the special case when $d = 2$ and all convex bodies are lines [4]. The algorithm relies on the observation that two different greedy algorithms are not constant-competitive for complementary instances: The first greedy algorithm always moves to the current line the cheapest way possible, and the second algorithm always moves to the intersection point of the current with the previous line (if it exists). Our algorithm is a combination of these two: In the first step, our algorithm always moves the cheapest way possible to the current line, say, a distance of d. In the second step, it then moves a distance of d towards the intersection point of the current and previous line (if it exists), but never over it. Using a potential-function argument, we show that the algorithm is constant-competitive.

We are able to extend the $\mathcal{O}(1)$-competitiveness result to the so-called lazy variant of this problem: In this problem, each line comes with a specific slope in [0,1]. In contrast to convex-body chasing, the algorithm can decide to move to any point within the entire space, but it needs to pay its emerging distance to the convex body, discounted by the associated slope. The $\mathcal{O}(1)$-competitive algorithm is achieved by interpreting the slope as an (independent) probability for the associated convex body to appear. Using a reduction from [11], which we present in a more rigorous way, we can then further extend this result to an $\mathcal{O}(1)$-competitive algorithm for the special case when $d = 3$ and all convex bodies are planes.

This approach turns out to even work more generally: We are able to extend this result even further to a $2^{\mathcal{O}(d)}$-competitive algorithm when d is arbitrary and all convex bodies are affine subspaces of the full space [4]. This is the first constant-competitive algorithm in this setting with fixed $d > 2$. However, a gap to the lower bound of $\sqrt{d}$ remains.

We also consider a more general special case of metrical task systems called convex-function chasing. Here, the algorithm is presented a sequence of functions mapping from d-dimensional space to the real numbers. In response, the server can then move to any position in that space, but pays the function value at that position in addition to the movement costs. While convex-body chasing is trivial for $d = 1$, convex-function chasing is not: We improve the previously known bound of $e/(e-1) \approx 1.58$ [13] on the achievable competitive ratio to 1.86 [5], thus almost matching the upper bound of 2 [2, 3, 5].

Apart from the aforementioned gaps, it is still not known whether $\mathcal{O}(1)$-competitive algorithms for general convex-body chasing exist in fixed dimension larger than 2.

References

1. Anand, S., Garg, N., & Megow, N. (2011). Meeting deadlines: How much speed suffices? *International Colloquium on Automata, Languages and Programming (ICALP)* (pp. 232–243).
2. Andrew, L. L. H., Barman, S., Ligett, K., Lin, M., Meyerson, A., Roytman, A., & Wierman, A. (2013). A tale of two metrics: Simultaneous bounds on competitiveness and regret. In *Conference on Learning Theory (COLT)*, 741–763.
3. Andrew, L. L. H., Barman, S., Ligett, K., Lin, M., Meyerson, A., Roytman, A., & Wierman, A. (2015). A tale of two metrics: Simultaneous bounds on competitiveness and regret. *CoRR*. arXiv:1508.03769.
4. Antoniadis, A., Barcelo, N., Nugent, M., Pruhs, K., Schewior, K., & Scquizzato, M. (2016). Chasing convex bodies and functions. *Latin American Theoretical Informatics Symposium (LATIN)* (pp. 68–81).
5. Bansal, N., Gupta, A., Krishnaswamy, R., Pruhs, K., Schewior, K., & Stein, C. (2015). A 2-competitive algorithm for online convex optimization with switching costs. *Workshop on approximation algorithms for combinatorial optimization problems (APPROX)* (pp. 96–109).
6. Borodin, A., Linial, N., & Saks, M. E. (1992). An optimal on-line algorithm for metrical task system. *Journal of the ACM, 39*(4), 745–763.
7. Chan, H., Lam, T. W., & To, K. (2005). Nonmigratory online deadline scheduling on multi-processors. *SIAM Journal on Computing, 34*(3), 669–682.
8. Chen, L., Megow, N., & Schewior, K. (2016). An $\mathcal{O}(\log m)$-competitive algorithm for online machine minimization. *ACM-SIAM Symposium on Discrete Algorithms (SODA)* (pp. 155–163).
9. Chen, L., Megow, N., & Schewior, K. (2016). The power of migration in online machine minimization. *ACM Symposium on Parallelism in Algorithms and Architectures (SPAA)* (pp. 175–184).
10. Dertouzos, M. L., & Mok, A. K. (1989). Multiprocessor on-line scheduling of hard-real-time tasks. *IEEE Transactions on Software Engineering, 15*(12), 1497–1506.
11. Friedman, J., & Linial, N. (1993). On convex body chasing. *Discrete & Computational Geometry, 9*, 293–321.
12. Kalyanasundaram, B., & Pruhs, K. (2001). Eliminating migration in multi-processor scheduling. *Journal of Algorithms, 38*(1), 2–24.
13. Karlin, A. R., Manasse, M. S., Rudolph, L., & Sleator, D. D. (1988). Competitive snoopy caching. *Algorithmica, 3*, 77–119.
14. Lam, T. W., & To, K.-K. (1999). Trade-offs between speed and processor in hard-deadline scheduling. *CM-SIAM Symposium on Discrete Algorithms (SODA)* (pp. 623–632).
15. Phillips, C. A., Stein, C., Torng, E., & Wein, J. (2002). Optimal time-critical scheduling via resource augmentation. *Algorithmica, 32*(2), 163–200.
16. Schewior, K. (2016). Handling critical tasks online: Deadline scheduling and convex-body chasing. *Dissertation*, Technische Universität Berlin.

Methodological Advances and New Formulations for Bilevel Network Design Problems

Pirmin Fontaine

1 Motivation

Transportation networks are a key element of our society. Many commuters use a car, bike or the public transportation system to get to work. And due to globalization, even more goods are shipped through the city. This results in congested and overutilized highways. City centers are dealing with traffic jams, especially during rush hour. Because of the high utilization, deterioration further stresses the already congested cities. But globalization also stresses the highways outside of the city. More goods are shipped throughout the country and worldwide. Thus, the improvement and regulation of traffic networks is very important for society.

In 2016, the Federal Ministry of Transport and Digital Infrastructure of Germany published the 2030 Federal Transport Infrastructure Plan [5]. The major goals of this plan are to select and prioritize projects for the federal trunk roads, the federal railway infrastructure and the federal waterway sector to improve the mobility in passenger traffic and to guarantee freight transportation. In addition to achieving these goals, the government further aims at reducing emissions, improving safety, limiting the use of nature, conserving the nature and improving the quality of life in general. Hereby a main constraint is the very restrictive budget of a total investment of 264.5 billion Euro. Already about 69% is used for preserving the current network. Only 31% can be used to build new or upgrade existing infrastructure to avoid and reduce bottlenecks.

The basis for this infrastructure plan is the forecast of transport interconnectivity 2030 [4]. According to this report, the transport performance of passenger transport will rise by about 12.2% from 1,184 billion pkm (passenger-kilometer) in 2010 to

---

P. Fontaine (✉)
CIRRELT - Interuniversity Research Centre on Enterprise Networks,
Logistics and Transportation, and School of Management,
Université du Québec à Montréal, C.P. 8888, Succ.Centre-ville, Montreal,
QC H3C 3P8, Canada
e-mail: pirmin.fontaine@cirrelt.ca

1,329 billion pkm in 2030. Hereby, the traffic volume is only increasing by 1.2% to 103 billion passengers, meaning that the largest driver is the increase in long-distance traveling.

In freight transportation, the expected growth of demand is even higher. Germany is expecting an increase of 38% in transportation performance. This means it is increasing from 607.1 billion tkm (ton-kilometer) to 837.6 billion tkm. Again the distance is the main driver, however also volume is supposed to increase by 18% (3,704.7 million tons to 4,358.4 million tons) between 2010 and 2030.

The transport performance within the territory of Germany will rise by 38% (from 607.1 billion tkm (ton-kilometer) to 837.6 billion tkm) and the traffic volume by 18% (3,704.7 million tons to 4,358.4 million tons) between 2010 and 2030. In particular, the transport performance of rail transportation is expected to increase by 42.9%.

The Federal Transport Infrastructure Plan consists of more than 2,000 projects. Since cities are already facing congestion and the traffic is supposed to further increase the selection of the most improving projects is crucial. But also in the local networks, the efficient use of budget and the scheduling of maintenance works is very important. In 2015 and 2016, Munich had more than 600 construction zones each year in the road network [16, 19].

These construction sites will further reduce the capacity of the network during maintenance phases, which can lead to even more congestion. This will especially be true if too many projects are scheduled in one area and a city struggles with congestion in general.

A special focus in freight traffic is the transportation of hazardous materials (hazmat). In Germany, 14% of the in 2013 transported volume were dangerous goods. 47% of those used the road network, 20% were shipped by train and about 16.5% each were transported on inland waterways and the sea [18]. The consequences of hazardous accidents are truly fear-evoking. Therefore, the risk of hazardous accidents should be reduced as much as possible. Despite the reduction of risk through technical advances, also the selection of lower risk paths can reduce the risk. Moreover, besides risk minimization, a fair distribution of risk among the population is requested more and more. Therefore, a second goal should be a better distribution such that not one part of the population takes all the risk.

2 Problem Statement

Because of the high increase in traffic and since most authorities have a tight budget for improving their network, it is even more important to use this budget efficiently when expanding the network and, further, to regulate the transport of dangerous goods in order to reduce risks. These problems have a hierarchical decision structure and are summarized under network design problems in the literature. On the one side, the users of the network (also called followers) want to minimize either their own travel time or the transportation costs. On the other side, the government or the responsible authorities (also called leader) want to regulate the traffic with the goal to minimize the overall congestion (e.g., [15]) or to reduce the risk of dangerous good accidents (e.g., [14]). Since the objective of the leader and the follower are not the same, the leader has to anticipate the reaction of the follower. These hierarchical decision problems are then modeled as bilevel problems [17]. In bilevel programming, the leader objective is optimized subject to a nested optimization problem: the follower optimization problem. But, most optimal solution approaches date back to the beginning of bilevel programming (e.g., [1, 13]). Most recent publications focused either on heuristics or metaheuristics.

In the literature of traffic network design problems, the focus is still mainly on the Discrete Network Design Problem (e.g., [15]), which only considers the extension of existing networks. The high share of the budget for maintenance in the Federal Transport Infrastructure Plan [5] shows that maintenance planning is becoming more and more important. However, in the literature these models are still scarce.

The problem of hazardous material transportation is known in the literature on the Hazmat Transport Network Design Problem (e.g., [14]). The focus of these models is still mainly the minimization of total risk in the network. To achieve this goal, the leader decides if a road of the network is allowed for the shipment of dangerous goods or not. Although, several publications motivated the consideration of risk equilibration (e.g., [6]), only a few researchers considered this problem in their model (e.g., [2, 3]). The risk is equilibrated fairly among all arcs and only road networks are considered. This distribution of risk is, however, not fair with respect to the population.

The dissertation of the author contributes to the literature by addressing these methodological challenges:

1. How to solve linear bilevel problems with discrete leader variables efficiently to optimality?
2. How to approximate the non-linear Discrete Network Design Problem to a linear bilevel problem without additional binary variables?
3. How to model a maintenance problem as a bilevel problem and solve it?
4. How to use the multiple follower structure in the Hazmat Transport Network Design Problem to solve it efficiently?
5. How to model hazardous material risk for fair distribution?

Using these methodological advances, the following research questions were addressed and shall support the responsible authorities in their decision:

1. How can a bilevel model for maintenance planning improve the use of the budget to reduce congestion compared to practical heuristics?
2. Why is it important to consider different modes in the Hazmat Transport Network Design Problem?
3. What is the trade-off between risk minimization and risk equilibration?

3 Results

The dissertation of the author introduces a Benders decomposition algorithm for solving discrete-continuous linear bilevel problems. To apply Benders decomposition, the lower level problem of the bilevel formulation is replaced by its Karush–Kuhn–Tucker conditions and a bilinear term is linearized without introducing auxiliary binary variables. In the resulting mixed-integer linear program, the binary leader variables are the complicating variables and the problem can be decomposed in the classical master and slave problem according to Benders decomposition. The method is tested on four different problems: the Discrete Network Design Problem, the Dynamic Discrete Network Design Problem, the Decentralized Capacitated Facility Selection Problem and the Hazmat Transport Network Design Problem. For all problems the numerical results show the efficiency of the method. Problem specific acceleration techniques are used to further accelerate the Benders decomposition.

In the first problem, the Discrete Network Design Problem, the non-linear time functions in the objective function are linearly approximated. Compared to existing approaches, we avoid introducing binary auxiliary variables by using the convexity of the functions. Besides that, the slave problem is divided into two subproblems for a fast calculation of the dual variables for the Benders decomposition. The numerical results show run time improvements of more than 60% compared to the mixed-integer linear programming formulation solved on a commercial solver. This chapter is based on [8].

In [11], the Discrete Network Design Problem is extended to a multi-period model to derive maintenance schedules in traffic networks. The multi-period structure, allows us to decompose the slave problem within the periods and apply a multi-cut version of Benders decomposition. Even though Benders decomposition does not reach convergence, it quickly finds good solutions and outperforms, the mixed-integer linear program, a genetic algorithm, and simple priority rules, which currently might be used in practice. Especially under tight budgets, our approach was the only approach among all tested methods which could find feasible solutions.

In the Decentralized Capacitated Facility Selection Problem [9] and the Hazmat Transport Network Design Problem [10], the calculation of pareto-optimal cuts significantly reduced the number of iterations and therefore improved the convergence. In the Hazmat Transport Network Design Problem, the multi-follower structure is again used to apply the multi-cut version of Benders decomposition. Compared to the mixed-integer linear programming formulation, our approach shows run time improvements of more than 90% for both problems. Furthermore, we analyze the benefits of using bilevel formulations. The results underline that, a good classification of dangerous goods, can reduce the risk.

The last chapter is based on [7] and is joined work with Teodor Gabriel Crainic (University of Quebec in Montreal), Michel Gendreau (Polytechnique Montreal), and Stefan Minner (Technical University of Munich). We introduce a new population-based risk definition to distribute the risk fairly among the population and extend the Hazmat Transport Network Design Problem to a multi-mode network design problem. The results show that pure risk equilibration leads to a very high total risk in the network and every population center in the network can end up with a higher risk. However, because of a convex correlation between these two objectives, already a small increase in total risk in the network can lead to a significantly better distribution of risk among the population. Therefore, it is important to consider the trade-off between risk minimization and risk equilibration and authorities need to select between the pareto-optimal solutions. Moreover, we show that multiple modes need to be considered and that the equilibration of risk over arcs, as it is done in the literature so far, does not distribute the risk among the population fairly.

Acknowledgements While working on this thesis, the author was doctoral student in the School of Management at the Technical University of Munich. The author also gratefully acknowledges a fellowship of Deutscher Akademischer Austauschdienst (DAAD), which helped to start the work of the last project.

References

1. Bard, J., & Moore, J. (1990). A branch and bound algorithm for the bilevel programming problem. *SIAM Journal on Scientific and Statistical Computing*, *11*(2), 281–292.
2. Bianco, L., Caramia, M., & Giordani, S. (2009). A bilevel flow model for hazmat transportation network design. *Transportation Research Part C: Emerging Technologies*, *17*(2), 175–196.
3. Bianco, L., Caramia, M., Giordani, S., & Piccialli, V. (2015). A game-theoretic approach for regulating hazmat transportation. *Transportation Science*.
4. Bundesministerium für Verkehr und digitale Infrastruktur (BMVI) (2014). Verkehrsverflechtungsprognose 2030.
5. Bundesministerium für Verkehr und digitale Infrastruktur (BMVI) (2016). Bundesverkehrswegeplan 2030.
6. Erkut, E., Tjandra, S. A., & Verter, V. (2007). Chapter 9 hazardous materials transportation. In C. Barnhart & G. Laporte (Eds.), *Transportation* (Vol. 14), Handbooks in Operations Research and Management Science. Elsevier.
7. Fontaine, P., Crainic, T. G., Minner, S., & Gendreau, M. (2016). Population-based risk equilibration for the multi-mode hazmat transport network design problem. Technical report CIRRELT-2016-63.
8. Fontaine, P., & Minner, S. (2014). Benders decomposition for discrete-continuous linear bilevel problems with application to traffic network design. *Transportation Research Part B: Methodological*, *70*, 163–172.
9. Fontaine, P., & Minner, S. (2016). Benders decomposition for the decentralized facility selection problem. Working paper.
10. Fontaine, P., & Minner, S. (2016). Benders decomposition for the hazmat transport network design problem. Working paper.
11. Fontaine, P., & Minner, S. (2017). A dynamic discrete network design problem for maintenance planning in traffic networks. *Annals of Operations Research*, *253*(2), 757–772.
12. Fontaine, P. S. R. (2016). *Methodological Advances and New Formulations for Bilevel Network Design Problems*. Ph.D. thesis, Dissertation, München, Technische Universität München.
13. Hansen, P., Jaumard, B., & Savard, G. (1992). New branch-and-bound rules for linear bilevel programming. *SIAM Journal on Scientific and Statistical Computing*, *13*(5), 1194–1217.
14. Kara, B. Y., & Verter, V. (2004). Designing a road network for hazardous materials transportation. *Transportation Science*, *38*(2), 188–196.

15. LeBlanc, L. J. (1975). An algorithm for the discrete network design problem. *Transportation Science*, 9(3), 183–199.
16. Schmidt, P. T., & Karowski, S. (2016). Hier droht Stau! Der Bauplan für 2016. http://www.merkur.de/lokales/muenchen/stadt-muenchen/muenchen-bauplan-moegliche-stauquellen-jahr-2016-meta-6021787.html.
17. Schneeweiß, C. (2003). *Distributed Decision Making* (2nd ed.). Berlin: Springer.
18. Statistisches Bundesamt Wiesbaden. (2015). Verkehr: Gefahrguttransporte 2013. *Fachserie 8 Reihe 1.4*.
19. Völklein, M. (2015). Baustellen in München. http://www.sueddeutsche.de/muenchen/baustellen-in-muenchen-hier-stehen-sie-im-stau-1.2287759.

Stable Clusterings and the Cones of Outer Normals

Felix Happach

1 Introduction

Informed decision-making based on large data sets is one of the big challenges in operations research. We are interested in one of the fundamental tasks in data analytics: The *clustering* of a data set into disjoint clusters. Data is often represented as a finite set $X \subseteq \mathbb{R}^d$ in d-dimensional Euclidean space. A clustering $C = (C_1, \ldots, C_k)$ then is a partition of X such that $\bigcup_{i=1}^{k} C_i = X$ and $C_i \cap C_j = \emptyset$.

There is a huge amount of clustering algorithms, a popular one being the k-means algorithm. It exhibits an interesting discrepancy between its excellent behaviour in practice and its known worst-case behaviour. In a theoretical worst-case, it may take exponentially many iterations, and it is easy to construct artificial examples for which its results do not capture the structure of the underlying data at all. In practice, however, it typically terminates after a few iterations and produces human-interpretable results. We use methods from polyhedral theory to better understand this difference and gain insights on "good" clusterings.

The polytope we are investigating was first introduced in [1] and it encodes all possible clusterings of the data point set X. Its vertices encode certain favorable clusterings [1], one property being that they are minimizers of the least-squares functional among all clusterings of the same cluster sizes. Most importantly, these vertices have strong separation properties: They allow for the construction of a *power diagram*, a generalized Voronoi diagram in $\mathbb{R}^d$ with one polyhedral cell for the data points for each cluster [3], see Fig. 1.

F. Happach (✉)
Technische Universität München, Munich, Germany
e-mail: felix.happach@tum.de

Fig. 1 A power diagram with one cell for the blue, red, and black cluster. The data points of each cluster lie in the interior of their respective cells. The small dots indicate the sites

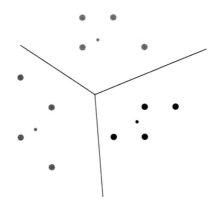

2 Preliminaries

Let $n, d, k \in \mathbb{N} := \{1, 2, \ldots\}$ be fixed. Let $X := \{x_1, \ldots, x_n\} \subseteq \mathbb{R}^d$ be a set of n distinct non-zero data points and for $m \in \mathbb{N}$ define $[m] := \{1, \ldots, m\}$.

2.1 Clusterings

A partition $C := (C_1, \ldots, C_k)$ of X is called **clustering**. For $i \in [k]$, we call C_i the *i*th **cluster** of C and $|C_i|$ its **size**. Let $s^- := (s_1^-, \ldots, s_k^-), s^+ := (s_1^+, \ldots, s_k^+) \in \mathbb{N}^k$ such that $0 \leq s_i^- \leq s_i^+ \leq n$ for all $i \in [k]$ be the lower and upper bounds on the cluster sizes. We only consider clusterings satisfying $s_i^- \leq |C_i| \leq s_i^+$ for all $i \in [k]$.

In order to compare two clusterings $C := (C_1, \ldots, C_k), C' := (C'_1, \ldots, C'_k)$, we use the **clustering difference graph**, c.f. [3], which is the labeled directed multigraph $CDG(C, C') := (V, E)$ with node set $V := [k]$ and edge set E constructed as follows: For each $x_j \in C_i \cap C'_l$ with $i, l \in [k]$ and $i \neq l$, there is an edge $(i, l) \in E$ with label x_j. We can derive the clustering C' from C by applying operations corresponding to the edges of $CDG(C, C')$:

Let $(i_1, i_2) - (i_2, i_3) - \cdots - (i_t, i_1)$ be a cycle in $CDG(C, C')$ with labels $x_{j_1}, \ldots, x_{j_t}$. Applying the **cyclical exchange**

$$CE: \quad C_{i_1} \xrightarrow{x_{j_1}} C_{i_2} \xrightarrow{x_{j_2}} \cdots \xrightarrow{x_{j_{t-1}}} C_{i_t} \xrightarrow{x_{j_t}} C_{i_1}$$

to C means deriving the clustering $\bar{C} = (\bar{C}_1, \ldots, \bar{C}_k)$ by setting $\bar{C}_{i_l} := (C_{i_l} \setminus \{x_{j_l}\}) \cup \{x_{j_{l-1}}\}$ for all $l \in \{1, \ldots, t\}$ (cyclical indexing) and $\bar{C}_r := C_r$ for all $r \in [k] \setminus \{i_1, \ldots, i_t\}$. Clearly, all cluster sizes remain the same.

For an edge path $(i_1, i_2) - (i_2, i_3) - \cdots - (i_t, i_{t+1})$ in $CDG(C, C')$ with labels $x_{j_1}, \ldots, x_{j_t}$, we consider the **movement**

$$M: \quad C_{i_1} \xrightarrow{x_{j_1}} C_{i_2} \xrightarrow{x_{j_2}} \cdots \xrightarrow{x_{j_t}} C_{i_{t+1}}.$$

Applying M to C means deriving the clustering $\bar{C} = (\bar{C}_1, \ldots, \bar{C}_k)$ by setting $\bar{C}_{i_l} := (C_{i_l} \setminus \{x_{j_l}\}) \cup \{x_{j_{l-1}}\}$ for all $l \in \{2, \ldots, t\}$, $\bar{C}_{i_1} := C_{i_1} \setminus \{x_{j_1}\}$, $\bar{C}_{i_{t+1}} := C_{i_{t+1}} \cup \{x_{j_t}\}$ and $\bar{C}_r := C_r$ for all $r \in [k] \setminus \{i_1, \ldots, i_{t+1}\}$.

One can obtain any clustering from any other clustering by greedily decomposing their clustering difference graph into paths and cycles and applying the corresponding movements and cyclical exchanges to C.

2.2 Power Diagrams

Let $a := (a_1^T, \ldots, a_k^T)^T \in \mathbb{R}^{d \cdot k}$ be a **site vector** with distinct **sites** $a_1, \ldots, a_k \in \mathbb{R}^d$ and let $w_1, \ldots, w_k \in \mathbb{R}$ be k **weights**. For $i \in [k]$, we call

$$P_i := \{x \in \mathbb{R}^d \mid \|x - a_i\|_2^2 - w_i \leq \|x - a_j\|_2^2 - w_j \text{ for all } j \in [k] \setminus \{i\}\}$$

the ith **(power) cell** of the **power diagram** $(P_1, \ldots, P_k)$. We are interested in power diagrams **inducing** C that satisfy $C_i \subseteq P_i$ for all $i \in [k]$, see Fig. 1. We want to stress the strength of this separation property, which is stronger than just the existence of separating hyperplanes. The constructed cells also partition the underlying space.

2.3 Bounded-Shape Partition Polytope

Let $N_P(v) := \{a \in \mathbb{R}^d \mid a^T v \geq a^T x \ \forall x \in P\}$ be the **normal cone** or **cone of outer normals** of a polytope P at $v \in P$. Figure 2 illustrates 2D and 3D examples. For a clustering $C = (C_1, \ldots, C_k)$ and $i \in [k]$, let $\sigma_i := \sum_{x \in C_i} x \in \mathbb{R}^d$. The **clustering vector** of C is $w(C) := (\sigma_1^T, \ldots, \sigma_k^T)^T \in \mathbb{R}^{d \cdot k}$ and $\mathcal{P} := \text{conv}(\{w(C) \mid s_i^- \leq |C_i| \leq s_i^+, \ \forall i \in [k]\})$ is called the **bounded-shape partition polytope**.

In [1], Barnes et al. gave a complete characterization of the vertices of $\mathcal{P}$. The authors proved that a clustering vector $w(C)$ is a vertex of $\mathcal{P}$ if and only if there exists a site vector $a \in \mathbb{R}^{d \cdot k}$ and k scalars fulfilling some extraordinary separation properties for C [1]. One can show that these conditions imply the existence of a power diagram with site vector a and weights only depending on a that induces C [4]. Moreover, any $a \in N_{\mathcal{P}}(w(C))$ can be chosen as a site vector to construct these power diagrams. Further, we define the **vector of a cyclical exchange** $w(CE)$ and **movement** $w(M)$ to be the difference vector of the respective clusterings vectors.

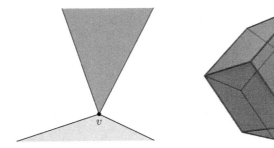

Fig. 2 A 2-dimensional polytope with vertex v in black and normal cone at v in blue (left) and a 3-dimensional polytope (right)

3 Main Results

3.1 Edge Structure and Construction of the Normal Cones

First, we extend the results of Barnes et al. to obtain a characterization of the edges of $\mathcal{P}$. This generalizes results of [3, 5] on some special cases.

Theorem 1 *Let $C := (C_1, \ldots, C_k)$, $C' := (C'_1, \ldots, C'_k)$ be two clusterings such that $w(C)$ and $w(C')$ are adjacent vertices of $\mathcal{P}$. Further, let no three points in X lie on a single line. Then C and C' either differ by a single cyclical exchange, a single movement or by two movements and there are distinct $i, j \in [k]$ such that both are of the form $C_i \to C_j$.*

Theorem 1 enables us to construct the cone of outer normals of a vertex $w(C)$ explicitly. For a cone $K \subseteq \mathbb{R}^{d \cdot k}$, its polar cone is defined as $K^\circ := \{y \in \mathbb{R}^{d \cdot k} \mid y^T x \leq 0, \, \forall x \in K\}$. If $F \subseteq K$ is a l-dimensional face of K, then its **polar face** $[F]^\circ := F^\perp \cap K^\circ$ is a $(d \cdot k - l)$-dimensional face of K°.

Consider a vertex $w(C) \in \mathcal{P}$ whose incident edges all correspond w.l.o.g. to cyclical exchanges $CE_1, \ldots, CE_t$ (Theorem 1). Otherwise, just add the corresponding movements M and consider their vectors as well. It is well-known from linear optimization that the normal cone then can be written as a polar cone:

$$N_\mathcal{P}(w(C)) = \left\{ \sum_{j=1}^{t} \lambda_j w(CE_j) \mid \lambda_j \geq 0 \text{ for all } j \in [t] \right\}^\circ.$$

So the facets of $N_\mathcal{P}(w(C))$ are polar faces to the edges of $\mathcal{P}$: $F_j = [w(CE_j)]^\circ$ for all $j \in [t]$, yielding an explicit representation of the normal cone via its facets.

3.2 The Volume of the Cones of Outer Normals

An important observation through polyhedral theory is that any positive multiple $\lambda a \in \mathbb{R}^{d \cdot k}$ ($\lambda > 0$) of a site vector $a \in N_\mathcal{P}(w(C))$ is contained in the normal cone and defines the same power diagram (all sites are scaled by the same factor). Therefore, we can identify every site vector by its unit norm vector.

We are interested in measuring the volume of the normal cones of $\mathcal{P}$ in order to characterize "good" clusterings. We follow the notation of Bonifas et al. [2] who used the volume of normal cones in a different context. Let $\mathbb{S}^{d \cdot k} := \{x \in \mathbb{R}^{d \cdot k} \mid \|x\|_2 = 1\}$ be the Euclidean unit sphere. For a cone $K \subseteq \mathbb{R}^{d \cdot k}$, let $B(K) := K \cap \mathbb{S}^{d \cdot k}$ be its **base**. The **volume of** K is defined as the $(d \cdot k - 1)$-dimensional area of $B(K)$ and is denoted by vol(K).

In simple terms, popular clustering algorithms such as the k-means iteratively compute sites of power diagrams and return the induced clustering. This can be interpreted as repeated linear optimization over $\mathcal{P}$. Informally we can say: The larger the area of the base $B(N_\mathcal{P}(w(C)))$ the higher the probability of a site vector $a \in \mathbb{R}^{d \cdot k}$ being in $N_\mathcal{P}(w(C))$. Therefore, the ratio

$$\frac{\text{vol}(N_\mathcal{P}(w(C)))}{\text{vol}(\mathbb{R}^{d \cdot k})} = \frac{\text{vol}(N_\mathcal{P}(w(C)))}{\text{area}(\mathbb{S}^{d \cdot k})} \quad (1)$$

encodes the probability with which a randomly chosen site vector $a \in \mathbb{R}^{d \cdot k}$ induces C. Hence, there is a direct relation between large cones and "good" clusterings explaining why many clustering algorithms work so well in practice.

3.3 Stable Clusterings and Cones of Different Shapes

In the previous section, we argued that large cones are likely to be found. They also encode "stable clusterings": clusterings which are robust w.r.t. perturbation of the site vector of the inducing power diagram, i.e. which do not change for small changes of the site vector. Geometrically, a "stable" site vector for a clustering can be described by informally dropping a p-norm unit ball into the normal cone $N_\mathcal{P}(w(C))$ with a gravity center at $0 \in \mathbb{R}^{d \cdot k}$ and computing where it "gets stuck" due to being blocked by the facets of the cone. The center of this unit ball then is a vector that lies "most centrally" within the normal cone. This gives the following optimization problem

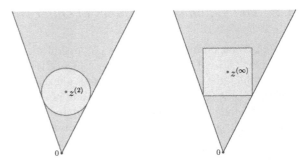

Fig. 3 Unit balls (red) blocked by facets of the normal cone (blue) (color figure online)

$$\min \|z\|_2$$
$$\text{s.t. } z \in N_{\mathcal{P}}(w(C)),$$
$$dist(z, F_j)_p \geq 1 \ \forall j \in [t], \quad (2)$$
$$z \in \mathbb{R}^{d \cdot k}.$$

Here, $dist(z, F_j)_p$ denotes the distance w.r.t. the p-norm of z and the facet F_j, i.e. $dist(z, F_j)_p = \inf\{\|z - f^j\|_p \mid f^j \in F_j\}$. Figure 3 illustrates two 2D examples of this approach with different p-norms.

For an optimal solution $z^{(p)}$ of (2), any vector in $\{z \in \mathbb{R}^{d \cdot k} \mid \|z - z^{(p)}\|_p \leq 1\}$ is still contained in the normal cone, i.e. its corresponding power diagram induces C as well. Projecting this p-norm ball around $z^{(p)} \in \mathbb{R}^{d \cdot k}$ onto the components of the k sites corresponds to smaller d-dimensional p-norm balls within which we can perturb the sites $z_1^{(p)}, \ldots, z_k^{(p)} \in \mathbb{R}^d$ without changing the clustering.

We can generalize (2) by not just considering p-norm unit balls, but their image under an invertible linear map $A : \mathbb{R}^{d \cdot k} \mapsto \mathbb{R}^{d \cdot k}$ with $|\det(A)| = 1$. This map (matrix) A contorts the p-norm unit ball without changing its volume and, by this, we can model any possible shape of the cones. Informally, this contortion corresponds to weighting some directions more than others, so – depending on the underlying structure – we might be able to perturb some sites more than others or in different directions with different amount. However, considering the generalization of (2), one can show that cones of equal volume, no matter which shape they have, encode equally stable ("good") clusterings.

4 Follow-Up Work

In the paper *Good Clusterings Have Large Volume* with Steffen Borgwardt, we improve and extend these results by studying the structure of the normal cones extensively. Besides improving notation, we formalize the notions of the **quality of**

clusterings and the **stability of site vectors** as well-defined measures and provide an optimization problem for computing a "best" site vector, as well as an approximation algorithm. We perform some proof-of-concept computations and exhibit how to use their results for informed decisions.

Acknowledgements I am grateful to my master's thesis advisor Steffen Borgwardt for introducing me to this exciting topic and for the continuous support.

References

1. Barnes, E. R., Hoffman, A. J., & Rothblum, U. G. (1992). Optimal partitions having disjoint convex and conic hulls. *Mathematical Programming, 54*(1), 69–86. https://doi.org/10.1007/BF01586042.
2. Bonifas, N., Di Summa, M., Eisenbrand, F., Hähnle, N., & Niemeier, M. (2014). On subdeterminants and the diameter of polyhedra. *Discrete and Computational Geometry, 52*(1), 102–115. https://doi.org/10.1007/s00454-014-9601-x.
3. Borgwardt, S. (2010). A Combinatorial Optimization Approach to Constrained Clustering. Ph.D. thesis. Technische Universität München.
4. Brieden, A., & Gritzmann, P. (2012). On optimal weighted balanced clusterings: gravity bodies and power diagrams. *SIAM Journal on Discrete Mathematics, 26*(2), 415–434. https://doi.org/10.1137/110832707.
5. Fukuda, K., Onn, S., & Rosta, V. (2003). An adaptive algorithm for vector partitioning. *Journal of Global Optimization, 25*(3), 305–319. https://doi.org/10.1023/A:1022417803474.

The Two Dimensional Bin Packing Problem with Side Constraints

Markus Seizinger

1 Introduction

In the *two dimensional bin packing problem with side constraints* (2DBP-SC) we are given a set of rectangular items $i \in I$, each defined by its height h_i, width w_i and type t_i. A bin consists of S sides with dimensions H and W, respectively. The goal of 2DBP-SC is to assign every item a concrete position such that all items are packed without overlapping and using as few bins as possible. Furthermore, items are packed orthogonally and the newly introduced *side constraint* has to be satisfied, meaning that no two items of different type may be placed face-to-face on the same bin but on different sides. We assume items may not be rotated. This problem is an extension of the well-studied *two dimensional bin packing problem* (2DBP), where only one bin side exists ($S = 1$). 2DBP-SC is strongly NP-hard, since 2DBP is known to be such [5, 6].

The example depicted in Fig. 1 gives an idea of the restrictions implied by the side constraint. This small instance contains bins with two sides and two item types. The side constraint reduces the available space on different sides of the bin. In contrast to the traditional 2DBP, where items simply occupy bin capacity for their respective shape, in 2DBP-SC items additionally block this capacity on all other sides of the bin. Only items of identical type may use this blocked area (indicated by hatched area). This means that the concrete capacity consumption of a single item depends on the actual packing of the bin. Notice that another item of type A (green) with dimensions $(w, h) = (4, 6)$ could be placed in the bottom right corner on the rear side of the bin, whereas an identically shaped item of type B cannot be placed in this position, because it would overlap with area blocked from the front side for type A.

M. Seizinger (✉)
Department of Health Care Operations/Health Information Management, Faculty of Business and Economics, University Center of Health Sciences at Klinikum Augsburg (UNIKA-T), University of Augsburg, Universitätsstraße 16, 86159 Augsburg, Germany
e-mail: markus.seizinger@unikat.uni-augsburg.de

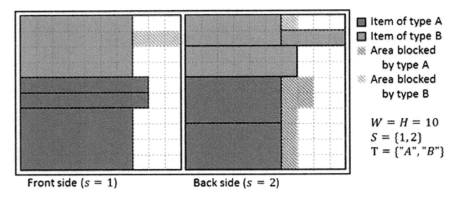

Fig. 1 Example Packing

The motivation for this problem originates from a real-world application. We analyzed a bottleneck resource in a paint shop. Items have to be placed on racks with two faces (front and rear). Additionally, two types of items exist. These types are not allowed to be placed face-to-face for quality reasons. Due to these restrictions the utilization of racks was very low and this production step became a limiting factor. 2DBP-LC corresponds directly to this problem. The same problem must be solved when packing a multi-temperature compartment truck. These trucks consist of flexible departments, dividing them into different temperature zones. The cargo has to be packed into shelves, while every item has to be in its respective temperature zone.

2 Lower Bounds

We introduce already known lower bounds for 2DBP and 3DBP and adapt them to our new class of 2DBP-SC. In general, two ideas exist: The geometric bound sums up volumes of all items and divides it by the area available in a single bin [1, 6]. In contrast, the bound of large items focusses on items fulfilling the conditions $w_i > \frac{W}{2}$ and $h_i > \frac{H}{2}$. We combine both ideas as in [5] to bound L_2^{SC}. By adapting the bound to 2DBP-SC, we can account for item types. This allows us to further improve existing bounds.

3 Upper Bound

For the 2DBP several solution heuristics exist: Lodi et al. [2] describe well-known heuristics Next-Fit Decreasing Height, First-Fit Decreasing Height, and Best-Fit Decreasing Height. The latter dominates the remaining two regarding solution quality [3].

We adapt the *Best-Fit Decreasing Height* (BFDH) algorithm, first introduced by [1]. Our algorithm, called *Best-Fit Decreasing Height with side constraints* (BFDH-SC), creates feasible packings but does not guarantee to find an optimal one. We first order all items according to (a) type and (b) non-increasing height. The first item in the list—the active item—is packed on a level containing only items with identical type and sufficient remaining space. According to the best-fit rule we select that level, where the remaining horizontal space is minimized. If no such level exists, a new level is initialized on top of the latest one, if the bin has enough remaining height. If not, a new bin is initialized as well. Finally, the active item is removed from the list. The algorithm terminates, if the list of items is empty. The heuristic has a runtime complexity of O (n^3).

4 Solution Algorithm

We decompose the problem straightforward according to Dantzig-Wolfe and solve the resulting set covering formulation with column generation. In particular, we form a *(Restricted) Master Problem* (RMP) and a *two dimensional packing problem* (2DPP), acting as the subproblem.

We will solve the RMP in every iteration of the column generation procedure to obtain the dual variables π_i associated with every item. We can then use this information to find a new feasible packing that can improve the current solution of the RMP by solving the subproblem. Thereafter, the next iteration starts over again by solving the slightly enlarged RMP. If all sufficient columns are added to the RMP, its solution is optimal, i.e. the LP-relaxation of RMP has been solved. From theory the bound improves the LP-relaxation of the original problem.

Initial columns are created by BFDH-SC, L_2^{SC} acts as initial lower bound. During the course of the algorithm, new lower bounds are available in every iteration [4], using optimal solution values of RMP and subproblem $L^{CG} = \left\lceil \frac{z_{RMP}}{1-z_{sub}} \right\rceil$.

4.1 Subproblem Decomposition

In our approach the subproblem has to find a feasible packing pattern p with negative reduced costs $c_p^\pi < 0$. If no negative reduced cost column exists, the column generation algorithm terminates with the optimal LP solution of RMP.

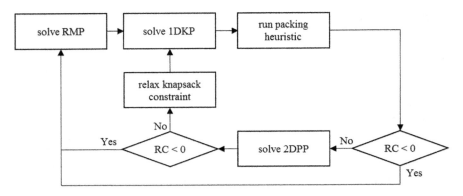

Fig. 2 Iteration scheme of column generation process

To solve this complex subproblem, we follow the approach of [7]. There, the pricing problem is decomposed into a *one dimensional knapsack problem* (1DKP) and a relaxed 2DPP. 1DKP selects a subset of items $I' \subseteq I$ prior to the packing problem. We then solve the 2DPP with this reduced set, what leads to an increase in speed for the latter problem. Notice that this comes at the cost of decreasing solution quality, since we cut a quite large region of the solution space. If any item, which is part of the optimal packing in this iteration, is not an element of I', then 2DPP is not able to find this optimal solution, so it will be necessary to solve 2DPP for the original set of items I in some iterations (Fig. 2).

First, we extend a MIP formulation for a traditional 2DPP first published in [7] to take item types into account. Decision variables indicate whether a pair of items is part of the packing $(a_i, a_j = 1)$ and if so, the relative position of two items (f_{ij}, l_{ij}, b_{ij}). Two non-overlapping items may ether be in front, behind, left, right, above, or below of each other. The side-constraint is implemented as (III). The problem's goal is to create a new pattern with minimal reduced costs.

Second, we additionally introduce a packing heuristic inspired by an algorithm by [5] to speed up the process of creating a new column. This greedy algorithm starts with an empty bin and aligns items along edges of already placed items until no more items fit into the bin. It takes the side constraint explicitly into account but does not guarantee to find the minimum reduced-cost packing, though it is much faster than solving the subproblem to optimality with the above formulation.

The algorithm works as follows. Items are initially sorted by non-increasing relative value, so $\frac{\pi_i}{w_i h_i}$. The first item is then placed on the bottom left corner of the front side. Now, items are placed iteratively by selecting the best valid position. These positions are generated based on the current packing and the shape of the current item. Valid positions in the sense of this heuristic are all corner points of already placed items. These points are copied to every side of the bin, no matter on which side the actual item is positioned. The points are reduced to those, which allow placing the current item without violating any constraint. We consider the point maximizing the distance between item and borders of the bin the best point. Any other selection-

z^{2DPP}	$=$	$\min 1 - \sum_{i \in I} \pi_i a_i$		(I)
s.t.:				
$f_{ij} + f_{ji} + l_{ij} + l_{ji} + b_{ij} + b_{ji} + (1-a_i) + (1-a_j)$	$\geq$	1	$\forall i, j \in I : i < j$	(II)
$l_{ij} + l_{ji} + b_{ij} + b_{ji} + (1-a_i) + (1-a_j)$	$\geq$	1	$\forall i, j \in I : i < j \wedge t_i \neq t_j$	(III)
$x_i - x_j + W l_{ij}$	$\leq$	$W - w_i$	$\forall i, j \in I$	(IV)
$y_i - y_j + H b_{ij}$	$\leq$	$H - h_i$	$\forall i, j \in I$	(V)
$s_i - s_j + S f_{ij}$	$\leq$	$S - 1$	$\forall i, j \in I$	(VI)
x_i	$\leq$	$W - w_i$	$\forall i \in I$	(VII)
y_i	$\leq$	$H - h_i$	$\forall i \in I$	(VIII)
s_i	$\leq$	$D - 1$	$\forall i \in I$	(IX)
x_i, y_i, s_i	$\geq$	0	$\forall i \in I$	(X)
l_{ij}, b_{ij}, f_{ij}	binary		$\forall i, j \in I$	(XI)
a_i	binary		$\forall i \in I$	(XII)

rule can be used as well, but trying to place items of same type behind of each other as soon as possible leads to good results.

We use this heuristic in every subiteration and only if it fails to find a promising packing, will we relax the knapsack constraint and apply an exact algorithm for 2DPP-SC. This combination of the two approaches proved to be very efficient. But the exact method is still computationally expensive, due to the geometrical structure of the problem.

5 Experimental Results

For a numerical study, we use test instances described in [6]. Originally, items were member of one of four groups ('wide', 'tall', 'large', small'), defining a range for width and height of the individual item. Groups were extended with subclasses ('A', 'B') determining types of items. The first subclass assigns types independently of an item's dimensions. In instances of subclass 'B', item types are predetermined by dimensions.

We defined bins to be of size $W = 10$, $H = 10$ with sides $S = 2$ and different item types $|T| = 2$. For this setup, the algorithm was able to solve 179 of 400 instances to proven IP optimality, although its goal is to solve the LP-relaxation only. Initial lower bounds were improved in 89 cases, and initial solutions in 77 cases, so a significant number of instances—especially but not only small ones—could be solved to optimality by BFDH-SC. The absolute gap could be reduced to 0.78 (initially 1.2) bins, which corresponds to a relative gap of 6.8% over all instances. Results prove, the used algorithm is able to generate many columns in short time. Its biggest weakness is to meet the termination criterion, so to prove that no new negative reduced-cost column exists.

6 Conclusion

This thesis is the first work to introduce a new class of bin packing problems side constraint. One of its main contributions is to formally describe this new problem class. We proved such problems to be NP-hard and that the usage of a compact MIP-formulation is practically unsolvable. We applied several lower bounds of related bin packing problems and showed in a computational study that these bounds are quite tight. Additionally, we developed a new best-fit algorithm for our new problem to obtain fast and good solutions heuristically.

The development of a column generation procedure and extensive computational experiments on a set of extended, standard benchmark instances is another main contribution of this thesis. First, we decomposed the problem according to Danzig-Wolfe. Due to the enormous complexity, we further decomposed the subproblem process and introduced an additional knapsack problem, as well as a heuristic packing algorithm.

References

1. Berkey, J. O., & Wang, P. Y. (1987). Two-dimensional finite bin-packing algorithms. *The Journal of the Operational Research Society, 38*(5), 423.
2. Lodi, A., Martello, S., & Vigo, D. (2002). Recent advances on two-dimensional bin packing problems. *Discrete Applied Mathematics, 123*(1–3), 379–396.
3. Lodi, A., Martello, S., & Monaci, M. (2002a). Two-dimensional packing problems. A survey. *European Journal of Operational Research, 141*(2), 241–252.
4. Lübbecke, M. E., & Desrosiers, J. (2005). Selected topics in column generation. *Operations Research, 53*(6), 1007–1023.
5. Martello, S., Pisinger, D., & Vigo, D. (2000). The three-dimensional bin packing problem. *Operations Research, 48*(2), 256–267.
6. Martello, S., & Vigo, D. (1998). Exact solution of the two-dimensional finite bin packing problem. *Management Science, 44*(3), 388–399.
7. Pisinger, D., & Sigurd, M. (2007). Using decomposition techniques and constraint programming for solving the two-dimensional bin-packing problem. *INFORMS Journal on Computing, 19*(1), 36–51.

Part II
Business Analytics, Artificial Intelligence and Forecasting

A First Derivative Potts Model for Segmentation and Denoising Using ILP

Ruobing Shen, Gerhard Reinelt and Stephane Canu

1 Introduction

Segmentation is a fundamental task for extracting semantically meaningful regions from an image. In this paper we consider the problem of partitioning a given image into an unknown number of segments, i.e., we assume that no prototypical features about the image are available, it is a so-called *unsupervised image segmentation problem*. In a general setting this problem is NP-hard. Exact optimization models such as the *multicut problem* [1, 2] are based on *integer linear programming* (ILP) and solved using branch-and-cut methods.

Another aspect of image processing is *denoising*. Main tools for denoising are the variational methods like the approach with Potts priors which was designed to preserve sharp discontinuities (edges) in images while removing noises. Given n signals, denote their intensities $y = (y_1, y_2, \ldots, y_n)$ (e.g. grey scale or color values) and define $w = (w_1, w_2, \ldots, w_n)$ as the vector of denoised values. The classical (discrete) Potts model (named after R. Potts [3]) has the form

$$\min_{w} \ \|w - y\|_k + \lambda \|\nabla^1 w\|_0, \tag{1}$$

where the first part measures the ℓ_k norm difference between w and y, and the second part measures the number of oscillations in w. Recall that the *discrete first derivative* $\nabla^1 x$ of a vector $x \in \mathbb{R}^n$ is defined as the $n-1$ dimensional vector $(x_2 - x_1, x_3 - x_2, \ldots, x_n - x_{n-1})$ and the ℓ_0 norm of a vector gives its number of nonzero entries. The scalar λ is a parameter for regularization. Recently, various modifications and improvements have been made for the Potts model, see [4] for an overview.

R. Shen (✉) · G. Reinelt
Institute of Computer Science, Heidelberg University, 69120 Heidelberg, Germany
e-mail: ruobing.shen@informatik.uni-heidelberg.de

S. Canu
LITIS, Normandie University, INSA Rouen, 76800 Rouen, France

In general, solving the discrete Potts model (1) is also NP-hard. In [5] local greedy methods are used to solve it. Recently, [6] uses an ILP formulation to deal with the ℓ_0 norm for a similar problem in statistics called the *best subset selection problem*.

Motivated by the above mentioned two models, we are interested in simultaneously segmentation and denoising. We assume that the input is an image given as grey scale values (RGB images can be easily transformed) for pixels located on an $m \times n$ grid. Let $V = \{p_{1,1}, \ldots, p_{m,n}\}$ denote this set of pixels. For representing relations between neighboring pixels, we define the corresponding grid graph $G = (V, E)$ where E contains edges between pixels which are horizontally or vertically adjacent. A general segmentation is a partition of V into sets $\{V_1, V_2, \ldots, V_k\}$ such that $\cup_{i=1}^{k} V_i = V$, and $V_i \cap V_j = \emptyset$, $i \neq j$. So in graph-theoretical terms the image segmentation problem corresponds to a graph partitioning problem.

The paper is organized as follows. In Sect. 2 we introduce our *mixed integer programming* (MIP) formulation of problem (1) for the 1D signal case. We then review the multicut problem in Sect. 3. Section 4 presents our main ILP formulation for 2D images and introduces two types of redundant constraints. Computational experiments of 3 instances are presented in Sect. 5. Finally, we conclude and point to future work in Sect. 6.

2 The First Derivative Potts Model: 1D

Given n signals $p = (p_1, \ldots, p_n)$ in some interval $D \subseteq \mathbb{R}$ with intensities $y = (y_1, \ldots, y_n)$. We call a function f piecewise constant over D if there is a partition of D into subintervals $D_1, \ldots, D_k$ such that $D = \cup_{i=1}^{k} D_i$, where $D_i \cap D_j = \emptyset$, and f is constant when restricted to D_i. Throughout the paper, we assume the input images or signals contain noises. The task of segmentation and denoising then becomes piecewise constant fitting. The fitting value for signal p_i is denoted $w_i = f(p_i)$.

In 1D, the associated graph $G(V, E)$ is simply a chain, where $V = \{p_i \mid i \in [n]\}$ and $E = \{e_i = (p_i, p_{i+1}) \mid i \in [n-1]\}$. Here, $[n]$ denotes the discrete set $\{1, 2, \ldots, n\}$. We propose to formulate problem (1) as an MIP by introducing $n-1$ binary variables x_{e_i}, where $x_{e_i} = 1$ if and only if the end nodes of e_i are in different segments. If so, the edge is called *active*, otherwise it is *dormant*. Since w is restricted to be constant within the same segment, it follows that $w_{i+1} - w_i \neq 0$ if and only if p_i and p_{i+1} are on the boundary (i.e., $x_{e_i} = 1$). Thus the signals between two active edges define one segment and the number of segments is $\sum_{i=1}^{n-1} x_{e_i} + 1$. See the left part of Fig. 1 for an example, where there are two active edges and three segments.

An MIP formulation for (1) is

$$\min \sum_{i=1}^{n} |w_i - y_i| + \lambda \sum_{i=1}^{n-1} x_{e_i} \qquad (2)$$

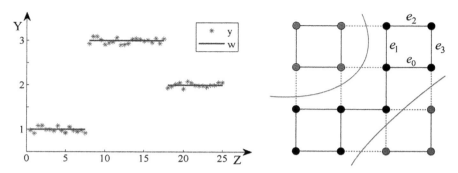

Fig. 1 Left: 1D-fitting, 3 segments and 2 active edges. Right: multicut in a 4 × 4-grid

$$|w_{i+1} - w_i| \leq M x_{e_i}, \qquad i \in [n-1], \qquad (2a)$$
$$w_n \in \mathbb{R}, \qquad i \in [n], \qquad (2b)$$
$$x_{e_i} \in \{0, 1\}, \qquad i \in [n-1], \qquad (2c)$$

where λ is the penalty parameter for the number of segments to prevent over-fitting, and M is usually called the "big M" constant in MIP to ensure that the constraints (2a) are always valid. It enforces that the pixels corresponding to the end nodes of a dormant edge ($x_e = 0$) have the same fitting value. Note that we use the ℓ_1 norm because it can be easily modeled with linear constraints. Namely, constraint (2a) is replaced by the two constraints $w_{i+1} - w_i \leq M x_{e_i}$ and $-w_{i+1} + w_i \leq M x_{e_i}$, and the term $|w_i - y_i|$ is replaced by $\varepsilon_i^+ + \varepsilon_i^-$ where $w_i - y_i = \varepsilon_i^+ - \varepsilon_i^-$ and $\varepsilon_i^+, \varepsilon_i^- \geq 0$. Moreover, it is more robust to noise than ℓ_2.

The solution of (2) gives the fitting value w_i for the signal p_i and the boundaries of two segments are given by the active edges ($x_e = 1$). From now on, for simplicity, we will just specify models in form (2).

3 The Multicut Problem

The multicut problem [1] formulates the graph partitioning problem as an *edge labeling* problem. For a partition $\mathcal{V} = \{V_1, V_2, \ldots, V_k\}$ of V, the edge set $\delta(V_1, V_2, \ldots, V_k) = \{uv \in E \mid \exists i \neq j \text{ with } u \in V_i \text{ and } v \in V_j\}$ is called the *multicut* induced by $\mathcal{V}$. We introduce binary edge variables x_e and represent the multicut by a set of active edges.

With the edge weight $c : E \to \mathbb{R}$ representing the absolute differences between two pixels' intensities, the multicut problem [2] can be formulated as the following ILP

$$\min \sum_{e \in E} -c_e x_e + \sum_{e \in E} \lambda x_e \qquad (3)$$

$$\sum_{e \in C \setminus \{e'\}} x_e \geq x_{e'}, \quad \forall \text{ cycles } C \subseteq E, e' \in C, \tag{3a}$$

$$x_e \in \{0, 1\}, \quad \forall e \in E, \tag{3b}$$

Constraints (3a) are called the *multicut constraints* and they enforce the consecutiveness of the active edges, and in turn the connectedness of each segment. Thus each maximal set of vertices induced only by dormant edges corresponds to a segment. The right part of Fig. 1 shows a partition of a 4×4-grid graph into 3 segments where the dashed active edges form the multicut.

Problem (3) is NP-hard in general and while the number of inequalities (3a) can be exponentially large, violated constraints can be found efficiently using shortest path algorithms. They are added iteratively until the solution is feasible [1, 2].

4 The First Derivative Potts Model in 2D

The main formulation is modeled as a discrete first derivative Potts model and is obtained by formulating (2) per row and column. We also talk about adding redundant constraints to speed up computation in Sect. 4.2.

4.1 Main Formulation

Given a 2D image, following notation from Sect. 2, we further divide $E = E^r \cup E^c$ into its row (horizontal) edge set E^r and column (vertical) edge set E^c. Denote $e^r_{i,j} \in E^r$ as the row edge $(p_{i,j}, p_{i,j+1})$ and $e^c_{i,j} \in E^c$ as the column edge $(p_{i,j}, p_{i+1,j})$. Our main formulation is

$$\min \sum_{i=1}^{m} \sum_{j=1}^{n} |w_{i,j} - y_{i,j}| + \lambda \sum_{e \in E} x_e \tag{4}$$

$$|w_{i,j+1} - w_{i,j}| \leq M x_{e^r_{ij}}, \quad i \in [m], \ j \in [n-1], \tag{4a}$$

$$|w_{i+1,j} - w_{i,j}| \leq M x_{e^c_{ij}}, \quad j \in [n], \ i \in [m-1], \tag{4b}$$

$$w_{i,j} \in \mathbb{R}, \quad i \in [m], \ j \in [n], \tag{4c}$$

$$x_e \in \{0, 1\}, \quad e \in E. \tag{4d}$$

4.2 Redundant Constraints

It is common practice to add redundant constraints to an MIP for computational efficiency. A constraint is redundant if it is not necessarily needed for a formulation to be valid. However, they may be useful because they forbid some fractional solutions

A First Derivative Potts Model for Segmentation and Denoising Using ILP

during the branch-and-bound approach, where the MIP solver iteratively solves the linear programming (LP) relaxation, or because they impose a structure that help shrink the search space.

It is well known that if a cycle $C \in G$ is chordless, then the corresponding constraint (3a) is facet-defining for the multicut polytope [2], thus providing the tightest LP relaxation. Inspired by the multicut problem (3), in a grid graph, although the number of such constraints is still exponential, it may be advantageous to add the following 4-edge chordless cycle constraints

$$\sum_{e \in C \setminus \{e'\}} x_e \geq x_{e'}, \quad \forall \text{ cycles } C \subseteq E, |C| = 4, e' \in C \tag{5}$$

to (4). Meanwhile, if the user has some prior knowledge or good guesses on the number of active edges, it might be beneficial to add the following cardinality constraints

$$\sum x_e \leq k \tag{6}$$

per row and column in a given image.

We show detailed experiments in Sect. 5 on how adding the above two types of constraints affects the computation.

5 Computational Experiments

Computational tests are performed using Cplex 12.6.1, a standard MIP solver, on a Intel i5-4570 quad-core desktop with 16GB RAM. We compare three different models where Model 1 is the multicut problem (3), Model 2 refers to our main formulation (4), with and without the 4-edge cycle constraints (5), and Model 3 is the main formulation (4) with both (5) and (6). We take two images from [7], and resize them to 40×40 and 41×58. We add Gaussian and salt and pepper noise, and set a time limit of 100 s.

Parameter setting. We first compute the average intensity of each 4×4 pixels block in the image, and then calculate the absolute difference of its maximum and minimum value, denoted Y^*. So Y^* somehow represents the global contrast of the image. We set the constant M to Y^*, and λ to $\frac{1}{4}\sigma_1 Y^*$, where σ_1 is a user defined parameter. When there exists an extreme outlier, model (4) tends not to treat the single outlier as a separate segment, since doing so would incur a penalty of 4λ. Denote $Y_i^r = (y_{i,1}, \ldots, y_{i,n})$, the constant k_i^r in (6) of row i is set to the number of elements in $\nabla^1 Y_i^r$ that are greater than $\sigma_2 Y^*$, where $0 < \sigma_2 < 1$ is some suitably chosen parameter. Constant k_j^c is computed similarly for each column j.

Figure 2 shows the input images, detailed setting of the parameters, and the segmentation results for the 3 models.

With and without (5). We first report that Model 2 with (5) saves 0.9 s in the second instance, and narrows 23.5% of the Cplex optimality gap on average, compared to without (5). In later comparison, we denote Model 2 as the formulation (4) with (5).

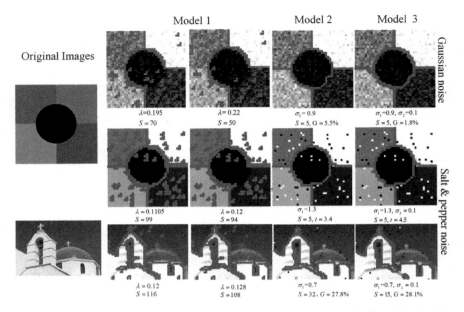

Fig. 2 Segmentation results of three models. S: number of segments. t: running time. G: optimality gap when it hits the time limit of 100 s

Running time and optimality gap. Model 1 is very fast to solve, takes less than 0.1 s in all three instances. Model 2 and 3 take 2.4 and 4.5 s in the second instance and hit the time limit in the other two. The optimality gap for Model 2 and 3 on the first and the third instance, are 5.5, 27.8, 1.8 and 28.1% respectively.

Model 2 versus 3. We keep σ_1 the same when comparing the effects of adding (6). There is no clear advantage of adding the cardinality constraints (6), since for example, it enlarges the optimality gap in the third instance while the solution is visually better.

As we can see from Fig. 2, Model 1 is sensitive to noise and the parameter λ. As a result, it is over partitioned, and is hard to control the desired number of segments. On the other hand, although requiring more computational time, Model 2 and 3 are robust to noise, less sensitive to parameters, and give better segmentation results. In addition, we found it beneficial to add the 4-edge cycle constraints (5), while there is no clear conclusion on whether to add the cardinality constraints (6) to (4).

6 Conclusions and Future Work

We present an ILP formulation of a discrete first derivative Potts model with ℓ_1 data term for simultaneously segmenting and denoising. The model is quite general, firstly, it can use any heuristic method like [5] as an initial solution and provide a guarantee (lower bound) by solving an LP. Secondly, it could improve the initial solution by finding a better solution within the branch-and-bound framework using any MIP solver.

Decomposition algorithms such as superpixel lattice algorithms [8] could be used as preprocessing towards larger images. We will also explore the possibilities of applying our model to 3D images. Finally, since the underlying problem is piecewise constant fitting, applications beyond the scope of computer vision are also of interest.

References

1. Andres, B., Kappes, J. H., Beier, T., Köthe, U. & Hamprecht, F. A. (2011). Probabilistic image segmentation with closedness constraints. In *International Conference on Computer Vision* (pp. 2611–2618).
2. Kappes, J. H., Speth, M., Andres, B., Reinelt, G. & Schnörr, C. (2011). Globally optimal image partitioning by multicuts. In *International Workshop on Energy Minimization Methods in Computer Vision and Pattern Recognition* (pp. 31–44).
3. Potts, R. B., & Domb, C. (1952). Some generalized order-disorder transformations. *Proceedings of the Cambridge Philosophical Society, 48*, 106–109.
4. Chan, T., Esedoglu, S., Park, F. & Yip, A. (2006). Total variation image restoration: Overview and recent developments. In N. Paragios, Y. Chen, & O. Faugeras (Eds.), *Handbook of Mathematical Models in Computer Vision* (pp. 17–31). Springer.
5. Nguyen, R. M. H. & Brown, M. S. (2015). Fast and effective l0 gradient minimization by region fusion. In *International Conference on Computer Vision* (pp. 208–216).
6. Bertsimas, D., King, A., & Mazumder, R. (2016). Best subset selection via a modern optimization lens. *Annals of Statistics, 44*(2), 813–852.
7. Storath, M., & Weinmann, A. (2014). Fast partitioning of vector-valued images. *SIAM Journal on Imaging Sciences, 7*, 1826–1852.
8. Moore, A. P., Prince, S. J. D. & J. Warrell (2010). Constructing superpixels using layer constraints. In *International Conference on Computer Vision* (pp. 2117–2124).

Consumer's Sport Preference as a Predictor for His/Her Response to Brand Personality

Friederike Paetz and Regina Semmler-Ludwig

1 Introduction

The increasing competition in saturated markets drives companies to adopt highly sophisticated communication mix strategies. This is the case especially for categories of products that possess similar physical attributes, e.g., sport shoes. For these products, the brand attribute constitutes a key driver affecting consumers' final purchase decision. However, what affects consumers' choice of a specific branded product? Wright [11] has claimed that consumers often tend to choose a product that arouses the most positive brand affect rather than taking certain product attributes into account. Hence, recent literature has focused on central drivers for (positive) brand affects, i.e., constructs that arouse positive brand emotions such as joy or happiness. In this context, brand personality has been identified as a central driver for affective brand loyalty, which encompasses the abovementioned brand-related emotions. Brand personality could be described according to a person's personality and therefore refers to personality characteristics, e.g., trustworthiness and activity, that are associated with the brand. Recent studies have revealed that consumers prefer brands that are aligned with their own personality traits (e.g., Mulyanegara et al. [7]). Because brand personality is created by a company's communication mix, companies may explore the personality traits of their target market segments and attach the corresponding personality traits to their brand. However, focusing on consumers' personality traits for market segmentation (i.e., market evaluation) is challenging since personality traits are not directly observable.

In the context of unobservable segmentation variables/bases, Wedel and Kamakura [10] stated that three of six criteria for successful market segmentation variables, i.e.,

---

F. Paetz (✉) · R. Semmler-Ludwig
Clausthal University of Technology, Clausthal-Zellerfeld, Germany
e-mail: friederike.paetz@tu-clausthal.de

R. Semmler-Ludwig
e-mail: regina.semmler@tu-clausthal.de

stability, accessibility and responsiveness, are not clearly supported in the relevant literature (p. 14), while identifiability, substantiality and actionability are supported. However, linking the unobservable variables underlying personality traits to observable variables may solve these drawbacks. In particular, the criterion accessibility, which refers to the degree to which the target market segment is reachable through a company's communication efforts, is a key variable identified in the research of drivers of affective brand loyalty. The existence of an easily observable variable, which could be linked with consumers' personality traits and may serve as a predictor of consumer response to brand personality, may therefore simplify and reduce a company's communication efforts.

In this contribution, we search for such an easily observable variable. We explore whether a consumer's preference for/choice of a specific sport and, correspondingly, his/her membership in a specific sport cluster, may serve as an appropriate observable variable in this context and, hence, as a predictor of the consumer's response to brand personality, i.e., affective brand loyalty.

In the next section, we review the theoretical bases of consumer and brand personality and discuss relations between these two constructs. In the third section, we use empirical data to explore the personality traits of several sport clusters. Furthermore, we give recommendations on how these results may be used within communication mix strategies. Finally, we conclude our results by explicitly pointing out the appropriateness of a consumer's sport preference as a predictor of the consumer's response to brand personality, i.e., affective brand loyalty.

2 Consumer Personality and Brand Personality

Since the 1980s, there has been a consensus about five (independent) personality traits that determine a person's personality. This Big Five model constitutes of the following five factors: agreeableness, conscientiousness, extraversion, neuroticism and openness to experience (cp. McCrae and John [6]).While agreeableness is linked to facets such as altruism and modesty, conscientiousness is connected with traits such as efficiency and dutifulness. Neuroticism summarizes facets such as vulnerability and impulsiveness, and extraversion is described by traits such as warmth and talkativeness. Openness could be described by fantasy and wide interests. To measure the degree of each of these five factors in a person, several rating-based tests have been developed. The most popular test constitutes the NEO-PI-R test of Costa and McCrae [2], which lays the ground for most subsequent tests. Almost all tests rely on Likert scales, where respondents rate their self-application to different personality statements. For example, "I am efficient" is a facet of consciousness. These facet results are subsequently pooled to yield an overall result for the focal Big Five factor.

In accordance with consumer personality, brand personality could be defined as the "set of human characteristics associated with a brand" (Aaker [1], p. 347). Aaker [1] worked out five factors that determine a brand's personality, i.e., sincerity, excitement,

Consumer's Sport Preference as a Predictor for His/Her Response to Brand Personality

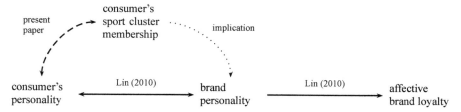

Fig. 1 Relations between consumer personality, sport cluster membership, brand personality and affective brand loyalty

competence, sophistication and ruggedness. These factors could be described by different facets, e.g., sincerity refers to cheerfulness, excitement to imaginativeness, sophistication to glamor and ruggedness to toughness (p. 351). Brand personality is created by company's communication mix, e.g., an advertisement that aligns the brand with certain personality traits. For example, Thomas and Sekar [9] found that the sport brand Nike is associated with ruggedness.

Aaker [1] has found parallels between three of the Big Five factors and brand personality factors: agreeableness and sincerity, extraversion and excitement, and consciousness and competence. Several studies identified relations between consumer and brand personality: Lin [5] found positive relationships between the Big Five factors agreeableness and extraversion and the brand personality trait excitement as well as a positive relation between the Big Five factor agreeableness and the brand personality traits of sincerity and competence. Geuens et al. [3] found positive correlations between consciousness resp. extraversion and competence as well as between agreeableness resp. consciousness and sincerity.

Brand personality has been identified as a central driver for affective brand loyalty. Affective brand loyalty constitutes one part of brand loyalty as it refers to a consumer's preference for a specific brand. Affective brand loyalty could be seen as a predictor of action loyalty, which constitutes another part of brand loyalty and could be measured by actual purchases (Lin [5]). Since consumers prefer brands that are aligned with their own personality traits, a consumer's personality may be used as a predictor for his/her affective brand loyalty (Lin [5]). The bottom part of Fig. 1 illustrates these relations. However, a consumer's personality traits are rarely directly observable. Hence, the search for a more easily observable variable (here: consumer's sport cluster membership), which may finally serve as a predictor for affective brand loyalty, is appropriated to reduce a company's communication efforts. In this contribution, we therefore check whether customers in different sport clusters differ in their personality traits. In this case, a consumer's membership in a specific sport cluster would mirror his/her personality traits (cp. dashed double arrow in Fig. 1), which would subsequently predict his/her affective brand loyalty as a result of brand personality (cp. dotted arrow in Fig. 1).

3 Empirical Study

To answer our research question, we conducted an empirical study at a German university. A total of 153 students of different study courses participated. Most of the respondents were male (67%), and the mean age was 26 years. The respondents were asked to answer two questionnaires. The first questionnaire was related to the respondents' sports activities. The respondents were asked about their favored sports. The second questionnaire encompassed a Big Five self-test adopted from Saum-Aldehoff [8], pp. 190–198. Respondents' self-reports on different facets of the Big Five factors were collected by using rating-scales. The score for each factor ranges from −20 to 20, where a high/low score reflects a high/low level of the focal Big Five factor. To handle the high number of reported sports activities, we followed the advice of Hartmann et al. [4], p. 43, and built seven sport clusters, i.e., sport games (39 members); fitness sports (36 members); endurance-trained athletes such as swimmers, long-distant runners and triathletes (21 members); adventure/nature sports (17 members); sport fighting (14 members); dancing/gymnastics (13 members); and others, such as riders (10 members). Furthermore, a special cluster was built for students who refused to state their preferred sports (three members). To check for personality differences between those clusters, we pooled respondents' individual scores of the Big Five factors for each sport cluster and conducted one-way ANOVAs. Extraversion ($p = 0.077$) and openness to experience ($p = 0.033$) turned out to differ significantly across clusters. While conscientiousness ($p = 0.121$) was on the threshold at a 90% significance, neuroticism ($p = 0.751$) and agreeableness ($p = 0.532$) showed no significant cluster-specific differences.

To gain further insight, we subsequently conducted pairwise t-tests between sport clusters for the Big Five factors that turned out to differ at least weakly significantly. Fig. 2 depicts bar charts, which plot the means and standard deviations for extraversion, conscientiousness and openness to experience in each sport cluster.

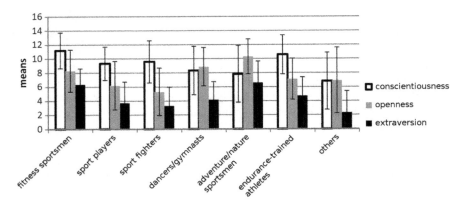

Fig. 2 Means of conscientiousness, openness and extraversion in sport clusters

On average, sport players (3.692), sport fighters (3.286) and others (2.333) rate themselves as weakly extraverted. These sport clusters are (at least by trend) less extraverted than fitness sportsmen (6.333, $p < 0.05$) and adventure/nature sportsmen (6.588, $p \leq 0.122$). Fitness sportsmen are the most conscientious (11.167). Their results differ (at least by trend) from the means of sport players (9.307, $p = 0.105$), dancers/gymnasts (8.307, $p = 0.121$) and significantly from adventure/nature sportsmen (7.824, $p = 0.073$) and others (6.800, $p = 0.042$). Adventure/nature sportsmen exhibit the highest mean (10.294) of openness to experience. They score significantly higher than sport players (6.180, $p = 0.031$), sport fighters (5.286, $p = 0.024$) and endurance-trained athletes (7.095, $p = 0.081$). Obviously, differences in personality traits across sport clusters exist. In accordance with Fig. 1, a consumer's membership in a specific sport cluster mirrors the consumer's personality traits. Hence, the consumer's sport cluster membership predicts affective brand loyalty resulting from brand personality. From a practical point of view, companies may use this result to simplify their communication strategies for creating brand personality. To harmonize brand personality and the personality of the target market and thereby achieve affective brand loyalty of consumers, companies may infer a consumer's personality from his or her sport cluster membership. For example, if a company wants to sell hiking boots, it could use information on highly extraverted and open adventure/nature sportsmen and attach the corresponding personality trait to its brand. Hence, rather than exploring the personality traits of their target market by cost-intensive market surveys, companies may simply focus on the (easily inferable/observable) product-corresponding sport cluster to draw conclusions for a harmonious brand personality. This approach saves communication costs and therefore contributes to the company's profit.

4 Conclusions

This study aimed to explore whether a consumer's preference for a specific sport - measured by the consumer's membership in a specific sport cluster - predicts the consumer's affective brand loyalty as a result of brand personality. To answer this research question, we used empirical data on personality traits - measured by the popular Big Five approach - and sport cluster memberships of respondents. Using one-way ANOVA and pairwise t-tests, we found evidence that different sport clusters differ significantly with respect to certain personality traits. Since the recent literature has identified a consumer's personality traits as predictors for his/her response to brand personality, our results imply that a consumer's sport cluster membership is a key predictor for his/her affective brand loyalty. From a practical point of view, consumers' sport cluster membership - compared to their personality traits - is easily observable. Hence, companies may use our results to simplify and reduce the communication efforts associated with creating brand personality.

References

1. Aaker, J. (1997). Dimensions of brand personality. *Journal of Marketing Research, 34*(3), 347–356.
2. Costa, P. T., & McCrae, R. R. (1992). *Revised NEO Personality Inventory (NEO-PI-R) and NEO Five-Factor Inventory (NEO-FFI) professional manual*. Odessa, FL: Psychological Assessment Ressources.
3. Geuens, M., Weijters, B., & Wulf, K. D. (2009). A new measure of brand personality. *International Journal of Research in Marketing, 26*(2), 97–107.
4. Hartmann, C., Minow, H. J., & Senf, G. (2011). *Sport verstehen - Sport erleben*. Berlin: Lehmanns.
5. Lin, L.-Y. (2010). The relationship of consumer personality trait, brand personality and brand loyalty: An empirical study of toys and video games buyers. *Journal of Product & Brand Management, 19*(1), 4–17.
6. McCrae, R. R., & John, P. O. (1992). An introduction to the five-factor model and its applications. *Journal of Personality, 60*, 175–215.
7. Mulyanegara, R., Tsarenko, Y., & Anderson, A. (2009). The big five and brand personality: Investigating the impact of consumer personality on preferences towards particular brand personality. *Journal of Brand Management, 16*(4), 234–247.
8. Saum-Aldehoff, T. (2012). *Big Five: Sich selbst und andere erkennen*. Ostfildern: Patmos.
9. Thomas, B. J., & Sekar, P. C. (2008). Measurement and validity of Jennifer Aaker's brand personality scale for Colgate brand. *VIKALPA, 33*(3), 49–61.
10. Wedel, M., & Kamakura, W. (2000). *Market Segmentation*. Norwell (MA): Kluwer Academic Publishers.
11. Wright, P. L. (1975). Consumer choice strategies: simplifying vs optimising. *Journal of Marketing Research, 12*, 60–67.

Detecting Changes in Statistics of Road Accidents to Enhance Road Safety

Katherina Meißner, Cornelius Rüther and Klaus Ambrosi

1 Introduction

When trying to detect changes in car accident statistics, police analysts are faced with a large amount of incidents on the one hand and many attributes with several attribute values leading to multitudinous possible combinations on the other. Of course, not all combinations and changes therein are essential to decide upon police actions to enhance road safety. But defining potentially interesting combinations to track in advance can lead to a narrow perspective on the actual situation. If there was an increase in the frequency of a particular combination, which had not manually been predefined, this increase would remain unrecognized and therefore untreated by the police for some periods. Tracking changes manually in these numerous operating figures is not possible.

We propose an automated approach based on Frequent Itemset Mining to detect significant changes in the statistical figures. It is based on the known Apriori algorithm which we apply to monthly slices of accident data to retain a sequence of monthly support values for each itemset. With these sequences, we try to classify the itemsets according to their appearance frequencies in each month. One major question to be answered within our framework is how to find changing itemsets that are worth being presented to the police analyst.

This paper is organized as follows: In Sect. 2, the related work is reviewed. The data set and the preparations made are then introduced in Sect. 3 before the algorithm and its parameters are presented in Sect. 4. With the frequent itemsets found for each month, we show how to detect changes in these data structures in Sect. 5. Finally, we discuss our conclusions and suggest future work in Sect. 6.

K. Meißner (✉) · C. Rüther · K. Ambrosi
Department of Economics and Information Systems, University of Hildesheim,
Universitätsplatz 1, 31141 Hildesheim, Germany
e-mail: meissner@bwl.uni-hildesheim.de

2 Related Work

There are some interesting approaches in change mining on the one hand and in association rule mining on accident data on the other hand. Song et al. [9] and Chen et al. [3] showed how to mine changes in customer behavior. They both focused on pattern mining in a marketing application. In a more general approach, Liu et al. [6] presented a method to distinguish between stable and trend rules. This approach is used for change detection in our framework.

Böttcher et al. [2] built a framework for change mining and defined the term itself. Baron et al. [1] divided the data mining process in two parts, mining the model and mining the changes, to speed up the mining process on evolving data.

As to the analysis of road accident data, Geurts et al. [4] made use of association rule mining to evaluate accident causes in so-called black spots, i.e. places where accidents regularly happen, and in contrast, the causes of accidents in other places in Belgium. Based on accident data from Florida, Pande et al. [8] conducted a market basket analysis to find associations between the accidents' characteristics. In 2014, Moradkhani et al. [7] did the same for UK-accident data, which is the data used for this research. The main focus of all of these approaches lies on finding the root causes for accidents. None of the above evaluated the change in accident statistics.

3 Data Preparation

The GB-accident data is openly available for the years 2005–2015. All accidents with personal injuries are provided with statistical information. The data set consists of 1.8 million accidents with 3.5 million vehicles and 2.6 million casualties, both having a one-to-many-relationship to accidents which has to be dissolved during data preparation. Some of the 55 variables utilized are e.g. date and time of the accident, weather conditions, type of vehicle, and age of casualty.

We decided to focus on the years 2014 and 2015 to build the analytics framework. With no other filters applied, we have a data set D consisting of 285,000 transactions with more than 300,000 different items (attribute-value-pair) after applying the following reduction methods. A single accident consists of 19–85 different items. Attributes containing location information were removed, as they were too detailed to find relevant frequencies. Attribute values like 'data missing' or 'none' were also not considered in order to prevent the mining algorithm from evaluating these uninteresting items. Moreover, attribute values with an occurrence level above 95% were pruned in advance.

Most of the attributes are provided as categorical data. The ones that are not, for example 'hour of accident' or 'age of vehicle', had to be discretized first. This is done automatically by building clusters with equal frequencies.

To analyze changes over time, D is finally separated into monthly data sets D_i and transformed to transactional data in order to find frequent itemsets.

4 Finding Frequent Itemsets

The Apriori algorithm for detecting frequent itemsets is performed on each D_i $\forall i = 1, \ldots, m$, resulting in $m = 24$ different sets of itemsets I_i. An itemset is considered frequent in our framework if at least 3% of the monthly data D_i support, i.e. contain, its combination of attribute values ($minsupp = 0.03$). For months with about 12,000 accidents, the algorithm detects about 55 million frequent itemsets. Because this amount is too high to find any interesting changing patterns, I_i must be condensed to a representative level. Therefore, the lossless representation of *closed frequent itemsets* is chosen. By removing supersets of itemsets with exactly the same support as the itemset itself, the amount is drastically reduced. The resulting itemsets are more general and therefore more applicable in practice than the pruned superset.

Xiong et al.'s [10] approach of a hyperclique pattern miner is used to only keep potentially interesting patterns, even when using a quite low *minsupp*. By applying this approach, all itemsets in I_i with an all-confidence value below 15% (*minAconf*) are pruned, as the items within these sets tend have a poor correlation. All-confidence is defined as all-confidence$(X) = \frac{\text{supp}(X)}{\max_{x \in X}\{\text{supp}(x)\}}$, where $\max_{x \in X}$ is the maximum support of all items x within itemset X. For association rule induction this would imply that all rules derived from this itemset X have a minimum confidence of all-confidence at least.

Association rules were not considered in the final framework. The dependence of items which the rules seemed to illustrate was contradictory, since many rules with similar confidence were found having the shape $A \Rightarrow B$ and $B \Rightarrow A$. Sorting them by confidence and removing the duplicates led to difficulties when joining the rules of two different intervals, because it could not be ensured that from one itemset the same rule was kept for all months. Hence, changes within the rules support or confidence could hardly have been detected that way.

The thresholds for *minsupp* and *minAconf* are determined by trading off the huge amount of itemsets returned with low parameter values and the possible interesting itemsets being pruned when using values that are too high. The data structure with an immense amount of items but only a relatively small number of transactions per month is optimal for a depth-first search algorithm like Eclat. Surprisingly, experiments on the data sets with different parameter combinations showed that Eclat was significantly slower than Apriori in finding closed frequent itemsets while the task of finding all frequent itemsets was performed faster.

5 Detecting Changes

The preparation for the change detection process is conducted in accordance with Liu et al.'s [6] approach. The itemsets I_i found for each month i are joined to one set of itemsets I to obtain support sequences of length m for each itemset. Missing support values for parts of the sequence, which occur when the respective itemset is infrequent

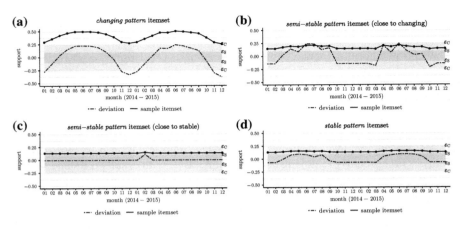

Fig. 1 Monthly progression of changing, semi-stable and stable itemsets

in one month's accident data D_i or has an all-confidence value below the threshold, are filled with the itemset's support generated on the entire transactional data D (*overall support*). This way, unintended breaks in the sequential data are avoided. Itemsets with an *overall support* or an overall all-confidence below the thresholds defined in Sect. 4 are then pruned to further reduce the amount of itemsets.

Change detection for each itemset $X_j \in I \; \forall \, j = 1, \ldots, |I|$ is initiated by calculating the relative deviation $\mathrm{dev}_i(X_j)$ between the itemset's support $\mathrm{supp}_i(X_j) \; \forall \, i = 1, \ldots, m$ for each month and the corresponding $\mathrm{mean}(X_j)$ of the monthly support values as shown in Eq. (1).

$$\mathrm{dev}_i(X_j) = \frac{\mathrm{supp}_i(X_j) - \mathrm{mean}(X_j)}{\mathrm{mean}(X_j)}, \quad i = 1, \ldots, m \qquad (1)$$

Based on this computation we define two thresholds ε_s and ε_c to classify all itemsets X_j according to their change level unambiguously.

Stable itemsets I_s with $|\mathrm{dev}_i| \leq \varepsilon_s \; \forall \, i = 1, \ldots, m$.
Semi-stable itemsets I_{ss} with $|\mathrm{dev}_i| \leq \varepsilon_c \; \forall \, i = 1, \ldots, m$ and $\exists \, i : |\mathrm{dev}_i| > \varepsilon_s$.
Changing pattern itemsets I_c with $|\mathrm{dev}_i| > \varepsilon_c \; \forall \, i = 1, \ldots, m$.

The thresholds ε_s and ε_c are determined by evaluating the itemset progresses visually using graphs. In particular, we examined the itemsets within these classes that have either a very high or very low sum of deviations, as the probability of misclassification is severe for these itemsets. In Fig. 1, we show some characteristic sequences for itemsets within the classes. Due to the boundary to both other classes, the class of *semi-stable* itemsets has to be evaluated for both thresholds ε_s and ε_c (cf. Fig. 1b, c).

We get similar class sizes for the *changing* and *stable* class with about 70,000 itemsets each, while the *semi-stable* class contains nearly twice the number of itemsets. Since the *changing* itemsets are most important for our purpose, these will be

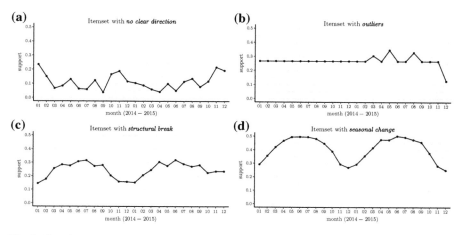

Fig. 2 Sample changing itemsets with particular progresses

presented to a police analyst first, while *stable* and *semi-stable* itemsets do not need to be investigated in the first place. The remaining class size is still too large to monitor all changes. Therefore, we propose to rank the *changing* itemsets by their amplitude of support values.

Based on our purpose to detect changes in accident characteristics, Fig. 2 displays some typical sequences for *changing* itemsets. Itemsets with no clear direction as in Fig. 2a or providing outliers in Fig. 2b are not following any trend and can therefore hardly be predicted. They have to be detected by measuring the support growth between two subsequent months for example, and presented to police analysts for further investigation. Itemsets with seasonal change as in Fig. 2d, are mostly depending on weather conditions and are therefore not surprising, which is why they can be neglected. Itemsets with structural breaks, as Fig. 2c shows, should be detected as fast as possible, since they point out a major change in the underlying data.

6 Conclusion and Future Work

Conclusion We were able to present a basic framework to detect change patterns in road accident statistics. Our assumptions were evaluated using the road safety data set for Great Britain. With a low *minsupp* and a condensed representation of itemsets, we were able to find the most interesting itemsets. We then divided the itemsets in three classes according to their dispersion from the mean support over the whole sequence and ranked the itemsets by their amplitude.

Research Agenda To utilize our framework in police practice the approach requires further research. For instance, the classification of change levels could not only be based on basic thresholds for the deviation from mean but also on growth rates for different time intervals. A time series analysis for each itemset sequence would also be conceivable to detect seasonal changes as well as linear trends. As can be seen in Fig. 1, many itemsets have the same shape of progression of the monthly support. Here, the approach of fundamental rule changes [5] could further reduce the number of itemsets without any information loss. With an approach to cluster these sequences, we could however refrain from using manual thresholds for detecting changing sequences at all.

The geographical aspect has not been considered yet. Taking the accident location into account, e. g. by geographical clustering, will lead to even more useful results for police forces, as they will be enabled to act preventative on local black spots and, even more important, on geographically shifting black spots.

References

1. Baron, S., Spiliopoulou, M. & Günther, O. (2003). Efficient monitoring of patterns in data mining environments. In *7th East-European Conference on Advances in Databases and Informations Systems* (pp. 253–265).
2. Böttcher, M., Höppner, F., & Spiliopoulou, M. (2008). On exploiting the power of time in data mining. *ACM SIGKDD Explorations Newsletter, 10*(2), 3.
3. Chen, M. C., Chiu, A. L., & Chang, H. H. (2005). Mining changes in customer behavior in retail marketing. *Expert Systems with Applications, 28*(4), 773–781.
4. Geurts, K., Thomas, I., & Wets, G. (2005). Understanding spatial concentrations of road accidents using frequent item sets. *Accident Analysis & Prevention, 37*(4), 787–799.
5. Liu, B., Hsu, W.& Ma, Y. (2001). Discovering the set of fundamental rule changes. In *Proceedings of the Seventh ACM SIGKDD International Conference on Knowledge Discovery and Data Mining (KDD-2001)*.
6. Liu, B., Ma, Y. & Lee, R. (2001). Analyzing the interestingness of association rules from the temporal dimension. In *Proceedings 2001 IEEE International Conference on Data Mining* (pp. 377–384).
7. Moradkhani, F., Ebrahimkhani, S., & Sadeghi Begham, B. (2014). Road accident data analysis: a data mining approach. *Indian Journal of Scientific Research, 3*(3), 437–443.
8. Pande, A., & Abdel-Aty, M. (2009). Market basket analysis of crash data from large jurisdictions and its potential as a decision support tool. *Safety Science, 47*(1), 145–154.
9. Song, H. S., Kim, J., & Kim, S. H. (2001). Mining the change of customer behavior in an internet shopping mall. *Expert Systems with Applications, 21*(3), 157–168.
10. Xiong, H., Tan, P. N., & Kumar, V. (2006). Hyperclique pattern discovery. *Data Mining and Knowledge Discovery, 13*(2), 219–242.

A Mixed Integer Linear Program for Election Campaign Optimization Under D'Hondt Rule

Evren Güney

1 Introduction

Prior to elections political parties try to influence voters by investing their resources in marketing campaigns [9]. The voters who do not have a clear decision about their choice, which are called swing voters are the primary targets of these campaigns. We define the Election Campaign Optimization Problem (ECOP) as determining the best way to allocate a political party's resources to maximize the seats or member of parliaments (MPs) won. The effect of advertising to convince voters has been studied extensively in designing the best advertising strategy or determining the marketing mix [11]. Most of the studies focus on computing the magnitude and significance of the effect of spending by using various regression analysis methods. Jacobson [8] is one of the first researches who tests the significance of monetary support for success in elections using regression analysis, which indicate a positive relation between the two and the significance of this relation is higher for the challengers than for the incumbents. Many other studies also focus on the significance of spending money for gaining votes reporting a positive correlation between the two, but some studies claim that the effect is statistically insignificant [4]. Also there is an on going discussion about the higher significance of financial support for challengers rather than the incumbents [2].

From the aspect of optimization several studies focus on different issues of the problem. Fleck [5] presents how voting effects the allocation of governmental resources by an optimization model which maximizes the probability of re-election under a limiting budget constraint. Belenky [1] develops a knapsack model for approximately calculating the minimal fraction of the popular vote needed to elect a US President in the Electoral College. Ostapenko et al. [10] formulates the problem of choosing an optimal strategy for allocating the scarce resources among regions.

E. Güney (✉)
Istanbul Arel University, Tepekent, Kumburgaz, Turkey
e-mail: evrenguney@arel.edu.tr

A game theoretic approach is proposed and the existence of a unique equilibrium is proven.

In this study, the objective is to determine the election regions to focus or to market heavily so that the party wins extra MPs in those regions. To achieve this, we provide mathematical relations to compute the minimum amount of votes needed to pass the opponent to win extra seat(s). We base our studies on the D'Hondt election rule, which is one the most widely used multiple-winners election rule [3]. Next, we develop a mixed integer linear program (MILP) for the ECOP and test it on the Turkish Parliamentary elections data.

The organization of the paper is as follows: In Sect. 2 basic assumptions and the mathematical methodology are presented. In Sect. 3, the details of the ECOP is provided as a MILP. In Sect. 4 experimental analysis and results are presented. The last section concludes the paper.

2 Election Model, D'Hondt Rule and Calculating the Amount of Necessary Votes

D'Hondt rule is one of the most widely used multiple-winners election method which determines the allocation of seats in the countries having a parliamentary and more than 40 countries are actively using it [3]. According to the D'Hondt rule, in an election region which has N seats, the seats are distributed among the parties according to the following steps. First, the votes of each party are divided into N consecutive numbers from 1 to N to determine the so called "quotients". Next, all quotients are ranked in a descending order to construct a sorted list. Finally, N seats are allocated one-by-one to the parties whose quotients are in the top N ranks of the sorted list.

To compute the minimum amount of necessary votes to win an extra seat under the D'Hondt rule, we need to develop certain mathematical relations. First we will focus on the simple case where there are only two competing parties and then we are going to extend our results for the general case. Let $\mathcal{I}$ show the set of parties attending to the election with $|\mathcal{I}| = I$. Assume that according to the poll results the votes of the participating parties are estimated and the vote for party i is v_i, $i \in \mathcal{I}$. Also let v_s be the amount of swing votes according to the polls again. Then the total amount of valid votes is $v_T = v_s + \sum_{i \in I} v_i$.

Assume that there are K seats to be distributed among two political parties with expected votes v_1 and v_2, respectively. Hence, the total votes for this election region is $v_T = v_1 + v_2 + v_s$. Let q_{1k} and q_{2k} shows the set of quotients computed by dividing the votes of each party to the D'Hondt divisors $k = 1, 2, \ldots, K$. So $q_{11} = v_1, q_{12} = v_1/2, \ldots, q_{1K} = v_1/K$. Also let's assume that out of a total of K seats the first party is taking n of them and the second party is taking $K - n$, where $n \geq K - n$. Therefore the nth quotient of first party should be greater than the $(K - n + 1)$-th quotient, otherwise the nth seat would go to the second party. Mathematically, $q_{1n} > q_{2(K-n+1)}$, or equivalently $v_1/n > v_2/K - n + 1$. Let's define $\lambda = (K - n + 1)/n$, showing the ratio between the next integer greater than the number of seats won by the challenger and the number of seats of the incumbent party.

Before the marketing campaign by the challenger party, the swing votes v_s are split among the two parties proportional to their estimated votes. After the split of swing votes, the total votes of parties are $v_1 + v_1/(v_1 + v_2)$ and $v_2 + v_2/(v_1 + v_2)$. Now when the challenger performs the extra marketing activities (by shifting some of the campaign budget that is previously allocated to some other region) and persuades v_x voters, where $v_x < v_s$ then only the remaining swing votes $(v_s - v_x)$ are split between two parties proportional to their initial votes. Given this setting we can state the following proposition. If the amount of swing votes v_s in a region is greater than v_x, then there is definitely an opportunity for the challenger party to win an extra seat. Conversely, for those election regions with $v_s < v_x$, there is no chance of winning an extra seat even if all the swing voters are convinced to choose the challenger party.

Proposition 1 *Given there are a total of K seats and the incumbent winning n seats, than the minimum number of votes v_x necessary for the challenger party to win one extra seat, i.e., $K - n + 1$ seats under the D'Hondt rule is $v_x > v_T \frac{(\lambda v_1 - v_2)}{v_1(1+\lambda)}$.*

In the multiple party case with $I > 2$, the ranked quotients of parties are mixed, so to compute the necessary votes required for an additional seat, one has to determine the first non-winning quotient q^{NW} and its ranking r^{NW} for the challenger party and the last winning quotient just ranked before q^{NW}, i.e. q^{LW} with rank r^{LW} of the corresponding party. Here r^{NW} and r^{LW} show the rank of the seat within the corresponding party's list. Let $\lambda = (r^{NW})/r^{LW}$. Also let v_1 and v_2 show the votes of the last winning party and the challenger (first non-winning) party, respectively. Lastly, let $v_c = v_T - v_1 - v_2 - v_s$, which is the total expected votes of all the remaining parties, before distributing the slack votes.

Proposition 2 *Given there are a total of K seats and I competing parties with a total vote of $v_T = v_s + \sum_{i \in I} v_i$, the minimum number of votes v_x necessary for the challenger party to win an extra seat under the D'Hondt rule is $v_x > v_T \frac{(\lambda v_1 - v_2)}{v_1(1+\lambda)+v_c}$.*

The proofs of the propositions are available in our previous work [6] and due to lack of space they are skipped.

3 A MILP Formulation for ECOP

ECOP seeks for the optimal resource allocation strategy for a single political party and aims to maximize to total number of seats won. Let the unit cost of persuading a swing voter be c, which is constant and equal for any party and region. Determining c is a hard task as it is difficult to materialize the monetary value to persuade an individual by a political party's marketing campaign. However, there are many studies that focus on how to find the unit cost of a vote [2]. Let $j \in J$ represent the set of election regions and $m_j = cv_{xj}$ represents the minimum amount of budget to win one more MP in region j. They are computed by the mathematical formulae presented in the previous section. Similar to m_j, one can compute n_j, the maximum amount of budget a party can cut in region j so that it still gets the same number of seats even when n_j/c votes are lost. Finally, let $s_j = cv_{sj}$, that is the required budget to cover all the swing votes in region j.

We assume that, given the level of swing votes and the voter population of the election regions, all parties allocate their budgets proportional to the amount of swing voters. Namely, if the number of swing votes in region j_1 is twice as much as of region j_2, then the budget allocated to region j_1 is also twice of the budget of region j_2. Another assumption is the budgets of parties are considered to be proportional to their expected votes. This rough assumption is again reasonable, because the government subsidy for political parties is usually proportional to the number of current seats in the parliament and the amount of sponsorship revenues of the parties are again usually parallel to their popularity and thus their estimated votes. Therefore, when the poll results are unleashed, each party allocates its budget among regions proportional to the swing votes of the region. Then the amount of swing votes they win are exactly proportional to their estimated votes. This assumption is also reasonable, because in the public polls most of the time the swing votes are distributed among the parties proportional to their original vote rates and since we assumed the budgets to be proportional to their expected votes, we will get the same ratios. Shortly, if the polls show that party A has twice the votes of party B, then party A will again win twice more swing votes than party B.

Observe that shifting some of the budget from various regions may be beneficial, only if the extra budget can result in winning extra seats. In other words if $m_{j1}/c < n_{j2}/c$, then the party can shift the required budget m_{j1} from region $j2$ to region $j1$ to win one more seat in region $j1$ without losing a seat in region $j2$. By exploiting this strategy for all possible regions a party can try to win as many seats as possible. Under a parliamentary system the number of seats that can be selected from a region is usually more than one. Especially in populous regions a party can aim to win more than one additional seats when the number of swing votes is respectively higher than the average votes necessary for a single seat. Thus a second index k is introduced representing the kth additional seat in a region. Let K_j be the maximum possible number of seats that can be won in region j. K_j are simply the last index k where $\sum_k m_{jk} \leq v_{sj}$ holds, meaning that the amount of swing votes in a region are large enough to win K_j more seats.

Two sets of continuous decision variables u_j and w_j are introduced to represent the amount of budget allocated to or removed from region j, respectively. Also let x_{jk} and y_{jk} be binary variables showing if the party wins or loses the kth seat in region j. Then the following MILP represents the ECOP:

ECOP:

$$\max z = \sum_{j \in \mathcal{J}} \sum_{k \in \mathcal{K}_j} x_{jk} - y_{jk} \tag{1}$$

$$\text{s.t.} \sum_{j \in J} u_j = \sum_{j \in J} w_j \tag{2}$$

$$x_{jk} \leq \frac{u_j}{m_{jk}} \quad j \in J, k \in K_j \tag{3}$$

$$\frac{w_j}{n_{jk}} - 1 \leq My_{jk} \quad j \in J, k \in K_j \tag{4}$$

$$u_j \leq MI_j \quad j \in J \tag{5}$$

$$w_j \leq M(1 - I_j) \quad j \in J \tag{6}$$

$$x_{jk}, y_{jk}, I_j \in \{0, 1\} \quad j \in J, k \in K_j \tag{7}$$

$$0 \leq u_j, 0 \leq w_j \leq s_j^p \quad j \in J \tag{8}$$

The objective function (1) maximizes the total number of seats won. First constraint (2) balances the amount of budget shifted (since c is constant it is equivalent to the amount of votes shifted) among regions. Constraints (3) and (4) are the threshold constraints. In other words, to win the kth seat in region j the budget u_j allocated should be at least m_{jk}. Similarly, a party will only lose a seat if the total amount of budget removed is at least n_{jk}. Constraints (5) and (6) disallow to both invest and save budget at the same region together with the auxiliary binary variables I_j. Lastly, constraints (7) and (8) are bounds and binary requirements on the decision variables. Notice that s_j^p is the estimated amount of swing votes that the party will win with its normal campaigns.

4 Computational Results

The computational results on the performance of our mathematical model and some insights on the effect of certain parameters on the optimal solution are given in this section. The 2015 (June) Turkish Parliamentary Election data is used in our analysis, where 4 major parties compete. There are some other small parties as well as independent candidates but to simplify the computations, they are discarded from the data. There are 85 election regions in Turkey with a total of 550 seats. m_{ik} and n_{ik} values are computed using the propositions developed in Sect. 2 and then they are used as parameters of the MILP. All the analysis are carried out for the second party, but similar results are obtained for the remaining parties as well. We run our tests by varying the following two parameters: swing vote rate percentage ($s = 1\%$ to $s = 20\%$) and unit vote cost ($c = 1$ to $c = 50$). The integer programs are coded in C# environment and CPLEX 12.6 callable library [7] is used.

The results are displayed in Fig. 1. Observe that by shifting budget among regions a significant number of additional seats can be won. Also the effects of swing votes or unit vote cost can be be easily tracked. As the unit costs decrease or swing vote rate increases, the gain increases. Finally, notice the diminishing returns behavior, where the effects of per increase in swing vote rate or decrease in unit cost decreases gradually.

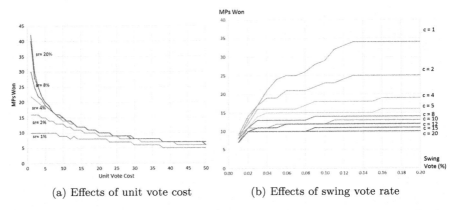

(a) Effects of unit vote cost (b) Effects of swing vote rate

Fig. 1 a Effects of unit vote cost. b Effects of swing vote rate Effects of unit vote cost and swing vote rates

5 Conclusion

In this study the election campaign optimization problem is formulated and solved. Various mathematical relations are proposed for easily computing the necessary amount of votes to identify potential gains according to poll estimates prior to the elections. Computational analysis show that a party can significantly increase its number of seats in the parliament without an increase in the total votes or marketing budget, but rather carefully allocating its budget in the best way. As a future work one can focus on the problem with a game theoretic approach, where the competitors will also benefit from similar optimizations.

References

1. Belenky, A. (2008). A 0–1 knapsack model for evaluating the possible electoral college performance in two-party us presidential elections. *Mathematical and Computer Modelling, 48*, 665–676.
2. Benoit, K., & Marsh, M. (2010). Incumbent and challenger campaign spending effects in proportional electoral systems:the irish elections of 2002. *Political Research Quarterly, 63*(1), 159–173.
3. di Cortona, P., CManzi Pennisi, A., Ricca, F. & Simeone, B. (1999). *Evaluation and Optimization of Electoral Systems*. Philadelphia, USA: SIAM - Society for Industrial and Applied Mathematics.
4. Fisher, J. (1999). Party expenditure and electoral prospects: a national level analysis of britain. *Electoral Studies, 18*, 519–532.
5. Fleck, R. (1999). The value of the vote: A model and test of the effects of turnout on distributive policy. *Economic Inquiry, 37*(4), 609–623.
6. Guney, E. (2017). Efficient election campaign optimization using integer programming. In *Proceedings of 11th International Conference on Industrial Engineering and Industrial Management, Valencia, SPAIN*. https://goo.gl/RqnuMC.

7. ILOG-CPLEX. (2013). Cplex 12.6 User's Manual. ILOG.
8. Jacobson, G. (1978). The effects of campaign spending in congressional elections. *The American Political Science Review*, *72*(2), 469–491.
9. Mueller, D. (2003). *Public Choice III*. UK: Cambridge University Press.
10. Ostapenko, V., Ostapenko, O., Belyaeva, E., & Stupnitskaya, Y. (2012). Mathematical models of the battle between parties for electorate or between companies for markets. *Cybernetics and Systems Analysis*, *48*(6), 814–822.
11. Steenburg, E. (2015). Areas of research in political advertising: A review and research agenda. *International Journal of Advertising*, *34*(2), 195–231.

Long-Term Projections for Commodity Prices—The Crude Oil Price Using Dynamic Bayesian Networks

Thomas Schwarz, Hans-Joachim Lenz and Wilhelm Dominik

1 Introduction

The commodity exploration and production industry has project life cycles of ten, twenty or up to 60 years. The scenario technique is a methodology for allowing such long lead times. It enables making assumptions and statements about the future system state with subject, time and region well defined. A "what-if"-analysis allows generating various future world states of interest.

In contrast to econometric models or time series analyses, scenario technique is able to make predictions on forecast horizons larger than 10 years. This is inevitable as the entire project development and production period (full project life cycle) may last for up to 60 years. This paper proposes a new mathematical approach for future projections of prices for those long horizons using a Dynamic Bayesian Network (DBN). The DBN approach is verified at the crude oil price example.

The advantage of using DBN over econometric models is twofold. First, the forecast horizon may be very long. A comprehensive overview on forecasting specifically the various prices of oil give [1]. However, they do not provide any detail on very long-term forecasts.

Second, a DBN allows analyses of different future states simply by changing assumptions towards for example extreme case scenarios. In order to approach the variability of the input variables, the possible input range may be divided into intervals. Whereas, the uncertainty of future developments is modeled by probabilities for the transition from one state to another. Additionally, in DBN linguistic terms can replace real numbers, which may make it easier to describe possible future states.

T. Schwarz (✉) · W. Dominik
Technische Universität Berlin, Berlin, Germany
e-mail: t.schwarz@campus.tu-berlin.de

H.-J. Lenz
Freie Universität Berlin, Berlin, Germany

2 Crude Oil Price and Oil Field Life Cycle

For the application of the scenario technique in order to project commodity prices, crude oil is chosen for two reasons. First, crude oil is the most important commodity as it is the basis for many industries. It is a global product, which is produced in various regions around the globe. This is combined with a global market for crude oil and its derivates. Furthermore, there are financial products on oil and oil can be shipped to any place around the globe.

Second, the oil market is rather liquid. The trading volume is reasonable as well as the number of players at the market. This holds in comparisons for example to the rare-earth elements.

There are various hydrocarbons and within them various crude oils with different characteristics. The two most important crude oils are Brent from the North Sea and WTI from the US. They are the most traded crude oils and often the price setters for other crude oils and hydrocarbons and their further derivatives.

2.1 Brief Overview of Oil Price History

In history, the price of crude oil is volatile. That means, that the forecast and especially the long-term forecast of the crude oil price is hardly possible using common econometric models, which do need a sufficiently long supporting area. However, having those long supporting area may yield to an inclusion of different price regimes. Analysing the crude oil price for horizons up to a century, which is necessary for forecasting 60 years, shows, that there are multiple changes in the crude oil price regime.

Figure 1 shows the annual crude oil price from 1861 to 2015 [2, p. 14]. The Nominal Values curve shows no significant changes after a small peak in the 1860s until the mid-1970s. That is, there is no change of crude oil prices at all for around 100 years. Beginning in the mid-1970s, the curve shows a higher volatility with a strong increase in prices from 2000 onwards.

In contrast, the 2015 Prices, which are the nominal values deflated using the Consumer Price Index for the US, show a strong volatility already in the first 20 years until 1880. Afterwards, there are again peaks, however, the price always returns to a USD-20-level. Likewise the Nominal Values curve, the 2015 Prices curve shows a stronger volatility with three major peaks in 1980, 2008 and 2011 with a price of more than USD 100.

2.2 Oil Field Life Cycle

From a user perspective being e.g. an oil company, it is necessary to project the future development of crude oil prices. For any development and production project, it is

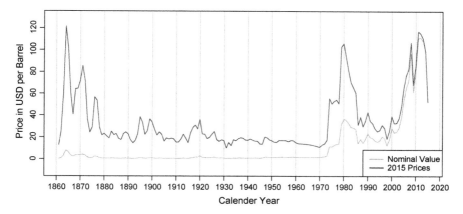

Fig. 1 Annual Crude Oil Price from 1861 to 2015. 2015 Prices: Nominal Values deflated using the Consumer Price Index for the US Data for 1861–1944: US average. Data for 1945–1983: Arabian Light posted at Ras Tanura. Data for 1984–2015: Brent dated. [p. 14 and corresponding data base] [2]

necessary to forecast the costs as well as the revenues. For those development and production projects the most important source for revenues is the price of crude oil sold. Thus, the decision makers need to project the future price of oil for the entire "field life cycle". The field life cycle describes the different phases of an oil field from gaining access to the oil field to exploration, development, production and, finally, decommissioning the oil field.

Figure 2 shows a detailed picture of an oil field life cycle, which is based on [6]. Focussing on the forecast of the price of oil, this life cycle shows only the periods until end of production or the economic cut-off. Also the phases of gaining access and appraisal are not considered here.

In total, there are three main stages. However, the production phase may be divided into two phases. First, the phase of conventional production from oil wells, which is equivalent to natural flow production. This phase may also be known as primary oil recovery. Second, the phase of production from improved oil recovery (IOR) and enhanced oil recovery (EOR). This second phase also comprises the so-called secondary oil recovery, which includes waterflooding and gas injection in order to keep the pressure in the reservoir constant. The EOR is also known as tertiary oil recovery.

The total length of an oil exploration and exploitation project may be up to 60 years.

Following this consideration, there are five main decision points, $t_1, \ldots, t_5$, in an oil field life cycle. At each point, a decision has to be made in order to conduct successfully an exploitation project. However, it happens that during an oil field life one passes the same decision point multiple times. Let t_0 being the date of modelling. At t_0 a forecast on the possible revenues of the project needs to be included in a first feasibility study. Thus, a forecast on the price of crude oil is needed.

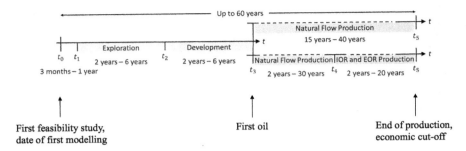

Fig. 2 The oil field life cycle and typical decision points (own illustration)

The length of each interval between two time points differ from phase to phase. Furthermore, each project is unique and, therefore, the length of each period may vary strongly from one project to another.

3 Scenario Technique and Dynamic Bayesian Networks

Scenario technique can be formalized by the tuple (G, I, E) with $G = (V, K, P)$ being the model graph, V being a finite, not empty set of variables (knots) with finite $Range\,(v)$, $v \in V$. K is a finite, not empty set of linked pairs of variables (directed edges) $v, w \in V$. P represents a set of measures of uncertainty on V and is a finite, not empty set of marginal, subjective probability distributions over V. I describes the inference mechanism, which derives in DBN for directed acyclic graphs (DAG) and for fixed $e \in E$ (evidences) the distribution $p_v \in P$ (compare [3, 5]). Under the first-order Markov assumption, P can be decomposed in $\prod p_{v|i}$ ($v \in parents\,(i)$).

Using scenario technique assumes at least two different points in time (today and in the future), multi-causal influence ($v \to (x, y, z)$) with interdependencies or interactions and cause and effect relationships without feedback (compare with [4]). In our application we use up to five points in time. Therefore, we need to define multiple time slices, which consists of V, K and I.

3.1 Dynamic Bayesian Networks

We apply the DBN framework in order to forecast the price of crude oil. The advantage is to model an observable variable by a number of causal unobservable variables. The assumption is, that the values of the unobservable variables may cause the value of the observable variable of interest. The quantification of this "causal" effect lies in the respective conditional probabilities.

Table 1 Unobservable variables of the DBN grouped by topic

Global demand	Global supply	Industrial development
Global consumption	Politics oil prod. countries	Oil technology
Global warming politics	Peak oil	Renew. technology
Substitution and competition	OPEC Cartel	Other technology

Additionally, we simplify the DBN by assuming, that the underlying data generating process follows a Markov-process of order 1. That means, that the current value of each variable only depends on the previous value but not on prior values.

$$P(X_t|X_{0:t-1}) = P(X_t|X_{t-1})$$

with X_t being the set of non-observable variables at time point t.

Another advantage of the DBN is the fact, that the different time points do not need to be equidistant. Furthermore, no two-way interaction is allowed, that is two-way interaction between variables at the same point in time.

3.2 Variables for Modelling Crude Oil Prices

The choice of variables is based on our considerations on the various factors influencing the price of oil. For oil price variables we use value-per-unit variables, whereas all other are stock variables. The variability of the input is determined by a pre-defined range.

The observable variable of interest is the price of *Brent* crude oil. The Brent spot price is publicly available on a daily basis. All other variables are unobservable.

In order to reflect the oil price at a global market, we introduce a variable *Global Oil Price*. This variable drives the *Brent* price in an asymmetric way. This occurs due to the fact, that *Brent* shows one of the highest qualities and, thus, is always at the higher end of possible prices. Additionally, the Brent oil field will be exhausted in near future. *Brent* and *Global Oil Price* are both value-per-unit variables.

The main driver in terms of modelling is the demand-supply-relationship. Every variable chosen affects the demand or supply side. We choose nine variables determining the *Global Oil Price*. Additionally, we introduce a specific variable addressing especially the scarcity of Brent crude oil. *Peak Oil* theory comes into play with the corresponding variable. Though, it describes a specific demand-supply-relationship problem. Substitutional effects will be modelled with the corresponding variable. The following table names all variables, which enter the DBN model as unobservable stock variables (Table 1).

The *Global Consumption* variable describes the global consumption of hydrocarbons based on the status of the global economic cycle. The *Global Warming Politics* variable describes the global political situation with respect to global warming, whereas the variable *Substitution & Competition* describes substitutional effects

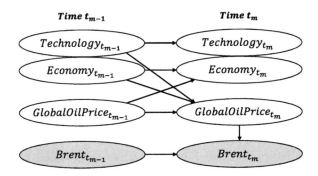

Fig. 3 Example of Dynamic Bayesian Network graph with one time slice t_m

Technology t_{m-1}	
No Development	0.2
Evolution	0.8

Economy t_{m-1}	
Demand High	0.25
Supply High	0.75

		Technology t_{m-1}	
		No Development	Evolution
Technology t_m	No Development	0.4	0.8
	Evolution	0.6	0.2

Global Oil Price t_{m-1}	
Low	0.8
High	0.2

Brent t_{m-1}	
Low	0.8
High	0.2

	Global Oil Price t_{m-1}	Low		High	
	Economy t_{m-1}	Demand High	Supply High	Demand High	Supply High
Economy t_m	Demand High	0.8	0.7	0.3	0.5
	Supply High	0.2	0.3	0.7	0.5

	Economy t_{m-1}	Demand High				Supply High			
	Technology t_{m-1}	No Development		Evolution		No Development		Evolution	
	Global Oil Price t_{m-1}	Low	High	Low	High	Low	High	Low	High
Global Oil Price t_m	Low	0.3	0.1	0.5	0.3	0.8	0.5	0.95	0.6
	High	0.7	0.9	0.5	0.7	0.2	0.5	0.05	0.4

	Brent t_{m-1}	Low		High	
	Global Oil Price t_{m-1}	Low	High	Low	High
Brent t_m	Low	0.85	0.1	0.4	0.05
	High	0.15	0.9	0.6	0.95

Fig. 4 Conditional probabilities tables for the example of Dynamic Bayesian Network

and the competition to other energy resources. Those three variables affect the global demand of crude oil.

The global supply of crude oil is affected by other three effects. The *Politics* of the *Oil Producing Countries* together with the fact whether the *OPEC* acts as a *Cartel* cause the supply side. Furthermore, the variable *Peak Oil* tackles the fact of crude oil being an exhaustive resource.

Finally, the developments or disruptions in the various technologies are modelled by three variables. The developments in the *Oil Technology* influence directly the price of oil. The developments in *Renewable Technologies* as well as in all *Other Technologies* affect the price of oil indirectly and depends also on substitutional effects.

3.3 An Introductory Example

Figure 3 shows an example of a Dynamic Bayesian Network. On the one hand, it illustrates the complexity of the approach. On the other hand, it points out how the model approach can be simplified assuming identically characterized time slices. In this example, there is one observable variable *Brent* and three unobservable variables *Technology*, *Economy* and *Global Oil Price*.

The value of the variable *Brent* at time t_m depends on its own past value at time t_{m-1} and on the value of the variable *Global Oil Price* at time t_m. The *Global Oil Price* at t_m depends on its own past value at time t_{m-1} and on past values (t_{m-1}) of the variables *Technology* and *Economy*. The variable *Technology* depends only on its own past, whereas *Economy* is dependent on its own past as well as the past of the *Global Oil Price*.

Figure 4 shows as an illustration the conditional probability tables for the given example. This example should give a sense on how those conditional probabilities have to be chosen and how large the number increases with an increasing number of variables as well as input range or intervals. For each variable, we choose only two different states "Low" versus "High", or "No Development" versus "Evolution".

References

1. Alquist, R., Kilian, L. & Vigfusson, R. J. (2013). Forecasting the price of oil. In G. Elliott & A. Timmermann (Eds.), *Handbook of Economic Forecasting*. Amsterdam: Elsevier (pp. 427 - 507).
2. BP p.l.c. (2016). BP Statistical Review of World Energy. Retrieved May 16, 2017 from http://www.bp.com/statisticalreview.
3. Lauritzen, S. L., & Spiegelhalter, D. J. (1988). Local computation with probabilities on graphical structures and their applications to expert systems. *Journal of the Royal Statistical Society B*, *50*(2).
4. Müller, R. M., & Lenz, H.-J. (2013). *Business Intelligence*. Heidelberg: Springer.
5. Pearl, J. (1986). Fusion, propagation, and structuring in belief networks. In Artificial Intelligence (vol. 29).
6. Wintershall Holding GmbH. (2017). The "lifecycle" of oil & gas fields. Retrived April 4, 2017 from https://www.wintershall.com/company/about-us/value-chain.html.

Prognosis of EPEX SPOT Electricity Prices Using Artificial Neural Networks

Johannes Hussak, Stefanie Vogl, Ralph Grothmann and Merlind Weber

1 Introduction

In the recent years, two major trends can be observed within the European electricity market: a growing share of renewable energy supply and an increased market interconnection. Simultaneously, a steadily growing trading volume and a high price volatility are observed at the EPEX SPOT day-ahead market [1]. Providing an accurate price forecast creates a strategic as well as an economic advantage, which is important to all participants of the EPEX SPOT day-ahead market. Current forecasting approaches mainly apply linear regression, GARCH, ARMA and ARIMA models or artificial neural networks [2]. Especially artificial neural networks are able to capture real world market processes mathematically and market dynamics can be transferred into a market model [3]. In terms of neural network approaches, three-layer networks, known as multilayer perceptrons (MLPs), are state of the art. However, deep neural networks have proven in many cases, that they can approximate complex dynamics better and with fewer units [4]. In this paper, we propose a market modeling approach applying deep neural networks in order to secure a holistic and robust market model with accurate predictions. The forecasts are given each day at 11:30 am, in order to have sufficient time to place the orders at the EPEX SPOT day-ahead market, where order books close at 12:00 pm. Only data, which are available until 11:30 am are considered in the modeling process. This makes the whole system real-time capable.

J. Hussak (✉) · M. Weber
Technical University Munich, Arcisstraße 21, 80333 Munich, Germany
e-mail: HussakJohannes@gmail.com

J. Hussak · S. Vogl · R. Grothmann
Siemens AG, Corporate Technology, Otto-Hahn-Ring 6, 81739 Munich, Germany
URL: https://www.siemens.com

2 Empiric Market Modeling

2.1 Neural Network Approach

Originally, artificial neural networks were developed to model the biological processes within the human brain. Their characteristic to capture highly non-linear and complex systems makes them well suited for econometric modeling. The internal information processing within an artificial neural network can be interpreted as the mathematical model of the real world decision-making process of a trader at the stock exchange. Here, the trader needs to rate all information according to their relevance, aggregate the weighted information and derive a final decision. In the artificial neuron, the first two steps, weighting and aggregation of relevant information, are mathematically captured by multiplying the numerical input information with a certain weight and summing them up: $\sum_{i=1}^{n} w_i x_i$. In a mathematical model, the decision-making process is described by a step function f. Since step functions are not continuously differentiable, sigmoid functions are applied in the artificial neuron: $f\left(\sum_{i=1}^{n} w_i x_i\right)$. These three steps form the information process within an artificial neuron (see Eq. 1 and Fig. 1 left). An additional threshold w_0 is considered here, which can be used as certain stimulus threshold for the decision making.

Mathematical information process within an artificial neuron:

$$y = f\left(\sum_{i=1}^{n} w_i x_i - w_0\right) \quad (1)$$

Since each artificial neuron can be interpreted as an individual trader, a network of artificial neurons can thus be seen as a whole market model. In the right side of Fig. 1, a three layer neural network with four hidden neurons is displayed. All hidden neurons are connected to each input neuron (numerical input information) and each

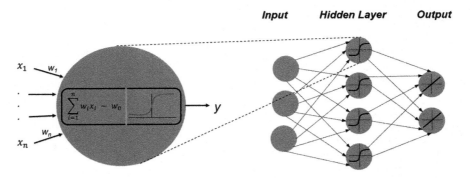

Fig. 1 Left: Information process within an artificial neuron, right: Structure of an artificial neural network

output neuron. To any link, a certain weight is attached, which filters the information. Whereas in reality, the trader weights the available information according to his gut feeling or his experience, in artificial neural networks a mathematical algorithm adjusts the weights such, that the outcome assimilates the target function. This is achieved by applying the backpropagation algorithm, which computes the gradient of the error function with respect to each weight. Afterward, a suitable optimization algorithm searches for the optimal weights.

2.2 Deep Neural Networks

According to [5], a sufficiently large three-layer neural network is actually able to capture any kind of continuous function on a compact domain [5]. However, deep neural networks often show better results using fewer units to approximate complex functions [4]. Therefore, we apply three-layer neural networks as well as deep neural networks. In this context, a neural network model is called "deep", if it consists more than one hidden layer. This can be achieved by simply adding additional hidden layers to the three-layer model. However, this has some major drawbacks. As described in [6], it is not ensured, that the lower layers contribute to the final output at all. Additionally, through a large tower-like construction, relevant input information might get lost on the long forward path, whereas the error signal in the backward path decays, while it propagates through the large number of hidden layers. Therefore, the topology depicted in Fig. 2 is applied within this study. Each hidden layer is separately connected to the input layer. By the use of the shared weight matrix A, each hidden layer will get the same input information. Moreover, all hidden layers are connected with a separate output layer. Thus, on the backward path learning is applied on each single intermediate layer. The information is transferred from

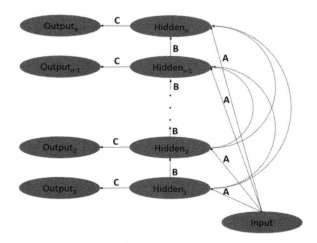

Fig. 2 Topology of the applied deep neural network

one hidden layer to the following through the backbone of the model. Applying additional highway connections, here visualized with dashed lines, the information can also surpass intermediate layers. The error, on the other hand, is not propagated through the backbone.

In the empirical study, the optimal values for the meta-parameters training set, validation set, pattern learning sequence, activation function, training epochs as well as the best topology needs to be determined.

2.3 Quantile Base Bias Correction

Just as per mathematical definition, artificial neural networks are not able to extrapolate. In most cases, it can be observed, that higher values are more likely to be underestimated whereas smaller values are often overestimated. We propose a quantile-based scaling method (QBS) in order to further reduce the prognosis error, especially for the rare events. In the quantile-based scaling process, the cumulative distribution functions (CDFs) are computed and their percentiles compared. The difference between the mean of the model quantile $\overline{x_{model_q}}$ and the corresponding target quantile $\overline{x_{target_q}}$ is computed and added to the corresponding model output x_{model_q}. Since model and target CFDs only show larger deviations at very large and very low values, only these areas are adjusted using this method. This results in more accurate predictions, especially for the rare events of high or low prices.

$$QBS: \quad \widehat{x_q} = \left(\overline{x_{model_q}} - \overline{x_{target_q}}\right) + x_{model_q} \tag{2}$$

3 Results and Conclusion

As described above, at first, optimal values for all meta-parameters need to be distinguished within an empirical study. In the case of modeling the EPEX SPOT day-ahead market, the underlying dynamics can be captured best applying the following setup:

Trainingset :	4368 pattern (182 days)
Validationset :	336 pattern (14 days)
Generalizationset :	24 pattern (1 day)
ActivationFunction :	tanh
Patternselection :	Permute
Trainingepochs :	50
Statedimension :	30

The subsequent modeling study shows, that a setup with four hidden layers combined with the use of shared weights matrices A, B, C and highway connections results

Table 1 Final performance errors (MAE [€/MWh])

	Last Day	FFNN	Deep	Deep + QBS
Total	6.47	2.62	2.56	**2.51**
Season				
Winter	8.51	3.25	3.31	**3.22**
Spring	5.95	2.43	2.41	**2.31**
Summer	4.85	1.65	1.66	**1.61**
Autumn	6.60	3.17	**2.89**	2.93
Day				
Sunday	6.89	3.44	**3.36**	3.37
Monday	11.15	**3.39**	3.45	3.45
Tuesday	6.22	2.74	2.75	**2.62**
Wednesday	5.30	2.16	2.07	**1.92**
Thursday	4.42	2.24	**2.09**	2.09
Friday	4.69	2.18	2.07	**2.02**
Saturday	6.66	2.23	2.17	**2.15**

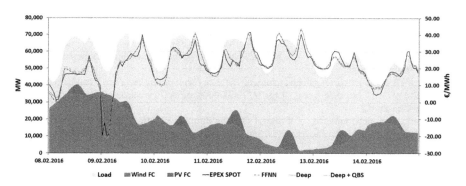

Fig. 3 Comparison of final results in an exemplary winter week in 2016

in the lowest mean absolute error (MAE). In Table 1 it can be seen, that the neural network approach by far outperforms the simple estimate of applying the previous days' prices. Compared to the best three-layer neural network setup, the optimal setup of the deep neural network shows superior results, especially in autumn. Here, the deep neural network reduces the remaining error by 9%. The quantile based bias correction further reduces the error. This effect can be seen in Fig. 3. Whilst the three models perform quite similar in the medium price range, larger deviations occur in sharp price peaks as on the 9th of February 2016. For these events, the deep neural network architecture and especially the QBS approach reduce the remaining residual errors significantly. As the QBS mostly affects the tales of a distribution (i.e. rare events), the overall error is only reduced slightly. However, this method helps to identify extreme values and can be effectively combined with a trigger function for rare events.

Through the systematic process of understanding and capturing real world information, extracting relevant parameter in the sensitivity analysis, a large empirical neural network study and the statistical bias correction step, we obtain accurate results from a robust model setup. The findings within this paper can either be directly used for an improved trading at the EPEX SPOT day-ahead market or can be used as the basis for further research. The proposed approach can easily be transferred for the modeling of other markets, other products or different forecasting horizons. Moreover, it can be applied to analyze the impact of future market developments on the EPEX SPOT price.

References

1. SE, E.S. (2017). Facts and figures. Technical report, EPEX SPOT SE (2017). https://www.epexspot.com/document/36931/2017-01_EPEX_SPOT_Facts.
2. Aggarwal, S. K., Saini, L. M., & Kumar, A. (2009). Electricity price forecasting in deregulated markets: A review and evaluation. *International Journal of Electrical Power & Energy Systems*, *31*(1), 13–22. https://doi.org/10.1016/j.ijepes.2008.09.003.
3. Zimmermann, H. & Neuneier, R. (2000). Modeling dynamical systems by recurrent neural networks. *Data Mining 2*. https://www.witpress.com/Secure/elibrary/papers/DATA00/DATA00054FU.pdf.
4. Bengio, Y. (2009). Learning deep architectures for ai. *Foundations and Trends in Machine Learning*, *2*(1), 1–127. https://doi.org/10.1561/2200000006. https://www.iro.umontreal.ca/~bengioy/papers/ftml.pdf.
5. Hornik, K., Stinchcombe, M., & White, H. (1989). Multilayer feedforward networks are universal approximators. *Neural Networks*, *2*(5), 359–366. https://doi.org/10.1016/0893-6080(89)90020-8.
6. Zimmermann, H. G. (2016). Analysis of complex dynamical systems with neural networks. https://indico.cern.ch/event/561247/attachments/1382489/2102391/ZimmermannCERN1.pdf.

Part III
Control Theory and Continuous Optimization

Computing the Splitting Preconditioner for Interior Point Method Using an Incomplete Factorization Approach

Marta Velazco and Aurelio R. L. Oliveira

1 Introduction

The interior point methods search directions [10] are computed by solving one or more linear systems with the same coefficient matrix. The performance of the implementations using an iterative solution depend upon the choice of an appropriate preconditioner. In particular, for interior point methods, the linear system becomes highly ill-conditioned as an optimal solution is approached. Recently, a hybrid preconditioner was proposed [1]. This approach assumes that the optimization occurs in two phases, and different preconditioners are used for each phase. In the initial phase, a preconditioner obtained by incomplete factorization [3] of the matrix is used. In the second phase, a splitting preconditioner specific for interior point systems is used [7]. This paper presents a new splitting preconditioner for the second phase and is organized as follows. Section 2 introduces the primal-dual interior point methods. The linear systems arising from these methods are presented in Sect. 3. Sections 4 and 5 describe the hybrid preconditioner and the new preconditioner, respectively. Numerical experiments are also carried out. Finally, Sect. 7 concludes the paper.

2 The Predictor-Corrector Primal-Dual Method

Consider the primal-dual pair of linear programming problems in standard form:

$$\begin{aligned} \text{minimize } & c^T x \quad \text{s.t} \quad Ax = b, \quad x \geq 0, \text{ Primal,} \\ \text{maximize } & b^T y \quad \text{s.t} \quad A^T y + z = c, \; z \geq 0. \text{ Dual,} \end{aligned}$$

M. Velazco (✉)
Faculty of Campo Limpo Paulista, Campo Limpo Paulista, São Paulo, Brazil
e-mail: marta.velazco@gmail.com

A. R. L. Oliveira
Applied Mathematics Department, University of Campinas, Campinas, São Paulo, Brazil

where $A \in \mathbb{R}^{m \times n}, rank(A) = m, c, x, z \in \mathbb{R}^n, b, y \in \mathbb{R}^m$. The Karush–Kuhn–Tucker optimality conditions for the primal and dual problems are:

$$\begin{aligned} Ax - b &= 0, \\ A^t y + z - c &= 0, \\ XZe &= 0, \\ (x, z) &\geq 0, \end{aligned} \qquad (1)$$

where $X = diag(x)$, $Z = diag(z)$ and $e \in \mathbb{R}^n$ is the vector of all ones. In interior point methods, the nonlinear equations system of the optimality conditions (1) disregarding the nonnegativity constraint is solved by the Newton method. The predictor-corrector primal-dual method computes the search direction in two steps: the predictor direction and the corrector direction. For each step, a linear system (2) is solved with the same coefficient matrix but different right-hand sides. A new point is computed from the directions, and the stepsize ensures that the point is interior $((x, z) > 0)$.

$$\begin{bmatrix} A & 0 & 0 \\ 0 & A^T & I \\ Z & 0 & X \end{bmatrix} \begin{bmatrix} \Delta x \\ \Delta y \\ \Delta z \end{bmatrix} = r. \qquad (2)$$

3 Linear System Solution

The solution of the linear systems to compute the search directions is the most expensive step for such methods. The linear system (2) can be reduced to a normal equation system, by eliminating the variables Δx and Δz, as shown below:

$$(ADA^T)\Delta y = AD(r_d - X^{-1}r_a) + r_p. \qquad (3)$$

where $D = Z^{-1}X$, $d_{ii} > 0$ is a diagonal matrix and r_p, r_d e r_a are the residuals: $r_p = b - Ax$, $r_d = c - A^T y - z$, e $r_a = -XZe$ of the system (1). When iterative methods are used for the solution of system (3), the success of the method depends on a good preconditioner. The elements of the diagonal matrix $D = Z^{-1}X$ have an undesirable characteristic; on final iterations of the interior point methods, the values of d_{ii} are either very large or very small. Consequently, the matrix ADA^T is very ill-conditioned and makes the system (3) very difficult to solve without an appropriate preconditioner.

4 The Hybrid Preconditioner

In 2007, Bocanegra et al. [1] introduced a hybrid preconditioner to solve the linear system from interior point methods. Bocanegra et al. [1] assume that the optimization process takes place in two different phases and different preconditioners are used in each phase. The first phase occurs at the beginning of the optimization, and the controlled Cholesky factorization preconditioner [3] is used. In the second phase, the splitting preconditioner [7] is used, which is specific for the ill-conditioning matrix from interior point method. A heuristic determines the change of phase [9].

4.1 The Controlled Cholesky Factorization Preconditioner

The controlled Cholesky factorization [3] (CCF) is an incomplete Cholesky factorization of the matrix ADA^T with controlled fill-in. Consider $ADA^T = LL^t = \widetilde{L}\widetilde{L}^t + R$, where L is the factor of the complete Cholesky factorization, $\widetilde{L}$ is the factor of the incomplete factorization, and R is a remainder matrix. By defining $E = L - \widetilde{L}$, the preconditioned coefficient matrix is: $\widetilde{L}^{-1}ADA^T\widetilde{L}^{-T} = (I_m + \widetilde{L}^{-1}E)(I_m + \widetilde{L}^{-T}E)^T$. From this formulation, when $\widetilde{L} \approx L \implies E \approx 0 \implies \widetilde{L}^{-1}(ADA^T\widetilde{L}^{-T} \approx I_m$. The CCF is built considering the minimization of the Frobenius norm of E: $minimize \quad \|E\|_F^2 = \sum_{j=1}^{m} c_j$, where $c_j = \sum_{i=1}^{m} |l_{ij} - \widetilde{l}_{ij}|^2$. Consider $c_j = \sum_{k=1}^{t_j+\eta} |l_{i_k j} - \widetilde{l}_{i_k j}|^2 + \sum_{k=t_j+\eta+1}^{m} |l_{i_k j}|^2$, where t_j is the number of nonzero elements below the diagonal in the jth column of matrix ADA^T and η is the number of extra element allowed for each column. The CCF can be computed by the following heuristics: (1) Choosing the $t_j + \eta$ elements of $\widetilde{L}$ with largest absolute value; (2) Increasing η, allowing more fill-in for the $\widetilde{L}$ factor. The preconditioner $\widetilde{L}$ is built by columns. During the computation of the column, small values or even negatives can appear in the diagonal; CCF uses an exponential shift to avoid loss of positive definiteness.

4.2 The Splitting Preconditioner

The splitting preconditioner [7] has a better performance near the solution of the optimization problem where the matrices are ill-conditioned. Consider $P \in \mathbb{R}^{n \times n}$, a permutation matrix, such that $A = [BN]P$, where $B \in \mathbb{R}^{m \times m}$ is nonsingular and $N \in \mathbb{R}^{m \times (n-m)}$; thus, we have: $ADA^T = BD_BB^T + ND_NN^T$. The splitting preconditioner for normal equations is given by $D_B^{-\frac{1}{2}}B^{-1}$, and the preconditioned matrix is defined as: $D_B^{-\frac{1}{2}}B^{-1}(ADA^T)B^{-T}D_B^{-\frac{1}{2}} = I_m + D_B^{-\frac{1}{2}}B^{-1}ND_NN^TB^{-T}D_B^{-\frac{1}{2}}$. The choice of A columns to form matrix B has a great influence on the performance of the preconditioner. A strategy to form B is to minimize $\|D_B^{-\frac{1}{2}}B^{-1}ND_N^{\frac{1}{2}}\|$. Results presented in [9] show that the best performance is obtained when the first m linearly independent columns of AD with the largest 2-norm are chosen. The columns of AD are found by LU factorization, and this is the most expensive step in the construction of the preconditioner. One column is treated at a time, and the linearly dependent columns are discarded. The factorization continues with the next column in the ordering. During the factorization, excessive fill-in can occur; in which case the independent columns found until then are reordered by sparsity and the factorization is restarted using the new ordering. The process is repeated until m independent columns are found. An advantage of the splitting preconditioner is that the same set of columns can be used for some iterations. Thus, the preconditioner will be very cheap and can be quickly compute for such iterations since B is kept and only the diagonal matrix D changes. Finding the m independent columns is the most expensive step in the construction of the splitting preconditioner. In this search, the order of the columns by the 2-norm may not necessarily be maintained. In the following section, a new splitting preconditioner will be presented where the order of the columns is maintained instead of its linear independence.

5 A New Splitting Preconditioner

In this work, a new splitting preconditioner for the second phase of optimization is proposed. In the construction of the splitting preconditioner, the most expensive step is to find the m linearly independent columns with the largest 2-norm via the LU factorization. In the new splitting preconditioner, the first k columns of AD with the largest 2-norm will be taken, thus eliminating the need to obtain a nonsingular matrix. Next, with the new B, the new preconditioner $\widetilde{L}_B$ will be computed by the CCF of the matrix $BD_B B^T$. For $k \geq m$ columns of AD, B may not be full rank, and therefore, the matrix $BD_B B^T$ may not be positive definite, and very small or even negative pivots may appear. When the failure occurs, an exponential increase in α is globally applied to $BD_B B^T$, and the incomplete factorization of $BD_B B^T + \alpha I$ is restarted. This process is repeated until the factorization is successfully completed.

6 Numerical Experiments

For the numerical experiments, a modified version of the code PCx was used. The original code PCx [4] solves linear programming problems using the predictor-corrector method, and the linear systems are solved with the Cholesky factorization. In the modified version of PCx, the linear systems are solved by the preconditioned gradient method with the hybrid preconditioner [1, 8]. The computational tests were performed on a 2.80 GHz Intel Core i7 platform with 6 GB RAM running 64-bit Linux and the GNU *gcc* and *gfortran* compilers. The preconditioner was tested on 26 problems from public libraries [2, 5, 6]. Table 1 describes the test problems. The columns are: name of the problem (**PROBLEM**), library (**LIBRARY**), size (**M** × **N**) and the sparsity of the restriction matrix (**SPARSITY**) after preprocessing.

Table 1 also presents the computational results with the two preconditioners. Columns 5 and 6 show the total time required to solve the problem by the predictor-

Table 1 Test problems and computational results. **N** means fail to converge

Problem	Library	M×N	Sparsity (%)	Time HP (s)	Time NHP (s)
AA01	MESZAROS	712 × 8904	99,0043	6,28	17,70
AA03	MESZAROS	690 × 8572	99,0053	7,40	7,39
AIR04	MESZAROS	712 × 8904	99,0034	6,27	17,69
AIR05	MESZAROS	367 × 7195	98,2761	2,6	3,97
AIR06	MESZAROS	690 × 8572	99,0053	7,77	7,38
CHR22B	QAP	5587 × 10417	99,9372	33,93	103
CHR25B	QAP	8149 × 15325	99,9564	53,71	324,98
DEGEN2	MISC	442 × 757	98,7546	0,15	0,34
INDATA	MESZAROS	2152 × 7440	98,8329	22,41	24,86
I09A13L1D	MESZAROS	244 × 1483	98,7124	0,18	0,33
NEMSCEM	MESZAROS	479 × 1540	99,5290	0,08	0,13
NUG05	MISC	148 × 225	97,7777	0,02	0,02
NUG05-3RD	MISC	1208 × 1425	99,6791	N	0,58
NUG06	MISC	280 × 486	98,7654	0,05	0,08
NUG06-3RD	MISC	3540 × 4686	99,8814	N	7,9
NUG07	MISC	474 × 931	99,2481	0,32	0,43
NUG07-3RD	MISC	8594 × 12691	99,9484	N	47,58
NUG08	MISC	742 × 1632	99,5098	0,60	0,76
NUG08-3RD	MISC	18270 × 29856	99,9747	N	316,30
NUG15	MISC	5698 × 22275	99,9326	N	856,85
P0040	MESZAROS	23 × 63	90,8212	0,0	0,0
PS	MESZAROS	5698 × 22275	99,9326	N	907,18
QAP08	QAP	742 × 1682	99,5243	0,54	0,84
QAP12	QAP	2794 × 8856	99,8644	N	138,87
QAP15	QAP	5698 × 22275	99,9326	N	798,93
T0331-4L	MESZAROS	664 × 46915	98,6164	126,68	73,56

corrector method when the hybrid preconditioner and the new hybrid preconditioner are used, respectively. The comparison of the two approaches can be summarized as follows:

- When the two approaches reach an optimal solution, the previous preconditioner requires less computational time in most cases: The new splitting preconditioner does not need to select the first m linearly independent columns; the CCF of BD_BB^T matrix is computed with the first k columns of A with the largest 2-norm. The CCF of the matrix is expensive to compute, because when small or even negative diagonal elements are found, the CCF discards the factor, increases the diagonal by a shift, and computes the factor again. Additionally, at each iteration of the interior point method, a new CCF is computed, and if the number of iterations

of the preconditioned conjugate gradient increase, the fill-in in the factorization increases too and more time is needed. The previous splitting preconditioner can use the same matrix B for some interior point iterations, thus computational time is saved.
- When the previous hybrid preconditioner does not complete the optimization process (the letter **N** represents this status on the table), the new preconditioner concludes the optimization process and an optimal solution is found. In the previous splitting preconditioner, during the LU factorization, an excessive fill-in may occur; in this case, the independent columns found until then are reordered by sparsity and the factorization is restarted using the new ordering. For these cases, the previous splitting preconditioner fails and the optimization process does not finish. On the other hand, the new preconditioner does not present these problems. Note that this occurs in problems where the coefficient matrix is sparser.

7 Conclusions

This paper presents a new splitting preconditioner, computed applying the CCF to the first k columns of the matrix AD ordered by the 2-norm. The new matrix can be singular, and in the CCF an exponential shift on the diagonal is used to compute the factorization. The new preconditioner exhibits good performance in the tested problems. The best performance was obtained for the problems where the factorization of matrix B is dense. This is the most expensive step in the splitting preconditioner because when the factorization is very dense, it must be restarted. In these cases, the new preconditioner provides a good solution of the systems, and the optimal solution is reached.

Acknowledgements This work was supported by the Foundation for the Support of Research of the State of São Paulo (FAPESP-2010/06822-4), the National Council for Scientific and Technological Development (CNPq) and Faculty of Campo Limpo Paulista (FACCAMP).

References

1. Bocanegra, S., Campos, F. F., & Oliveira, A. R. L. (2007). Using a hybrid preconditioner for solving large-scale linear systems arising from interior point methods. *Computational Optimization and Applications*, *36*(2), 149–164. https://doi.org/10.1007/s10589-006-9009-5.
2. Burkard, R. S., Karisch, S., & Rendl, F. (1991). QAPLIB a quadratic assignment problem library. *European Journal of Operational Research*, *55*, 115–119.
3. Campos, F. F., & Birkett, N. R. C. (1998). An efficient solver for multi-right hand side linear systems based on the CCCG(η) method with applications to implicit time-dependent partial differential equations. *SIAM Journal on Scientific Computing*, *19*(1), 126–138. https://doi.org/10.1137/S106482759630382X.
4. Czyzyk, J., Mehrotra, S., Wagner, M., & Wright, S. J. (1999). PCx: An interior-point code for linear programming. *Optimization Methods and Software*, *11*(1–4), 397–430.

5. Miscellaneous LP models. Hungarian Academy of Sciences OR Lab. Online at http://www.sztaki.hu/meszaros/public_ftp/lptestset/misc
6. Mittelmann LP models. Miscellaneous LP models collect by Hans D. Mittelmann. Online at http://plato.asu.edu/ftp/lptestset/pds/
7. Oliveira, A. R. L., & Sorensen, D. C. (2005). A new class of preconditioners for large-scale linear systems from interior point methods for linear programming. *Linear Algebra and Its Applications*, *394*, 1–24. https://doi.org/10.1016/j.laa.2004.08.019.
8. Silva, D., Velazco, M., & Oliveira, A. (2017). Influence of matrix reordering on the performance of iterative methods for solving linear systems arising from interior point methods for linear programming. *Mathematical Methods of Operations Research*, *85*(1), 97–112. https://doi.org/10.1007/s00186-017-0571-7.
9. Velazco, M. I., Oliveira, A. R. L., & Campos, F. F. (2010). A note on hybrid preconditions for large scale normal equations arising from interior-point methods. *Optimization Methods and Software*, *25*, 321332. https://doi.org/10.1080/10556780902992829.
10. Wright, S. J. (1997). *Primal-dual interior-point methods*. Philadelphia: SIAM Publications.

Quadratic Support Functions in Quadratic Bilevel Problems

Oleg Khamisov

1 Statement of the Problem

We consider bilevel programming problem in the following form [1]

$$F(x, y) \to \min_{x,y}, \tag{1}$$

$$A_1 x + B_1 y \leq r_1, \quad f(x, y) - \psi(x) \leq 0, \quad A_2 x + B_2 y \leq r_2, \tag{2}$$

$$\psi(x) = \min_{y}\{f(x, y) : A_2 x + B_2 y \leq r_2\}, \tag{3}$$

where $F : R^{n_1} \times R^{n_2} \to R$ is a quadratic function, A_1 is an $(m_1 \times n_1)$ matrix, B_1 is an $(m_1 \times n_2)$ matrix, A_2 is an $(m_2 \times n_1)$ matrix, B_2 is an $(m_2 \times n_2)$ matrix, $r_1 \in R^{m_1}$, $r_2 \in R^{m_2}$,

$$f(x, y) = x^T P y + y^T G y + w^T y, \tag{4}$$

P is a $(n_1 \times n_2)$ matrix, G is a positive definite $(n_2 \times n_2)$ matrix $(G \succ 0)$, $w \in R^{n_2}$.

The approach suggested in the paper differs from others applicable to problem (1)–(4) (see [3, 4, 11, 12]). We do not elaborate optimality conditions (see, for example [2]). The idea of our approach is based on using so called nonlinear support functions [8, 9] in combination with outer approximation technique in global optimization [7].

Below we will assume that linear inequalities in (2) are consistent and determine a bounded polyhedron in $R^{n_1+n_2}$. This assumption is not in general strict, we just concentrate our attention on the idea description and omit here some technical details related to a more general case. The same can be said about condition $G \succ 0$. It is used also to simplify the below description and can be relaxed to $G \succeq 0$.

O. Khamisov (✉)
Melentiev Energy Systems Institute SB RAS, Lermontov str. 130, Irkutsk, Russia
e-mail: khamisov@isem.irk.ru

2 Support Functions

Since the function ψ in (3) is given implicitly we can not directly use different global optimization techniques. For approximating implicit inequality (3) upper and lower support functions can be used.

Problem (3) is a convex programming problem which has a unique solution for a given x. From the duality theory we have another representation for ψ,

$$\psi(x) = \max_{\lambda \geq 0}\{-\frac{1}{4}(P^T x + B_2^T \lambda + w)^T G^{-1}(P^T x + B_2^T \lambda + w) + (A_2 x - r_2)^T \lambda = \tag{5}$$

$$= \frac{1}{4} \max_{\lambda \geq 0}\{-\lambda^T B_2 G B_2^T \lambda - (2 B_2 G^{-1} P^T x + A_2 x + 2 B_2 G^{-1} w - r_2)^T \lambda\} - \tag{6}$$

$$-\frac{1}{4} x^T P G^{-1} P^T x - \frac{1}{2} w^T G^{-1} P^T x + w^T G^{-1} w. \tag{7}$$

It is not difficult to see, that expression (6) represents a convex in x function and expression (7) represents a concave in x function. Hence, function ψ is a so called d.c. (difference of convex) function [7], in which convex part is still implicit, but concave part is already explicit.

Let some $\tilde{x}$ be given. By solving problem (3) or, equivalently, problem (5) we find the corresponding dual solution $\tilde{\lambda}$. Define function

$$\varphi(x) = -\frac{1}{4}(P^T x + B_2^T \tilde{\lambda} + w)^T G^{-1}(P^T x + B_2^T \tilde{\lambda} + w) + (A_2 x - r_2)^T \tilde{\lambda}. \tag{8}$$

Then, $\psi(\tilde{x}) = \varphi(\tilde{x})$ and $\psi(x) \geq \varphi(x)$ for $x \neq \tilde{x}$, i.e. function φ is a lower support function of ψ. If we substitute implicit inequality $f(x, y) - \psi(x) \leq 0$ by explicit inequality $f(x, y) - \varphi(x) \leq 0$ in (2), we obtain inner explicit approximation of the implicit feasible set of problem (1)–(4). Then we can solve this explicit inner approximation problem and obtain an improvement of the given $\tilde{x}$ in the sense of the objective function F. However, quite often the only feasible point of the inner approximation problem is point $\tilde{x}$ itself, so no progress w.r.t. objective F. That is the reason why we suggest to use upper support functions.

Denote by $\lambda^*(x)$ dual solution of (5) for a given x. Assuming that x is bounded and using different techniques (see [6, 10]) we can compute a bound γ such, that $\lambda_i^*(x) \leq \gamma \ \forall x$, $i = 1, \ldots, m_2$. Then for given $\tilde{x}$ we have

$$\psi(x) = \max_{\lambda \in \Lambda} \min_y \{f(x, y) + \lambda^T(A_2 x + B_2 y - r_2)\} = \tag{9}$$

$$= \min_y \max_{\lambda \in \Lambda}\{f(x, y) + \lambda^T(A_2 x + B_2 y - r_2)\} \leq \tag{10}$$

$$\leq f(x, \tilde{y}) + \max_{\lambda \in \Lambda}\{\lambda^T(A_2 x + B_2 \tilde{y} - r_2)\} = \tag{11}$$

$$= f(x, \tilde{y}) + \gamma \max\{0, (A_{21}x + B_{21}\tilde{y} - r_{21}), \ldots, (A_{2m_2}x + B_{2m_2}\tilde{y} - r_{2m_2})\} = \eta(x), \quad (12)$$

where $\Lambda = \{\lambda \geq 0 : \sum_{i=1}^{m_2} \lambda_i \leq \gamma\}$, A_{2i}, B_{2i} are ith rows of matrices A_2 and B_2 respectively, r_{2i} is ith element of vector r_2. Hence, $\psi(\tilde{x}) = \eta(\tilde{x})$ and $\psi(x) \leq \eta(x)$ for $x \neq \tilde{x}$, i.e. η is an nonconvex but explicit upper support function of ψ. Substituting inequality $f(x, y) - \psi(x) \leq 0$ by inequality $f(x, y) - \eta(x) \leq 0$ in (2) we obtain explicit outer approximation problem for problem (1)–(4). Optimal value of the outer approximation problem gives a lower bound for an optimal value of the initial bilevel problem. Upper support functions is the base for outer approximation method for solving (1)–(4).

3 The Outer Approximation Algorithm

Detailed description of the algorithm is as follows.

Step 0. Solve the problem
$$F(x, y) \to \min_{x,y}, \quad (13)$$
$$A_1 x + B_1 y \leq r_1, \quad A_2 x + B_2 y \leq r_2. \quad (14)$$

Let (x^0, y^0) be a solution of (13)–(14). Set $k = 0$.

Step 1. Solve the problem
$$f(x^k, y) \to \min_y, \quad (15)$$
$$A_2 x^k + B_2 y \leq r_2. \quad (16)$$

Let $\tilde{y}^k$ be a solution of (15)–(16).

Step 2. If $f(x^k, y^k) = f(x^k, \tilde{y}^k)$ then stop: (x^k, y^k) is a solution of the initial bilevel problem (1)–(4).

Step 3. Solve the outer approximate problem
$$F(x, y) \to \min_{x,y}, \quad (17)$$
$$A_1 x + B_2 y \leq r_1, \quad A_2 x + B_2 y \leq r_2, \quad (18)$$
$$f(x, y) - f(x, \tilde{y}^j) - \gamma \max\{0, (A_{2i}x + B_{2i}\tilde{y}^j - r_{2i}), i = 1, \ldots, m_2\} \leq 0, \quad (19)$$
$$j = 0, \ldots, k. \quad (20)$$

Let (x^{k+1}, y^{k+1}) be a solution of (17)–(20).

Step 4. Set $k = k + 1$ and goto Step 1.

Algorithm generates sequence of functions which are used at the Step 3. From the boundedness assumption and the problem data structure we can conclude that this sequence of functions is equicontinuous and uniformly bounded. Then, from the theory of global outer approximation algorithms [7] it follows that every accumulation point of sequence (x^k, y^k) is a solution of the initial implicit global optimization problem (1)–(4).

At each iteration we have to solve one convex optimization problem (15)–(16) and one multiextremal problem (17)–(20). The main advantage of the multiextremal problem is that it is explicit, so one can use different global optimization techniques for its solution. Outer Approximation Algorithm substitute solution of one implicit global optimization problem by a sequence of explicit global optimization problems.

From the above consideration we can see that properties of the objective function F and function f w.r.t. x are not used. So, the performance of the algorithm can be improved by using this fact. For example, if functions F and f are convex, then the approximate problem (17)–(20) breaks up into a number of convex optimization problems, which essentially improves the effectiveness of the algorithm. On the other hand it follows that the outer approximation algorithm can be used in a more general situation.

4 Examples

In order to show that the outer approximation algorithm is practically implementable we consider two examples.

Example 1 Consider the following bilevel problem

$$(x - 3)^2 + (y - 2)^2 \to \min,$$

$$0 \le x \le 6,$$

$$(y - 5)^2 \to \min_y,$$

$$-2x + y - 1 \le 0, \quad x - 2y + 2 \le 0, \quad x + 2y - 14 \le 0,$$

$$0 \le y \le 6.$$

Its solution $(x^*, y^*) = (1, 3)$, $F(x^*, y^*) = 5$. Outer approximation algorithm finds an approximate solution in 15 iteration. Absolute error in the objective $\varepsilon = 0.01$. This example shows us the useless of the lower support function (8). Take $\tilde{x} = 0$, then the corresponding primal and dual solutions of the follower problem are $\tilde{y} = 1$ and $\tilde{\lambda} = (8, 0, 0, 0, 0)$. Point $(\tilde{x}, \tilde{y})$ is feasible and $F(0, 1) = 10$. The corresponding inequality $f(x, y) - \varphi(x) = (y - 5)^2 - 16 + 16x \le 0$ in combination with all linear constraints gives only feasible point $(\tilde{x}, \tilde{y}) = (0, 1)$, so using lower support function gives no progress in the objective F. We just stuck at $(0, 1)$.

Example 2 ([5]) This example demonstrates that the suggested algorithm can be extended to a nonquadratic case. The bilevel problem is

$$F(x, y) = x^3 y_1 + y_2 \to \min_{x,y},$$

$$0 \leq x \leq 1,$$

$$-y_2 \to \min,$$

$$g_1(x, y) = xy_1 - 10 \leq 0,$$

$$g_2(x, y) = y_1^2 + xy_2 - 1 \leq 1,$$

$$-1 \leq y_1 \leq 1, 0 \leq y_2 \leq 100.$$

Solution $(x^*, y_1^*, y_2^*) = (1, 0, 1)$, $F(x^*, y^*) = 1$. Outer approximation problem at the Step 3 of the algorithm has now the following form

$$F(x, y) \to \min_{x,y},$$

$$0 \leq x \leq 1, \ g_1(x, y) \leq 0, \ g_2(x, y) \leq 0,$$

$$f(x, y) - f(x, \tilde{y}^j) - \gamma \max\{0, g_1(x, \tilde{y}^j), g_2(x, \tilde{y}^j)\} \leq 0, j = \overline{0, k},$$

$$y \in Y = \{y : -1 \leq y_1 \leq 1, 0 \leq y_2 \leq 100\}.$$

After 5 iterations the following approximate solution was obtained $(x^5, y_1^5, y_2^5) = (1, -0.001, 1)$, $F(x^5, y^5) = 0.9985$.

Acknowledgements This work is supported by the Russian Science Foundation (project 17-11-01021).

References

1. Dempe, S. (2002). *Foundations of bilevel programming*. Dordrecht: Kluwer Academic Publishers.
2. Dutta, J., & Dempe, S. (2006). Bilevel programming with convex lower level problem. In S. Dempe & V. Kalashnikov (Eds.), *Optimization with multivalued mappings* (pp. 51–71). LLC: Springer Science + Business Media.

3. Etoa Etoa, J. B. (2010). Solving convex quadratic bilevel programming problems using an enumeration sequential quadratic programming algorithm. *Journal of Global Optimization, 47*, 615–637.
4. Etoa Etoa, J. B. (2011). Solving quadratic convex bilevel programming problems using a smoothing method. *Applied Mathematics and Computation, 217*, 6680–6690.
5. Gümüs, Z. H., & Floudas, C. A. (2001). Global optimization of nonlinear bilevel programming problems. *Journal of Global Optimization, 20*, 1–31.
6. Hiriart-Urruty, J.-B., & Lemaréchal, C. (1993). *Convex analysis and minimization algorithms II*. Berlin: Springer.
7. Horst, R., & Tuy, H. (1996). *Global optimization. (Deterministic approaches)*. Berlin: Springer.
8. Khamisov, O. V. (1999). On optimization properties of functions with a concave minorant. *Journal of Global Optimization, 14*, 79–101.
9. Khamisov, O. V. (2016). Optimization with quadratic support functions in nonconvex smooth optimization. *AIP Conference Proceedings, 1776*, 050010. https://doi.org/10.1063/1.4965331.
10. Mangasarian, O. L. (1985). Computable numerical bounds for Lagrange multipliers of stationary points of non-convex differentiable non-linear programs. *Operations Research Letters, 4*, 47–48.
11. Muu, L. D., & Quy, N. V. (2003). A global optimization method for solving convex quadratic bilevel programming problems. *Journal of Global Optimization, 26*, 199–219.
12. Strekalovsky, A. S., Orlov, A. V., & Malyshev, A. V. (2010). On computational search for optimistic solution in bilevel problems. *Journal of Global Optimization, 48*, 159–172.

Part IV
Decision Theory and Multiple Criteria Decision Making

Valuation of Crisp and Intuitionistic Fuzzy Information

Olga Metzger, Thomas Spengler and Tobias Volkmer

1 Introduction

Information decision problems are decision problems in which it is necessary to clarify whether additional information (and if so, which) should be obtained in order to make good decisions. In this regard, the decision maker is confronted with two decision problems: on one hand with the choice of an action from the original decision problem, and on the other hand with the meta-problem of making a rational information decision. The information valuation in the standard model using Bayesian statistics is limited to the processing of additional information on prior state probabilities. This article provides an extended view to the problem of obtaining and evaluating additional information. Therefore, we extend the standard approach of Emery [1] in the version of Laux [2] and present different variations of the standard case. In this context, we consider situations where a decision maker obtains information in order to improve prior assessments of consequences, identify new actions or new decision-relevant states of nature. Additionally, we discuss their economic effects on information valuation. We also address the role of a possible negative information value and its importance for rational information decisions. Subsequently we present another variation of the standard model, where we use intuitionistic fuzzy logic to handle the presence of vague information within information decision problems.

O. Metzger (✉) · T. Spengler · T. Volkmer
Otto von Guericke University, 39106 Magdeburg, Germany
e-mail: olga.metzger@ovgu.de
URL: http://www.ufo.ovgu.de/

T. Spengler
e-mail: bwl-uo@ovgu.de

2 Extentions of Crisp Information Valuation

Let $S := \{s | s = 1, 2, \ldots, \overline{S}, \overline{S}+1, \overline{S}+2, S^*\}$ be a set of states of nature with corresponding probabilities $w(s)$ and let $A := \{a | a = 1, 2, \ldots, \overline{A}, \overline{A}+1, \overline{A}+2, A^*\}$ be a set of actions. In addition, let E^{pr} be a set of all prior consequences e_{as}, $E^{pos}_{(*)}$ a set of posterior consequences e_{as} given information result $(*)$ and e_{as}, $E^{pra}_{(*)}$ a set of prior consequences e_{as} which a posteriori become irrelevant due to information result $(*)$. The subsets $A^0 = \{a | a = 1, 2, \ldots, \overline{A}\} \subset A$ and $S^0 = \{s | s = 1, 2, \ldots, \overline{S}\} \subset S$ represent a priori known actions and states of nature. Combined with the corresponding $w(s)$ and the set E^{pr} they characterize the standard case. Within this standard case the decision maker may obtain information in order to improve his or her prior probability judgements. The gross value $(VI = EVI - EV)$ of this kind of additional information (before subtracting information costs) is calculated as follows:

$$VI = \sum_{i \in I} w(i) \cdot \max_{a \in A^0} \sum_{s \in S^0} w(s|i) \cdot e_{as} - \max_{a \in A^0} \sum_{s \in S^0} w(s) \cdot e_{as} \quad (1)$$

with $\quad EVI = \sum_{i \in I} w(i) \cdot \max_{a \in A^0} \sum_{s \in S^0} w(s|i) \cdot e_{as} \quad$ and $\quad EV = \max_{a \in A^0} \sum_{s \in S^0} w(s) \cdot e_{as}$ (2)

By obtaining information, the prior probabilities $w(s)$ alter into conditional posterior probabilities $w(s|i)$ of state s given $i \in I$, where I is defined as a set of information results leading to a revision of decision maker's prior estimates on state probabilities. The probability update results formally from the application of the Bayes' theorem and depends on the probabilities of receiving a particular information result $(w(i))$ as well as the decision maker's estimates on the prediction quality of the information source $(w(i|s))$. We now extend the standard case to cases where additional information regarding new consequences (case 1), new actions (case 2) and new decision-relevant states (case 3) is obtained. $J := \{j | j = 1, \ldots, \overline{J}\}$ is defined as a set of information results for case 1, $K := \{k | k = 1, \ldots, \overline{K}\}$ as a set of information results for case 2 and $L := \{l | l = 1, \ldots, \overline{L}\}$ as a set of information results for case 3. Case 4 combines these three decision situations (see Table 1). The probabilities for receiving information results j, k and l are denoted by $w(j), w(k)$ and $w(l)$. It should be noted that the value EVI in the extended cases differs from the value EVI in the standard model in terms of content and formulation, depending on the extension. But EV is always the reference point for determining the information value in all cases. There is also a distinction concerning prior and posterior components in the model variations. Whereas the standard model solely focuses on prior and posterior probabilities and expected values, in the extended cases we also deal with prior and posterior consequences (cases 1 and 4), actions (cases 2 and 4) as well as states of nature (cases 3 and 4). So far, these types of information decisions have been identified as significant in the literature, but yet no attempt has been made to present a formalization of these types of decision situations.

Valuation of Crisp and Intuitionistic Fuzzy Information

Table 1 (Components of) posterior pay-off matrices

	$w(s=1)$	...	$w(s=\overline{S})$	$w(s=\overline{S}+1)$	...	$w(s=S^*)$
	$s=1$	...	$s=\overline{S}$	$s=\overline{S}+1$	...	$s=S^*$
$a=1$	e_{11}	...	$e_{1\overline{S}}$	$e_{1,\overline{S}+1}$	...	e_{1,S^*}
$\vdots$	$\vdots$	$\ddots$ (Q1)	$\vdots$	$\vdots$	$\ddots$ (Q2)	$\vdots$
$a=\overline{A}$	$e_{\overline{A}1}$	...	$e_{\overline{A}\overline{S}}$	$e_{\overline{A},\overline{S}+1}$	...	$e_{\overline{A},S^*}$
$a=\overline{A}+1$	$e_{\overline{A}+1,1}$	...	$e_{\overline{A}+1,\overline{S}}$	$e_{\overline{A}+1,\overline{S}+1}$	...	$e_{\overline{A}+1,S^*}$
$\vdots$	$\vdots$	$\ddots$ (Q3)	$\vdots$	$\vdots$	$\ddots$ (Q4)	$\vdots$
$a=A^*$	$e_{A^*,1}$	...	$e_{A^*,\overline{S}}$	$e_{A^*,\overline{S}+1}$	...	e_{A^*,S^*}

In **Case 1 (Q1)**, we consider decision situations where information is obtained to improve judgements regarding prior consequences. Therefore, in this case a posteriori available consequences e'_{as} have to be considered in order to determine EVI. It is assumed that the respective information results do not lead to a probability update ($w(s)$ replaces $w(s|i)$). Under these circumstances the information value VI is calculated by

$$VI = \sum_{j \in J} w(j) \cdot \max_{a \in A^0} \sum_{s \in S^0} w(s) \cdot e'_{as} - \max_{a \in A^0} \sum_{s \in S^0} w(s) \cdot e_{as} \quad (3)$$
$$\text{with } e'_{as} \in \{E^{pr} \setminus E^{pra}_j\} \cup E^{pos}_j.$$

Case 2 (Q1, Q3) depicts decision situations where information about new actions is obtained. In addition to the previously known actions $a \in A^0$, a posteriori new actions $a \in \{\overline{A}+1, \overline{A}+2, \ldots, A^*\}$ become available to the decision maker. The index sets A_k contain the posterior actions which are discovered after receiving information result k. In analogy to case 1, the prior probabilities $w(s)$ are used to determine EVI. This results in

$$VI = \sum_{k \in K} w(k) \cdot \max_{a \in A^0 \cup A_k} \sum_{s \in S^0} w(s) \cdot e'_{as} - \max_{a \in A^0} \sum_{s \in S^0} w(s) \cdot e_{as} \quad (4)$$
$$\text{with } e'_{as} \in E^{pr} \cup E^{pos}_k.$$

Case 3 (Q1, Q2) maps situations where the decision maker acquires information in order to discover previously unknown states $s \in \{\overline{S}+1, \overline{S}+2, \ldots, S^*\}$. The index sets S_l capture those posterior states added through information result l. Due to stochastic dependencies between s and l, conditional probabilities $w(s|l)$ are now to be considered here. This results in

$$VI = \sum_{l \in L} w(l) \cdot \max_{a \in A^0} \sum_{s \in S^0 \cup S_l} w(s|l) \cdot e'_{as} - \max_{a \in A^0} \sum_{s \in S^0} w(s) \cdot e_{as} \tag{5}$$

$$\text{with } e'_{as} \in E^{pr} \cup E_l^{pos}.$$

In **Case 4 (Q1–Q4)** the first three decision situations are combined. Taking $\overline{J} \cdot \overline{K} \cdot \overline{L}$ posterior pay-off matrices into account, for case 4 we determine the information value VI by

$$VI = \sum_{j \in J} w(j) \sum_{k \in K} w(k) \sum_{l \in L} w(l) \cdot \max_{a \in A^0 \cup A_k} \sum_{s \in S^0 \cup S_l} w(s|l) \cdot e'_{as} - \max_{a \in A^0} \sum_{s \in S^0} w(s) \cdot e_{as}$$

$$\text{with } e'_{as} \in \{E^{pr} \setminus E_j^{pra}\} \cup E_j^{pos} \cup E_k^{pos} \cup E_l^{pos}. \tag{6}$$

It can easily be shown that the gross information value can never become negative in the standard model using the Bayes' theorem. The same applies to case 2. If new actions become available to the decision maker, with expected values lower than EV he or she will still choose the initial action. This results in a gross information value of 0. It also can easily be shown that in cases 1, 3 and 4, however, negative gross information values may result due to certain data constellations. Whereas the standard model recommends to obtain any information that has a positive net information value at the time of the information decision, now a new question arises: Should the decision be made for or against obtaining information, when the information value becomes negative? Despite the fact, that the negative information value results from a lower posterior expected value compared to the expected value of the initially chosen action, it represents an information gain. Given reliability of the information source used, the EVI corresponds with the "true" expected value. Therefore, a rational decision maker should not strictly reject a negative information value. He or she has to consider its relevance before making the particular information decision individually. It may also be useful to take the absolute value of the information into account.

3 Vague Information Valuation

In the context of the extensions described so far, cases with crisp information structures are considered; in the following, we discuss possibilities for a formal handling of vague information within information decisions. In principle, all elements of the decision field as well as the objective function can be vague. A decision maker could, for example, have only vague judgements about state probabilities $w(s)$, consequences e_{as}, or the prediction quality $w(i|s)$ of an information source. In the following, we particularly examine the latter case. For this purpose, we present a further extension of the standard model based on intuitionistic fuzzy logic (see [3]). To illustrate our approach, we formulate a numerical example with three actions and three states. The decision maker assumes crisp prior probabilities, where $w(s=1) = 0, 7$, $w(s=2) = 0, 2$

Valuation of Crisp and Intuitionistic Fuzzy Information   117

Table 2 Pay-off matrix (example)

	$s=1$	$s=2$	$s=3$
$a=1$	180	100	−120
$a=2$	150	150	20
$a=3$	80	80	80

Table 3 $w(i|s)$ - intervals

	$i=1$	$i=2$	$i=3$
$s=1$	[0, 65; 0, 7]	[0, 05; 0, 1]	[0, 15; 0, 2]
$s=2$	[0, 05; 0, 1]	[0, 75; 0, 8]	[0, 05; 0, 1]
$s=3$	[0, 25; 0, 3]	[0, 05; 0, 1]	[0, 55; 0, 6]

Table 4 Intuitionistic fuzzy values

	$i=1$	$i=2$	$i=3$
$s=1$	(0, 65; 0, 3; 0, 05)	(0, 05; 0, 9; 0, 05)	(0, 15; 0, 8; 0, 05)
$s=2$	(0, 05; 0, 9; 0, 05)	(0, 75; 0, 2; 0, 05)	(0, 05; 0, 9; 0, 05)
$s=3$	(0, 25; 0, 7; 0, 05)	(0, 05; 0, 9; 0, 05)	(0, 55; 0, 4; 0, 05)

and $w(s=3) = 0, 1$. The consequences e_{as} can also be precisely determined by the decision maker (according to Table 2). The inaccuracy regarding the prediction quality of the information source is represented by probability intervals of the form $[\underline{w}(i|s); \overline{w}(i|s)]$. The corresponding sample values for three information results are given in Table 3.

The probability intervals are transformed into intuitionistic fuzzy values of the form $\alpha(i|s) = (\mu_\alpha(i|s), \nu_\alpha(i|s), \pi_\alpha(i|s))$ according to the method proposed by Metzger and Spengler [4] (see Table 4). The particular elements of the $\alpha(i|s)$ are defined as subjectively perceived degrees of realizability($\mu_\alpha(i|s)$), non-realizability ($\nu_\alpha(i|s)$) and indeterminacy ($\pi_\alpha(i|s)$) concerning the (non-)realizability of the respective information result i under the assumption that state s occurs in the future. Formally the following interdependencies apply: $\mu_\alpha(i|s) \in [0, 1]$, $\nu_\alpha(i|s) \in [0, 1]$, $\mu_\alpha(i|s) + \nu_\alpha(i|s) \leq 1$ and $\pi_\alpha(i|s) = 1 - \mu_\alpha(i|s) - \nu_\alpha(i|s)$.

As Table 4 shows, in our example $\pi_\alpha(i|s)$ has a value of 0.05, which is equivalent to the width of the corresponding intervals $[\underline{w}(i|s); \overline{w}(i|s)]$. It can be interpreted as a relatively low degree of the decision makers indeterminacy concerning his or her judgement about the information source prediction quality. Basically, $\pi_\alpha(i|s)$ can be redistributed ex post fully or in part to the degree $\mu_\alpha(i|s)$ (realizability of information result i is assessed higher than before) or to the degree $\nu_\alpha(i|s)$ (non-realizability of i is assessed higher than before). After this redistribution, π equals 0 and the corresponding probability $w(i|s)$ then again corresponds to a point (crisp) value within the interval boundaries. The decision maker is not able to observe this redistribution even after the state of nature s has occurred. Therefore, a precise information value cannot be determined at the time of the information valuation. In this regard we have to identify the set of information values which anticipates all potential cases of the redistribution of $\pi_\alpha(i|s)$. For the formal handling of the lower bound of the expected value with information ($\underline{EVI}$), we propose to take the degree $\mu_\alpha(i|s)$ into account, to which the decision maker considers $i|s$ to be necessarily feasible. This seems to be plausible since this value represents the lower bound of the prediction quality, the decision maker assigns to the information source. It is easy to show that VI can also be determined by modifying Eq. (1) and replacing $w(i|s)$ by $\mu_\alpha(i|s)$. Using this modification, the lower bound of the information value ($\underline{VI}$) can be determined as follows:

$$\underline{VI} = \sum_{a \in A^0} \max_a \sum_{s \in S^0} \mu_\alpha(i|s) \cdot w(s) \cdot e_{as} - \max_{a \in A^0} \sum_{s \in S^0} w(s) \cdot e_{as} \qquad (7)$$

The decision maker cannot observe the redistribution of $\pi_\alpha(i|s)$ to $\mu_\alpha(i|s)$, or to $\nu_\alpha(i|s)$, neither at the time of the information decision nor after the occurrence of state s. Nevertheless, the best case is a full reallocation of $\pi_\alpha(i|s)$ to $\mu_\alpha(i|s)$. Therefore, we suggest determining the upper bound of the information value ($\overline{VI}$) as follows:

$$\overline{VI} = \sum_{a \in A^0} \max_a \sum_{s \in S^0} ((\mu_\alpha + \pi_\alpha)(i|s)) \cdot w(s) \cdot e_{as} - \max_{a \in A^0} \sum_{s \in S^0} w(s) \cdot e_{as} \qquad (8)$$

Accordingly, the information value is element of the interval $[\underline{VI}; \overline{VI}]$. Applying (7) and (8) to the example illustrated in Tables 2, 3 and 4, we get an information value interval of $VI \in [-10, 9; 9, 5]$. It turns out that indeterminacy, even to a small extent, can have a strong impact on the information value. In this example, we get among others a negative $\underline{VI}$. We recommend not to neglect this fact when making the corresponding information decision.

4 Conclusions

In this article, we present various (crisp and fuzzy) extensions of the standard model of information valuation. For one thing, these do not focus on the prior probabilities of states considered, but on other elements of the decision field. For another thing, vague information is taken into account (at least partially) using intuitionistic fuzzification. Here, we consider the intuitionistic fuzzy logic in terms of intuitionistic fuzzy values according to [4]. In further research, it is necessary to clarify how vagueness concerning consequences, actions and states of nature can be integrated in the information valuation process based on the intuitionistic fuzzy set theory.

References

1. Emery, J. T. (1969). *Organization planning and control systems: Theory and technology*. London: Macmillan Pub Co.
2. Laux, H. (1979). *Grundfragen der Organisation: Delegation Anreiz und Kontrolle*. Berlin: Springer.
3. Atanassov, K. T. (1986). Intuitionistic fuzzy sets. *Fuzzy Sets and Systems, 20*(1), 87–96. https://doi.org/10.1016/S0165-0114(86)80034-3.
4. Metzger, O., & Spengler, T. (2017). Subjektiver Erwartungsnutzen und intuitionistische Fuzzy Werte. In T. Spengler, W. Fichtner, M. J. Geiger, H. Rommelfanger, & O. Metzger (Eds.), *Entscheidungsunterstuetzung in Theorie und Praxis* (pp. 109–137). Wiesbaden: Springer Gabler. https://doi.org/10.1007/978-3-658-17580-1_6.

Multiobjective Spatial Optimization: The Canadian Coast Guard

H. A. Eiselt, Amin Akbari and Ron Pelot

1 Introduction

The concern of this short paper is the improvement of the operations of the Canadian Coast Guard by mathematical methods. At present the Canadian Coast Guard operates 140 vessels of different kinds for a variety of missions, see [1]. About one third of these vessels operates in Newfoundland and Labrador and the Maritime Provinces. This piece deals exclusively with search and rescue missions. Services are provided to a diverse clientele, ranging from large tanker and container vessels to smaller commercial fishing vessels, as well as recreational boats. Similar to the standard triage system in hospitals, incidents are classified into three categories from distress calls to potential distress and non-distress calls. The years 2005–2012 (except 2007, which was excluded for technical reasons) saw 8,033 incidents on the Atlantic Coast [2]. These incidents are not distributed evenly throughout the year as Fig. 1 clearly demonstrates.

There are a number of decision variables that planners can use to optimize the service provided by the Coast Guard. Among them are the choice of the equipment (in the long run), the location of the existing vessels, and operational procedures. This paper will investigate and optimize the location of the vessels. In order to do so, we need to define objectives. An obvious customer-centered objective is the

---

H. A. Eiselt (✉)
Faculty of Business Administration, University of New Brunswick, 4400, Fredericton, NB E3B 5A3, Canada
e-mail: haeiselt@unb.ca

A. Akbari · R. Pelot
Department of Industrial Engineering, Dalhousie University, 15000, Halifax, NS B3H 4R2, Canada
e-mail: Amin.Akbari@dal.ca

R. Pelot
e-mail: Ronald.Pelot@dal.ca

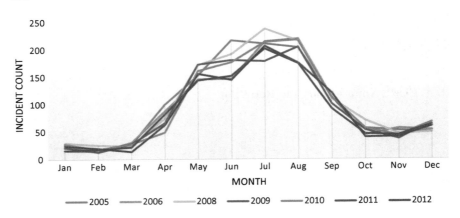

Fig. 1 Incidents distribution by month in the Atlantic region

time it takes to reach the site of an incident with the appropriate vessel in order to render assistance. The Coast Guard in Atlantic Canada has a total of 24 vessels, which include lifeboats, multi-tasking ships and offshore patrol vessels. Since the available data do not distinguish between different types of calls and their respective requirements regarding equipment, we will locate the existing vessels based on their speed and other features. A second criterion concerns the number of vessels that are "within reach" of the Coast Guard and, by extension, help. In order to operationalize this measure, we will define an acceptable response time, i.e., the time between receiving a distress call and the time that the vessel is reached, as six hours. This figure was chosen in consultation with the Coast Guard, who use it as a benchmark. Given that, we will then attempt to maximize the number of vessels that can be reached within that time. The third criterion is based on the possibility of congestion. While past experience indicates that congestion is not really a problem, this may become an issue at peak times. For that reason, we have decided to also include backup coverage as an objective to be maximized.

Given that vessels are not restricted to specific routes (in contrast to commercial traffic that is restricted to established shipping lanes), suitably modified Euclidean distances are an appropriate tool for the task. They should be modified, though, in order to include land avoidance for some trips close to the shoreline. The use of Euclidean distances will make the problem nonlinear and as such presents computational problems. In order to avoid these, we have chosen to digitize the model by defining cells, whose size ranges from ¼ × ¼ degree close to the shoreline, four times this size farther from shore, and sixteen times that size in outlying areas that had very low incident rates in the past. The digitization will allow us to formulate our model in integer terms. Another concern relates to the historic data. While the past is often a good guide to the future, it is realistic to assume that incidents do not generally occur at the same sites at which they occurred in the past. Instead, we make the weaker assumption that incidents occur with the same probability they did in the

Fig. 2 Number of past incidents in cells

past. For that purpose, we determine the relative frequency of incidents happening in the individual cells and use those frequencies in different scenarios via kernel density estimation. In each of these scenarios, we simulate specific incidents and determine the solution that best satisfy our objectives.

In order to provide a visualization of the problem, Fig. 2 shows the past incidents in the cells. Darker colors indicate more incidents. It is apparent that concentrations of incidents are found in the Grand Banks east of Newfoundland (the area also includes the Hibernia oil fields), the southwestern shore of Nova Scotia, and areas in the Northumberland Strait and areas close to the shore of Newfoundland.

2 The Model

This section will develop the main features of the model. For further details, readers are referred to [2]. The objective function includes:

(1) primary coverage at each possible cell location (defined as the expected number of customers that are covered, i.e., are located in the predefined coverage radius centered at each of the cells,
(2) backup coverage, defined as the expected number of incidents in a scenario that are covered by at least two vessels, and
(3) the mean access time between a customer and the vessel that is located closest to him.

Note that the first two objectives are of the "maximization" type, while the third objective is of the "minimization" type.

The formulation of the model then integrates the three objectives as follows. First there are constraints that require each of the objectives not to fall short of a value that represents a minimal requirement. This is an absolute constraint that cannot violated. Secondly, the decision maker defines target values. These are optional and the optimizer will attempt to minimize the weighted deviation of a solution's consequence (i.e., its primary coverage, backup coverage, and average access time) from the target values. Clearly, the deviations only measures the differences between the target values and the undesirable outcomes, i.e., the coverage that falls short of the target value and the access times that exceed the target values. Among the constraints are also capacity restrictions as well as limitations of pre-positioning rescue vessels at offshore locations. The latter constraints are necessary as some small boats do not allow overnight stays of the crew.

This model is a weighted goal programming problem of the type discussed in many texts and research papers; see, e.g., [3–5]. This type of model was chosen, as it allows maximal flexibility. In particular, while the different scenarios allow us to avoid explicit inclusion of probabilistic features in the model, the reduction of the multiple objectives to a single weighted objective—while we acknowledge the typical problems with commensurability and interpretation—allows flexibility in modeling and provides insight into the sensitivity of the solutions.

3 Optimization and Computational Results

As applied to the Coast Guard, Atlantic Region, our model has 242,761 variables and 248,849 constraints. We used Gurobi 6.0.4 as a solver with Intel Core i7 CPU and 8 GB of RAM. The solution times were typically below 6 min, with only one instance requiring about 24 min. All different scenarios resulted in 93–95% of primary coverage, 71–83% of backup coverage, and access times of 2.5–2.7 h. In contrast, the present arrangement has a primary coverage of 89.4%, backup coverage of 60.2%, and a mean access time of 3.14 h. In other words, the optimization has produced solutions that represent improvements of 5% over the present primary coverage, 25% better than the present backup coverage, and have 17% better access time.

In order to present further insight into the results, we compare the results of our model with weights of 0.5, 0.2, and 0.3 on the respective objective function components with the solutions of single-objective models that optimize the coverage (the usual max cover problem first formulated in [6]) and the p-median model (first envisaged in [7] and first formulated in [8]). Table 1 compares the results, in which the rows represent the model that is solved, while the columns measure the achievements regarding the three criteria used to evaluate solutions. As an example, solving the max cover problem, the mean access time in the solution is 2.79 h. The other entries in the table are interpreted similarly.

It is apparent that the multiobjective solution has a quality somewhere between that of the max cover problem and the p-median model on primary coverage and mean access time, but it shines when it comes to backup coverage, which is more than ten percentage points better than that of either of the other two single-objective models.

Table 1 Comparison of our multiobjective model with single-objective formulations

	Solution (objective values)		
	Primary coverage (%)	Backup coverage (%)	Mean access time (hrs)
Maximal covering	95.2	60.1	2.793
p-Median	94.3	63.1	2.534
Multiobjective model	94.1	74.8	2.611

4 Outlook

This paper has examined a location problem that finds sites for Coast Guard vessels in the Northwest Atlantic along the Canadian coast, so as to optimize three criteria, which are different expressions of the quality of service. The results demonstrate that there is room for improvement and one of the advantages of our solutions is that they can easily be implemented, as they use the resources that are already available and simply move them to different locations.

An extension and a long-term view was taken in [9]. More specifically, the authors assumed that the present resources were all to be sold and new vessels would be purchased with the resulting revenue plus any additional amount, which may become available. The model's objective then minimized the average access time to any of the customers who were covered. Given the Coast Guard's wishes to hold on to their large multitasking vessels, the model ensured that they would be kept. The constraint included a budget constraint, and a constraint that ensured that at least a given proportion α of all customers would be covered. Coverages beyond 95% proved to be impossible with the given budget and average access times were stable at about 2.3 h. What is more interesting, though, is the fact that the optimized mix of vessels is very different from the equipment that is presently at the Coast Guard's disposal. The optimized types of equipment are very stable until the required coverage is high. At that point the solutions become very "jumpy" and unstable.

There are a number of research directions that could be followed in the future. One issue is to model the impacts of cooperation of the Coast Guard with other Government agencies, e.g., air force and navy. In this context, it would also be desirable to include helicopters and fixed-wing aircraft in the model. However, this would require more detailed historical data on their usage in order to incorporate their operating characteristics and the types of incidents that they typically respond to.

References

1. Canadian Coast Guard 2010–2011. *Fleet Annual Report*. Retrieved June 13, 2017, from http://www.ccg-gcc.gc.ca/folios/00888/docs/FAR-RAF-2010-2011-eng.pdf.
2. Akbari, A., Pelot, R., & Eiselt, H. A. (2017). A modular capacitated multi-objective model for maritime search and rescue location. *Annals of Operations Research*, July 2017. Retrieved from https://link.springer.com/content/pdf/10.1007%2Fs10479-017-2593-1.pdf.
3. Romero, C. (2014). *Handbook of critical issues in goal programming*. Oxford, UK: Pergamon Press.
4. Jones, D., & Tamiz, M. (2010). *Practical goal programming*. International Series in Operations Research and Management Science, vol. 141, Springer Science+ Business Media, New York.
5. Aouni, B., & Kettani, O. (2001). Goal programming model: a glorious history and a promising future. *European Journal of Operational Research, 133*(2), 225–231.
6. Church, R. L., & ReVelle, C. S. (1974). The maximal covering location problem. *Papers of the Regional Science Association, 32*, 101–118.
7. Hakimi, S. L. (1964). Optimal locations of switching centers and the absolute centers and medians of a graph. *Operations Research, 12*, 450–459.
8. ReVelle, C. S., & Swain, P. W. (1970). Central facilities location. *Geographical Analysis, 2*(1), 30–42.
9. Akbari, A., Eiselt, H. A., MacMackin, W. D., & Pelot, R. P. (2017). Determining the optimal mix and location of search and rescue vessels for the Canadian Coast Guard. *International Journal of Operations and Quantitative Management, 23*(2), 131–146.

Part V
Discrete and Integer Optimization

A Linear Program for Matching Photogrammetric Point Clouds with CityGML Building Models

Steffen Goebbels, Regina Pohle-Fröhlich and Philipp Kant

1 Introduction

CityGML is an XML-based description standard for city models (see [7]). Such models are used for cadastral, planning and simulation purposes [2]. Each CityGML polygon has a semantic meaning. Thus it represents either a wall or a roof facet or a door, etc. On the other hand, textured (photo) meshes are often used for 3D visualization. They just represent triangulated surfaces without considering the types of objects they show. Their vertices can be taken from photogrammetric point clouds, and triangles can be textured using the photogrammetric input data.

Most current CityGML models are given in a level of detail that requires buildings to only have walls, roofs, and a ground plane. To add detailed facades, we want to map textured meshes onto CityGML polygons, see Fig. 1. Based on such textures, detection of windows and doors can be performed.

To generate point clouds and textured meshes, one can use overlapping photos or videos. Depending on the material's source there might or might not be georeferencing. In general, it is possible to manually do a coarse registration with, for example, the UTM (ETRS89) coordinate system. But precision might be not adequate to directly map textured meshes to city models. One needs an automated adjustment of the given coarse point cloud registration.

The Iterative Closest Point algorithm (ICP) is the standard non-feature-based approach to align a roughly calibrated point cloud P with a calibrated cloud Q. It

---

S. Goebbels (✉) · R. Pohle-Fröhlich · P. Kant
iPattern Institute, Niederrhein University of Applied Sciences, 47805 Krefeld, Germany
e-mail: steffen.goebbels@hs-niederrhein.de

R. Pohle-Fröhlich
e-mail: regina.pohle@hs-niederrhein.de

P. Kant
e-mail: philipp.kant@stud.hn.de

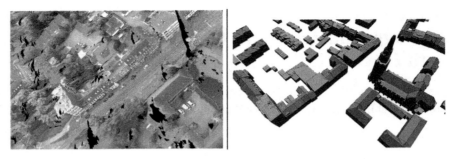

Fig. 1 The textured mesh to the left has to be aligned with the city model to the right

estimates a transformation matrix A iteratively by greedily assigning each point of cloud P to its nearest neighbor in cloud Q, measured in Euclidean l^2 norm, cf. [9].

We generate point clouds and corresponding textured meshes from internet UAV videos. Resulting clouds have large gaps but cover multiple building corners. For principal, a direct ICP registration is possible with a point cloud Q that is sampled from the city model and enriched with points of a digital terrain model. In our case, Point Cloud Library's ICP based on Singular Value Decomposition does not yield reliable results or converges slowly in point-to-point mode. It performs better in point-to-plane mode. However, running times exceed those of Linear Programs by powers of ten. Therefore, we try to detect a set P of building corners in the photogrammetric cloud. Then we align it with a set Q of vertices taken from a CityGML model. Unfortunately, P and Q are both small and not all elements of P correspond with model vertices in Q. For the data shown in Fig. 1, Point Cloud Library's ICP, applied to align feature set P with Q, does not find enough correspondences. Instead, we propose to use a Mixed Integer Linear Program (MIP) or a Linear Program (LP) relaxation of this MIP. To obtain an LP description, we measure absolute distances in l^1 instead of l^2 norm, see [4] for implications.

LP and MIP are established means in 3D modeling. For example, in [3] a MIP is used to reconstruct surfaces from point clouds. An LP is used in [6] to heal non-planarity in 3D city models. Coarse registration of point clouds based on MIP is described in [10]. Also, a MIP is used to compute non-rigid matchings between 3D shapes based on small surface patches [12]. Wang et al. [8, 11] use a MIP that is a linearized min-max version of the quadratic assignment problem. They align two sets so that distances between points in one set correspond to distances between matched points in the second set. But as in the definition of the quadratic assignment problem, they match all points of the first set. In our application there might exist no counterpart of a point of one set within the other.

We use an LP relaxation of a MIP to match a largest subset P' of the cloud's building corners P with cadastral vertices Q. Finally, we compute a transformation matrix A that adjusts P' with correspondences in Q by executing another LP.

2 Detection of Corner Points and Linear Programs

To detect building corners, we first rotate the point cloud so that walls become vertical. We do this based on RANSAC estimates of wall planes. Then we project points to a density greyscale picture that represents the x-y-plane. Thereby, we exclude green points because they most likely belong to vegetation and not to buildings. The number of points projected to the same pixel determines the grey value of this pixel so that, because of the initial rotation, vertical walls become clearly visible dense lines, see Fig. 2. After thresholding this picture to a binary mask, Harris corner detector finds candidates for building corners, see circles in Fig. 2. For each corner we select both its probable intersection point with the ground and its probable intersection point with a building's roof. To this end we look for the smallest and the greatest z-coordinates of all points within a surrounding (with radius $\frac{1}{3}$ m) of each candidate. This results in two building corner points. Let $P \subset \mathbb{R}^4$, $P = \{p_1, \ldots, p_m\}$ be the set of all corner points in homogeneous coordinates, i.e. the fourth component of each point is set to one. Thus, each corner point p_i is given as a column vector with coordinates $p_i.x$, $p_i.y$, $p_i.z$, 1. We will align P with a set $Q = \{q_1, \ldots, q_n\}$ of vertices from a city

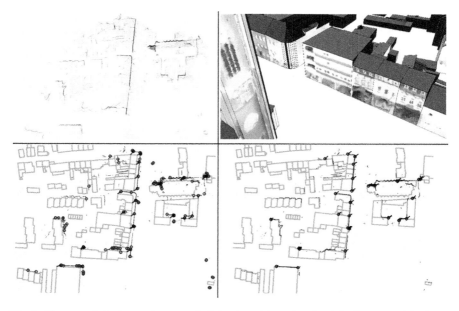

Fig. 2 The upper left picture shows vertical density of the point cloud. In the upper right figure, walls are textured with points that coincide with model facades after registration. Both pictures in the second row show footprints of buildings (grey), candidates of walls detected from density representation of point cloud (black) and their detected corners (circles). To the left, the situation before transformation is shown. Walls of the point cloud differ significantly from footprints. Using LP relaxation we find correspondences between detected corners and vertices of footprints. These correspondences are marked with short black lines. Based on such pairs, the transformation matrix is computed using an LP. The right picture shows the result of transformation. Short black lines visualize corner movements

model that are also given in homogeneous coordinates. To be more concise, we put vertices of CityGML intersection lines between walls and terrain into Q. For each such vertex, we also add the highest roof vertex with same x- and y-coordinates to Q. Therefore, we get points with different height values, i.e. z-coordinates, even for flat terrains. This will allow to compute z-coordinate scaling and translation. Before any computation, we shift coordinates so that P's center of gravity becomes the origin. Small coordinates support numerical stability.

To find a transformation matrix $A := (a_{k,l}) \in \mathbb{R}^{4\times 4}$ with $a_{4,1} = a_{4,2} = a_{4,3} = 0$, $a_{4,4} = 1$, that optimally aligns P with Q, we first define a MIP to detect a maximum set of matching pairs of points $p_i \in P$ and $q_j \in Q$. This is different to the 2D point registration approach of Baird [1] that uses an LP (within a pruned search) to check if a given set of pairs fulfills a registration property.

Since we require a coarse registration, distances between corresponding points can be assumed to be shorter than a threshold value d (we use $d = 6$ m). Therefore, we do not have to consider all pairs of points but only those within $R := \{(i, j) : |p_i - q_j| < d \text{ for } i \in [m] := \{1, 2, \ldots, m\}, j \in [n]\}$.

Binary variables $x_{i,j}$, $(i, j) \in R$, indicate whether points $p_i \in P$ and $q_j \in Q$ match. Then $x_{i,j} = 1$, otherwise $x_{i,j} = 0$. A MIP determines an initial version of matrix A that maps matching points to each other within a certain error bound. In a second step, we then fine-tune the matrix A using an LP.

The MIP's task is to find a largest number of matchings such that a transformation matrix A exists so that for matching pairs (p_i, q_j) the coordinates of $A \cdot p_i$ and q_j are within a small threshold distance $\varepsilon > 0$. Let M be a large number, for example twice the greatest distance between points of P and Q. We find correspondences with the following MIP for $x_{i,j} \in \{0, 1\}, d_{i,j}^+, d_{i,j}^- \in (\mathbb{R}^{\geq 0})^4, (i, j) \in R, a_{k,l} \in \mathbb{R}, k \in [3], l \in [4]$:

$$\text{Max} \sum_{(i,j)\in R} x_{i,j}, \text{ s.t.} \sum_{i\in[m]:(i,j)\in R} x_{i,j} \leq 1 \text{ for } j \in [n], \text{ and } \sum_{j\in[n]:(i,j)\in R} x_{i,j} \leq 1 \text{ for } i \in [m],$$

$$d_{i,j}^+ - d_{i,j}^- = q_j - A \cdot p_i, \max\{d_{i,j}^+.x, d_{i,j}^+.y, d_{i,j}^+.z, d_{i,j}^-.x, d_{i,j}^-.y, d_{i,j}^-.z\} + Mx_{i,j} \leq \varepsilon + M.$$

A maximum might be obtained for a matrix A that cannot be described as a product of matrices for scaling, rotation, and translation. For example, we have to avoid mirroring. Thus, we seek a matrix A that is defined with small angles α, β, and γ near to zero, scaling factors s_1, s_2, s_3 near to one, and offsets d_1, d_2, and d_3 for translations:

$$\begin{pmatrix} s_1(\cos\alpha\cos\gamma - \sin\alpha\sin\beta\sin\gamma) & -s_1(\sin\alpha\cos\gamma + \cos\alpha\sin\beta\sin\gamma) & -s_1\cos\beta\sin\gamma & d_1 \\ s_2\sin\alpha\cos\beta & s_2\cos\alpha\cos\beta & -s_2\sin\beta & d_2 \\ s_3(\cos\alpha\sin\gamma + \sin\alpha\sin\beta\cos\gamma) & s_3(\cos\alpha\sin\beta\cos\gamma - \sin\alpha\sin\gamma) & s_3\cos\beta\cos\gamma & d_3 \\ 0 & 0 & 0 & 1 \end{pmatrix}$$

$$\approx \begin{pmatrix} s_1 & -s_1\alpha & -s_1\gamma & d_1 \\ s_2\alpha & s_2 & -s_2\beta & d_2 \\ s_3\gamma & s_3\beta & s_3 & d_3 \\ 0 & 0 & 0 & 1 \end{pmatrix}.$$

Taylor expansion and omission of even smaller products of small angles lead to the approximate version of A. As a heuristics to reduce the set of feasible matrices we use threshold values $\delta = 0.3$ and $\mu = 0.1$ in connection with

$$1 - \delta \leq a_{i,i} \leq 1 + \delta \text{ for } i \in [3], \quad -\delta \leq a_{i,j} \leq \delta \text{ for } i, j \in [3], i \neq j, \quad (1)$$

$$-\mu \leq a_{1,2} + a_{2,1} \leq \mu, \quad -\mu \leq a_{1,3} + a_{3,1} \leq \mu, \quad -\mu \leq a_{3,2} + a_{2,3} \leq \mu. \quad (2)$$

Instead of calling a MIP solver, it turned out to be sufficient to approximately solve the MIP using LP relaxation. Thus we allow $x_{i,j} \in [0, 1]$. Based on an optimal LP solution, we define the set R' of pairs $(i, j) \in R$ for which $x_{i,j} \geq 1 - 4\frac{\varepsilon}{M}$ (instead of selecting pairs with $x_{i,j} = 1$ in a MIP solution). There exists a linear mapping A that maps p_i to a point that is close to p_j for all $(i, j) \in R'$, i.e. $0 \leq d^{\pm}_{i,j}.x, d^{\pm}_{i,j}.y, d^{\pm}_{i,j}.z \leq \varepsilon + M(1 - x_{i,j}) \leq 5\varepsilon$. There might still be a systematic error up to the magnitude of 5ε. Therefore, we now use a second LP that minimizes distances between coordinates of matching pairs in R'. The LP relaxation also computes a transformation matrix A that can serve as an initial configuration for this second LP. We use the same variable names as for the MIP, i.e. $d^+_{i,j}, d^-_{i,j} \in (\mathbb{R}^{\geq 0})^4$, $(i, j) \in R'$, and $a_{k,l} \in \mathbb{R}$, $k \in [3], l \in [4]$. To compute matrix A we solve

$$\text{Min} \sum_{(i,j) \in R'} (d^+_{i,j}.x + d^+_{i,j}.y + d^+_{i,j}.z + d^-_{i,j}.x + d^-_{i,j}.y + d^-_{i,j}.z)$$

$$\text{s.t.} \quad d^+_{i,j} - d^-_{i,j} = q_j - Ap_i \text{ and conditions (1), (2).}$$

3 Results

We execute LPs with the GNU Linear Programming Kit library. The approach works if a couple of significant corners can be detected. Then results depend on the error bound ε and on the resolution of the density image that is used to detect corners. Data shown in Fig. 1 belong to a point cloud with 5,577,195 points. The cloud has to be matched with 1,440 vertices of the city model. Figure 2 illustrates the outcome for resolution 9 dots/m^2 and $\varepsilon = 0.6$ m. Figure 3 shows distances between transformed corners and model vertices of matching pairs in R'. Overall running time on one i5 processor core is less than one second. Without relaxation, the running time does not change significantly (ε replaced by 5ε for consistency). With increasing resolution also accuracy increases. However, the higher the resolution is, the less visible become walls, and the number of detected corners decreases.

Our approach works for scenes that cover multiple significant building corners. This might not be the case if videos are taken from street level. But then one can detect (few) lines in the density picture and match them with corresponding lines of the city model's building footprints using a MIP, see [5].

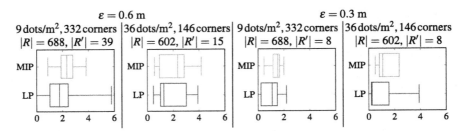

Fig. 3 Distances between transformed corners and CityGML vertices of matching pairs in meters: Boxplots show the results of LP relaxation of the MIP (denoted MIP) and subsequent LP optimization (denoted LP) for density images with 9 and 36 dots/m². Median 0.25 for parameter $\varepsilon = 0.3$ m and resolution 36 dots/m² is within metering precision

References

1. Baird, H. S. (1984). Model-based image matching using location. *ACM Distinguished Dissertation*. MIT Press, Cambridge, Mass.
2. Biljecki, F., Stoter, J., Ledoux, H., Zlatanova, S., & Çöltekin, A. (2015). Applications of 3D city models: State of the art review. *ISPRS International Journal of Geo-Information, 4*, 2842–2889.
3. Boulch, A., de La Gorce, M., & Marlet, R. (2014). Piecewise-planar 3D reconstruction with edge and corner regularization. *Computer Graphics Forum, 33*(5), 55–64.
4. Flöry, S., & Hofer, M. (2010). Surface fitting and registration of point clouds using approximations of the unsigned distance function. *Computer Aided Geometric Design, 27*(1), 60–77.
5. Goebbels, S. & Pohle-Fröhlich, R. (2018) Line-based registration of photogrammetric point clouds with 3D city models by means of mixed integer linear programming. In: *International Conference on Computer Vision Theory and Applications (VISAPP 2018), Funchal*.
6. Goebbels, S., Pohle-Fröhlich, R. & Rethmann, J. (2017). Planarization of CityGML models using a linear program. In: *Operations Research (OR 2016 Hamburg)* (pp. 591–597). Berlin: Springer.
7. Gröger, G., Kolbe, T. H., Nagel, C. & Häfele, K. H. (2012). OpenGIS City Geography Markup Language (CityGML) Encoding Standard. Version 2.0.0. Open Geospatial Consortium.
8. Maiseli, B., Gu, Y., & Gao, H. (2017). Recent developments and trends in point set registration methods. *Journal of Visual Communication and Image Representation, 46*, 95–106.
9. Rusinkiewicz, S. & Levoy, M. (2001). Efficient variants of the ICP algorithm. In: *Third International Conference on 3-D Digital Imaging and Modeling* (pp. 145–152).
10. Sakakubara, S., Kounoike, Y., Shinano, Y., & Shimizu, I. (2007). Automatic range image registration using mixed integer linear programming. In Y. Yagi, S. B. Kang, I. S. Kweon, & H. Zha (Eds.), *Computer Vision - ACCV 2007: 8th Asian Conference on Computer Vision, Tokyo, Japan, November 18–22, 2007, Part II* (pp. 424–434). Berlin, Heidelberg: Springer.
11. Wang, Y., Moreno-Centeno, E., & Ding, Y. (2017). Matching misaligned two-resolution metrology data. *IEEE Transactions on Automation Science and Engineering, 14*(1), 222–237.
12. Windheuser, T., Schlickewei, U., Schmidt, F. R., & Cremers, D. (2011). Large-scale integer linear programming for orientation preserving 3D shape matching. *Computer Graphics Forum, 30*(5), 1471–1480.

A Modified Benders Method for the Single- and Multiple Allocation P-Hub Median Problems

Hamid Mokhtar, Mohan Krishnamoorthy and Andreas T. Ernst

1 Introduction

Hubs are employed in several network design contexts that involve flow interchange between nodes. Hubs are used in the design of, for example, airline networks, parcel delivery networks, and telecommunications networks. Flow between nodes (referred to as *access nodes*) is routed via hubs, which acts as a consolidator and forwarder. The (volume) flow between the hubs is discounted because of the large volumes that accrues from flow consolidation. Hub location design problems then determine the location of the hubs and the allocations of access nodes to the hubs. Thus, through the use of hubs, origin-destination flows/demands can be fulfilled using a smaller number of links, while delivering economies of scale [4]. Usually it is assumed that the hubs do not have capacity/flow restrictions, and the hubs are fully connected. Then, given a positive integer p, we either get the uncapacitated single allocation p-hub median problem (USApHMP) if each access-node is allocated to exactly one hub, or the uncapacitated multiple allocation p-hub median problem (UMApHMP) if access nodes can be allocated to multiple hubs.

After seminal works of O'Kelly [12, 13] on the hub location problem (HLP), a few hub median problems were introduced and formulated by Campbell [2, 3]. Ernst and Krishnamoorthy [7–9] developed and presented a compact 3-index formulation for USApHMP and UMApHMP and provided exact solution approaches for these problems. Recently, modified Benders decomposition methods have been developed, with remarkable success, for solving some classes of HLPs (see [5, 6]). In many cases degeneracy in the subproblems of the method could lead to slow convergence of the method. This issue was noticed and addressed by Magnanti and Wong

---

H. Mokhtar (✉) · M. Krishnamoorthy
Dept. Mech. and Aerospace Eng., Monash University, Clayton, VIC 3800, Australia
e-mail: h.mokhtar@uq.edu.au

A. T. Ernst
School of Math. Sciences, Monash University, Clayton, VIC 3800, Australia

[11] who introduced the generation of 'pareto optimal' cuts using some 'core point'. This improvement is employed in many implementations of the Benders method, including the one by Contreras et al. [5] for HLP with multiple allocation. In the current paper we take advantage of these pareto optimal cuts. We then remarkably enhance this approach by choosing better core points and generating stronger cuts. We also come up with a more efficient approach for solving subproblems to generate cuts by converting the subproblems to minimum cost network flow problems. Through the use of more effective Benders cuts and more efficient solution of subproblems, our method results in fewer iterations and faster running times. We believe this paper is the first implementation of Benders method to solve USApHMP and UMApHMP.

2 Problem Statement

We are given a set of n nodes $N = \{1, 2, \ldots, n\}$, distances between each pair of nodes, d_{ij}, and a positive integer p with $n \geq p$. We consider a complete digraph $G = (N, A)$, where $A = N \times N$ so that the weight of each link is the distance of its endpoints. We suppose that hubs are connected through a complete graph on the set of hubs, and non-hub nodes are only connected to hubs. For every pair of nodes (i, j), $W_{ij} \geq 0$ denotes the amount of flow demand from i to j.

In practice, flow between hubs is discounted by a *transfer* coefficient α, and flow from a non-hub to a hub, and flow from a hub to a non-hub have *collection* (χ) and *distribution* (δ) cost coefficients respectively. Usually $\alpha \leq 1$, $\chi \geq \alpha$ and $\delta \geq \alpha$ in practical applications. The problem of locating p hubs among $n \in N$ nodes, and allocating each non-hub node to exactly one hub with minimum total cost of fulfilling flow demands is USApHMP. When each non-hub node can be allocated to arbitrary number of hubs, the problem is UMApHMP.

We may assume that all flow must be routed through at most two (not necessarily distinct) hubs since using two hubs is always cheaper than using three or more hubs because of the triangular inequality assumption. Therefore, any path between i and j must contain three links, (i, k), (k, l), and (l, j), where i and j are connected to hubs k and l respectively. We denote such a path by i-k-l-j. Let x_{ijkl} be the fraction of flow request W_{ij} that is sent on the i-k-l-j path, for all $i, j, k, l \in N$. Let binary decision variable $h_k = 1$ if node k is chosen as hub, and $h_k = 0$ otherwise, for each $k \in N$. Let binary decision variable $z_{ik} = 1$ if node i is connected to hub k, and $z_{ik} = 0$ otherwise, for all $i, k \in N$. Then the cost of using i-k-l-j path, considering the cost coefficients of different link types, is $C_{ijkl} = \chi d_{ik} + \alpha d_{kl} + \delta d_{lj}$. Furthermore, the establishment of node $k \in N$ as a hub corresponds to a fixed cost F_k. An integer linear programming formulation of USApHMP is presented below:

$$\min \quad \sum_{k \in N} F_k h_k + \sum_{i \in N} \sum_{j \in N} \sum_{k \in N} \sum_{l \in N} C_{ijkl} W_{ij} x_{ijkl} \qquad (1)$$

$$\text{s.t.} \quad \sum_{k \in N} h_k = p, \qquad (2)$$

$$\sum_{k \in N} z_{ik} = 1, \qquad \forall i \in N \qquad (3)$$

$$z_{ik} \leq h_k, \qquad \forall i, k \in N \qquad (4)$$

$$\sum_{k \in N} \sum_{l \in N} x_{ijkl} = 1, \qquad \forall i, j \in N \qquad (5)$$

$$\sum_{l \in N} x_{ijkl} \leq z_{ik}, \qquad \forall i, j, k \in N \qquad (6)$$

$$\sum_{k \in N} x_{ijkl} \leq z_{jl}, \qquad \forall i, j, l \in N \qquad (7)$$

$$h_k, z_{ik} \in \{0, 1\}, x_{ijkl} \geq 0 \qquad \forall i, j, k, l \in N. \qquad (8)$$

By dropping allocation variables z_{ik} and constraints (3)–(4) from the above formulation of USApHMP, and substituting (6)–(7) with the following constraints, we obtain a formulation for UMApHMP:

$$\sum_{l \in N} x_{ijkl} \leq h_k, \qquad \forall i, j, k \in N$$

$$\sum_{k \in N} x_{ijkl} \leq h_l, \qquad \forall i, j, l \in N.$$

Both USApHMP and UMApHMP are NP-hard in general. However, UMApHMP with fixed location of hubs can be solved polynomially [9], but USApHMP for $p \geq 3$ is NP-hard even when the location of hubs are fixed [10]. The large number of variables and constraints corresponding to the flows between pairs of nodes leads us to the idea of solving the problems using Benders decomposition.

3 Benders Decomposition

In Benders decomposition method [1], the original problem is decomposed into a master problem MP, which may consist of integer variables and corresponding constraints, and a subproblem SP, which consists of the remaining variables and constraints. MP and SP are solved iteratively in a dependant manner. In the case that SP is feasible for any feasible solution of MP (such as the problems in this paper), only *optimality* Benders cuts will be added to MP to improve the feasibility of the current MP solution, until no further improvement is needed.

In order to apply Benders decomposition method to USApHMP, in each iteration, the location and allocation variables, $h = (h_k)_{k \in N}$ and $z = (z_{ik})_{i,k \in N}$ respectively, are fixed to some $\hat{h}$ and $\hat{z}$. We then obtain a linear programming subproblem in the iteration, which is the problem of optimal routing between n^2 pairs for specified hubs and allocations by $\hat{h}$ and $\hat{z}$. The subproblem can be further decomposed into n^2 subproblems because we can find an optimal routing for each pair of nodes separately/independently. This decomposition results in n^2 Benders cuts, which may provide tighter cuts for MP, and faster convergence. The dual of SP for any pair of nodes $(i, j) \in N^2$ is as follows:

DS_{ij} : max $f_{ij} - \sum_{k \in N} \hat{z}_{ik} u_{ijk} - \sum_{l \in N} \hat{z}_{jl} v_{ijl}$

s.t. $f_{ij} - u_{ijk} - v_{ijl} \leq C_{ijkl} W_{ij},$ $\forall k, l \in N$ (9)

$u_{ijk}, v_{ijl} \geq 0, f_{ij} \in \mathbb{R}$ $\forall k, l \in N.$ (10)

By an optimal solution $(\hat{f}_{ij}, \hat{u}_{ij}, \hat{v}_{ij})$ of DS_{ij} for each $(i, j) \in N^2$, a Benders cut is generated. Thus, in each iteration we obtain n^2 Benders cuts:

$$\eta_{ij} \geq \hat{f}_{ij} - \sum_{k \in N} z_{ik} \hat{u}_{ijk} - \sum_{l \in N} z_{jl} \hat{v}_{ijl} \quad \forall i, j \in N, \quad (11)$$

where η_{ij} is a real non-negative variable. Then the master problem is:

MP: min $\sum_{k \in N} F_k h_k + \sum_{i \in N} \sum_{j \in N} \eta_{ij}.$

s.t. $(2) - (4), (11), \quad h_k, z_{ik}, \eta_{ij} \geq 0 \quad \forall k, i, j \in N.$

The optimal solution of DS_{ij} is not unique since SP is degenerate. The strength of Benders cuts (11) is dependent on the choice of optimal solutions of DS_{ij}. We maximise a weighted summation of dual variables among optimal solutions of DS_{ij} by defining a slope for the objective function of a second LP to generate different cuts. Let z'_{ik}, z'_{jl} for $k, l \in N$ be non-negative real parameters and m'_0 be a real parameter. Define

DL_{ij} : max $m'_0 f_{ij} - \sum_{k \in N} z'_{ik} u_{ijk} - \sum_{l \in N} z'_{jl} v_{ijl}$

s.t. $f_{ij} - \sum_{k \in N} \hat{z}_{ik} u_{ijk} - \sum_{l \in N} \hat{z}_{jl} v_{ijl} = \hat{\delta}_{ij}$

$(9) - (10),$

where $\hat{\delta}_{ij}$ is the optimal value of DS_{ij}.

The strength of our generated Benders cuts is directly related to the slope of the objective function of the above SP. When $(h'_k, z'_{ik})_{i,k \in N}$ is an interior point of the convex hull of MP (called a 'core point'), and $m'_0 = 1$, we obtain an acceleration of the Benders method proposed by Magnanti and Wong [11], which is shown to generate 'pareto optimal' cuts. Generated cuts by their method, however, might not be the strongest cuts in general. We observed in most cases (across a few test implementations we ran) that a modification of their method – in which the objective function is minimised – results in stronger cuts, fewer number of branch and bound iterations, and faster convergence. So pareto optimality is not sufficient to measure the strength of Benders cuts.

Here we set $z'_{ij} = (z'_{i1}, \ldots, z'_{in}, z'_{j1}, \ldots, z'_{jn})$ for DL_{ij} which may not be equivalent to a core point, but results in stronger cuts for USApHMP. We set

$$z'_{ik} = \Gamma_i/(n-p) \text{ if } \hat{z}_{ik} = 0, \quad z'_{jl} = \Gamma_j/(n-p) \text{ if } \hat{z}_{jl} = 0, \quad (12)$$

where $\Gamma_i, \Gamma_j > 0, z'_{ik} = 1$ if $\hat{z}_{ik} = 1$, and $z'_{jl} = 1$ if $\hat{z}_{jl} = 1$. By this combination, we control the number of non-zero coefficients of allocation variables in Benders cuts which are associated to i-j paths through non-hubs shorter than the shortest path through allocations specified by $\hat{z}$, i-$\hat{k}$-$\hat{l}$-j. We show in next section that, on average, our approach for some choice of z'_{ij} substantially reduces the number of iterations and the computational time.

Another issue we address in this paper is the computationally expensive generation of Benders cuts. By exploiting the structure of SP, we reformulate SP as a minimum cost network flow problem on an auxiliary network with $2n + 2$ nodes and $n^2 + 2n$ arcs, and the capacity of arcs are determined by z'_{ik} for $k \in N$ and z'_{jl} for $l \in N$. The amount of flow for pair (i, j) in this network is $\Gamma_i + \Gamma_j$. This results in generating Benders cuts more efficiently.

With a similar discussion for UMApHMP and a given feasible solution of MP, the corresponding DL$_{ij}$ has the same set of constraints as that of USApHMP, but its objective function is $m'_0 f_{ij} - \sum_{k \in N} h'_k u_{ijk} - \sum_{l \in N} h'_l v_{ijl}$. Then for an optimal solution $(f'_{ij}, \boldsymbol{u}'_{ij}, \boldsymbol{v}'_{ij})$ of DL$_{ij}$, the Benders cuts for UMApHMP are:

$$\eta_{ij} \geq f'_{ij} - \sum_{k \in N} h_k u'_{ijk} - \sum_{l \in N} h_l v'_{ijl} \quad \forall i, j \in N.$$

For UMApHMP we set $h'_k = \Gamma_i/(n - p)$ if $\hat{h}_k = 0$, and $h'_k = 1$ if $\hat{h}_k = 1$.

4 Computational Results

We compare the computational results when SPs are solved by simplex, and minimum cost network flow (MCNF) methods, and also our modified Benders methods for USApHMP and UMApHMP. We suffice to present our computational results since in the literature of the two problems, there is no implementation of Benders method, and not many recent results using exact methods.

Tables 1 and 2 present the computational results of, respectively, USApHMP and UMApHMP tests on the problem instances with n nodes and p hubs in the Australia Post data sets [7]. All methods were coded in C/C++ using CPLEX 12.7, and performed on a PC with 8 cores of 3.6 GHz speed and 32G RAM with time limit of 2 hours. Columns SPLX and MCNF, respectively, show results of computations when SPs are solved using simplex, and MCNF algorithms for some 'core-point' ($h'_k = p/n$ and $z'_{ik} = 1/n$ for $i, k \in N$). The results of our judicious choice of objective function slope of DL$_{ij}$ is presented in column z'-DL$_{ij}$, in which we set $\Gamma_i = 2.4$, $\Gamma_j = 1.6$ in (12). For each node pair we generated two cuts with $m'_0 = 1$ and $m'_0 = -0.2$.

As Tables 1 and 2 show, solving SPs by MCNF significantly improves the computational times (shown in sec.), although in a few cases the number of Benders iterations (B-itr) and branch&bound nodes (BBN) are increased. Note that different methods with the same parameters may get different cuts due to the degeneracy of SPs. On the other hand, SPLX method encounters memory shortage (indicated by 'm') so that it is unable to solve problems with $n \geq 60$. In contrast, MCNF did

Table 1 USApHMP on Australia Post data sets

n	p	SPLX			MCNF			z'-DL$_{ij}$		
		B-itr	BBN	Time	B-itr	BBN	Time	B-itr	BBN	Time
25	3	7	0	59.78	9	11	1.58	4	0	0.37
25	4	14	27	119.46	13	37	3.11	6	2	0.99
50	3	6	0	2773.01	6	14	24.10	6	0	3.26
50	4	7	0	3522.72	9	42	41.58	5	0	3.76
75	3			m	17	57	330.02	6	4	31.32
75	4			m	17	140	534.28	9	28	53.80
100	3			m	19	114	1918.67	8	14	191.97
100	4			m	22	203	1712.07	8	8	106.41
125	3			m	15	125	5122.10	11	23	745.47
125	4			m	20	184	5349.48	14	16	362.12
150	3			m			t	13	8	1759.09
150	4			m			t	32	22	2618.44

Table 2 UMApHMP on Australia Post data sets

n	p	SPLX			MCNF			z'-DL$_{ij}$		
		B-itr	BBN	Time	B-itr	BBN	Time	B-itr	BBN	Time
25	3	12	0	47.50	7	22	0.49	6	24	0.27
25	4	10	0	38.54	10	12	0.59	10	24	0.41
50	3	10	10	2502.39	8	64	11.63	8	54	5.30
50	4	7	12	1980.66	9	74	15.56	8	66	6.77
75	3			m	14	68	112.59	8	52	21.05
75	4			m	10	56	49.95	10	102	41.61
100	3			m	19	76	973.85	10	154	195.21
100	4			m	14	164	634.43	19	184	720.72
125	3			m	16	122	2200.58	15	138	537.01
125	4			m	18	196	914.29	15	132	613.04
150	3			m			t	19	72	1083.39
150	4			m			t	28	190	4721.91
200	3			m			t	10	72	2427.48
200	4			m			t	27	94	5021.77

not have memory issue for tested problems, but was unable to solve instances with $n \geq 150$ in 2 h (indicated by 't').

Our modified Benders method outperforms the traditional accelerated Benders method in the computational times for all cases, and also in B-itr and BBN for most cases. In particular, it solved a few instances which other methods were not able to solve. Thus our approach enables us to solve larger problems.

References

1. Benders, J. F. (1962). Partitioning procedures for solving mixed-variables programming problems. *Numer. Math.*, *4*(1), 238–252.
2. Campbell, J. F. (1994). Integer programming formulations of discrete hub location problems. *Eur. J. Oper. Res.*, *72*(2), 387–405.
3. Campbell, J. F. (1996). Hub location and the p-hub median problem. *Oper. Res.*, *44*(6), 923–935.
4. Campbell, J. F., & O'Kelly, M. E. (2012). Twenty-five years of hub location research. *Transp. Sci.*, *46*(2), 153–169.
5. Contreras, I., Cordeau, J.-F., & Laporte, G. (2011). Benders decomposition for large-scale uncapacitated hub location. *Oper. Res.*, *59*(6), 1477–1490.
6. de Camargo, R. S., Miranda, Gd, & Luna, H. (2008). Benders decomposition for the uncapacitated multiple allocation hub location problem. *Comput. Oper. Res.*, *35*(4), 1047–1064.
7. Ernst, A. T., & Krishnamoorthy, M. (1996). Efficient algorithms for the uncapacitated single allocation p-hub median problem. *Locat. Sci.*, *4*(3), 139–154.
8. Ernst, A. T., & Krishnamoorthy, M. (1998a). An exact solution approach based on shortest-paths for p-hub median problems. *INFORMS J. Comput.*, *10*(2), 149–162.
9. Ernst, A. T., & Krishnamoorthy, M. (1998b). Exact and heuristic algorithms for the uncapacitated multiple allocation p-hub median problem. *Eur. J. Oper. Res.*, *104*(1), 100–112.
10. Love, R. F., Morris, J. G., & Wesolowsky, G. O. (1988). *Facilities location*.
11. Magnanti, T. L., & Wong, R. T. (1981). Accelerating Benders decomposition: Algorithmic enhancement and model selection criteria. *Oper. Res.*, *29*(3), 464–484.
12. O'Kelly, M. E. (1986). The location of interacting hub facilities. *Transp. Sci.*, *20*(2), 92–106.
13. O'Kelly, M. E. (1987). A quadratic integer program for the location of interacting hub facilities. *Eur. J. Oper. Res.*, *32*(3), 393–404.

The Ubiquity Generator Framework: 7 Years of Progress in Parallelizing Branch-and-Bound

Yuji Shinano

1 Introduction

This paper deals with the *Ubiquity Generator (UG) framework* [20, 24], a software package that allows to parallelize branch-and-bound (B&B) solvers—in particular solvers for mixed integer linear programming (MILP) problems. The standard algorithm used to solve MILP is an LP-based branch-and-bound with many advanced procedures such as primal heuristics, preprocessing and conflict analysis, which implicitly enumerates the whole solution space to find an optimal solution. The reader is referred to [10] for details about these procedures and the latest survey of parallel MILP solvers. This paper presents the ground design and general features of UG, current development based on it, and discusses 7 years of progress in parallelizing branch-and-bound solvers with UG.

2 Towards a General Branch-and-Bound Parallelization

Standardization of the message passing interface started in the mid-90s. In the same period of time, general parallel branch-and-bound software framework/library development started [2, 21, 23]. Comparing between a sequential sophisticated B&B implementation and a naive parallel B&B one for solving an optimization problem,

---

This work has been supported by the Research Campus MODAL *Mathematical Optimization and Data Analysis Laboratories* funded by the Federal Ministry of Education and Research (BMBF Grant 05M14ZAM). All responsibility for the content of this publication is assumed by the authors.

Y. Shinano (✉)
Zuse Institute Berlin, Takustr. 7, 14195 Berlin, Germany
e-mail: shinano@zib.de
URL: http://www.ug.zib.de/

the former is overwhelmingly high-performance in terms of solvability. In order to investigate effectiveness of parallelization for a sophisticated B&B implementation, the CPLEX solver was parallelized by using PUBB2 [19]. However, it soon turned out that parallelizing a black-box solver with a general parallel B&B framework does not easily lead to a significantly enhanced performance. Therefore, the development of ParaLEX [18] was started, which was specialized for the CPLEX solver and could run on distributed memory environments. Yet, when ParaLEX was redesigned in 2008 [13] by the author of this article, the idea of developing a general software framework to exploit state-of-the-art MILP solvers re-emerged and subsequently gave rise to the UG framework described in the following.

2.1 The Ubiquity Generator (UG) Framework

UG is a generic framework to parallelize any existing state-of-the-art B&B based solver, subsequently referred to as *base solver*. UG is composed of a collection of base C++ classes, which define interfaces that can be customized for any base solver and allow descriptions of subproblems and solutions to be translated into a solver independent form. Additionally, there are base classes that define interfaces for different message-passing protocols. Implementations of ramp-up, dynamic load balancing, and check-pointing and restarting mechanisms are available as a generic functionality (see details in [20]). The B&B tree is maintained as a collection of subtrees by the base solvers, while UG only extracts and manages a small number of subproblems from the base solvers for load balancing.

The concept of UG is thus to abstract from a base solver and parallelization library and to provide a framework that can be used, in principle, to parallelize any powerful state-of-the-art base solver on any computational environment. For a particular base solver, only the interface to UG in form of specializations of base classes needs to be implemented. Similarly, for a particular parallelization library, a specialization of an abstract UG class is necessary.

A particular instantiated parallel solver is referred to as ug [a specific solver name, a specific parallelization library name]. Here, the specific parallelization library is used to realize the message-passing based communications. In [10], recent parallel MILP solvers are summarized in terms of aspects such as load coordination mechanisms, or granularity of working unit. According to the term defined in [10], UG employs a Supervisor-Worker coordination mechanism with subtree-level parallelism (the unit of work is a subtree). One of the most important characteristics of UG is that it makes algorithmic changes to that of the base solver, such as multiple presolving, and performs very adaptive algorithms, such as racing ramp-up [20] and distributed domain propagation [5].

2.2 Instantiated Parallel Solvers by UG

The following parallel solvers are instantiated by UG. The current distribution of UG has the capability to use the parallelization libraries MPI (Message Passing Interface) and pthreads (POSIX Threads).

Academic solver SCIP as the base solver Two solvers have been developed for the academic SCIP solver [11], ParaSCIP (= ug [SCIP, MPI]) [14] and FiberSCIP (= ug [SCIP, Pthreads]) [20]. Algorithmically, both solvers are identical, since they are parallelized by the same software framework UG. The run-time behavior has been investigated in detail for the MIPLIB2010 benchmark instances by using FiberSCIP. ParaSCIP successfully solved 14 previously unsolved instances from MIPLIB2003 and MIPLIB2010 as of writing this document [15, 16]. The longest and the biggest scale computation conducted to solve an open instance by ParaSCIP is presented in Fig. 1. The rmine10 instance was solved for the first time with 48 restarted runs from checkpoint files that were generated by previous runs using between 6144 and 80,000 cores. In total, it took about 75 days and 6,405 years of CPU core hours.

Commercial solver FICO Xpress as the base solver Two solvers have been developed for the commercial Xpress, the solvers ParaXpress (= ug [Xpress, MPI]) and FiberXpress (= ug [Xpress, Pthreads]). Xpress itself is a shared memory parallel MILP solver. Therefore, FiberXpress can be viewed as a multi-level threaded parallel shared memory MILP solver. When there is more than one core, it is necessary to decide how cores are assigned to UG threads and how many to the Xpress threads. The assignment also changes the solving behavior of the algorithm. ParaXpress has the same assignment issue in between UG processes and FICO Xpress internal threads. The difference in assignments was investigated in [17].

Distributed memory parallel solver PIPS-SBB as the base solver UG has also been used to parallelize the PIPS-SBB [8] solver for two-stage stochastic programming problems (ug [PIPS-SBB,MPI]) [9]. PIPS-SBB can solve large-scale LPs on distributed memory computing environments. Therefore, this parallel solver instantiation shows that UG is capable of parallelizing distributed memory parallel base solvers.

2.3 Instantiated Parallel Solvers by ug [SCIP,*] Libraries

UG has been developed mainly in concert with SCIP. Therefore, ug [SCIP,*] is the most mature and also has user-customizable libraries. By using these libraries with the plug-in architecture of SCIP, a customized parallel solver can be developed with minimal effort.

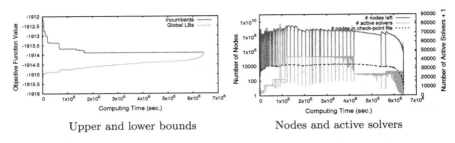

Fig. 1 Evolution of computation for solving **rmine10** by using up to 80,000 cores

ug [SCIP-Jack, *] One of the most successful results of using this development mechanism is the SCIP-Jack solver for Steiner tree problems and its variants: ug [SCIP-Jack, MPI] solved three open instances from the SteinLib [22] benchmark set [4]. The SCIP Optimization Suite contains all source codes of this parallel solver. Only one file with 116 lines of code (without comments) in the source code of ug is required for the parallelization of SCIP-Jack.

ug [SCIP-Scheduler, *] The SCIP applications moreover contain a Scheduler, which is a solver for resource-constrained project scheduling problems [1]. Also this solver can be parallelized by ug [SCIP,*] libraries.

ug [SCIP-SDP, *] The Mixed Integer Semidefinite Programming (MISDP) solver has been developed in a project of SCIP-SDP [12] at TU Darmstadt. The solver is realized as plugin for SCIP. Therefore, this solver can be parallelized by ug [SCIP,*]; libraries and the code will be published in the near future.

3 UG Synthesizer (UGS)

The strategy of composing multiple heuristic algorithms within a single solver that chooses the best suited one for each input is called *algorithm portfolio*. In order to exploit performance variability [3, 6] for MILP solving, a solver may solve an instance in parallel with several different configurations of parameters (including parameter for permutation of columns and rows of input data). This procedure is called *racing* [20]. It is a natural idea to run several (parallel) heuristic solvers together with several B&B based solvers in parallel to share a good primal solution to cut-off search trees in the B&B based solvers. *UG synthesizer (UGS)* is a software framework to realize this strategy on a distributed memory computing environment; it is a general framework to realize any combinations of algorithm portfolio and racing.

Although a parallel solver instantiated by UG has a single executable file and runs as a SPMD (Single Program, Multiple Data) model MPI program, a parallel solver configured by UGS has several executable files and it runs as MPMD (Multiple Program, Multiple Data) model MPI program. In UGS solvers, a heuristic solver or a B&B solver has a separate executable file and is referred to as a UGS solver that can be a distributed memory parallel solver. Currently, shared memory parallel B&B UGS solvers ugs_Xpress, ugs_CPLEX, and ugs_Gurobi have been developed. As for these solvers, the corresponding commercial solvers are extended to run as UGS solvers. And also, distributed memory heuristic UGS solvers ugs_PAC_CPLEX and ugs_PAC_Xpress have been developed. These are implementations of alternative criteria search [7] using different MILP solvers. Any ug [*,*] solver can run as a UGS solver. UGS provides one special executable file ugs, which mediates incumbent solutions among the UGS solvers.

A parallel UGS solver can be configured at run-time flexibly. For example, it runs with ugs, ugs_Xpress1, ugs_Xpress2, ugs_CPLEX1, ugs_PAC_CPLEX1 and ug [Xpress,MPI]1. The different numbers at the end of the same solver name denote the multiple solvers for one UGS solver can run in parallel with different parameter settings. The solver configuration is specified by a special file and is passed to each solver at run-time. Therefore, the configuration can be decided flexibly depending on the computing environment used to solve an instance. On top of that, whenever a new promising algorithm implementation has appeared, a new UGS solver can be added without any modification of the other existing solvers, since the executable file is separate.

4 Concluding Remarks

Some of the instances solved by ParaSCIP for the first time are currently solvable by commercial solvers on a common desktop machine in a reasonable amount of time. This can be taken as an indication that algorithmic improvements are more crucial than parallelizations. Nevertheless, by providing a way to apply large-scale parallelization to the latest algorithm implementations, UG has in many cases succeeded to "look ahead in time" and in some cases helped to guide sequential solver development.

Besides this fact, a solver instantiated by UG causes algorithmic changes to that of the base solver by adding more cores, though the program code is the same. An open question is whether this kind of algorithmic change can fundamentally help to increase the solvability of problems or not.

References

1. Berthold, T., Heinz, S., Lübbecke, M. E., Möhring, R. H., & Schulz, J. (2010). A constraint integer programming approach for resource-constrained project scheduling. In A. Lodi, M. Milano, & P. Toth (Eds.), *Integration of AI and OR techniques in constraint programming for combinatorial optimization problems* (Vol. 6140, pp. 313–317). Lecture notes in computer science. Berlin: Springer.

2. Cun, B. L., & Roucairol, C. (1995). The PNN team: BOB: A unified platform for implementing branch-and-bound like algorithms. Rapports de Recherche 95/16, PRiSM.
3. Danna, E. (2008). Performance variability in mixed integer programming. In: *Presentation, workshop on Mixed Integer Programming (MIP 2008)*, Columbia University, New York. http://coral.ie.lehigh.edu/~jeff/mip-2008/talks/danna.pdf.
4. Gamrath, G., Koch, T., Maher, S. J., Rehfeldt, D., & Shinano, Y. (2017). SCIP-Jack–a solver for stp and variants with parallelization extensions. *Math. Program. Comput.*, 9(2), 231–296.
5. Gottwald, R. L., Maher, S. J., & Shinano, Y. (2016). Distributed domain propagation. ZIB-Report 16-71, Zuse Institute Berlin, Takustr. 7, 14195 Berlin; *Leibniz international proceedings in informatics SEA 2017* (to appear).
6. Koch, T., Achterberg, T., Andersen, E., Bastert, O., Berthold, T., Bixby, R. E., et al. (2011). MIPLIB 2010. *Math. Prog. Comp.*, 3, 103–163.
7. Munguía, L. M., Ahmed, S., Bader, D. A., Nemhauser, G. L., & Shao, Y. (2016). Alternating criteria search: A parallel large neighborhood search algorithm for mixed integer programs. Submitted for publication.
8. Munguía, L. M., Oxberry, G., & Rajan, D. (2016). PIPS-SBB: A parallel distributed-memory branch-and-bound algorithm for stochastic mixed-integer programs. In: *2016 IEEE IPDPSW* (pp. 730–739).
9. Munguía, L. M., Oxberry, G., Rajan, D., & Shinano, Y. (2017). Parallel pips-sbb: Multi-level parallelism for stochastic mixed-integer programs. ZIB-Report 17-58, Zuse Institute Berlin.
10. Ralphs, T., Shinano, Y., Berthold, T., & Koch, T. (2016). Parallel solvers for mixed integer linear programming. Technical Report 16-74, ZIB, Takustr.7, 14195 Berlin.
11. SCIP: Solving Constraint Integer Programs. http://scip.zib.de/.
12. SCIP-SDP: A mixed integer semidefinite programming plugin for SCIP. http://www.opt.tu-darmstadt.de/scipsdp/.
13. Shinano, Y., Achterberg, T., & Fujie, T. (2008). A dynamic load balancing mechanism for new ParaLEX. *Proc. ICPADS, 2008*, 455–462.
14. Shinano, Y., Achterberg, T., Berthold, T., Heinz, S., & Koch, T. (2012). ParaSCIP – a parallel extension of SCIP. In: C. Bischof, H. G. Hegering, W. E. Nagel & G. Wittum (Eds.), *Competence in high performance computing 2010* (pp. 135–148). Springer.
15. Shinano, Y., Achterberg, T., Berthold, T., Heinz, S., Koch, T., & Winkler, M. (2014). Solving hard MIPLIB2003 problems with ParaSCIP on supercomputers: An update. In: *2014 IEEE International Parallel Distributed Processing Symposium Workshops (IPDPSW)* (pp. 1552–1561).
16. Shinano, Y., Achterberg, T., Berthold, T., Heinz, S., Koch, T., & Winkler, M. (2016). Solving open MIP instances with ParaSCIP on supercomputers using up to 80,000 cores. In: *2016 IEEE International Parallel and Distributed Processing Symposium (IPDPS), IEEE Computer Society, Los Alamitos, CA, USA* (pp. 770–779).
17. Shinano, Y., Berthold, T., & Heinz, S. (2016). A first implementation of ParaXpress: Combining internal and external parallelization to solve MIPs on supercomputers (pp. 308–316). Cham: Springer International Publishing. https://doi.org/10.1007/978-3-319-42432-3_38.
18. Shinano, Y., & Fujie, T. (2007). ParaLEX: A parallel extension for the CPLEX mixed integer optimizer. In: F. Cappello, T. Herault, J. Dongarra (Eds.), *Proceedings of Recent advances in parallel virtual machine and message passing interface* (pp. 97–106). Berlin, Heidelberg: Springer. https://doi.org/10.1007/978-3-540-75416-9_19.
19. Shinano, Y., Fujie, T., & Kounoike, Y. (2003). Effectiveness of parallelizing the ILOG-CPLEX mixed integer optimizer in the PUBB2 framework. In H. Kosch, L. Böszörményi, & H. Hellwagner (Eds.), *Proceedings of Euro-Par 2003 Parallel Processing* (pp. 451–460). Berlin Heidelberg: Springer.
20. Shinano, Y., Heinz, S., Vigerske, S., & Winkler, M. (2018). FiberSCIP – a shared memory parallelization of SCIP. *INFORMS J. Comput.*, 30(1), 11–30. https://doi.org/10.1287/ijoc.2017.0762.
21. Shinano, Y., Higaki, M., & Hirabayashi, R. (1995). A generalized utility for parallel branch and bound algorithms. In: *Proceedings of seventh IEEE symposium on parallel and distributed processing* (pp. 392–401). https://doi.org/10.1109/SPDP.1995.530710.

22. SteinLib Testdata Library. http://steinlib.zib.de/steinlib.php.
23. Tschöke, S., & Polzer, T. (1996). Prortabl Parallel Branch-and-Bound Library PPBB-Lib. User manual version 2.0, University of Paderborn.
24. UG: Ubiquity Generator framework. http://ug.zib.de/.

Exploring the Numerics of Branch-and-Cut for Mixed Integer Linear Optimization

Matthias Miltenberger, Ted Ralphs and Daniel E. Steffy

1 Introduction

The branch-and-cut algorithm for mixed integer linear optimization problems (MILPs) combines aspects of the branch-and-bound algorithm with the cutting plane algorithm to strengthen the initial LP relaxation (see [4] for a complete description of these operations and the definitions of these terms). While branching increases the number of subproblems to be solved and should thus be avoided in principle, the addition of too many cutting planes often results in an LP relaxation with undesirable numerical properties. Recent research into the viability of solving MILPs using a pure cutting plane approach has provided some insight into how and why this happens and has explored techniques to generate a sequence of valid inequalities whose addition to the LP relaxation is less likely to cause difficulties [5, 9].

In general, branching and cutting must be used carefully in concert with each other to maintain numerical stability. The effect of these operations on numerics is not well understood, however, and is difficult to control directly. There exists a number of approaches to effectively combine the branching and cutting operations. In some solvers, cutting is only done at the root node, while in others, cuts are added throughout the tree. As with any numerical process, implementations of these solution algorithms use floating-point arithmetic and are subject to accumulation of roundoff errors within the computations. Without appropriate handling of these

---

M. Miltenberger (✉)
Zuse Institute Berlin, 14195 Berlin, Germany
e-mail: miltenberger@zib.de

T. Ralphs
Department of Industrial and System Engineering, Lehigh University,
Bethlehem, PA 18015, USA
e-mail: ted@lehigh.edu

D. E. Steffy
Mathematics and Statistics, Oakland University, Rochester, MI 48309, USA
e-mail: steffy@oakland.edu

errors, the algorithms may return unreliable results, failing to behave or terminate as expected.

Modern MILP solvers use a wide range of techniques to mitigate the difficulties associated with numerical errors. For example, it is standard practice to discard or modify cuts whose coefficients differ significantly in magnitude, since these inequalities are likely to degrade the conditioning of the LP relaxation. This and other techniques help to ensure that the LP relaxation will have better numerical properties and increases the computational stability of the algorithm.

It is well understood that the addition of cutting planes has the potential to negatively impact the numerical properties of the LP relaxation, even after steps have been taken to improve their reliability. On the other hand, branching may counteract this effect to some extent, leading to a more stable algorithm overall. In this paper we seek to carefully investigate the impact of both branching and cutting on the numerical properties of the LP subproblems solved in the branch-and-cut algorithm. The purpose of this work is both to confirm existing folklore, namely that branching improves condition and cutting degrades it, as well as to explore the potential for directly controlling numerical properties through judicious algorithmic choices.

In Sect. 2 we discuss the choice and computation of the basis matrix condition numbers as a measure of numerical stability. In Sect. 3 we describe computational results regarding how branching and cutting affect the condition numbers. Section 4 discusses some implications of our findings and ongoing work.

2 Condition Numbers

The *condition number* of a numerical problem is a bound on the relative change (in terms of a given norm) in the solution to a problem that can occur as a result of a change in the input (see [3] for formal definitions). For example, the condition number of a matrix A is $\kappa(A) = \|A\|_2 \|A^{-1}\|_2$ and yields a bound on how much the solution to the linear system of equations $Ax = b$ might change, relative to a change in the right hand side vector b. For LPs, a handful of different condition numbers have also been defined; a comprehensive treatment of condition numbers for LPs, along with much more general discussion regarding the concept of problem condition, is given in [3].

When LPs are solved by the simplex method, a sequence of basis matrices are encountered (see [4]), each corresponding to a square system of linear equations. Although condition numbers can be defined for LPs themselves, it is the condition number of the basis matrices encountered during a simplex solve (particularly the optimal basis) that is the most relevant measure of numerical stability of the branch-and-cut algorithm. A primary reason for this is that the solution to the LP relaxation is obtained by solving a system of equations involving the basis matrix so that the condition number of this matrix determines the multiplicative effect of numerical errors in the computed cuts.

After applying cutting planes or branching at a node, the resulting modified LP is re-solved. In general, we expect that the newly added cuts or branching inequalities will be binding at the new basic solution, which means that these additional constraints are a factor in determining the conditioning of the basis. Thus, measuring the condition number of these linear systems and how they change as a result of the added cuts or branching inequalities should give some insight into the numerical behavior of the simplex algorithm and ultimately the branch-and-cut algorithm. In this paper, we are looking for overall trends (how much does the addition of cuts *generally* degrade the conditioning), so we consider these numbers in the aggregate and provide some suggestions for visualizing this data.

Since we are interested in an accurate picture, we use the 2-norm power iteration method to determine condition numbers. This method provides an accurate answer, though it is unlikely to be efficient enough for practical use. An excellent discussion on algorithms for condition number estimation is given in [6, Chap. 15].

3 Experiments

To study the effect of cuts on conditioning, we solved a subset of instances from MIPLIB 3 [2], MIPLIB 2003 [1], and MIPLIB 2010 [7] test sets, collecting detailed statistics. The solver used was SCIP 4.0 with the LP solver SoPlex 3.0 [8] (with slight modifications to allow access to the condition number information). We used a time limit of one hour and a node limit of 10,000.

To get a clearer picture, we deactivated many advanced features, such as primal heuristics, domain propagation, and conflict analysis. Furthermore, we only generated Gomory cuts and disabled all other cutting plane generators. While SCIP only applies cutting planes at the root by default, we enabled cut generation at all nodes in order to study how this affects conditioning. Note that although cuts are generated throughout the tree, SCIP still uses a scoring strategy to determine which inequalities should actually be added.

In what follows, we first study how the condition number of the basis matrix evolves at the root node, where the initial LP relaxation is solved and initial rounds of cuts are added, and then study how the condition number of the basis matrices are affected by branching and cutting as the algorithm progresses.

3.1 Root Node Analysis

In general, we expect the condition number of the basis matrix to degrade as a result of operations performed in the root node and our initial computations are aimed at confirming this. Figure 1 shows the condition number of each basis matrix encountered during each iteration of the solution of the initial LP relaxation and during each iteration of the re-solve occurring after adding each round of cuts for

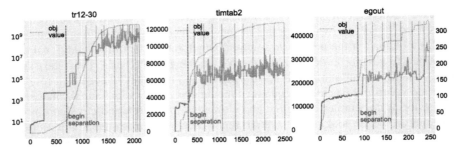

Fig. 1 Condition number development (vertical axis, in log scale) for every simplex iteration in the root node (horizontal axis) including re-optimizations after adding cutting planes in multiple rounds (vertical lines). A plot of objective values at each iteration is overlaid as a dashed gray line with the scales given to the right of each plot

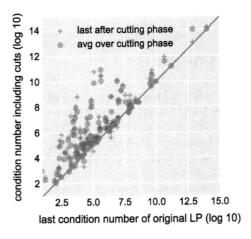

Fig. 2 Root node: Comparison of condition numbers of the original LP and including cutting planes

selected instances from our test set. One can observe that during the early iterations—especially of the initial relaxation in the root node—the condition numbers of the basis matrices grow quickly. This is expected, as more structural variables are pivoted into the basis, while slack variables are pivoted out. Since the initial basis is always the identity matrix, which has condition number 1, the conditioning can only degrade at first. After the initial optimization, the MILP solver tries to generate Gomory cuts. This computation involves the basis matrix itself, so an ill-conditioned basis matrix can prevent precise calculation of the coefficients of the new constraint. Moreover, adding these new rows to the LP often deteriorates its condition number even further as can be seen in Fig. 1. This sample of instances clearly shows the expected behavior.

Figure 2 is a visualization of the difference between the condition number of the optimal basis of the original LP and two other numbers: (1) the average over all bases encountered during the cutting procedure and (2) the condition number of the final optimal basis. While for some instances there is a slight improvement after

Fig. 3 Effects of branching and cutting

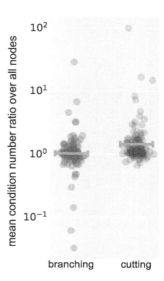

adding cuts, in most cases addition of cuts leads to an increased condition number, as expected.

3.2 Tree Analysis

One way in which the addition of cuts can cause basis matrices to become poorly conditioned is if the associated hyperplanes are nearly parallel; addition of many such cuts may lead to a *tailing off* of the cutting plane algorithm as many similar cuts are generated and the process stalls. Although branching also involves imposing a special kind of "cut" to the resulting subproblems, these branching constraints have a simple form (the coefficient vector is a unit vector), which makes them quite attractive from a numerical point of view. In particular, they are mutually orthogonal and unlikely to degrade the conditioning much in general. As such, we may be tempted to hope that the addition of this special kind of inequality may even improve conditioning.

Despite the apparent plausibility of this hypothesis, our experiments do not fully support it, though they do show a significant difference between the effect of branching versus cutting, as expected. In Fig. 3, we show how branching and cutting impact the numerical stability. The left plot shows the average relative change in the condition number as a result of the addition of the branching constraints. Similarly, the right plot shows the average relative change in conditioning resulting from the addition of cuts. In each case, we took the difference between the condition numbers of the optimal basis matrices before and after either branching or cutting. Each dot then represents the average across all nodes for a given instance. The bar represents the mean over all instances. While branching does not seem to have a significant

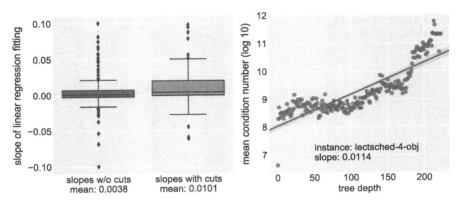

Fig. 4 Condition number development in the tree. Left: Distribution of linear regression slopes of all instances in the test set. Right: Single instance example

effect on average, adding cutting planes clearly leads to an increase in the condition number. Thus, despite the observation that branching does not appear to degrade the condition number in the same way as cut generation, it does not appear to help it either.

In Fig. 4 we visualize how condition numbers degrade generally as a function of the depth of a given node. The idea is to determine whether conditioning generally degrades consistently as the tree gets deeper. The right figure plots the average condition number across all nodes at a given depth, along with a regression line showing the average degradation in the log of the condition number per level in the tree for a single instance. The left figure shows the distribution of slopes of this same linear regression across all instances both with cuts and with a pure branch-and-bound.

It appears that in general, the condition number often has a strong positive correlation with the tree depth if cuts are added throughout the solving process. When cutting is disabled this effect is much less strong. One has to be aware that the behavior of a single instance might be much different from what the trend predicts.

4 Outlook

In this paper, we presented a preliminary exploration of the numerical behavior of SCIP, a state-of-the-art MILP solver. In the future, we hope to do similar explorations with other solvers to determine what the overall behavior is and where additional control of the numerical stability might have an impact. The eventual goal is to determine whether it is possible to more directly estimate the impact of certain algorithmic choices on numerical behavior and whether this could lead to improved control mechanisms.

Acknowledgements The work for this article has been partly conducted within the *Research Campus Modal* funded by the German Federal Ministry of Education and Research (fund number 05M14ZAM). The support of Lehigh University is also gratefully acknowledged.

References

1. Achterberg, T., Koch, T., & Martin, A. (2006). MIPLIB 2003. *Operations Research Letters*, *34*(4), 1–12.
2. Bixby, R. E., et al. (1998). An updated mixed integer programming library: MIPLIB 3.0. *Optima*, *58*, 12–15.
3. Bürgisser, P., & Cucker, F. (2013). *Condition—The geometry of numerical algorithms* (Vol. 349)., Grundlehren der math. Wissenschaften Heidelberg: Springer.
4. Conforti, M., Cornuéjols, G., & Zambelli, G. (2014). *Integer programming*. Berlin: Springer.
5. Fischetti, M., & Salvagnin, D. (2011). A relax-and-cut framework for Gomory mixed-integer cuts. *Mathematical Programming Computation*, *3*(2), 79–102.
6. Higham, N. J. (2002). *Accuracy and stability of numerical algorithms*. Philadelphia: Society for Industrial and Applied Mathematics.
7. Koch, T., et al. (2011). MIPLIB 2010. *Mathematical Programming Computation*, *3*(2), 103–163.
8. Maher, S. J. et al. (2017). *The SCIP optimization suite 4.0*. Technical report 17-12. ZIB.
9. Zanette, A., Fischetti, M., & Balas, E. (2011). Lexicography and degeneracy: can a pure cutting plane algorithm work? *Mathematical Programming*, *130*(1), 153–176.

Four Good Reasons to Use an Interior Point Solver Within a MIP Solver

Timo Berthold, Michael Perregaard and Csaba Mészáros

1 Introduction: MIP and the Analytic Center

Mixed integer programming (MIP) is one of the most important techniques in Operations Research and Discrete Optimization. A mixed integer program is an optimization problem of the form:

$$\min\{c^t x : Ax = b,\ x \geq 0,\ x_I \in \mathbb{Z}^I\}, \tag{1}$$

with matrix $A \in \mathbb{R}^{m \times n}$, vectors $b \in \mathbb{R}^m$ and $c \in \mathbb{R}^n$, and a subset $I \subseteq N := \{1, \ldots, n\}$. The LP relaxation of a MIP is the continuous optimization problem which we get by dropping the integrality requirements of (1). The feasible region of the LP relaxation is a polyhedron. For an introduction to MIP, see [19].

The analytic center x^{ac} of a bounded polyhedron given in equality form ($Ax = b, x \geq 0$) has been introduced by Sonnevend [21] and is defined as

$$x^{ac} = \operatorname{argmin}\{-\sum_{j \in I} \ln x_j : Ax = b\}. \tag{2}$$

The analytic center can be efficiently computed by using a barrier algorithm, see next section. Note that the strong convexity of the logarithm implies that the analytic center of a bounded polyhedron is indeed uniquely defined. Furthermore, it maximizes the

---

T. Berthold (✉)
FICO (Fair Isaac Corporation), Takustr 7, 14195 Berlin, Germany
e-mail: timoberthold@fico.com

M. Perregaard
FICO, Birmingham, UK
e-mail: michaelperregaard@fico.com

C. Mészáros
FICO, Budapest, Hungary
e-mail: csabameszaros@fico.com

distance to the boundary due to the logarithm going towards minus infinity when going to zero. If the polyhedron is a simplex, the analytic center is also the barycenter of the polyhedron [22].

The FICO Xpress Optimization Suite is a toolbox for mathematical optimization [6, 15]. It features software tools used to model and solve linear, integer, quadratic, nonlinear, and robust optimization problems. The core solver of this suite is the FICO Xpress-Optimizer (from here on: Xpress), a state-of-the-art MIP solver which combines ease of use with speed and flexibility. All computational results mentioned in this paper have been conducted with Xpress.

MIP solvers like Xpress feature a variety of algorithmic components which extend the basic branch-and-bound search and which are the reason that modern MIP solvers can solve some of the most complex optimization problems. These include primal heuristics to find feasible solutions, presolving techniques to reduce the problem size and branching strategies to efficiently split the problem into disjoint subproblems. All of the named techniques will be addressed in the present paper.

2 Impact of Barrier for LP Solving

Unlike the simplex algorithm which iterates over extremal vertex solutions, the Newton barrier method iterates through solutions in the interior of the feasible region of the LP [13]. The name barrier comes from replacing the non-negativity constraints in the LP by logarithmic penalty terms, as in (2), and solving the problem

$$\min\{c^t x - \mu \sum_{j \in N} \ln x_j : Ax = b\}. \tag{3}$$

When μ converges to zero, the solution of the barrier problem converges to an optimal solution of the original LP (the barrier solution x^{bar}). The set of solutions for different μ describes the so-called central path which connects x^{ac} with x^{bar}. Note that x^{bar} is the analytic center of the optimal face of the LP. The barrier algorithm can be generalized to convex programming and thereby in particular to convex quadratic programming. Besides the nice property of being of polynomial complexity, the barrier method excels by its practical running time in particular on sparse problems. For an overview on interior point methods, see [20]. The barrier algorithm in Xpress is based on the solver described in [16]. It is a primal-dual algorithm extended by predictor-corrector and target-following techniques.

The observation that there is no clear dominance between simplex and barrier algorithm in terms of solving speed leads to the first, probably most obvious, application of an interior point solver within a MIP solver. For solving the initial LP relaxation, it is the default behavior of Xpress to run primal simplex, dual simplex and barrier (with a subsequent crossover) side-by-side in separate threads. Primal and dual each use one thread, all other threads are occupied by barrier. This method is known as concurrent LP solving. It will be interrupted when one of the algorithm solves the LP relaxation to proven optimality.

For MIP solving, the impact of concurrent LP solving on the overall running time is limited, about 2% speedup on MIPLIB [14]. However, the time for solving the initial LP relaxation improves by 36%. As a side effect of solving the initial LP faster, the time to find a first MIP solution improves by about 4%.

Note that even if barrier plus crossover is not the concurrent winner, barrier alone might have finished before the winning simplex algorithm. In this case, we might use x^{bar} without the need to compute it from scratch.

3 Using the Analytic Center for Presolving

The barrier solution is in the center of the optimal face, hence it minimizes the number of variables which are at their bound. In particular, it maximizes the number of fractional binaries. Similar to simplex solutions, the barrier solution comes with a dual solution, from which a set of reduced costs can be computed. These can be used for reduced cost fixing, a bound tightening algorithm.

The analytic center of the optimal face provides a maximal set of non-zero dual values and hence of non-zero reduced costs. If a variable has a zero reduced cost in this solution, it must have a zero reduced cost in any optimal LP solution. Note that this is different from, but related to a recent work by Bajgiran et al. [3] who compute a set of reduced costs s.t. a maximal set of variables can be fixed w.r.t. a given primal MIP solution value. Conversely, if we wanted to find an integer solution in the optimal face, we could safely fix all variables with non-zero duals to their current bounds. This could, also be used in the context of pump-reduce [1] to reduce the size of the auxiliary LP.

Besides using the dual values, there is a direct, primal way to use the analytic center for fixing variables. Assume for the remainder of this section, that we have an analytic center solution x^{ac} for the whole feasible region. By definition, this solution is strictly in the relative interior of the feasible region. Hence, any variable that is at its bound in this solution must therefore be at its bound in any feasible solution. This allows us to fix and remove such variables from any further consideration during a MIP solve. Principally, the same holds for slack variables. This is potentially beneficial for ranged rows, where fixing the slack will result in the ranged row to be tightened to an equation.

Note that the fixed variables would have been at their bound in any LP or MIP solution anyway. So the direct impact of this presolving step is limited. There is a certain benefit from the sheer reduction in the problem size if many variables can be fixed that way. Additional benefit might occur indirectly from extra presolving that has been enabled by the analytic-center-based fixings. In our computational experiments, we observed about 2% improvement from fixing variables w.r.t. the analytic center of the whole problem. The impact of using reduced costs from barrier solutions was performance neutral.

4 An Analytic Center Heuristic

Naoum-Sawaya [18] introduced *(recursive) central rounding*, which is based on the idea to round the analytic center to the nearest integer vector. Their intuition was that for general integer variables, a point in the "middle" of the relaxation's feasible region is more likely to have an integer feasible solution in its vicinity as compared to an extremal solution of the relaxation.

We suggest a different way to interpret the analytic center of the whole problem in a heuristic context. It can be seen as an indicator for the direction into which a variable is likely to move when going towards feasibility. This is particularly interesting for variables that are close to zero or one in a pure binary problem. However, not all of them might be simultaneously set to their corresponding bound value; thus a pure rounding approach might not be sufficient. We propose to rather apply a soft rounding in a large neighborhood search fashion, compare, e.g., proximity search [9].

Thus, we set up an auxiliary objective function, whose coefficients are proportional to the analytic center solution values. That is, the closer a binary is to one in the analytic center solution, the more the objective will try to push it towards one. Further, we tentatively fix some variables that are very close to one of their bounds and finally apply an auxiliary MIP solve with strict working limits on the modified problem. Note that this heuristic, similar to the feasibility pump [8], completely disregards the original objective function of the MIP. Thus, it makes most sense as a start heuristic to find a first feasible solution, not so much as an improvement heuristic.

This heuristic is relatively expensive, as it involves computing the analytic center x^{ac} and solving at least some nodes of a MIP of similar size as the original. Thus, Xpress only uses it in rare cases. For instances for which it is particularly cumbersome to find a feasible solution, however, it makes a big difference. We observed an overall speedup of 40% on the Feasibility benchmark of Hans Mittelmann [17]. This big difference is due to the fact that there are a few instances for which this heuristic is the only one finding a solution within the time limit of that benchmark.

5 Branching w.r.t. Analytic Centers

Finally, we present a branching strategy that makes use of the analytic center of the whole problem. As argued in the previous section, the analytic center can be understood as an indicator in which direction variables are easiest to move while maintaining feasibility. The analytic center branching in Xpress branches on binary variables that are close to one. Additionally, it searches the subtree resulting from the up-branch first.

This is applied only for extremely dual degenerate MIPs and only on the top levels of the branch-and-bound tree. This follows the idea that when the analytic center has only a few binaries close to one, then it is likely that those, or at least most of those, should be one in any optimal MIP solution. While for the variable fixing procedure in Sect. 3 it is important that we use the actual analytic center, for the heuristic in Sect. 4 or for branching, other interior points might work similarly well.

As said, this strategy is only applied in a few cases, but relatively efficient on those. Overall, branching w.r.t. analytic centers gives 2% speedup on MIPLIB. This comes from few instances on which this branching strategy improves performance by orders of magnitude.

6 Conclusion

Taking all the uses of the barrier solver and analytic center solutions together, having a barrier available makes up for 10% speedup in solving MIPs to proven optimality and three more MIPLIB2010 problems being solved by Xpress. The number of branch-and-bound nodes reduces by 6% and the primal-dual integral [5] by 9%.

Within Xpress, the proposed presolving, heuristic, and branching strategy all improve performance, but come with the computational burden of having to compute the analytic center first. For each of the individual procedures, this is a rather big overhead. However given that the analytic center only needs to be computed once to enable the application of all of them, it seems worthwhile to consider further applications of the analytic center within MIP, even if a single application will not justify the computational cost. This includes, e.g., extensions of the feasibility pump which make use of the analytic center [2, 7]. In the present paper, we did not discuss possibilities to use the analytic center for generating cutting planes, as it is done in convex programming [11, 12] or for filtering cuts. Compare also [4, 10] for the use of interior points for cutting plane separation.

References

1. Achterberg, T. (2010). LP basis selection and cutting planes.
2. Baena, D., & Castro, J. (2011). Using the analytic center in the feasibility pump. *Operations Research Letters*, *39*(5), 310–317.
3. Bajgiran, O. S., Cire, A. A., & Rousseau, L. -M. (2017). A first look at picking dual variables for maximizing reduced cost fixing. In *International Conference on AI and OR Techniques in Constraint Programming for Combinatorial Optimization Problems* (pp. 221–228). Springer.
4. Ben-Ameur, W., & Neto, J. (2007). Acceleration of cutting-plane and column generation algorithms: Applications to network design. *Networks*, *49*(1), 3–17.
5. Berthold, T. (2013). Measuring the impact of primal heuristics. *Operations Research Letters*, *41*(6), 611–614.
6. Berthold, T., Farmer, J., Heinz, S., & Perregaard, M. (2017). Parallelization of the FICO xpress-optimizer. *Optimization Methods and Software*, 1–12.
7. Boland, N. I., Eberhard, A. C., Engineer, F. G., Fischetti, M., Savelsbergh, M. W. P., & Tsoukalas, A. (2014). Boosting the feasibility pump. *Mathematical Programming Computation*, *6*(3), 255–279.
8. Fischetti, M., Glover, F., & Lodi, A. (2005). The feasibility pump. *Mathematical Programming*, *104*(1), 91–104.

9. Fischetti, M., & Monaci, M. (2014). Proximity search for 0–1 mixed-integer convex programming. *Journal of Heuristics*, *20*(6), 709–731.
10. Fischetti, M., & Salvagnin, D. (2010). An in-out approach to disjunctive optimization. In *International Conference on Integration of Artificial Intelligence (AI) and Operations Research (OR) Techniques in Constraint Programming* (pp. 136–140). Springer.
11. Goffin, J.-L., & Vial, J.-P. (2002). Convex nondifferentiable optimization: A survey focused on the analytic center cutting plane method. *Optimization Methods and Software*, *17*(5), 805–867.
12. Gondzio, J., Du Merle, O., Sarkissian, R., & Vial, J.-P. (1996). ACCPM–a library for convex optimization based on an analytic center cutting plane method. *European Journal of Operational Research*, *94*(1), 206–211.
13. Karmarkar, N. (1984). A new polynomial-time algorithm for linear programming. In *Proceedings of the sixteenth annual ACM symposium on Theory of computing* (pp. 302–311). ACM.
14. Koch, T., Achterberg, T., Andersen, E., Bastert, O., Berthold, T., Bixby, R. E., et al. (2011). MIPLIB 2010. *Mathematical Programming Computation*, *3*(2), 103–163.
15. Laundy, R., Perregaard, M., Tavares, G., Tipi, H., & Vazacopoulos, A. (2009). Solving hard mixed-integer programming problems with Xpress-MP: a MIPLIB 2003 case study. *INFORMS Journal on Computing*, *21*(2), 304–313.
16. Mészáros, C. (1999). The bpmpd interior point solver for convex quadratic problems. *Optimization Methods and Software*, *11*(1–4), 431–449.
17. Mittelmann, H. Benchmarks for optimization software: Feasibility benchmark. http://plato.asu.edu/bench.html.
18. Naoum-Sawaya, J. (2013). Recursive central rounding for mixed integer programs. *Computers and Operations Research*.
19. Nemhauser, G. L., & Wolsey, L. A. (1988). *Integer and combinatorial optimization*. New York: Wiley.
20. Nesterov, Y., & Nemirovski, A. (1994). *Interior-point polynomial algorithms in convex programming.*, Studies in applied and numerical mathematics Philadelphia: Society for Industrial and Applied Mathematics.
21. Sonnevend, G. (1986). An "analytical centre" for polyhedrons and new classes of global algorithms for linear (smooth, convex) programming. In A. Prékopa, J. Szelezsáan, & B. Strazicky (Eds.), *System modelling and optimization* (Vol. 84, pp. 866–875)., Lecture notes in control and information sciences Berlin: Springer.
22. Sonnevend, G. (1989). Applications of the notion of analytic center in approximation (estimation) problems. *Journal of Computational and Applied Mathematics*, *28*, 349–358.

Measuring the Impact of Branching Rules for Mixed-Integer Programming

Gerald Gamrath and Christoph Schubert

1 Introduction

All state-of-the-art solvers for mixed-integer programs (MIPs) are based on the linear programming (LP) based branch-and-bound method [9]. One of the key components of the algorithm is the *branching rule*, which splits the current problem into two or more disjoint sub-problems. How this is done can have a large impact on the solving process and has been subject to intensive research over the last decades, see [3, 4, 6, 7] among many others. When new ideas are presented in publications, they are typically evaluated on a set of benchmark instances and compared to other common rules. One of the most common criteria for comparison is the solving time to optimality, complemented by the number of instances solved to optimality within the given time limit.

In this paper, we focus on another measure that is often used to describe the impact of branching rules: the size of the branch-and-bound tree needed to prove optimality. This gives a good estimate of the effectiveness of branching rules, as those are responsible for building the tree. This estimate, however, can be flawed by side-effects of a branching rule that artificially decrease the tree size. An extreme case of such a branching rule would just solve the problem corresponding to the current node as a sub-MIP. Then, it transfers all solutions back and installs the dual bound computed by the sub-MIP as the local dual bound of the current node. If no limits are applied to the sub-MIP solving process, this branching rule would solve every instance at the root node and clearly dominate all other branching rules in this regard! The reader will probably agree that this is an unfair comparison, because this had nothing to do with branching and the rule did not even do any branching.

G. Gamrath (✉) · C. Schubert
Zuse Institute Berlin, Takustr. 7, 14195 Berlin, Germany
e-mail: gamrath@zib.de

C. Schubert
e-mail: schubert@zib.de

However, it is just exaggerating the flaw present in many comparisons of branching rules with inherent side-effects.

In the next section, we discuss the usefulness of the tree size as a performance indicator for branching and illustrate side-effects of common branching rules. Based on this, we propose the *fair node number*, a new measure for the quality of the decisions taken by a branching rule. Section 3 presents an evaluation of some common branching rules, before we close with concluding remarks.

2 Measuring the Impact of Branching Rules

When evaluating new features of a MIP solver, important insights can be achieved by complementing the solving time with another criterion that is tailored more to the algorithmic change being investigated. In the case of branching rules, the canonical candidate for this is the number of branch-and-bound nodes needed to solve an instance to optimality. This is a direct indicator for the quality of the branching decisions taken, which build the tree and naturally aim at keeping it small. The node number has several positive attributes in this context. It allows to measure the potential of a branching rule even with a first prototype that is not necessarily implemented very efficiently. This allows to evaluate research ideas at an early stage and might motivate further investigation of ideas that show some potential without already reducing the solving time. From a practical viewpoint, it is not depending on reliable time measurements, which allows to run multiple experiments in parallel on the same machine. Finally, the tree size reduction may become more important than sequentially measured running times when switching to a massive parallel environment, where distributing nodes of the tree causes a message passing overhead. Summing up, there is a strong incentive to use the size of the branch-and-bound tree created by a branching rule as a measure for its performance in addition to the solving time.

When comparing node numbers between different branching rules, we must only take into account instances solved to optimality within the time limit by all of the rules. Otherwise, final node numbers are unknown and cannot be compared. If a solver timed out, a smaller node count might be an indicator for better decisions creating a smaller tree but could as well point to a slowdown caused by the branching rule, resulting in a larger solving time with a similar or even higher node count.

Many publications follow this approach. However, side-effects of branching rules are often disregarded, but can have a huge impact, see the motivation in the introduction. In common branching rules, side-effects are often encountered when the branching rule uses some form of *strong branching* (SB) [7]. SB evaluates a potential branching decision by solving an auxiliary LP for each of the potential child nodes. In the first place, this provides very reliable predictions of the dual bounds improvement in both child nodes, which most branching rules try to predict and maximize. However, there are more potential implications which are not directly related to the final branching decision. First, if SB on a variable detects that both auxiliary

Measuring the Impact of Branching Rules for Mixed-Integer Programming

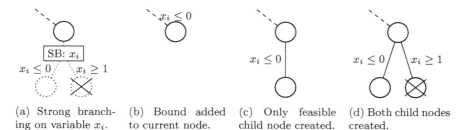

(a) Strong branching on variable x_i.
(b) Bound added to current node.
(c) Only feasible child node created.
(d) Both child nodes created.

Fig. 1 Strong branching proves infeasibility of a potential child node (**a**) and three different ways to apply this information in branching

LPs are infeasible, infeasibility of the current node can be deduced and the node is just cut off without any branching. Analogously, if exactly one of the two auxiliary LPs is infeasible, this node does not need to be investigated by the branch-and-bound algorithm anymore. The way this information is exploited differs among solvers, see Fig. 1. Some solvers will just branch on the variable but omit the infeasible child node as illustrated in Fig. 1c. Others just add the branching bound change of the feasible child directly to the current node and iterate the processing of that node, cf. Fig. 1b. Clearly, this leads to inconsistencies between solvers when evaluating the impact of branching rules and "hides" nodes from the evaluation that could not have been omitted by a branching rule that just returns the best branching variable. Other side-effects exceeding the pure branching decision include the knowledge of dual bounds for both child nodes created by branching as well as an improved dual bound for the current node based on the minimum of the dual bounds of any pair of SB LPs. This information might be used to discard them later in case a better primal solution was found. We also get better estimates for the best primal solution contained in the sub-trees rooted at the child nodes and more accurate pseudo-costs—history information about the dual bound improvement after branching on a variable [4]. Finally, conflict constraints can be extracted from infeasible SB LPs [1]. All this introduces a bias towards SB based branching rules, because it decreases the number of nodes reported by the solver additionally to the effect of better branching decisions.

This motivates us to propose a new measure to cover the quality of the branching decision better: the *fair node number*. It is based on the notion of a *branching oracle*, which, given the current LP solution, does nothing else than returning a variable to branch on. The valuable information that one or both of the two potential child nodes for a branching candidate are infeasible can then only be returned indirectly by selecting this variable for branching. This branching adds two nodes to the branch-and-bound tree to obtain the same information otherwise learned directly from SB, as the node(s) will be found infeasible when being processed, cf. Fig. 1d. If all compared branching rules can be interpreted as such an oracle, they are on a level field and their branch-and-bound nodes allow for a fair comparison. There are two means to reach this fair comparison for other branching rules. First, node numbers should be adjusted by mimicking that two branch-and-bound nodes were created for every cutoff or domain change identified. This means that the number of nodes is increased by two for each SB cutoff and by one or two for each domain change identified by SB, depending on whether the solver already created one child node in such a case or directly applied the reduction at the current node (see Fig. 1).

Other algorithmic features that cause side-effects, e.g., conflict analysis for infeasible SB children, should be disabled if possible. This typically includes providing the optimal objective value as an upper bound at the beginning of the solving process and disabling primal heuristics, cf. [6]. By this, the variability introduced by finding solutions at a different point of time is removed, while focusing on the main task of branching rules, which is improving the dual bound and proving optimality. This helps to reduce most side-effects. Since the upper bound is already optimal, improved dual bounds obtained by SB for child nodes as well as the current node either directly cause a reduction that will lead to an adjusted node count or will never allow to prune that node. They might influence node selection, as do the changed estimates, but due to the optimum being known upfront, the tree size is not influenced by the node selection, except for rare side-effects of the selection. Finally, pseudo costs are updated as a part of the branching rule and used later for branching. However, they also have other uses which might cause side-effects, e.g., in primal heuristics and node selection. Since these main other applications are disabled or without effect due to the installed upper bound, the side-effects should be negligible.

3 Computational Results

We performed computational experiments to evaluate common branching rules implemented in SCIP [1] with respect to time, node count, and fair node number. All regarded rules perform variable branching, i.e. create the sub-problems by splitting the domain of an integer variable with fractional value in the current LP solution into two disjoint parts. However, the methodology could as well be applied to general constraint branching, an active field of research that did not make it into state-of-the-art rules for general MIP so far. All computations were performed on a cluster of 2.5 GHz Intel Xeon E5-2670v2 CPUs with 64 GB main memory, running only one job per node at a time. We used the MMMC test set consisting of all benchmark instances from the last three MIPLIB [8] versions as well as the COR@L test set [5]. As suggested in the previous section, we installed the optimal objective value as an upper bound upfront and disabled primal heuristics as well as conflict analysis for temporary SB children. We restrict the comparison to the 156 instances that needed some branching to be solved and were solved by all branching rules within the time limit of 8 h.

Table 1 summarizes the results. For each branching rule, we list the average solving time to optimality, the number of nodes reported by SCIP and the fair node number introduced in the previous section, as well as the number of domain reductions and cutoffs which are needed to compute the fair node number. All averages are computed using the shifted geometric mean [1] with a shift of 10.

Table 1 Aggregated results over the 156 instances solved by all branching rules

Settings	Time (s)	Nodes	Fair nodes	DomReds	Cutoffs
MIB	134.77	10957.54	10957.54	0.00	0.00
PCB	45.37	3192.05	3192.05	0.00	0.00
FSB	80.86	203.44	1301.82	534.50	85.15
FSBDP	82.40	190.97	1287.02	535.19	90.61
RB	37.58	993.94	1748.13	167.21	21.60
HB	34.20	857.97	1582.86	156.40	25.97

The first two rows show results for *most infeasible branching* (MIB) and the *pseudo-cost branching rule* [4] (PCB), two rules without any side-effects, so that the fair nodes equal the number of nodes reported by SCIP. While MIB selects an integer variable with fractional part of the solution value closest to 0.5, PCB uses pseudo-costs to predict the lower bounds obtained in both child nodes and chooses a variable which maximizes the lower bound improvements. This helps to decrease both time and nodes by a factor of about three.

More accurate, but also more expensive, is *full strong branching* (FSB) which applies SB at every node for each fractional variable. The node reduction by a factor of 15 is in line with experiments in the literature [3]. However, we now see that this is largely caused by side-effects: SB proves infeasibility for more than every third node, and identifies on average 2.5 bound changes per node. Thus, the fair node number is more than six times higher than the node count. This means that the better branching decisions reduce the tree size by "only" 59% compared to PCB. The side-effects, on the other hand, further reduce the node number by another 84%. Full strong branching with domain propagation [6] (FSBDP) improves SB predictions by applying domain propagation techniques during SB. This mainly identifies more domain changes and cutoffs per node, so that a node decrease of 5% diminishes to about 1% in the fair node number.

The *reliability pseudo-cost branching rule* [3] (RB) is a combination of PCB and SB and uses SB only a few times on each variable to obtain reliable predictions. This reduces the average solving time significantly while increasing the number of reported nodes by a factor of almost five compared to FSB. However, this is mainly due to fewer SB calls which result in fewer domain reductions and cutoffs being identified. Consequently, the difference to FSB is smaller in the fair node number, where reliability branching needs only 34% more nodes.

Finally, *hybrid branching* [2] extends reliability branching by using additional statistics about infeasible nodes and domain changes inferred by domain propagation as tie breakers. The implementation we used also makes use of SBDP as well as statistical methods to filter out unpromising candidates. All this together gives the fastest variant and a fair node number that is only 23% higher than that of FSBDP, the branching rule performing best with respect to this criterion. This proves how important it is to support early branching decisions by SB, while using pseudo costs later does not deteriorate the quality of branching decisions very much anymore but just generates fewer side-effects.

4 Conclusions

We presented the *fair node number*, a new measure for the quality of a branching rule. It distinguishes between the quality of the branching decisions themselves and additional reductions learned, e.g., by strong branching. Both help to reduce the tree size, but investigating those effects individually can provide valuable insights. The fair node number, which focuses on the former effect, can be read from the statistics of SCIP if a few parameters are adjusted. It can be used to fairly assess the potential of branching rules with different side-effects. Thereby, it does not replace other measures like the branch-and-bound tree size reported by the solver or the solving time, but rather complements them, allowing for a better analysis of the impact of branching rules.

Acknowledgements The work for this article has been conducted within the *Research Campus Modal* funded by the German Federal Ministry of Education and Research (fund number 05M14ZAM).

References

1. Achterberg, T. (2007). Constraint Integer Programming. Ph.D. thesis. Technische Universität, Berlin.
2. Achterberg, T., & Berthold, T. (2009). Hybrid Branching. In W. J. van Hoeve & J. N. Hooker (Eds.), *CPAIOR 2009* (Vol. 5547, pp. 309–311). LNCS, Springer.
3. Achterberg, T., Koch, T., & Martin, A. (2005). Branching rules revisited. *Operations Research Letters*, *33*, 42–54.
4. Benichou, M., et al. (1971). Experiments in mixed-integer linear programming. *Mathematical Programming*, *1*, 76–94.
5. COR@L MIP Instances. Accessed June 2017. http://coral.ise.lehigh.edu/data-sets/mixedinteger-instances/.
6. Gamrath, G. (2014). Improving strong branching by domain propagation. *EURO Journal on Computational Optimization*, *2*(3), 99–122.
7. Gauthier, J. -M., & Ribière, G. (1977). Experiments in mixed-integer linear programmingusing pseudo-costs. *Math Prog*, *12*(1), 26–47.
8. Koch, T., et al. (2011). MIPLIB 2010. *Mathematical Programming Computation*, *3*(2), 103–163.
9. Land, A. H., & Doig, A. G. (1960). An automatic method of solving discrete programming problems. *Econometrica*, *28*(3), 497–520.

The Multiple Checkpoint Ordering Problem

Philipp Hungerländer and Kerstin Maier

1 Introduction

In this paper we introduce and analyze the multiple Checkpoint Ordering Problem (mCOP), which is a new variant of a row layout problem. An instance of the mCOP consists of n one-dimensional departments $D := \{1, 2, \ldots, n\}$ with given positive lengths $\ell_1, \ldots, \ell_n$, m checkpoints $C := \{n+1, \ldots, n+m\}$ with given positions and pairwise weights w_{ij}, $i \in D$, $j \in C$. We are looking for a non-overlapping placement of the departments without gaps between them, where the weighted sum of the distances of all departments to all checkpoints is minimal. The corresponding optimization problem can be written down as $\min_{\pi \in \Pi_n} \sum_{i \in D, j \in C} w_{ij} z_{ij}^\pi$, where Π_n is the set of permutations of the departments D and z_{ij}^π is the distance between the center of Department i and Checkpoint j with respect to a particular permutation $\pi \in \Pi_n$.

Hungerländer [3] introduced and analyzed the Checkpoint Ordering Problem (COP), which is the special case of the mCOP with $m = 1$. The COP is weakly NP-hard, while the complexity of the mCOP is still open. The mCOP has connections to other combinatorial optimization problems like the Single-Row Facility Layout Problem (SRFLP) and scheduling on m identical parallel machines with the objective of minimizing the sum of weighted completion times that is defined as follows: We are given a set of jobs $\mathcal{J}$ that have to be scheduled on m identical parallel machines. Each Job $j \in \mathcal{J}$ is specified by its processing time $p_j \geq 0$ and by its weight $w_j \geq 0$. Every machine can process at most one job at a time, and every job has to be processed on one machine in an uninterrupted fashion. The completion time of Job j is denoted by C_j. The goal is to minimize the total weighted completion time

---

P. Hungerländer · K. Maier (✉)
Alpen-Adria-Universität Klagenfurt, Klagenfurt, Austria
e-mail: kerstin.maier@aau.at

P. Hungerländer
e-mail: philipp.hungerlaender@aau.at

$\sum_{j \in \mathcal{J}} w_j C_j$. In the standard classification scheme of Graham et al. [2] this scheduling problem is denoted by $P || \sum w_j C_j$ for m part of the input, and by $Pm || \sum w_j C_j$ for constant m. The key differences between the mCOP and the weakly NP-hard [5] $Pm || \sum w_j C_j$ are the following ones:

1. For the mCOP the sum of the lengths of the departments that are placed to the left and to the right of the checkpoints are predetermined through the positions of the checkpoints as there are no spaces allowed between the departments. E.g. for the COP with a centered checkpoint the sums of the lengths of the departments to the left and to the right of the checkpoint have to be equal. Contrary to that for the $Pm || \sum w_j C_j$ there are typically no capacity restrictions imposed on the machines.
2. The checkpoints must not lie exactly at a splitting point of two departments, but they can also be covered by departments. I.e. the checkpoints do not necessarily define a partition of the departments. When considering a scheduling set-up, the COP can be described as follows: We are given two machines and it is allowed to split an arbitrary job into two parts at any point and then the two parts have to be scheduled first on the two machines. The mCOP with $m \geq 2$ cannot be formulated in this scheduling set-up anymore because the distances of the departments to all checkpoints are relevant in the objective.

Due to these differences it is not possible to directly carry over complexity and polyhedral results, dynamic programming (DP) algorithms and integer linear programming (ILP) models and their corresponding approximation results from scheduling on identical parallel machines [4] to the mCOP.

In this paper we propose two solution approaches for the mCOP and compare them in a computational study, where we observe that the mCOP seems much harder to solve in practice than the related SRFLP and $Pm || \sum w_j C_j$. There is no clear winner between our two methods. While the ILP approach is hardly influenced by the department lengths and number of checkpoints considered, the performance of the DP algorithm, which is only exact for one checkpoint, deteriorates for increasing department lengths and an increasing number of checkpoints.

The paper is structured as follows. In Sects. 2 and 3 we suggest an ILP approach and a DP algorithm for solving the mCOP and in Sect. 4 we conduct computational experiments, indicating the practical applicability and limitations of the approaches suggested. For future research it would be interesting to design more sophisticated exact approaches and heuristics for the mCOP. Furthermore it is still an open question if the mCOP is weakly or strongly NP-hard.

2 An ILP Formulation for the mCOP

In this section we propose an integer linear programming (ILP) approach for solving the mCOP with an arbitrary but fixed number of checkpoints that is a generalization of the ILP for the COP suggested in [3]. First we define S as the sum of the lengths of all departments

$$S = \sum_{i \in D} \ell_i. \tag{1}$$

The locations of the m checkpoints are defined by $p_j \in [0, 1]$, $j \in C$, where $S \cdot p_j$ gives the position of Checkpoint j.

Next we introduce binary ordering variables x_{ij}, $i \in D$, $j \in D \cup C$, $i < j$,

$$x_{ij} = \begin{cases} 1, & \text{if Department } i \text{ lies to the left of} \\ & \text{Department respectively Checkpoint } j, \\ 0, & \text{otherwise,} \end{cases}$$

to relate the positions of the n departments to each other and to the m checkpoints. To ensure transitivity on these variables, we use the 3-cycle inequalities

$$0 \leq x_{ij} + x_{jk} - x_{ik} \leq 1, \quad i, j \in D, \ k \in D \cup C, \ i < j < k, \tag{2}$$

which are sufficient for guaranteeing that there is no directed cycle.

Now we are able to express the distances of the departments from the m checkpoints as quadratic terms in ordering variables. The position d_i of the center of Department $i \in D$ is given as the sum of the lengths of the departments left of i plus $\ell_i/2$. The difference $d_j - d_i$, $i \in D$, $j \in C$, gives the distance of the center of Department i to Checkpoint j, if Department i is located to the left of the checkpoint. If Department i is located to the right of the checkpoint, this difference is minus the distance of the center of Department i from Checkpoint j. Therefore we multiply $d_j - d_i$, $i \in D$, $j \in C$, by the term $(2x_{ij} - 1)$ that is 1, if the center of Department i lies to the left of Checkpoint j and -1 otherwise:

$$z_{ij} = (2x_{ij} - 1)(d_j - d_i), \ i \in D, \ j \in C, \tag{3}$$

$$d_i = \frac{\ell_i}{2} + \sum_{k \in D, \ k < i} \ell_k x_{ki} + \sum_{k \in D, \ k > i} \ell_k (1 - x_{ik}), \ i \in D, \quad d_j = S \cdot p_j, \ j \in C,$$

Expanding and simplifying (3) yields

$$z_{ij} = (2x_{ij} - 1)\left(S \cdot p_j - \frac{\ell_i}{2} - \sum_{\substack{k \in D \\ k < i}} \ell_k x_{ki} - \sum_{\substack{k \in D \\ k > i}} \ell_k (1 - x_{ik})\right), \ i \in D, \ j \in C. \tag{4}$$

The multiplication of $(d_j - d_i)$ with $(2x_{ij} - 1)$ ensures a correct calculation of all distances through the following constraints:

$$z_{ij} \geq 0, \quad i \in D, \ j \in C. \tag{5}$$

To model the mCOP as an ILP, we apply standard linearization and introduce new variables for all products of ordering variables in (4):

$$y_{ijki} = x_{ij}(1 - x_{ik}), \ i < k, \qquad y_{ijki} = x_{ij}x_{ki}, \ i > k,$$

where $i, k \in D, \ j \in C$. Now (4) can be further rewritten as:

$$\begin{aligned} z_{ij} = & (2x_{ij} - 1)\left(S \cdot p_j - \frac{\ell_i}{2}\right) + \sum_{\substack{k \in D \\ k < i}} \ell_k x_{ki} \\ & + \sum_{\substack{k \in D \\ k > i}} \ell_k(1 - x_{ik}) - 2\sum_{\substack{k \in D \\ k \neq i}} \ell_k y_{ijki}, \quad i \in D, \ j \in C. \end{aligned} \quad (6)$$

Moreover we use the following standard constraints to relate the ordering variables and their products:

$$\begin{aligned} y_{ijki} &\leq x_{ij}, & y_{ijki} &\leq 1 - x_{ik}, \ i < k, & y_{ijki} &\leq x_{ki}, \ i > k, \\ y_{ijki} &\geq x_{ij} - x_{ik}, \ i < k, & y_{ijki} &\geq x_{ij} + x_{ki} - 1, \ i > k, \end{aligned} \quad (7)$$

where $i, k \in D, \ j \in C$. Overall we obtain the following ILP model for the mCOP:

$$\begin{aligned} \min \ & \sum_{i \in D, \ j \in C} w_{ij} z_{ij} \\ \text{s.t.} \ & (1), (2), (5) - (7), \\ & x_{ij} \in \{0, 1\}, \quad i \in D, \ j \in D \cup C, \ i < j, \\ & y_{ijki} \in \{0, 1\}, \quad i, k \in D, \ j \in C, \ i \neq j. \end{aligned}$$

3 A Dynamic Programming Algorithm for the mCOP

In [3] an exact dynamic programming (DP) algorithm for solving the COP was proposed. In this section we suggest how to extend this algorithm to the mCOP. As our extension is not exact for $m \geq 2$, it is still an open question if the mCOP with $m \geq 2$ is weakly or strongly NP-hard. Note that $Pm||\sum w_j C_j$ is weakly NP-hard as it can be solved in pseudopolynomial time by a DP approach [5].

Now let us give a brief outline of our DP algorithm. In an optimal layout departments that are positioned to the left or to the right of *all* checkpoints adhere to the well-known V-shaped property [1], i.e. they are arranged in non-increasing order from the leftmost or rightmost checkpoint to the border of the layout with respect to their relative weights $\left(\sum_{j \in C} w_{ij}\right)/\ell_i, \ i \in D$. Contrary to that departments with a high relative weight that are located between two checkpoints should not nec-

The Multiple Checkpoint Ordering Problem 175

Table 1 Results obtained by our ILP approach using Gurobi 6.5 restricted to one thread with a time limit of 12 h. The running times are given in hh:mm:ss

# Checkpoints	1			2		
Instance	Best solution	Gap (%)	Time	Best solution	Gap (%)	Time
AnKeVa80set4dep15	2042.00	0.0	00:10:21	7280.50	0.0	00:53:25
AnKeVa80set4dep20	2280.50	20.3	12:00:00	10802.00	34.6	12:00:00
AnKeVa80set4dep25	3663.00	60.9	12:00:00	19273.00	65.6	12:00:00
AnKeVa80set4dep30	5594.00	96.2	12:00:00	25929.00	96.3	12:00:00
HuRe40set4dep15	259.50	0.0	00:01:31	894.75	0.0	00:06:22
HuRe40set4dep20	303.00	0.0	00:02:56	1169.50	0.0	01:53:40
HuRe40set4dep25	324.50	0.0	00:05:15	2084.50	15.9	12:00:00
HuRe40set4dep30	1116.00	37.9	12:00:00	3529.75	64.2	12:00:00
# Checkpoints	3			4		
Instance	Best solution	Gap [%]	Time	Best solution	Gap [%]	Time
AnKeVa80set4dep15	7662.25	0.0	05:01:40	976.20	0.0	07:20:14
AnKeVa80set4dep20	14017.00	61.7	12:00:00	22402.50	73.2	12:00:00
AnKeVa80set4dep25	25523.50	78.1	12:00:00	34161.00	86.7	12:00:00
AnKeVa80set4dep30	34290.00	99.9	12:00:00	44552.50	97.4	12:00:00
HuRe40set4dep15	1176.50	0.0	00:35:10	1718.50	0.0	02:17:54
HuRe40set4dep20	1486.50	27.3	12:00:00	3757.30	53.4	12:00:00
HuRe40set4dep25	2685.50	44.1	12:00:00	4655.80	81.2	12:00:00
HuRe40set4dep30	3952.50	97.3	12:00:00	9037.90	94.5	12:00:00

essarily be positioned close to a checkpoint. This is why we arrange departments between two checkpoints k and $k+1$ in non-increasing order regarding the ratio $\left(\sum_{j=n+1}^{k} w_{ij}\right) / \left(\sum_{j=k+1}^{n+m} w_{ij}\right)$, $i \in D$. Unfortunately the obtained arrangements may not be optimal for $m \geq 2$. In fact the approach is not even guaranteed to find a feasible solution. Nonetheless our DP algorithm proves to be a good heuristic for the mCOP, in particular if m is small.

At the heart of our approach lies a recursive relation that is used to decide where Department j should be placed with respect to the checkpoints. For one checkpoint the recursion tells us whether to assign Department j to the left or to the right of the checkpoint:

$$F_j(s) = \frac{\ell_j w_j}{2} + \min\left\{F_{j-1}(s + \ell_j) + (s + \ell_c^1)w_j;\ F_{j-1}(s) + (M - s + \ell_c^2)w_j\right\},$$

where s indicates the remaining free space to the left of the checkpoint, M gives the overall remaining free space either to the left or to the right of the checkpoint, c is

Table 2 Results obtained by our DP algorithm with a time limit of 12 h. The running times are given in hh:mm:ss. In column DP versus ILP we compute $\frac{\text{Solution of DP} - \text{Solution of ILP}}{\text{Solution of DP}}$, hence in case of a negative entry the DP provided a better feasible layout than the ILP

# Checkpoints	1			2		
Instance	Best solution	DP versus ILP	Time	Best solution	DP versus ILP	Time
AnKeVa80set4dep15	2280.50	0.00	00:00:02	7280.50	0.00	12:00:00
AnKeVa80set4dep20	2042.00	−10.46	00:00:01	10899.00	0.89	12:00:00
AnKeVa80set4dep25	3663.50	0.00	00:00:06	20482.00	6.27	12:00:00
AnKeVa80set4dep30	5584.00	−0.17	00:00:08	25885.00	−0.17	12:00:00
HuRe40set4dep15	259.50	0.00	00:00:01	894.75	0.00	00:00:14
HuRe40set4dep20	304.00	0.00	00:00:01	1187.50	1.54	00:02:22
HuRe40set4dep25	324.50	0.00	00:00:01	2118.50	1.66	00:07:26
HuRe40set4dep30	1115.00	−0.09	00:00:01	3544.25	0.41	00:20:44
# Checkpoints	3			4		
Instance	Best solution	DP versus ILP	Time	Best solution	DP versus ILP	Time
AnKeVa80set4dep15	8651.75	12.91	12:00:00	–	–	12:00:00
AnKeVa80set4dep20	15419.00	10.00	12:00:00	–	–	12:00:00
AnKeVa80set4dep25	–	–	12:00:00	–	–	12:00:00
AnKeVa80set4dep30	–	–	12:00:00	–	–	12:00:00
HuRe40set4dep15	1176.50	0.00	01:08:56	1719.50	0.06	12:00:00
HuRe40set4dep20	1517.50	2.09	12:00:00	4074.90	8.45	12:00:00
HuRe40set4dep25	2681.50	−0.01	12:00:00	5226.60	12.26	12:00:00
HuRe40set4dep30	4022.50	1.77	12:00:00	8859.30	−1.98	12:00:00

the center department covering the checkpoint and ℓ_c^1 (ℓ_c^2) is the length of the part of the center department left (right) to the checkpoint.

A detailed description of our DP algorithm, including in particular a discussion of the general recursive relation for $m \geq 2$ checkpoints, is omitted in this short paper due to space limitations and will be provided in a forthcoming paper.

4 Computational Experiments

All experiments were performed on a Linux 64-bit machine equipped with Intel(R) Xeon(R) CPU e5-2630 v3@2.40 GHz and 128 GB RAM. The algorithms were implemented in C (DP) and Gurobi 6.5 (ILP) respectively. To generate mCOP instances, we utilized benchmark instances from row layout literature by simply randomly choosing $m + n$ departments from these instances and using them as our n

departments and m checkpoints. Accordingly we took the corresponding pairwise connectivities in these instances as our mCOP weights w_{ij}, $i \in D$, $j \in C$.

In our computational study we consider two different instance sets from the literature. AnKeVa80 consists of 80 departments with department lengths between 1 and 60. HuRe40 contains 40 departments with department lengths between 1 and 10. Each of our mCOP instances consists of 10–30 departments and has 4 checkpoints. We choose the checkpoint positions dependent on the number of checkpoints considered, but independent from the number of departments. All instances can be downloaded from http://tinyurl.com/layoutlib.

In Tables 1 and 2 we respectively state the results of our ILP approach and our DP algorithm. We observe that the mCOP with $m \geq 2$ is already very hard to solve to optimality for instances of moderate size. In particular the mCOP seems much harder to solve in practice than the closely related strongly NP-hard Single-Row Facility Layout Problem and the weakly NP-hard $Pm||\sum w_j C_j$.

The performance of the ILP approach is hardly influenced by increasing the number of checkpoints or the length of the departments. Contrary to that the performance of our DP algorithm deteriorates both for an increasing number of checkpoints and for larger department lengths. Note that on the AnKeVa80 instances the DP algorithm does not provide any *feasible* solution when considering all 4 checkpoints. Nonetheless for each number of checkpoints there are instances for which the DP algorithm provides better solutions than the ILP approach.

References

1. Eilon, S., & Chowdhury, I. G. (1977). Minimising waiting time variance in the single machine problem. *Management Science*, *23*(6), 567–575.
2. Graham, R. L., Lawler, E. L., Lenstra, J. K., & Kan, A. H. R. (1979). Optimization and approximation in deterministic sequencing and scheduling: a survey. In P. L. Hammer, E. L. Johnson, & B. H. Korte (Eds.), *Discrete optimization II* (Vol. 5, pp. 287–326). Annals of discrete mathematics. Amsterdam: Elsevier.
3. Hungerländer, P. (2017). The checkpoint ordering problem. *Optimization*, 1–14.
4. Queyranne, M., & Schulz, A. S. (2004). Polyhedral approaches to machine scheduling. Technical report.
5. Rothkopf, M. H. (1966). Scheduling independent tasks on parallel processors. *Management Science*, *12*(5), 437–447.

An Improved Upper Bound for the Gap of Skiving Stock Instances of the Divisible Case

John Martinovic and Guntram Scheithauer

1 Introduction and Preliminaries

The 1D skiving stock problem (SSP) is strongly related to the dual bin packing problem (DBPP) and can be formulated as follows: how many (large) objects of length not less than L can at most be build by connecting $m \in \mathbb{N}$ given (small) item types of length l_i and quantity b_i, $i \in I := \{1, \ldots, m\}$? Without loss of generality, we may assume all input data to be positive integers.

Originating from applications in paper recycling [3], such objectives are of high interest in industrial production [2, 9] and wireless communications [4] as well. After having been introduced as the *dual bin packing problem* by Assmann et al. [1], a generalization for larger availabilities was firstly considered in [9] and termed *skiving stock problem*. A detailed survey on different modelling approaches and their computational behaviour has recently been published in [5]. Throughout this paper, we will use the abbreviation $E := (m, l, L, b)$ for an *instance* of the SSP with $l = (l_1, \ldots, l_m)^\top \in \mathbb{Z}_+^m$ and $b = (b_1, \ldots, b_m)^\top \in \mathbb{Z}_+^m$. Any feasible arrangement of items leading to an object of length not less than L is called *(packing) pattern* of E, and is represented by a nonnegative vector $a = (a_1, \ldots, a_m)^\top \in \mathbb{Z}_+^m$ where $a_i \in \mathbb{Z}_+$ denotes the number of contained items of type $i \in I$. Note that considering the set $P^\star(E)$ of *minimal patterns* (where each appearing item is indeed necessary to ensure $l^\top a \geq L$) is sufficient. Let $x_j \in \mathbb{Z}_+$ count how often the minimal pattern $a^j = (a_{1j}, \ldots, a_{mj})^\top \in \mathbb{Z}_+^m$ ($j \in J^\star$) of E is used where $J^\star = \{1, \ldots, n\}$ represents an index set of $P^\star(E)$. Then the standard model of the skiving stock problem [5, 9] can be formulated as

---

J. Martinovic (✉) · G. Scheithauer
Institut für Numerische Mathematik, Technische Universität Dresden, 01062 Dresden, Germany
e-mail: john.martinovic@tu-dresden.de

G. Scheithauer
e-mail: guntram.scheithauer@tu-dresden.de

$$z^{\star}(E) = \max \left\{ \sum_{j \in J^{\star}} x_j \;\middle|\; \sum_{j \in J^{\star}} a_{ij} x_j \leq b_i,\, i \in I,\, x_j \in \mathbb{Z}_+,\, j \in J^{\star} \right\}. \quad (1)$$

A common (approximate) solution approach consists in considering the continuous relaxation

$$z_c^{\star}(E) = \max \left\{ \sum_{j \in J^{\star}} x_j \;\middle|\; \sum_{j \in J^{\star}} a_{ij} x_j \leq b_i,\, i \in I,\, x_j \geq 0,\, j \in J^{\star} \right\} \quad (2)$$

and/or the application of appropriate heuristics. Then, the difference $\Delta(E) := z_c^{\star}(E) - z^{\star}(E)$ is called *gap* (of E).

Definition 1 A set $\mathscr{T}$ of instances has the *integer round-down property* (IRDP), if $\Delta(E) < 1$ holds for all $E \in \mathscr{T}$, and it has the *modified integer round-down property* (MIRDP), if $\Delta(E) < 2$ holds for all $E \in \mathscr{T}$.

It is conjectured, see [9], that the skiving stock problem possesses the MIRDP. Indeed, the currently largest known gap is given by $\Delta(E) = 325/276 \approx 1.1775$. But, since a general verification for arbitrary instances (or the presentation of a counterexample) is very difficult, frequently only some special subclasses, e.g. the so-called *divisible case*[1] [6], are considered. For this special case, we may assume without loss of generality that E does not contain any *exact pattern*,[2] i.e., a pattern $a \in \mathbb{Z}_+^m$ with $a \leq b$ and $l^\top a = 1$. Moreover, upper bounds for the gap of arbitrary instances can be obtained from those of the divisible case [7].

In the next section, we describe the best fit decreasing heuristic (for the SSP) and some of its most important properties. Afterwards, we show how a detailed analysis of this algorithm leads to an improved upper bound for the gap of the divisible case.

2 The Best Fit Decreasing Heuristic and Basic Properties

In [6], the MIRDP of the divisible case has been proved by a quite deep study of a two-phase *first fit decreasing* algorithm. In this section, we apply an alternative heuristic of much easier description that will lead to an improved upper bound. Let $t \in \mathbb{N}$ and empty bins $B_1, \ldots, B_t$ be given. In a particular step of the algorithm, let $C(j)$ describe the current total length of items that have been allocated to B_j

---

[1] An instance $E = (m, l, L, b)$ of the SSP belongs to the *divisible case* (for short: $E \in \mathscr{DC}$) if $l_i \mid L$ holds for all $i \in I$. Then, E can be described by $E = (m, l, 1, b)$ with $l_i \in \{1/2, 1/3, \ldots\}$ for all $i \in I$.

[2] Each upper bound of the gap that holds for such instances can easily be transferred to unrestricted instances of the divisible case, too (see [6, Theorem 9]).

Algorithm 1 Best Fit Decreasing with Parameter $t \in \mathbb{N}$
―――
1: Consider a number $t \in \mathbb{N}$ of bins $B_1, \ldots, B_t$. Define $T := \{1, \ldots, t\}$, $\tilde{n} := e^\top b$ and sort all items according to decreasing lengths, i.e. $1 > l'_1 \geq l'_2 \geq \ldots \geq l'_{\tilde{n}}$.
2: **for all** $i \in \{1, \ldots, \tilde{n}\}$ **do**
3:      Allocate item i to bin B_τ where τ is given by $\tau := \tau(i) := \min \operatorname{argmin}_{j \in T}\{C(j)\}$.
4: **end for**

As an introductory observation, note that also this heuristic leads to the MIRDP of the divisible case, see [8] for the details. In the following, we will show that the currently best upper bound of $\Delta(E) < 3/2$ (for the divisible case, see [8]) can be improved by a more detailed analysis of this algorithm. Moreover, this new upper bound will also contain the problem-specific input data (i.e., the lengths l and availabilities b) of a given instance.

Definition 2 Let $K \in \mathbb{N}$ be given. Then we define $\mathscr{DC}(K)$ as the set of all instances of the divisible case satisfying $\lfloor l^\top b \rfloor = K$.

Let us consider an instance $E = (m, l, 1, b) \in \mathscr{DC}(K)$ with

$$l^\top b - K \geq \frac{1}{2} - \frac{1}{2K}. \tag{3}$$

Our aim is to show that, in this particular setting, Algorithm 1 with $t := \lfloor l^\top b \rfloor (= K)$ leads to a feasible (integer) solution of the skiving stock problem with t patterns. As a direct consequence, this observation implies the improved upper bound $\Delta(E) < 3/2 - 1/(2K)$ for instances $E \in \mathscr{DC}(K)$, since $z_c^\star(E) = l^\top b$ always holds for $E \in \mathscr{DC}$, see [6].

Lemma 1 *At the end of Algorithm 1 with $t := \lfloor l^\top b \rfloor$, there is some $j \in T$ with $C(j) > 1$.*

Proof This result is an immediate consequence of the pigeonhole principle and the absence of exact patterns. □

Let $j^\star \in T$ denote the bin that is filled first during Algorithm 1, and let item $i^\star \in \{1, \ldots, \tilde{n}\}$ of length $l_{i^\star} = 1/q$ for some $q \in \mathbb{N}, q \geq 2$ be responsible for that. In order to ease the notation, we define $W(q) = \sum_{d=1}^{i^\star - 1} l'_i$, i.e., $W(q)$ denotes the total length of all items that have been allocated prior to that moment when $B_{j^\star}$ is filled. Note that $W(q)$ can be bounded above by the optimal objective value of the discrete optimization problem

$$W^\star(q) := \max \left\{ \kappa(q)^\top x \,\middle|\, x \in \mathbb{Z}_+^q, \forall y \in \mathbb{Z}_+^q, y \leq x : \kappa(q)^\top y \neq 1 \right\}, \tag{4}$$

where $\kappa(q) := (1, 1/2, 1/3, \ldots, 1/q)^\top \in \mathbb{Q}_+^q$ holds[3]. In order to ease the notation, we define $Q := Q(q) := \{1, \ldots, q\}$. As an introductory result, note that the feasible region of this problem can obviously be considered to be finite:

―――――――――――
[3]Due to the absence of exact patterns, $x_1 = 0$ has to hold in any feasible solution of (4). This means that, actually, it is not necessary to consider the length $\kappa_1 = 1$ in the corresponding maximization problem. Nevertheless, we will work with the given definition of the vector κ for the sake of an easier presentation, since then $\kappa_d = 1/d$ is true for all $d \in \{1, \ldots, q\}$.

Lemma 2 *Let x be a feasible solution of (4), then we have $x_j \leq j - 1$ for all $j \in Q$.*

Proof Assuming $x_j > j - 1$ for some $j \in Q$, we can choose $y = j \cdot e^j$ as a feasible vector with $y \leq x$ and $\kappa(q)^\top y = 1$ which gives the contradiction. □

In a more sophisticated approach this observation can be extended to:

Lemma 3 *Let x be a feasible solution of (4), then there is a feasible solution x' with the same objective value such that $x'_j \leq t(j) - 1$ holds for all $j \in Q \setminus \{1\}$, where $t(j) := \min\{\rho \in \mathbb{P} \mid \rho \text{ divides } j\}$ denotes the smallest prime divisor of $j \in Q \setminus \{1\}$.*

Proof Let x be a feasible solution of (4) with $x_j \geq t(j)$ for some $j \in Q \setminus \{1\}$. Without loss of generality, let j be chosen maximal with respect to this property. Due to $t(j) \mid j$ we have $t(j)/j = 1/k$ for some $k \in \{2, \ldots, j-1\}$. But then, we can define $x'_i = x_i$ for $i \in Q \setminus \{j, k\}$, $x'_i = x_i - t(i)$ for $i = j$, and $x'_i = x_i + 1$ for $i = k$. Note that x' is feasible for (4). In particular, we obtain that x' provides the same objective value as x, and $x'_j < x_j$ as well as $x'_i \leq t(i) - 1$ for all $i > j$ are satisfied. After a finite number of such steps we are done. □

3 An Improved Upper Bound

These preliminaries can be used to obtain the following upper bound.

Theorem 1 *The inequality $W^*(q) \leq \frac{q+1}{2} \cdot \left(1 - \frac{1}{q}\right)$ holds.*

Proof Let x be a solution of (4). Note that every denominator k from the set $\{1/1, 1/2, \ldots, 1/q\}$ possesses a unique representation as $k = 2^u \cdot v$ with $u \in \mathbb{Z}_+$ and v odd. Let $\mathbb{O} := \mathbb{O}(q)$ denote the set of odd numbers not larger than q. Then we obtain

$$W^*(q) = \sum_{j=1}^{q} \frac{x_j}{j} = \sum_{v \in \mathbb{O}} \left(\sum_{k = 2^u \cdot v,\, u \in \{0, \ldots, \lfloor \log_2(q/v) \rfloor\}} \frac{x_k}{k} \right)$$

$$\leq \sum_{v \in \mathbb{O}} \left(\frac{v-1}{v} + \sum_{k = 2^u \cdot v,\, u \in \{1, \ldots, \lfloor \log_2(q/v) \rfloor\}} \frac{1}{k} \right)$$

$$= \sum_{v \in \mathbb{O}} \left(\frac{v-1}{v} + \sum_{u=1}^{\lfloor \log_2(q/v) \rfloor} \frac{1}{2^u \cdot v} \right) = \sum_{v \in \mathbb{O}} \left(\frac{v-1}{v} + \frac{1}{v} \cdot \sum_{u=1}^{\lfloor \log_2(q/v) \rfloor} \left(\frac{1}{2}\right)^u \right)$$

where $x_k \leq 1$ has been used for all $k \in \mathbb{N}$ with $2 \mid k$ in the second line.

For all $v \in \mathbb{O}$, the corresponding geometric sum leads to

$$\sum_{u=1}^{\lfloor \log_2(q/v) \rfloor} \left(\frac{1}{2}\right)^u = 1 - \left(\frac{1}{2}\right)^{\lfloor \log_2(q/v) \rfloor} \leq 1 - 2^{-\log_2(q/v)} = 1 - \frac{v}{q}$$

Plugging in this result in our first calculation gives

$$W^\star(q) \leq \sum_{v \in O}\left(\frac{v-1}{v} + \frac{1}{v} \cdot \left(1 - \frac{v}{q}\right)\right) = \sum_{v \in O}\left(1 - \frac{1}{q}\right) \leq \frac{q+1}{2} \cdot \left(1 - \frac{1}{q}\right)$$

since there are exactly $\lceil q/2 \rceil \leq (q+1)/2$ odd numbers not larger than q. □

Consequently, we can obtain the following result by using Theorem 1.

Corollary 1 *The inequality $q \geq 2t$ holds.*

Proof This is an immediate consequence of

$$\frac{q+1}{2} \cdot \left(1 - \frac{1}{q}\right) \geq W^\star(q) \geq W(q) > t \cdot \left(1 - \frac{1}{q}\right),$$

where the last inequality has been shown in [8, Proof of Lemma 2]. □

Now we are able to prove the main result of this section.

Theorem 2 *Let $E \in \mathscr{DC}(K)$ denote an instance of the skiving stock problem that satisfies inequality (3). Then, for the choice $t = \lfloor l^\top b \rfloor$, all bins are filled at the end of Algorithm 1.*

Proof Let us assume that $C(k) < 1$ holds for some $k \in T$ after applying Algorithm 1. Then, no item has been added to any filled bin (since they would not have been minimal with respect to line 3 of Algorithm 1). Hence, we have $C(j) < 1 + 1/q$ for all $j \neq k$ and $C(k) < 1$ leading to

$$l^\top b - t = \sum_{j=1}^{t} C(j) - t < (t-1) \cdot \left(1 + \frac{1}{q}\right) + 1 - t = \frac{t-1}{q} \leq \frac{t-1}{2t},$$

where Corollary 1 has been used. Due to $t = K$, this gives the contradiction to the initial assumption (3). □

Consequently, an instance of $\mathscr{DC}(K)$ possesses the IRDP whenever $l^\top b - K \geq 1/2 - 1/(2K)$ holds. Thus we can state:

Theorem 3 *Let $E = (m, l, 1, b) \in \mathscr{DC}(K)$. Then the inequality $\Delta(E) < 3/2 - 1/(2K)$ holds.*

Remark 1 By means of an extensive theoretical study, it can be shown that $W^\star(q) = \mathscr{O}(q/\ln(q))$ holds for $q \to \infty$. Hence, better upper bounds for $W(q)$ (and also better upper bounds for $\Delta(E)$) can be found if q is assumed to be sufficiently large.

Note that better upper bounds for the divisible case are also helpful to obtain improved bounds for the gap of arbitrary instances, see [7].

4 Conclusions

In this paper we investigated the gap of the one-dimensional skiving stock problem with respect to the well known divisible case. To this end we applied the best fit decreasing heuristic and showed how an improved upper bound can be obtained by means of a detailed theoretical analysis. Due to $W^\star(q) = \mathcal{O}(q/\ln(q))$ (for $q \to \infty$) even better bounds are possible for special subclasses of the divisible case.

References

1. Assmann, S. F., Johnson, D. S., Kleitman, D. J., & Leung, J. Y.-T. (1984). On a dual version of the one-dimensional bin packing problem. *Journal of Algorithms, 5*, 502–525.
2. Chen, Y., Song, X., Ouelhadj, D., & Cui, Y. (2017). A heuristic for the skiving and cutting stock problem in paper and plastic film industries. To appear in: International Transactions in Operational Research, https://doi.org/10.1111/itor.12390.
3. Johnson, M. P., Rennick, C., & Zak, E. J. (1997). Skiving addition to the cutting stock problem in the paper industry. *SIAM Review, 39*(3), 472–483.
4. Martinovic, J., Jorswieck, E., Scheithauer, G., & Fischer, A. (2017). Integer linear programming formulations for cognitive radio resource allocation. *IEEE Wireless Communication Letters, 6*(4), 494–497.
5. Martinovic, J., & Scheithauer, G. (2016). Integer linear programming models for the skiving stock problem. *European Journal of Operational Research, 251*(2), 356–368.
6. Martinovic, J., & Scheithauer, G. (2016). Integer rounding and modified integer rounding for the skiving stock problem. *Discrete Optimization, 21*, 118–130.
7. Martinovic, J., & Scheithauer, G. (2016). New Theoretical Investigations on the Gap of the Skiving Stock Problem. Preprint MATH-NM-03-2016, Technische Universität Dresden
8. Martinovic, J., & Scheithauer, G. (2017). An upper bound of $\Delta(E) < 3/2$ for skiving stock instances of the divisible case. *Discrete Applied Mathematics, 229*, 161–167.
9. Zak, E. J. (2003). The skiving stock problem as a counterpart of the cutting stock problem. *International Transactions in Operational Research, 10*, 637–650.

Closed Almost Knight's Tours on 2D and 3D Chessboards

Michael Firstein, Anja Fischer and Philipp Hungerländer

1 Introduction

The closed knight's tour problem is a well-studied problem [3, 6, 7]. Given a rectangular two-dimensional (2D) chessboard of size $m \times n$ does there exist a tour (a Hamiltonian cycle or closed knight's tour) over all cells of the chessboard such that each move is a knight's move? A knight's move is a step of length $\sqrt{5}$ where we move one cell in one direction and two cells in the other direction. There also exist extensions of the closed knight's tour problem to generalized chessboards in arbitrary dimension [3, 4]. It is well-characterized for which 2D and 3D chessboards a closed knight's tour exists. For the 2D case Schwenk [7] proved the following result:

Theorem 1 ([7]) *An $m \times n$ chessboard with $m \leq n$ has a closed knight's tour if $(m, n) \notin \{(3, 4), (3, 6), (3, 8)\} \cup \{(m, n): m \in \{1, 2, 4\}\}$ and if m, n are not both odd.*

DeMaio and Mathew [3] have extended Theorem 1 to 3D chessboards:

Theorem 2 ([3]) *An $m \times n \times \ell$ chessboard with $m, n, \ell \geq 2$ has a closed knight's tour if $(m, n, l) \neq (2, 3, 3)$, $(m, n) \neq (2, 2)$ and at least one of m, n, ℓ is even.*

For chessboards that do not have a closed knight's tour we define a *closed almost knight's tour* as a Hamiltonian cycle of minimal length if only moves of at length at least $\sqrt{5}$ are allowed. The problem of determining closed (almost) knight's tours

---

M. Firstein · P. Hungerländer (✉)
Alpen-Adria-Universität Klagenfurt, Klagenfurt, Austria
e-mail: philipp.hungerlaender@aau.at

M. Firstein
e-mail: michaelfir@edu.aau.at

A. Fischer
Georg-August-Universität Göttingen, Göttingen, Germany
e-mail: anja.fischer@mathematik.uni-goettingen.de

on 2D and 3D chessboards is equivalent to the so called traveling salesman problem with forbidden neighborhoods (TSPFN) with radius two on regular 2D and 3D grids. Given points in the Euclidean space and some radius $r \in \mathbb{R}_+$, the TSPFN asks for a shortest Hamiltonian cycle, where connections between points with distance at most r are forbidden. The TSPFN was originally motivated by an application in laser beam melting where a workpiece is built in several layers. By excluding the heating of positions that are too close during this process, one hopes to reduce the internal stresses of the workpiece. We refer to [5] and the references therein for further details on this application and for results for regular 2D grids and $r \in \{0, 1, \sqrt{2}\}$. In this paper we present construction schemes for closed almost knight's tour on 2D and 3D chessboards that are based on the ideas for constructing knight's tours suggested by Lin and Wei [6].

The paper is structured as follows. In Sect. 2 we consider 2D chessboards of size $m \times n$ with $m, n \geq 5$. For m and n odd, a closed almost knight's tour uses only knight's moves except for one move of length $\sqrt{8}$. In Sect. 3 we consider 3D chessboards of size $m \times n \times \ell$, $m, n \geq 5$, $\ell \geq 3$. For m, n, ℓ odd a closed almost knight's tour uses only knight's moves except for one move of length $\sqrt{6}$. In Sect. 4 we conclude the paper and give suggestions for future work.

2 Closed Almost Knight's Tours on 2D Chessboards

We consider $m \times n$ chessboard where m is the number of rows and n is the number of columns. Each cell of the chessboard is denoted by a tuple (i, j), $i \in \{1, \ldots, m\}$, $j \in \{1, \ldots, n\}$.

Lemma 1 *Given an $m \times n$ chessboard with $m, n \geq 5$ and m and n odd, then $(mn - 1)\sqrt{5} + \sqrt{8}$ is a lower bound on the length of a closed almost knight's tour.*

This result follows directly from Theorem 1 because there does not exist a closed knight's tour for the considered chessboard sizes and the shortest move longer than $\sqrt{5}$ has length $\sqrt{8}$. Next we show that there always exists a closed almost knight's tour with this length. For $m \times m$ chessboard the existence of such an almost knight's tour follows from a result in [1] on the existence of s-t-knights path on quadratic chessboard between cells s and t.

Theorem 3 *Given an $m \times n$ chessboard with $m, n \geq 5$ and m and n odd, then there always exists a closed almost knight's tour with length $(mn - 1)\sqrt{5} + \sqrt{8}$.*

Proof We prove this result algorithmically by providing a construction scheme for building closed almost knight's tours. Explicit closed almost knight's tours for $m \times n$ chessboards with $m, n \in \{5, 7, 9\}$ are depicted in Fig. 1. These basic tours for our construction scheme can, e. g., be derived by solving an integer linear program that is obtained by fixing certain variables to zero in the classical formulation of the traveling

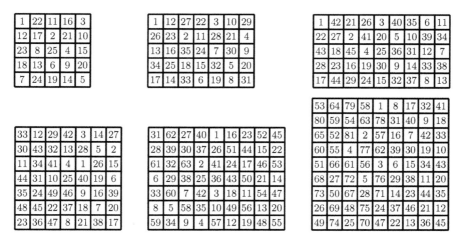

Fig. 1 Closed almost knight's tours on $m \times n$ chessboards with $m, n \in \{5, 7, 9\}$

Fig. 2 Open knight's tours for 5×6, 6×6, 7×6, 9×6 chessboards needed (sometimes mirrored at the diagonal) in the proof of Theorem 3. They start at $(1, 1)$ and end in $(2, 1)$

salesman problem by Dantzig et al. [2], see [5] for details. We want to emphasize that our closed almost knight's tours in Fig. 1 have some special structure. Indeed, they all contain the edges $\{(m-1, 1), (m, 3)\}$ and $\{(1, n-1), (3, n)\}$.

In our construction scheme, which is similar to the approach for constructing knight's tours suggested in [6], we now extend our basic tours from the right and from below in order to derive a closed almost knight's tour for the whole chessboard. This is done iteratively, increasing the size in horizontal or vertical direction by six in each step. For these extensions we use open knight's tours on the 5×6, the 6×6, the 7×6 and the 9×6 chessboard, see Fig. 2, and also [6], for a visualization. It is easy to check that the open tours and our basic tours are built in such a way such that the construction depicted in Fig. 3, where we delete the dotted edges and connect the single tours and paths by knight's moves, is feasible.

In summary starting with one of the basic tours from Fig. 1 we can construct a closed almost knight's tour for all $m \times n$ chessboards $m, n \geq 5$ and m and n odd by iteratively adding open knight's tours from Fig. 2. □

Fig. 3 Combining an $m_1 \times n_1$ basic tour with $m_1 \times 6$, $6 \times n_1$ and 6×6 open knight's tours to obtain a closed almost knight's tour on an $m \times n$ chessboard with $m, n \geq 5$ and m and n odd. For large values of m, n several open knight's tour have to be included. To improve visibility we separated the single building blocks by some space

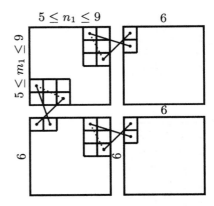

Corollary 4 *Optimal TSPFN tours with $r = 2$ on regular $m \times n$ grids with $m, n \geq 5$ have length $mn\sqrt{5}$ for m or n even and length $(mn - 1)\sqrt{5} + \sqrt{8}$ for m and n odd.*

3 Closed Almost Knight's Tours on 3D Chessboards

In this section we consider $m \times n \times \ell$ chessboards where ℓ is the number of layers. For an illustration of the $3 \times 3 \times 3$ chessboard we refer to Fig. 4.

Lemma 2 *Given an $m \times n \times \ell$ chessboard with $m, n, \ell \geq 3$ and m, n, ℓ odd, $(mn\ell - 1)\sqrt{5} + \sqrt{6}$ is a lower bound on the length of a closed almost knight's tour.*

This result follows directly from Theorem 2 because there does not exist a closed knight's tour for the considered chessboard sizes and the shortest move longer than $\sqrt{5}$ has length $\sqrt{6}$. Next we show that there always exists a closed almost knight's tour with this length. The proof idea is, similar to [4], to use slightly adapted 2D tours in each layer and to delete some specific edges so that nodes from different layers can be connected by some new edges.

Theorem 5 *Given an $m \times n \times \ell$ chessboard with $m, n \geq 5$, $\ell \geq 3$ and m, n, ℓ odd, there always exists a closed almost knight's tour with length $(mn\ell - 1)\sqrt{5} + \sqrt{6}$.*

Proof We prove this result algorithmically by providing a construction scheme for building closed almost knight's tours. First, we use specific knight's tours for every layer and then we connect different layers via new edges.

We build an $m \times n$ closed almost knight's tour using the construction scheme described in the proof of Theorem 3 and then modify the upper left area that belongs to the $5 \leq m, n \leq 9$ building blocks, see Fig. 1, as follows. We delete the edge of length $\sqrt{8}$ connected to Cell 1 and one further edge and adapt the cell sequence of the new tours (subtours for the 3D chessboard) as follows:

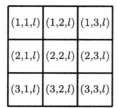 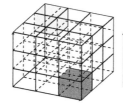 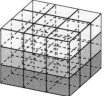

Fig. 4 Visualizations of the $3 \times 3 \times 3$ chessboard. The left picture shows the numbering of layers $l \in \{1, 2, 3\}$. In the middle picture, the dark gray cell is $(3, 3, 1)$ and the light grey one is $(1, 1, 3)$. In the right picture, each layer has one color, where layers with smaller numbers are darker

- $(1, \ldots, 21, 22, 25, 24, 23, 1)$ for the 5×5 chessboard,
- $(1, \ldots, 12, 35, 34, \ldots, 13, 1)$ for the 5×7 chessboard,
- $(1, \ldots, 42, 45, 44, 43, 1)$ for the 5×9 chessboard,
- $(1, \ldots, 40, 49, 48, \ldots, 41, 1)$ for the 7×7 chessboard,
- $(1, \ldots, 40, 63, 62, \ldots, 41, 1)$ for the 7×9 chessboard and
- $(1, \ldots, 78, 81, 80, 79, 1)$ for the 9×9 chessboard,

where the numbering refers to Fig. 1. After the first modification, each such 2D tour has length $(mn - 1)\sqrt{5} + 2$ instead of $(mn - 1)\sqrt{5} + \sqrt{8}$.

Now we use these modified tours for all layers except for Layer ℓ and connect Layers k and $k + 1$, k odd, by replacing the edges $\{(i, j, k), (i + 2, j, k)\}$ and $\{(i, j, k + 1), (i + 2, j, k + 1)\}$ of length 2 that lie directly above each other by the edges $\{(i, j, k), (i + 2, j, k + 1)\}$ and $\{(i, j, k + 1), (i + 2, j, k)\}$ that switch the layer.

Next we connect Layers k and $k + 1$, k even, by using the same exchange as in [4]. We delete the two edges of a so called *bi-site* (see [4]) where two knight's moves of our subtours cross each other. A bi-site can be found in the lower left corner of each grid in Fig. 1, e. g. $\{23, 24\}$ and $\{7, 8\}$ for the 5×5 chessboard or $\{35, 36\}$ and $\{23, 24\}$ for the 7×7 chessboard. In general let the bi-sites be $\{(i, j, k), (i + 2, j + 1, k)\}$ and $\{(i, j + 1, k), (i + 2, j, k)\}$ for $k = 1, \ldots, \ell - 1$, then we use afterwards edges $\{(i, j, k), (i + 2, j, k + 1)\}$ and $\{(i + 2, j + 1, k), (i, j + 1, k + 1)\}$.

It remains to connect Layer ℓ where we use an original closed almost knight's tour to the other layers. To do so we delete the edge of length $\sqrt{8}$ in Layer ℓ and one appropriate edge in Layer $\ell - 1$ (depending on the position of the edge of length 2) such that the exchange visualized in Fig. 5 is possible. The squares highlighted light gray are connected by a move of length $\sqrt{6}$ and the squares highlighted dark gray are connected by a move of length $\sqrt{5}$. It is easy to check that the respective edges exist in each of these tours and that we do not create subtours. □

For illustration purposes we depict a closed almost knight's tour for the $5 \times 5 \times 5$ chessboard in Fig. 6.

Corollary 6 *Optimal TSPFN tours with $r = 2$ on regular $m \times n \times \ell$ grids with $m, n \geq 5$ and $\ell \geq 3$ have length $mn\ell\sqrt{5}$ for m or n or ℓ even and length $(mn\ell - 1)\sqrt{5} + \sqrt{6}$ for m, n, ℓ odd.*

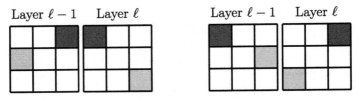

Fig. 5 Moves connecting Layers $\ell - 1$ and ℓ in closed almost knight's tours

1	22	11	16	3
12	17	2	21	10
25	8	23	4	15
18	13	6	9	20
7	24	19	14	5

26	47	36	41	28
37	42	27	46	35
125	33	48	29	40
43	38	31	34	45
32	49	44	39	30

118	64	53	58	120
54	59	119	63	52
67	50	65	121	57
60	55	123	51	62
124	66	61	56	122

68	114	78	108	70
104	109	69	113	77
117	75	115	71	107
110	105	73	76	112
74	116	111	106	72

79	100	89	94	81
90	95	80	99	88
101	86	103	82	93
96	91	84	87	98
85	102	97	92	83

Fig. 6 Visualization of a closed almost knight's tour on a $5 \times 5 \times 5$ chessboard

4 Conclusion and Future Work

In this paper we introduced the concept of closed almost knight's tours and proposed construction schemes for such tours on 2D and 3D chessboards that do not have a classical closed knight's tour. We restricted our analysis to the case with $m, n \geq 5$ in 2D and $m, n \geq 5$ and $\ell \geq 3$ in 3D. It remains to determine closed almost knight's tours for the remaining chessboard sizes. Furthermore, one can think of extending the results to arbitrary chessboard dimensions. For the closely related traveling salesman problem with forbidden neighborhoods it seems to be worthwhile to study also other values for r.

References

1. Conrad, A., Hindrichs, T., Hussein, M., & Wegener, I. (1994). Solution of the knight's Hamiltonian path problem on chessboards. *Discrete Applied Mathematics*, *50*(2), 125–134.
2. Dantzig, G., Fulkerson, R., & Johnson, S. (1954). Solution of a large-scale traveling-salesman problem. *Operations Research*, *2*, 393–410.
3. DeMaio, J., & Mathew, B. (2011). Which chessboards have a closed knight's tour within the rectangular prism?. *The Electronic Journal of Combinatorics 18*.
4. Erde, J., Golénia, B., Golénia, S. (2012). The closed knight tour problem in higher dimensions. *The Electronic Journal of Combinatorics*, *16*(4).
5. Fischer, A., & Hungerländer, P. (2017). The traveling salesman problem on grids with forbidden neighborhoods. *Journal of Combinatorial Optimization*.
6. Lin, S., & Wei, C. (2005). Optimal algorithms for constructing knight's tours on arbitrary $n \times m$ chessboards. *Discrete Applied Mathematics*, *146*, 219–232.
7. Schwenk, A. J. (1991). Which rectangular chessboards have a knight's tour? *Mathematics Magazine*, *64*(5), 325–332.

SCIP-Jack—A Solver for STP and Variants with Parallelization Extensions: An Update

Daniel Rehfeldt and Thorsten Koch

1 Introduction

The Steiner tree problem in graphs (STP) is a classical $\mathcal{NP}$-hard problem [1] entailing a wealth of research articles. Given an undirected, connected graph $G = (V, E)$, costs $c : E \to \mathbb{Q}_{\geq 0}$ and a set $T \subseteq V$ of *terminals*, the problem is to find a tree $S \subseteq G$ of minimum cost that includes T.

While Steiner tree problems can be found in various applications, these problems are usually one of the many variants of the STP, such as for instance the rectilinear Steiner tree problem [2] or the prize-collecting Steiner tree problem [3]. The 2014 DIMACS Challenge, dedicated to Steiner tree problems, marked a revival of research on the STP and related problems: Both at and in the wake of the Challenge several new Steiner problem solvers were introduced and many articles were published. One of these new solver is *SCIP-Jack*, which was by far the most versatile solver participating in the DIMACS Challenge, being able to solve the STP and 10 of its variants (note that in the current version one more variant can be handled). Moreover, SCIP-Jack was able to win two categories of the Challenge.

SCIP-Jack is described in detail in the article [4], but already in an updated version that vastly outperforms its predecessor participating in the DIMACS Challenge. However, the development of SCIP-Jack did not stop with [4]. In the following we will report on recent improvements and provide current results that again demonstrate a significant speed-up of SCIP-Jack (Table 1).

Internally, all problems are transformed into the Steiner arborescense problem (SAP), the directed version of the STP [4]. In only two cases it is necessary to add

---

D. Rehfeldt · T. Koch (✉)
Zuse Institute Berlin, Takustr. 7, 14195 Berlin, Germany
e-mail: koch@zib.de
URL: http://www.zib.de/rehfeldt

D. Rehfeldt
e-mail: rehfeldt@zib.de

Table 1 SCIP-Jack can solve the STP and 11 related problems

Abbreviation	Problem Name
STP	Steiner tree problem in graphs
SAP	Steiner arborescence problem
RSMT	Rectilinear Steiner minimum tree problem
OARSMT	Obstacle-avoiding rectilinear Steiner minimum tree problem
NWSTP	Node-weighted Steiner tree problem
PCSTP	Prize-collecting Steiner tree problem
RPCSTP	Rooted prize-collecting Steiner tree problem
MWCSP	Maximum-weight connected subgraph problem
RMWCSP	Rooted maximum-weight connected subgraph problem
DCSTP	Degree-constrained Steiner tree problem
GSTP	Group Steiner tree problem
HCDSTP	Hop-constrained directed Steiner tree problem

Table 2 Transformations, heuristics, and preprocessing according to problem type

Problem	Special constraints	Virtual vertices	Virtual arcs	Special preprocessing	Special heuristics
STP	–	–	✓	✓	✓
SAP	–	–	–	✓	✓
RSMT	–	✓	✓	–	–
OARSMT	–	✓	✓	–	–
NWSTP	–	–	✓	–	–
PCSTP	–	✓	✓	✓	✓
RPCSTP	–	✓	✓	✓	✓
MWCSP	–	✓	✓	✓	✓
RMWCSP	–	✓	✓	–	✓
DCSTP	✓	–	✓	–	✓
GSTP	–	✓	✓	–	–
HCDSTP	✓	–	–	✓	✓

specific constraints. The transformations are a distinct feature of SCIP-Jack and allow for a generic solving approach with a single branch-and-cut algorithm. Descriptions on some of the transformations can be found in [5]. While in principle it would be possible to only employ solving routines for the SAP (since each problem variant can be transformed to it), this approach falls far short of being competitive as it fails to utilize special properties of particular problem types. Therefore, SCIP-Jack includes a plethora of specialized heuristics and preprocessing routines. Table 2 indicates for which problem variants specialized algorithms are used. Detail on the heuristics is given in [4, 5] (and in Sect. 2), while detail on the preprocessing can be found in [6].

2 Recent Improvements

Apart from general improvements, particular progress has been made with the MWCSP and the PCSTP. Therefore, these two variants will be discussed separately in Sect. 2.2.

2.1 General Improvements

The general improvements of SCIP-Jack include a change in the default propagator, see [4], which now additionally employs reduction techniques to fix variables (of the underlying IP formulation) to zero. These variables correspond to arcs in the SAP—to which all Steiner tree variants including the STP are transformed. Whenever ten percent of all arcs have been newly fixed during the branch-and-cut procedure, the underlying, directed, SAP graph D is (re-) transformed into a graph G for the respective Steiner tree problem variant. All edges (or arcs) in G that correspond to arcs that have been fixed to 0 in D are removed. Thereupon, the default reduction techniques of SCIP-Jack are used to further reduce G and the changes are retranslated into arc fixings in D.

A further important development is the reimplementation of the separation algorithm of SCIP-Jack, which is based on the warm-start preflow-push algorithm described in [7]. The new separation algorithm is for many instances more than ten times faster than the old one. The cause of this speed-up lies both with an improved, cache-optimized implementation and the use of new heuristics. Notably, the underlying maximum-flow routine also vastly outperforms the algorithm described in [8], which is commonly used as a benchmark for maximum-flow algorithms.

2.2 Improvements for MWCSP and PCSTP

Maximum-weight connected subgraph problem. Given an undirected graph $G = (V, E)$ and node weights $p : V \to \mathbb{Q}$, the objective is to find a connected subgraph $S = (V_S, E_S) \subseteq G$ such that $\sum_{v \in V_S} p_v$ is maximized.

Prize-collecting Steiner tree problem. Given an undirected graph $G = (V, E)$, edge-weights $c : E \to \mathbb{Q}_+$, and node-weights $p : V \to \mathbb{Q}_{\geq 0}$, a tree $S = (V_S, E_S) \subseteq G$ is required that minimizes

$$C(S) := \sum_{e \in E_S} c_e + \sum_{v \in V \setminus V_S} p_v. \tag{1}$$

The most important component for accelerating exact solving of both PCSTP and MWCSP is graph reduction—which can for instance be employed in preprocessing.

A simple reduction routine for the MWCSP is for example to contract all adjacent vertices of positive weight. However, several reduction techniques are considerably more sophisticated—in [9], for instance, three techniques for the MWCSP are described that involve NP-hard subproblems. By adding these preprocessing techniques to SCIP-Jack, not only most MWCSP problems can be solved during preprocessing, but also several instances could be solved for the first time to optimality [6, 9].

Another important component is constituted by heuristics. The current version of SCIP-Jack includes for instance a straightforward greedy heuristic for the PCSTP that starts with a single vertex tree $S_0 = v$ with $v \in V$ and repeatedly connects the current tree S_i to another vertex $w \in V$ with $p_w > 0$ such that this extension leads to a tree S_{i+1} with $C(S_{i+1}) \leq C(S_i)$. This procedure is implemented by a modification of Dijkstra's algorithm. More refined heuristics in SCIP-Jack for both PCSTP and MWCSP are described in [5, 9].

3 Computational Results

The computational experiments described in the following were performed on a cluster of Intel Xeon X5672 CPUs with 3.20 GHz and 48 GB RAM. SCIP 4.0.0 was used and CPLEX 12.6[1]) was employed as the underlying LP solver. Moreover, the overall run time for each instance was limited by two hours. If an instance was not solved to optimality within the time limit, the gap is reported, which is defined as $\frac{|pb-db|}{\max\{|pb|,|db|\}}$ for final primal bound (pb) and dual bound (db). The average gap is obtained as an arithmetic mean. The averages of the number of nodes and the solving time are computed by taking the shifted geometric mean with a shift of 10.0 and 1.0, respectively. For reasons of space we only provide results for STP, PCSTP, and MWCSP.

The results in Table 3 show that the majority of STP instances can be solved within short time. The new version of SCIP-Jack can solve several more instances to optimality than the previous version described in [4]. Also, the run time has been more than halved for the majority of instances.

As can be seen in Tables 4 and 5, with the combination of the new reduction techniques, heuristics and transformations most of the PCSTP and MWCSP instances can be solved easily. Only the PUCNU test set has unsolved instances left. Notably, the results not only demonstrate a speed-up of more than 200% for many instances as compared to the previous version of SCIP-Jack, but also mark a demarcation from other state-of-the-art PCSTP or MWCSP solvers. For example for several problems from the SHINY test, SCIP-Jack outperforms the best run-times reported in the literature [10] by three orders of magnitude and solves problems in less than 0.1 s that are intractable for other solvers.

---

[1] http://www-01.ibm.com/software/commerce/optimization/cplex-optimizer/

Table 3 Computational results for steiner tree problem in graphs

Test set	#	Solved	Optimal		Timeout	
			⌀ nodes	⌀ time [s]	⌀ nodes	⌀ gap [%]
X	3	3	1.0	0.1	–	–
E	20	20	1.8	0.3	–	–
I640	100	80	23.3	6.6	980.1	0.7
ALUE	15	13	1.3	22.9	1.0	1.9
Vienna-i-advanced	85	83	1.8	70.2	1.8	0.0

Table 4 Computational results for (rooted) prize-collecting Steiner tree problem ((R)PCSTP)

Test set	#	Solved	Optimal		Timeout	
			⌀ nodes	⌀ time [s]	⌀ nodes	⌀ gap [%]
Cologne1	14	14	1.0	0.0	–	–
Cologne2	15	15	1.0	0.1	–	–
JMP	34	34	1.0	0.0	–	–
CRR	80	80	1.0	0.2	–	–
PUCNU	18	11	28.0	26.2	865.6	1.8

Table 5 Computational results for maximum-weight connected subgraph problem (MWCSP)

Test set	#	Solved	Optimal		Timeout	
			⌀ nodes	⌀ time [s]	⌀ nodes	⌀ gap [%]
JMPALMK	72	72	1.0	0.0	–	–
SHINY	39	39	1.0	0.0	–	–
ACTMOD	8	8	1.0	0.1	–	–

4 Conclusions and Outlook

The computational results of SCIP-Jack demonstrate that improved preprocessing and transformation techniques can have a dramatic effect on performance when solving Steiner tree variants. In many cases it is possible to solve problems to optimality even before it is necessary to employ the branch-and-cut kernel.

In the future we will continue on this path, adding more specific routines, while at the same time improving those sections that apply to all problem variants. The aim is to both improve the run-time of SCIP-Jack, in particular for the STP, and, equally important, tackle additional previously unsolved instances.

Acknowledgements The work for this article has been conducted within the *Research Campus Modal* funded by the German Federal Ministry of Education and Research (fund number 05M14ZAM). It was supported by the Federal Ministry for Economic Affairs and Energy within the BEAM-ME project (ID: 03ET4023A-F). It has been further supported by a Google Faculty Research Award.

References

1. Karp, R. (1972). Reducibility among combinatorial problems. In R. Miller, & J. Thatcher (Eds.), *Complexity of computer computations* (pp. 85–103). Plenum Press.
2. Warme, D., Winter, P., & Zachariasen, M. (2000). Exact algorithms for plane Steiner tree problems: A computational study. In D.Z. Du, J. Smith, & J. Rubinstein (Eds.), *Advances in steiner trees* (pp. 81–116). Kluwer.
3. Ljubić, I., Weiskircher, R., Pferschy, U., Klau, G. W., Mutzel, P., & Fischetti, M. (2006). An algorithmic framework for the exact solution of the prize-collecting steiner tree problem. *Mathematical Programming, 105*(2), 427–449. Feb.
4. Gamrath, G., Koch, T., Maher, S., Rehfeldt, D., & Shinano, Y. (2017). SCIP-Jack–a solver for STP and variants with parallelization extensions. *Mathematical Programming Computation, 9*(2), 231–296.
5. Rehfeldt, D., & Koch, T. (2016). Transformations for the Prize-Collecting Steiner Tree Problem and the Maximum-Weight Connected Subgraph Problem to SAP. Technical Report 16–36, ZIB, Takustr.7, 14195 Berlin.
6. Rehfeldt, D., Koch, T., & Maher, S. (2016). Reduction Techniques for the Prize-Collecting Steiner Tree Problem and the Maximum-Weight Connected Subgraph Problem. Technical Report 16–47, ZIB, Takustr.7, 14195 Berlin.
7. Hao, J., & Orlin, J. B. (1994). A faster algorithm for finding the minimum cut in a directed graph. *Journal of Algorithms, 17*(3), 424–446.
8. Cherkassky, B. V., & Goldberg, A. V. (1997). On implementing the push–relabel method for the maximum flow problem. *Algorithmica, 19*(4), 390–410.
9. Rehfeldt, D., & Koch, T. (2017). Combining NP-Hard Reduction Techniques and Strong Heuristics in an Exact Algorithm for the Maximum-Weight Connected Subgraph Problem. Technical Report 17–45, ZIB, Takustr.7, 14195 Berlin.
10. Loboda, A. A., Artyomov, M. N., & Sergushichev, A. A. (2016). *In Solving generalized maximum-weight connected subgraph problem for network enrichment analysis* (pp. 210–221). Cham: Springer International Publishing.

Extended Formulations for Column Constrained Orbitopes

Christopher Hojny, Marc E. Pfetsch and Andreas Schmitt

1 Introduction

Modeling combinatorial optimization problems as mixed integer optimization problems often leads to formulations which contain symmetry, i.e., solutions can be permuted to obtain new solutions with the same objective value. This slows solving via branch-and-bound down, since equivalent solutions are repeatedly inspected in different nodes of the branching scheme.

We illustrate this setting with the aid of the so-called *balanced partitioning problem* (BPP), a variant of the graph partitioning problem. Given an undirected graph (V, E) with $m = |V|$ nodes and a positive divisor n of m, the task is to find a partition of the nodes into exactly n equal sized parts that minimizes the number of edges between different parts, see, e.g., Lisser and Rendl [11] and Karisch and Rendl [9]. Applications include solving sparse matrix-vector multiplications with iterative-parallel algorithms, see Karypis and Kumar [10]. The problem can be formulated using binary variables $X_{v,i}$ to model whether node v belongs to part i.

The problem exhibits obvious symmetry: Permuting the labels of the parts of a feasible solution yields another feasible solution. If the X-variables are considered as an $m \times n$-matrix, this permutation can be expressed by permuting columns of X.

A class of techniques to handle symmetry considers the removal of all but one lexicographically maximal representative of the set of equivalent solutions from the problem by cutting planes. In the above case, these hyperplanes would cut off all solutions whose columns in X are not sorted lexicographically non-increasing.

C. Hojny · M. E. Pfetsch · A. Schmitt (✉)
Department of Mathematics, Technische Universität Darmstadt, Darmstadt, Germany
e-mail: aschmitt@mathematik.tu-darmstadt.de

C. Hojny
e-mail: hojny@mathematik.tu-darmstadt.de

M. E. Pfetsch
e-mail: pfetsch@mathematik.tu-darmstadt.de

To understand the polyhedral properties of these cutting planes, the notion of *full orbitopes* $O_{m,n}$ has been introduced in [8]. The polytope $O_{m,n}$ is the convex hull of all binary $m \times n$-matrices whose columns are sorted lexicographically non-increasing. Adding its facet-defining inequalities to the problem at hand destroys the symmetry.

Utilizing further problem structure of orbitopes can give stronger cuts. In (BPP), for example, each row of X contains exactly one 1-entry. This additional structure can be handled by *partitioning orbitopes*, see [8] or Faenza and Kaibel [3]. Besides the row constraints, (BPP) consists of further bounds on X, namely exactly $\frac{m}{n}$ 1-entries are allowed in each column. For this reason, we consider the *column constrained (partitioning) orbitope* $C_{m,n}^{=k}$ with bound $k \in [m] := \{1, \ldots, m\}$. This polytope is the convex hull of all binary $m \times n$-matrices contained in $O_{m,n}$ that have exactly k 1-entries in every column, i.e.,

$$C_{m,n}^{=k} := \mathrm{conv}\Big(\Big\{ X \in O_{m,n} \cap \{0,1\}^{m \times n} : \sum_{i=1}^{m} X_{i,j} = k, j \in [n] \Big\}\Big).$$

In this paper, we will shortly sketch an efficient algorithm to optimize over $C_{m,n}^{=k}$. Afterwards, an extended formulation for the case of two columns is derived via a disjunctive programming approach. A numerical comparison of symmetry handling with this extended formulation and the partitioning orbitope concludes our work.

2 Column Constrained Orbitopes

Loos [12] presented a polynomial time algorithm to optimize over full orbitopes. It uses dynamic programming and relies on the vertex structure of orbitopes. A binary matrix $X \in \{0, 1\}^{m \times n}$ is a vertex of $O_{m,n}$ if and only if there exists a column index $t \in [n] \cup \{0\}$ such that the first row of X consists of t 1-entries followed by $n - t$ 0-entries and the two submatrices spanning all rows below the first row and columns 1 to t and $t + 1$ to n, respectively, are vertices of orbitopes with suitable dimensions. Thus, an optimal vertex can be computed recursively and a Bellman equation for a dynamic programming algorithm can be derived.

The vertices of $C_{m,n}^{=k}$ have a similar structure. The first row of a vertex again divides into a 1-part of size t and a 0-part of size $n - t$. The two submatrices below the first row must belong to $C_{m-1,t}^{=k-1}$ and $C_{m-1,n-t}^{=k}$, respectively. By adjusting Loos' algorithm to regard the different bounds k and $k - 1$, optimization over column constrained orbitopes is possible in time $\mathcal{O}(mn^3k)$. Thus, there are no obstacles from complexity theory preventing us to find a compact linear description of $C_{m,n}^{=k}$. For the bound $k = 1$ the description of $C_{m,n}^{=k}$ is provided by the following system:

$$\sum_{i=1}^{m} X_{i,j} = 1, \qquad j \in [n], \tag{1a}$$

$$\sum_{i=1}^{s} X_{i,j+1} - \sum_{i=1}^{s} X_{i,j} \leq 0, \qquad (s,j) \in [m] \times [n-1], \tag{1b}$$

$$X_{i,j} \geq 0, \qquad (i,j) \in [m] \times [n]. \tag{1c}$$

These inequalities force feasible binary matrices to contain exactly one 1-entry in each column and their columns to be lexicographically non-increasing. Consequently, these matrices are exactly the vertices of $C_{m,n}^{=1}$. In fact, System (1) is a complete linear description of $C_{m,n}^{=1}$, because the coefficient matrix of (1) is a network matrix, and thus totally unimodular.

For general bounds k, computational experiments with polymake [5], however, show that $C_{m,n}^{=k}$ has a much more complicated facet structure. But since handling symmetry associated with $C_{m,n}^{=k}$ is completely possible by forcing each pair of adjacent columns of X to belong to $C_{m,2}^{=k}$, cf. [6, Proposition 29], it suffices to only consider the column constrained orbitope $C_{m,2}^{=k}$ with two columns. The latter polytope has an easier vertex structure than $C_{m,n}^{=k}$: The first column of a vertex X of $C_{m,2}^{=k}$ has to be lexicographically not smaller than the second column by definition. Thus, either the first column coincides with the second or there exists a row $\ell \in [m]$ such that the entries coincide in each row above ℓ and $(X_{\ell,1}, X_{\ell,2}) = (1, 0)$. This leads to the definition of the critical row of a vertex X of $C_{m,2}^{=k}$, cf. Kaibel and Loos [7]:

$$\mathrm{crit}(X) := \min(\{\ell \in [m] : (X_{\ell,1}, X_{\ell,2}) = (1,0)\} \cup \{m+1\}).$$

We can partition the vertices of $C_{m,2}^{=k}$ based on their critical row and define the polytopes P_ℓ for $\ell \in [m+1]$ by

$$P_\ell := \mathrm{conv}(\{X \in C_{m,2}^{=k} \cap \{0,1\}^{m \times 2} : \mathrm{crit}(X) = \ell\}).$$

These polytopes are completely described by the following constraints

$$\sum_{i=1}^{m} X_{i,j} = k, \qquad j \in [2], \tag{2a}$$

$$X_{i,1} - X_{i,2} = 0, \qquad i \in [\ell - 1], \tag{2b}$$

$$(X_{\ell,1}, X_{\ell,2}) = (1, 0), \tag{2c}$$

$$0 \leq X_{i,j} \leq 1, \qquad (i,j) \in [m] \times [2]. \tag{2d}$$

Lemma 1 *For $\ell \in [m]$ the polytope P_ℓ consists of all those matrices $X \in \mathbb{R}^{m \times 2}$ which satisfy constraints (2a) to (2d). The polytope P_{m+1} is given by the constraints (2a), (2b) and (2d) with $\ell = m + 1$.*

Proof Our arguments above show that every vertex of $C_{m,2}^{=k}$ with critical row ℓ satisfies the constraints. Also every integer point satisfying them is a vertex of $C_{m,2}^{=k}$, since there are exactly k 1-entries in each column by (2a) and the second column is lexicographically smaller or equal to the first by (2b) and (2c). Via elementary operations, we can transform the coefficient matrix of constraints (2a) and (2b) into a node-arc incidence matrix of a directed graph. Thus, the matrix is totally unimodular and system (2a)–(2d) defines an integral polytope. □

With the descriptions of the P_ℓ, we derive an extended formulation of $C_{m,2}^{=k}$.

Theorem 1 *Let $m \in \mathbb{N}$ and $k \in [m]$. A matrix $X \in \mathbb{R}^{m \times 2}$ lies within the column constrained orbitope $C_{m,2}^{=k}$ if and only if there exist matrices $Y^\ell \in \mathbb{R}^{m \times 2}$ and scalars $\lambda^\ell \in \mathbb{R}$ for $\ell \in [m+1]$ satisfying the constraints*

$$X_{i,j} = \sum_{\ell=1}^{m+1} Y_{i,j}^\ell, \qquad (i,j) \in [m] \times [2], \tag{3a}$$

$$\sum_{i=1}^{m} Y_{i,j}^\ell = k\lambda^\ell, \qquad (\ell, j) \in [m+1] \times [2], \tag{3b}$$

$$Y_{i,1}^\ell - Y_{i,2}^\ell = 0, \qquad \ell \in [m+1], \; i \in [\ell-1], \tag{3c}$$

$$(Y_{\ell,1}^\ell, Y_{\ell,2}^\ell) = (\lambda^\ell, 0), \qquad \ell \in [m], \tag{3d}$$

$$0 \leq Y_{i,j}^\ell \leq \lambda^\ell, \qquad (\ell, i, j) \in [m+1] \times [m] \times [2], \tag{3e}$$

$$\sum_{\ell=1}^{m+1} \lambda^\ell = 1. \tag{3f}$$

Proof Since $C_{m,2}^{=k} = \operatorname{conv}\left(\cup_{\ell=1}^{m+1} P_\ell\right)$ and we know a complete linear description of each P_ℓ, we can apply disjunctive programming, see Balas [1, Theorem 2.1], to derive an extended formulation of $C_{m,2}^{=k}$ which is given by (3). □

Theorem 1 shows that $C_{m,2}^{=k}$ admits a compact extended formulation with $\mathcal{O}(m^2)$ variables and constraints. Note, however, that a complete linear description of $C_{m,2}^{=k}$ in the original space is unknown.

These results can be modified for the *column constraint packing orbitope* with two columns $C_{m,2}^{\leq k} := \operatorname{conv}(\{X \in O_{m,2} \cap \{0,1\}^{m \times 2} : \sum_{i=1}^{m} X_{i,j} \leq k, j \in [2]\})$. Replacing equality in Constraint (2a) by inequality preserves totally unimodularity. Thus, changing equality in Constraint (3b) to inequality yields an extended formulation for $C_{m,2}^{\leq k}$. A similar modification results in an extended formulation for the column constrained covering orbitope with two columns $C_{m,2}^{\geq k}$, which has at least k 1-entries in each column.

Coming back to the initial example (BPP), we observe that not only the number of 1-entries in each column of a solution is bounded, but additionally exactly one entry in each row must be 1. We can incorporate this further information by considering the polytope

$$\text{conv}\left(\left\{X \in C_{m,2}^{=k} \cap \{0,1\}^{m \times 2} : X_{i,1} + X_{i,2} \leq 1,\ i \in [m]\right\}\right) \quad (4)$$

to handle symmetry. Adding this polytope for each pair of adjacent columns, the number of 1-entries in each row is forced to be at most one. By a slight modification we can adjust the extended formulation from Theorem 1 to comply with Polytope (4): For each vertex X, the rows $(X_{i,1}, X_{i,2})$ with $i < \text{crit}(X)$ must be zero, since these rows are either $(0,0)$ or $(1,1)$, but are only allowed to contain at most one 1-entry. These rows can thus be neglected in P_ℓ. Also Constraint (2b) is obsolete; instead $X_{i,1} + X_{i,2} \leq 1$ for $i \in \{\ell+1, \ldots, m\}$ needs to be added. Furthermore, the vertices can only have a critical row $\ell \leq m - 2k$, otherwise there would not be enough 1-entries in each column. Thus, we do not use P_ℓ for $\ell > m - 2k$ in the extended formulation.

3 Numerical Results and Conclusions

We investigated the practical use of the derived extended formulation of $C_{m,2}^{=k}$ to handle the symmetry of (BPP) using the framework SCIP 3.2.1 [4] on a Linux cluster with Intel Xeon E5 CPUs with 3.50 GHz, 10 MB cache, and 32 GB memory. The adjacency matrices of the graphs of our instances are from the University of Florida Sparse Matrix Collection [2]. We considered matrices with at least 6 and less than 100 rows, which are non-diagonal and pattern symmetric, i.e., only square matrices A with $A_{i,j} \neq 0$ if and only if $A_{j,i} \neq 0$. This results in 40 instances, which are to be partitioned into $n \in \{5, 10, 20\}$ parts. We added isolated nodes to the graph if n did not divide the number of nodes and disregard calculation when n exceeds the number of nodes. We compared two settings to estimate the strength of the extended formulation. The first setting restricts the variables to lie inside the partitioning orbitope via an already existing SCIP plugin. This plugin fixes the variables in the upper right triangle of X to zero and separates the facets of the partitioning orbitope. In the second setting, we added for each pair of adjacent columns of X the variables and constraints belonging to the extended formulation of the Polytope (4) with $k = \frac{m}{n}$.

We are most interested in the improvement of the LP relaxation by adding (constrained) orbitopes. Thus, we compared for each setting the value of the first solved LP relaxation.

To evaluate the improvement of adding the extended formulation, we calculated for each instance the quotient of the value of the extended formulation's LP relaxation d_{ef} and the value of the LP relaxation obtained by the orbitope setting d_{orbi}, i.e., the value $d_{\text{ef}}/d_{\text{orbi}}$. Since we consider a minimization problem a value greater than 1 indicates a better LP relaxation of the extended formulation.

Table 1 summarizes our results. Column "#instances" shows the number of considered instances for a fixed value of n (recall that some instances are disregarded), whereas the last three columns collect the arithmetic mean together with the min and max over the relative values for instances with the same partition size. We see

Table 1 Statistics on the improvement ratio $d_{\text{ef}}/d_{\text{orbi}}$ for partition sizes $n \in \{5, 10, 20\}$

n	#instances	Arithmetic mean	Min	Max
5	40	1.0276	1.00	1.1931
10	37	1.0270	1.00	1.1382
20	34	1.0218	0.85	1.1149

that on average the LP relaxation of the extended formulation setting performs about 2% better than the orbitope. Except for one instance with $n = 20$, the values are always at least 1. With an increasing number of parts the advantage of the extended formulation decreases. Nevertheless, improvements of nearly 20% for $n = 5$ to 11% for $n = 20$ can be observed for some instances.

Conclusions. In this paper, we introduced orbitopes with additional requirements and incorporated these properties to orbitopes via an extended formulation. In a computational study, we have seen that this extended formulation improves the first LP relaxation, on some instances even substantially. However, this effect decreases if we consider orbitopes with more columns. Furthermore the blow up of the variable space makes adding the whole extended formulation impractical. To benefit from the positive effect on the LP relaxation, schemes to separate the extended formulation as well as descriptions of $C_{m,n}^{=k}$ with fewer variables should be considered. In particular, a complete description in the original space would be desirable, but could not be achieved so far.

Acknowledgements The research of the second and third author was supported by the German Research Foundation (DFG) as part of the Collaborative Research Centre 666 and 805, respectively.

References

1. Balas, E. (1998). Disjunctive programming: Properties of the convex hull of feasible points. *Discrete Applied Mathematics, 89*(1), 3–44.
2. Davis, T. A., & Hu, Y. (2011). The University of Florida sparse matrix collection. *ACM Transactions on Mathematical Software, 38*(1), 1–25.
3. Faenza, Y., & Kaibel, V. (2009). Extended formulations for packing and partitioning orbitopes. *Mathematics of Operations Research, 34*(3), 686–697.
4. Gamrath, G., Fischer, T., Gally, T., Gleixner, A. M., Hendel, G., Koch, T., Maher, S.J., Miltenberger, M., Müller, B., Pfetsch, M. E., Puchert, C., Rehfeldt, D., Schenker, S., Schwarz, R., Serrano, F., Shinano, Y., Vigerske, S., Weninger, D., Winkler, M., Witt, J. T., Witzig, J. (2016). The SCIP Optimization Suite 3.2. Technical Report 15-60, ZIB, Takustr. 7, 14195 Berlin.
5. Gawrilow, E., & Joswig, M. (2000). Polymake: A framework for analyzing convex polytopes. In *Polytopes – combinatorics and computation* (pp. 43–74).
6. Hojny, C., & Pfetsch, M. E. (2017). Polytopes associated with symmetry handling. www.optimization-online.org/DB_HTML/2017/01/5835.html.
7. Kaibel, V., & Loos, A. (2011). Finding descriptions of polytopes via extended formulations and liftings. In: A. R. Mahjoub (Ed.), *Progress in combinatorial optimization*. New Jersey: Wiley.

8. Kaibel, V., & Pfetsch, M. E. (2008). Packing and partitioning orbitopes. *Mathematical Programming*, *114*(1), 1–36.
9. Karisch, S. E., & Rendl, F. (1998). Semidefinite programming and graph equipartition. *Topics in Semidefinite and Interior-Point Methods*, *18*, 77–95.
10. Karypis, G., & Kumar, V. (1998). A fast and high quality multilevel scheme for partitioning irregular graphs. *SIAM Journal on Scientific Computing*, *20*(1), 359–392.
11. Lisser, A., & Rendl, F. (2003). Graph partitioning using linear and semidefinite programming. *Mathematical Programming*, *95*(1), 91–101.
12. Loos, A. (2011). Describing orbitopes by linear inequalities and projection based tools. Ph.D. thesis, Universität Magdeburg.

The k-Server Problem with Parallel Requests and the Corresponding Generalized Paging Problem

R. Hildenbrandt

1 The Formulation of the Model

Firstly, we want to describe the generalized k-server problem. Let $k \geq 1$ be an integer, and $\mathcal{M} = (M, d)$ be a finite metric space where M is a set of points with $|M| = N$. An algorithm controls k mobile servers, which are located on points of M. Several servers can be located on one point. Requests r^t for service at several points come in over time. In online computation, an algorithm must decide how to act on incoming requests without any knowledge of future inputs. In contrast, an offline procedure would be allowed to know the entire sequence of requests in advance, before it makes any decisions. Now, let $\sigma = r^1, r^2, \ldots, r^n$ be such a sequence of requests. A request r is defined as an N-ary vector of integers with $r_i \in \{0, 1, \ldots, k\}, i = 1, 2, \ldots, N$ ("parallel requests"). The request means that r_i servers are needed on point i, $i = 1, 2, \ldots, N$.

Principally, two cases of requests have to distinguished: $\sum_{i=1}^{N} r_i \leq k$ describes the surplus-situation. The request can be completely fulfilled. We say a request r is served if at least r_i servers lie on $i, i = 1, 2, \ldots, N$. In contrast, $\sum_{i=1}^{N} r_i \geq k$ means the scarcity-situation. The request cannot be completely met, however it should be met as much as possible. The request r is served if at most r_i servers lie on $i, i = 1, 2, \ldots, N$.

By moving servers, the online algorithm must serve the requests $r^1, r^2, \ldots, r^n$ sequentially. For any request sequence σ and any k-server algorithm ALG, ALG(σ) is defined as the total distance (measured by the metric d) moved by the ALG's servers in servicing σ.

R. Hildenbrandt (✉)
Ilmenau Technical University, PF 10 06 65, 98684 Ilmenau, Germany
e-mail: r.hildenbrandt@tu-ilmenau.de

Analogous to (Borodin and El-Yaniv, p. 152) working with lazy algorithms ALG is sufficient. This means, servers are not moved in a step if they are not needed to fulfil requests in this step. For that reason we define the set of feasible servers' positions with respect to the previous servers' positions s and the request r in the following way

$$\hat{A}_{N;k}(s,r) = \{s' \in S_N(k) \,|\, r_i \leq s'_i \leq \max\{s_i, r_i\},\ i = 1, \ldots, N\} \quad (1)$$

$$\text{where } S_N(k) := \left\{s \in \mathbb{Z}_+^N \,\Big|\, \sum_{i=1}^{N} s_i = k\right\} \quad (2)$$

in the case of the surplus-situation and

$$\hat{A}_{N;k}(s,r) = \{s' \in S_N(k) \,|\, \min\{s_i, r_i\} \leq s'_i \leq r_i,\ i = 1, \ldots, N\} \quad (3)$$

in the case of the scarcity-situation. The metric d implies that $(S_N(k), \hat{d})$ is also a finite metric space where $\hat{d}$ are the optimal values of the classical transportation problems with availabilities s and requirements $s' \in S_N(k)$.

We want online algorithms whose cost compares favorably to the cost of an optimal offline algorithm. The competitive ratio is a measure of how much better an algorithm could do if it knew the future (in more detail, see e.g. [2], pp. 3, 46).

In the specific case that all distances between two points of M are equal the above described problem could be seen as a generalized paging problem. An other interpretation of a paging problem includes a two-level memory system, where each level can store a number of pages. The first level, the slow memory, hereby stores a fixed set of N pages. The second level, the fast memory can store any k pages. Given a request for certain pages, then the system must make these pages available in the fast memory (or as much pages as possible in the case of the scarcity-situation). If these pages are already in the fast memory, the system does not do anything. (See also [2], pp. 32, 33.)

2 Generalized Paging

A well-known optimal offline algorithm for usual paging is the algorithm LFD (Longest-Forward-Distance), see e.g. [2], p. 33. A natural generalization of this algorithm, adapted to the generalized paging problem, is the following:

Surplus-situation: If it is necessary, replace pages in the fast memory, whose next requests are latest.
Scarcity-situation: Do not copy requested pages in the fast memory, whose next requests are latest.

Theorem 1 *LFD is an optimal offline algorithm for generalized paging.*

This Theorem has been proved by E. Jäger[1].

Well-known competitive online algorithms for usual paging are, for example, the deterministic algorithm LRU (Least-Recently-Used) and the randomized algorithms RAND and MARK, see e.g. [2], pp. 33, 46, 50. These algorithms can also be generalized in a natural way. It seems that these generalizations of the algorithms are competitive in the cases of the surplus-situation with the same bounds of the ratio as in the cases of the usual problems. In the case of the scarcity-situation these generalizations are also competitive, but additional conditions may be necessary. Proofs of the corresponding statements for the generalized algorithms LRU and MARK are ongoing research.

The natural generalization of RAND can be stated as follows:

Surplus-situation: If it is necessary evict pages, which are chosen randomly and uniformly among all not requested fast memory pages.

Scarcity-situation: Copy requested pages, which are chosen randomly and uniformly in the fast memory.

Theorem 2 *RAND attains a competitive ratio of $c(k) = max\{k, R(k) - k + 1\}$ against an adaptive online adversary if $\sum_{i \in M} r_i^t \leq R(k)$ $\forall t$ for given $R(k) > k$ ([4]).*

3 The Generalized k-Server Problem

Well-known competitive online algorithms for usual k-server problems are the deterministic work function algorithm and the randomized Harmonic k-server algorithm. The proofs to show the competitiveness of these algorithms are extensive, see [2], pp.164–174 and [1].

Natural generalizations of the algorithms are not competitive in the case of the scarcity-situation (see Example 1 in [5] and Example 1 in [6]). Until now, to answer the question, whether these algorithms are competitive or not in the case of the surplus-situation are difficult open problems. Thats why, we suggest new "compound algorithms". We have proved that these algorithm are competitive with the bounds of the ratio as in the cases of the usual problems. The compound algorithms are derived from surrogate problems, where each step of the original problem will be replaced by a number of steps in the surrogate problem.

3.1 The Compound Harmonic k-Server Algorithm

A natural generalization of the Harmonic algorithm would send a sufficient set of servers with probability proportional to the inverse of its distance from the current request locations. In more detail,

---

[1] Diploma thesis, in preparation.

$$P_H(s'|s,r) = \frac{\frac{1}{d(s,s')}}{\sum\limits_{s'':s''\in \hat{A}_{N;k}(s,r)} \frac{1}{d(s,s'')}} \quad \text{for } s' \in \hat{A}_{N;k}(s,r). \quad (4)$$

A surrogate problem is created in the following way: Let s denote the online servers' positions at the beginning of a step t, let r be the request in the tth step and $\bar{N} \subseteq \{1, 2, \ldots, N\}$ the set such that $r_l > 0$ if and only if $l \in \bar{N}$. Furthermore, let $\bar{n} = |\bar{N}|$ and i a bijection from $\{1, 2, \ldots, \bar{n}\}$ to $\bar{N}$. If $\bar{n} > 1$ then we replace the tth step of the generalized k-server problem with the request r by a number of steps $t_{1,1}, t_{1,2}, \ldots, t_{1,\bar{n}}, t_{2,1}, t_{2,2}, \ldots, t_{2,\bar{n}}, \ldots$ with the request sequence $(\bar{r}^{1,1}, \bar{r}^{1,2}, \ldots, \bar{r}^{1,\bar{n}}, \bar{r}^{2,1}, \bar{r}^{2,2}, \ldots, \bar{r}^{2,\bar{n}}, \ldots)$, where $\bar{r}^{j,f}_{i(f)} = r_{i(f)}$ and $\bar{r}^{j,f}_{\bar{f}} = 0$ for $j \in \{1, 2, \ldots\}, \bar{f} \in \{1, 2, \ldots, N\}, \bar{f} \neq i(f)$. Probabilities $P_M(\bar{s}'^{j,f}|\bar{s}^{j,f}, \bar{r}^{j,f})$, for $\bar{s}'^{j,f} \in \hat{A}_{N;k}(\bar{s}^{j,f}, \bar{r}^{j,f})$ are be computed as m−step transition probabilities:

$$P_M(\bar{s}'^{j,f}|\bar{s}^{j,f}, \bar{r}^{j,f}) := \sum_{\{(s'^1, s'^2, \ldots, s'^{m-1}, \bar{s}'^{j,f})\}} P_H(s'^1|\bar{s}^{j,f}, r'^1) \cdot P_H(s'^2|s'^1, r'^2) \cdots P_H(s'|s'^{m-1}, r'^m) \quad (5)$$

where $m = r_{i(f)}, r'^{f'}$ with $r'^{f'}_{i(f)} = f', r'^{f'}_{\bar{f}} = 0$ for $f' \in \{1, 2, \ldots, m\}$, $\bar{f} \in \{1, 2, \ldots, N\}, \bar{f} \neq i(f), s'^{f'} \in \hat{A}_{N;k}(s'^{f'-1}, r'^{f'})$ and

$$P_H(s'^{f'}|s'^{f'-1}, r'^{f'}) = \frac{1/d(l_0, i(f))}{\sum\limits_{l:s_l^{f'-1} > 0, l \neq i(f)} 1/d(l, i(f))}, \quad l_0 : s'^{f'}_{l_0} = s^{f'-1}_{l_0} - 1. \quad (6)$$

If s' denotes the online servers' positions at the end of step t in case of the generalized k-server problem then several sequences $(\bar{s}'^{1,0}, \bar{s}'^{1,1}, \bar{s}'^{1,2}, \ldots, \bar{s}'^{1,\bar{n}}, \bar{s}'^{2,1}, \bar{s}'^{2,2}, \ldots, \bar{s}'^{2,\bar{n}}, \ldots, \bar{s}'^{\bar{j},1}, \bar{s}'^{\bar{j},2}, \ldots, \bar{s}'^{\bar{j},\bar{l}})$ with $s'^{j,f} \in \hat{A}_{N;k}(\bar{s}^{j,f}, \bar{r}^{j,f})$ for $j \in \{1, 2, \ldots, \bar{j}\}, f \in \{1, 2, \ldots, \bar{n}\}$ exist, where $\bar{s}'^{1,0} = s, s'^{\bar{j},\bar{l}} = s'$ and $\bar{s}^{j,f} = s'^{j,f-1}$ ($f > 1$ or $j = 1$) or $\bar{s}^{j,1} = s'^{j-1,\bar{n}}$, respectively. (If $\bar{s}^{j,f}_{i(f)} \geq \bar{r}^{j,f}_{i(f)} > 0$ then the corresponding surrogate step could be also omitted.)

Such sequences represent realizations of a time-homogeneous Markov chain with transient states $(\bar{s}^{j,f}, \bar{r}^{j,f})$, absorbing states $s' \in \hat{A}_{N;k}(s, r)$ and transition probabilities $P_M(\bar{s}'^{j,f}, \bar{r}^{j,f+1}|\bar{s}^{j,f}, \bar{r}^{j,f}) := P_M(\bar{s}'^{j,f}|\bar{s}^{j,f}, \bar{r}^{j,f})$ ($f < \bar{n}$) or $P_M(\bar{s}'^{j,\bar{n}}, \bar{r}^{j+1,1}|\bar{s}^{j,\bar{n}}, \bar{r}^{j,\bar{n}}) := P_M(\bar{s}'^{j,\bar{n}}|\bar{s}^{j,\bar{n}}, \bar{r}^{j,\bar{n}})$, respectively.

The probabilities $P_C(s'|s, r), s' \in \hat{A}_{N;k}(s, r)$, which are used by the compound Harmonic algorithm, are defined as absorbing probabilities. Absorbing probabilities can be computed by means of linear systems (see e.g. [3], Theorem 6.6). For this purpose all states of the above mentioned Markov chains must be known and the corresponding transition probabilities $P_M(\cdot|\cdot, \cdot)$ are the coefficients of these linear systems. The number of these states is finite. Furthermore $\sum\limits_{s' \in \hat{A}_{N;k}(s,r)} P_C(s'|s, r) = 1$ is valid and the solutions of the linear systems are unique.

Clearly, if $\bar{n} = 1$ then the computation of more-step transition probabilities is sufficient in this special case.

In general, **the compound Harmonic algorithm** uses absorbing probabilities P_C instead of the probabilities P_H. In the following Example probabilities P_C, used by the compound Harmonic algorithm, are compared with probabilities P_H used by the Harmonic algorithm.

Example 1 Let $k = 4$ and let the metric space M consist of 6 points $p_1, p_2, \ldots, p_6$ of the two-dimensional Euclidean space with the distances
$d(p_3, p_1) = 5, d(p_4, p_1) = 3, 85, d(p_5, p_1) = 1, 6, d(p_6, p_1) = 4, 5, d(p_2, p_1) = 2, 4$ and $d(p_3, p_2) = 4, d(p_4, p_2) = 5, d(p_5, p_2) = 2, 1, d(p_6, p_2) = 4, 55$.

The current online servers' positions are given by $s = (0, 0, 1, 1, 1, 1)^T$ and the current request by $r = (1, 1, 0, 0, 0, 0)^T$.

Then we have 6 feasible online servers' positions with respect to s and r:
$s'^{(1)} = (1, 1, 0, 0, 1, 1)^T, s'^{(2)} = (1, 1, 0, 1, 0, 1)^T, s'^{(3)} = (1, 1, 0, 1, 1, 0)^T,$
$s'^{(4)} = (1, 1, 1, 0, 0, 1)^T, s'^{(5)} = (1, 1, 1, 0, 1, 0)^T, s'^{(6)} = (1, 1, 1, 1, 0, 0)^T.$

Corresponding distances $\hat{d}(s, s'^{(i)})$, probabilities $P_H(s'^{(i)}|s, r)$ and $P_C(s'^{(i)}|s, r)$ can be found in the following Table.

i	1	2	3	4	5	6
$\hat{d}(s, s'^{(i)})$	7,85	5,60	8,50	5,95	8,40	6,15
$P_H(s'^{(i)}|s, r)$	0,1459	0,2045	0,1347	0,1924	0,1363	0,1862
$P_C(s'^{(i)}|s, r)$	0,0836	0,2504	0,0781	0,2582	0,0829	0,2466

We can observe that $P_C(s'^{(i)}|s, r) < P_H(s'^{(i)}|s, r)$ for greater distances $\hat{d}(s, s'^{(i)})$ and $P_C(s'^{(i)}|s, r) > P_H(s'^{(i)}|s, r)$ for smaller $\hat{d}(s, s'^{(i)})$.

Theorem 3 *The compound Harmonic algorithm applied to the generalized k-server problems with parallel requests is $((k + 1)(2^k - 1) - k)$-competitive against an adaptive online adversary in the case of the surplus-situation ([5]).*

3.2 The Compound Work Function Algorithm

Firstly, we want to generalize the definition of the work functions, which are used in the work function algorithm applied to the usual k-sever problem (see e. g. [2], pp. 164, 165). A work functions $w_{t+1}(s)$ is defined as the optimal offline cost (sequentially) servicing all requests in σ_t, starting from the initial configuration s^0 and ending at configuration s. Work functions can be computed recursively (by means of dynamic programming) as follows:

$$w_\emptyset(s) = \hat{d}(s^0, s) \quad \text{(There } \emptyset \text{ denotes the empty request sequence.)} \quad (7)$$

Assume that the value $w_t(s)$ is known for any configuration s. Given the next request r^{t+1} and a configuration s,

$$w_{t+1}(s) = \min_{\tilde{s} \in \hat{A}_{N;k}(s, r^{t+1})} \{w_t(\tilde{s}) + \hat{d}(\tilde{s}, s)\}. \quad (8)$$

Note, that these generalized work functions are quasi-convex in the cases of the surplus- and the scarcity-situation ([2], Lemma 1).

The following algorithm represents a natural generalizations of the deterministic work function algorithm (WFA): Let σ_t be the request sequence thus far and let $s^{\prime t}$ be the configuration of the WFA algorithm after servicing σ_t. Then, given the next request r^{t+1}. WFA serves r^{t+1} with s^{*W} satisfying

$$s^{*W} = \arg \min_{\tilde{s} : \in \hat{A}_{N;k}(s^{\prime t}, r^{t+1})} \{w_t(\tilde{s}) + \hat{d}(\tilde{s}, s^{\prime t})\}. \tag{9}$$

The cost of algorithm WFA to serve the request r^{t+1} is $\hat{d}(s^{*W}, s^{\prime t})$.

Now, we create a surrogate problem, where at most one server must be moved in servicing the request in each step. For this we replace the steps of the original problem by a number of steps in the surrogate problem.

In more detail, let $r^t = (r_1^t, r_2^t, \ldots, r_N^t)$ be the request in the tth step. Then we set $\bar{N} := \{i \mid r_i^t > 0\}$, $\bar{n} := |\bar{N}|$ and $\rho_i := r_i^t$, $\bar{\rho}_i := \sum_{l=1}^{i} \rho_l$ for $i = 1, \ldots, N$, $\bar{\rho}_0 := 0$, $\bar{\rho} := \bar{\rho}_N$. (Note that $\bar{N}$ and $\bar{\rho}, \ldots$ depend on t.)

Furthermore, let $\bar{j}$ be an integer with $\bar{j} = 1$ if $\bar{n} = 1$ and with

$$\delta_* \cdot (\bar{j} - 1) > \bar{\rho} \cdot \delta^* = (\sum_{i=1}^{N} r^t) \cdot \delta^*, \text{ if } \bar{n} > 1, \text{ respectively,} \tag{10}$$
$$\text{where } \delta_* = \min_{\{i_1, i_2\} \subseteq \bar{N}, i_1 \neq i_2} d(i_1, i_2), \delta^* = \max_{i \in \bar{N}, s \in M} d(i, s).$$

We replace a step t of the generalized k-server problem by steps $t_{1,1}, \ldots, t_{1,\bar{\rho}}, t_{2,1}, \ldots, t_{2,\bar{\rho}}, \ldots, t_{\bar{j},1}, \ldots, t_{\bar{j},\bar{\rho}}$ with requests $\bar{r}^{1,1}, \ldots, \bar{r}^{1,\bar{\rho}}, \bar{r}^{2,1}, \ldots, \bar{r}^{2,\bar{\rho}}, \ldots, \bar{r}^{\bar{j},1}, \ldots, \bar{r}^{\bar{j},\bar{\rho}}$ in the surrogate problem, where

$$\bar{r}_i^{j,f} = \begin{cases} f - \bar{\rho}_{i-1} & \text{for } i \text{ with } r_i^t \neq 0, f \in \{\bar{\rho}_{i-1} + 1, \bar{\rho}_{i-1} + 2, \ldots, \bar{\rho}_i\}, \\ & \text{and } j = 1, \ldots, \bar{j} \\ 0 & \text{otherwise} \end{cases} \tag{11}$$

Since $\bar{r}^{j,f}$ are independent of j we set $\bar{r}^f := \bar{r}^{j,f}$ for $f = 1, 2, \ldots, \bar{\rho}$.

For example, $r^t = (0, 2, 3)$ implies that $\bar{n} = 2$, $\bar{\rho} = 5$ and $\bar{r}^1 = (0, 1, 0)$, $\bar{r}^2 = (0, 2, 0)$, $\bar{r}^3 = (0, 0, 1)$, $\bar{r}^4 = (0, 0, 2)$, $\bar{r}^5 = (0, 0, 3)$.

The compound work function algorithm (compound WFA) consists of the following two steps:
1. Apply the usual WFA to the surrogate problem.
2. Skip the movements of servers which are unnecessary for the original problem.

Theorem 4 *The compound WFA algorithm is $(2k - 1)$-competitive for any k and any metric space in the case of the surplus-situation ([6]).*

References

1. Bartal, Y., & Grove, E. (2000). The Harmonic k-Server Algorithm Is Competive. *Journal of the ACM, 47*(1), 1–15.
2. Borodin, A., & El-Yaniv, R. (1998). *Online computation and competitive analysis.* Cambrigde: University Press.
3. Langrock, P., & Jahn, W. (1979). *Einführung in die Theorie der Markovschen Ketten und ihre Anwendungen.* Leipzig: Teubner.
4. Hildenbrandt, R. (2014). A k-server problem with parallel requests and unit distances. *Information Processing Letters, 114*(5), 239–246. https://doi.org/10.1016/j.ipl.2013.12.011.
5. Hildenbrandt, R. (2016). The k-server problem with parallel requests and the compound harmonic algorithm. *Baltic Journal of Modern Computing, 4*(3), 607–629.
6. Hildenbrandt, R. (2017). *The k-Server Problem with Parallel Requests and the Compound Work Function Algorithm* (pp. 17–04). Preprint No: Inst. f. Mathematik.

The Traveling Salesperson Problem with Forbidden Neighborhoods on Regular 3D Grids

Anja Fischer, Philipp Hungerländer and Anna Jellen

1 Introduction

In this paper we study the Traveling Salesperson Problem with Forbidden Neighborhoods (TSPFN) on regular three-dimensional (3D) grids. The task of the TSPFN is to determine a shortest Hamiltonian cycle over given points in the Euclidean plane, such that the distance between consecutive points along the tour is larger than some given $r \in \mathbb{R}_+$.

The TSPFN was originally motivated by an application in laser beam melting, where a workpiece is built in several layers. By excluding the heating of positions that are too close during this process, one hopes to reduce the internal stresses of the workpiece, see [3] and the references therein for further details. Furthermore, the TSPFN has connections to the maximum scatter TSP (msTSP), where the length z of a shortest edge in a tour is maximized. Clearly for optimal z, there exists a TSPFN tour for all $r < z$. For details on the msTSP see, e. g., [1] for the general case and [4] for a version on 2D grids.

In this work we determine optimal TSPFN tours on regular 3D grids with arbitrary grid sizes for the smallest reasonable forbidden neighborhoods $r = 0$ and $r = 1$. We consider $m \times n \times \ell$ grids, where m is the number of rows, n is the number of columns and ℓ is the number of layers. So each cell/vertex of the grid is represented by three coordinates. For a visualization we refer to Fig. 1. In the 3D case we always assume

---

A. Fischer
Tehnische Universität Dortmund, Dortmund, Germany
e-mail: anja2.fischer@tu-dortmund.de

P. Hungerländer · A. Jellen (✉)
Alpen-Adria-Universität Klagenfurt, Klagenfurt, Austria
e-mail: anna.jellen@aau.at

P. Hungerländer
e-mail: philipp.hungerlaender@aau.at

© Springer International Publishing AG, part of Springer Nature 2018
N. Kliewer et al. (eds.), *Operations Research Proceedings 2017*, Operations Research Proceedings, https://doi.org/10.1007/978-3-319-89920-6_30

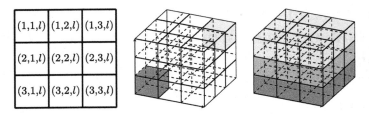

Fig. 1 Visualizations of the $3 \times 3 \times 3$ grid. The left picture shows the numbering of layers $l \in \{1, 2, 3\}$. In the middle cube, the dark gray cell has the coordinates $(3, 1, 1)$ and the light gray one has the coordinates $(1, 3, 3)$. In the right cube, each layer has one color, where layers with larger numbers are lighter

Fig. 2 Optimal TSP tours on $m \times n$ grids for m even, n even and both m and n odd [3]. The two tours on the left are denoted as "rook tours"

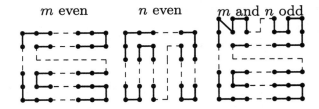

$m, n, \ell \geq 2$. If $1 \in \{m, n, \ell\}$, we are in the 2D case of the TSPFN. Results for this case can be found in [3].

The TSPFN on grids can easily be described as an integer linear program (ILP). In comparison to the classic formulation of the TSP by Dantzig et al. [2], we additionally forbid connections that are too short by setting the respective variables to zero, see [3] for the ILP formulation of the TSPFN in the 2D case.

2 Results for $r = 0$

If $r = 0$, then the TSPFN is equivalent to the Euclidean TSP. Since our constructions in the 3D case are an extension of the ones for the 2D case, first we recall some construction schemes of optimal TSP tours on regular 2D grids in Fig. 2.

Considering now the 3D case, we divide the grid cells into *odd* and *even vertices*, where a vertex is *odd (even)*, if the sum of its indices is odd (even). We denote the subgrid of the odd (even) vertices as o-grid (e-grid).

For $r = 0$, the shortest connections have length 1, running between an even and an odd vertex. Hence a trivial lower bound on the optimal tour length is $mn\ell$. If m, n, ℓ are odd, there is one more odd than even vertex. Thus in this case $mn\ell - 1 + \sqrt{2}$ is a lower bound. Next we show that these lower bounds are in fact the optimal lengths of TSPFN tours with $r = 0$ on regular 3D grids.

Theorem 1 *An optimal TSP tour on an $m \times n \times \ell$ grid has length $mn\ell$, if m or n or ℓ is even, and length $(mn\ell - 1) + \sqrt{2}$, if m, n, ℓ are odd.*

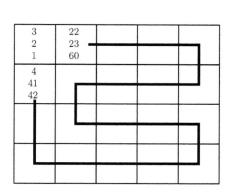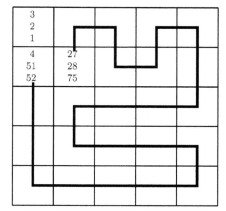

Fig. 3 Optimal TSP tours on the $4 \times 5 \times 3$ grid and the $5 \times 5 \times 3$ grid, respectively. The three numbers in the cells correspond to the three layers with Layer 1 at the bottom

Proof We prove this by presenting tour construction schemes, where the tour lengths equal the lower bounds given above. The construction always starts in $(1, 1, 1)$ and we go with steps of length 1 to $(1, 1, 2), (1, 1, 3), \ldots, (1, 1, \ell)$. When we reach Layer ℓ, we apply the appropriate construction schemes from the 2D case, see Fig. 2, of course not yet closing the tour. We distinguish two cases:

- **m or n or ℓ is even**: W. l. o. g. we assume that m or n is even. Hence we can apply one of the rook tours that end in $(1, 2, \ell)$. Now we change the layer by taking a step of length 1 to $(1, 2, \ell - 1)$. Here we apply the open rook tour again, but in the other direction, such that it ends in $(2, 1, \ell - 1)$. We repeat using these construction schemes until we reach Layer 1. Depending on the parity of ℓ, the last rook tour ends in $(2, 1, 1)$ or $(1, 2, 1)$. Both cells have distance 1 to the start vertex $(1, 1, 1)$.
- **m, n, ℓ are odd**: We proceed in a similar way as above, applying the construction scheme of the 2D case with both m, n odd. The alternating start and end cells of these open rook tours are $(2, 1, i)$ and $(2, 2, i)$, $i \in \{1, \ldots, \ell\}$. Finally, in Layer 1 we end in $(2, 2, 1)$ that has a distance of $\sqrt{2}$ to $(1, 1, 1)$. □

To further clarify the construction schemes of optimal TSP tours on regular 3D grids, we depict some optimal tours on the $4 \times 5 \times 3$ and $5 \times 5 \times 3$ grids in Fig. 3.

3 Results for $r = 1$

For the TSPFN with $r = 1$ the shortest possible step has length $\sqrt{2}$, see Fig. 4 for a visualization of the corresponding forbidden neighborhood. A step of length $\sqrt{2}$ is either possible between two even or between two odd vertices. For determining a

Fig. 4 Illustration of the forbidden neighborhood for the TSPFN with $r = 1$ on regular 3D grids. The current grid cell is the black one in the middle, the forbidden cells are the gray ones around it and the cells that we are allowed to visit next are white

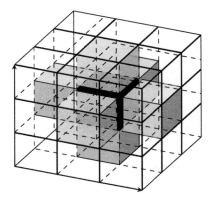

lower bound we assume the existence of Hamiltonian paths on the o-grid and the e-grid that consist only of steps of length $\sqrt{2}$. The shortest feasible step for connecting the paths on the subgrids has length $\sqrt{3}$. Hence $(mn\ell - 2)\sqrt{2} + 2\sqrt{3}$ is a lower bound for the length of optimal TSPFN tours with $r = 1$ on an $m \times n \times \ell$ grid. In the following we present construction schemes for tours whose lengths coincide with this lower bound. We start with $\ell = 2$, then we consider $\ell = 3$ and eventually we extend our construction schemes to an arbitrary number of layers.

Construction scheme for $\ell = 2$: We construct tours on each subgrid and connect them with steps of length $\sqrt{3}$. Putting two layers of a subgrid upon each other, we obtain a full 2D grid, where every step is allowed, because vertices belonging to the same subgrid have a distance of at least $\sqrt{2}$. Now we can again apply the construction schemes for $r = 0$ of the 2D case, this time for determining tours over two layers, where every step of length 1 in the original construction scheme corresponds to a step of length $\sqrt{2}$. Slightly adapting the 2D solution by using steps of length $\sqrt{2}$ within one layer, we can choose different start- and end-cells and derive a tour of the desired length. To further clarify this construction scheme we depict the optimal TSPFN tour with $r = 1$ on the $4 \times 5 \times 2$ grid in Fig. 5.

Construction scheme for $\ell = 3$: As above we put layers of the same subgrid upon each other so that they look like a full 2D grid and connect the two subgrids by steps of length 3. Note that in this case vertices of the 2D visualization that are in Layer 1 or 3 have to be visited twice and vertices that are in Layer 2 have to be visited once. Our construction scheme starts on the o-grid in $(1, 1, 1)$. Initially we visit all cells in the first two columns and continue with covering the columns pairwise until three columns are left. Then we go to $(1, n - 3, 1)$ and visit the first two remaining rows. Again we cover the rows pairwise and stop when five rows are left. It remains a 5×3 grid on which we apply an explicitly determined subpath. With a step of length $\sqrt{3}$ we change to the e-grid. The construction scheme on the e-grid is a slightly adapted and rotated (by 180 degrees) version of the construction of the o-grid. A detailed description of the construction scheme over three layers is given in Fig. 6. To further

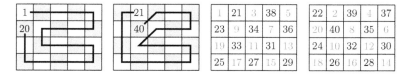

Fig. 5 Illustration of an optimal TSPFN tour with $r = 1$ on the $4 \times 5 \times 2$ grid. The tour consists of two Hamiltonian paths on the o-grid (left) and the e-grid (Picture 2) that are connected by steps of length $\sqrt{3}$. The two right pictures display the tour explicitly (Level 1 left)

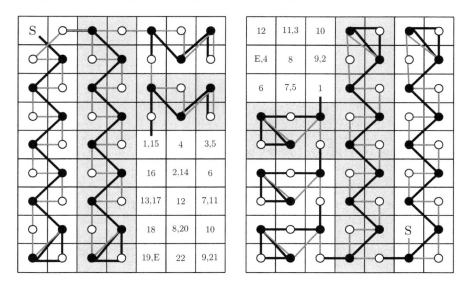

Fig. 6 Illustration of drawing patterns for $\ell = 3$. The pictures show the construction scheme for the odd vertices (left) and the even vertices (right). For larger m, n the gray parts can be repeated. Black vertices correspond to Layers 1 and 3 and hence are visited twice. White vertices correspond to Layer 2 and hence are visited once. "S" and "E" indicate the start and end vertices

clarify this construction scheme we depict an optimal TSPFN tour with $r = 1$ on the $7 \times 5 \times 3$ grid in Fig. 7.

Construction scheme for an arbitrary ℓ: Finally in the proof of Theorem 2 we describe how the above construction schemes for $\ell = 2$ and $\ell = 3$ can be used to obtain a construction scheme for an arbitrary number of layers ℓ.

Theorem 2 *An optimal TSPFN tour on the $m \times n \times \ell$ grid with $r = 1$ has length $(mn\ell - 2)\sqrt{2} + 2\sqrt{3}$.*

Proof We prove this by presenting an explicit construction scheme, where the length of the tours equals the lower bound derived above. The optimal solutions for the $3 \times 3 \times 3$ and the $3 \times 5 \times 3$ grids were derived explicitly by the ILP formulation discussed in the introduction. So let us consider the other cases.

Fig. 7 Illustration of an optimal TSPFN tour with $r = 1$ on the $7 \times 5 \times 3$ grid. The odd (even) vertices are depicted in the left (right) picture. The gray cells belong to Layers 1 and 3 and hence are visited twice

1,20	21	22,29	27	24,26
19	2,18	30	23,28	25
3,16	17	31,45	34	33,35
15	4,14	46	32,44	36
5,12	13	43,47	42	37,41
11	6,10	48	38,50	40
7,9	8	49,53	52	39,51

104	103,95	102	68,65	67
105,96	100	101,94	64	69,66
98	99,97	93	70,63	62
90,87	91	92,85	60	71,61
88	89,86	84	72,59	58
81,78	82	83,76	54	73,57
79	80,77	75	74,55	56

First note that for all grid sizes there exists a construction scheme over two layers that visits all odd vertices such that $(1, 1, i)$ and $(1, 2, i + 1)$ for i odd are the start- and end-cells, see Fig. 8. We start in $(1, 1, 1)$, apply this construction scheme and hence terminate in $(1, 2, 2)$. Then we take a step of length $\sqrt{2}$ to $(1, 1, 3)$ and apply again the same construction scheme.

Now we distinguish two cases:

1. **m or n or ℓ is even:** We assume, w.l.o.g., that ℓ is even. Hence we can continue with the application of the construction scheme over two layers until we reach $(1, 2, \ell)$. Now we have covered all odd vertices and need to take a step of length $\sqrt{3}$ to change to the e-grid. We go to $(2, 1, \ell - 1)$ and apply a slightly adapted construction scheme that is depicted in Fig. 8, Picture 5. Using this drawing pattern we visit all even vertices on the Layers ℓ and $\ell - 1$ and terminate in $(1, 2, \ell - 1)$. For the even vertices on the remaining layers, except for Layers 1 and 2, we again iteratively apply the construction schemes illustrated in Pictures 1 to 3 of Fig. 8. Finally on Layers 1 and 2 we use a slightly adapted construction depicted in Fig. 8, Picture 4, in order to end in $(2, 2, 2)$. From $(2, 2, 2)$ we go to $(1, 1, 1)$ with a step of length $\sqrt{3}$.

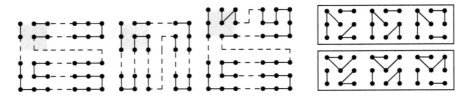

Fig. 8 Construction schemes for tours on either the o-grid or the e-grid over two layers using only steps of length $\sqrt{2}$. The three left construction schemes are applied for subgrids with different parities, i.e. m even, n even or m and n odd (from left to right). For the e-grid over Layers 1 and 2 (Picture 4) as well as Layers $\ell - 1$ and ℓ (Picture 5) slight adaptions are needed. The pictures show the interesting part of the upper 9 highlighted nodes, the remaining connections do not change

2. m, n, ℓ **are odd**: We apply the construction scheme over two layers from Fig. 8 on the o-grid until we reach $(1, 2, \ell - 3)$. We make a step of length $\sqrt{2}$ to $(1, 1, \ell - 2)$ and then use the construction scheme from Fig. 6 on the last three layers and end in $(2, 1, \ell - 2)$. We make a step of length $\sqrt{2}$ to $(1, 1, \ell - 3)$. Finally, we visit the remaining even vertices as in Case 1. □

It remains for future work to extend the presented results to larger values of r or to grids, where the cells are cuboids. Looking at the application in laser beam melting it is not only interesting to enforce some minimum distance between consecutive points in the tour, but also of points that are close in the tour so that, e. g., some areas of the workpiece can cool down over a longer time period.

References

1. Arkin, E. M., Chiang, Y.-J., Mitchell, J. S. B., Skiena, S. S., & Yang, T.-C. (1999). On the maximum scatter traveling salesperson problem. *SIAM Journal on Computing, 29*(2), 515–544.
2. Dantzig, G., Fulkerson, R., & Johnson, S. (1954). Solution of a large-scale traveling-salesman problem. *Operations Research, 2*, 393–410.
3. Fischer, A., & Hungerländer, P. (2017). The traveling salesman problem on grids with forbidden neighborhoods. *Journal of Combinatorial Optimization, 34*(3), 891–915.
4. Stock, I. (2017). *The Maximum Scatter TSP on a Regular Grid: How to Avoid Heat Peaks in Additive Manufacturing.* Ph.D thesis, University of Bayreuth.

Traveling Salesman Problems with Additional Ordering Constraints

Achim Hildenbrandt

1 Introduction and Notation

The Traveling Salesman Problem (TSP) is probably one of the most studied problems of combinatorial optimization. In this paper we consider four variants of the TSP which each contain an additional type of ordering constraint. Applications of such problems often appear in the context of disaster logistics.

The first variant we consider is the Clustered Traveling Salesman Problem (CTSP). In this problem the set of nodes is partitioned into k clusters. Each feasible tour must then visit the nodes of one cluster successively. We denote the clusters with $C_i \subseteq V$, $i = 1, \ldots, k$.

A special case of this problem is the Ordered Clustered Traveling Salesman Problem (OCTSP), where an additional order in which the clusters must be visited is specified. This problem is also sometimes called Hierarchical TSP. When there are only two clusters this problem is equivalent to the Traveling Salesman Problem with Backhauls (TSPB). We denote the start node of the tour by d. The clusters are numbered according to the order in which they have to be visited.

The third problem we study in this paper is the Precedence Constrained Traveling Salesman Problem (PCTSP). In this variant each feasible tour must fulfill a list of precedence constraints. We denote these constraints with the ordered set $P \subseteq V \times V$ where (i, j) is in P when i has to be visited before j. This problem is called Sequential Ordering Problem in the Hamiltonian Path version.

Last but not least we consider the Target Visitation Problem (TVP) which is a connection of the TSP and the Linear Ordering Problem (LOP). So for each two nodes $i, j \in V$ there exists a value p_{ij} which we get as bonus when we visit i before j in the tour.

A. Hildenbrandt (✉)
INF 205, 69120 Heidelberg, Germany
e-mail: achim.hildenbrandt@informatik.uni-heidelberg.de

The aim of this paper is to study the connection between these four problems and present a fast method to solve them to optimality.

2 Literature-Review

In this section we give a short overview of existing research considering these four problems. We focus on literature which is concerned with exact approaches for solving these problems. The CTSP was introduce by Chisman in [4]. In this article the CTSP is solved exactly by applying the well know branch-and-bound algorithm by Little et al. [7]. Jongens and Volgant propose an Lagrangean Relaxation for the symmetric variant of the CTSP in [8] which seems to work very well. For the OTCP, there exists not much literature. Aramgiatisiris describes an algorithm in [2] which solves the TSPB exactly using decomposition techniques. This algorithm could also be used for the OCTSP. For the PCTSP a branch-and-cut algorithm could be found in [1]. Valid inequalities for the associated polytope are described in [3]. The TVP was introduced by Grundel et al. in [5]. An extensive survey on different solving methods for the TVP can be be found in [6]. In addition we also want to mention an article by Lokin [9] concerning a branch-and-bound algorithm (based on the algorithm of Little) which is applied to the CTSP, PCTSP und OCSTP.

3 Connections

Obviously all these problems are connected with each other. In the following we explain these connections in more detail.

Every instance of an OCTSP can be transformed to an instance of the TVP which has the same optimal solution. This transformation is done as follows: The distance cost matrix is copied while the preference matrix is constructed in the following way (M hereby denotes a very large value compared to the traveling cost of the TSP):

$$p_{ij} := \begin{cases} -M & \text{if } i \in C_l \text{ and } j \notin \{C_l, C_{l+1}\} \text{ with } l \in \{1, \ldots, k-1\}, \\ 0 & \text{otherwise.} \end{cases}$$

Also every instance of PCTSP can be transformed to an instance of the TVP with a similar big-M construction. Again the distance matrix is copied and this time p_{ij} is defined as follows:

$$p_{ij} := \begin{cases} M & \text{if } (i, j) \in P, \\ 0 & \text{otherwise.} \end{cases}$$

Last but not least every instance of an OCTSP could be transformed to an instance of a PCTSP by defining P as follows:

$(i, j) \in P$ if and only if $i \in C_l$ and $j \in C_m$ with $l < m$.

4 IP Modeling

In this section we examine how these four problems can be modeled as integer programs. To make things easier we transform our TSPs to Hamiltonian path problems (HPP). That means instead of a tour which visits all nodes we are looking from now for a path that visits all nodes. So we can forget about the start node of the tour. Such transformation have been introduced by Queyranne et al. in [10] and been used for the TVP before (see [6]).

So from now on we treat our problems as HPPs with additional ordering constraints. So we first present an IP model for the HPP. There are existing many different approaches. The most common one uses variables x_{ij} defined by;

$$x_{ij} := \begin{cases} 1 & \text{if node } i \text{ is directly followed by node } j \text{ in the path,} \\ 0 & \text{otherwise.} \end{cases}$$

which leads to the objective function

$$\min \sum_{i=1}^{n} \sum_{\substack{j=1 \\ j \neq i}}^{n} d_{ij} x_{ij},$$

where d_{ij} denotes the traveling costs from node i to node j. To obtain a valid solution it is sufficient to include degree constraints:

$$\sum_{\substack{i=1 \\ i \neq j}}^{n} x_{ij} \leq 1, \ j \in V \text{ and } \sum_{\substack{j=1 \\ j \neq i}}^{n} x_{ij} \leq 1, \ i \in V$$

as well an as the subtour elimination constraints

$$\sum_{i \in S} \sum_{\substack{j \in S \\ i \neq j}} x_{ij} \leq |S| - 1, \ S \subset V, \ 2 \leq |S| < n$$

and the constraint that a path through n nodes must contain $n - 1$ edges

$$\sum_{i=1}^{n} \sum_{\substack{j=1 \\ j \neq i}}^{n} x_{ij} = n - 1.$$

To model our problems we will extend this model and add additional constrains (and in some cases also new variables). We start with the TVP which is the only problem where the added constraints are not depending on the given instance. The following integer programming model can be found in [6]. It combines the IP model of the HPP with the IP model for the Linear Ordering Problem. That means we have to introduce some additional variables

$$w_{ij} := \begin{cases} 1 & \text{if node } i \text{ is visited before node } j \text{ in the path,} \\ 0 & \text{otherwise.} \end{cases}$$

With this the objective function must be modified as follows:

$$\max \sum_{\substack{i=1 \\ j \neq i}}^{n} \sum_{\substack{j=1 \\ j \neq i}}^{n} p_{ij} w_{ij} - \sum_{\substack{i=1 \\ j \neq i}}^{n} \sum_{\substack{j=1 \\ j \neq i}}^{n} d_{ij} x_{ij}$$

We also have to add three new classes of constrains to the HPP model:

$$w_{ij} + w_{jk} + w_{ki} \leq 2,\ 1 \leq i, j, k \leq n,\ i < j,\ i < k,\ j \neq k,$$

$$x_{ij} - w_{ij} \leq 0,\ 1 \leq i, j \leq n,\ i \neq j,$$

$$w_{ji} + w_{ij} = 1,\ 1 \leq i < j \leq n.$$

For the case of the CTSP an IP model has been stated in [4]. It consists of adding the following constraints to the HPP model:

$$\sum_{i \in C_u} \sum_{j \in C_u} x_{ij} = |C_u| - 1\ \forall u = 1, \ldots, k \text{ where } |C_u| \geq 1.$$

These equations describe the fact that a path that visits all nodes of a cluster C_u successively must contain $|C_u| - 1$ edges. Alternative: Add variables w_{ij} which are analogously defined as in the TVP-model and constraints $w_{ij} + w_{jl} \leq 1$, $\forall i, l \in C_u, j \in C_v$.

For modeling the OCTSP we simply add the constraints:

$$x_{ij} = 0\ \forall i \in C_u \text{ and } j \in C_v \text{ with } C_u > C_v$$

to the HPP model. Optionally the following equations could also be added:

i) $x_{ij} = 0\ \forall i \in C_u$ and $j \in C_v$ with $C_v > C_u + 1$
ii) $\sum_{i \in C_u} \sum_{j \in C_{u+1}} x_{ij} = 1\ \forall u \in \{1, \ldots, k-1\}$

Table 1 Results on the PCTSP problems

Instance	Nodes	10%	20%	30%	40%	50%
p43	43	0:00:10	0:00:05	0:00:07	0:00:02	0:00:04
ry48	48	Mem.	Mem.	0:00:24	0:00:05	0:00:06
ftv55	55	55:52:28	0:05:04	0:00:51	0:00:07	0:00:03
ft70	70	Mem.	0:34:51	0:01:23	0:00:47	0:00:12
kro124p	100	Mem.	Mem.	Mem.	Mem.	Mem.

Table 2 Results on the OCTSP problems

Instance	Nodes	3 Clu.	4 Clu.	5 Clu.	6 Clu.	8 Clu.	10 Clu.
p43	43	1:50:42	0:20:42	0:01:14	0:07:31	0:00:54	0:00:03
ry48	48	0:00:22	0:00:10	0:00:17	0:00:05	< 1	< 1
ftv55	55	0:00:02	0:00:04	0:00:02	< 1	< 1	< 1
ft70	70	3:15:28	0:13:47	0:02:50	0:00:04	0:00:02	0:00:02
kro124p	100	4:38:15	1:04:59	4:03:26	0:21:54	0:01:21	0:01:41
ftv70	170	4:22:33	1:46:06	0:21:07	0:10:03	0:03:59	0:03:24

For the PCTSP an IP model can be found for example in [1]. This formulation contains an exponential number of constraints, but they can be separated in polynomial time. However we try a new approach and use the IP model of the TVP. There we only have to add the constraints $w_{ij} = 1 \ \forall (i,j) \in P$ and we have to leave out the ordering part of the objective function.

5 Branch-and-Cut

In this section we present a branch-and-cut algorithm which could solve PCTSP, OCTSP and TVP to optimality. For this purpose we adopted code written by Reinelt and Rinaldi. The algorithm is written in C and uses CPLEX, especially its opportunity of using callback functions. We also use the TVP facet classes 25, 29, 31, 39, and 41 which have been stated in [6].

The test instances for the TVP were also taken from [6]. For the other two problems we took asymmetric TSP instances from the TSP-library [11] and extended them in the following way: In case of the OCTSP we put the first k_1 nodes in cluster C_1, the next k_2 nodes in cluster C_2 and so on. For the PCTSP we generated a random tour, and then we randomly included a defined percentage of constraints implied by this tour.

We execute our algorithm on an Intel(R) Core(TM) i7-2600 CPU @ 3.40GHz with 16GB Ram. As one can see the TVP is the hardest of these problems since at the moment it is not possible to solve instances with more than 50 nodes, even if

Table 3 Results for the TVP instances

Instance	n	B.-and-C. in [6]	New B.-and-C.
ER_CFO_30_2	30	0:23:49	0:05:14
ER_MCO_30_2	30	0:26:08	0:10:01
LD_CFO_35_1	35	0:24:19	0:46:06
LB_CFO_35_1	35	0:17:37	0:01:40
ER_CFO_40_2	40	3:44:44	0:45:51
ER_CFO_40_4	40	Mem.	12:01.38
LD_MCO_40_1	40	3:24:02	0:01:00

they have a metric order on the distances and a complete ordering on the preferences (see Table 3 for results). Nevertheless our algorithm is faster than the branch-and-cut algorithm presented in [6]. As one can see in Table 2 in case of the OCTSP it seems that an increasing number of clusters makes the problem much easier. So we were able to solve instances where clusters contain at most 10 nodes in at most 3 minutes. In case of the PCTSP it depends on the number of constraints how difficult a instance is, even if there are more than 50 % of all possible constraints included in P, it is not possible to solve instances with 100 or more nodes (see Table 1).

6 Conclusion and Further Research

In this paper we considered several TSPs with additional ordering constraints. We showed that they are all connected to each other. That why it was possible to use results valid for the TVP to solve the other problems as well. We also present a branch-and-cut algorithm which shows good results. In the future we want to extend our branch-and-cut algorithm so it can also solve CTSP instances and we want to examine associated polytopes to determine some additional cutting planes to speed up the algorithm.

References

1. Ascheuer, N., Jnger, M., & Reinelt, G. (2000). A branch-and-cut algorithm for the asymmetric traveling salesman problem with precedence constraints. *Computational Optimization and Applications*, *17*, 61–84.
2. Aramgiatisiris, T. (2004). An exact decomposition algorithm for the traveling salesman problem with backhauls. *Journal of Research in Engineering and Technology*, (1), 151164.
3. Balas, E., Fischetti, M., & Pulleyblank, W. (1995). The precedence-constrained asymmetric traveling salesman polytope. *Mathematical Programming*, *68*(1), 241–265.
4. Chisman, J. A. (1975). The clustered traveling salesman problem. *Computers and Operations Research*, *2*(2), 115–119.

5. Grundel, D. A., & Jeffcoat, D. E. (2004). Formulation and solution of the target visitation problem. In *Proceedings of the AIAA 1st Intelligent Systems Technical Conference*.
6. Hildenbrandt, A. (2015). The target visitation problem. Ph.D thesis.
7. Sweeney, D., Little, J., Murty, K. & Karel, C. (1963). An algorithm for the traveling salesman problem. *Operations Research*, (11), 972–989.
8. Jongens, K., & Volgenant, T. (1985). The symmetric clustered traveling salesman problem. *European Journal of Operational Research*, *19*(1), 68–75.
9. Lokin, F. C. J. (1979). Procedures for travelling salesman problems with additional constraints. *European Journal of Operational Research*, *3*(2), 135–141.
10. Queyranne, M., & Wang, Y. (1993). Hamiltonian path and symmetric travelling salesman polytopes. *Mathematical Programming*, *58*, 89–110.
11. Reinelt, G. Tsplib. http://comopt.ifi.uni-heidelberg.de/software/TSPLIB95/index.html.

Part VI
Energy and Environment

Benefits and Limitations of Simplified Transient Gas Flow Formulations

Felix Hennings

1 Introduction

For the past years, the mathematics of gas transport have been intensively studied, mainly focusing on the stationary (time-independent) case, which can be applied to planning scenarios for example [2, 6]. However, when aiming to optimize the actual short-term control of the technical gas network elements, we have to consider the time dependent so-called transient case. Here, research is still in the early stages and current state-of-the-art approaches cannot solve instances of large real-world network size [7].

One difficulty are the Euler Equations [2] describing the one dimensional gas flow in a cylindric pipeline, a set of nonlinear hyperbolic partial differential equations. For the isothermal case they can be stated as

$$\frac{\partial \rho}{\partial t} + \frac{\partial (\rho v)}{\partial x} = 0$$

$$\frac{\partial (\rho v)}{\partial t} + \frac{\partial (p + \rho v^2)}{\partial x} + \lambda \frac{|v| v}{2D} \rho + g \rho h' = 0,$$

where x denotes the position in the pipe, t the time, ρ the density of the gas, v the velocity of the gas, p the pressure of the gas, λ the friction factor of the pipeline, D the diameter of the pipeline, g the gravitational acceleration and h' the constant slope of the pipe. Note that ρ, v and p depend on x and t. The second equation can be further simplified by assuming the terms $\partial_t(\rho v)$ and $\partial_x(\rho v^2)$ to be small as in [1]. We finally reduce the number of variables by rewriting the constraints using the equation of state for real gases $p = R_s \rho T z(p)$ and the definition of the mass flow $q = A\rho v$ with $A = D^2 \pi / 4$ being the cross sectional area of the pipe as

---

F. Hennings (✉)
Zuse Institute Berlin, Takustr. 7, 14195 Berlin, Germany
e-mail: hennings@zib.de

$$\frac{A}{R_s T z} \frac{\partial p}{\partial t} + \frac{\partial q}{\partial x} = 0 \tag{1}$$

$$\frac{\partial p}{\partial x} + \frac{\lambda R_s T z}{2DA^2} \frac{|q|q}{p} + gh' \frac{p}{R_s T z} = 0. \tag{2}$$

Here R_s denotes the specific gas constant, T the constant gas temperature and $z(p, T)$ the compressibility factor, which is often assumed to be constant and hence just stated as z.

This model of gas flow in pipelines still contains non-convex terms, which introduce a lot of complexity to any model aiming to solve these equations. For this reason, we will present an additional simplification of the constraints and investigate the resulting theoretical properties as well as evaluate the caused errors based on historic flow data of real pipelines.

2 A Linearization Approach

The non-convexity of the stated equations is based in the second term in (2)

$$f := \frac{\lambda R_s T z}{2DA^2} \frac{|q|q}{p}$$

describing the *friction-based pressure difference per meter* on a pipeline. Using the definition of mass flow and the equation of state above we get a definition of the velocity in terms of pressure and mass flow as

$$v = \frac{R_s T z}{A} \frac{q}{p} \quad \Rightarrow \quad |v| = \frac{R_s T z}{A} \frac{|q|}{p}. \tag{3}$$

We can now rewrite (2) as

$$\frac{\partial p}{\partial x} + \frac{\lambda |v|}{2DA} q + gh' \frac{p}{R_s T z} = 0$$

and observe that the equation becomes linear if we assume the absolute velocity in the friction term to be constant, that is $|v| = v_c$. Note that we do not restrict ourselves to one flow direction, since we fix the *absolute* velocity. Furthermore, we only fix the absolute velocity in the friction term. Hence, the *actual* absolute velocity calculated from p and q might be different from v_c.

If the proposed simplification can be verified to be reasonable, the overall modeling complexity would decrease drastically. However, since the friction-based pressure difference scales linearly with the velocity, the assumption of fixed velocity might easily lead to large errors in terms of pressure differences. On the other hand, both the friction-induced pressure difference and the absolute velocity increase with increasing absolute flow values. As a consequence, overestimating the absolute velocity should in general be more favorable for minimizing the error in terms of pressure differences.

3 Analysis of Real-World Data

In order to see if the approximation of the friction term by using a constant absolute velocity is reasonably close to the actual friction-induced pressure differences, we use real pipeline data in the network of our project partner OGE [4], which is the biggest gas network operator in Germany. For this network, a history of states is given, measured every three minutes over a period of two years. There are two types of pipelines we consider here: (a) four large pipes A to D, which are used to transport gas between large network intersection areas, and (b) two small pipelines E and F, which are part of the network section connecting customers nodes with bigger pipelines. An overview of the properties of the six pipes can be found in Table 1.

One idea to choose a fixed absolute velocity value for each pipe is to set this value to a mean velocity computed over a given time period. The fixed velocity error would be small if a major part of absolute velocity values were to lie within a small interval and hence near the mean value. To investigate this, we plotted the absolute velocity values occurring on all pipelines over the two years in Fig. 1.

We can observe that the pipes A, B and C share a distinct characteristic with a high population of similar values and steep tails of the distributions. For example, for pipe A, the absolute difference between the 10th percentile and the 90th percentile is 3.10. Thus, only a relative error of less than 36% has to be taken into account for these values, when assuming a mean value of 4.290 m/s. For B and C the same percentiles yield a relative error of less than 44% resp. 54% for mean values of 3.445 resp. 2.610 m/s.

Table 1 Properties of the analyzed pipelines. The averages are computed over each pipe for each state and afterwards over all states of the two years. For the pressure averages a stationary formula (Lemma 2.3 from [2]) is used. The calculation of the velocities (3) uses for the compressibility factor z the formula of Papay [5, 8] and as an aggregated gas mixture computed from the mixtures at entries using a formula for mixtures at junctions from [2] Chap. 2 column denotes the percentage of times when gas was flowing into the main direction

Pipe	L [km]	D [mm]	Avg p [bar]	Avg $|v|$ [m/s]	Flow in main direction (%)
A	16	1000	56	4.2	100
B	16	900	63	3.4	99
C	15	1100	70	2.5	99
D	20	1100	71	1.4	76
E	3	400	54	2.7	93
F	2	300	16	4.4	100

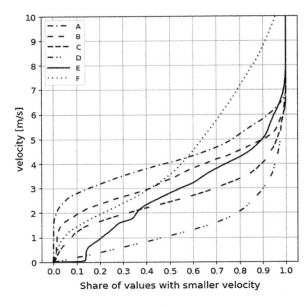

Fig. 1 Sorted absolute velocity values of each pipe over two years (cumulated distribution functions)

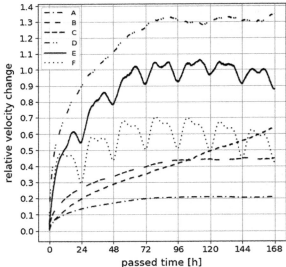

Fig. 2 Relative velocity change over time. Average values over two years ignoring abs. velocities below 0.02 m/s

In contrast, the other pipes have a much larger span of velocity values, even when ignoring the first 13% of values smaller than 0.02 m/s of pipe E. We can conclude that for the bigger pipes with unique flow directions, namely A, B and C, a precalculated constant mean velocity value should lead to relatively small errors in terms of friction loss.

A second idea to approximate the velocity term, would be to fix the velocity to some value known from the recent history, such as the velocity value given in the initial state of the gas network control problem for example. To estimate a possible error here, we compute for each time step the relative velocity changes over one week and average these over the whole time period of two years. The results are given in Fig. 2. In our discussion we focus on the value at 48 h since this is a typical time horizon for short-term gas network control problems.

In the picture we see, that for the pipes A, B and C the velocities change only slowly over short time periods, i.e. on average less than 0.35 m/s in 48 h for each of the pipes. Pipe A and C even seem to reach some constant level of relative velocity change over time. For the other three pipes we have higher relative velocity changes in the first 48 h, especially for pipe D. However, the smaller pipes E and F seem to have some daily pattern with a local minimum every 24 h, which should be taken into account when substituting the velocity with historical values.

4 Determining Fixed Velocity Values

After studying the historical velocity values on the six pipes, we will now compute concrete predefined velocity values and examine the actual errors in terms of pressure difference. We use the two approaches already briefly mentioned above: Calculating a constant velocity for each pipe based on a large set of historical data (approach $\mathcal{A}$) and taking the velocity, which has been measured on the pipe exactly 48 h before (approach $\mathcal{B}$).

The constant velocity of $\mathcal{A}$ is calculated as the velocity that minimizes the sum over time of squared pressure difference errors on the whole pipe length fL. These errors are derived based on (2) by replacing the derivative with the corresponding difference quotient and using v_c. Again we use Papay for the compressibility factor z and the formula of Nikuradse [3] for the friction factor λ. To have an unbiased evaluation, we use only the first year of the time period for calculating the constant velocity and compare the results of the two approaches on the basis of the second year data.

The results of approach $\mathcal{A}$ can be found in Table 2. We observe, that the constant velocity values are slightly above the average values given in Table 1. This supports our claim from Sect. 2 that overestimating the velocity is more favorable to reduce the pressure difference error. The actual average error values are quite small in terms of absolute values. However, when comparing to the actual average friction values, the values are quite high with an error to friction ratio above 20%. For pipe D the error is even higher than the friction-based difference. When looking at the maximum values, the relative errors are mainly at a level of about 50%, which is also rather high.

In contrast to the results expected in Sect. 3, the values for pipes A to C are not significantly better than the ones of pipes D to F. The only clear difference between the pipes in the results is the bad average error of pipe D in relation to the friction values.

Table 2 Results of approach $\mathcal{A}$ including the calculated constant absolute velocity v_c, the average pressure difference error, the corresponding maximum error, the average and maximum real friction-based pressure differences and the ratio of errors to friction values for the average and the maximum case

Pipe	v_c in m/s	Avg err in bar	Max err in bar	Avg $|fL|$ in bar	Max $|fL|$ in bar	Avg err/avg $|fL|$	Max err/max $|fL|$
A	4.969	0.130	0.801	0.621	1.519	0.210	0.527
B	5.165	0.103	0.355	0.451	1.247	0.228	0.285
C	4.355	0.084	0.330	0.183	0.926	0.458	0.356
D	5.021	0.120	0.406	0.100	0.826	1.202	0.491
E	4.669	0.064	0.976	0.207	1.788	0.309	0.546
F	8.603	0.052	0.526	0.133	1.124	0.390	0.468

Table 3 Results of approach $\mathcal{B}$ including the average pressure difference error, the corresponding maximum error, the average and maximum real friction-based pressure differences and the ratio of error to friction values for the average and maximum case

Pipe	Avg err in bar	Max err in bar	Avg $|fL|$ in bar	Max $|fL|$ in bar	Avg err/avg $|fL|$	Max err/max $|fL|$
A	0.104	0.938	0.621	1.519	0.167	0.618
B	0.100	0.675	0.451	1.247	0.222	0.542
C	0.052	0.399	0.183	0.926	0.284	0.431
D	0.047	0.781	0.100	0.826	0.471	0.945
E	0.048	0.878	0.207	1.788	0.234	0.491
F	0.032	0.613	0.133	1.124	0.238	0.545

For approach $\mathcal{B}$ the results are shown in Table 3. For all pipes, the average values are smaller for approach $\mathcal{B}$. The best improvement is made on pipe D, where the error could be more than halved. However, in relation to the average friction, pipe D has still by far the highest ratio.

For the maximum values, $\mathcal{B}$ performs even worse than $\mathcal{A}$ on all pipes except for pipe E. On pipe B and D the error value nearly doubled, which leads to a maximum error nearly as big as the actual friction on pipe D.

Regarding the expected performance due to the investigated velocity changes in Sect. 3, we observe that the values for pipe D are better than expected, but are clearly still the worst among all pipes, which is consistent with Fig. 2. Despite the different values for hour 48 in the graphic, the other five pipes have quite similar results. One reason could be the different evaluation periods: two years for Fig. 2 and only the second year for the results of Table 3.

5 Conclusion

We linearized the isothermal Euler Equations by fixing the velocity in the friction term to a constant value for each pipe. The results of approach $\mathcal{B}$ for determining the constant value, where we fixed the velocity to the historic values of two days before, indicate that the average errors made are not too large compared to the overall friction-induced pressure drop on the pipelines. However, the maximal error values turned out to be quite high, even in relation to the maximal friction values. Hence, we can conclude that the presented fixed velocity approaches can only be considered as rough approximations.

For future research, the results on our six pipes have to be verified on the complete network. Especially the bad values of pipe D have to be analyzed to find potential structural problems, maybe due to the change in flow direction. Furthermore, more sophisticated approaches to determine the fixed velocity might be possible. One option is to examine if compressor configurations at network intersection points have an impact on the velocity of the adjacent pipelines.

Acknowledgements The work for this article has been conducted within the Research Campus MODAL funded by the German Federal Ministry of Education and Research (BMBF) (fund number 05M14ZAM).

References

1. Ehrhardt, K., & Steinbach, M. C. (2003). Nonlinear Optimization in Gas Networks. ZIB Report.
2. Koch, T., Hiller, B., Pfetsch, M. E., Schewe, L. (Eds.) (2015). *Evaluating gas network capacities* (Vol. 21). Philadelphia: SIAM.
3. Nikuradse, J. (1950). Laws of flow in rough pipes. National Advisory Committee for Aeronautics Washington.
4. Open Grid Europe GmbH. www.open-grid-europe.com
5. Papay. (1968). A termeléstechnológiai paraméterek változása a gáztelepek muvelése során. OGIL Musz. Tud. Kozl.
6. Pfetsch, M. E., et al. (2014). Validation of nominations in gas network optimization: Models, methods, and solutions. *Optimization Methods and Software*.
7. Ríos-Mercado, R. Z., & Borraz-Sánchez, C. (2015). Optimization problems in natural gas transportation systems: A state-of-the-art review. *Applied Energy*, *147*, 536–555.
8. Saleh, J. M. (2002). *Fluid flow handbook*. McGraw-Hill Professional.

An Optimization Model to Develop Efficient Dismantling Networks for Wind Turbines

Martin Westbomke, Jan-Hendrik Piel, Michael H. Breitner, Peter Nyhius and Malte Stonis

1 Introduction and Research Background

Today, the dismantling of onshore wind turbines is generally conducted completely on-site. The rotor blades are cracked, the tower segments are separated, and the nacelle is cutted into smaller pieces. This undistributed dismantling of wind turbines is very time-consuming, inefficient, and implies ecological and economic risks. The dismantling of a single wind turbine takes around two weeks and entails costs of more than €130,000 [1]. The entire dismantling infrastructure needs to be transported to the wind farm and thus its operating capacity is not fully utilized. Further, in case of a repowering project, the time-consuming undistributed dismantling substantially delays the construction of new wind turbines and leads to lost revenues. An option to the undistributed dismantling is the transportation of only partly dismantled wind turbines to specialized dismantling sites for further handling. These specialized sites permit a better refining and thus higher revenues from selling the raw materials. Further, the specific costs for the necessary dismantling steps can be significantly reduced due to the specialization in the dismantling sites. However, this distributed

M. Westbomke (✉) · P. Nyhius · M. Stonis
Institut für Integrierte Produktion Hannover, Hollerithallee 6,
30419 Hannover, Germany
e-mail: westbomke@iph-hannover.de

P. Nyhius
e-mail: nyhuis@ifa-hannover.de

M. Stonis
e-mail: stonis@iph-hannover.de

J.-H. Piel · M. H. Breitner
Information Systems Institute, Leibniz University Hanover,
Königsworther Platz 1, 30167 Hannover, Germany
e-mail: piel@iwi.uni-hannover.de

M. H. Breitner
e-mail: breitner@iwi.uni-hannover.de

dismantling implies higher costs for the complex transportation of large-scale components and the initialization of dismantling sites. Consequently, when planning a distributed dismantling of wind turbines, assigned companies face the challenge of determining the optimal dismantling depth for each component as well as the optimal location of specialized dismantling sites.

Although such location and allocation problems are extensively investigated in various studies in the field of reverse logistics, studies presenting solutions for the dismantling of large-scale products, as for example wind turbines, are rare. Most of the studies focus on the end-of-life handling of conventional products, such as batteries [2] and vehicles [4]. However, in contrast to conventional products, large-scale products are typically stationary such that their dismantling needs to begin on-site [4]. Hence, the application of existing solutions to the dismantling of wind turbines is highly limited. One rare example focusing on the network design for end-of-life wind turbines is the study of Cinar and Yildirim [5]. They present a mixed integer linear programming model proposed to determine a long-term strategy for the dismantling of wind turbines. Their objective function minimizes the transportation and operation costs of the network by determining optimal locations for recycling and re-manufacturing sites. Nonetheless, their optimization model is based on the assumption of fixed component sizes. Variable dismantling depths are not taken into account, although both transportation and dismantling costs of large-scale products highly depend on the dismantling depths. Given the trade-off between these two types of costs, the consideration of variable depths is, however, essential for the network optimization.

In order to address this research gap, we present an optimization model for the development of efficient dismantling networks for wind turbines that optimizes both the location of specialized dismantling sites and the dismantling depths of specific wind turbine components. Our model generates the network design by optimally allocating every necessary dismantling step to a dismantling site, including either the wind farm itself or a specified dismantling factory.

2 Optimization Model

Different approaches exist for the optimal design of material streams in reverse logistic networks. If the locations within such networks are not fixed, location planning problems arise. For the efficient design of dismantling networks for wind turbines, the decision to establish potential dismantling sites also requires an allocation of dismantling steps to the respective sites. Planning problems of this kind belong to the class of Koopmans-Beckmann problems and are predominantly formulated as quadratic assignment problems [6]. Consequently, we transfer and adapt the Koopmans-Beckmann modelling approach to fit to the problem of designing efficient dismantling networks for wind turbines. To do so, several model assumptions need to be established: (1) The sites of the wind turbines and the potential dismantling sites and sinks are known; (2) the dismantling process begins at the location of the wind turbines; (3) a second market for the complete sale of decommissioned wind turbines is not considered; (4) dismantling sites can be established according to demand at predefined potential locations, which entails initialization costs; (5) the transportation and disposal costs depend on the dismantling depth; (6) the dismantling costs depend on the dismantling sites.

$$Min\ Z = \underbrace{\sum_{m=1}^{M} T_m \cdot \left(\sum_{w=1}^{W} \sum_{d=1}^{D} e_{wd} y_{wdm} + \sum_{w=1}^{W} \sum_{s=1}^{S} e_{ws} y_{wsm} + \sum_{d=1}^{D} \sum_{s=1}^{S} e_{ds} y_{dsm} \right)}_{transport\ costs} + \quad (1)$$

$$\underbrace{\sum_{s=1}^{S} \sum_{m=1}^{M} E_m y_{sm}}_{disposal\ costs} + \underbrace{\sum_{m=1}^{M} \left(\sum_{w=1}^{W} C_{wm} y_{wm} + \sum_{d=1}^{D} C_{dm} y_{dm} \right)}_{dismantling\ costs} + \underbrace{\sum_{d=1}^{D} I_d y_d}_{initialization\ costs} \quad (2)$$

$$\sum_{m=2}^{M} y_{wm} \geq 1 \qquad \forall w \in W \quad (3)$$

$$\sum_{m=1}^{M} y_{dm} \leq y_d \cdot B \qquad \forall d \in D \quad (4)$$

$$y_{wdm} + y_{wsm} = y_{wm} \qquad \forall d \in D,\ w \in W,\ m \in M \quad (5)$$

$$y_{sdm} = y_{dm} \qquad \forall m \in M,\ s \in S,\ d \in D \quad (6)$$

$$y_{sdm} + y_{wsm} = y_{sm} \qquad \forall m \in M,\ s \in S,\ d \in D \quad (7)$$

Our optimization model consists of the following indices, sets, parameters, and variables. The dismantling tasks are defined as $m \in (1, \ldots, M)$. The indices $w \in (1, \ldots, W)$, $d \in (1, \ldots, D)$ and $s \in (1, \ldots, S)$ determine the locations. The corresponding sets are W for wind turbines, D for dismantling sites and S for disposal sites. The parameters e_{wd}, e_{ws} and e_{ds} describe the distance between the locations, while T_m describes the specific transportation costs after dismantling task m. C_{wm} and C_{dm} define the dismantling costs of dismantling task m at location w and d. The disposal costs after dismantling task m are described by E_m. I_d are the initialization costs of a dismantling site d and B is a sufficiently large number. Z describes the dismantling network costs. The binary variable y_d ensures that the initialization costs are incurred only if at least one task m is allocated at location d. The binary variables y_{wm}, y_{dm} and y_{sm} assume a value of one if a dismantling task m is assigned to the respective location w, d or s. If a transportation takes place after dismantling task m at location w to location d, the binary variable y_{wdm} takes a value of one. The same logic applies to the binary variables y_{wsm} and y_{dsm}. The objective function (1) of our optimization model minimizes the total costs Z of a dismantling network for wind turbines, including total transportation costs in the first term, total disposal costs in the second term, total dismantling costs in the third term and total initialization costs for dismantling sites in the fourth term. Constraint (2) determines that further dismantling tasks can be allocated at the locations of each wind turbine. Constraint (3) ensures that initialization costs are incurred for each implemented dismantling site. Constraints (4) and (5) secure that each transport is carried out from a location where a dismantling task was performed. The constraint (6) ensures that the disposal costs are taken into account.

3 Case-Study and Results

In order to demonstrate the applicability of our optimization model, we present a proof-of-concept based on a case-study of an exemplary wind farm in Northern Germany. We obtained the case-study characteristics in corporation with a German recycling company operating in this region. The case-study includes the dismantling of a wind farm with six Nordex N117/2400 wind turbines [7] and two sinks. We consider two different scenarios with the aim of investigating whether a distributed dismantling can be more cost efficient than the undistributed alternative. The first scenario represents the undistributed dismantling, while the second scenario represents the distributed dismantling in a dismantling network. The second scenario considers three potential dismantling sites in addition to the wind turbines and sinks. Figure 1 presents the locations of the wind farm, the potential dismantling sites and the sinks. Table 1 shows the given dismantling, transportation, and disposal costs depending on the dismantling depth of the blades, tower, and nacelle. The dismantling depth is indicated in different stages. The higher the stages, the more dismantling tasks have been performed. While the first stage indicates that no dismantling is performed, the last stage represents a complete dismantling. Proportionate initialization costs are €1,000 per established dismantling site and the disposal of the foundations entails costs of €21,500 per wind turbine. In order to permit the application of our optimization model to the case-study, we implemented the optimization model in GAMS and utilized a CPLEX solver.

Figure 1 presents the resulting allocation of the necessary dismantling tasks to the potential dismantling sites for both scenarios, which finally determines the dismantling depth at the wind farm site. Dismantling sites and sinks that are not taken into account by the solver are grayed out. The arrows show the established transport connections. For the undistributed and distributed dismantling, the solver allocated the entire dismantling of tower and foundation to the wind farm site. The solver determines that the blades are sawn to 8 m pieces (stage 2) and that the drive train is cut off from the nacelle (stage 3) at the wind farm site. Afterwards, both components are transported to the sink. This optimal solution results in minimal costs of €30,986 per wind turbine for the undistributed dismantling. For the distributed dismantling, two out of three potential dismantling sites are established in the optimal solution. The blades are sawn to 8 m pieces at the wind farm site. The dismantling of the nacelle

An Optimization Model to Develop Efficient Dismantling ...

Table 1 Dismantling (DM), transportation, and disposal costs depending on the dismantling depth of each component

Component	Blades					Tower			Nacelle				
DM depth	1	2	3	4	5	1	2	3	1	2	3	4	5
DM costs on-site (€/t)	–	150	250	250	450	–	20	30	–	35	40	45	65
DM costs at DM site (€/t)	–	30	60	70	100	–	10	15	–	30	35	40	50
Transport costs (€/km)	5.23	0.25	0.1	0.08	0.05	3.92	1	0.06	1.12	0.25	0.25	0.25	0.01
Disposal costs (€/t)	450	300	270	250	200	–180	–200	–220	–180	–200	–200	–200	–220

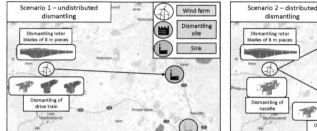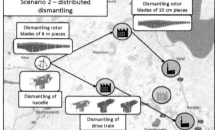

Fig. 1 Optimal network designs for the undistributed and distributed dismantling

differs, as it is only roughly dismantled to 8 m pieces (stage 2) at the wind farm site before it is transported to a dismantling site. At the dismantling site, the drive train is cut off from the nacelle (stage 3) and the partly dismantled nacelle is then transported to the sink. Another dismantling site is established for further handling of the blades, which are reduced up to a maximum size of 10 cm pieces (stage 3). The resulting minimal costs for the distributed dismantling in the described dismantling network are €20,320 per wind turbine, which is a cost reduction of €10,666 relative to the undistributed alternative.

4 Discussion and Conclusions

The aim of this paper was to present an optimization model for the development of cost-efficient dismantling networks for wind turbines. Due to high similarities with our optimization problem, we adapted an existing modelling approach for the optimization of the allocation of tasks to machines and factory layouts, known as Koopmans-Beckmann problem, to the dismantling of wind turbines. In order to proof our concept, we applied our optimization model to a real-world case-study and analyzed the two extrema of network designs: an undistributed and a distributed dismantling. Our results indicate that dismantling companies can realize high cost reductions (of up to 34.42% in the analyzed case-study) when selecting a distributed dismantling in a cost-efficient dismantling network instead of an undistributed dismantling entirely on-site. Consequently, depending on the conditions in each individual case, a distributed dismantling can be economically reasonable. In order to further confirm this hypothesis and to identify important input factors, future research should focus on extensive sensitivity and robustness analyses. Furthermore, as our current optimization model depicts only a part of the factors influencing the dismantling of wind turbines, future research should, alongside economic aspects, also implement ecological and logistical aspects in the optimization model.

References

1. Masherova, A. (2015, 12/2015). Rätselraten um Rückbaukosten. Eine Kalkulation für neue Windpark problematisch. 4initia Newsletter, 2–4.
2. Kannan, G., Sasikumar, P., & Devika, K. (2010). A genetic algorithm approach for solving a closed loop supply chain model. A case of battery recycling. *Applied Mathematical Modelling, 34*(3), 655670. https://doi.org/10.1016/j.apm.2009.06.021.
3. Cruz-Rivera, R., & Ertel, J. (2009). Reverse logistics network design for the collection of End-of-Life Vehicles in Mexico. *European Journal of Operational Research, 196*(3), 930939. https://doi.org/10.1016/j.ejor.2008.04.041.
4. Behrens, B.-A., Nyhuis, P., Overmeyer, L., Bentlage, A., Rüther, T., & Ullmann, G. (2014). Towards a definition of large scale products. *Production Engineering, 8*(1–2), 153–164.
5. Cinar, S., & Yildirim, B. (2017). Reverse logistic network design for end-of-life wind turbines. *Optimization and Dynamics with their Applications: Essays in Honor of Ferenc Szidarovszky*. Singapore: Springer.
6. Edwards, C. S. (1980). A branch and bound algorithm for the koopmans-beckmann quadratic assignment problem. *Mathematical Programming Study* (pp. 35–52)
7. Nordex. (2011). Rückbauaufwand für Windenergieanlagen. No. K0801025550DE Hamburg. Nordex Energy GmbH.

Model Generator for Water Distribution Systems

Corinna Hallmann and Stefan Kuhlemann

1 Motivation

In recent years, the optimization of water distribution systems has gained more and more attention. In the literature, there can be found different optimization problems, such as water network design [6], pipe optimization [8], tank optimization [7], energy minimization [2] or pump scheduling [4]. All those problems have in common that there exist many mathematical optimization models and a variety of different solution methods for each of them. Most research work is evaluated either on few realistic networks that are only available for one special use case or on those few test networks that are available in the literature. These networks are mostly very small and often lack in realistic properties, cf. [3]. In order to evaluate models and solution methods and compare them to other research work, it is inevitable to have many different water network models. These models require to be of realistic size and to have realistic properties. In this paper, we present a model generator for water distribution systems. With this generator it is possible to create network models with realistic properties and size. The generator can also control the structure of the network to guarantee a realistic reflection of a water distribution system. To ensure the hydraulic properties in the network we use a hydraulic simulation tool in our generator. Details about the generation process and the application of different use cases will be shown in this paper.

C. Hallmann (✉) · S. Kuhlemann
Paderborn University, 33098 Paderborn, Germany
e-mail: hallmann@dsor.de
URL: http://www.dsor.de

2 Literature Review

There are a few research papers that consider the generation of water network models. The generation approaches can be divided into manual and automatic ones. A manually generated network is the network *EXNET*, cf. [5]. It is a very large network model with several realistic properties and can be used as a test model to evaluate different use cases. Brumbelow et al. [1] developed two different network models *Micropolis* and *Mesopolis* representing a small and a large city with 5000 and 100,000 inhabitants, respectively. They use the models mainly for simulating cases of incidence such as electrical power failure. Those manually generated networks have very detailed and realistic properties, but it is a very time-consuming process to generate such networks. Thus, there are some research papers considering automatic generation of network models. Muranho et al. [10] developed the EPANET extension *WaterNetGen* for automatically generating network models. With this generator it is possible to generate network models with hundreds of nodes within a few minutes. The generator does not consider valves, pumps and reservoirs. Thus, the generated network models are not very realistic. Möderl et al. [9] present the graph-theory-based Modular Design System (MDS). This system generates networks depending on the demand of the nodes, the number of sources and the connectivity of the system. The authors present a set of 2280 virtual networks. In their approach the length of the pipes, the roughness and the elevation of the nodes are considered to be constant and pumps, valves and tanks are not integrated. De Corte and Sörensen [3] state that the generation of the networks is therefore not realistic. Sitzenfrei et al. [12] use the MDS to generate more realistic network models. Therefore, they use GIS data and the structure of the landscape and connect different graph structures. With these techniques it is possible to achieve more realistic structures. The tool HydroGen, cf. [3], can generate water network models of arbitrary size. It was developed to extend existing water network models.

3 Generator for Water Distribution Network Models

This section describes the process implemented in the generator for network models. The process can be divided into six different steps, which are described in the following. The application of these steps will be visualized in Fig. 1, where an example network is shown in Fig. 1a.

1. **Import Network**: The generator is able to import any network model in the *inp* format, which is commonly used when simulating and modeling water networks, cf. [11].
2. **Adding components for basic structure**: It is decided which components should be added to the network. It is possible to add pipes and nodes, which also can be tanks. Nodes are randomly added to one of the four edges or one of the four corners of the existing network model and then connected to the network via

Model Generator for Water Distribution Systems

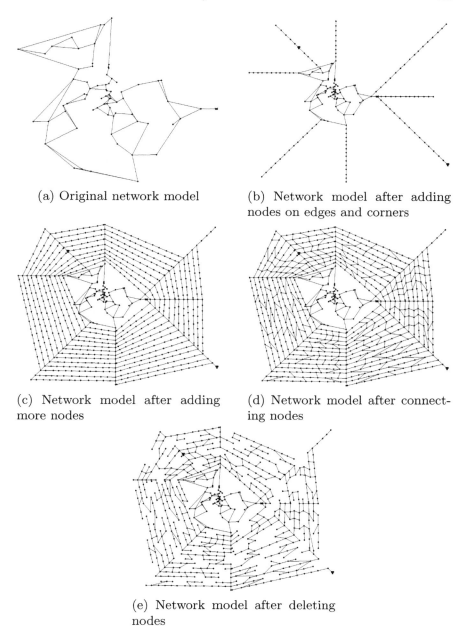

(a) Original network model

(b) Network model after adding nodes on edges and corners

(c) Network model after adding more nodes

(d) Network model after connecting nodes

(e) Network model after deleting nodes

Fig. 1 Different steps of the generation process

pipes. To determine the parameters of the new components, values of the existing components are used and are varied by predefined parameters. After adding a node with a corresponding pipe the network is simulated via a hydraulic simulation tool to check the modified network for infeasibilities. The simulation tool was recently developed by our industry partner *Rechenzentrum für Versorgungsnetze Wehr GmbH* and is based on EPANET [11]. The infeasibilities are for example a too low pressure on a node or numerical warnings of the algorithms. If numerical warnings occur, the node and the link that causes the problem will be deleted. If there are some infeasibilities, the parameters of the new node and link are modified until the network is feasible. If the modification of the parameters failed five times in a row, the node and the link will be deleted from the network as well. The user can decide how many of the new nodes should represent tanks. The implementation of tanks is important to ensure that there is enough water in the system also for the new components. Figure 1b shows the network after applying this step.
3. **Adding more nodes**: In this section more nodes are added to the network. These nodes are added to connect the nodes that were added in the previous step. The parameters of the pipes that are added to connect the nodes differ in such a manner that the resulting network is diversified and realistic. After adding new nodes and pipes, the network is simulated again to verify the correctness of the network. If some infeasibilities occur the parameters of the pipes are modified. If this fails, the new components are deleted from the network. The structure of the network is shown in Fig. 1c.
4. **Connect nodes**: This step connects the nodes of the previous step to each other. The nodes are chosen randomly but it is ensured that the new pipes do not intersect. The user can specify the maximal number of connections in order to vary the generated network. This step is displayed in Fig. 1d.
5. **Deleting Nodes**: In the previous steps a dense network with new nodes and pipes was generated around the original network. To get a more realistic network model, some of the new nodes and pipes are deleted in this step. With that, the network model gets a more realistic structure as the nodes are no longer aligned in the same way. The user can specify the rate of deleted nodes from the network. The resulting network is visualized in Fig. 1e.
6. **Export Network**: In this section the network is exported into the *inp* format.

4 Results

With the generator, any network model can be extended to an arbitrary size subject to the feasibility within a few minutes. In this section we present different versions of one network. Therefore, we use the network model *HG-SP-1-4*, which was generated by Hydrogen, cf. [3]. Figure 2 shows 6 different versions of the model and Table 1 shows the properties of those models. It can be seen, that it is possible to generate a

Model Generator for Water Distribution Systems

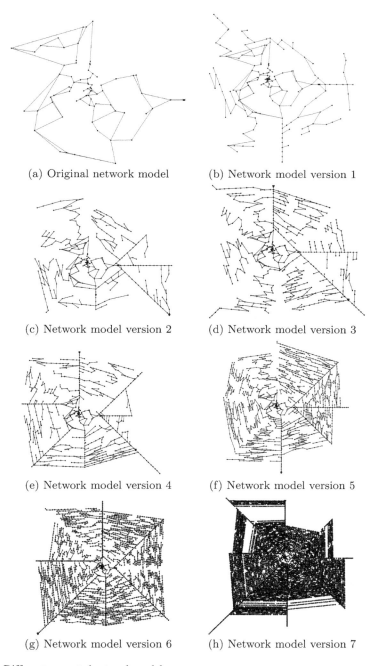

(a) Original network model (b) Network model version 1
(c) Network model version 2 (d) Network model version 3
(e) Network model version 4 (f) Network model version 5
(g) Network model version 6 (h) Network model version 7

Fig. 2 Different generated network models

Table 1 Properties of the generated network models

	Junctions	Pipes	Tanks
Original	73	102	0
Net1	163	198	1
Net2	311	355	2
Net3	515	565	5
Net4	555	627	3
Net5	1060	1171	4
Net6	2262	2357	6
Net7	6194	6796	10

variety of different network models with different sizes and different structures out of one original model. This shows the usefulness and the power of the presented generator.

5 Summary and Outlook

In this paper, we presented a generator for water network models. With this generator we can construct network models of arbitrary size. A simulation tool guarantees that the generated network model is feasible. The generator adds components in a structured manner and afterwards deletes some of those components to ensure a realistic structure of the model. In future work we would like to implement the adding of valves and pumps to the generation of network models.

References

1. Brumbelow, K., Torres, J., Guikema, S., & Bristow, E. (2007). Virtual cities for water distribution and infrastructure system research. *World Environmental and Water Resources Congress* (pp. 15–19).
2. Burgschweiger, J., Gnädig, B., & Steinbach, M. C. (2004). *Optimization Models for Operative Planning in Drinking Water Networks* (pp. 04–48). Berlin: Konrad-Zuse-Zentrum für Informationstechnik. ZIB-Report.
3. De Corte, A., & Sörensen, K. (2014). HydroGen: An artificial water distribution network generator. *Water Resources Management*, 28(2), 333–350.
4. De La Vega, J., & Alem, D. (2014). An improved stochastic optimization model for water supply pumping systems in urban networks. *CEP*(vol. 18052).
5. Farmani, R., Savic, D. A., & Walters, G. A. (2004). HydroGen: Exnet benchmark problem for multi-objective optimization of large water systems. *Modelling and Control for Participatory Planning And Managing Water Systems*.

6. Farmani, R., Walters, G. A., & Savic, D. A. (2005). Trade-off between total cost and reliability for anytown water distribution network. *Journal of Water Resources Planning and Management*, *131*(3), 161–171.
7. Hallmann, C., & Suhl, L. (2015). Optimizing water tanks in water distribution systems by combining network reduction, mathematical optimization and hydraulic simulation. *OR Spectrum*, *38*(3), 577–595.
8. Marques, J., Cunha, M., Savić, D., & Giustolisi, O. (2014). Dealing with uncertainty through real options for the multi-objective design of water distribution networks. *Procedia Engineering*, *89*(3), 856–863.
9. Möderl, M., Sitzenfrei, R., Fetz, T., Fleischhacker, E., & Rauch, W. (2011). Systematic generation of virtual networks for water supply. *Water Resources Research*, *47*(2).
10. Muranho, J., Ferreira, A., Sousa, J., Gomes, A., & Marques, A. (2012). WaterNetGen: An EPANET extension for automatic water distribution network models generation and pipe sizing. *Water Science and Technology: Water Supply*, *12*(1), 117–123.
11. Rossman, L. A. (2000). EPANET 2 - users manual. *Water Supply and Water Resources Division, National Risk Management Research Laboratory*. Cincinnati.
12. Sitzenfrei, R., Möderl, M., & Rauch, W. (2013). Automatic generation of water distribution systems based on GIS data. *Environmental Modelling & Software*, *47*, 138–147.

Effects of Constraints in Residential Demand-Side-Management Algorithms—A Simulation-Based Study

Dennis Behrens, Cornelius Rüther, Thorsten Schoormann, Thimo Hachmeister, Klaus Ambrosi and Ralf Knackstedt

1 Problem Identification

Climatic changes, urbanization, changing in living patterns and new technologies are only few of the current challenges occurring in energy grids. Hence, improvements in the management of energy grids, especially in managing and controlling appliances, are required. New developments contribute to face these challenges by, for example, considering sustainable but volatile energy resources (e.g., photovoltaic, wind or water), installing Smart Meters (e.g., [7, 9]) or managing Electric Vehicles (EVs). Demand-Side-Management (DSM) is one possibility that contributes to this by controlling loads, saving energy or reducing peaks in energy grids (e.g., [8]). Often not the whole energy grid is managed, as this is a very complex and challenging task. Instead, a microgrid consisting of different living units (houses, flats, etc.), is regarded. For implementing DSM (overviews can be found for example in [1–3]), different frameworks are available such as radius optimization, centralized and decentralized controlled optimization.

However, most of these algorithms make assumptions regarding load characteristics, which are not meeting the reality [6]. Behrens et al. [6] identified five different constraints, which are regarded by (some) DSM algorithms. Nevertheless, not all algorithms consider all constraints. *Accordingly, our study adresses the following goal: analysing the effects of constraints on DSM algorithms.*

Our paper is structured as follows: First, the individual constraints and the mathematical model to represent them are described (Sect. 2). In order to analyse the effects of these constraints, we choose a simulation. The setting and further selections, for

---

D. Behrens (✉) · T. Schoormann · T. Hachmeister · R. Knackstedt
Department of Information Systems, University of Hildesheim,
Universitätsplatz 1, 31141 Hildesheim, Germany
e-mail: dennis.behrens@uni-hildesheim.de

C. Rüther · K. Ambrosi
Department of Economics, University of Hildesheim, Universitätsplatz 1,
31141 Hildesheim, Germany

example, the data used, are described first (Sect. 3.1). Afterwards, the results are presented (Sect. 3.2). By discussing these results, it can be seen that the constraints have an effect to the results regarding the quality of the flattened load profile and savings. As the possibility of being able to manage appliances is increasing and even more shiftable devices will be implemented in the future, these effects will be more important (Sect. 4).

2 Constraints

2.1 Description

Behrens et al. [6] introduced a taxonomy of five constraints (C). Our focus lies in C1 (horizontal separability), C2 (vertical separability) and C3 (time interval of use). We excluded C4 (environmental effects, too much user information needed) and C5 (dependencies between loads, rated most unimportant) (see Fig. 1).

Horizontal separability of loads (C1). This means, for example, to pause the load after starting it. The load profile itself remains unchanged. Examples are the washing process of a washing machine or a dishwasher.

Vertical separability of loads (C2). A load can be separated vertically according its intensity. The load profile is therefore variable. An example is the charging process of an EV, which often offers a fast charging mode.

Time interval of use (C3). Some loads can only be turned on during specific time intervals or need to be finished at a specific time. Most of these deadlines or intervals are connected to the user's habits and requirements (e.g., an EV needs to be charged in the morning because of the user's travel to work).

2.2 Mathematical Model

Based on [4, 6], we derive a mathematical model. Let N be all considered living units, A_n be the appliances of living unit $n \in N$ and ω be the sample rate of the

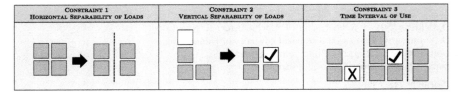

Fig. 1 Schematic representation of DSM constraints (cf. [6])

discrete model (time periods) over one day. Moreover, let $x^h = \sum_{n \in N} \sum_{a \in A_n} x_{n,a}^h$ with $h \in T = \{0, 1, \ldots, \omega\}$ be the sum of the consumption $x_{n,a}^h$ for all appliances $a \in A_n$ and all living units $n \in N$ in the timeslot h.

Let $l_{n,a}^k$ be the load profile in a local time interval $k \in T_l = \{0, 1, \ldots, \delta_{n,a}\}$, $l_{n,a} = \sum_{k \in T_l} l_{n,a}^k$ the load sum and $\delta_{n,a}$ the length of the load of $a \in A_n$. In doing so, we can transform a given horizontal and inseparable load profile from its local time interval T_l to the global one T (C1) through shifting the whole T_l by an appropriate constant $m_{n,a}$, i.e. $h_k = k + m_{n,a}$ with $0 \leq m_{n,a} \leq \omega - \delta_{n,a}$.

Furthermore, let $\gamma_{n,a}^{h,\min}$ be the min and $\gamma_{n,a}^{h,\max}$ be the max borders for a load $x_{n,a}^h$ with $h \in T, n \in N, a \in A_n$, so we can specify, in which borders the intensity of the load of a can be shifted, i.e. $\gamma_{n,a}^{h,\min} \leq x_{n,a}^h \leq \gamma_{n,a}^{h,\max}$ (C2). We note that the given load profiles have to satisfy the inequality $\gamma_{n,a}^{h,\min} \leq l_{n,a}^{h_k} \leq \gamma_{n,a}^{h,\max}$ for all $k \in T_l$ to get a feasible solution.

Let $\alpha_{n,a}$ be the starting and $\beta_{n,a}$ be the ending time slot of a load for an appliance a, then we can restrict time interval T to $[\alpha_{n,a}, \beta_{n,a}]$ (C3). We note that the interval length between $\alpha_{n,a}$ and $\beta_{n,a}$ has to be at least the length of the load profile $\delta_{n,a}$ to get a feasible solution, i.e. $\beta_{n,a} - \alpha_{n,a} \geq \delta_{n,a}$.

To turn on and off the constraint i for each appliance a let $c_a^i \in \{0, 1\}$ be a binary variable that shows if a constraint is turned on ($c_a^i = 1$) or not ($c_a^i = 0$).

The objective function describes the total cost of the given load profiles, while the cost in a time slot h is a function depending on h and the total load x^h, i.e. $c^h = c(h, x^h) \cdot x^h$. Generally, c is a concave function with regard to the load x^h, which causes that the optimal load profile is smoothed, which is an important property for energy providers.

The resulting mathematical model that minimizes the energy costs and holds the described contraints can be formulated as follows.

$$\min_{x_{n,a}^h} C = \sum_{h=0}^{\omega} c^h = \sum_{h=0}^{\omega} c(h, x^h) \cdot x^h \tag{1}$$

$$\sum_{h=0}^{\omega} x_{n,a}^h = l_{n,a} \qquad \forall n \in N, a \in A_n \tag{2}$$

$$(x_{n,a}^{h_k} - l_{n,a}^k) \cdot c_a^1 = 0 \qquad \forall k = 0, \ldots, \delta_{n,a}, \forall n \in N, a \in A_n \tag{C1}$$

$$\left.\begin{array}{c}(x_{n,a}^h - \gamma_{n,a}^{h,\max}) \cdot c_a^2 \leq 0 \\ (\gamma_{n,a}^{h,\min} - x_{n,a}^h) \cdot c_a^2 \leq 0\end{array}\right\} \forall n \in N, a \in A_n \tag{C2}$$

$$\left(\sum_{h=\alpha_{n,a}}^{\beta_{n,a}} x_{n,a}^h - l_{n,a}\right) \cdot c_a^3 = 0 \qquad \forall n \in N, a \in A_n \tag{C3}$$

$$x_{n,a}^h \geq 0, \quad c_a^i \in \{0, 1\} \qquad \forall h \in T, n \in N, a \in A_n, i = 1, 2, 3 \tag{3}$$

3 Simulation

3.1 Setting

In order to analyse the effects of the constraints, a simulation is the method of choice here. To do so, we need to choose a suitable dataset (scenario), give these data to several algorithms and compare the results (different impact factors) with each other. The data need to be extended with additional data (e.g., the information which intervals of usage are suitable for a user to meet his behaviour patterns, etc.). Moreover, the dependency between different loads and many environmental effects is difficult to provide, because they are highly user-specific. We therefore decided to implement C1, C2 and C3. C1 and C2 can be realized without any additional assumptions (e.g., a washing machine cannot be paused and the load profile is fixed).

Data selection. As described in [5], two types of data can be selected: artificial or real data. As we want to cover a wide range of possible scenarios, and real datasets often lack in additional information, we decided to use artificial data, for example generated by the LoadProfileGenerator (urlwww.loadprofilegenerator.de). Here, different predefined load profiles can be chosen. We decided to choose the following seven (predefined) load profiles: (1) family, 3 children, both work, (2) single woman, 2 children, with work, (3) multigenerational home (working couple, 2 children, 2 seniors), (4) single woman under 30 with work, (5) single man under 30 years with work, (6) couple under 30 years with work and (7) family with 2 children, man at work. All of these living units possess a washing machine, a dryer and a dishwasher as shiftable appliances. Out of these seven living units we formed one microgrid, which tries to reach a global optimum. We simulated each household over one complete year, including holidays, seasons, etc. Besides the habits, which are already included in the data, we needed to add several time intervals, where certain loads could be run or has to be finished. We proposed the washing machine, dryer and dishwasher could run between 8 and 24 o'clock.

Algorithm selection. In order to implement two DSM algorithms, we chose a Greedy-based algorithm (e.g., [10]) and a Multi-Agent-System-based algorithm (e.g., [4]), as they embody two contrary classes of DSM algorithms: a centralized (Greedy) and a decentralized (MAS) decision making.

Factor selection. For comparing the results, several impact factors can be selected. By analysing existing DSM algorithms, several factors can be identified. In order to measure the results, we chose the following factors: (a) peak-to-average-ratio (PAR, measure the highest peak), (b) (root-)mean-squared-error ((R)MSE, measure flattening) and (c) savings (compared to no DSM is used).

3.2 Results

By simulating the scenario described above (seven living units over one complete year), we were able to derive several results. The mean values over 365 days of the stated factors can be found in Fig. 2. We excluded C2 from the figure, as it has no effect. The reason is therefore quite easy to find: We have not included any appliances in our data (such as an EV) that fulfills this constraint. Regarding the remaining constraints C1 and C3 and also C1+C3, we can see that the effects varied in accordance with the individual day. Even the savings showed negative numbers during our simulation during a few days. Comparing the individual constraints to each other, we can state that C1 has a bigger effect on the result (in a negative way) than C3 does. However, C3 is strongly dependent on the set intervals from the user. If we shortened these, C3 might be more restrictive. The "worst" results are achieved

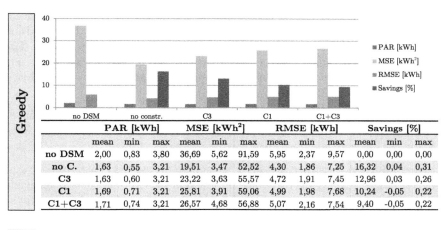

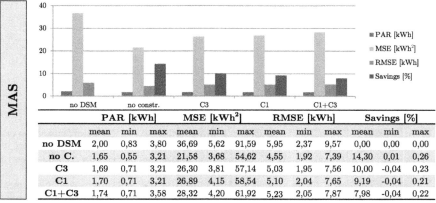

Fig. 2 Results for 365 days with the mean, the max and min value for each factor

by considering C1 and C3, as for example the savings dropped from over 16% to under 10% (Greedy) and from 14% to under 8% (MAS). The other factors, i.e. PAR, MSE and RMSE, show the same picture only with rising values (lower values are more advantageous). However, all combinations of simulated constraints on average achieved a benefit compared to not using DSM. Comparing both algorithms, we can state that Greedy algorithm performs better then MAS. Nevertheless, there is a deterioration of around 7% (Greedy) to 6.3% (MAS). On the basis of the results in our scenario, the MAS seems to generatea better outcome.

4 Discussion

Contribution. We conducted a simulation with a microgrid consisting of seven different living units (e.g., with certain consumption patterns, appliances). For simplicity, we did not add further infrastructure (e.g., PV, batteries). As a result, we can state that the reviewed constraints have effects on savings, PAR and (R)MSE. Thus, we conclude that DSM algorithms should take these constraints into account. This gets even more important nowadays, as EVs rush into the market and so do new electric appliances, which will cause more consumption and possibly peaks. Hence, constraints will have bigger effects in the future.

Limitations. We used artificial data that is built with regard to real and observed consumption patterns, behaviour, etc. Thus, we claim to have met the reality. Additionally, we made assumptions about the time intervals of use.

Research agenda. For future research, we plan to (a) use field data, (b) review all five constraints, (c) vary the additional information such as intervals and behaviour patterns, (d) add more infrastructure and EVs to the microgrid, (e) take the utility of the user into account, (f) investigate how the acceptance might be affected by (not) considering these constraints and (g) investigate the effects in different contexts such as the industrial and commercial context.

References

1. Al-Sumaiti, A. S., Ahmed, M. H., & Salama, M. M. A. (2014). Smart home activities: A literature review. *Electric Power Components and Systems, 42*(3–4), 294–305.
2. Balijepalli, V. M., Pradhan, V., Khaparde, S., & Shereef, R. (2011). Review of demand response under smart grid paradigm. *Innovative Smart Grid Technologies.*
3. Barbato, A., & Capone, A. (2014). Optimization models and methods for demand-side management of residential users: A survey. *Energies, 7*(9), 5787–5824.
4. Behrens, D., Gerwig, C., Knackstedt, R., & Lessing, H. (2014). Selbstregulierende Verbraucher im smart grid: Design einere Infrastruktur mit Hilfe eines Multi-Agenten-Systems. *Proceedings of the Multikonferenz Wirtschaftsinformatik 2014.*

5. Behrens, D., Schoormann, T., & Knackstedt, R. (2016). In Heinrich, C. & Pinzger, M. (Eds.), *Datensets für Demand-Side-Management Literatur-Review-Basierte Analyse und Forschungsagenda.*, Lecture notes in informatics.
6. Behrens, D., Schoormann, T., & Knackstedt, R. (2017). Towards a taxonomy of constraints in demand-side-management-methods for a resedential context. *Lecture Notes in Business Information Processing (LNBIP), 288*, 283–295.
7. Feuerriegel, S., Bodenbenner, P., & Neumann, D. (2016). Value and granularity of ICT and smart meter data in demand response systems. *Energy Economics, 54*, 1–10.
8. Gellings, C. W., & Chamberlin, J. H. (1993). *Demand-Side Management: Concepts and Methods* (2nd ed.). GA: Prentice Hall.
9. Schoormann, T., Behrens, D., Kolek, E., & Knackstedt, R. (2016). Sustainability in business models - a literature-review-based design-science-oriented research agenda. *Proceedings of the European Conference on Information Systems.*
10. Sianaki, O. A., Hussain, O., & Tabesh, A. R. (2010). A knapsack problem approach for achieving efficient energy consumption in smart grid for endusers' life style. *2010 IEEE Conference on Innovative Technologies for an Efficient and Reliable Electricity Supply* (pp. 159–164).

Production Process Modeling for Demand Management

Stefanie Kabelitz and Martin Matke

1 Introduction

The electricity supply is growing increasingly dependent on the weather as the share of renewables increases. Nevertheless, different measures can maintain grid reliability and quality. These include the usage of storage technologies, grid expansion and options for responsiveness of supply and demand. The latter is known as demand-side management. In the private sector, this idea is gaining in importance with the emergence of smarthome-technologies [3]. In the industrial sector, potential for demand-side management arises from the occurence of processes that are not directly linked with the production. In practice this means, that quantities of electricity can be purchased spontaneously at the stock market for energy (EEX), or potentials to shift the load can be offered at the German Electricity Balancing Market.

The theoretical and economical potentials of demand-side management have been determined in [2, 7]. However, the economic potential has been studied solely from the perspective of the energy supply side. The effects of the current legal situation on the willingness of the demand side to participate have been neglected so far. Furthermore, studies that investigate incentives for companies to look for flexibility potential among production processes are still missing. These are part of the aims of the investigations within the innovation cluster for Smart, Energy Efficient Regional Value Chains in Industry (ER-WIN®) [6] which our study is part of.

S. Kabelitz (✉) · M. Matke
Fraunhofer IFF, 39106 Magdeburg, Germany
e-mail: Stefanie.Kabelitz@iff.fraunhofer.de
URL: https://www.iff.fraunhofer.de/en.html

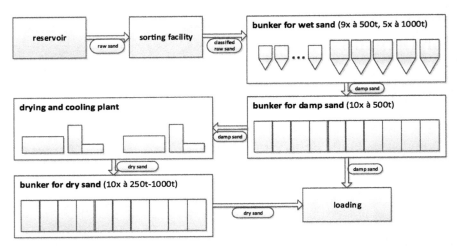

Fig. 1 Production scheme

2 Problem Description

We investigated a production-planning problem for a real sand processing facility in order to determine the economic potential for load shifting. At the facility, raw sand is extracted from a lake and processed in several stages to obtain sand for the construction industry. The main steps consist of classifying the raw sand into different grain sizes, draining the wet sand, drying the damp sand, and cooling the warmed up sand. At the end, the raw sand is processed into eight main products and ready for loading onto truck or train containers. Figure 1 illustrates the general procedure.

The production site aroused our interest through its potential for demand management. From a mathematical point of view, the problem at hand is interesting for its relatively large-scaled production scheme and the characteristic that it includes both continuous processes and discontinuous batch processes.

3 Problem Modeling

The aim of our model is to reflect the planning process of the sand processing facility for a period of 48 hours. In particular, we designed our model in order to obtain a production plan that safeguards the demand requirements while considering the integration of a cogeneration unit to the energy supply. Additionally, we sought to formulate our model in such a way, that it enables us to investigate potential effects of changed conditions, for instance the introduction of time-dependent energy prices or the option to participate at the German Electricity Balancing Market and the day-ahead spot market.

3.1 The MILP in General

In a nutshell our model represents the facility as a combinatorial graph based on a mixed integer linear program (MILP). We discretized our planning horizon into 192 periods of 15 min and declared our decision variables (for storage, flow, etc.) for these periods. The underlying graph consists of 66 nodes and 102 edges, which represent the production processes and map the connections among them. The size of the graph and the number of periods considered led to a relatively large-scaled problem. Nevertheless, the actual complexity of our problem is due to the introduction of the discontinuous batch processes. Needless to say, it was also the aspect that kept the problem being interesting.

The objective function of our model consists of costs for the electricity usage. In addition, we tested the integration of balancing energy into the objective and manipulated the cost vector to extract information about potential effects of different pricing schemes.

The constraints of our model can be divided into five main groups. Namely constraints modeling the nature of the input (raw sand) and the output (demands); constraints mapping production, storage and energy consumption for continuous processes and discontinuous batch processes, and, finally, constraints representing the interactions among the various processes, which primarily consist of constraints on the edges.

3.2 Assumptions

To put our model into work we first assumed that all data is given a priori. We further simplified the production facility by not considering a distinction between train and truck loading. These initial assumptions can be justified by creating a stock to compensate inexact forecasts for the composition of the raw sand and the extent of the demand.

We integrated a stockpile into our model by initializing the storage variables $x_{v,0}^S$ at the start of the planning horizon as non-zero for some of the processes v, and forced our optimization problem to attempt to maintain the initial storage volume at the end of the planning horizon by penalizing the offset x_v^O in the objective function. This can be achieved via:

$$\forall v \in \mathcal{V}: \\ x_{v,t_{\max}}^S - x_{v,0}^S \leq x_v^O \\ x_v^O \geq 0 \quad (1)$$

where $t_{\max}$ represents the last period of the planning horizon.

The integration of the cogeneration unit proofed to be a minor challenge in both modeling and computational terms. If we assume that the maximum output of the cogeneration unit is a priori known for all periods t to be x_t^{CE} and the energy needed in our system to produce accumulates to x_t^{PE} for each period t, then the energy consumption our system needs to cover in addition to the cogeneration unit x_t^{EE} can be obtained via:

$$\forall t \in \mathcal{T}:$$
$$x_t^{PE} - x_t^{CE} \leq x_t^{EE} \qquad (2)$$
$$x_t^{EE} \geq 0.$$

The formulation in (2) is essentially the same as in (1). A greater challenge occurred with the integration of both continuous and discontinuous processes into our model.

3.3 Continuous Process Modeling

To model the continuous processes of our production scheme we essentialy formulated the following constraints.

We linked the output $x_{v,t}^P$ of a continuous process v to the previous inflow $I(v, t-1)$ via:

$$x_{v,t}^P = I(v, t-1), \qquad (3)$$

and linked the storage variable $x_{v,t}^S$ at each process v for each period t with the production $x_{v,t}^P$ and the outflow $O(v, t)$ via:

$$x_{v,t}^S = x_{v,t-1}^S + x_{v,t}^P - O(v, t). \qquad (4)$$

Note that $I(v, t)$ and $O(v, t)$ denote the sums of the flow variables at t on the edges that enter and leave v, respectively. For the source node (raw sand processing) and the final nodes (loading) the equations in (3) and (4) require minor changes. The time and energy consumption are in the continuous case directly proportional to the production output. Detailed information on the concrete implementation can be found in [5].

3.4 Batch Process Modeling

To model the discontinuous batch processes of the bunkers for wet sand and the drying and cooling plants we introduced a binary decision variable $\delta_{v,t}$ for each of these processes v in each period t, with $\delta_{v,t} = 1$ denoting that a batch process starts in v at t and $\delta_{v,t} = 0$ that it does not. To represent the nature of the discontinuous processes in our model we included the following constraints.

First, in each discontinuous process no more than one batch can be processed at a time. Given the time $T(v)$ a batch takes in process v we can express this restriction via:

$$\forall t : \sum_{i=t}^{t+T(v)-1} \delta_{v,i} \leq 1. \tag{5}$$

To avoid that completed charges mix up with untreated sand we introduce:

$$x_{v,t}^L / L(v) + \delta_{v,t}^B \leq 1, \tag{6}$$

where $L(v)$ is the maximum charge volume of a given process v (e.g. the volume of a bunker) and $x_{v,t}^L$ is the volume contained at a given period t. The inequality (6) is a linear formulation of the expression that $\delta_{v,t}^B = 0$ or $x_{v,t}^L = 0$ and ensures, that a batch process can only start when already processed sand has been forwarded completly.

By assuming a continual use of energy over the period of a batch process, we integrated the energy consumption of the discontinuous processes into our existing model via:

$$x_{v,t}^{PE} = E(v)/T(v) * \sum_{i=t-T(v)+1}^{t} \delta_{v,t}, \tag{7}$$

with $E(v)$ being the total energy consumption to process a single charge in v. Constraint (7) could be easily customized for energy consumption patterns different to the assumed continual consumption.

The modeling of the flow and the storage variables for the discontinuous processes basically follows the formulations in 3.3. Further details can be found in [5].

4 Results and Discussion

For the results we used real input data. Our industrial partner provided the overall electric consumption of the facility accumulated for 15 min periods and information on the supply and demand of the different sand types. Furthermore, we had information available for the processes, such as the installed load of the machines.

We were able to successfully describe the nature of both the continuous and discontinuous processes within our MILP model. Figure 2 illustrates the developement of the storage and energy decision variables for a discontinuous batch process, namely a sand drying plant, within our model. For the planning horizon two charges are processed in the illustrion. The characteristics of batch processes, mapped by the constraints (5)–(7) as well as the fulfillment of the storage constraints in (4) are clearly visible.

We could show that it is possible to create a production plan that reacts to volatile energy prices. Hence, it is possible to reduce the energy costs in our model assuming real-time pricing. However, such purchasing strategies are uncommon due to the inherent risks and the considerable effort for individual enterprises.

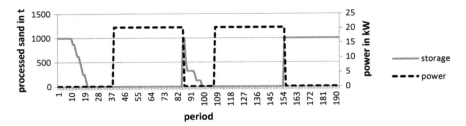

Fig. 2 Illustration of a batch process

5 Outlook

We have been able to model the production processes of our case example. The benefit is, that we gained the possibility to investigate potential effects of different law concepts that are designed to make the electricity price components more dynamic [1]. In further studies, we seek to investigate how existing laws promote flexibility schemes and which legal situation could be favourable for the demand side.

Nevertheless, it is to question whether there is a simpler procedure. One such approach is the evaluation of the energy flexibility of the production processes, as for instance in [4, 8]. These approaches examine, whether the energy flexibility of the modeled processes match the energy flexibility of the real world experiences and to which extent these can be substituted. It is possible, that the question of economical sustainability to offer energy flexibility options can be determined by the sole knowledge about the energy flexibility of the production or individual production processes.

Furthermore, energy flexibility would provide a measurement to classify a production. The investigations provided a rather limited view for the case study. However, the production sites energy flexibility might represent a variety of other facilities. This approach will be further investigated in the course of the innovation cluster ER-WIN®.

References

1. Bundesministerium für Wirtschaft und Energie (Hg.) (2015): Ein Strommarkt für die Energiewende. Ergebnispapier des Bundesministeriums für Wirtschaft und Energie (Weißbuch).
2. Gils, Hans Christian. (2014). Assessment of the theoretical demand response potential in Europe. *Energy, 67*, 1–18. https://doi.org/10.1016/j.energy.2014.02.019.
3. Hoogsteen, G., Molderink, A., Hurink, J. L., & Smit, G. J. M. (2016). Generation of flexible domestic load profiles to evaluate demand side management approaches. *IEEE International Energy Conference (ENERGYCON)* (pp. 1–6). Belgium: Leuven.
4. Kabelitz, S., & Streckfuß, U. (2014). ZWF Artikel Energieflexibilität 2014. Ein analytischer Identifikationsansatz. *ZWF Zeitschrift für wirtschaftlichen Fabrikbetrieb, 109*((1–2)), 43–45.

5. Matke, Martin. (2015). *Energieorientierte Produktionsplanung am Beispiel eines Sandaufbereitungswerkes mithilfe von IBM ILog CPLEX*. Magdeburg: Otto-von-Guericke Universität.
6. Neugebauer, Reimund (2016): Ressourceneffizienz. Schlüsseltechnologien für Wirtschaft & Gesellschaft. [Place of publication not identified]: Springer Science and Business Media; Springer.
7. Paulus, Moritz, & Borggrefe, Frieder. (2011). The potential of demand-side management in energy-intensive industries for electricity markets in Germany. *Applied Energy, 88*(2), 432–441. https://doi.org/10.1016/j.apenergy.2010.03.017.
8. Reinhart, G., Graßl, M., & Datzmann, S. (2014). Methode zur Bewertung der Energieflexibilität. *wt Werkstattstechnik online* (Vol. 5, pp. 313–319)

Part VII
Finance

Decoupled Net Present Value—An Alternative to the Long-Term Asset Value in the Evaluation of Ship Investments?

Philipp Schrader, Jan-Hendrik Piel and Michael H. Breitner

1 Introduction and Research Background

The uproarious times of the financial crisis in 2008/09 presented a critical juncture for the practice of ship valuation, which had hitherto been mainly conducted according to comparable transactions in the market. The LTAV, introduced in 2009 by the Hamburg Shipbroker Association VHSS e.V., allowed banks to assess ship values based on a DCF approach. Designed in particular for valuations in distressed markets, the LTAV quickly gained in popularity among banks, as it allowed to postpone necessary impairments on shipping loans and to smooth asset values. As with other DCF models, the selection of an appropriate discount rate is most critical. The LTAV mandates the use of a risk-adjusted discount rate (RADR), comprised of the risk-free rate and a risk premium to account for both the time value of money and risk at the same time.

However, the bundling of time preference and risk distorts the evaluation of investment risks, in general by assuming time and risk to be interchangeable [1]. In particular, [8] for instance discuss the problem of valuing negative cash flows. In this case, increasing the discount rate may result in a higher overall Net Present Value (NPV), especially if cash outflows occur late in the life cycle of the investment. This also creates the incentive to transfer liabilities into the future, which in turn encourages short-term thinking [3]. The seemingly obvious solution to apply a lower RADR to negative cash flows, on the other hand, further disconnects risk from its actual

---

P. Schrader (✉) · J.-H. Piel · M. H. Breitner
Information Systems Institute, Leibniz University Hannover,
Königsworther Platz 1, 30167 Hannover, Germany
e-mail: philipp.schrader@stud.uni-hannover.de

J.-H. Piel
e-mail: piel@iwi.uni-hannover.de

M. H. Breitner
e-mail: breitner@iwi.uni-hannover.de

sources and assumes that negative cash flows are less uncertain, which might not be the case [1]. The shortcomings of RADRs are exacerbated if company-wide rates are used to evaluate single projects. Investments riskier than the company as a whole seem more favourable when evaluated using the lower company wide RADR and vice versa, leading to over-investment in relatively riskier projects [6].

The Decoupled Net Present Value (DNPV), introduced by [2, 3] presents itself as an alternative to traditional DCF models, promoting the segregation of risk and time preference as well as a thorough analysis of investment risks. This is achieved by first acknowledging that investors in real projects face both systematic and unsystematic risk and have to be compensated for both. The DNPV framework then defines investors as insurance providers for risks that are not transferred to third parties. In exchange for holding onto these risks, they are compensated with so-called synthetic insurance premiums (SIP), which represent the price of individual risks and are treated as costs to the investment. Capturing risks in these SIPs renders the investment's cash flows risk-free and hence allows discounting with the risk-free rate. The pricing of risks via SIPs requires the identification, segmentation and quantification of all risks affecting the investment. As a result, the DNPV can support a systematic approach to risk and a comprehensive basis for investment decisions. In contrast to aggregating risks in the discount rate, individual modeling also enables an accurate communication of the investment risk profile [3].

Shipping has always been a high-risk business, given the fluctuations in charter revenues (CR) and its capital intensive, highly leveraged assets. Understanding the factors that drive the profitability of an investment in a vessel means understanding its risks. Incorporating these in an investment evaluation is of utmost importance for shipowners and banks alike. However, despite being developed as an alternative to the market approach in distressed and therefore highly volatile markets, the LTAV itself is ill-equipped to account for the full range of risks in a consistent manner. The DNPV as an approach focused towards risk has been applied to infrastructure and renewable energy projects [3, 4]. However, to the best of our knowledge, an application to a cargo ship investment case is missing from the literature. In order to address this research gap, we provide a proof of concept illustrating the applicability of the DNPV in maritime investment evaluation. To this end, we develop a prototype in Python and document its application to a real-world vessel investment simulation study.

2 Vessel Investment Case

Our investment case is designed akin to the exemplary LTAV valuation presented in [7]. It takes the perspective of a ship financing bank that is approached with a loan request for the acquisition of the Conti Emden, a container vessel with 2.700 TEU. Determining a reasonable purchase price going into negotiations and judging the profitability of the investment, the LTAV and the DNPV are to be applied alongside each other. The remaining operative live time of the vessel is 15 years, at the end of

Table 1 Free cash flows analysis of the investment case in thousand $

Parameter	Year 1	Year 4	Year 11	...	Year 15	PV	Distribution[a]
Charter revenues	2,427	4,413	4,308	...	10,020		
OPEX	2,330	2,547	3,132	...	5,482		
EBITDA	**96.29**	**1,866**	**1,176**	**...**	**4,537**	**9,801**	**=LTAV**
GCR SIP	126.98	230.89	225.43	...	233.78	1,875	T(11,600, 80%, 140%)
OPEX SIP	46.82	51.16	62.92	...	110.13	550	BP(6,000, 95%, 125%)
TS SIP	42.04	76.43	74.62	...	80.77	616	BIN(20%, 31 days)
PS SIP	178.29	165.46	84.83	...	36.11	1,167	BIN(0.3%, $ 2m clean up)
RV SIP	22.82	33.47	145.23	...	128.52	723	Stochastic Approach
Total SIPs	**416.95**	**557.40**	**593.03**	**...**	**589.32**	**4.903**	
Decoupled FCF	**-320.65**	**1,308**	**583.31**	**...**	**3,948**	**8,947**	**=DLTAV**

[a] Triangular T(mode, min in %, max in %); betaPERT BP(mode, min in %, max in %); binomial BIN(μ, comment)

which it is scheduled to be scrapped at a steel price of $ 300 per lightweight tonne, adjusted for inflation. Operating days are 358 in regular years and 343 in years requiring survey. Inflation rates of 2 and 3% apply to Gross Charter Rate (GCR) and Operating Expenditure (OPEX), respectively.

For the following DNPV analysis, we identified the most important risk categories as GCR and OPEX risk, risk of permanent shutdown (PS) and temporary shutdown (TS) of the vessel and residual value risk (RV). The type of distribution to model each risk is determined in accordance with [3, 4]. The choice of distributions and their shapes in the last column of Table 1 have been matched in consultation with industry experts. From the expected values of the distributions in Table 1, the revenues and OPEX for the cash flow model are derived. For the detailed planning period of three years, the GCR equal $ 7.109, the average of the rates suggested by [5]. The 10-year average of $ 11.600 serves as the mode for the triangular distribution. Fees and commissions of 6.5% apply to the GCR. OPEX include costs for repairs and maintenance, insurance, stores, crew and administration and are set at $ 6,000 per day in the BetaPERT distribution. Drydocking and special survey costs sum up to $ 1.25M every five years [5].

3 Decoupled Net Present Value Analysis

Equation 1 illustrates the central equation in the DNPV framework. In the equation, $\tilde{R}_{V,t}$ and $\tilde{R}_{I,t}$ represent the costs of risks, i.e. the SIPs associated with revenues V_t and expenditures I_t, respectively. Revenues from the vessel's residual value are included in V_T. After accounting for risks as costs in the numerator of Eq. 1, the resulting net cash flows are considered risk-free and can be discounted at the risk-free rate. The separate treatment of revenues and expenditures highlights the understanding of risk underlying the DNPV. According to [2], risk is defined as a deviation from the expected value that negatively affects profitability, i.e. revenues being lower than expected or expenditures being higher than expected. This notion of risk is also central to the actual pricing of risks via the concept of SIPs.

$$DNPV = \sum_t \sum_{i,j} \frac{(\tilde{V}_{t,i} - \tilde{R}_{t,i}) - (\tilde{I}_{t,j} + \tilde{R}_{t,j})}{(1+r_f)^t} \quad (1)$$

The DNPV framework offers three distinct methods to derive the price of risk. Whereas the heuristic approach uses expert opinion about the impact of risks, the stochastic and probabilistic approaches feature the use of stochastic processes and probability distributions, respectively. Focusing on the latter, we highlight the versatility of this approach by illustrating the calculation of SIPs with two different distributions. When dealing with distributions to calculate SIPs, a distinction has to be made between revenue and expenditure risk. For the former, Eq. 2 shows that the price of revenue risk is comprised of the probability of revenues being lower than expected ($PR\left[\tilde{V}_t > V_t\right]$) and the average amount lost in such a case, i.e. the mean shortfall ($\tilde{V}_t - \tilde{V}_t$). As shown in Eq. 3, the difference between actual and expected cash flows is multiplied with the probability that costs exceed their expected value.

$$\tilde{R}_{V,t} = (V_t - \tilde{V}_t) \cdot Pr\left[\tilde{V}_t > V_t\right] = L_t \cdot Pr\left[\tilde{V}_t > V_t\right] \quad (2)$$

$$\tilde{R}_{I,t} = (\tilde{I}_t - I_t) \cdot Pr\left[I_t > \tilde{I}_t\right] = L_t \cdot Pr\left[I_t > \tilde{I}_t\right] \quad (3)$$

To further illustrate the calculation of SIPs, Fig. 1 shows the Monte-Carlo simulation with 10,000 iterations of the triangular distribution for revenue risk as well as the pricing of PS risk with a binomial distribution. The latter assumes a total hull loss at a certain period with probability Θ. The cash flow loss in this case equals the value of the ship in that period plus costs for environmental clean up. The ship value is determined by the remaining cash flows, discounted at a probability-adjusted risk-free rate $(1 + r_f + \Theta)$. Calculating the expected value of the cash flow loss distribution gives the SIP [2].

Table 1 depicts the cash flow model of the investment case as well as the SIPs calculated based on the distributional settings in the last column. Looking at the present values of the SIPs gives a sense of the risk structure of the investment. Unsurprisingly, the risk associated with charter revenue fluctuation is the most important risk category, followed closely by PS risk, which decreases with the value of the ship. OPEX risk is limited as the uncertainty surrounding OPEX forecasts is far less than for charter rates. Discounting the EBITDA at the discount rate of 7.3% published

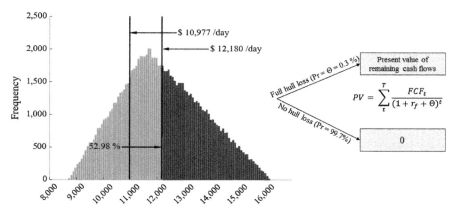

Fig. 1 SIP calculation for CR and TS in year one. Applying Eq. 3 to the CR distribution results in an SIP of ($ 12,180–$ 10,977) · 52.98% = 5.2% of CR or $ 126.984. Applying it to the binomial distribution for PS gives a SIP of 0.3% · (PV of remaining cash flows + Liability) = $ 178, 294

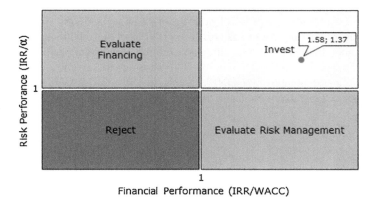

Fig. 2 Though profitable from both perspectives, risk performance could still be improved. Easing the impact of PS risk could be achieved by purchasing loss of hire and/ or vessel pollution insurance, as long as the premiums do not exceed the increase in value resulting from the reduced risk

by [9] as the official rate suggested for LTAV applications yields a value of $ 9.8M. Deducting the SIPs from EBITDA and discounting the resulting decoupled free cash flows at the risk-free rate of 1.58% yields a (decoupled) LTAV of $ 8.9M.

At a purchase price of $ 7M, both approaches signal a viable investment. However, the lower DNPV suggests that the LTAV underestimates the risks of the investment. In order to reconcile both approaches and ease preconceptions among LTAV proponents, [2] advise to evaluate the investment's financial and risk performance separately in the matrix of Fig. 2. The ratio of Internal Rate of Return (IRR) to the applied RADR, i.e. the Weighted Average Cost of Capital (WACC) in Fig. 2 is used to measure financial performance only and represents the standard NPV decision rule, while risk performance is measured by the ratio of IRR to implied risk-adjusted discount rate (iRADR), the RADR that sets the NPV equal to the DNPV and therefore incorporates the same amount of risk as suggested by the DNPV.

4 Conclusion and Outlook

In the current market environment, the feasibility of ship investments depends on a thorough understanding of their risks. We demonstrated that the DNPV can be applied to ship investments and can support investment decisions by framing risks as costs to the investment. Modeling of risks in SIPs allows for the decoupling of time preference and risk. This not only alleviates some of the shortcomings of traditional DCF methods, but also provides a new and comprehensive perspective on the risk profile of the investment. However, the advantages of the DNPV depend on the ability to model risks in shipping accurately, which confronts investors with additional efforts compared to the LTAV. We contributed to the existing research on the DNPV by providing a proof of concept in shipping and ship finance. Further research has to be conducted to establish the DNPV in these fields and in order for it to rival incumbent valuation techniques. Certainly, important areas to address are how specific risks can be modeled in the DNPV framework and the accuracy of risk modeling. The acceptance of the DNPV as an addition to the LTAV in practice would benefit from comparative studies of time series of ship values resulting from both methods.

References

1. Carmichael, D. G. (2016). Adjustments within discount rates to cater for uncertainty - Guidelines. *The Engineering Economist*, 1–14.
2. Espinoza, D. (2014). Separating project risk from the time value of money: a step toward integration of risk management and valuation of infrastructure investments. *International Journal of Project Management, 32*, 1056–1072.
3. Espinoza, D., & Morris, J. W. (2013). Decoupled NPV: A simple, improved method to value infrastructure investments. *Construction Management and Economics, 31*(5), 471–496.
4. Espinoza, D., & Rojo, J. (2015). Using DNPV for valuing investments in the energy sector: A solar project case study. *Renewable Energy, 75*, 44–49.
5. Hartwig, T. (2016). Maritime Overview 11/2016. Ernst Russ Shipbroker GmbH & Co.KG
6. Krüger, P., Landier, A., & Thesmar, D. (2015). The WACC fallacy: The real effects of using a unique discount rate. *The Journal of Finance, 70*(3), 1253–1285.
7. Mayr, D. (2015). Valuing vessels. In O. Schinas, C. Grau, & M. Johns (Eds.), *HSBA Handbook on Ship Finance* (pp. 141–163)
8. Robichek, A., & Myers, S. (1966). Conceptual problems in the use of risk-adjusted discount rates. *The Journal of Finance, 21*(4), 727–730.
9. Vereinigung Hamburger Schiffsmakler und Schiffsagenten e.V.: Long-Term Asset Value Parameter Diskontierungssatz zum 1. Oktober 2016. Private communication, 28.06.2017

Fast Methods for the Index Tracking Problem

Dag Haugland

1 Introduction

Passive fund managers aim to compose a portfolio that brings approximately the average market return. A common strategy to accomplish this, referred to as *index tracking*, is to construct funds that simulate a chosen benchmark, typically a stock market index. To save administration costs, the index tracking portfolios consist of only a small subset of the assets present in the index.

It is crucial for the fund manager to select the portfolio assets carefully, and to determine how large proportion of the portfolio each asset is to constitute. A measure of difference between a portfolio and the benchmark which it mimics is referred to as the *tracking error*, and the problem is to find a portfolio minimizing this measure. Only portfolios of a given (small) size are to be considered.

Building on recent theoretical results and computational procedures for cardinality constrained index tracking, the contributions of the current work are fast methods for near-optimal solutions. Answering the need to reduce the running time of previously suggested construction and improvement methods, we develop new methods, by which we prove that the running time can be reduced by one order of magnitude.

2 Notation and Definitions

For finite sets $M \subseteq N$ and $i \in N$, $M + i$ and $M - i$ denote $M \cup \{i\}$ and $M \setminus \{i\}$, respectively. Denote by x_i the component of $\mathbf{x} \in \mathbb{R}^N$ corresponding to i, and let $\mathbf{x}_M$ denote the vector with components x_j ($j \in M$). For finite sets $M_j \subseteq N_j$ ($j = 1, 2$) and a matrix $\mathbf{A} \in \mathbb{R}^{N_1 \times N_2}$, $\mathbf{A}_{M_1 M_2}$ denotes the submatrix of $\mathbf{A}$ consisting of rows

---

D. Haugland (✉)
University of Bergen, Bergen, Norway
e-mail: dag@ii.uib.no

corresponding to $M_1 \subseteq N_1$ and columns corresponding to $M_2 \subseteq N_2$. If $M_1 = \{k\}$ ($M_2 = \{k\}$), then notation $\mathbf{A}_{kM_2}$ ($\mathbf{A}_{M_1 k}$) is used, while the element in row $i \in N_1$ and column $j \in N_2$ is denoted a_{ij}. Vector $\mathbf{e}_i \in \mathbb{R}^n$ is the unit vector with a 1-entry in position i.

Henceforth, N is the set of assets in a benchmark index, and the index tracking portfolio is denoted $M \subseteq N$. Vectors $\mathbf{w}, \mathbf{x} \in [0, 1]^N$ consist of asset weights in the index and the portfolio, respectively, and m denotes the imposed portfolio size. Jansen and van Dijk [3] suggest the tracking error definition $f(\mathbf{x}) = (\mathbf{x} - \mathbf{w})^\mathsf{T} \mathbf{Q}(\mathbf{x} - \mathbf{w})$, where $\mathbf{Q} \in \mathbb{R}^{N \times N}$ is the estimated covariance matrix of the stock returns. Adopting this definition, the minimum tracking error corresponding to portfolio M is $z(M) = \min_{\mathbf{x} \in \mathbb{R}_+^N} \{f(\mathbf{x}) : \sum_{i \in M} x_i = 1, \mathbf{x}_{N \setminus M} = \mathbf{0}_{N \setminus M}\}$, and the INDEX TRACKING PROBLEM is defined as: $\min \{z(M) : M \subseteq N, |M| = m\}$.

For an overview of alternative error definitions, index tracking problems, and associated algorithms, readers are referred to recent literature [1, 2]. It has been shown [4] that the INDEX TRACKING PROBLEM is strongly NP-hard, and that mixed integer programming solvers are unsuitable for the problem. The construction and improvement heuristics developed in [4] are also time-consuming when the index has more than a thousand assets. In the sequel, we develop alternative methods, and prove their more favorable running times.

3 Construction Heuristic - Best Extension by One Asset

Consider a portfolio $M \subset N$ and an asset $k \in N \setminus M$, and the corresponding matrix $\mathbf{A}(M, k) \in \mathbb{R}^{M \times M}$ and vector $\mathbf{b}(M, k) \in \mathbb{R}^M$ defined as

$$\mathbf{A}(M, k) = \mathbf{Q}_{MM} - \mathbf{Q}_{Mk}\mathbf{1}_M^\mathsf{T} - \mathbf{1}_M \mathbf{Q}_{kM} + q_{kk}\mathbf{1}_M \mathbf{1}_M^\mathsf{T}, \text{ and} \quad (1)$$

$$\mathbf{b}(M, k) = q_{kk}\mathbf{1}_M + \mathbf{Q}_{MN}\mathbf{w} - \mathbf{1}_M \mathbf{Q}_{kN}\mathbf{w} - \mathbf{Q}_{Mk}. \quad (2)$$

It was recently proved [4] that

- $\mathbf{A}(M, k)$ is positive definite, implying that $\mathbf{x}_M = \mathbf{A}^{-1}(M, k)\mathbf{b}(M, k)$ exists.
- If $\mathbf{x}_M \geq \mathbf{0}$ and $\sum_{i \in M} x_i \leq 1$, then the optimal asset weights corresponding to portfolio $M + k$ are x_i ($i \in M$), $x_k = 1 - \sum_{i \in M} x_i$, and $x_i = 0$ ($i \in N \setminus M - k$).
- If $\mathbf{x} \geq \mathbf{0}$, then $z(M + k) = \mathbf{w}^\mathsf{T} \mathbf{Q} \mathbf{w} - 2\mathbf{w}^\mathsf{T} \mathbf{Q}_{Nk} + q_{kk} - \mathbf{b}^\mathsf{T}(M, k)\mathbf{x}_M$.

The construction heuristic in [4] keeps track of $\mathbf{A}(M, k)$ for all possible extensions k. Extension by asset k is then evaluated by solving an $|M| \times |M|$-system of linear equations. If $k^* \neq k$ is added to M, the system $\mathbf{A}(M + k^*, k)\mathbf{x}_{M+k^*} = \mathbf{b}(M + k^*, k)$ is solved in the next iteration. Whereas $\mathcal{O}(|M|^3)$ operations are needed to solve $\mathbf{A}(M, k)\mathbf{x}_M = \mathbf{b}(M, k)$ from scratch, only $\mathcal{O}(|M|^2)$ operations are needed if the QR-factorization $\mathbf{P}(M, k)\mathbf{R}(M, k)$ of $\mathbf{A}(M, k)$ is available. Updating $\mathbf{P}(M, k)$ and $\mathbf{R}(M, k)$ to $\mathbf{P}(M + k^*, k)$ and $\mathbf{R}(M + k^*, k)$, respectively, runs in $\mathcal{O}(|M|^2)$ time. Thus, Algorithm 1 runs in $\mathcal{O}\left(n^2 + nm^3\right)$ time ($n = |N|$), whereas a straightforward implementation needs $\mathcal{O}\left(n^2 + nm^4\right)$ time. The term n^2 stems from the computation of $\mathbf{w}^T\mathbf{Qw}$, needed in the calculation of $z(M+k)$.

Algorithm 1 Best extension by one

$M \leftarrow \emptyset, \mathbf{v} \leftarrow \mathbf{Qw}, \text{wQw} \leftarrow \mathbf{w}^T\mathbf{v}, \text{for } k \in N \text{ do } \mathbf{x}(k) \leftarrow \mathbf{e}_k$
for $t \leftarrow 1, \ldots, m$ **do**
    for $k \in N \setminus M$ **do** $z(M+k) \leftarrow \text{wQw} - 2v_k + q_{kk} - \sum_{i \in M} b_i(k) x_i(k)$
    Choose $k^* \in \arg\min \{z(M+k) : k \in N \setminus M, \mathbf{x}(k) \geq \mathbf{0}\}$
    $\mathbf{x} \leftarrow \mathbf{x}(k^*), z \leftarrow z(M+k^*), M \leftarrow M + k^*$
    for $k \in N \setminus M$ **do**
        $(\mathbf{P}(k), \mathbf{R}(k)) \leftarrow$ QR-decomposition of $\mathbf{A}(M, k)$
        $\mathbf{x}_M(k) \leftarrow \mathbf{R}^{-1}(k)\mathbf{P}^T(k)\mathbf{b}(k), x_k(k) \leftarrow 1 - \sum_{i \in M} x_i(k), \mathbf{x}_{N \setminus M}(k) \leftarrow \mathbf{0}$
return $(M, \mathbf{x}, z)$

---

4 Construction Heuristic - Fast Extension by One Asset

In each of m iterations of Algorithm 1, the most error-reducing extension by one asset is made. By sacrificing this quality, a faster method is achievable. Consider an $\ell \in N$ for which the tracking error $\mathbf{w}^T\mathbf{Qw} - 2\mathbf{Q}_{\ell N}\mathbf{w} + q_{\ell\ell}$ of single-asset portfolios is minimized. Asset ℓ, referred to as the *root*, is the first to be included in M.

Algorithm 2 Fast extension by one

Compute $\mathbf{v} \leftarrow \mathbf{Qw}$ and $\{c_{k0} \leftarrow \mathbf{w}^T\mathbf{v} - 2v_k + q_{kk} : k \in N\}$
Choose a root $\ell \in \arg\min_{k \in N} \{c_{k0}\}$, let $M \leftarrow \{\ell\}$ and $\mathbf{x} \leftarrow \mathbf{e}_\ell$
while $|M| < m$ **do**
    $\text{xQx} \leftarrow \mathbf{x}_M^T \mathbf{Q}_{MM} \mathbf{x}_M$
    for $k \in N \setminus M$ **do**
        $c_{k1} \leftarrow q_{kk} + \mathbf{v}_M^T \mathbf{x}_M - \mathbf{Q}_{kM}\mathbf{x}_M - v_k, c_{k2} \leftarrow \text{xQx} + q_{kk} - 2\mathbf{Q}_{kM}\mathbf{x}_M$
    Choose $k^* \in \arg\min_{k \in N \setminus M} \{c_{k0} - c_{k1}^2/c_{k2} : 0 \leq c_{k1} \leq c_{k2}\}, M \leftarrow M + k^*$
    repeat
        Update $\mathbf{P}$ and $\mathbf{R}$ such that $\mathbf{PR} = \mathbf{A}(M - \ell, \ell)$
        $\mathbf{x}_{M-\ell} \leftarrow \mathbf{R}^{-1}\mathbf{P}^T\mathbf{b}(M-\ell, \ell), x_\ell \leftarrow 1 - \sum_{i \in M-\ell} x_i, \mathbf{x}_{N \setminus M} \leftarrow \mathbf{0}$
        $M \leftarrow M \setminus \{i \in M : x_i < 0\}, \text{if } \ell \notin M \text{ then choose a new root } \ell \in M$
    until $\mathbf{x} \geq \mathbf{0}$
return $\left(M, \mathbf{x}, \mathbf{x}_M^T \mathbf{Q}_{MM} \mathbf{x}_M + \mathbf{w}^T\mathbf{v} - 2\mathbf{x}_M^T\mathbf{v}_M\right)$

Algorithm 2 evaluates a currently excluded asset $k \in N \setminus M$ by computing the minimum of f on the straight line segment between the current solution $\mathbf{x}$, where $\mathbf{A}(M-\ell, \ell)\mathbf{x}_{M-\ell} = \mathbf{b}(M-\ell, \ell)$, and $\mathbf{e}_k$. Only vectors where the relative proportions $x_j/\sum_{i \in M} x_i$ remain unchanged are considered. At $\lambda \in [0, 1]$, we have $f(\lambda\mathbf{x} + (1-\lambda)\mathbf{e}_k) = c_{k0} - 2c_{k1}(\mathbf{x})\lambda + c_{k2}(\mathbf{x})\lambda^2$, where $c_{k0} = (\mathbf{e}_k - \mathbf{w})^T \mathbf{Q}(\mathbf{e}_k - \mathbf{w}), c_{k1}(\mathbf{x}) = (\mathbf{e}_k - \mathbf{w})^T \mathbf{Q}(\mathbf{e}_k - \mathbf{x})$, and $c_{k2}(\mathbf{x}) = (\mathbf{e}_k - \mathbf{x})^T \mathbf{Q}(\mathbf{e}_k - \mathbf{x})$. The minimum value $c_{k0} - c_{k1}^2(\mathbf{x})/c_{k2}(\mathbf{x})$ of f is obtained for $\lambda = c_{k1}(\mathbf{x})/c_{k2}(\mathbf{x})$. When x has negative components, Algorithm 2 backs up by removing all corresponding assets from M. If $x_\ell < 0$, a new root in M is picked arbitrarily.

In the case of no removals, Algorithm 2 requires $\mathcal{O}\left(n^2 + nm^2\right)$ operations, $\mathcal{O}\left(n^2\right)$ for the computation of $\mathbf{v}$ and $\{c_{k0} : k \in N\}$, and $\mathcal{O}(n|M|)$ for the updates of $\{c_{k1}, c_{k2} : k \in N\}$. This dominates the updating of **PR** and $\mathbf{x}$ ($\mathcal{O}(|M|^2)$). For each removal, it takes $\mathcal{O}(|M|^2)$ operations to update **PR** and $\mathbf{x}$. Root removals require computation of **PR** from scratch, which amounts to $\mathcal{O}(|M|^3)$ operations.

5 Local Search - Best Exchange by One

In portfolio M, Algorithm 3 replaces an asset $j^* \in M$ by a $k^* \in N \setminus M$ until no reduction in z is achieved, while maximizing the reduction in each iteration. It is shown [4] that each iteration runs in $\mathcal{O}(n|M|^3)$ time.

Algorithm 3 Best exchange by one $(M, \mathbf{x}, z)$

$\mathbf{v} \leftarrow \mathbf{Qw}, \text{wQw} \leftarrow \mathbf{w}^T \mathbf{v}$
repeat
  done←true
  for $(j, k) \in M \times (N \setminus M)$ **do**
    $\mathbf{y}_{M-j}(j, k) \leftarrow \mathbf{A}^{-1}(M-j, k)\mathbf{b}(M-j, k)$
    $y_k(j, k) \leftarrow 1 - \sum_{i \in M-j} y_i(j, k), \mathbf{y}_{N \setminus M + j - k}(j, k) \leftarrow \mathbf{0}$
    $z(j, k) \leftarrow \text{wQw} - 2v_k + q_{kk} - \mathbf{b}^T(M, k)\mathbf{y}_{M-j}(j, k)$
  Choose $(j^*, k^*) \in \arg\min_{j,k} \{z(j, k) : \mathbf{y}(j, k) \geq 0\}$
  if $z(j^*, k^*) < z$ **then**
    $M \leftarrow M + k^* - j^*, \mathbf{x} \leftarrow \mathbf{y}(j^*, k^*), z \leftarrow z(j^*, k^*)$, done←false
until done
return $(M, \mathbf{x}, z)$

---

6 Local Search - Fast Exchange by One

Faster asset swapping is accomplished by assuming no change in relative weights among remaining assets. Letting $\mathbf{y}(j) = (1 - x_j)^{-1} (\mathbf{x} - x_j \mathbf{e}_j)$, this corresponds to minimizing $f(\lambda \mathbf{y}(j) + (1 - \lambda)\mathbf{e}_k)$ with respect to $\lambda \in [0, 1]$ and $(j, k) \in M \times (N \setminus M)$. From Sect. 4, we get $f(\lambda \mathbf{y}(j) + (1 - \lambda)\mathbf{e}_k) = c_{k0} - c_{k1}^2(\mathbf{y}(j))/c_{k2}(\mathbf{y}(j))$ for $\lambda = c_{k1}(\mathbf{y}(j))/c_{k2}(\mathbf{y}(j))$. In Algorithm 4, (j^*, k^*) is consequently chosen as a minimizer of $c_{k0} - c_{k1}^2(\mathbf{y}(j))/c_{k2}(\mathbf{y}(j))$. Portfolio $M + k^* - j^*$ is accepted if its tracking error is smaller than the one of M, otherwise the search terminates.

Like Algorithms 2 and 4 stores the QR-factorization **PR** of only one matrix $\mathbf{A}(M, \ell)$. If $j^* = \ell$, a new root $\ell \in M$ with corresponding $\mathbf{A}(M, \ell)$ and $\mathbf{b}(M, \ell)$ are chosen, and $\mathbf{PR} = \mathbf{A}(M, \ell)$ is computed ($\mathcal{O}(|M|^3)$). Otherwise, only an update of **PR** is needed ($\mathcal{O}(|M|^2)$). If the weight vector **x** has negative entries, it is replaced by $\lambda \mathbf{y}(j^*) + (1 - \lambda)\mathbf{e}_{k^*} \geq \mathbf{0}$.

Algorithm 4 Fast exchange by one (M, **x**, z)

Compute $\mathbf{v} \leftarrow \mathbf{Qw}$, $\mathbf{wQw} \leftarrow \mathbf{w}^\mathsf{T}\mathbf{v}$, $\{c_{k0} = \mathbf{w}^\mathsf{T}\mathbf{v} - 2v_k + q_{kk} : k \in N\}$
Choose a root $\ell \in M$, and compute $\mathbf{A}(M, \ell)$ and $\mathbf{b}(M, \ell)$ using (1)–(2)
repeat
  done←true
  for $(j, k) \in M \times (N \setminus M)$ **do**
    $\mathbf{y}(j) \leftarrow (1 - x_j)^{-1}\left(\mathbf{x} - x_j\mathbf{e}_j\right), c_0[j, k] \leftarrow (\mathbf{e}_k - \mathbf{w})^\mathsf{T} \mathbf{Q}(\mathbf{e}_k - \mathbf{w})$
    $c_1[j, k] \leftarrow (\mathbf{e}_k - \mathbf{w})^\mathsf{T} \mathbf{Q}(\mathbf{e}_k - \mathbf{y}(j)), c_2[j, k] \leftarrow (\mathbf{e}_k - \mathbf{y}(j))^\mathsf{T} \mathbf{Q}(\mathbf{e}_k - \mathbf{y}(j))$
  Choose $(j^*, k^*) \in \arg\min_{j,k} \{c_0[j, k] - c_1^2[j, k]/c_2[j, k]\}$
  $\widehat{M} \leftarrow M + k^* - j^*, \lambda \leftarrow c_1[j^*, k^*]/c_2[j^*, k^*]$
  if $\ell = j^*$ **then** choose a new root $\ell \in M + k^* - j^*$
  Compute/update the QR-factorization $\mathbf{PR} = \mathbf{A}(\widehat{M}, \ell)$
  $\hat{\mathbf{x}}_{\widehat{M}-\ell} \leftarrow \mathbf{R}^{-1}\mathbf{P}^\mathsf{T}\mathbf{b}(\widehat{M}, \ell), \ \hat{x}_\ell = 1 - \sum_{j \in \widehat{M}-\ell} \hat{x}_j, \ \hat{\mathbf{x}}_{N \setminus \widehat{M}} \leftarrow \mathbf{0}$
  if $\hat{\mathbf{x}} \not\geq \mathbf{0}$ **then** $\hat{\mathbf{x}} \leftarrow \lambda\mathbf{y}(j^*) + (1 - \lambda)\mathbf{e}_{k^*}$
  if $f(\hat{\mathbf{x}}) \leftarrow \mathbf{wQw} + \hat{\mathbf{x}}_{\widehat{M}}^\mathsf{T} \mathbf{Q}_{\widehat{M}\widehat{M}} \hat{\mathbf{x}}_{\widehat{M}} - 2\mathbf{w}^\mathsf{T}\mathbf{Q}_{N\widehat{M}}\hat{\mathbf{x}}_{\widehat{M}} < z$ **then**
    $\mathbf{x} \leftarrow \hat{\mathbf{x}}, z \leftarrow f(\mathbf{x}), M \leftarrow \widehat{M}$, done←false
until done
return (M, **x**, z)

Once $\mathbf{Q}_{NM}\mathbf{x}_M$ ($\mathcal{O}(nm)$) and $\mathbf{v}^\mathsf{T}\mathbf{x}$ ($\mathcal{O}(n)$) are available, c_1 and c_2 are computed in constant time. In iterations where $j^* = \ell$, the calculations of **P** and **R** take $\mathcal{O}(m^3)$ floating point operations. When $j^* \neq \ell$, only $\mathcal{O}(m^2)$ operations are needed for updating the factorization. In the worst case ($j^* = \ell$), one iteration of Algorithm 4 hence takes $\mathcal{O}(nm + m^3)$ time. Under the assumption that $j^* = \ell$ occurs at a frequency $1/m$ or less, average running time of one iteration is $\mathcal{O}(nm)$.

7 Computational Experiments

Application of Algorithm 1 (Algorithm 2) to construct M, followed by Algorithm 3 (Algorithm 4) to improve it, is denoted Algorithm 1+3 (Algorithm 2+4). To assess the practical performance of the methods, we report results from experiments with Algorithms 1+3 and 2+4. All experiments are run on a computer with an Intel Core i5–2400 processor with a frequency of 3.10 GHz and 3.8 GB memory. All test instances, a subset of those in [4], correspond to real stock market indices. They are denoted S&P 500 (n = 482), S&P 600 (n = 569), Russell 1000 (n = 852), Russell 2000 (n =1301), and Russell 3000 (n = 1920). For $m \in \{20, 40, \ldots, 100\}$, Table 1 shows tracking errors (z) obtained and the CPU-times (cpu) in seconds spent by the two algorithms.

Observe from the table that the reduction in tracking error obtained by applying the slower Algorithm 1+3 is considerable. On the other hand, the amount of

Table 1 Performance of Algorithms 1+3 and 2+4 in instances S&P 500 and 600

m	S&P 500 z 1+3	2+4	cpu 1+3	2+4	S&P 600 z 1+3	2+4	cpu 1+3	2+4
20	0.3638	0.6740	2	0.0	0.0427	0.0595	2	0.0
40	0.0927	0.1533	35	0.0	0.0143	0.0193	19	0.0
60	0.0492	0.0676	64	0.0	0.0066	0.0100	78	0.1
80	0.0295	0.0353	284	0.1	0.0040	0.0062	310	0.1
100	0.0203	0.0247	548	0.1	0.0026	0.0043	828	0.1

m	Russell 1000 z 1+3	2+4	cpu 1+3	2+4	Russell 2000 z 1+3	2+4	cpu 1+3	2+4	Russell 3000 z 1+3	2+4	cpu 1+3	2+4
20	0.2464	0.8153	4	0.0	0.0674	0.3559	5	0.1	0.1779	0.9305	14	0.1
40	0.0683	0.1286	65	0.1	0.0238	0.1101	57	0.1	0.0524	0.2442	147	0.2
60	0.0349	0.0603	122	0.1	0.0127	0.0313	247	0.3	0.0267	0.0928	355	0.4
80	0.0203	0.0360	501	0.2	0.0086	0.0216	506	0.3	0.0169	0.0465	925	0.6
100	0.0140	0.0261	1411	0.3	0.0065	0.0128	1365	0.5	0.0124	0.0298	1381	0.9

computational work is correspondingly big. While the faster Algorithm 2+4 concludes in less than a second even in the largest instance with $m = 100$, the running time of Algorithm 1+3 is, at worst (Russell 3000, $m = 100$), 23 minutes. This is only partly explained by the running time analysis in Sects. 3, 4, 5 and 6. The more accurate improvement method (Algorithm 3) proves to identify, on average, about four times as many error reducing moves as does Algorithm 4. This reflects the risk of neglecting error-reducing moves when applying an inexact measure of the tracking error.

8 Conclusions

Fast construction and improvement methods for the problem of computing cardinality constrained portfolios with minimum tracking error are developed in the current work. It is reviewed how QR-factorization of a matrix, derived from the covariance matrix of stock returns, can be exploited in order to speed up the selection of assets. It is further demonstrated how construction and improvement moves can be made faster if the exact computation of the resulting tracking error is replaced by an approximate one. This is accomplished by assuming constant relative proportions of assets remaining in the portfolio. When applied to real-life stock market indices, computational experiments prove that the accelerated solution techniques run faster than the original versions. The reduced effort does however come at the cost of portfolios with significantly larger tracking errors.

References

1. Canakgoz, N. A., & Beasley, J. E. (2009). Mixed-integer programming approaches for index tracking and enhanced indexation Eur. *The Journal of the Operational Research, 196*(1), 384–399.
2. Guastaroba, G., & Speranza, M. G. (2012). Kernel search: An application to the index tracking problem. *European Journal of Operational Research, 217*(1), 54–68.
3. Jansen, R., & van Dijk, R. (2002). Optimal benchmark tracking with small portfolios. *The Journal of Portfolio Management, 28*(2), 33–39.
4. Mutunge, P., & Haugland, D. (2018). Minimizing the tracking error of cardinality constrained portfolios. *Computers and Operations Research, 90*, 33–41.

Part VIII
Game Theory and Experimental Economics

Non-acceptance of Losses—An Experimental Study on the Importance of the Sign of Final Outcomes in Ultimatum Bargaining

Thomas Neumann, Stephan Schosser and Bodo Vogt

1 Introduction

Consider the standard ultimatum game [1]: two players have to share a pie between them. One player, the proposer, suggests an allocation of the pie. The second player, the responder, decides whether to accept or reject this offer. In the event of the latter, both players receive nothing. If the responder accepts the proposer's offer, the allocation of the pie is paid. The game theoretic solution concept for an ultimatum game is that of a subgame perfect Nash equilibrium [2]. The responder's best choice is to accept any positive share, since it is (strictly) better than nothing. Foreseeing the responder's behavior, it is optimal for the proposer to offer the smallest possible share.

Typically, the ultimatum game is played in a gains domain; i.e., if both players reach an agreement, their payoffs increase in comparison to the initial situation. One very common finding of studies that have focused on ultimatum bargaining over gains is that the actual behavior of the players deviates from the subgame perfect equilibrium prediction (e.g., [3–6]). On average, proposers offer between 40 and 50% of the pie and that these offers are almost always accepted by the responders [3]. Another finding is that distributions that result in the responder receiving less than 40% of the pie are frequently rejected.

T. Neumann (✉) · S. Schosser · B. Vogt
Faculty of Economics and Management, Otto-von-Guericke-University Magdeburg,
Chair in Empirical Economics, Postbox 4120, 39016 Magdeburg, Germany
e-mail: t.neumann@ovgu.de

S. Schosser
e-mail: stephan.schosser@ovgu.de

B. Vogt
Faculty of Medicine, Otto-von-Guericke-University Magdeburg, Institute of Social Medicine
and Health Economics, Leipziger Str. 44, 39120 Magdeburg, Germany
e-mail: bodo.vogt@ovgu.de

The literature that specifically focuses on ultimatum bargaining over losses is sparse. Only a few studies have investigated players' bargaining behavior in an ultimatum game over losses. Corresponding to the findings of research that focused on games over gains, a high fraction of players in games over losses split the pie equally. In addition, Camerer et al. [7] reported that the average offer was similar in the game over losses as it was in the game over gains. However, Buchan et al. [5] found that offers and demands were higher in games over losses than in games over gains. Lusk and Hudson [8] found that players made more aggressive offers when bargaining over losses. In line with this finding, Camerer et al. [7] found that players more frequently rejected offers when bargaining over losses.

All the studies described above compare the results from ultimatum games over gains with those of games over losses. In this experimental study, we combined both aspects within the bargaining task. We introduced a game in which the players bargained over the distribution of a pie; however, the strategic advantages (implemented through different initial positions) of the players were different.

In contrast to existing work on ultimatum bargaining, we found that while the behavior of proposers does not change, responders focused on breaking even, i.e., they rejected offers which yield a negative payoff for themselves and accepted any higher offer. The wish to break-even was even stronger than the desire to reach an outcome close to the equal split. We argue that this result can help to better understand ultimatum game behavior under losses: Deviations from the subgame perfect equilibrium in ultimatum game behavior can be justified with other regarding preferences (e.g., [9, 10]). As we observed, the wish to break-even is a strong motive, too. Hence, the more aggressive behavior under losses, described in related work, might be driven by the desire to break-even.

2 Game Design

We found preliminary evidence of bargaining behavior based on two modified ultimatum games, as presented in Table 1. These ultimatum games differ from previous work as the payoffs for not reaching an agreement were negative for both players. In addition, we framed the games in a sense that proposers perceived all offers as a loss, while responders saw it as a gain.

Table 1 Ultimatum games over mixed outcomes

	Payoffs, if responder...			
	...**accepts** the proposer's offer		...**rejects** the proposer's offer	
	Proposer	Responder	Proposer	Responder
Game 1	$75 - x_P$	$-25 + x_P$	-25	-25
Game 2	$50 - x_P$	$-50 + x_P$	-50	-50

In both games, the players who acted as proposers were asked to distribute a pie of size $s = 100$ by choosing how much of the pie, x_p, they would give to the responder. Simultaneously, the responders specified the minimum share, x_R, they wanted to receive. Varying the initial positions of the players imposed different bargaining motives and goals. A piece of the pie $x_p > 0$ reduced the proposer's final payoff and, therefore, served as a loss. From the perspective of the responder, each offer of $x_p > 0$ increased his final payoff and, therefore, served as a gain. Both games represented bargaining sets, which we refer to as mixed outcomes.

According to prospect theory [11], players are loss averse, meaning that losses loom larger than gains. Therefore, one could expect the proposer to offer a lower share of the pie than he would in the standard ultimatum game. This expectation is supported by the consideration that the responder is also loss averse and, therefore, should have a strong interest in reducing his loss (his initial position).

On the other hand, offering an equal split might be covered by common fairness considerations within the players. From the perspective of the responder, the costs of not reaching an agreement are much lower than the costs of the proposer. Thus, it is a credible threat that the responder will reject unfair offers. Another possible assumption is that the responder will accept any offer that compensates him for his initial loss. If the proposer assumes this, he will offer the exact initial loss of the responder and the smallest possible amount in addition.

3 The Experiment and Procedure

We played the ultimatum games (Table 1) using the strategy vector method [12]. That is, each player made decisions as proposer and responder, and each player participated in two treatments: Game 1 and Game 2.

We ran the experiment in the MaXLab, the experimental laboratory of the University of Magdeburg. The participants consisted of students from various faculties of the university. We ran our experiment over two sessions with 25 participants in the first session and 26 participants in the second. Participants were recruited using hroot [13] and were randomly assigned to seats in the laboratory.

The participants were not permitted to communicate with each other at any point during the experiment. They did not receive any information about the identity of their matched partners. To ensure that no reciprocity occurred, the players also did not get to know the outcomes of the first treatment until the end of the experiment.

The participants played both roles, proposer and responder, in the two different ultimatum games; i.e., they made four decisions, of which one was randomly selected and paid off. Prior to the experiment, all participants received a show-up fee of 5.00 EUR and were told that this payoff could increase or decrease depending on the decisions they make during the experiment.

4 Results

In the remainder of this section, we will first focus on the behavior of the proposers and, thereafter, on the behavior of the responders. In both games, the proposer offered 45% of the pie on average (see Table 2), which was in line with previous findings in the literature.

In Game 1, offers below 25% were not made; i.e., all proposers compensated the responder's initial loss at a minimum. The vast majority of proposers offered an equal split of the pie in both games.

Since we used a within-subject deign, we were able to analyze the proposer's behavior during the two games in more detail. To analyze the difference in behavior, we used the formula $d_P = x_P^{G_1} - x_P^{G_2}$, where $x_P^{G_1}$ ($x_P^{G_2}$) was the offer of the proposer in Game 1 (Game 2).

As Table 3 shows, 34 players offered the same share of the pie in Game 1 as they did in Game 2, 8 (9) participants offered less (more) in the second game. The behavioral difference between both games was insignificant (Wilcoxon test, two-sided, $p = 0.7436$).

Let's now focus on the responders. We found that the responder's average demand was 33% in Game 1 and 40% in Game 2 (see Table 4). Only 8 participants demanded an equal split in Game 1 in comparison to 28 participants in Game 2.

Table 2 Comparison of proposers' offers in both games

No. of proposers who offered...	Game 1	Game 2	*Difference*
...more than 50%	0	0	*0*
...exactly 50% (equal split)	34	39	*5*
...less than 50%	17	12	*5*
Average Offer	0.4489	0.4527	*0.0038*
Standard Deviation	0.0858	0.1046	*0.0816*

Table 3 Classification of differences (d_P) in proposers' offers

Classification of Differences (d_P) in Proposers' Offers				
<–more–		equal	–less–>	
$d_P \leq -0.10$	$-0.10 < d_P < 0.00$	$d_P = 0.00$	$0.00 < d_P < 0.10$	$0.10 \leq d_P$
6	3	34	2	6
(−0.1583)	(−0.0500)	(0.0000)	(0.0150)	(0.1458)

Averages of differences in parenthesizes

Table 4 Comparison of responders' demands in both games

No. of responders who demanded...	Game 1	Game 2	*Difference*
...more than 50%	0	1	*1*
...exactly 50% (equal split)	8	28	*20*
...less than 50%	43	22	*21*
Average Offer	0.3323	0.4014	*0.0691*
Standard Deviation	0.1359	0.1588	*0.1257*

Table 5 Classification of differences (d_R) in responders' demands

Classification of Differences (d_R) in Responders' Demands

<–more–		equal	–less–>	
$d_R \leq -0.10$	$-0.10 < d_R < 0.00$	$d_R = 0.00$	$0.00 < d_R < 0.10$	$0.10 \leq d_R$
24	7	15	0	5
(−0.1725)	(−0.0443)	(0.0000))	(—)	(0.1850)

Averages of differences in parenthesizes

We analyzed the behavioral differences of the responders in more detail using the formula $d_R = x_R^{G_1} - x_R^{G_2}$, where $x_R^{G_1}$ ($x_R^{G_2}$) was the demand of the responder in Game 1 (Game 2).

As Table 5 shows, 15 participants demanded the same share of the pie in both games. While 5 participants demanded less in Game 2 than Game 1, 31 participants demanded more. The behavioral difference between both games was significant (Wilcoxon-test, two-sided, $z = 3.3999$, $p = 0.0007$).

5 Discussion and Conclusion

We presented the preliminary results of bargaining in a modified ultimatum game that incorporated asymmetric strategic advantages. While, for one player, the bargaining problem represented a distribution of a loss, the same problem represented a distribution of a gain for the counterparty.

In line with previous research findings, we found that the majority of proposers offered an equal split of the pie. The decision to offer an equal split may have been driven by two possible motives: (1) a common fairness consideration, and (2) the desire to compensate for the initial loss of the other player, which could be assumed to represent a minimum requirement.

Contrary to the findings related to the behavior of responders in an ultimatum game over losses, we found that responders accepted offers below 40% of the pie in Game 1. They need to be awarded a compensation for their initial loss, i.e., the responders' desire to break-even seemed to be stronger than their longing for fairness.

Of course, one could think of other motives and factors that could be at play in this scenario; e.g., changes in the players' beliefs of the other players behavior, differences in the players' risk attitudes, etc. Future research should investigate these factors in more depth.

Acknowledgements This research is funded by the DFG (German Science Foundation) under grant number BE 2373/4-1 as part of the research project "The fair division of losses".

References

1. Güth, W., Schmittberger, R., & Schwarze, B. (1982). An experimental analysis of ultimatum bargaining. *Journal of Economic Behavior and Organization, 3*(4), 367–388.
2. Selten, R. (1975). Reexamination of the perfectness concept for equilibrium points in extensive games. *International Journal of Game Theory, 4*(1), 25–55.
3. Güth, W., & Kocher, M. G. (2014). More than thirty years of ultimatum bargaining experiments: Motives, variations, and a survey of the recent literature. *Journal of Economic Behavior and Organization, 108*, 396–409.
4. Camerer, C. (2003). *Behavioral Game Theory: Experiments in Strategic Interaction.* Princeton University Press.
5. Buchan, N., Croson, R., Johnson, E., & Wu, G. (2005). Gain and loss ultimatums. *Experimental and Behavorial Economics* (pp. 1–23). Emerald Group Publishing Limited.
6. Neumann, T., Schosser, S., & Vogt, B. (2017). Ultimatum bargaining over losses and gains - an experimental comparison. *Social Science Research, 67*, 49–58.
7. Camerer, C. F., Johnson, E., Rymon, T., & Sen, S. (1993). Cognition and framing in sequential bargaining for gains and losses. *Frontiers of Game Theory* (pp. 27–47)
8. Lusk, J. L., & Hudson, M. D. (2010). Bargaining over losses. *International Game Theory Review, 12*(01), 83–91.
9. Fehr, E., & Schmidt, K. M. (1999). A theory of fairness, competition, and cooperation. *The Quarterly Journal of Economics, 114*(3), 817–868.
10. Bolton, G. E., & Ockenfels, A. (2000). Erc: A theory of equity, reciprocity, and competition. *The American Economic Review*, 166–193.
11. Kahneman, D., & Tversky, A. (1979). Prospect theory: An analysis of decision under risk. *Econometrica: Journal of the Econometric Society*, 263–291.
12. Rauhut, H., & Winter, F. (2010). A sociological perspective on measuring social norms by means of strategy method experiments. *Social Science Research, 39*(6), 1181–1194.
13. Bock, O., Baetge, I., & Nicklisch, A. (2014). hroot: Hamburg registration and organization online tool. *European Economic Review, 71*, 117–120.

Part IX
Graphs and Networks

A Graph Theoretic Approach to Solve Special Knapsack Problems in Polynomial Time

Carolin Rehs and Frank Gurski

1 Introduction

Let $G = (V, E)$ be a graph. An *independent set* of G is a subset V' of V such that there is no edge in G between two vertices from V'. A *maximum independent set* is an independent set of largest size. A *maximal independent set* is an independent set that is not a proper subset of any other independent set. A *clique* of G is a subset V' of V such that there is an edge in G between every two different vertices from V' and a *maximum clique* is clique set of largest size. As usual let $\alpha(G)$ be the size of a maximum independent set and $\omega(G)$ be the size of a maximum clique in G.

The family of all maximal independent sets of some graph G is denoted by $MIS(G)$ and its cardinality is denoted by $mis(G)$. Enumerating and counting maximal independent sets in graphs is an often studied problem in the field of special graph classes, see [6, 10]. We will solve these problems for k-threshold graphs and apply the results to knapsack problems.

An extended version of this paper can be found in [3].

C. Rehs (✉) · F. Gurski
University of Düsseldorf, Institute of Computer Science,
Algorithmics for Hard Problems Group, 40225 Düsseldorf, Germany
e-mail: carolin.rehs@hhu.de

F. Gurski
e-mail: frank.gurski@hhu.de

2 Knapsack Problem and Threshold Graphs

Name: MAX KNAPSACK (MAX KP)
Instance: A set $A = \{a_1, \ldots, a_n\}$ of n items, where for every a_j there is a size s_j and a profit p_j. Further there is a capacity c of the knapsack. The parameters n, p_j, s_j, and c are assumed to be positive integers.
Task: Find some subset $A' \subseteq A$, such that the total profit of X is maximized and the total size of A' is at most c.

Let I be an instance for MAX KP. Every subset A' of A such that $\sum_{a_j \in A'} s_j \leq c$ is a *feasible solution* of I. A feasible solution which is not the subset of another feasible solution is called *maximal*. An instance I for MAX KP and a graph $G = (V, E)$ are *equivalent*, if there is a bijection $f : A \to V$ such that $A' \subseteq A$ is a feasible solution of I if and only if $\{f(a_j) \mid a_j \in A'\}$ is an independent set of G.

In general not every instance for MAX KP has an equivalent graph and not every graph has an equivalent instance for MAX KP. This situation changes if we restrict to threshold graphs.

2.1 Threshold Graphs

Threshold graphs were introduced by Chvátal and Hammer in the 1970s [2] as a graph class which allows to distinguish between independent and non-independent sets in a very simple way. Formally $G = (V, E)$ is a *threshold graph* if there exist non-negative integers w_v, $v \in V$, and T such that for every $U \subseteq V$ it holds $\sum_{v \in U} w_v \leq T$ if and only if U is an independent set of G.

Threshold graphs have many known characterizations, see [7, Theorem 1.2.4]. We next mention two of them. A graph G is a threshold graph if and only if G contains no cycle C_4, no path P_4, and no matching $2K_2$ as induced subgraph. Further a graph $G = (V, E)$ with $V = \{v_1, \ldots, v_n\}$ is a threshold graph if and only if it can be constructed from the one-vertex graph by repeatedly adding an isolated vertex or a dominating vertex. A *creation sequence* for G (cf. [4]) is a binary string $t_1, \ldots, t_n$ of length n such that there is a bijection $v : \{1, \ldots, n\} \to V$ with $t_i = 1$ if $v(i)$ is a dominating vertex for the graph induced by $\{v(1), \ldots, v(i)\}$ and $t_i = 0$ if $v(i)$ is an isolated vertex for the graph induced by $\{v(1), \ldots, v(i)\}$. W.l.o.g. we define $t_1 = 1$. Using the linear time recognition algorithm of [7, Figure 1.4] a creation sequence can be found in linear time for some given threshold graph. Further there is linear time recognition algorithm which also gives a forbidden induced subgraph from $\{2K_2, P_4, C_4\}$ if the input is not a threshold graph [5].

By definition every threshold graph has an equivalent instance of MAX KP. It even holds that for every knapsack instance with equivalent graph, this graph is threshold [8]. The maximal independent sets in threshold graphs represent maximal feasible solutions of corresponding knapsack instances. Every optimal solution is among one of these sets. Therefore we show how to count and enumerate these sets.

```
A = ∅; MIS(G) = ∅;
for( i = n, ..., 1 )
    if (t_i = 1)
        MIS(G) = MIS(G) ∪ {A ∪ {v(i)}};
    else                ▷ t_i = 0
        A = A ∪ {v(i)};
```

Fig. 1 Enumerating all maximal independent sets in a threshold graph

2.2 Maximal Independent Sets in Threshold Graphs

We give a method to enumerate all maximal independent sets in a threshold graph.

Theorem 1 *For every threshold graph G on n vertices which is given by a creation sequence MIS(G) can be enumerated in time $\mathcal{O}(\omega(G) \cdot n)$ and it holds that $mis(G) = \omega(G)$.*

Proof Let G be a threshold graph and $t = t_1 \ldots t_n$ be a creating sequence for G. By the method given in Fig. 1 we generate all maximal independent sets in G. Since $\{v(i) \mid t_i = 1\}$ leads a maximum clique, it holds that $mis(G) = \omega(G)$. □

A creation sequence can be found in linear time by [7, Figure 1.4].

Corollary 1 *For every threshold graph G on n vertices and m edges MIS(G) can be enumerated in time $\mathcal{O}(\omega(G) \cdot n + m)$ and it holds that $mis(G) = \omega(G)$.*

Theorem 2 *Let I be an instance for* MAX KP *on n items which has an equivalent graph. Then I can be solved in time $\mathcal{O}(n^2)$.*

Proof Let I be some instance for MAX KP on n items which has an equivalent graph. Then I is equivalent (proof in [3]). to graph $G(I) = (V(I), E(I))$ with $V(I) = \{v_j \mid a_j \in A\}$ and $E(I) = \{\{v_j, v_{j'}\} \mid s_j + s_{j'} > c\}$, which can be constructed from I in time $\mathcal{O}(n^2)$. Graph $G(I)$ is a threshold graph.

Thus the $\omega(G) \le n$ maximal independent sets in $G(I)$ can be found in $\mathcal{O}(n^2)$ by Corollary 1 and correspond to the maximal feasible solutions of I. For every of these solutions we can compute its profit in time $\mathcal{O}(n)$. □

3 Multidimensional Knapsack Problem and k-Threshold Graphs

Name: MAX d- DIMENSIONAL KNAPSACK (MAX d- KP)
Instance: Given is a set $A = \{a_1, \ldots, a_n\}$ of n items and a number d of dimensions. Every item a_j has a profit p_j and for dimension i the size $s_{i,j}$. Further there is a capacity c_i for every dimension i. The parameters $n, d, p_j, s_{i,j}$, and c_i are assumed to be positive integers.

Task: Find some subset $A' \subseteq A$, such that the total profit of A' is maximized and for every dimension i the total size of A' is at most the capacity c_i.

Let I be an instance for MAX D- KP. Every subset A' of A such that $\sum_{a_j \in A'} s_{i,j} \leq c_i$ for every $i \in [d]$ is a *feasible solution* of I. A feasible solution which is not the subset of another feasible solution is called *maximal*. An instance I for MAX D- KP and a graph $G = (V, E)$ are *equivalent*, if there is a bijection $f : A \to V$ such that $A' \subseteq A$ is a feasible solution of I if and only if $\{f(a_j) \mid a_j \in A'\}$ is an independent set of G.

For some MAX D- KP instance I and $i \in [d]$ we define by I_i the MAX KP instance on the same item set A with profits $p_1, \ldots, p_n$, sizes $s_{i,1}, \ldots, s_{i,n}$, and capacity c_i.

3.1 k-Threshold Graphs

A graph $G = (V, E)$ on n vertices is a *k-threshold graph*, if there are at most k linear inequalities

$$a_{j,1}x_1 + \cdots + a_{j,n}x_n \leq T_j \tag{1}$$

such that $X \subseteq V$ is an independent set in G if and only if the characteristic vector $(x_1, \ldots, x_n)$ of X satisfies for $j = 1, \ldots, k$ the inequalities of type (1). Equivalently, a graph $G = (V, E)$ has shown to be a k-threshold graph if there are at most k threshold graphs $G_i = (V, E_i)$, $1 \leq i \leq k$, such that $E = E_1 \cup \ldots \cup E_k$, see [2].

Thus 1-threshold graphs correspond to threshold graphs and can be recognized in linear time. The set of 2-threshold graphs can be recognized in polynomial time [9]. For every fixed $k \geq 3$ it is NP-complete to determine whether a given graph is a k-threshold graph, see [11].

3.2 Maximal Independent Sets in k-Threshold Graphs

Next we show how to enumerate and count the maximal independent sets in a k-threshold graph G using these sets for k covering threshold graphs of G.

Theorem 3 *For every k-threshold graph G on n vertices and m edges whose edge set can be covered by k threshold graphs $G_i = (V, E_i)$, $1 \leq i \leq k$, MIS(G) can be enumerated in time $\mathcal{O}(\sum_{i=1}^{k} m_i + n \cdot (\Pi_{i=1}^{k} \omega(G_i))^2) \subseteq \mathcal{O}(n^{2k+1})$ and it holds that $mis(G) \leq \Pi_{i=1}^{k} \omega(G_i)$.*

Proof Let G be a k-threshold graph on n vertices and m edges. Further let $G_i = (V, E_i)$, $1 \leq i \leq k$, be a covering by k threshold graphs for G and m_i denote the number of edges in G_i. By the method given in Fig. 2 we generate all maximal independent sets in G.

```
MIS(G) = ∅; 𝒥 = ∅;
for (i = 1,...,k)
    compute MIS(G_i)                    ▷ see Fig. 1
    for each (M_1,...,M_k) ∈ MIS(G_1) × ... × MIS(G_k)
        𝒥 = 𝒥 ∪ {M_1 ∩ ... ∩ M_k}       ▷ compute 𝒥 = {M_1 ∩ ... ∩ M_k | M_i ∈ MIS(G_i)}
    for each (I,J) ∈ 𝒥 × 𝒥              ▷ remove all sets which are subsets
        if (I is a subset of J)
            𝒥 = 𝒥 - I
MIS(G) = 𝒥;
```

Fig. 2 Enumerating all maximal independent sets in a k-threshold graph

The running time for computing $MIS(G_i)$ for $1 \leq i \leq k$ can be bounded using Corollary 1 by $\mathcal{O}(\sum_{i=1}^{k} \omega(G_i) \cdot n + m_i) = \mathcal{O}(n \cdot \sum_{i=1}^{k} \omega(G_i) + \sum_{i=1}^{k} m_i)$.

For every tuple $(M_1, \ldots, M_k)$ the intersection $M_1 \cap \ldots \cap M_k$ can be computed in time $\mathcal{O}(k \cdot n)$ and the set of all intersections $\mathcal{J}$ can be computed in time $\mathcal{O}(k \cdot n \cdot (\Pi_{i=1}^{k} \omega(G_i)))$.

Then we have to eliminate non-maximal subsets in $\mathcal{J}$, which takes time $\mathcal{O}(n \cdot (\Pi_{i=1}^{k} \omega(G_i))^2)$.

By assuming $\omega(G_i) \geq 2$ for $1 \leq i \leq k$ the overall running time is in

$$\mathcal{O}(\sum_{i=1}^{k} m_i + n \cdot (\Pi_{i=1}^{k} \omega(G_i))^2) \subseteq \mathcal{O}(n^{2k+1}).$$

The correctness holds as follows. Every independent set S in G is also an independent set in graph G_i for every $1 \leq i \leq k$. Thus every independent set S in G is a subset of some *maximal* independent set M_i in graph G_i for every $1 \leq i \leq k$. Thus every independent set S in G is a subset of the intersection $M_1 \cap \ldots \cap M_k$ for some maximal independent sets M_i in graph G_i for every $1 \leq i \leq k$. Further since every such intersection $M_1 \cap \ldots \cap M_k$ is an independent set in G and we remove the non-maximal independent sets from the set of all these intersections in the last step of our method, we create exactly the set of all maximal independent sets of G. □

Theorem 4 *Let I be an instance for* MAX D-KP *on n items such that for every dimension $i \in [d]$ instance I_i has an equivalent graph. Then I can be solved in time* $\mathcal{O}(n^{2d+1})$.

Proof Let I be some instance for MAX D-KP on n items such that for every dimension $i \in [d]$ instance I_i for MAX KP has an equivalent graph G_i. For every dimension $i \in [d]$ instance I_i is equivalent [3] to graph $G(I_i) = (V(I_i), E(I_i))$ which was defined in the proof of Theorem 2 and which can be constructed in time $\mathcal{O}(n^2)$ from I_i. Further graph $G = (V, E)$ where $V = V(I_1)$ and $E = E(I_1) \cup \ldots \cup E(I_d)$ leads an equivalent graph for instance I. Graph G is a d-threshold graph.

Thus the $\Pi_{i=1}^{d} \omega(G_i) \leq n^d$ maximal independent sets in G can be found in time $\mathcal{O}(n^{2d+1})$ by Theorem 3 and correspond to the maximal feasible solutions of I. For every of these solutions we can compute its profit in time $\mathcal{O}(n)$. □

The theorem is even true for instances I of MAX D- KP having an equivalent graph, it is not necessary that there is an equivalent graph for each dimension $i \in [d]$. The much longer proof of this theorem can be found in our paper [3].

4 Conclusions

The presented methods allow us to solve special instances of the NP-hard knapsack problem in polynomial time using a method to list all maximal independent sets in a corresponding threshold graph in polynomial time. This results extend an approach to solve special instances for MAX D- KP suggested by Chvátal and Hammer in [2].

Since every maximum independent set is a maximal independent set, our results also can be applied to list all maximum independent sets in threshold graphs. By the method given in Fig. 1 and removing non-maximum sets we obtain a method of running time $\mathcal{O}(\omega(G) \cdot n)$ for listing all maximum independent sets in a threshold graph G. We can compute $\alpha(G)$ in the same time. The related problem of finding *one* maximum independent set in a threshold graph was solved in [1] in time $\mathcal{O}(n \log n)$. This problem can also be solved in $\mathcal{O}(n)$ along a given creation sequence.

Further by omitting the last step of the method given in Fig. 2 and removing non-maximum sets we obtain a method of running time $\mathcal{O}(\sum_{i=1}^{k} m_i + k \cdot n \cdot (\Pi_{i=1}^{k} \omega(G_i))) \subseteq \mathcal{O}(k \cdot n^{k+1})$ for listing all maximum independent sets in a k-threshold graph G. We can compute $\alpha(G)$ in the same time. The related problem of finding *one* maximum independent set in a k-threshold graph was solved in [1] in time $\mathcal{O}(n \log n + n^{k-1})$.

Comparing our solutions and these of [1] we observe that we require graph representations and the authors of [1] use the coefficients occurring in knapsack instances. Each of these versions can transformed into the other. Especially when we can bound the vertex degree of the threshold graphs our results are much better.

References

1. Caprara, A., Lodi, A., & Rizzi, R. (2004). On d-threshold graphs and d-dimensional bin packing. *Networks, 44*(4), 266–280.
2. Chvátal, V., & Hammer, P. (1977). Aggregation of inequalities in integer programming. *Annals of Discrete Mathematics, 1*, 145–162.
3. Gurski, F., & Rehs, C. (2017). Counting and Enumerating Independent Sets with Applications to Knapsack Problems. arXiv: 1710.08953
4. Hagberg, A., Swart, P., & Schult, D. (2006). Designing threshold networks with given structural and dynamical properties. *Physical Review E, 056116*, 00.
5. Heggernes, P., & Kratsch, D. (2007). Linear-time certifying recognition algorithms and forbidden induced subgraphs. *Nordic Journal of Computing, 14*(1–2), 87–108.
6. Leung, J. Y. T. (1984). Fast algorithms for generating all maximal independent sets of interval, circular-arc and chordal graphs. *Journal of Algorithms, 5*(1), 22–35.

7. Mahadev, N., & Peled, U. (1995). Threshold Graphs and Related Topics. *Annals of Discrete Mathematics, 56*. Elsevier, North-Holland.
8. Robinson, T. (1997). Knapsack graphs. *New Zealand Journal of Mathematics, 26*, 107–123.
9. Sterbini, A., & Raschle, T. (1998). An $O(n^3)$ time algorithm for recognizing threshold dimension 2 graphs. *Information Processing Letters, 67*(5), 255–259.
10. Vadhan, S. (2001). The complexity of counting in sparse, regular, and planar graphs. *SIAM Journal on Computing, 31*(2), 398–427.
11. Yannakakis, M. (1982). The complexity of the partial order dimension problem. *SIAM Journal on Algebraic Discrete Methods, 3*(3), 351–358.

Bootstrap Percolation on Degenerate Graphs

Marinus Gottschau

1 Introduction

An r-neighbor bootstrap percolation process on some given graph $G = (V, E)$ with vertex set V and edge set E is a discrete time infection process. Initially at time zero there is some set of infected vertices. Then, at every time step, every vertex that has at least r infected neighbors becomes infected in the next time step, too. The process was first introduced by Chalupa, Leath and Reich in 1979 in [1] and is a simple example for a cellular automaton. It is also closely related to the Glauber dynamics which represent the Ising model at zero-temperature (see [2]). Another application one can think of is rumor spreading in a social network, where individuals start spreading a rumor to all their friends once they have heard that rumor from a number of other friends. Instead of speaking of infection, the literature also uses the term activation but we shall stick to the term infection and write bootstrap percolation or simply process instead of r-neighbor bootstrap percolation process. We now shortly introduce the process formally before giving some known results. Call the set of initially infected vertices A_0 and the vertices that are infected at the end of the process A_f. More formally, let A_t be the vertices which are infected at time t, where $A_t := A_{t-1} \cup \{v \in V : |N(v) \cap A_{t-1}| \geq r\}$ and $N(v)$ denotes the neighborhood of some vertex $v \in V$ and thus $A_f = \bigcup_{t>0} A_t$. Usually, the set A_0 is a random set of vertices, where each vertex is initially infected independently with a given probability p, or some other reasonable probabilistic choice function is used. The dynamics of bootstrap percolation shall be depicted on an example graph in the following figure (Fig. 1).

For several graph classes there is already much known about the behavior of this process and there are a quite few things of interest. First of all one can study the probability of percolation, i.e. the probability that $A_f = V$, depending on p, which

M. Gottschau (✉)
TUM School of Management and Department of Mathematics,
Technical University of Munich, Arcisstraße 21, 80333 Munich, Germany
e-mail: marinus.gottschau@tum.de

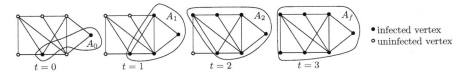

Fig. 1 Dynamics of the 2-neighbor bootstrap percolation process on an example graph where $|A_0| = 2$ and $A_f = V$

is the probability with which each vertex independently is initially infected. Several authors surveyed the threshold for the infection probability at which percolation is more likely to occur than not. For example, if the underlying graph is the d-dimensional cube graph $[n]^d$, the exact threshold function for $d = r = 2$ was shown by Holroyd in [3] to be $\frac{\pi^2}{18\log n} + o(\frac{1}{\log n})$. Later, Balogh et al. gave the exact threshold function for all $d \geq r \geq 2$ in [4].

Additionally, bootstrap percolation was studied on trees, like periodic trees in [5] and Galton–Watson trees in [6] and [7].

Some papers also study the size of sets of vertices that infect the whole graph, so called percolating sets. There is also a special class of percolating sets, namely minimal percolating sets, which are sets of vertices where any proper subset does not infect the whole graph. Riedl showed in [8] that for a tree on n vertices with l vertices of degree less than r, a minimal percolating set A_0 is of size $\frac{(r-1)n+1}{r} \leq |A_0| \leq \frac{rn+l}{r+1}$. Riedl also gave an algorithm in his paper that computes the size of a minimum percolating set for $r \geq 2$ as well as the size of a maximum minimal percolating set for $r = 2$ in linear time. In another paper he extended the studies of minimal percolating sets to hypercubes under 2-bootstrap percolation (see [9]).

Also, other graphs like the well known Erdős-Rényi random graph have been studied, for example in [10] by Janson et al. There, the authors give a function for the edge probabilities when percolation occurs with high probability, depending on the size of A_0.

Another quite interesting parameter is the running time of such a process, which is the time until no new vertex becomes infected, i.e. the least t such that $A_t = A_{t+1} = A_f$. This parameter has been studied for several graphs like the grid $[n]^2$, where Benevides and Przykucki [11] showed that for $r = 2$ the running time is bounded by $\frac{13}{18}n^2 + \mathcal{O}(n)$. In [12], Przykucki considered bootstrap percolation on the d-dimensional hypercube and proved the time to be at most $\lfloor \frac{d^2}{3} \rfloor$ again for $r = 2$. Bollobás et al. [13] analyzed the time of bootstrap percolation on the discrete torus while Janson et al. [10] gave a time bound for bootstrap percolation on the Erdős-Rényi random graph.

In this paper we focus on the size of the infected set A_f at the end of the process on degenerate graphs. We give a result for r-neighbor bootstrap percolation when the underlying graph is a degenerate graph and the set A_0 is any subset of the vertices. Furthermore, our theorem has implications on the size of minimum percolating sets as well as the running time of the process on degenerate graphs. Finally, it is noteworthy that, due to bounded degeneracy of several known graph classes, like trees or planar graphs, our results also covers more commonly studied graph classes.

2 Bootstrap Percolation on Degenerate Graphs

Our result gives a bound on the size of the set A_f of vertices that are infected at the end of the process on degenerate graphs, so let us define degeneracy first.

Definition 1 A finite graph $G = (V, E)$ is called d-degenerate if every subgraph contains a vertex of degree at most d.

There are many graph classes that have a bounded degeneracy. For example forests are 1-degenerate graphs, planar graphs are 5-degenerate while outerplanar graphs are 2-degenerate. Also, scale free networks generated by the Barabási-Albert model using a preferential attachment mechanism, have bounded degeneracy.

The definition of degeneracy has a useful equivalence as stated in the following lemma, which we are going to make use of later.

Lemma 1 *A graph $G = (V, E)$ is d-degenerate if and only if it has an ordering of the vertices on a line such that each vertex has at most d neighbors to its left, which is formally there exists a permutation $\pi : V \to V$ such that $|\{w \in N(v) : \pi(w) < \pi(v)\}| \leq d$ for all $v \in V$.*

The proof of this lemma is straightforward and left as an exercise to the reader.

Let us now state our main theorem.

Theorem 1 *Consider bootstrap percolation on a d-degenerate graph G with $r \geq d + 1$ and a set of initially infected vertices A_0. Then the set A_f of vertices that are infected at the end of the process fulfills*

$$|A_0| \leq |A_f| \leq \left(1 + \frac{d}{r-d}\right)|A_0|.$$

Proof We introduce a potential Ψ_i, which, after each timestep i of the infection process, bounds the number of vertices that might be infected in the next step from above. Due to Lemma 1 there exists an ordering $\pi : V \to V$ of the vertices, where each vertex has at most $d < r$ neighbors to its left, i.e. $|\{w \in N(v) : \pi(w) < \pi(v)\}| \leq d$. Since $d < r$, every vertex that becomes infected at some point must have at least one already infected vertex to its right. Now let $\Psi_i = \sum_{v \in V} \Psi_i^v$, where

$$\Psi_i^v := \begin{cases} 0, & \text{if } v \notin A_i \\ |\{w \in N(v) : \pi(w) < \pi(v)\} \setminus A_i|, & \text{else} \end{cases}$$

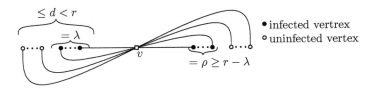

Fig. 2 Left and right neighborhood of v at the time of its infection

is the number of uninfected vertices in the left neighborhood of vertex v if it is infected and is 0 otherwise. Note that an uninfected vertex might be counted more than once in our potential if it has more than one infected vertex in its right neighborhood. Clearly, $\Psi_0 \leq d|A_0|$ since every initially infected vertex has at most d uninfected neighbors to its left.

Claim The potential decreases for each infection at time i by at least $r - d$, i.e. we have that $\Psi_{i-1} - \Psi_i \geq (r - d)|A_i \setminus A_{i-1}|$.

Proof (Proof of claim) Consider the set of vertices infected at time i and let vertex v be such a vertex, i.e. $v \in A_i \setminus A_{i-1}$. Due to the degeneracy of the graph, v has at most d uninfected neighbors in its left neighborhood and therefore v can increase the potential by at most d. Observe next that for vertex v to become infected it must have at least r infected neighbors. Now there are two kinds of such infecting vertices, namely those that lie in the right neighborhood of v and those that lie in the left neighborhood of v as depicted in Fig. 2.

Let λ be the number of infected vertices in the left neighborhood and ρ be the number of infected vertices in the right neighborhood. For each infected vertex w that lies in the right neighborhood, the potential decreases by one, since vertex v was accounted for in Ψ_{i-1}^w but is no longer uninfected and therefore does not contribute to Ψ_i^w. Each of the λ infected vertices in the left neighborhood of v does not add to the potential either, since it is already infected, so $\Psi_i^v \leq (d - \lambda) + \Psi_{i-1}^v$. If v is the only vertex infected at time i, we have that

$$\Psi_i \leq (d - \lambda) - \rho + \Psi_{i-1} \leq (d - \lambda) - (r - \lambda) + \Psi_{i-1} \leq \Psi_{i-1} - (r - d),$$

which, since $r > d$, means that the potential decreases by $r - d$. Finally note that additional, simultaneous infections do not increase Ψ_i^v and the above argumentation can independently be done for each vertex that is infected at time i, and thus each infection decreases the potential by the aforementioned amount and the claim follows.

Once the potential is less than 1 there does not exist an uninfected vertex that has an infected vertex in its right neighborhood and hence could be infected next, as $r > d$, and the process stops. Using the claim and the fact that $\Psi_0 \leq d|A_0|$ we get

$$|A_f| \leq |A_0| + \frac{\Psi_0}{r - d} \leq |A_0| + \frac{d|A_0|}{r - d} = \left(1 + \frac{d}{r - d}\right)|A_0|.$$

It is obvious that $|A_0| \leq |A_f|$ which finishes the proof of Theorem 1.

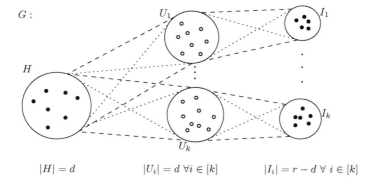

Fig. 3 Schematic picture of the constructed graphs that get arbitrarily close to the given upper bound

It is also remarkable that our bound on the size of A_f is sharp in a sense: For every $d \geq 1$, $r \geq 2$ and $\varepsilon > 0$ there exists a d-degenerate graph with an initially infected set A_0 such that the set A_f at the end of the process fulfills $|A_f| \geq (1 - \varepsilon)\left(1 + \frac{d}{r-d}\right)|A_0|$. For the construction of such a graph take a set H of d vertices. Next, take k pairs (U_i, I_i) of sets of vertices with $|U_i| = d$ and $|I_i| = r - d$ where every vertex in I_i is adjacent to all vertices in U_i for all $i \in [k]$. Furthermore, every vertex in H is adjacent to all vertices in $\bigcup_{i \in [k]} U_i$. Finally, choose $A_0 = H \cup (\bigcup_{i \in [k]} I_i)$. It is easy to see that this graph is a d-degenerate graph by construction. A schematic picture of the graph can be seen in Fig. 3.

The set A_0 is of size $d + k(r - d)$ while the set A_f, since every vertex will be infected at the end, is of size $d + k(r - d) + kd = d + kr$. Therefore

$$\frac{|A_f| - |A_0|}{|A_0|} = \frac{d + kr - (d + k(r-d))}{d + k(r-d)} = \frac{kd}{d + k(r-d)} = \frac{d}{\frac{d}{k} + r - d},$$

which tends to $\frac{d}{r-d}$ for k tending to infinity. Thus, one can choose k large enough such that $|A_f|$ is arbitrarily close to the given upper bound.

Let us now state two corollaries.

Corollary 1 *Consider bootstrap percolation on a d-degenerate graph with n vertices and let $r \geq d$. Then the size of a percolating set A_0 fulfills $\frac{r-d}{r}n \leq |A_0|$.*

This is simply obtained by setting $|A_f| = n$ in Theorem 1 and observing that $\frac{r-d}{r} \leq 0$ for $r \leq d$.

Next, we draw a connection to the results given by Riedl in [8]. First of all, note that every tree is a 1-degenerate graph, since every subgraph of a tree is a forest and thus contains at least one leaf, which is a vertex of degree one. As mentioned in the introduction, Riedl considered the size of minimal percolating sets on trees and proved that a minimal percolating set is of size at least $\frac{r-1}{r}n + \frac{1}{r}$. Our bound yields due to integrality of $|A_0|$ the same bound unless $n \equiv 0 \mod r$, in which case our bound is one less. Also the construction given in Fig. 3 can be used to show that there exist d-degenerate graphs for which the smallest percolating set A_0, which is of course also a minimal percolating set, is of size $(\frac{r-d}{r} + \varepsilon)n$.

Next, recall that the running time τ of a bootstrap percolation process is the least t such that $A_t = A_{t+1}$.

Corollary 2 *Given a d-degenerate graph, then the running time τ of the r-bootstrap percolation process with a given set A_0 and $r \geq d + 1$ is bounded by $\tau \leq \frac{d}{r-d}|A_0|$.*

This again follows immediately from Theorem 1, as every additional infection takes at most one time step.

Acknowledgements This work was supported by the Alexander von Humboldt Foundation with funds from the German Federal Ministry of Education and Research (BMBF).

References

1. Chalupa, J., Leath, P . L., & Reich, G . R. (1979). Bootstrap percolation on a Bethe lattice. *Journal of Physics C: Solid State Physics*, *12*(1).
2. Levin, A. D., Łuczak, M. J., & Peres, Y. (2010). Glauber dynamics for the mean-field ising model: cut-off, critical power law, and metastability. *Probability Theory and Related Fields*, *146*(1–2), 223–265.
3. Holroyd, A. E. (2003). Sharp metastability threshold for two-dimensional bootstrap percolation. *Probability Theory and Related Fields*, *125*(2), 195–224.
4. Balogh, J., Bollobás, Ba, Duminil-Copin, H., & Morris, R. (2012). The sharp threshold for bootstrap percolation in all dimensions. *Transactions of the American Mathematical Society*, *364*(5), 2667–2701.
5. Bradonjić, M., & Saniee, I. (2014). Bootstrap percolation on periodic trees. In *2015 Proceedings of the Twelfth Workshop on Analytic Algorithmics and Combinatorics (ANALCO), SIAM*, (pp. 89–96).
6. Bollobás, B., Gunderson, K., Holmgren, C., Janson, S., & Przykucki, M. (2014). Bootstrap percolation on Galton-Watson trees. *Electronic Journal of Probability*, *19*.
7. Balogh, J., Peres, Y., & Pete, G. (2006). Bootstrap percolation on infinite trees and non-amenable groups. *Combinatorics, Probability and Computing*, *15*(5), 715–730.
8. Riedl, E. (2012). Largest and smallest minimal percolating sets in trees. *The Electronic Journal of Combinatorics*, *19*(1), P64.
9. Riedl, E. (2010). Largest minimal percolating sets in hypercubes under 2-bootstrap percolation. *The Electronic Journal of Combinatorics*, *17*(1).
10. Janson, S., Łuczak, T., Turova, T., & Vallier, T. (2012). Bootstrap percolation on the random graph $G_{n,p}$. *Annals of Applied Probability*, *22*(5), 1989–2047.
11. Benevides, F., & Przykucki, M. (2015). Maximum percolation time in two-dimensional bootstrap percolation. *SIAM Journal on Discrete Mathematics*, *29*(1), 224–251.
12. Przykucki, M. (2012). Maximal percolation time in hypercubes under 2-bootstrap percolation. *The Electronic Journal of Combinatorics*, *19*, 1–13.
13. Bollobás, B., Holmgren, C., Smith, P., & Uzzell, A. J. (2014). The time of bootstrap percolation with dense initial sets. *The Annals of Probability*, *42*(4), 1337–1373.

A Hypergraph Network Simplex Algorithm

Isabel Beckenbach

1 Directed and Graph-Based Hypergraphs

A directed hypergraph is usually defined as a pair $(V, \mathcal{A})$ where V is a finite set of vertices and $\mathcal{A}$ is a set of pairs $a = (T(a), H(a))$ where $T(a), H(a)$ are disjoint subsets of V. The set $T(a)$ is called the tail and $H(a)$ the head of hyperarc a. A good survey on directed hypergraphs can be found in [1]. Inspired by an application to railway rotation planning Borndörfer et al. [2] defined directed hypergraphs slightly differently. Namely, they start with an ordinary directed graph and define a hyperarc to be a set of pairwise disjoint arcs.

Definition 1 Let $D = (V, A)$ be a directed graph. A directed hypergraph based on D is a pair $H = (V, \mathcal{A})$ where V is the vertex set of D and $\mathcal{A}$ is a set of non-empty subsets $E \subseteq A$ consisting of vertex-disjoint arcs. In this setting we call H a graph-based hypergraph.

A graph-based hypergraph can be seen as a special kind of directed hypergraph by setting $T(E) = \{v \in V : \exists w \in V, (v, w) \in E\}$ and $H(E) = \{v \in V : \exists w \in V, (w, v) \in E\}$ for $E \in \mathcal{A}$. For $v \in V$ we set $\delta^{in}(v) := \{E \in \mathcal{A} : v \in H(E)\}$ and $\delta^{out}(v) := \{E \in \mathcal{A} : v \in T(E)\}$. Using linear programming terminology the minimum cost flow problem can be stated as follows.

Definition 2 Given a graph-based hypergraph $H = (V, \mathcal{A})$, and functions $b : V \to \mathbb{R}$, $c : \mathcal{A} \to \mathbb{R}_{\geq 0}$, $u : \mathcal{A} \to \mathbb{R} \cup \{\infty\}$ the minimum cost hyperflow problem is the following linear optimization problem

---

I. Beckenbach (✉)
Zuse Institute Berlin, Takustr. 7, 14195 Berlin, Germany
e-mail: beckenbach@zib.de

$$\min \sum_{E \in \mathcal{A}} c(E) f(E)$$

$$\sum_{E \in \delta^{in}(v)} f(E) - \sum_{E \in \delta^{out}(v)} f(E) = b(v) \; \forall v \in V \quad (1)$$

$$0 \le f(E) \le u(E) \; \forall E \in \mathcal{A}. \quad (2)$$

For integral input data b, c, u there always exists an integral min-cost flow in directed graphs. For the special case that every hyperarc has at most one vertex in its head and b is nonnegative, the integrality of b implies the existence of an integral min-cost hyperflow which can be found by a combinatorial primal-dual algorithm, see [3]. However, this is not true for the min-cost hyperflow problem in general. In particular, it is $\mathcal{NP}$-hard to find an integral min-cost hyperflow (e.g. by reduction to 3D-Matching); even if all hyperarcs consist of at most two arcs (see [4] where the $\mathcal{NP}$-hardness of the hyperassignment problem which can be formulated as an integral min-cost hyperflow problem is proven).

In the remainder we only consider the uncapacitated minimum cost flow problem ($u \equiv \infty$) to make the presentation less technical and focus more on the underlying algorithmic ideas. However, the network simplex type algorithm described in the next section can also be adjusted to the capacitated case (details are deferred to a full version of the paper).

2 Min-Cost Hyperflow on Graph-Based Hypergraphs

In this section we characterize the basis matrices in the min-cost hyperflow problem on graph-based hypergraphs and show how most of the simplex operations can be done combinatorially. We do not specify any particular simplex rule, and leave any issues on the number of pivot iterations open for future research. Convergence can be guaranteed by usual methods.

In the remainder of this section let $H = (V, \mathcal{A})$ be a hypergraph based on the directed graph $D = (V, A)$ and let $M \in \{0, 1, -1\}^{V \times \mathcal{A}}$ be its incidence matrix, i.e.,

$$M_{v,E} = \begin{cases} 1 & v \in H(E) \\ -1 & v \in T(E) \\ 0 & v \notin H(E) \cup T(E) \end{cases}. \quad (3)$$

With this definition, all inequalities of type (1) can be written as $Mf = b$. We assume without loss of generality that D is connected and $\{a\} \in \mathcal{A}$ for all $a \in A$. The column $M_{\cdot E}$ corresponding to hyperarc $E = \{a_1, \ldots, a_k\}$ equals the sum of the columns $M_{\cdot a_k}$. This implies that the rank of M is the same as the rank of the vertex-arc incidence matrix of D which is $|V| - 1$ as D is connected.

In the following we will denote the submatrix of M restricted to the columns in some set $B \subseteq \mathcal{A}$ by M_B. For a set $B \subseteq \mathcal{A}$ we denote by $B_1 = \{E \in B : |E| = 1\}$

the set of all standard arcs and by $B_2 := B \setminus B_1$ the set of all "proper" hyperarcs. An easy observation shows that if B is a basis, then $D[\{a \in A : \{a\} \in B_1\}]$ is a forest having $|B_2| + 1$ connected components, see for example [5]. If $B_2 \neq \emptyset$ this condition is not sufficient. In this case, we choose a root $r \in V$ for each tree of the forest $D[\{a \in A : \{a\} \in B_1\}]$, denote this tree by T_r, and let R be the set of roots. We define a matrix $M_R \in \mathbb{Z}^{R \times B_2}$ by

$$M_R(r, E) = |V(T_r) \cap H(E)| - |V(T_r) \cap T(E)|.$$

M_R is independent of the concrete choice of the roots for the trees T_r. Furthermore, the columns of M_R have the following useful property.

Theorem 1 *Let B be a basis with B_1, B_2, R, $\{T_r\}_{r \in R}$ and M_R as defined above. Given $E \in B_2$ there exists a unique function $f : B \to \mathbb{R}$ such that $f(E) = 1$, $f(E') = 0$ for all $E' \in B_2 \setminus \{E\}$ and $f(\delta^{in}(v)) = f(\delta^{out}(v))$ for all $v \in V \setminus R$. Furthermore, the demand $f(\delta^{in}(r)) - f(\delta^{out}(r))$ at a root vertex $r \in R$ is given by $M_R(r, E)$.*

Proof We first show that a function f with $f(E) = 1$, $f(E') = 0$ for all $E' \in B_2 \setminus \{E\}$ and $f(\delta^{in}(v)) = f(\delta^{out}(v))$ for all $v \in V \setminus R$ exists. Therefore we set $b(v) = 1$ for all $v \in T(E)$, $b(v) = -1$ for all $v \in H(e)$ and $b(r) := |V(T_r) \setminus \{r\} \cap H(E)| - |V(T_r) \setminus \{r\} \cap T(E)|$. With this definition we have $\sum_{v \in V(T_r)} b(v) = 0$ for all trees T_r in particular $\sum_{v \in V} b(v) = 0$. This implies that there exists $f' : \{a : \{a\} \in B_1\} \to \mathbb{R}$ such that $f'(\delta^{in}(v)) - f'(\delta^{out}(v)) = b(v)$ for every $v \in V$. The uniqueness follows from the fact that f' is uniquely determined on every tree T_r. Setting $f(\{a\}) := f'(\{a\})$ for all $\{a\} \in B_1$, $f(E) = 1$, and $f(E') = 0$ for all $E' \in B_2 \setminus \{E\}$ gives a unique function satisfying the requirements of Theorem 1.

Now, we look at the demand induced by f on the roots. If $r \notin T(E) \cup H(E)$, then $f(\delta^{in}(r)) - f(\delta^{out}(r)) = b(r) = M_R(r, E)$. If r is a head vertex of E, then $f(\delta^{in}(r)) - f(\delta^{out}(r)) = b(r) + 1 = |V(T_r) \setminus \{r\} \cap H(E)| - |V(T_r) \setminus \{r\} \cap T(E)| + 1 = M_R(r, E) - 1 + 1 = M_R(r, E)$. The case $r \in T(E)$ is similar. □

Theorem 1 shows that M_R has the same properties as the matrix Cambini et al. [6] defined. In contrast to us, they used matrix operations and assumed that M has full rank which is not the case in our setting. The matrix M_R enables us to characterize the basis matrices for the min-cost hyperflow problem.

Theorem 2 *Let $B \subseteq A$ be a subset of size $|V| - 1$. M_B is a basis matrix for the linear program defined by (1)–(2) if and only if*

(a) $D[a \in A : \{a\} \in B]$ is a forest with $|B_2| + 1$ connected components.
(b) M_R has rank $|B_2|$.

Proof Let M_B be a basis matrix. (a) is easy to show. If (b) does not hold, then there exists a non-zero vector $y \in \mathbb{R}^{B_2}$ with $M_R \cdot y = 0$. For every $E \in B_2$, let $f^E \in \mathbb{R}^B$ be a vector with the properties described in Theorem 1, and set $f = \sum_{E \in B_2} y(E) f^E$. For every $v \in V \setminus R$ we have

$$f(\delta^{in}(v)) - f(\delta^{out}(v)) = \sum_{E \in B_2} y(E) \cdot \left(f^E(\delta^{in}(v)) - f^E(\delta^{out}(v))\right)$$
$$= \sum_{E \in B_2} y(E) \cdot 0 = 0,$$

and for $r \in R$ we get

$$f(\delta^{in}(r)) - f(\delta^{out}(r)) = \sum_{E \in B_2} y(E) \cdot \left(f^E(\delta^{in}(r)) - f^E(\delta^{out}(r))\right)$$
$$= \sum_{E \in B_2} y(E) \cdot M_R(r, E) = 0.$$

Furthermore, $f^E(E) = 1$ and $f^{E'}(E) = 0$ for all $E' \in B_2 \setminus \{E\}$ imply that $f(E) = y(E)$ for all $E \in B_2$. Thus, f is a non-zero vector with $M_B \cdot f = 0$ which is impossible as the columns of M_B are linearly independent.

Now, suppose (a) and (b) hold. The rows of M_B sum to zero, thus its rank is at most $|B|$. By this fact and basic linear algebra, the rank of M_B equals $|B|$ if and only if for every $b \in \mathbb{R}^V$ with $\sum_{v \in V} b(v) = 0$ the system $M_B \cdot f = b$ has a unique solution. By (a) we can find a flow f' on B such that $f'(E) = 0$ for all $E \in B_2$ and $f'(\delta^{in}(v)) - f'(\delta^{out}(v)) = b(v)$ for all $v \in V \setminus R$. Next, we set $\delta(r) := b(r) - (f'(\delta^{in}(r)) - f'(\delta^{out}(r)))$ for all $r \in R$ and solve $M_R \cdot y = \delta$. Again, let f^E be the unique flow with the properties of Theorem 1. We set $f = \sum_{E \in B_2} y(E) f^E + f'$. For $v \in V \setminus R$ we have

$$f(\delta^{in}(v)) - f(\delta^{out}(v))$$
$$= \sum_{E \in B_2} y(E) \cdot \left(f^E(\delta^{in}(v)) - f^E(\delta^{out}(v))\right) + f'(\delta^{in}(v)) - f'(\delta^{out}(v))$$
$$= 0 + b(v),$$

and for $r \in R$

$$f(\delta^{in}(r)) - f(\delta^{out}(r))$$
$$= \sum_{E \in B_2} y(E) \cdot \left(f^E(\delta^{in}(r)) - f^E(\delta^{out}(r))\right) + f'(\delta^{in}(r)) - f'(\delta^{out}(r))$$
$$= \sum_{E \in B_2} y(E) \cdot M_R(r, E) + b(r) - \delta(r) = b(r).$$

This shows that $M_B \cdot f = b$ holds. The uniqueness follows from the fact that the function values at B_2 are uniquely determined by (b), and given the values on B_2 the function f on B_1 is uniquely determined by property (a). □

Algorithm 1 describes a network simplex type algorithm for the min-cost hyperflow problem on graph-based hypergraphs. By our assumption, there exists a

Algorithm 1

Input: Digraph $D = (V, A)$, Hypergraph $H = (V, \mathcal{A})$ based on $D, b : V \to \mathbb{R}$ with $\sum_{v \in V} b(v) = 0$, and $c : \mathcal{A} \to \mathbb{R}_{\geq 0}$, a feasible flow x on D, corresponding basis B, initial spanning tree $T = D[B]$.
Output: A min-cost hyperflow $x : \mathcal{A} \to \mathbb{R}_{\geq 0}$.
1: Choose a root r arbitrarily, set $T_r = T$, $R = \{r\}$.
2: Solve $\pi^T M_B = c_B^T$ (Dual).
3: Compute reduced cost $c^\pi(E) = c(E) - \sum_{v \in H(E)} \pi(v) + \sum_{v \in T(E)} \pi(v)$ for all non-basic hyperarcs $E \in \mathcal{A} \setminus B$.
4: **if** $c^\pi \geq 0$ **then**
5:    Output x (x is optimal).
6: **else**
7:    Choose a hyperarc $E^{in} \in \mathcal{A} \setminus B$ with $c^\pi(E^{in}) < 0$.
8:    Solve the system $M_B f = -M_{E^{in}}$ (Primal).
9:    Choose a hyperarc $E^{out} = \arg\min\{x(E)/ - f(E) : f(E) < 0, E \in B\}$.
10:    Set $B \leftarrow B \setminus \{E^{out}\} \cup \{E^{in}\}$ update x, R, trees $\{T_r\}_{r \in R}$, and matrix M_R.
11:    Goto 1.
12: **end if**

feasible min-cost hyperflow if and only if there exists a feasible flow on the underlying digraph. Our algorithm receives such a feasible flow together with the corresponding Basis B as an input.

Algorithm 2 Flow

1: **procedure** FLOW($B, \{T_r\}_{r \in R}, d_N, f_2$)
2:    $d(r) \leftarrow 0$ for all $r \in R$ and $d(v) \leftarrow d_N(v)$ for all $v \in V \setminus R$.
3:    **for all** $E \in B_2, v \in V$ **do**
4:      **if** $v \in T(E)$ **then** $d(v) \leftarrow d(v) + f_2(E)$.
5:      **if** $v \in H(E)$ **then** $d(v) \leftarrow d(v) - f_2(E)$.
6:    **end for**
7:    **for all** trees T_r **do**
8:      **for** $j = |V(T_r)| - 1$ to 1 **do**
9:         **if** $a_j = (v, v_{j+1})$ **then** $f_1(a_j) \leftarrow d(v_{j+1})$.
10:        **if** $a_j = (v_{j+1}, v)$ **then** $f_1(a_j) \leftarrow -d(v_{j+1})$.
11:        $d(v) \leftarrow d(v) + d(v_{j+1})$
12:      **end for**
13:    **end for**
14:    $d_R(r) \leftarrow -d(r)$ for all $r \in R$.
15:    **return** d_R, f_1
16: **end procedure**

In the remainder we show how to solve problems of the type $M_B f = b$ (Primal) and $\pi^T M_B = c_B^T$ (Dual) where M_B is a basis matrix. We always assume that the trees $\{T_r\}_{r \in R}$ have its vertices and arcs ordered such that v_1 is the root, v_j is a leaf in $T[\{v_1, \ldots, v_j\}]$ and a_{j-1} is the arc v_j is incident to. We start with the primal problem $M_B f = b$ for which we basically use the algorithm described in the proof of Theorem 2. As a subroutine we need Algorithm 2 which given the demand d_N on the non-root vertices $N := V \setminus R$, and flow f_2 on the non-tree hyperarcs B_2 computes the unique flow f_1 on the tree arcs B_1 and demand d_R on the roots R such that

$$M_B \cdot \begin{pmatrix} f_1 \\ f_2 \end{pmatrix} = \begin{pmatrix} d_N \\ d_R \end{pmatrix}$$

where the rows and columns of M_B are arranged accordingly.

Using the algorithm above we can solve $M_B f = b$ as follows:

1. Compute Flow$(B, \{T_r\}_{r \in R}, b_N, 0)$.
2. Solve $M_R \cdot y = (b_R - d_R)$.
3. Compute Flow$(B, \{T_r\}_{r \in R}, b_N, y)$.

In the first step we compute a flow with value zero on all hyperarcs $E \in B_2$ which induces the right demands on the non-root vertices. In the second step we calculate the flow needed on B_2 to correct the demand at the root vertices, and finally in step 3 we adjust the flow on the tree arcs.

For the dual problem $\pi^T M_B = c_B^T$ we need Algorithm 3 as a subroutine. Given the cost c_1 of all tree arcs B_1, and the potential π_R at the root vertices it computes a cost vector e_2 on B_2 and potential π_N on the non-root vertices such that $(\pi_N^T, \pi_R^T) M_B = (c_1^T, e_2^T)$, i.e., the reduced cost of every basic hyperarc is zero.

Algorithm 3 Potential

1: **procedure** POTENTIAL$(B, \{T_r\}_{r \in R}, c_1, \pi_R)$
2:     $\pi(v) \leftarrow \pi_R(v)$ for all $v \in R$ and $\pi(v) = 0$ for all $v \in V \setminus R$.
3:     **for all** tress T_r **do**
4:        **for** $j = 1$ to $|V(T_r)| - 1$ **do**
5:           **if** $a_j = (v, v_{j+1})$ **then** $\pi(v_{j+1}) \leftarrow \pi(v) + c_1(a_j)$.
6:           **if** $a_j = (v_{j+1}, v)$ **then** $\pi(v_{j+1}) \leftarrow \pi(v) - c_1(a_j)$.
7:        **end for**
8:     **end for**
9:     **for all** $E \in B_2$ **do**
10:       $e_2(E) \leftarrow \sum_{v \in H(E)} \pi(v) - \sum_{v \in T(E)} \pi(v)$.
11:     **end for**
12:     **return** e_2, π.
13: **end procedure**

As the rank of M_B is $|V| - 1$ the system $\pi^T M_B = c_B^T$ has no unique solution. Thus, we can fix the value of one vertex, for example we can choose one of the roots $r_1 \in R$ and set $\pi(r_1) = 0$.

Now, we can solve Dual as follows.

1. Compute Potential$(B, \{T_r\}_{r \in R}, c_1, 0)$.
2. Find a solution to $y^T M_R = (c_2^T - e_2^T)$ with $y(r_1) = 0$.
3. For all $r \in R$ set $\pi(r) \leftarrow y(r)$ and $\pi(v) \leftarrow \pi(v) + \pi(r)$ for all $v \in V(T_r)$.

First, the potential on the roots is set to zero, and we compute a potential on the non-root vertices such that the reduced cost of every tree arc is zero. In the second step the correct potential of the root vertices is calculated, and in step 3 the potential on the non-roots is adjusted. In contrast to the primal problem, we do not have to call Algorithm 3 a second time. It suffices to add the potential of the root vertex to the potential of the other vertices in the tree.

References

1. Gallo, G., Longo, G., Pallottino, S., & Nguyen, S. (1993). Directed hypergraphs and applications. *Discrete Applied Mathematics*, *42*(2–3), 177–201.
2. Borndörfer, R., Reuther, M., Schlechte, T., & Weider, S. (2011). A hypergraph model for railway vehicle rotation planning. In *OASIcs-OpenAccess Series in Informatics*, *20*.
3. Jeroslow, R. G., Martin, K., Rardin, R. L., & Wang, J. (1992). Gainfree Leontief substitution flow problems. *Mathematical Programming*, *57*(1), 375–414.
4. Borndörfer, R., & Heismann, O. (2011). Minimum Cost Hyperassignments with Applications to ICE/IC Rotation Planning. In *OR* (pp. 59–64).
5. Heismann, O. (2014). The hypergraph assignment problem, Doctoral dissertation, Technische Universität Berlin.
6. Cambini, R., Gallo, G., & Scutellà, M. G. (1997). Flows on hypergraphs. *Mathematical Programming*, *78*(2), 195–217.

Rapid Mathematical Programming for Cooperative Truck Networks

Jörg Rambau

1 Cooperative Truck Networks

Cooperative Truck Networks (CTN) are an idea from [1]: In order to reduce idle times of the trucks and overnight stays of the drivers, cooperate among various (small) logistic companies in the following way: Any driver leaves the depot with a truck and a full trailer in the morning and returns with that same truck and (maybe) another trailer in the afternoon. We call such a one-day home-away-home truck tour a *depot commute*. The full truckload stays on its trailer, and the trailer is passed on from truck to truck until it arrives at its destination. For example: Instead of two full truckload transportations, one from Hamburg to Munich and one from Munich to Hamburg that take a whole day one-way one could exchange the trailers in Kassel to get home on the same day.

Of course, dispatching such a cooperative transportation system is a complicated task. Usually, not all transport routes will fit together. For example, if some transport traverses an edge that is not traversed in the opposite direction by any other transport route on the same day, then this transport cannot be carried out by a sequence of depot commutes.

The organizational task to maximize the number of transports that can be operated by the CTN for *Fixed Routing (FR)* (i.e., with given fixed scheduled routes for all transport requests) was algorithmically studied for the first time in [2]. Let us call this problem the *fixed-route CTN relay problem (FR-CTNRP)*. In order to evaluate the actual benefit, also monetary consequences must be assessed. Therefore, cost calculations have been provided in [3]. Using an ad-hoc optimization algorithm, it was observed in [2] that in many real-world cases the fraction of transport requests that can be cooperatively processed is too small to convince small companies to participate in a CTN. This problem was later confirmed by exact optimization calculations carried

---

J. Rambau (✉)
University of Bayreuth, 95440 Bayreuth, Germany
e-mail: joerg.rambau@uni-bayreuth.de
URL: http://www.rambau.wm.uni-bayreuth.de

out by the author. Thus, the idea of injecting more flexibility into the problem setting was born. The first attempt in this direction is *Multi Routing (MR)*, i.e., to allow for multiple routing alternatives for each transport request. The resulting problem is called the *multi-route CTN relay problem (MR-CTNRP)*. Detailed information concerning the business process and the data handling can be read in two other contributions to this volume [4, 5] and the thesis [6].

This paper contributes a *Rapid Mathematical Programming Approach* in the spirit of [7] based on mixed-integer linear programming (MILP). Based on the presented numerical results on real-world data some insights for the algorithm designer and for the manager are provided.

2 Formal Problem Definition

Let the set of *time slots* be $T = \{0, 1, \ldots N\}$, where one time slot extends to half the duration of a shift. *Morning time slots* are even, *afternoon time slots* are odd t's. Any driver can drive for the length of two time slots per day, first a morning time slot, then an afternoon time slot. Let $G = (V, E)$ be the *transportation network* of the CTN. Its nodes V are the possible trailer exchange points including all the home depots of the trucks. Its edges E connect two nodes whenever it takes no more than one time slot to go from one node to the other. Each truck belongs to a *depot*. The *set of depots* is denoted by $D \subseteq V$.

Moreover, there is an index set of transport requests Q. For each $q \in Q$ we have an index set of *routing alternatives* P_q with a default route $p(q) \in P_q$. We set $P = \bigcup_{q \in Q} (\{q\} \times P_q)$. The routing alternatives are specified by a *routing and scheduling function* $r \colon P_q \to 2^{T \times E}$ that assigns to each routing alternative $p \in P_q$ for transport request q a *scheduled route*, i.e., a set of time-stamped, chronologically adjacent edges that specify when and where the trailer loaded with transport request q shall go from one node to another. Origin and destination of a transport request q can be read off the common origin and destination nodes of its routing alternatives. We define a *depot commute* as a pair of scheduled edges (p, t, d, b) and $(p', t+1, b, d)$ where p, p' are routing alternatives, t is a morning time slot, d is a depot, and b is some node.

For some subset of routing alternatives $P' \subseteq P$ let $\mathcal{L}(P') = \bigcup_{p \in P'} (\{p\} \times r(p))$ be the *link collection of* P', i.e., the total set of route-labeled scheduled transport links traversed by the routes in P'. Each route $p \in P$ induces a *network transport cost* β_p if $p \in P'$ and a *direct transport cost* $\gamma_p > \beta_p$ if $p \in P \setminus P'$. The network transport cost is calculated assuming that the transport can be carried out along the route by depot commutes of trucks only, passing on the trailer at the end of a morning time slot. The direct transport cost is calculated assuming that the transport is carried out along the route by a single driver and a single truck keeping the same trailer throughout.

The task of the MR-CTNRP is to find a (by some criterion) optimal subset $P^* \subseteq P$ whose link collection $\mathcal{L}(P^*)$ can be partitioned into loaded depot commutes and for which at most one routing alternative is chosen for each transport request. For an MR-CTNRP-solution P^*, *nettrips* are the transport requests using a route from P^* cooperatively, and *dirtrips* are remaining transport requests using the default route directly. The FR-CTNRP is the special case of the MR-CTNRP where $|P_q| = 1$ for all $q \in Q$, i.e., there is only one routing alternative for each transport request.

The optimality criterion we investigated for this work are, first, the maximization of the number $|P^*|$ of nettrips and, second, the minimization of the total transportation cost incurred by nettrips and dirtrips.

3 A Mixed-Integer Linear Programming Model

We use the principle of "Rapid Mathematical Programming", first systematically discussed in [7]. Moreover, we use the modeling language zimpl introduced in the same work. Our goal is to provide a lean model for the MR-CTNRP with enough flexibility to evaluate model variants fast. We define binary selection variables z_p for each routing alternative $p \in P$. Moreover, we define binary depot assignment variables $x_{p,t,d,b}$ and $x_{p,t,a,d}$ indicating that p uses the departure link (d, b) or home link (a, d), respectively, of a depot commute at time slot t. We introduce additional non-negative measurement variables for the number of nettrips (u integer) and the total cost (v continuous) in order to be able to read off those values from the solution values. The resulting model reads as follows:

$$\max u \text{ or } \min v \qquad (1)$$

such that

$$u - \sum_{p \in P} z_p = 0 \qquad (2)$$

$$v - \sum_{p \in P} \beta_p z_p - \sum_{q \in Q} \gamma_{p(q)} (1 - \sum_{p \in P_q} z_p) = 0 \qquad (3)$$

$$x_{p,t,d,b} - z_p = 0 \quad \forall (p,t,d,b) \in \mathcal{L}(P)$$
$$\text{with } d \in D, \ t \text{ even} \qquad (4)$$

$$x_{p,t,a,d} - z_p = 0 \quad \forall (p,t,a,d) \in \mathcal{L}(P)$$
$$\text{with } d \in D, \ t \text{ odd} \qquad (5)$$

$$\sum_{\substack{p \in P: \\ (p,t,d,b) \in \mathcal{L}(P)}} x_{p,t,d,b} - \sum_{\substack{p \in P: \\ (p,t+1,b,d) \in \mathcal{L}(P)}} x_{p,t+1,b,d} = 0 \quad \forall d \in D, \forall t \in T \text{ even} \qquad (6)$$

$$\sum_{\substack{p \in P: \\ (p,t,a,d) \in \mathcal{L}(P)}} x_{p,t,a,d} - \sum_{\substack{p \in P: \\ (p,t-1,d,a) \in \mathcal{L}(P)}} x_{p,t-1,d,a} = 0 \quad \forall d \in D, \forall t \in T \text{ odd} \quad (7)$$

$$\sum_{p \in P_q} z_p \leq 1 \qquad \forall q \in Q \quad (8)$$

$$x_{p,t,d,b}, z_p \in \{0,1\} \quad \forall (p,t,d,b) \in \mathcal{L}(P). \quad (9)$$

We optimize in the objective function (1) one of the measurement variables u or v. Restrictions (2) and (3) compute these values from the independent variables. Restrictions (4) and (5) ensure that a path can be selected if and only all of its morning links and all of its afternoon links have been assigned to the appropriate part of some depot commute. Restrictions (6) and (7) guarantee that for each depot commute the number of paths assigned to the leaving part of the commute equals the number of paths assigned to the returning part. In restriction (8) we allow for at most one selected route per transport request. Whenever we want to compare the FR optimum, we can fix the selection variables to be one on the fixed route only. This way, we can easily test variants of possible CTN operations with only slight modifications.

4 Computational Results

In Table 1 we show the results of our tests. In numerical experiments we computed optimal solutions for model variants on the basis of real-world instances. The experiments with 6 start time slots (t6) are typical instance sizes for a daily operation on the moving horizon. The experiments with 14 start time slots (t14) have been carried out in order to evaluate the value of the additional future information. We applied two strategies to keep the number of routes under control: The instances l3 allow only routes with at most 3 nodes, the instances l5 allow routes with at most five notes. Moreover, we distinguish the case in which *shifting* is allowed: shifting (−s) provides for each single-link transport request a canonical routing alternative that uses the same single-link route but on the other time slot of the same day. For each solution we are interested in the fraction of transport requests that are carried out cooperatively and in the cost savings incurred by cooperative transportation. The instances **maxtrips** maximize the number of nettrips, the instances **mincost** minimize total costs.

All instances were provided by Bernd Nieberding from the ILAN project at the FH Erfurt. The cost for each route and transportation mode was estimated by the ILAN project team according to [3].

We used zimpl 3.3.1/cplex 12.5.0.0 on a MacBookPro (2012)/MacOS 10.11.6. The cplex parameter were changed to "set timelimit 3600" (timeout), "set mip tol mipgap 0.01" (increased optimality gap), "set mip `tol integrality 1e-7`" (decreased integrality gap), and "`set emph mip 1`" (search preferably for integer feasible solutions).

The following insights can be drawn from Table 1:

1. Our model scales well for alternative routes of length at most three nodes; the computation times explode if the maximal path length is increased to five nodes. Managerial insight: stick with routes of length at most three. Algorithmic insight: if longer routes need to be handled in the model for some reason, then more sophisticated solution methods are needed (most notably dynamic column generation for alternative routes). Possible explanation: with routes of at most three nodes (starting in the morning and arriving in the afternoon), our model reveals that then all decisions can be made independently for each day, so that the solution method scales well with an increasing number of time slots.
2. The idea to provide multiple routes for each transport request is effective; the optimal number of nettrips increases by around 50%. Managerial insight: motivate companies to provide alternative routes.
3. The idea to allow of shifting is very effective: it roughly yields another 60–70% increase of possible nettrips. Managerial insight: utilize both time slots of each day for single-link transports.
4. Maximizing nettrips usually yields low-cost solutions, and minimizing costs usually yields high-nettrips solutions. Managerial insight: use a suitable combination of both, e.g., if driver satisfaction is important in its own right.
5. Cost minimization seems to scale better in the MILP solution process with problem size (see MR-mincost-t14-l5, which yield a better fraction of nettrips than both MR-maxtrips-t14-l5 and MR-maxtrips-t14-l5-s). Algorithmic insight: for maximizing nettrips, perturb the objective by a cost term.
6. A longer horizon does not lead to a substantially larger fraction of nettrips. Managerial insight: the model is suitable for a rolling-horizon planning on a short planning horizon. Possible explanation: again, the model for short paths is uncoupled over days.

The rapid mathematical programming investigation with exact mathematical software yields useful information for the design and operation of a CTN. Even more (capacities, fair share of benefit, etc.) could be incorporated. We showed that almost half of the transports can be carried out as nettrips in a CTN with multirouting and shifting. Lifting additional potential of the CTN concept requires a more sophisticated machinery of Mathematical Programming – which lifts the entry hurdle for small companies maybe too high. Thus, there is evidence for the fact that multirouting plus shifting on single-day routes (length 3) yields a well-balanced CTN operation mode with high potential.

Table 1 Computational results: FR/MR: fixed route/multiple routes; maxtrips/mincost: maximize network trips/minimize costs; t6/t14: start at latest at time slot 6/14; 13/15: routes have at most 3/5 nodes; vars: no. of model variables; cons: no. of model constraints; nettrips: no. of transport requests operated cooperatively; dirtrips: total no. of transport requests; netcost: total cost for solution; dircost: total cost for individual transportation only; CPU/s: runtime with timeout set to 3600.00

Instance	Vars	Cons	Nettrips (%)	Dirtrips	Netcost (%)	Dircost	CPU/s
FR-maxtrips-t6-l3	90,982	95,932	2,641 (20)	13,111	7,703,501 (98.5)	7,818,695	0.56
MR-maxtrips-t6-l3	90,982	77,284	3,694 (28)	13,111	7,687,741 (98.3)	7,818,695	2.82
MR-maxtrips-t6-l3-s	99,588	81,813	**6,282 (48)**	13,111	7,627,221 (97.6)	7,818,695	2.55
[1ex] FR-mincost-t6-l3	90,982	95,932	2,629 (20)	13,111	7,692,226 (98.4)	7,818,695	0.65
MR-mincost-t6-l3	90,982	77,284	2,515 (19)	13,111	7,722,639 (98.8)	7,818,695	1.20
MR-mincost-t6-l3-s	99,588	81,813	5,808 (44)	13,111	**7,590,015 (97.1)**	7,818,695	1.97
[2ex] FR-maxtrips-t6-l5	2,059,303	2,068,444	2,031 (12)	17,120	11,513,675 (99.2)	11,601,729	4.54
MR-maxtrips-t6-l5	2,059,303	1,633,786	2,072 (12)	17,120	11,540,968 (99.5)	11,601,729	3600.00
MR-maxtrips-t6-l5-s	2,067,909	1,638,127	**8,103 (47)**	17,120	11,367,909 (98.0)	11,601,729	3600.00
[1ex] FR-mincost-t6-l5	2,059,303	2,068,444	2,021 (12)	17,120	11,504,815 (99.2)	11,601,729	4.76
MR-mincost-t6-l5	2,059,303	1,633,786	4,492 (26)	17,120	11,387,051 (98.1)	11,601,729	117.75
MR-mincost-t6-l5-s	2,067,909	1,638,127	7,128 (42)	17,120	**11,300,286 (97.3)**	11,601,729	307.60

(continued)

Table 1 (continued)

Instance	Vars	Cons	Nettrips (%)	Dirtrips	Netcost (%)	Dircost	CPU/s
[2ex] FR-maxtrips-t14-13	233,310	245,159	6,942 (20)	33,330	19,549,079 (98.5)	19,846,312	0.87
MR-maxtrips-t14-13	233,310	197,096	9,681 (29)	33,330	19,506,043 (98.3)	19,846,312	7.63
MR-maxtrips-t14-13-s	255,064	208,527	**16,258 (49)**	33,330	19,355,058 (97.5)	19,846,312	7.91
[1ex] FR-mincost-t14-13	233,310	245,159	6,905 (20)	33,330	19,519,537 (98.4)	19,846,312	0.89
MR-mincost-t14-13	233,310	197,096	9,206 (28)	33,330	19,424,448 (97.9)	19,846,312	3.51
MR-mincost-t14-13-s	255,064	208,527	15,004 (45)	33,330	**19,258,280 (97.0)**	19,846,312	5.27
[2ex] FR-maxtrips-t14-15	4,735,593	4,754,452	5,642 (13)	43,522	29,235,965 (99.2)	29,480,953	10.67
MR-maxtrips-t14-15	4,735,593	3,753,389	5,513 (13)	43,522	29,321,667 (99.5)	29,480,953	3600.00
MR-maxtrips-t14-15-s	4,757,347	3,764,328	**8,612 (20)**	43,522	29,248,792 (99.2)	29,480,953	3600.00
[1ex] FR-mincost-t14-15	4,735,593	4,754,452	5,592 (13)	43,522	29,211,036 (99.1)	29,480,953	10.97
MR-mincost-t14-15	4,735,593	3,753,389	13,870 (32)	43,522	**28,800,759 (97.7)**	29,480,953	1999.10
MR-mincost-t14-15-s	4,757,347	3,764,328	**8,612 (20)**	43,522	29,224,747 (99.1)	29,480,953	3600.00

References

1. Apfelstädt, A., & Gather, M. (2014). New design of a truck load network. In *Dynamics in Logistics: Proceedings of the 4th International Conference LDIC, 2014 Bremen, Germany*, Lecture Notes in Logistics (pp. 183–191). Berlin: Springer International Publishing.
2. Apfelstädt, A., Dashkovskiy, S., & Nieberding, B. (2016). Modeling, optimization and solving strategies for matching problems in cooperative full truckload networks. *IFAC-PapersOnLine, 49*(2), 18–23.
3. Dashkovskiy, S. N., & Nieberding, B. (2016). Costs and travel times of cooperative networks in full truck load logistics. In *Dynamics in Logistics* (pp. 193–201). Berlin: Springer International Publishing.
4. Apfelstädt, A. (2017). New production approach for European truckload cargo industry. In *This volume*.
5. Nieberding, B. (2017). Multi-routing for transport-matching in cooperative, full truckload, relay networks. In *This volume*.
6. Apfelstädt, A. (2017). Handlungsoptionen im euronationalen Ladungsverkehr. Dissertation, Bergische Universität Wuppertal Cuvillier, ISBN 9783736995635.
7. Koch, T. (2004). Rapid Mathematical Programming. Dissertation, Technische Universität Berlin

Multi-routing for Transport-Matching in Cooperative, Full Truckload, Relay Networks

Bernd Nieberding

1 Introduction

Full truckload (FTL) is described as so-called point-to-point or over-the-road dispatching, with standard curtain-sided or box trailers, where the smallest transportation unit is one trailer and a load is directly transported between origin and destination by a single driver, [1]. In Germany the segment of FTL, has a high lack of productivity due to small companies with non-industrialized transport processes, which leads to vehicle-workload of less than 30%, [2]. Approaches using truckload relay networks are known as so-called *Advanced Truckloading* and were introduced to reduce the high driver turnover of US truckload carriers, [1]. To fit the characteristics of the German FTL market these approaches were adapted and modified in [3]. The basic idea is a cooperative network of different carriers, with transport processes similar to the processes in the part load or less than truckload (LTL) segment, [4, 5]. Each location of a carrier serves as a relay depot with an assigned zone for pickup or delivery of loads, designated to be dispatched in the relay network. To increase the operation time of the truck, a restaffing of the truck is necessary. Therefore, direct transports between two depots are only executed, if the driving distance between them satisfies conditions such that a driver returns to his home depot by the end of his working period. Other transports are handled as a relay of direct transports in the network. A *parial matching* of transports on each relay link avoids empty runs and enhances the overall-costs.

Previous investigations, [6, 7], have shown that the use of only single routes for each transport leads to an unsatisfying number of transports handled within the network. This paper generalizes the model presented in [6, 7] to sets of routes, called multi-routes, which gives a higher flexibility in the matching of relay transports.

B. Nieberding (✉)
University of Applied Sciences, 99085 Erfurt, Germany
e-mail: bernd.nieberding@fh-erfurt.de

2 Multi-routes and Matching

The relay-network and transport routes are described by a directed graph $G = (V, E)$, where a node $v_i \in V$, $i = 1, \ldots, D$, represents a depot or exchange place in the network and an arc $(v_i, v_j) \in E \subseteq V \times V$ represents the connection between two depots $v_i, v_j \in V$. We define $\overline{(v_i, v_j)} := (v_j, v_i)$. To take transport distances or transport times into account we consider the weighted graph $(G, \omega))$, where the weight function $\omega : E \to \mathbb{R}_{\geq 0}$ assigns each arc $(v, w) \in E$ to a transport distance or time $\omega(v, w) \in \mathbb{R}_{\geq 0}$. The set of scheduled operation periods is defined as $T = \{t_n \in \mathbb{R}_{>0}\}_{n=1,\ldots,N}$, where $t_n < t_{n+1}$ holds for all $n = 1, \ldots, N$ and $N \in \mathbb{N}_{>0}$ is the total number of considered operation periods.

2.1 Transport Routes

Let $P \subset G$ be a directed path between two depots $v_{P_0} \in V$ and $v_{P_l} \in V$, with $V_P = \{v_{P_0}, \ldots, v_{P_l}\}$ and $E_P = \{(v_{P_0}, v_{P_1}), \ldots, (v_{P_{l-1}}, v_{P_l})\}$. Further, let $(t_n \in T)_{n=n_0,\ldots n_{l-1}}$, with $n_{l-1} - n_0 = l$, be a sequence of operation periods. Then, inside the relay-network a dispatching from a pickup-depot v_{P_0} to a delivery-depot v_{P_l}, along path P, beginning in period t_{n_0} and ending in period $t_{n_{l-1}}$, can be described as a sequence

$$R(v_{P_0}, v_{P_l}, t_{n_0}, t_{n_{l-1}}) = \left(\left((v_{P_i}, v_{P_{i+1}}), t_{n_i}\right)\right)_{i=0,\ldots,l-1}. \quad (1)$$

We call $R(v_{P_0}, v_{P_l}, t_{n_0}, t_{n_{l-1}})$ a transport route from depot v_{P_0} to depot v_{P_l} and the tuples $\left((v_{P_i}, v_{P_{i+1}}), t_{n_i}\right)$ are the so-called transport relay segments. For a transport $k = 1, \ldots, K$ all existing transport routes R_k^j, $j = 1, \ldots, J(k)$, between two nodes $v_k \in V$ and $w_k \in V$, i.e. there exists a path P from v_k to w_k satisfying constraint ω_{net} on each edge in $E(P)$, beginning earliestly in operation period t_{n_k} and ending latestly in period t_{m_k} are stored in the set

$$\mathcal{R}_k(v_k, w_k, t_{n_k}, t_{m_k}) := \left\{R_k^j(v_k, w_k, t_{n_0}^j, t_{n_{l-1}}^j)\right\}_{j=1,\ldots,J(k)}, \quad (2)$$

where $t_1 \leq t_{n_k} \leq t_{n_0}^j \leq t_{n_{l-1}}^j \leq t_{m_k} \leq t_N$ holds for all j. For a given set of K transports and their transport route sets, a compact notation of the finite set of all transport routes is given by $\mathcal{R} := \bigcup_{k=1}^{K} \mathcal{R}_k(v_k, w_k, t_{n_k}, t_{m_k})$.

2.2 Transport-Matching

Empty-runs play a crucial role with respect to transport costs in freight dispatching. The aim of our solution approach is to match two transport relay segments in a way such that each relay segment of a transport receives a backload given by a relay segment from another transport in a subsequent operation period. More precisely, for two given transport routes $R_{k_1}^{j_1}$ and $R_{k_2}^{j_2}$ in $\mathcal{R}$, with $k_1 \neq k_2 \in \{1, \ldots, K\}$, $j_1 \in \{1, \ldots, J(k_1)\}$, $j_2 \in \{1, \ldots, J(k_2)\}$, there exists a partial matching between a transport relay segments $((v_{k_1}, w_{k_1}), t_{n_{k_1}})$ of $R_{k_1}^{j_1}$ and $((v_{k_2}, w_{k_2}), t_{n_{k_2}})$ of $R_{k_2}^{j_2}$, iff $(v_{k_1}, w_{k_1}) = \overline{(v_{k_2}, w_{k_2})}$ and $t_{n_{k_1}} = t_{n_{k_2}} + 1$ holds, with $n_{k_1} = 2m_{k_1} - 1$ and $m_{k_1} \in \mathbb{N}_{>0}$. Due to small driver and truck capacities, the condition $n = 2m - 1$ is necessary to design a two-periodical process consisting of parallel network processes in one half (t_n and t_{n+1}) and parallel pick up and delivery processes in the other half (in the non-scheduled times before t_n and after t_{n+1}) of each day in our model, without overlapping. In general this condition can be dropped to allow both process types at all times.

The partial matching graph $\mathcal{G}$ is a tuple $(V_\mathcal{G}, E_\mathcal{G})$ consisting of a set $V_\mathcal{G} := \bigcup_{k=1}^{K} V_k := \bigcup_{k=1}^{K} \bigcup_{j=1}^{J(k)} V_k^j$ of nodes, where $v_k^{j,n} \in V_k^j$ iff there exists $((v, w), t_n) \in R_k^j : v, w \in V \wedge t_n \in T$ and a set $E_\mathcal{G}$ of edges, where for $k_1 \neq k_2$ $\left\{v_{k_1}^{j_1,n}, v_{k_2}^{j_2,n+1}\right\} \in E_\mathcal{G}$ iff there exists partial matching between $v_{k_1}^{j_1,n}$ and $v_{k_2}^{j_2,n+1}$.

If a set of transports is handled in the intended framework. Two conditions must be satisfied: Firstly, if a relay segment of a transport is used for a partial matching, then the transport has to be matched on all its segments. Secondly, each route segment of a transport can serve as a load or back load only one time, because the smallest transport unit is one trailer. This leads to the following definition of a matching: A matching $\mathcal{M} \subseteq \mathcal{G}$ is a tuple $(V_\mathcal{M}, E_\mathcal{M}) \subseteq (V_\mathcal{G}, E_\mathcal{G})$ such that if there exists $k \in \{1, \ldots, K\}$, $n \in \{1, \ldots, N\}$, $j \in \{1, \ldots, J(k)\}$ such that $v_k^{j,n} \in V_\mathcal{M}$, then $V_k^j \subset V_\mathcal{M}$ and $\bigcup_{\substack{l=1 \\ l \neq j}}^{J(k)} V_k^l \not\subseteq V_\mathcal{M}$, and for all $v \in V_\mathcal{M}$ holds $\deg(v) = 1$.

The problem to maximize the number of transports in the matching was considered with an ad-hoc method in [6], for two operation periods and using directed cycles in [7], and with an rapid mathematical programming approach with mixed-integer linear programming and further objective functions in [8].

3 Calculation of Multi-routes

The calculation of a route set (2) for each transport belongs to the field of *path enumeration*, i.e. the aim is not to find a specific solution, as for example given by *Dijkstra's algorithm*, instead we are searching for a set of feasible solutions, which are all satisfying specific constraints. Here, we are searching for the k shortest paths, with unknown k, satisfying constraints with respect to edge-weights, path-length, scheduling-time and costs, which in general is known as the *k-constrained shortest path problem*. A similar approach was used in [1], where path enumeration was used for a non-cooperative FTL network in the USA, with constraints regarding to path-length and circuity.

The main routing-process is divided into two steps: In a pre-processing step all paths between two different depots are calculated pairwisely under constraints to edge-weight, path-length and costs. This process has to be re-calculated only if the network structure is changed. Therefore, calculation time plays a minor role. In a post-processing each route given by the pre-processing is assigned to a transport, if constraints corresponding to pickup- and delivery depot, scheduling-time and costs are satisfied. This has to be performed for each new transport in the calculation, but does not require high computing ressources, see Table 2.

3.1 Pre-processing

In the desired framework a returning of the driver to his home-depot is an important precondition to restaff the vehicle. To this, we introduce a network operation constraint $\omega_{\max} \in \mathbb{R}_{\geq 0}$, which describes the maximally allowed driving distance or time of a relay segment such that a driver can return to his home depot. Under this constraint a route (1) is only valid, if $\max_{i=0,\ldots,l-1} \left\{ \omega\left((v_{P_i}, v_{P_{i+1}})\right), \omega\left((v_{P_{i+1}}, v_{P_i})\right) \right\} \leq \omega_{\max}$ is satisfied. In this context $\omega_{\max}$ puts a ceiling on the distance between two relay-depots that a driver is allowed to drive. It is assumed to be in the range of 4–4.5 h with respect to driving time and a variable, route-based transport-velocity or between 250–300 Km in case of a constant transport-velocity. In a dense network structure, i.e. a big number of depots distributed to a small region, where the radii for pickup- and delivery-processes of different depots overlap, it can be useful to bound the distances of the transport relay segments from below by $\omega_{\min} \in \mathbb{R}_{\geq 0}$ with $\min_{i=0,\ldots,l-1} \left\{ \omega\left((v_{P_i}, v_{P_{i+1}})\right), \omega\left((v_{P_{i+1}}, v_{P_i})\right) \right\} \geq \omega_{\min}$.

In realistic scenarios it is useful to restrict the length of paths in transport routes due to two main reasons. The first reason is related to the transport processes. Here, geographical data, especially the network-covered region and distances between depots in the network, may lead to a maximal number of trailer exchanges on the route, which are accepted by the forwarders in the network. On the other hand a dense network structure will drastically increase the calculation time to find all possible paths between to depots. In this case we have to find a good balance between the flexibility of transport routes with respect to the number of transports in the matching and the available ressources allowing to calculate and evaluate all routes in an acceptable time. Let $\lambda_{\max}$ be the maximal path-length, then a transport route (1) beginning in period t_{n_0} and ending in period $t_{n_{l-i}}$ is only valid if $n_{l-1} - n_0 = l \leq \lambda_{\max}$.

Similar to [1], it is not a requirement to the network to perform all transports within this framework. Especially transports with a better cost-situation in the conventional dispatching, without relays, can be excluded from the data set. To this, we restrict the costs of a path P. Let $C_{\text{dir}}^{\text{pre}}$ be the costs of a direct transport between two depots and $C_{\text{net}}^{\text{pre}}(P)$ the costs of a relay transport along P, where $C_{\text{dir}}^{\text{pre}}$ and $C_{\text{net}}^{\text{pre}}(P)$ are as proposed in [2]. Then, a path P is only valid if $1 - \frac{C_{\text{net}}^{\text{pre}}(P)}{C_{\text{dir}}^{\text{pre}}} \geq c_{\text{adv}}$, where c_{adv} is a parameter for a desired cost-advantage of network transports.

The computation of valid paths between two different depots $v \in V$ and $w \in V$ consists of the following steps: First, $\omega_{\max}$ and $\omega_{\min}$ are used to reduce the edge set of G such that the constraints $\omega_{\max}$ and $\omega_{\min}$ are satisfied for all egdes in the reduced edge set $E^* \subseteq E$. Then, a *depth-limited search (DLS)* algorithm is used to calculate all paths satisfying the determined constraints.

3.2 Post-processing

The scheduling of each transport to the set of operation periods T may be subjected to restrictions given by shippers or receivers. To this, each transport k has a minimal time period $t_{n_{\min}^k}$, which is the first operation period, where the transport can be handled in the relay network, and a maximal time period $t_{n_{\max}^k}$, which is the last period, in which a relay transport can be performed. For route sets (2) this leads to the time-constraint $t_{n_{\min}^k} \leq t_{n_0}^j \leq t_{n_{l-1}}^j \leq t_{n_{\max}^k}$ for all routes j. In the case $t_{n_{\min}^k} > t_{n_{\max}^k}$ the transport can not be handled as a relay transport. Beside the effect on transport scheduling, $t_{n_{\min}^k}$ and $t_{n_{\max}^k}$ implicitly give a maximal length $l_{\max}^k = n_{\max}^k - n_{\min}^k + 1$ of paths that can be used to route transport between two depots.

While in the pre-processing the costs of a transport along a path P in the network graph were only considered between two depots, the post-processing considers the total transport cost, i.e. including the distances from its origin and destination to the pickup and delivery depots, respectively. Let $C_{\text{dir}}^{\text{post}}$ be the total costs of a direct transport and $C_{\text{net}}^{\text{post}}(P)$ the total costs of a relay transport along P, where $C_{\text{dir}}^{\text{post}}$ and $C_{\text{net}}^{\text{post}}(P)$ are also as proposed in [2]. Then, a path P of a route (1) in a route set (2) is only valid, if $1 - \frac{C_{\text{net}}^{\text{post}}(P)}{C_{\text{dir}}^{\text{post}}} \geq c_{\text{adv}}$.

3.3 Computational Results

The nodes in the network graph consists of 87 depots in allmost all of the 95 postcode areas in Germany, which is similar to the situation of a part load network. Especially in the west, the network structure is dense. For $c_{\text{adv}} = 0$, Table 1 shows the results for the pre-processing.

Here, Rel$_{\text{iso}}$ is the number relations without any route, Rel$_{\text{con}}$ is the number of relations with at least one route, Rel$_{\text{av}}$ is the average number of routes on a relation, Rel$_{\text{min}}$ or Rel$_{\text{max}}$ are the minimal or maximal number of routes on a relation and t_{comp} is the computation time in minutes. While $\omega_{\min}$ has a small effect on the relation structure and computation time, the path-length $\lambda_{\max}$ increases the number of routes on these connection drastically with a high drawback to t_{comp}.

Table 1 Results of pre-processing for 87 depots and $c_{adv} = 0$

ω_{min}	ω_{max}	λ_{max}	Rel$_{iso}$	Rel$_{con}$	Rel$_{av}$	Rel$_{min}$	Rel$_{max}$	t_{comp}
0	300	3	2856	4626	1.7	0	25	1.04
0	300	5	1262	6220	25.3	0	972	120.35
100	300	3	3432	4050	1.3	0	15	0.65
100	300	5	1872	5610	9.6	0	420	28.39

Table 2 Results of post-processing for 100414 transports and $c_{adv} = 0$

Shift	ω_{min}	ω_{max}	λ_{max}	Tr$_{out}$	Tr$_{in}$	Tr$_{av}$	Tr$_{max}$	t_{comp}
0/1	0	300	3	67084	33330	5.6	22	0.2
0/1	0	300	5	56892	43522	35.7	972	1.8
0/1	100	300	3	70924	29490	5.0	13	0.1
0/1	100	300	5	60973	39441	14.7	420	0.7

In the post-processing we have evaluated a set of 100414 transports given by our project-partners. In the first scenario, Shift $= 0$, there is no degree of freedom with respect to scheduling, i.e. $t_{n_{min}^k} = t_{n_0}^j \leq t_{n_{l-1}}^j = t_{n_{max}^k}$. In the second scenario, Shift $= 1$, transport routes with only one relay segment will have a route alternative, where the transport route with only one relay segment is shifted to the subsequent operation period. This makes sense, due to the fact that network-processes cover two subsequent operation periods. For $c_{adv} = 0$, Table 2 shows the results for the post-processing, where Tr$_{out}$ is the number of transports with an empty route set, Tr$_{in}$ is the number of transport with at least one valid route, Tr$_{av}$ is the average number of routes of transports with non-empty route sets, Tr$_{max}$ is the maximal number of routes in a route set and t_{comp} is the computation time in minutes.

Compared to the pre-processing, the post-processing is much faster. The effects of the parameter ω_{min} and λ_{max} are the same as in the pre-processing.

References

1. Vergara, H. (2012). *Optimization Models and Algorithms for Truckload Relay Network Design.* Fayetteville: University of Arkansas.
2. Apfelstädt, A. (2017). Handlungsoptionen im Euronationalen Ladungsverkehr Cuvillier Verlag (Vol. 1).
3. Jäger, S. (2017). *Netzwerk-Design für LKW-Komplettladungsverkehre unter Berücksichtigung ökonomischer und sozialer Aspekte.* Wiesbaden: Springer Fachmedien Wiesbaden.
4. Apfelstädt, A., & Gather, M. (2016). New design of a truck load network. In *Dynamics in Logistics - Proceedings of the 4th International Conference LDIC, 2014 Bremen, Germany,* Berlin: Springer International Publishing.

5. Dashkovskiy, S., & Nieberding, B. (2016). Costs and travel times of cooperative networks in full truck load logistics. In *Dynamics in Logistics* (pp. 193–201). Berlin: Springer International Publishing.
6. Apfelstädt, A., Dashkovskiy, S., & Nieberding, B. (2016). Modeling, optimization and solving strategies for matching problems in cooperative full truckload networks. *IFAC-PapersOnLine*, *49*(2), 18–23. Amsterdam: Elsevier.
7. Apfelstädt, A., Dashkovskiy, S., & Nieberding, B. (2017). Modeling, cycles as a solving strategy for matching problems in cooperative full truckload networks. In *Proceedings of the 20th IFAC World Congress, Toulouse*.
8. Rambau, J. (2017). Rapid mathematical programming for cooperative truck networks. In *Operations Research Proceedings 2018*.

On a Technique for Finding Running Tracks of Specific Length in a Road Network

David Willems, Oliver Zehner and Stefan Ruzika

1 Introduction

Many sport events, especially endurance sport competitions, take place in urban spaces. Throughout the past years, fun runs or obstacle runs gained public attention with growing participant numbers. Labels like XLETIX, Tough Mudder, Strongman, Münz Sportkonzept or B2Run emerged from the popularity of such events. For example, the Münz Firmenlauf 2017 in Koblenz is expecting about 17,500 athletes [1]. For a city like Koblenz this constitutes an immense intervention into traffic for each afternoon the run is being carried out. Often, those competitions are reoccurring on an annual basis and have grown to an extent that their race track disrupts the local traffic situation. Traffic participants may encounter road closures and long waiting times. Up until now event organizers have to propose a potential track at the corresponding municipal administration office and order office including start point and finish point considering all requirements for the track. If the authorities approve the proposed track the event organizer is free to publish and advertise their venture. This planning step is not trivial as the event may be canceled because authorities consider the route as inadequate or the track may negatively affect the local traffic situation. Overcrowded roads and impatient drivers can lead to a bad reputation for the organizer, which may lead to cancellation in the future as well. In this paper, we present a combinatorial algorithm to find possible running tracks of specific length in a road network. Since also running tracks are often designed as cycles the presented method constitutes a valuable tool for organizers of running events.

D. Willems (✉) · O. Zehner
Mathematical Institute, University of Koblenz-Landau, 56070 Koblenz, Germany
e-mail: davidwillems@uni-koblenz.de

S. Ruzika
Optimization Research Group, Department of Mathematics, TU Kaiserslautern, 67653 Kaiserslautern, Germany

In this article, we provide a combinatorial algorithm to solve the problem of finding given-length cycles or given-length paths. This method of finding such paths and cycles especially benefits organizers of running events as this algorithm can be applied to road networks. As a result, this approach constitutes a simplification tool in the decision making process of finding adequate race tracks. Since the algorithm returns a number of possibilities from which organizers are able to choose the track that meets most requirements desired by the organizer, it is also thinkable to apply a second optimization phase to these solutions.

2 Preliminaries from Network Optimization

The algorithms used in this paper rely on basics of graph theory and network optimization. We will briefly recall the most important definitions in this section. For a more detailed overview, we refer the reader to the book of [2].

A *directed graph* $G = (V, A)$ consists of a set V of nodes (or sometimes called vertices) and a set A of edges whose elements are ordered pairs of distinct nodes. A *directed network* is a directed graph whose nodes or edges have associated numerical values (typically costs, lengths, travel times, capacities...).

In this paper we do not make a distinction between graphs and networks, so we use the terms "graph" and "network" synonymously. For convenience we set $n = |V|$ the number of nodes and $m = |A|$ the number of edges in the network.

For a directed graph $G = (V, A)$ with node set V, edge set A and *cost function* $c: A \to \mathbb{N}$ that associates costs c_{ij} with each edge $(i, j) \in A$ and a distinguished node $s \in V$, *Dijkstra's Algorithm* [2] can be used to compute shortest paths from the source node s to all other nodes in the graph.

Remark 1 Dijkstra's algorithm using a naïve implementation has a worst case time-complexity of $\mathcal{O}(n^2)$ [2]. Using Fibonacci heaps, the time complexity can be improved to $\mathcal{O}(m + n \log n)$ [3].

In the following, we assume that only *simple paths* are qualified as candidates for running tracks. A path is called simple if it contains no repeated vertices. The simpleness of a running track is an important property, since otherwise the path would contain repeated edges and thus intersections which may lead to interruptions in the operational flow.

3 Finding Paths or Cycles of Specific Lengths

In this paragraph, we describe the algorithm used to find simple paths or cycles in a network with prespecified length.

First, we address the computational complexity of the problem. For a given graph $G = (V, A)$, two distinguished nodes $s, t \in V$ and a target length $L \in \mathbb{N}$, the task is to find a simple path $P_{s,t}$ from s to t such that its cost $c\left(P_{s,t}\right)$ is as close as possible to L. If $L \leq c^*\left(P_{s,t}\right)$ where $c^*\left(P_{s,t}\right)$ is the cost of the shortest path between s and t, the problem is solvable in polynomial time since it suffices to compute the shortest path between s and t. Otherwise, it can be shown that the problem is NP-hard.

Theorem 1 *The general problem of finding a simple path P between two nodes s, t in a weighted directed graph whose cost $c(P)$ equals a given target value L is* NP-*hard.*

Proof The problem is clearly in NP. We construct a reduction from the subset-sum problem, which is known to be NP-complete [4].

Given an instance $(\{s_1, \ldots, s_n\}, L)$ of the subset-sum problem, construct a weighted graph $G_L = (V_L, A_L)$, where $V_L = \{v_0, \ldots, v_n, v'_0, \ldots, v'_n\}$ and there is an edge (v_{i-1}, v_i) with weight s_i. Additionally, add the edges (v_{i-1}, v'_i) and (v'_i, v_i) with a weight of zero. Obviously, there exists a simple path between v_0 and v_n of cost L if and only if there is a subset of $\{s_1, ..., s_n\}$ whose sum is equal to L, which concludes the proof. □

As a consequence of Theorem 1, there is no algorithm to solve the general problem of finding a path with specific length in a network in polynomial time, provided that $P \neq NP$.

Nevertheless, a combinatorial approach to solve the problem is now as follows: from a given starting point s use the K-shortest path algorithm by [5] to successively compute new shortest paths to the target point t until the length of the kth shortest path (for $1 \leq k \leq K$) is within a threshold of the desired target length L. A formal description of Yen's Algorithm is given in Algorithm 1.

Yen's Algorithm works in two phases, determining the first of the K-shortest path P^1 and subsequently determining all other K-shortest path for $K > 1$. The algorithm maintains a list A to save the k-shortest path and a heap B to hold the potential k-shortest paths. Using this notation, the first element of A is the shortest path from the starting point s to finish t. To determine this shortest path, we use Dijkstra's algorithm.

Lemma 1 *For a given graph $G = (V, A)$, Yen's Algorithm to find the K shortest loopless paths can be implemented in such a way that the time complexity is $\mathcal{O}(Kn(m + n \log n))$.*

Proof The time complexity of Yen's Algorithm mainly depends on the shortest path algorithm used in line 12 for the computation of the spur paths. We assume that Dijkstra's Algorithm is used. Yen's Algorithm makes Kl calls to the Dijkstra algorithm in computing the spur paths, where l is the length of spur paths. In the worst case, the spur path passes all other nodes in the graph, so it holds that $l = n$. This concludes the proof. □

Remark 2 The technique described in this section can be used to find paths of specific length. Due to the way how Yen's Algorithm works, Algorithm 1 cannot be used if we want to find a cycle, i.e., it holds that $s = t$. This can be fixed by a small modification

Algorithm 1: Yen's Algorithm, [5]

Input : A weighted graph $G = (V, A)$, a nonnegative cost function $c \colon A \to \mathbb{N}$, a source $s \in V$, a target $t \in V$ and some $K \in \mathbb{N}$.
Output: A list A with the K shortest paths from s to t in ascending order.

```
1  A[0] = SHORTESTPATH(G, s, t)       // Determine shortest path from s to t
2  B = []                             // Initialize heap to store candidates
3  for k = 1 to K do
4     for i = 0 to length(A[k − 1]-1) do
5        spurNode = A[k − 1].node(i)
6        rootPath = A[k − 1].nodes(0, i)
7        foreach path p ∈ A do
8           if rootPath == p.nodes(0, i) then
9              G.remove_edge(i, i + 1)
10       foreach rootPathNode ∈ rootPath \ {spurNode} do
11          G.remove_node(rootPathNode)
12       spurPath = SHORTESTPATH(G, spurNode, t)
13       totalPath = rootPath + spurPath    // Build the entire s-t-path
14       B.push(totalPath)                  // Push candidate onto the heap
15       G.restore_edges()      // Add back the edges that were removed
16       G.restore_nodes()      // Add back the nodes that were removed
17    if B.is_empty() then
18       break
19    B.sort()
20    A[k] = B.pop()                     // Add path with lowest cost to A
21 return A
```

of the original network: instead of using the identical nodes s and t as input for the algorithm, we introduce an artificial node $\tilde{t}$ with small distance ε to s to the network and use the nodes s and $\tilde{t}$ for the computations.

4 Application and Results

The core idea of finding a running track with prespecified length is now as follows: For a given starting point s and target t compute successively paths $P_{s,t}^k$ with Yen's Algorithm until the length $c(P_{s,t}^k)$ is within a feasible windows around the desired length L. Algorithm 2 shows the algorithmic framework of this approach.

To illustrate the application of our model, we compare our results to the actual track of the 5 km long *B2RUN Kaiserslautern* run from the year 2017. The map of the event is shown in Fig. 1a. One solution with the same starting and target point and similar length ($\approx$ 4993 m) found by our method is shown in Fig. 1b. Obviously, it is also possible to reconstruct the original running track for a fitting threshold parameter l.

Furthermore, the presented methodology can easily be adapted to to integrate "routes of interest" into the running track. This can be done in an iterative approach

Algorithm 2: Running track algorithm

Input : A weighted graph $G = (V, A)$, a nonnegative cost function $c: A \to \mathbb{N}$, a source $s \in V$, a target $t \in V$, the desired target length L and a threshold l.
Output: A list A with s-t-paths with lengths from $L - l$ to $L + l$ in ascending order.

```
1 A = [ ]                                    // Initialize empty list to store paths
2 for path in YEN(G, s, t, MAXINT) do        // Iteratively generate paths
3    if L − l <= length(path) <= L + l then
4        A.append(path)
5    if length(path) > L + l then
6        Break
7 return A
```

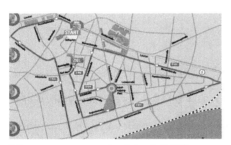

(a) The running track of the B2RUN event in Kaiserslautern, Source: [2]

(b) A solution found with our method. Start and target are at the same location as in the original.

Fig. 1 Application and comparison of the proposed method

by adding intermediate targets: to include a desired route in the running track, one uses Algorithm 2 from the starting point of the running track to the starting point of the route and additionally uses Algorithm 2 from the endpoint of the route to the target point. This approach may cause the algorithm to malfunction in the following way: since the algorithm operates on two separate parts of the running track, it is not assured that the overall concatenated path is simple, so an additional filtering step might be needed.

5 Conclusions and Further Research

In this paper we presented a real world problem of finding paths or cycles of prespecified lengths in a network. We showed that this problem cannot be solved in polynomial time until $P \neq NP$. However, we adapted a combinatorial algorithm to solve the problem in reasonable time.

In contrast to other methods like integer programming based models, our method is capable to find several solutions within the window around the desired target length. All the found solutions fulfill the length requirement by construction of the algorithm. In an additional step, those solutions can be classified further. A possible extension of our model would be (multiobjective) optimization over the solution set. For instance, the altitude difference that is traveled in the course of the running track is one criterion, that can be optimized. Alternatively, the shape of the route can be optimized. Since the original algorithm only takes the length of the track into account, the solutions may contain artifacts like zig-zags in such a way, that the track is not suitable as a running track.

Acknowledgements This work was partially funded by the German Federal Ministry of Education and Research through the project MultikOSi on assistance systems for urban events – multicriteria integration for openness and safety (Grant No. 13N12825).

References

1. Münz Sportkonzept. Münz Firmenlauf 2017. [accessed 10-April-2017]. 2017. https://www.muenz-sportkonzept.de/aktiv-event/firmenlauf/.
2. Ahuja, R. K., Magnanti, T. L., & Orlin, J. B. (1993). In Ahuja, R. K. (Ed.), *Network Flows: Theory, Algorithms, and Applications*. London: Pearson.
3. Fredman, M. L., & Tarjan, R. E. (1987). Fibonacci heaps and their uses in improved network optimization algorithms. *Journal of the ACM (JACM)*, *34*(3), 596–615.
4. Garey, M. R., & Johnson, D. S. (1979). In W. H. Freeman (Ed.), *Computers and Intractability: A Guide to the Theory of NP-Completeness* (1st ed.)., Series of Books in the Mathematical Sciences New York: W. H. Freeman & CO. ISBN: 9780716710455.
5. Yen, J. Y. (1971). Finding the K shortest loopless paths in a network. *Management Science*, *17*(11), 712–716.

Finding Maximum Minimum Cost Flows to Evaluate Gas Network Capacities

Kai Hoppmann and Robert Schwarz

1 Introduction

Recent regulation towards a market liberalization in the EU has led to the decoupling of gas trading and transport. Now the transport system operators (TSOs), who own and operate the gas networks, sell so-called transport capacity, which can be booked by the traders at entry and exit points independently. Within these booked capacities, the traders may then nominate any amount of gas that they want to insert into or withdraw from the network in the short term, that is hours or days. The TSOs have to ensure that transportation can be realized in all balanced situations, meaning that the amount of gas inserted at the entries is equal to the amount withdrawn at the exits during a certain time horizon.

Different strategies to estimate the overall transport capacity of gas networks have been developed, usually based on the evaluation of realistic and severe transport situations, see [1] or [2] for examples. An important measure for the severity of a scenario is the so-called transportmoment, i.e., the value of the induced Minimum Cost Flow Problem (MCF) [3]. The goal of the Uncapacitated Maximum Minimum Cost Flow Problem (UMMCF), which we introduce in this article, is to determine a scenario with maximum transportmoment.

In the following, we formulate two linear bilevel optimization models for UMMCF differing in the linear program used for the induced MCF. Additionally, we propose a greedy-style heuristic and present our first computational results.

K. Hoppmann (✉) · R. Schwarz
Zuse Institute Berlin, Takustraße 7, 10555 Berlin, Germany
e-mail: kai.hoppmann@zib.de

R. Schwarz
e-mail: schwarz@zib.de

2 Definitions and Notation

In this article, we consider directed flow networks $G = (V, A)$ with node set V and arc set $A \subseteq V \times V$, where each arc $a \in A$ has an associated nonnegative length $\ell_a \in \mathbb{R}_{\geq 0}$ and infinite capacity. $V^+ \subseteq V$ and $V^- \subseteq V$ denote the sources and sinks of the network and w.l.o.g. we assume that $V^+ \cap V^- = \emptyset$. Additionally, we demand that there exists at least one directed path from each source $u \in V^+$ towards each sink $w \in V^-$ in the network.

Further, for each source $u \in V^+$ a lower and an upper bound $\underline{b}_u, \overline{b}_u \in \mathbb{R}_{\geq 0}$ with $\underline{b}_u \leq \overline{b}_u$ on its supply are given. Similarly, for each sink $w \in V^-$ there is a lower and an upper bound $\underline{b}_w, \overline{b}_w \in \mathbb{R}_{\leq 0}$ with $\underline{b}_w \leq \overline{b}_w$ on its demand. At the inner nodes $V^0 := V \setminus (V^+ \cup V^-)$ flow conservation is assumed. Hence, we define $\underline{b}_v = \overline{b}_v = 0$ for each $v \in V^0$. Finally, $b \in \mathbb{R}^{|V|}$ is called demand and supply vector or scenario if $b_v \in [\underline{b}_v, \overline{b}_v]$ for all $v \in V$. It is called balanced if $\sum_{v \in V} b_v = 0$.

3 Bilevel Optimization Models

Next, we formulate the first linear bilevel optimization model for UMMCF. For an introduction to bilevel optimization and common notation and definitions we refer to [4].

$$\max_{b} \quad \sum_{a \in A} \ell_a f_a \qquad (1)$$

$$\text{s.t.} \quad b_v \in [\underline{b}_v, \overline{b}_v] \qquad \forall v \in V \qquad (2)$$

$$\min_{f} \quad \sum_{a \in A} \ell_a f_a \qquad (3)$$

$$\text{s.t.} \quad \sum_{a \in \delta^+(v)} f_a - \sum_{a \in \delta^-(v)} f_a = b_v \qquad \forall v \in V \qquad (4)$$

$$f_a \geq 0 \qquad \forall a \in A \qquad (5)$$

For each node $v \in V$ the variable b_v represents its supply or demand. Its value is chosen by the leader with respect to the upper and lower bounds (2). Given the resulting supply and demand vector, the follower solves the induced MCF problem with unlimited capacities on the arcs stated in (3)–(5). Here, the nonnegative f_a variables (5) describe the amount of flow on arc $a \in A$. Constraints (4) guarantee that the demands and supplies of all sources and sinks are satisfied and that flow conservation holds at all inner nodes. While the follower routes the flow through the network such that the cost $\sum_{a \in A} \ell_a f_a$ is minimized, it is the leaders goal to choose the supplies and demands in such a way that the cost is maximized, see (3) and (1). Since the arc flow formulation for the MCF problem [3] is used, we call this model the arc flow formulation (AFF) for UMMCF. Note that the demand and supply vector $b \in \mathbb{R}^{|V|}$ as it is chosen by the leader must be balanced. Otherwise the follower's MCF problem does not admit a feasible solution. If the bounds allow for no balanced scenarios, the problem is infeasible.

Another well-known formulation for the MCF problem, the path flow formulation [3], features flow variables for all directed paths from the sources towards the sinks instead. Since all arcs have infinite capacity in UMMCF, we can restrict ourselves to shortest paths (w.r.t. the arc lengths) here. Hence, let p_{uw} denote an arbitrary but fixed shortest path for each pair $u \in V^+$ and $w \in V^-$ in G. Additionally, we define $P := \bigcup_{u \in V^+} \bigcup_{w \in V^-} p_{uw}$, as well as $P_u := \bigcup_{w \in V^-} p_{uw}$ and $P_w := \bigcup_{u \in V^+} p_{uw}$ to simplify notation. With $\ell_p := \sum_{a \in p} \ell_a$ being the length of the path $p \in P$ and f_p denoting the flow on it, the path flow formulation (PFF) for UMMCF can then be stated as follows:

$$\max_{b} \quad \sum_{p \in P} \ell_p f_p \tag{6}$$

$$\text{s.t.} \quad b_v \in [\underline{b}_v, \overline{b}_v] \qquad \forall v \in V \tag{7}$$

$$\min_{f} \quad \sum_{p \in P} \ell_p f_p \tag{8}$$

$$\text{s.t.} \quad \sum_{p \in P_u} f_p = b_u \qquad \forall u \in V^+ \tag{9}$$

$$-\sum_{p \in P_w} f_p = b_w \qquad \forall w \in V^- \tag{10}$$

$$f_p \geq 0 \qquad \forall p \in P. \tag{11}$$

It is easy to verify that each optimal solution of PFF can be identified with an optimal solution of AFF. Thus, we can restrict ourselves to the reduced network $G' = (V, A')$ where $A' := \{a \in A \mid a \in p \text{ for some } p \in P\}$ when solving AFF.

4 Classical KKT Reformulation

A common way to solve bilevel optimization problems is to reduce them to single level problems. In this article we apply the so-called classical Karush-Kuhn-Tucker (KKT) transformation [4]. The linear programs of the follower are replaced by their KKT conditions: The primal and dual constraints together with the corresponding complementary slackness conditions. Applying it to AFF and PFF yields the following two non-linear single level problems:

$$\max_{b,f,\pi,\phi} \quad \sum_{a \in A} \ell_a f_a \tag{12}$$

$$\text{s.t.} \quad \sum_{a \in \delta^+(v)} f_a - \sum_{a \in \delta^-(v)} f_a = b_v \qquad \forall v \in V \tag{13}$$

$$\pi_v - \pi_u + \phi_a = \ell_a \qquad \forall (u,v) = a \in A \tag{14}$$

$$\phi_a f_a = 0 \qquad \forall a \in A \tag{15}$$

$$f_a \geq 0 \qquad \forall a \in A \tag{16}$$

$$\phi_a \geq 0 \qquad \forall a \in A \tag{17}$$

$$\overline{b}_v \geq b_v \geq \underline{b}_v \qquad \forall v \in V, \tag{18}$$

$$\max_{b,f,\lambda,\mu} \quad \sum_{p \in P} \ell_p f_p \tag{19}$$

$$\text{s.t.} \quad \sum_{p \in P_u} f_p = b_u \qquad \forall u \in V^+ \tag{20}$$

$$-\sum_{p \in P_w} f_p = b_w \qquad \forall w \in V^- \tag{21}$$

$$\lambda_w - \lambda_u + \mu_p = \ell_p \qquad \forall p_{uw} = p \in P \tag{22}$$

$$\mu_p f_p = 0 \qquad \forall p \in P \tag{23}$$

$$f_p \geq 0 \qquad \forall p \in P \tag{24}$$

$$\mu_p \geq 0 \qquad \forall p \in P \tag{25}$$

$$\overline{b}_v \geq b_v \geq \underline{b}_v \qquad \forall v \in V. \tag{26}$$

In the following, we denote these two models by KKT-AFF and KKT-PFF.

5 Greedy Minimum Cost Flow Heuristic

The Greedy Minimum Cost Flow Heuristic is based on the Greedy Minimum Cost Flow Method presented in [5]. To describe it here, we introduce some additional notation: For a balanced scenario $b \in \mathbb{R}^{|V|}$ we denote the optimal value of the induced MCF problem by $T(b)$. Furthermore, if we say two nodes are close to another or far away from another, this is always w.r.t. the length of a shortest path between them. The Greedy MCF Heuristic works as follows:

1. Choose a balanced scenario b_{init} and set $b_{\max} := b_{\text{init}}$ and $T_{\max} := T(b_{\text{init}})$.
2. For each source $x \in V^+$:

    (a) Set $u := x$, $b := b_{\text{init}}$, and $T := T(b_{\text{init}})$.
    (b) Choose $w \in V^-$ with $b_w > \underline{b}_w$ being farthest away from u.

(c) Increase b_u and decrease b_w simultaneously until one of the values hits a bound. Denote the resulting balanced scenario by b' and let $T' := T(b')$.
(d) If $T' < T$ go to (e). Otherwise set $b := b'$ and $T := T'$. If $b'_u = \overline{b}_u$ for all $u \in V^+$ or $b'_w = \underline{b}_w$ for all $w \in V^-$ go to (e). Otherwise, assign u an entry u' with $b_{u'} < \overline{b}_{u'}$ being closest to x and go to (b).
(e) If $T > T_{\max}$, set $b_{\max} := b$ and $T_{\max} := T$.

3. For each sink $y \in V^-$:
  (a) Set $w := y$, $b := b_{\text{init}}$, and $T := T(b_{\text{init}})$.
  (b) Choose $u \in V^+$ with $b_u < \overline{b}_u$ being farthest away from w.
  (c) Increase b_u and decrease b_w simultaneously until one of the values hits a bound. Denote the resulting balanced scenario by b' and let $T' := T(b')$.
  (d) If $T' < T$ go to (e). Otherwise set $b := b'$ and $T := T'$. If $b'_u = \overline{b}_u$ for all $u \in V^+$ or $b'_w = \underline{b}_w$ for all $w \in V^-$ go to (e). Otherwise, assign w an exit w' with $b_{w'} > \underline{b}_{w'}$ being closest to y and go to (b).
  (e) If $T > T_{\max}$, set $b_{\max} := b$ and $T_{\max} := T$.

4. Return $b_{\max}$.

The roles of entry x and exit y in the inner loops of the heuristic are to describe the direction from which flow is supposed to enter or leave the network in the created scenario, respectively. To do this, the supplies or demands of the entries or exits close to them are increased in a greedy fashion by using the corresponding farthest aways node for balancing.

Further, in the initialization phase (Step 1) the Greedy MCF Heuristic needs a scenario to start with. One way to generate it is the following: Start with the scenario where $b_u = \underline{b}_u$ for all $u \in V^+$ and $b_w = \overline{b}_w$ for all $w \in V^-$. If for example $\sum_{v \in V} b_v < 0$, increase the supply at one entry after the other until it hits its upper bounds using any consecutive order on V^+. Continue until the scenario is balanced. In case that $b_u = \overline{b}_u$ for all $u \in V^+$ at some point, but the scenario is still not balanced, the problem is infeasible. If $\sum_{v \in V} b_v > 0$ proceed analogously.

6 Computational Experiment

Next, we present a small computational experiment based on the data from the gaslib-582 network from the GasLib benchmark library [6]. The network topology and parameters are based on real data of a part of the German pipeline system, but slightly perturbed. In addition, it contains a collection of 4227 balanced scenarios, that were created with the methods described in [2]. For each source we used zero as the lower and the maximum supply value occuring in these scenarios as upper supply bound. Equivalently, for each sink we used zero as the upper and the minimum demand value occuring in these scenarios as the lower demand bound. The table below lists some important quantities concerning the reduced network.

instance	$|V|$	$|A'|$	$|V^+|$	$|V^-|$
gaslib-582	582	420	15	70

In our experiment we solve the KKT-AFF and KKT-PFF model of the gaslib-582 instance. All experiments were performed with the non-commercial MIP solver SCIP 4.0.0, using SoPlex 3.0.0 as LP solver [7]. The experiments were run on an Intel Core i7-5600U CPU with 2.6 GHz and 8GB of RAM and a time limit of 3600 s. All computations were run single-threaded.

It is important to note that for our experiment all non-linear constraints of type (15) and (23) were reformulated as SOS1 (Special Ordered Set of Type 1) constraints: Given such a set of variables, at most one of them is allowed to be non-zero in any feasible solution. In our case all the SOS1 sets contain two variables only, namely the f- and ϕ-variable for each $a \in A'$ in KKT-AFF and the f- and μ-variable for each $p \in P$ in KKT-PFF. For more information about SOS in general we refer to [8]. Important quantities of the two models and the computational results are shown below.

	vars	cons	sos1	LB	UB	time (in s)
KKT-AFF	1507	1002	420	–	–	3600
KKT-PFF	2270	1135	1050	1406674	1406674	338

The second column lists the number of variables, while the third column states the number of linear constraints of the models. In the fourth column the number of SOS1 sets can be found. In the fifth column the value of the best solution is given while the sixth column contains the best upper bound found by SCIP. Finally, the last column states the solving time for the models.

While the KKT-PFF is solved to optimality within 338 s, the KKT-AFF fails to find a feasible solution, even though it contains significantly less constraints of type SOS1. Additionally, SCIP is not able to determine any upper bound on the optimal solution within the time limit. A possible explanation for this behaviour is that SCIP detects useful variable bounds for KKT-PFF due to the sparsity of its coefficient matrix compared to KKT-AFF. The result is going to be the topic of future analysis.

We additionally ran the Greedy MCF Heuristic for the instance. For the MCF problems arising in the different steps of the heuristic, we solved the arc flow formulation for the reduced network. The Greedy MCF Heuristic provided a feasible solution with value 1379907 in 88 s. The heuristic should be used to provide an initial feasible solution for the two models in future implementations.

Acknowledgements The work for this article has been conducted within the Research Campus MODAL funded by the German Federal Ministry of Education and Research (BMBF) (fund number 05M14ZAM).

References

1. Steringa, J. J., Hoogwerf, M., Dijkhuis, H., et al. (2015). A systematic approach to transmission stress tests in entry-exit systems.
2. Koch, T., Hiller, B., Pfetsch, M. E., & Schewe, L. (2013). *Evaluating Gas Network Capacities*.
3. Ahuja, R. K, Magnanti, T. L., & Orlin, J. B. (1993). *Network Flows: Theory, Algorithms, and Applications*. Upper Saddle River: Prentice Hall.
4. Dempe, S., Kalashnikov, V., Perez-Valdes, G. A., & Kalashnykova, N. (2015). *Bilevel Programming Problems: Theory, Algorithms and Applications to Energy Networks*. Berlin: Springer.
5. Hennig, K., & Schwarz, R. (2016). Using bilevel optimization to find severe transport situations in gas transmission networks.
6. Schmidt, M., Aßmann, D., Burlacu, R., Humpola, J., Joormann, I., Kanelakis, N., Koch, T., Oucherif, D., Pfetsch, M. E., Schewe, L., Schwarz, R., Sirvent, M.: GasLib - A Library of Gas Network Instances (2017). gaslib.zib.de.
7. Maher, S. J., Fischer, T., Gally, T., Gamrath, G., Gleixner, A., Gottwald, R. L., et al. (2017). The SCIP Optimization Suite (Vol. 4).
8. Tomlin, J. A. (1988). Special ordered sets and an application to gas supply operations planning.

A Synthetic Model for Multilevel Air Transportation Networks

Marzena Fügenschuh, Ralucca Gera and Tobias Lory

1 Motivation and Related Work

The global transportation system is a very dynamic and intricate network. Optimizing travel through this network to efficiently transport goods and people via air travel, as well as analyzing its resilience to disruption, is highly desirable. Based on the real-world limitations of airports, aircrafts, financial and personnel resources as well as the unpredictability of weather and natural disasters, many variables must be taken into account. In order to effectively study the real world development of this complex network, methodical means of creating synthetic networks comparable in scope and behavior to real world data are needed. The natural development of air transportation networks is difficult to model because of the multilayered nature of the networks. Each airline independently creates routes based on market analysis for profit, competitor routes and available resources and destinations. On the other hand, each airport is separately developed by the municipalities it services with input and oversight from national and international governing bodies.

One way that this network has been studied in the past is through the analysis of multilayered networks. Multilevel or multilayered networks, frequently referred to as multiplexes, have been considered as a detailed extension of the single layered networks [1–3]. This structure is desirable in our case, as each airline company can easily be modeled by a layer, with the airports being captured by the nodes. While generating synthetic networks [4] has been very active research area, less has been done in synthetic multilayered network generation [3]. In the most common approach

---

M. Fügenschuh (✉) · T. Lory
Beuth University of Applied Sciences, Berlin, Germany
e-mail: fuegenschuh@beuth-hochschule.de

R. Gera
Department of Applied Mathematics, Naval Postgraduate School, Monterey, CA 93943, USA

growing multiplex network models are based on preferential attachment [5, 6] as they usually model relations in social networks.

Particular attention has been paid to the European Air Transportation Network (EATN), studied in [7]. A model for the network was introduced in [8], where the scale-free structure of airline networks is exploited and models simulating air traffic network based on preferential attachment are introduced. However, these models do not exploit the multilayered structure. In [9] the multilayer and the scale-free structure of EATN is exploited to design a generative model based on an enhanced preferential attachment method to imitate the EATN. As investigations of existing air transportation networks confirmed their scale-free nature [8], the approach of Barabási-Albert comes in handy to model the layers of this network. The preferential attachment method can indeed deliver a reliable multiplex network model [9]. However, the inter- and intra-layer structure has not been considered in detail.

In the current work, we build on the *BinBall* model using the Barabási-Albert approach to model the diversity of the layers within a multiplex network.

2 An Enhanced Synthetic Model for a Multiplex

A multiplex as a complex network consists of several layers (subnetworks), on the same set of nodes. As each layer is given by a different attribute (different airline in our case), the edges of the layers may duplicate each other. Thus, the multiplex M, is an undirected multigraph consisting of simple undirected graphs, the layers, $L_1, \ldots, L_\ell$, for some $\ell > 1$, i.e. $M = \bigcup_{k=1}^{\ell} L_k$. A node of a multiplex can be viewed within a single layer, or globally in the whole network. Thus one distinguishes between the *local degree* of a node u with respect to some layer L, $\deg_L(u)$, and the *global degree* with respect to the multiplex, $\deg_M(u)$.

In the *BinBall* model [9], an empty network on the node set shared across all layers is initialized. The node set is divided into possibly equally-sized subsets indicating the layers. Edges are added iteratively. For each edge, $e = (u, v)$, the layer L is chosen randomly. The selection of the end nodes is based on their local and global degrees. The probability of a node u being chosen as the first end-node of an edge, and a node v as the second end-node is:

$$\frac{\alpha \deg_L(u) + s}{\sum_{t \in V_L}(\deg_L(t) + s)} \quad \text{and} \quad \frac{\alpha \deg_M(v) + P(v) + s}{\sum_{t \in V}(\deg_M(t) + P(v) + s)},$$

respectively. Here, α, s and P are predefined values: α is a scaling factor mapping a node degree to a weight, s the *zero appeal* - a base value added to all nodes' weights when randomly choosing a node, and P a mapping from the nodes to positive reals indicating a node's global weight.

The *BinBall* model simplifies the multiplex structure, because a unified evolution manner is applied to all layers. As a result, layers of similar node and edge sizes contribute to the network. All layers evolve alike with respect to their degree distribution.

We introduce *StarGen*, a model summarized in Algorithm 1, that focuses on the diversity of the distinct layers within a multiplex. Inspired by *BinBall*'s preferential attachment we create an asynchronous growth of the layers in the multiplex. To do so, we allow different sizes of the layers based on a predefined distribution of layers' edge count. Furthermore, we decouple the scaling factor α by distinguishing between *local* and *global* α-values. We vary the local α-values to influence the variety of the intra-layer structure: to each layer L_k, $1 \leq k \leq \ell$, we assign α_k as the layer's own local exponent. We consider

$$\frac{(\deg_L(u))^{\alpha_k}}{\sum_{t \in V_L}(\deg_L(t))^{\alpha_k}} \qquad (1)$$

as the probability of a node u being chosen as the first end node, as well as,

$$\frac{\alpha \deg(v) + s}{\sum_{t \in V}(\alpha \deg(t) + s)}, \qquad (2)$$

the probability of a node being chosen as the second end node.

Algorithm 1 *StarGen*

Input
l, m, n - the total number of layers, edges and nodes in the multiplex, resp.
s - zero appeal, α - global α-value, $\alpha_1, \ldots \alpha_l$ - local α-values
$P_L^E = (p_1, \ldots, p_l)$ - layer edge sizes distribution
1: initialize multiplex M on n nodes, and empty layers $L_1, \ldots, L_l$
2: **for** each edge $e \in 1 \ldots m$ **do**
3:     select a layer, say L_i, with respect to P_L^E
4:     **if** $node\_size(L_i) \leq 0.25 \cdot n$ **then**
5:       select start node u according to the local preferential attachment (1)
6:       select end node v according to the global preferential attachment (2)
7:     **else**
8:       select start and end node u, v randomly from nodes in L_i
9:     **end if**
10:    add the edge $e = (u, v)$ to layer L_i and to multiplex M
11:    update local and global degree distribution of u and v according to (1) and (2)
12: **end for**
Output $M, L_1, \ldots, L_l$.

The layer's sizes evolve via the preferential attachment. To avoid very large layers we enforce a random selection of both nodes from the layer, if its node count exceeds 25% of the multiplex node size.

3 Data Analysis and Model Validation

Following [9] we validate our model with a real-world multiplex network data of [7]. In airline networks, nodes represent airports and edges represent flights between two airports on a given airline. A layer in this network represents the contribution of a particular airline to the network. As already reported in [7] the EATN consists of 450 distinct node labels, 37 layers, and 3588 edges (including duplicates from different layers). The layers, especially those corresponding to national airlines, tend to build a hub and spoke structure. The emergence of a hub in one layer makes it a good candidate for a spoke in another layer. As a result, the multiplex as the union of all layers has a power law degree distribution.

Our analysis of the inner, layered structure of the network revealed that the layers vary from 35 to 128 nodes, and from 34 to 601 edges. While the layer's sizes based on nodes are nearly uniformly distributed, the edge counts follow a power law distribution. Although almost all layers resemble hub and spoke structure, it shapes differently over the layers. We deduce it from the highly volatile percentage of one degree nodes across the layers, see the first chart on the left in Fig. 1. Each color represents the group of nodes of degree 1, followed by the ones of degree less than $t\%$ of local maximum degree, where $t \in \{10, 20, \ldots, 100\}$. For each x-value representing a layer, the y-value is the count of each color group, normalized by the layer's node count.

We measure the performance of the *StarGen*-model by comparing it to the *BinBall*-model and EATN. We sample 100 synthetic networks of both models with common input values for $\ell = 37$, $m = 3588$, $n = 450$, and $\alpha = 1.0$. In *BinBall*-model, the P-values represent node degrees of a random preferential attachment graph on the multiplex's node set, with incoming nodes attaching with one edge, and s is set to 0.9 as in [9]. In *StarGen*-model, we generated the probabilities P_L^E using the degree distribution of a random preferential attachment graph on the set of ℓ nodes, with incoming nodes attaching with one edge.

Based on our experiments, we chose local α-values in *StarGen* algorithm at random, uniformly distributed over the interval [1.1, 1.8]. Varying the types of distributions and the boundaries of the sampled interval, we observed that wider intervals

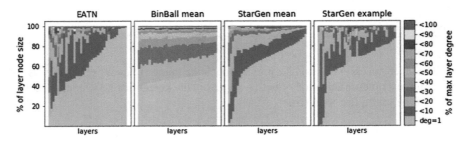

Fig. 1 The comparison of the layer degree structure of the multiplex models

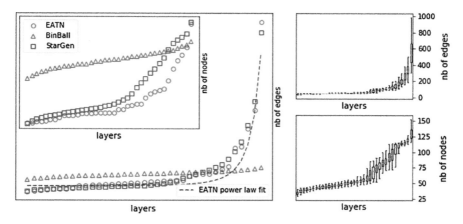

Fig. 2 Layer edge and node counts comparison: Average over *BinBall* and *StarGen* samples (left), statistics on *StarGen* sample (right)

lead to higher fluctuations of one-degree node count per layer, independently of the distribution. Additionally, the percentage of one-degree nodes increases with growing local α-values. Therefore we assign small local α-values to layers with big P_L^E-values. Furthermore, we noticed that the zero appeal (s-value) influences the number of zero degree nodes as well as the maximum degree value in the multiplex. In our setting the value $s = 1.1$ ascertained to perform best.

We refer once more to Fig. 1 showing four plots, the first being EATN, the next one is the average of 100 runs of *BinBall*, followed by the average of 100 runs of *StarGen*, and lastly one example of the analysis of a *StarGen* network. Particularly, the one-degree node count is very large overall and variable for different layers in EATN which we reproduced in *StarGen* due to the varying local α-value. The other color bands are also less uniform in the *StarGen* than in the *BinBall* samples, and match better the EATN's profile.

Figure 2 shows the edge and node (inset) count per layer for EATN, and the average of 100 runs of *BinBall* and *StarGen* algorithm. The right two figures show the boxplots of the *StarGen* samples. The appropriate choice of the distribution for layer edge counts in *StarGen*-model substantiates the good match of the layer sizes. Even the node sizes evolve adequately, although influenced only by the preferential attachment method and the limit on the maximum value. As seen in Fig. 3, the *StarGen*-model delivers a better model for the EATN-multiplex, based on the degree distribution, the average shortest path length per node, and the average centrality per node. Nevertheless, *StarGen*'s multiplexes tend to come out with higher values for the highest degree nodes.

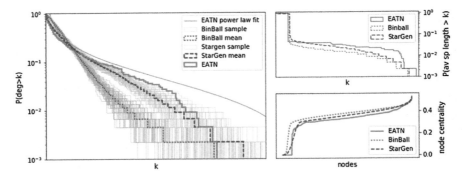

Fig. 3 Multiplex: Degree distribution (left), average shortest path length per node (upper right), average centrality per node (lower right)

4 Conclusion

Synthetic networks provide a valuable tool to generate replicas of real world networks or to predict their growth. To obtain reliable models, various characteristics of the modeled network have to be reproduced. The more complex the network is, the more challenging it is to design a straightforward procedure to emulate the network. In this work we shaped an easy-to-follow method to replicate a multiplex supporting the variety in the layers' structure. We were able to show that our model considerably outperforms its prototype *BinBall* and delivers a reliable replication of EATN, especially its intra-layer formation.

In our tests we set the interlayer structure out of scope. We observed however that it needs a further consideration as *StarGen*'s as well as *BinBall*'s layers overlap very poorly in comparison with those of EATN.

References

1. De Domenico, M., Solé-Ribalta, A., Cozzo, E., et al. (2013). Mathematical formulation of multilayer networks. *Physical Review, X3*(4). https://doi.org/10.1103/PhysRevX.3.041022.
2. Kivelä, M., Arenas, A., Barthelemy, M., Gleeson, J . P., Moreno, Y., & Porter, M. A. (2014). Multilayer networks. *Journal of Complex Networks, 2*(3), 203–271. https://doi.org/10.1093/comnet/cnu016.
3. Boccaletti, S., Bianconi, G., Criado, R., et al. (2014). The structure and dynamics of multilayer networks. *Physical Reports, 544*(1), 1–122. https://doi.org/10.1016/j.physrep.2014.07.001.
4. Barthélemy, M. (2011). Spatial networks. *499*, 1–101. https://doi.org/10.1016/j.physrep.2010.11.002.
5. Kim, J . Y., & Goh, K.-I. (2013). Coevolution and correlated multiplexity in multiplex networks. *Physics Review Letters, 111*(5). https://doi.org/10.1103/PhysRevLett.111.058702.
6. Nicosia, V., Bianconi, G., Latora, V., & Barthelemy, M. (2013). Growing multiplex networks. *Physical Review Letters, 111*(5). https://doi.org/10.1103/PhysRevLett.111.058701.

7. Cardillo, A., Gmez-Gardees, J., Zanin, M., Romance, M., et al. (2013). Emergence of network features from multiplexity. *Scientific Reports*, *3*(2). https://doi.org/10.1038/srep01344.
8. Guimer, R., & Amaral, L . A . N. (2004). Modeling the world-wide airport network. *The European Physical Journal*, *38*(2), 381–385. https://doi.org/10.1140/epjb/e2004-00131-0.
9. Basu, P., Sundaram, R., & Dippel, M. (2015). Multiplex networks: a generative model and algorithmic complexity. In *IEEE/ACM* (pp. 25–28). https://doi.org/10.1145/2808797.2808900.

On Finding Subpaths With High Demand

Stephan Schwartz, Leonardo Balestrieri and Ralf Borndörfer

1 Introduction

In this paper we consider the following *subpath load computation problem* (SLCP). Given a directed graph G and user demands in the form of weighted paths in G, compute the load of every (sub)path in G. The load of a subpath P is the sum of the weights of all user paths T which contain P as a subpath.

The problem has applications in toll billing where users of a given network are billed for certain subpaths, called segments, which they cover during their trip. This graph segmentation problem is described in detail in [1] where the problem is solved using a set-packing integer programming formulation. The information of all subpaths' loads serves as an input for the IP and is therefore crucial for the formulation.

The SLCP has connections to finding frequent subpaths. In [2], the problem of mining paths which are frequent subpaths of given trajectories is considered. There, as usual for mining frequent substructures of a graph, the term frequent is determined by a given threshold value. Consequently, the focus of the used algorithms is a bottom-up approach where frequent substructures are combined to larger substructures which are then pruned if they are not frequent themselves, see [3] for an overwiew. In contrast, we aim at computing the loads or frequencies of all possible subpaths, favoring a different approach.

While the SLCP can be solved in polynomial time, efficient computations become necessary with large networks and even larger numbers of user paths. We tackle the problem in two steps. First, we construct a *subpath-graph* to better handle duplicate subpaths. In the second step, we employ a recursive approach on the subpath-graph

---

S. Schwartz (✉) · L. Balestrieri · R. Borndörfer
Zuse Institute Berlin, Takustr. 7, 14195 Berlin, Germany
e-mail: schwartz@zib.de

R. Borndörfer
e-mail: borndoerfer@zib.de

to compute the loads of all subpaths. Our runtime analysis shows, that the presented approach compares very well against the theoretical minimum runtime.

2 The Subpath Load Computation Problem

Let $G = (V, E)$ be a directed graph with $|V| = n$ and let $\mathcal{P}$ denote the set of simple paths in G. Moreover, let $\mathcal{T} \subseteq \mathcal{P}$ be a set of *user trajectories* in G with $|\mathcal{T}| = t$ and a demand $d_T \in \mathbb{N}$ for every $T \in \mathcal{T}$. For a path $P \in \mathcal{P}$ we define the *load* of P as follows:

$$\ell(P) := \sum_{T \in \mathcal{T}:\, P \subseteq T} d_T.$$

In other words, the load of a path can be seen as the number of users covering the path during their trip. The subpath load computation problem (SLCP) is to compute the load of every possible path in G.

First, we can observe that $\ell(P) = 0$ if $P \nsubseteq T$ for all $T \in \mathcal{T}$. Consequently, we define

$$\mathcal{P}_\mathcal{T} := \{P \in \mathcal{P} \mid \exists T \in \mathcal{T} : P \subseteq T\}$$

and state that $|\mathcal{P}_\mathcal{T}| \leq t \binom{n}{2}$ since each user trajectory $T \in \mathcal{T}$ has at most $\binom{n}{2}$ subpaths. As a result, we only have to compute the loads for paths $P \in \mathcal{P}_\mathcal{T}$ and therefore avoid the exponential size of $|\mathcal{P}|$.

Now let us take a closer look at the size of $\mathcal{P}_\mathcal{T}$ and define $s := |\mathcal{P}_\mathcal{T}|$. While there are instances with $s \in \Theta(tn^2)$, e.g. with arc-disjoint user trajectories, in many cases we have $s \ll tn^2$ due to intersecting user trajectories. For example, consider a path graph on n nodes with every possible user trajectory $\bigl(\text{i.e. } t \in \Theta(n^2)\bigr)$. Therefore, we have $t \binom{n}{2} \in \Theta(n^4)$ while on the other hand, we have $s \in \Theta(n^2)$.

A natural first approach is to consider every user trajectory $T \in \mathcal{T}$ and every possible subpath of T to collect the demand for all subpaths. As pointed out above, this algorithm runs in $\mathcal{O}(tn^2)$ since we consider every subpath of every user trajectory.

In particular, every subpath in the intersection of two trajectories is considered multiple times. For example, consider the instance given in Fig. 1. Since the subpath $(2, 3)$ is part of every trajectory it is explored $|\mathcal{T}|$ times with the above algorithm. In particular, if trajectories share a longer subpath, e.g. $(1, 2, 3)$, all subpaths of this subpath are considered for each of those trajectories.

In order to avoid these multiple considerations we introduce a *subpath-graph* that ensures that every subpath is expanded only once.

(1,2,3,4,5)	(2,3,4,5)	(5,1,2,3)	(4,1,2,3)	(1,2,3)
(1,2,3,4)	(2,3,4)	(5,1,2)	(4,1,2)	(1,2)
(2,3,4,5)	(3,4,5)	(1,2,3)	(1,2,3)	(2,3)
(1,2,3)	(2,3)	(5,1)	(4,1)	
(2,3,4)	(3,4)	(1,2)	(1,2)	
(3,4,5)	(4,5)	(2,3)	(2,3)	
(1,2)				
(2,3)				
(3,4)				
(4,5)				

Fig. 1 User trajectories and considered subpaths for an instance of SLCP with $G = K_5$, $d \equiv 1$ and $\mathcal{T} = \{(1, 2, 3, 4, 5), (2, 3, 4, 5), (5, 1, 2, 3), (4, 1, 2, 3), (1, 2, 3)\}$

3 Constructing the *Subpath-Graph*

In the following we describe a problem-specific construction of what we call the *subpath-graph*. For a given instance $(G, \mathcal{T}, d)$ of the SLCP, the corresponding subpath-graph $D = (W, A)$ is a directed graph, where each node $w \in W$ represents a path $w = (v_1, \ldots, v_k)$ in G. More specifically, we have $W = \mathcal{P}_\mathcal{T}$, i.e. the nodes in D correspond to the subpaths of $\mathcal{T}$. For every node $w = (v_1, \ldots, v_k) \in W$ with $k \geq 3$ we introduce an arc (w, w_1) with $w_1 = (v_1, \ldots, v_{k-1})$ and another arc

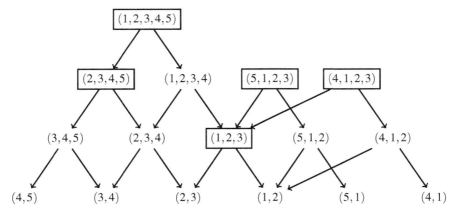

Fig. 2 Examplary subpath-graph for $\mathcal{T} = \{(1, 2, 3, 4, 5), (2, 3, 4, 5), (5, 1, 2, 3), (4, 1, 2, 3), (1, 2, 3)\}$ and $G = K_5$ as well as $d \equiv 1$

(w, w_2) with $w_2 = (v_2, \ldots, v_k)$. Figure 2 presents an exemplary subpath-graph on a small network.

First, we can observe that the subpath-graph is a directed acyclic graph, since the head of every arc represents a proper subpath of its tail. Moreover, every node in D not representing an arc in G has exactly two successors, namely the two subpaths obtained by removing the first and the last node, respectively, which implies that $|A| \leq 2|W|$.

Algorithm 3.1 specifies the construction of the subpath-graph. The set W contains nodes for which all outgoing arcs have been created while Q contains the candidates to be added to W. For every candidate w we check if it has already been considered (line 5). If it was not added to W before, we generate the successors of w as well as the corresponding arcs and add the successors as candidates to Q. The algorithm terminates if the set of candidates is empty.

Algorithm 3.1 construct subpath-graph

Require: user trajectories $\mathcal{T}$ of paths in G
Ensure: subpath-graph $D = (W, A)$
1: $W, A := \emptyset$
2: $Q := \mathcal{T}$
3: **while** $Q \neq \emptyset$ **do**
4:     $w := Q.\text{pop}()$ $// w = (v_1, \ldots, v_k)$
5:     **if** $w \notin W$ **then**
6:       $W := W \cup \{w\}$
7:       **if** $k \geq 3$ **then**
8:          $w_1 := (v_1, \ldots, v_{k-1})$
9:          $w_2 := (v_2, \ldots, v_k)$
10:         $A := A \cup \{(w, w_1), (w, w_2)\}$
11:         $Q := Q \cup \{w_1, w_2\}$
12: **return** (W, A)

With the observations above we can evaluate the runtime of this algorithm. First, note that in line 11 we only add node w to the candidate set Q if w is a head of an arc in the subpath-graph D. As we have $|A| \leq 2s$, the loop in line 3 is executed at most $2s + t$ times which lies in $\mathcal{O}(s)$. If the check in line 5 is implemented using a prefix tree with already added nodes (cf. [2]), the lookup can be done in $\mathcal{O}(n \, \Delta(G))$ where $\Delta(G)$ is the maximum degree of G. The total runtime of Algorithm 3.1 is then in $\mathcal{O}(s \, n \, \Delta(G))$.

4 Solving the SLCP Recursively

Now that we have constructed the subpath-graph, we will describe an algorithm to efficiently compute the loads of all nodes in D to solve the SLCP.

Algorithm 4.1 Compute loads for all subpaths

Require: subpath-graph D, user trajectories $\mathcal{T}$ with demands (d_T)
Ensure: loads $L = \bigl(\ell(w)\bigr)_{w \in W}$
1: Compute $W_m := \{w = (v_1, \ldots, v_m) \in W\}$ for $m = 2, \ldots, n$
2: $\ell(w) := 0 \quad \forall w \in W$
3: $\ell(w) := d_T \quad \forall w = T \in \mathcal{T}$
4: **for** $m \in \{n, \ldots, 2\}$ **do**
5:    **for** $(v_1, \ldots, v_m) \in W_m$ **do**
6:       **if** $m \geq 3$ **then**
7:          $\ell(v_1, \ldots, v_{m-1}) := \ell(v_1, \ldots, v_{m-1}) + \ell(v_1, \ldots, v_m)$
8:          $\ell(v_2, \ldots, v_m) \;\;\;\;:= \ell(v_2, \ldots, v_m) \;\;\;+ \ell(v_1, \ldots, v_m)$
9:       **if** $m \geq 4$ **then**
10:          $\ell(v_2, \ldots, v_{m-1}) := \ell(v_2, \ldots, v_{m-1}) - \ell(v_1, \ldots, v_m)$
11: **return** $L = \bigl(\ell(w)\bigr)_{w \in W}$

Algorithm 4.1 starts by partitioning the node set W into several level sets depending on the length of the path associated with each node (line 1). Afterwards, starting at the top level we descend the graph and for every node $w = (v_1, \ldots, v_m)$ that we consider, we add the load of the current node to the load of both of its successors, given that w is not a leaf, i.e. $m \geq 3$. We also subtract the current load from the load of "the inner path" $(v_2, \ldots, v_{m-1})$, if this is still a path, i.e. $m \geq 4$. We will see in a moment that this is necessary to respect the inclusion-exclusion principle (cf. Theorem 1) and that the algorithm indeed computes the loads of all subpaths.

Let us first analyze the runtime of Algorithm 4.1. Computing the level sets W_m can be done in $\mathcal{O}(s)$ if we start at the leafs ($m = 2$) and traverse the graph with reversed arcs. In the main part we consider every node in W exactly once and since all other operations can be performed in $\mathcal{O}(1)$ the total runtime of this algorithm is in $\mathcal{O}(s)$.

In the following we prove the recursion which is implemented in Algorithm 4.1. We start by introducing further notation to simplify the illustration of the recursion. For $P = (v_1, \ldots v_k) \in \mathcal{P}$ we define $N^-(P) := \{v_0 \in V \mid (v_0, v_1, \ldots, v_k) \in \mathcal{P}\}$ and for $v \in N^-(P)$ we set $v \boxplus P := (v, v_1, \ldots, v_k)$. This means that $N^-(P)$ are the predecessors of node v_1 which are not part of path P. Therefore, $(v, v_1, \ldots, v_k)$ with $v \in N^-(P)$ is a simple path in G which we denote by $v \boxplus P$. Analogously, we define $N^+(P) := \{v_{k+1} \in V \mid (v_1, \ldots, v_k, v_{k+1}) \in \mathcal{P}\}$ and set $P \boxplus v := (v_1, \ldots, v_k, v)$ for $v \in N^+(P)$. Finally, for $P \in \mathcal{P}$ we write $\mathcal{T}(P) := \{T \in \mathcal{T} : P \subseteq T\}$ and obtain the following result.

Lemma 1 *For $P \in \mathcal{P}$ we have*

$$\mathcal{T}(P) = (\{P\} \cap \mathcal{T}) \cup \bigcup_{u \in N^-(P)} \mathcal{T}(u \boxplus P) \cup \bigcup_{v \in N^+(P)} \mathcal{T}(P \boxplus v). \tag{1}$$

Moreover, for arbitrary $u \in N^-(P)$ and $v \in N^+(P)$ we have

$$\mathcal{T}(u \boxplus P) \cap \mathcal{T}(P \boxplus v) = \mathcal{T}(u \boxplus P \boxplus v).$$

Proof Let $P \in \mathcal{P}$ and $T \in \mathcal{T}$. We know that $P = T \iff (\{P\} \cap \mathcal{T}) = T$ and we can also observe that P is a proper subpath of T if and only if there is a node $u \in N^-(P)$ or $v \in N^+(P)$ such that $u \boxplus P \subseteq T$ or $P \boxplus v \subseteq T$. This proves the first equation. To prove the second equation let $P = (v_1, \ldots, v_k)$ and let $u \in N^-(P)$ and $v \in N^+(P)$. Now obviously $(u, v_1, \ldots, v_k) \subseteq T$ and $(v_1, \ldots, v_k, v) \subseteq T$ iff $(u, v_1, \ldots, v_k, v) \subseteq T$ which concludes the proof. □

Now we extend the user demand to all paths by defining $d_P := 0 \ \forall P \notin \mathcal{T}$ to formulate the following recursion.

Theorem 1 *Let $P \in \mathcal{P}$, then*

$$\ell(P) = d_P + \sum_{u \in N^-(P)} \ell(u \boxplus P) + \sum_{v \in N^+(P)} \ell(P \boxplus v) - \sum_{u \in N^-(P)} \sum_{v \in N^+(P)} \ell(u \boxplus P \boxplus v).$$

Proof We use Lemma 1 and the inclusion-exclusion principle. First note that for $u_1 \neq u_2 \in N^-(P)$ we have $\mathcal{T}(u_1 \boxplus P) \cap \mathcal{T}(u_2 \boxplus P) = \emptyset$. Analogously, for $v_1 \neq v_2 \in N^+(P)$ we have $\mathcal{T}(P \boxplus v_1) \cap \mathcal{T}(P \boxplus v_2) = \emptyset$. Inserting the identity from (1) into the definition of $\ell(P)$, the statement immediately follows using the inclusion-exclusion principle. □

Theorem 2 *Algorithm 4.1 is correct.*

Proof With Theorem 1 it is easy to prove the correctness of Algorithm 4.1. For any path $P \in \mathcal{P}_T$ and for arbitrary $u \in N^-(P)$ we know that either $u \boxplus P \notin \mathcal{P}_T$ or $u \boxplus P$ is a predecessor of P in the path-graph D. While the first implies that $\ell(u \boxplus P) = 0$, the latter ensures that the load is added to $\ell(P)$ in line 8 of the algorithm when the node $u \boxplus P$ and its successors are considered. Analogously, this holds for $v \in N^+(P)$ and the paths $P \boxplus v$ and $u \boxplus P \boxplus v$, proving the correctness of Algorithm 4.1. □

We conclude that the subpath-graph can be constructed in $\mathcal{O}(s \ n \ \Delta(G))$. In many networks we can assume that the maximum degree is bounded, leading to a runtime of $\mathcal{O}(s \ n)$. If the subpath-graph is constructed, our recursive algorithm to solve the SLCP runs in $\mathcal{O}(s)$. Given that considering every subpath at least once leads to a minimum runtime of $\mathcal{O}(s)$, the presented algorithms are very well suited for solving the SLCP.

References

1. Schwartz, S., Borndörfer, R., & Bartz, G. (2015). The graph segmentation problem. In *Proceedings of INOC 2017, ENDM* (to appear).
2. Guha, S. (2014). Finding frequent subpaths in a graph. *International Journal of Data Mining & Knowledge Management Process, 5*, 35.
3. Cook, D., & Holder, L. (Eds.). (2006). *Mining Graph Data*. New Jersey: John Wiley & Sons.

Part X
Health Care Management

Kidney Exchange Programs with a Priori Crossmatch Probing

Filipe Alvelos and Ana Viana

1 Introduction

Patients suffering from chronic kidney disease have three alternatives for transplant – find a compatible donor in a deceased donors waiting list, have a willing compatible living donor or join a kidney exchange program (KEP). In these programs patients with a willing incompatible donor join a pool of incompatible patient-donor pairs and, if compatibility between patient in one pair and donor in another is found, patient in one pair can receive an organ from the donor in another pair and vice-versa. The problem can be represented by a graph where each node P_i represents an incompatible pair i and an arc from P_i to P_j means that the donor in pair i is compatible with the patient in pair j. A feasible exchange plan corresponds to a set of disjoint cycles in the graph.

This work has been supported by COMPETE: POCI-01-0145-FEDER-007043 and FCT Fundação para a Ciência e Tecnologia within the Project Scope: UID/CEC/00319/2013 and by the ERDF European Regional Development Fund through the Operational Programme for Competitiveness and Internationalisation - COMPETE 2020 Programme, and by National Funds through the Portuguese funding agency, FCT - Fundao para a Cincia e a Tecnologia, within project POCI-01-0145-FEDER-016677.

F. Alvelos (✉)
Centro Algoritmi/Departamento de Produção e Sistemas, Universidade do Minho,
4710-057 Braga, Portugal
e-mail: falvelos@deps.uminho.pt

A. Viana
INESC TEC, Campus da FEUP, 4200-465 Porto, Portugal
e-mail: aviana@inesctec.pt

A. Viana
ISEP - School of Engineering, Polytechnic of Porto, 4200-072 Porto, Portugal

© Springer International Publishing AG, part of Springer Nature 2018
N. Kliewer et al. (eds.), *Operations Research Proceedings 2017*, Operations Research Proceedings, https://doi.org/10.1007/978-3-319-89920-6_49

Fig. 1 Compatibility graph and proposed solution (arcs in bold)

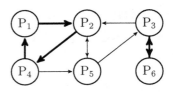

In Fig. 1 we present an example of a compatibility graph for a program with six pairs, and a possible set of exchanges (arcs in bold) corresponding to the disjoint cycles $1-2-4-1$ and $3-6-3$.

To assess preliminary pair compatibility besides comparing patient and donor blood types it is necessary to perform an additional test, *virtual* crossmatch, that detects whether patients have antibodies to donors specific antigens, or not. If the virtual crossmatch result is negative, patient and donor(s) are considered compatible. Based on this information, an exchange plan is proposed. However, a more accurate crossmatch test (hereby simply referred to as crossmatch as opposed to virtual crossmatch) is performed later for selected donor and receiving patient. This test can detect new incompatibilities that will prevent the actual transplant, as well as all transplants in the cycle that involves that transplant, from being performed. Because the bigger the cycle the more transplants will be cancelled due to new incompatibilities detection, in general KEPs define a limit k to the size of the cycle. This limit is also due to logistics reasons.

The problem of selecting the pairs that should be considered for transplant so that a given objective is optimised, was modeled as an Integer Program (IP) by several authors [1, 3]. Classically, those models do not consider in the decision process the possibility of new incompatibilities being detected and, as so, it can happen that the actual number of transplants performed is significantly reduced when compared to the planned number. The work in [6] addresses the problem by considering an objective where cycles that can be replaced by other (sub)cycles if an arc fails are preferred. In [5], probabilities of positive crossmatch are taken into account for maximizing the *expected* number of transplants. A robust optimization approach can be found in [7].

In this paper we propose and evaluate a new approach that also aims at increasing the number of actual transplants. We depart from a compatibility graph and, prior to proposing a solution, we consider the possibility of making crossmatch tests to a pre-defined number of arcs, possibly modifying the transplants plan each time an incompatibility is found. Two types of policies are proposed: one where exchanges that were already tested are fixed in the plan; another, where those exchanges can be removed from the plan if a plan with more (potential) transplants exists (flexible policies).

Computational tests on instances of realistic size show that the flexible policies allow for substantial improvement on the actual number of transplants with a small number of tests.

The paper is organized as follows. Following this introduction, in Sect. 2 we describe the problem and the proposed approaches. In Sect. 3, we report computational results. In Sect. 4, the main conclusions are drawn.

2 A Priori Crossmatch

If no a priori crossmatch tests are considered, the optimal number of transplants in a KEP can be obtained via integer programming using, most commonly, the cycle formulation (see [1]).

For each transplant in the plan corresponding to the optimal solution found, a more accurate test, consisting in physically mixing cells from both the donor and patient, is conducted. Some of these actual crossmatch tests may contradict the previous compatibility assumptions and preclude the tested transplants, as well as the others involved in affected exchanges. In the optimization model, an actual crossmatch corresponds to testing the existence of the corresponding arc. If the crossmatch is positive, the transplant is not feasible, the arc is removed from the graph, and the cycle including the arc must be removed.

In this paper, we study a priori crossmatch policies, where actual crossmatches are conducted before a definite transplants plan is defined. Given that an actual crossmatch requires resources (at least, money and time), a limit on the number of actual crossmatches to be conducted is imposed. The problem now is to decide on the (possibly temporary) solution to consider and on the arcs to test as, opposed to maximizing the number of planned transplants. In the next subsections we describe three policies that differ on how to select the arc to test and on the solution considered in each iteration.

2.1 Fixed Solution Policy

In this policy, in each iteration, one arbitrary arc of the current solution (that was not tested before) is tested. According to the general algorithm, if the arc exists (i.e. the crossmatch test is negative), the solution is kept and another arc is chosen to be tested. If the arc does not exist, a new solution is obtained by maximizing the number of transplants in the residual graph. In this policy all arcs already tested that exist are forced to be part of the solution (if it does not preclude feasibility).

For the example, in Fig. 1, 5 transplants will be performed if all arcs exist. Suppose now that arc $1 - 2$ is selected to be tested. If it exists, then another arc is chosen to be tested and arc $1 - 2$ will be forced to be part of the solution in the remaining procedure (unless no feasible solution exists with arc $1 - 2$). If incompatibility between 1 and 2 is found, another plan is obtained after removing arc $1 - 2$ from the graph.

It is important to note that this policy may not lead to the number of transplants that would be reached with complete information, even if all arcs could be tested.

An arc that exists but is not part of the optimal solution with complete information may be fixed in the solution. Nevertheless, this policy may be relevant in practise, as it assures that if a crossmatch is negative, the involved pairs are being considered for actual transplantation.

2.2 Flexible Solution Policies

In a flexible policy, the solution does not necessarily include existing tested arcs. If, in an iteration, the tested arc does not exist, lexicographic optimization is applied. The first objective is to maximize the number of transplants, assuring that no solution with a high potential is lost (and thus convergence to the complete information optimal number of transplants). Among the solutions with maximum number of transplants, a solution with arcs that were already tested (potentially allowing a higher actual number of transplants) and as similar as possible to the current one (avoiding large variations on the actual number of transplants) is preferable. These two criteria are treated in a single objective through weights (incentives are given to arcs tested that exist, and to cycles and arcs that belong to the current solution).

We consider two flexible policies that differ in the way the arc to be tested is selected. In the first policy (named arbitrary), that arc is chosen arbitrarily. In the second policy (named delta), the arc to be tested is the one whose failure implies the loss of more planned transplants. The rationale for the *delta* policy is that if the failure of one arc does not have an impact in the value of the maximum number of transplants (because there are solutions not including the arcs with the same value), conducting the corresponding crossmatch does not provide any additional useful information - independently of the crossmatch being positive or negative, a solution with the same number of planned transplants exists.

As an example, let us consider again the plan of Fig. 1. The *deltas* for the arcs in the plan are $\delta_{12} = 0$ (solution $2 - 4 - 5 - 2$ and $3 - 6 - 3$ also have 5 transplants), $\delta_{24} = 1$ (optimal plan without arc $2 - 4$ is $2 - 5 - 2$ and $3 - 6 - 3$ with value 4), $\delta_{41} = 0, \delta_{36} = 2$ and $\delta_{63} = 2$. Since they both have the larger delta, in this case one of the arcs $3 - 6$ and $6 - 3$ would be tested. If the arc fails, an optimization is conducted to obtain a new plan. However, in this policy arcs already tested are not forced to be part of the plan, although if there is more than one solution with the same (maximum) number of transplants, solutions with more tested arcs with will be preferred.

3 Computational Results

We performed a set of computational tests in order to assess the merits of the proposed policies and quantify the relation between the number of arcs tested and the number of actual transplants. In particular, it is of most practical relevance to address two questions: (i) How many additional transplants can be actually performed for a given

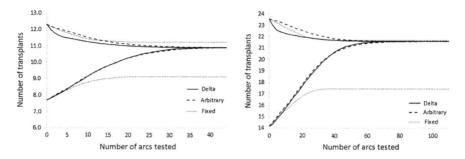

Fig. 2 Number of transplants versus number of arcs tested for the 30 (left) and 50 (right) nodes instances. Upper curves: planned transplants; lower curves: actual transplants

number of arcs tested and (ii) How many tests must be conducted to achieve a given number of additional actual transplants (in the limit, the complete information number of transplants)?

For that purpose we used two sets of 50 instances each, one with 30 nodes and the other with 50 nodes. The instances' generator takes into account probabilities of blood type and HLA as described in [4]. For each instance, 100 independent runs were conducted. In each run, the existence of each arc is determined according to the estimation of the probability of a positive crossmatch from [2]. All tests were conducted on a machine with a Intel Core i7 CPU @ 2.3 GHz and 6 GB RAM. The maximum length of a cycle was set to three (as common in practice). The computational times revealed very small: for the 50 nodes instances, in average, it took less than one second for testing all the arcs for all the three methods.

Figure 2 shows the average number of planned (upper curves) and actual (lower curves) number of transplants with respect to the number of tested arcs for the instances with 30 and 50 nodes. When no arcs are tested, the number of actual transplants is around 60% of the number of planned transplants (for the 30 and 50 nodes instances, in average, 12.3 and 23.5 transplants are planned and 7.7 and 14.2 are performed, respectively).

Both flexible policies behave similarly in both sets of instances, converging to the complete information number of actual transplants (10.9 and 21.6) when around 10% of the arcs are tested. For the 30 nodes instances, on average, after 29.4 arcs are tested the complete information solution is reached for both flexible policies. For the 50 nodes instances, on average, after 55.5 (delta policy) and 58.0 (arbitrary policy) arcs are tested the complete information solution is reached. The same does not happen with the fixed policy which does not converge to the complete information actual number of transplants.

Furthermore, for flexible policies, there is a approximately linear behaviour in the first part of the actual transplants curve for both sets of instances, which allows to conclude that for around 7 arcs tested, there is one additional transplant.

4 Conclusions

In this paper we proposed new policies for Kidney Exchange Programs that try to reduce the number of planned transplants that are cancelled due to last minute incompatibilities and increase the actual number of transplants. These policies allow for actual crossmatch tests to be performed in a set of selected pairs. Three policies were proposed: a fixed policy and two flexible policies. For the fixed policy, after a transplant plan is set, one arbitrary arc of the current solution (that was not tested before) is tested. If the test is positive, the current solution is updated to a solution that maximizes the number of transplants, excludes that arc and includes all arcs already tested and having negative crossmatch. For the flexible policies, if a test is positive, the current solution is updated to a solution that maximizes in lexicographic order two objectives. Two possibilities are considered when selecting the arc for probing: (1) the one that implies a larger deterioration in the number of transplants if the arc fails, and (2) arbitrarily.

Results show that the flexible policies have similar results, that are much better than the fixed policy. For the flexible policies only a small percentage (around 10%) of the total number of potential transplants needs to be tested to achieve the maximum possible number of transplants (that corresponds to the case where no new incompatibilities arise). For small absolute values of the number of arcs tested, the number of actual transplants grows approximately linearly (around one more transplant for every 7 arcs tested).

References

1. Constantino, M., Klimentova, X., Viana, A., & Rais, A. (2013). New insights on integer-programming models for the kidney exchange problem. *European Journal of Operational Research, 231*(1), 57–68.
2. Glorie, K. (2012). Estimating the probability of positive crossmatch after negative virtual cross-match (No. EI 2012-25), pp. 1–8.
3. Dickerson, J. P., Manlove, D. F., Plaut, B., Sandholm, T., & Trimble, J. (2016). Position-indexed formulations for kidney exchange. In *Proceedings of the 2016 ACM Conference on Economics and Computation* (pp. 25–42).
4. Saidman, S., Roth, A., Sonmez, T., Unver, M., & Delmonico, F. (2006). Increasing the opportunity of live kidney donation by matching for two- and three-way exchanges. *Transplantation, 81*, 773–782.
5. Klimentova, X., Pedroso, J. P., & Viana, A. (2016). Maximising expectation of the number of transplants in kidney exchange programmes. *Computers and Operations Research, 73*, 1–11.
6. Manlove, D., & OMalley, G. (2012). Paired and altruistic kidney donation in the UK: algorithms and experimentation. *Lecture Notes in Computer Science, 7276*, 271–282.
7. Glorie, K. M. (2014). Clearing barter exchange markets: kidney exchange and beyond, Chapter VI. Ph.D thesis, Erasmus University Rotterdam.

Preventing Hot Spots in High Dose-Rate Brachytherapy

Björn Morén, Torbjörn Larsson and Åsa Carlsson Tedgren

1 Introduction

High Dose-Rate Brachytherapy (HDR BT) is a modality of radiation therapy, used e.g. in prostate cancer treatment. Contrary to external beam radiation, in BT the radiation dose is delivered from within the body using catheters (hollow needles). The radiation source steps through the catheters between dwell positions, and at each of these it can dwell for some time. We consider the number of catheters and their placements to be predetermined in the dose planning. In addition to the tumour (Planning Target Volume, PTV), which should receive a high enough dose, there are healthy organs and tissue nearby (Organs At Risk, OAR), which should be spared if possible. The PTV and OAR are discretised into dose points, where the radiation doses (in Gray, Gy) are calculated and used for evaluating dose plans. For an introduction to radiotherapy and BT see [4].

Clinically, dose plans are constructed either manually, with graphical tools available in treatment planning software, or with mathematical optimization. The most used optimization model for BT dose planning is the Linear Penalty Model (LPM, see e.g. [5, pp. 33–35]), which we also used in this study. For each dose point there is a penalty if the dose is outside a specified interval, and this penalty increases

---

B. Morén (✉) · T. Larsson
Department of Mathematics, Linköping University, 58183 Linköping, Sweden
e-mail: bjorn.moren@liu.se

B. Morén
Department of Medical and Health Sciences, Linköping University, Linköping, Sweden

Å. C. Tedgren
Medical Radiation Physics and Nuclear Medicine,
Karolinska University Hospital, Solna, Sweden

Å. C. Tedgren
Department of Oncology Pathology, Karolinska Institute, Solna, Sweden

© Springer International Publishing AG, part of Springer Nature 2018
N. Kliewer et al. (eds.), *Operations Research Proceedings 2017*, Operations
Research Proceedings, https://doi.org/10.1007/978-3-319-89920-6_50

linearly with the deviation from the interval. The LPM is easily solved with Linear Programming (LP) methods.

The primary criteria for evaluating dose distributions are Dosimetric Indices (DIs). For the PTV, the portion of its volume that receives at least a specified prescription dose is of interest, while for an OAR it is the portion that receives at most a specified dose. For the PTV a DI is denoted V_x^{PTV}, where x is a percentage of the prescription dose. Note that all DIs are only aggregate measures of the dose. An optimization model that handles DIs explicitly is the dose-volume model. It has been studied in e.g. [2, 7]. For a thorough introduction to mathematical optimization in HDR BT dose planning see [5].

While it is important that the PTV receives a dose that is high enough, it is not good if the dose is too high. The concept of a hot spot refers to a noticeable contiguous volume that receives a dose that is much too high. When a dose distribution satisfies the treatment goals in terms of DIs, it is common clinical practice to visually inspect the spatial dose distribution before approving the plan. If there are too large hot spots, the dose plan is manually adjusted to reduce their volumes, while trying to maintain the levels of the DIs.

The purpose of our work is to study this adjustment process by means of mathematical optimization and introduce criteria that explicitly take the spatial distribution of the dose into account. The aim is twofold and we show that

1. there is some degree of freedom in the dose planning even though we keep the levels of the DIs close to the values of an initial acceptable plan, and
2. it is possible to use this degree of freedom to reduce the prevalence of hot spots, within a clinically feasible computing time.

We have designed a method to resemble the clinical planning process. After finding an initial acceptable dose plan with respect to the primary criteria (the DIs), the plan is adjusted, with constraints on the DIs, to improve it with respect to the spatial dose distribution. Thus we consider both aggregate and spatial measures, which is our contribution. This two-step approach was chosen since there is a clear priority of the two goals. It is also an advantage that the adjustment step is independent of the first step and therefore applicable to any given dose plan.

2 Models

The set of dose points in the PTV is T, the set of OAR is S, the set of dose points in organ at risk s is OAR_s, and the set of dwell positions is J. The value of a DI is ω for the PTV and τ^s, $s \in S$, for OAR. The prescribed dose for the PTV is L. For OAR the prescribed upper bounds are U^s, $s \in S$, and the maximal allowed doses are M^s, $s \in S$. The parameter d_{ij} denotes dose contribution from dwell position j to dose point i. Dwell time variables are denoted t_j. The received dose at a dose point i is $D_i = \sum_{j \in J} d_{ij} t_j$, $i \in T$. The contribution to a DI from a dose point is defined by a step function (Heaviside function, H) and can be modelled using binary variables.

Preventing Hot Spots in High Dose-Rate Brachytherapy 371

These are y_i, $i \in T$, for the PTV, and v_i^s, $i \in OAR_s$, $s \in S$, for OAR. The former variables take the value 1 if the dose is high enough and the latter take the value 1 if the dose is low enough.

The adjustment step is based on the following optimization model, with a general objective function, $f(t)$, where $t = (t_j)_{j \in J}$, which will be defined later.

$$
\begin{aligned}
\min \quad & f(t) \\
s.t. \quad & \sum_{j \in J} d_{ij} t_j \geq L y_i, & i \in T \\
& \sum_{i \in T} y_i \geq \omega |T| \\
& \sum_{j \in J} d_{ij} t_j \leq U^s + (M^s - U^s)(1 - v_i^s), & i \in OAR_s,\ s \in S \\
& \sum_{i \in OAR_s} v_i^s \geq \tau^s |OAR_s|, & s \in S \\
& t_j \geq 0, & j \in J \\
& v_i^s \in \{0, 1\}, & i \in OAR_s,\ s \in S \\
& y_i \in \{0, 1\}, & i \in T
\end{aligned} \quad (1)
$$

The first two constraints together model a requirement on the DI V_{100}^{PTV}, where the first constraint ensures that each binary variable takes the correct value and the second is forcing a high enough dose to a large enough portion of the PTV. The combination of the third and fourth constraints models requirements on DIs for OAR. The third constraint ensures that each binary variable takes the correct value and the fourth is forcing a large enough portion of the OAR to receive a dose that is low enough. Model (1) is based on a standard formulation of a dose-volume model, which includes the same constraints, with τ^s, $s \in S$, as parameters but with ω as a variable, and objective to maximise ω (i.e., V_{100}^{PTV}).

The prevalence of hot spots can be reduced, either by decreasing the number of dose points with a too high dose, if possible, or by redistributing them, to make the spatial dose distribution more uniform. To achieve this, we suggest an objective function that gives a penalty for each pair of dose points in the PTV if the dose is too high in both points. Further, this penalty is larger the closer the dose points are. The objective function is given by

$$
\sum_{i, j \in T: i \neq j} \frac{g(D_i) g(D_j)}{distance(i, j)^2}, \quad (2)
$$

where the denominator is the squared Euclidean distance between dose points i and j. We tried two formulations of the function g:

$$
g(D_i) = \{\max(0, D_i - b)\}^2, \quad (3)
$$
$$
g(D_i) = H(D_i - b). \quad (4)
$$

Here b is an upper bound on the dose (e.g. 200% of the prescription dose).

We also considered a third objective function. Here we represent potential hot spots by dividing the PTV into small subvolumes, where each subvolume is located around two adjacent dwell positions. These subvolumes can overlap. Letting K denote the set of all subvolumes we construct the objective function

$$100 \times \max_{k \in K} A_k + \sum_{k \in K} A_k, \qquad (5)$$

where A_k, is the mean dose in subvolume k, $k \in K$.

3 Experiments and Results

We have tested our models on a data set consisting of 10 patients, previously treated for prostate cancer, and all values given in this section are mean values for these 10 patients. The number of dose points was in the range 4 369–7 939, the number of dwell positions in the range 190–352, and the number of catheters in the range 14–20. In line with clinical practice, the PTV and OAR (urethra and rectum) were contoured on medical images and we added artificial healthy tissue surrounding the PTV. Prescription dose for the PTV was 8.5 Gy and the bounds on the DIs for OAR were 90%. To simulate problems in the catheter placement, which is a common cause for hot spots, we modified the patient data by removing three catheters and corresponding dwell positions for each patient.

We compare four models. Model I is the LPM, which gave an initial dose plan as input to the adjustment step, in which models II–IV were used. These are based on the model (1) with objective to minimise (3)–(5), respectively. In the adjustment step the lower bound on V_{100}^{PTV} was set to its initial value minus one percentage point.

We tried to solve model III as a Mixed Integer Program (MIP), which however showed to be intractable. Therefore, the binary variables were approximated (cf. [3]) and replaced with nonlinear functions $0.5(1 + \tanh(\beta (D_i - b)))$, where $\beta > 0$ is a constant. This approximation was also used in models II and IV.

Gurobi was used to solve LPs and MIPs, and the Matlab solver fmincon was used for nonlinear models, with computing times in the order of a few minutes.

First, to study the degrees of freedom in the adjustment step, we calculate the difference in dwell times between the adjusted plan and the initial plan. The mean total dwell time was 437 s for model I and 420 s for model II, and the 1-norm deviation of the dwell times was 401 s. Since this value includes both increased and decreased dwell times, half of it, 200.5 s, is more representative for the actual deviation. This is 46% of the total dwell time and indicates that a dose plan can be substantially changed while keeping the initial levels of DIs.

A large number of active dwell positions (that is, with $t_j > 0$) is considered advantageous because it can give a more homogeneous dose distribution and also a more robust plan with respect to uncertainties in the catheter placement. Restrictions on dwell times to get more active dwell positions and more evenly distributed dwell times have been studied in [1, 2]. Typically the LPM gives fewer active dwell positions and longer maximum dwell times compared to manual planning [6]. In Fig. 1a we

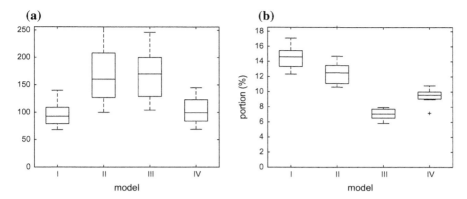

Fig. 1 a Shows the number of active dwell positions and **b** shows V_{200}^{PTV}

see that the number of active dwell positions from models II and III are significantly higher than from model I.

Figure 1b shows a decrease in mean V_{200}^{PTV}, from 15% for model I to 7% for model III. This indicates that the dose plan has indeed changed substantially. The decrease in V_{200}^{PTV} is probably beneficial in terms of prevalence of hot spots, although the DI does not take spatiality of the dose into account.

Second, for a comparison of dose plans with respect to spatial distribution, Fig. 2a is based on the division of the PTV into subvolumes. For each subvolume k, V_{200}^{k} is calculated and the figure shows the highest values for the 10 patients. Since each subvolume corresponds to a small, spatially connected volume, a high value indicates a hot spot. From model III, we got a mean result of 41%, compared to 62% from model I. Figure 2b shows the mean dose in the subvolume with the highest mean

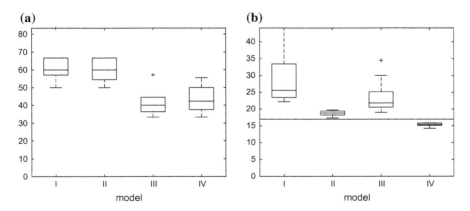

Fig. 2 a Shows the highest V_{200}^{k} in a subvolume. **b** Shows the highest mean dose in a subvolume, with the horizontal line at the 200% dose level

dose, with results of 28 and 15 Gy for models I and IV, respectively. Both these results indicate significant dose reductions for the hottest subvolumes, which is an important target for the adjustment step.

4 Conclusions and Future Research

The primary criteria for evaluating HDR BT dose distributions are DIs, which are aggregate measures. We study the possibility to take spatial aspects into account and find that it is possible to adjust dwell times and dose plans significantly in this respect while maintaining DIs on acceptable levels. Our models can be used clinically since they provide solutions within a short time.

We have presented criteria that evaluate the dose distribution in subvolumes of the PTV which indicate potential hot spots, and we show improvements for the hottest subvolumes. There might however be other aspects worth studying, possibly better related to treatment results. For evaluating treatment results with respect to prevalence of hot spots, more studies on the effect on complications for patients are needed. Such studies would also be helpful in designing objectives for spatial distribution that fit the clinical goals. We would also like to study patient data where hot spots are of higher concern, e.g. for head-and-neck cancer, and with clinically used dose plans as input.

Since models II–IV give non-convex optimization models, fmincon is not guaranteed to find the global optimum, and it is therefore possible to improve the results by using more advanced optimization algorithms.

Acknowledgements The project was funded by the Swedish Research Council, grant no VR-NT 2015-04543 and by the Swedish Cancer Foundation, grant no CAN 2015/618.

References

1. Baltas, D., Katsilieri, Z., Kefala, V., Papaioannou, S., Karabis, A., Mavroidis, P., et al. (2009). Influence of modulation restriction in inverse optimization with HIPO of prostate implants on plan quality: Analysis using dosimetric and radiobiological indices. *IFMBE Proceedings, 25*(1), 283–286.
2. Deist, T. M., & Gorissen, B. L. (2016). High-dose-rate prostate brachytherapy inverse planning on dose-volume criteria by simulated annealing. *Physics in Medicine and Biology, 61*(3), 1155–1170.
3. Guthier, C. V., Damato, A. L., Viswanathan, A. N., Hesser, J. W., & Cormack, R. A. (2017). A fast multi-target inverse treatment planning strategy optimizing dosimetric measures for high-dose-rate (HDR) brachytherapy. *Medical Physics, 44*(9), 4452–4462.
4. Halperin, E. C., Brady, L. W., Perez, C. A., & Wazer, D. E. (2013). *Perez and brady's principles and practice of radiation oncology*. Philadelphia: Wolters Kluwer Health.
5. Holm, Å. (2013). Mathematical optimization of HDR brachytherapy. Ph.D thesis, Linköping University.

6. Holm, Å., Larsson, T., & Carlsson Tedgren, Å. (2012). Impact of using linear optimization models in dose planning for HDR brachytherapy. *Medical Physics*, *39*(2), 1021–1028.
7. Siauw, T., Cunha, A., Atamtürk, A., Hsu, I. C., Pouliot, J., & Goldberg, K. (2011). IPIP: A new approach to inverse planning for HDR brachytherapy by directly optimizing dosimetric indices. *Medical Physics*, *38*(7), 4045–4051.

Part XI
Logistics and Freight Transportation

Active Repositioning of Storage Units in Robotic Mobile Fulfillment Systems

Marius Merschformann

1 Introduction

In today's increasingly fast-paced e-commerce an efficient distribution center is one crucial element of the supply chain. Hence, new automated parts-to-picker systems have been introduced to increase throughput. One of them is the Robotic Mobile Fulfillment System (RMFS). In a RMFS mobile robots are used to bring rack-like storage units (so-called pods) to pick stations as required, thus, eliminating the need for the pickers to walk and search the inventory. A task which can take up to 70% of their time in traditional picker-to-parts systems (see [6]). This concept was first introduced by [7] and an earlier simulation work by [2]. The first company implementing the concept at large scale was Kiva Systems, nowadays known as Amazon Robotics.

One of the features of RMFS is the continuous resorting of inventory, i.e. every time a pod is brought back to the storage area a different storage location may be used. While this potentially increases flexibility and adaptability, rarely used pods may block prominent storage locations, unless they are moved explicitly. This raises the question whether active repositioning of pods, i.e. picking up a pod and moving it to a different storage location, can be usefully applied to further increase the overall throughput of the system. In order to address this issue we focus on two approaches for active repositioning. First, we look at repositioning done in parallel while the system is constantly active and, second, we look at repositioning during system downtime (e.g. nightly down periods). While the assignment of storage locations to inventory is a well studied problem in warehousing (see [1]) the repositioning of inventory is typically not considered for other systems, because it is usually very expensive.

M. Merschformann (✉)
Paderborn University, Paderborn, Germany
e-mail: marius.merschformann@uni-paderborn.de

2 Repositioning in RMFS

In a RMFS *passive repositioning* of pods is a natural process, if the storage location chosen for a pod is not fixed. For example, in many situations using the next available storage location is superior to a fixed strategy, because it decreases the travel time of the robots and by this enables an earlier availability for their next tasks. However, this strategy might cause no longer useful pods to be stored at very prominent storage locations, which introduces the blocking problem discussed earlier. There are two opportunities to resolve this issue: on the one hand it is possible to already consider characteristics of the pod content while choosing an appropriate storage location, while on the other hand it is possible to actively move pods from inappropriate storage locations to better fitting ones. We call the latter approach *active repositioning* of pods. Both repositioning approaches are shown in Fig. 1. Additionally, the figure shows an excerpt of the basic layout used for the experiments, i.e. the replenishment stations (yellow circles), pick stations (red circles), pods (blue squares), the storage locations (blocks of 2 by 4) and the directed waypoint graph used for path planning. This layout is based on the work by [3].

In order to assess the value of storage locations and pods we introduce the following metrics. First the *prominence* F^{SL} of a storage location $w \in \mathcal{W}^{SL}$ is determined by measuring the minimum shortest path time to a pick station $m \in \mathcal{M}^O$ (see Eq. 1). The shortest path time $f^{A_t^*}$ is computed with a modified A* algorithm that considers turning times to achieve more accurate results. The storage location with the lowest $F^{SL}(w)$ is considered the most prominent one, since it offers the shortest time for bringing the pod to the next pick station. In order to assess the value of a pod b at time t we introduce the pod-speed (F^{PS}) and pod-utility (F^{PU}) measures. The *speed* of a pod (see Eq. 2) is calculated by summing up (across all SKUs) the units of an SKU contained (f^C) multiplied with the frequency of it (f^F). This frequency is a relative value reflecting the number of times a SKU is part of a customer order compared to all other SKUs. By using the minimum of units of an SKU contained and the demand for it (f^D), the *utility* of a pod (see Eq. 3) sums the number of potential picks when considering the customer order backlog. Thus, it is a more dynamic value. Both scores are then combined in the metric F^{PC} (see Eq. 4). For our experiments we consider the weights $w^S = w^U = 1$ to value both characteristics equally.

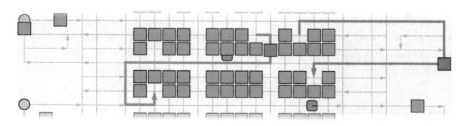

Fig. 1 Active repositioning move (green arrow) versus passive repositioning (red arrow)

Algorithm 1: CalculateWellsortednessCombined

1 $\mathcal{L} \leftarrow$ Sort ($\mathcal{W}^{SL}, i \Rightarrow F^{SL}(i)$) , $r' \leftarrow 1$, $f' \leftarrow \min_{i \in \mathcal{W}^{SL}} F^{SL}(i)$
2 **foreach** $i \in \{0, \ldots, \text{Size}(\mathcal{L}) - 1\}$ **do**
3   **if** $F^{SL}(i) > f'$ **then** $r' \leftarrow r' + 1$, $f' \leftarrow F^{SL}(i)$
4      $r_i \leftarrow r'$
5 $c \leftarrow 0, d \leftarrow 0$
6 **foreach** $i_1 \in \{0, \ldots, \text{Size}(\mathcal{L}) - 1\}$ **do**
7   **foreach** $i_2 \in \{i_1 + 1, \ldots, \text{Size}(\mathcal{L}) - 1\}$ **do**
8   **if** IsPodStored($\mathcal{L}[i_1]$) $\wedge$ IsPodStored($\mathcal{L}[i_2]$) $\wedge r_{\mathcal{L}[i_1]} \neq r_{\mathcal{L}[i_2]}$ **then**
9              $b_1 \leftarrow$ GetPod($\mathcal{L}[i_1]$), $b_2 \leftarrow$ GetPod($\mathcal{L}[i_2]$)
10   **if** $F^{PC}(b_1, t) < F^{PC}(b_2, t)$ **then** $c \leftarrow c + 1, d \leftarrow d + (r_{\mathcal{L}[i_2]} - r_{\mathcal{L}[i_1]})$
11 **return** $a \leftarrow \frac{d}{c}$

$$F^{SL}(w) := \min_{m \in \mathcal{M}^O} f^{A_t^*}(w, m) \tag{1}$$

$$F^{PS}(b, t) := \sum_{d \in \mathcal{D}} \left(f^C(b, d, t) \cdot f^F(d) \right) \tag{2}$$

$$F^{PU}(b, t) := \sum_{d \in \mathcal{D}} \left(\min \left(f^C(b, d, t), f^D(d, t) \right) \right) \tag{3}$$

$$F^{PC}(b, t) := \frac{F^{PS}(b, t)}{\max_{b' \in B} F^{PS}(b', t)} \cdot w^S + \frac{F^{PU}(b, t)}{\max_{b' \in B} F^{PU}(b', t)} \cdot w^U \tag{4}$$

For evaluation purposes we can use these measures to determine an overall "well-sortedness" score for the inventory. The procedure for calculating the well-sortedness score is described in Algorithm 1. At first we sort all storage locations by their prominence score in ascending order (see line 1 f.). Next, ranks r_i are assigned to all storage locations $i \in \mathcal{W}^{SL}$, i.e., the best ones are assigned to the first rank and the rank is increased by one each time the prominence value increases (see line 2 f.). Then, we assess all storage location two-tuples and count misplacements, i.e., both storage locations are not of the same rank and the score of the better placed pod at i_1 is lower than the worse placed pod at i_2 (see line 5 f.). In addition to the number of misplacements we track the rank offset. From this we can calculate the average rank offset of all misplacements, i.e., the well-sortedness. Hence, a lower well-sortedness value means a better sorted inventory according to the given combined pod-speed and pod-utility measures.

In this work we investigate the following repositioning mechanisms:

Nearest (N) For passive repositioning this mechanism always uses the nearest available storage location in terms of estimated path time ($f^{A_t^*}$). This mechanism does not allow active repositioning.

Cache (C) This mechanism uses the nearest 25% of storage locations in terms of estimated path time ($f^{A_t^*}$) as a cache. During passive repositioning pods with combined score (F^{PC}) above a determined threshold are stored at a cache storage location and others are stored at one of the remaining storage locations. In its active variant it swaps pods from and to the cache.

Utility (U) This mechanism matches the pods with the ranks of the storage locations (see Algorithm 1) on the basis of their combined score (F^{PC}). A nearby storage location with a close by rank is selected during passive repositioning. In its active variant pods with the largest difference between their desired and their actual storage location are moved to an improved one.

3 Computational Results

For capturing and studying the behavior of RMFS we use an event-driven agent-based simulation that considers acceleration / deceleration and turning times of the robots (see [5]). Since diverse decision problems need to be considered in an RMFS we focus the scope of the work by fixing all remaining mandatory ones to simple assignment policies and the FAR path planning algorithm described in [4]. A more detailed overview of the core decision problems of our scope are given in [5]. For all experiments we consider a simulation horizon of one week, do 5 repetitions to reduce noise and new customer and replenishment orders are generated in a random stream with a Gamma distribution ($k = 1$, $\Theta = 2$) used for the choice of SKU per order line from 1000 possible SKUs. Furthermore, we analyze repositioning for four layouts. The specific characteristics are set as follows:

Layout	Small	Wide	Long	Large
Stations (pick / replenish)	4/4	8/8	4/4	8/8
Aisles (hor. x vert.)	8x10	16x10	8x22	16x22
Pods	673	1271	1407	2658

For the evaluation of active repositioning effectiveness we consider two scenarios. At first, we look at a situation where the system faces a nightly down period (22:00–6:00) during which no worker is available for picking or replenishment, but robots can be used for active repositioning. In order to keep the replenishment processes from obscuring the contribution of nightly inventory sorting, replenishment orders are submitted to the system at 16:00 in the afternoon in an amount that is sufficient to bring the storage utilization back to 75% fill level. For pick operations we keep a constant backlog of 2000 customer orders to keep the system under pressure. Additionally, we generate 1500 orders per station at 22:00 in the evening to increase information for the pod utility metric about the demands for the following day. Secondly, we look at active repositioning done in parallel in a system that is continuously in action. For this, we consider three subordinate configurations distributing the robots per station as following:

R1P3A0: $\frac{1}{4}$ replenishment, $\frac{3}{4}$ picking, no active repos.
R1P2A1: $\frac{1}{4}$ replenishment, $\frac{2}{4}$ picking, $\frac{1}{4}$ active repos.
R1P3A1: $\frac{1}{5}$ replenishment, $\frac{3}{5}$ picking, $\frac{1}{5}$ active repos. (+1 robot per station)

This scenario is kept under continuous pressure by keeping a backlog of constant size for both: replenishment (200) and customer orders (2000).

The main performance metric for the evaluation is given by the unit throughput rate score (UTRS). Since we use a constant time of $T^P = 10s$ for picking one unit an upper bound for the number of units that can possibly be handled by the system during active hours can be calculated by $UB := |\mathcal{M}^O| \frac{3600}{T^P}$ with the set of all pick stations $\mathcal{M}^O$. Using this we can determine the fractional score by dividing the actual picked units per hour in average by this upper bound.

The results of the experiment are summarized in Table 1. For the comparison of resorting the inventory during the nightly down period (line: Activated) vs. no active repositioning at all (line: Deactivated) we can observe an advantage in throughput. However, for the parallel active repositioning it is not possible to observe a positive effect. When moving one robot per pick station from pick operations to active repositioning (lines: R1P3A0 and R1P2A1) we observe a loss in UTRS, because less robots bring inventory to the pick stations. Even with an additional robot per pick station (line: R1P3A1) we cannot observe a substantial positive effect. For most cases, the effect is rather negative as a result of the increased congestion potential for robots moving within the storage area.

In the following we take a closer look at the nightly down period scenario. If we keep the system sorted with the passive repositioning mechanism (C-C and U-U), nightly active repositioning does not have a noticeable positive effect, because the passive repositioning mechanism already keeps the inventory sorted for the most part. However, the Nearest mechanism which has a better overall performance, can benefit from a nightly active repositioning (N-C and N-U). Especially for the Large and Long layouts we can observe a reasonable boost in UTRS. The greater merit for layouts with more vertical aisles suggest that shorter trip times of the robots are the reason. This can also be observed when looking at the detailed results of a run with and without active repositioning (see Fig. 2). First, more orders can be completed per hour after the inventory was sorted over night (first graph). This boost is eliminated as soon as replenishment operations begin. Thus, when and how replenishment is done is crucial to the benefit of resorting during down times, because the effect may be lost quite quickly. In the third graph the shorter times for completing trips to the pick stations after sorting the inventory support the assumption that these are the the main reason for the boost. Lastly, the second graph provides the well-sortedness measure and shows that sorting the inventory can be done reasonably fast.

Table 1 Unit throughput rate score of the different scenarios, layouts and mechanisms (values in percent (%), read columns as [passive mechanism-active mechanism])

Layout setup/Mech.	Small						Wide						Long						Large					
	C-C	U-U	N-C	N-U			C-C	U-U	N-C	N-U			C-C	U-U	N-C	N-U			C-C	U-U	N-C	N-U		
Deactivated	52.9	53.6	55.0	55.0			50.6	50.5	52.7	52.7			43.3	46.0	47.8	47.8			42.6	44.8	46.3	46.3		
Activated	52.9	53.5	55.5	55.9			50.5	50.4	53.1	53.5			42.8	46.3	49.0	49.7			42.2	45.1	47.6	48.2		
R1P3A0	52.6	51.7	55.6	55.6			50.3	48.9	53.3	53.3			43.3	43.4	48.2	48.2			42.4	42.4	46.6	46.6		
R1P2A1	40.9	40.7	44.6	43.8			39.7	38.1	43.3	42.3			31.7	33.8	39.1	38.1			31.5	32.8	38.8	37.8		
R1P3A1	51.9	51.5	55.3	54.5			50.3	48.5	53.7	53.0			41.6	44.0	49.3	48.1			41.2	42.6	48.9	47.6		

Note: The columns for each layout setup are: C-C, U-U, N-C, N-U.

Layout setup/Mech.	Small				Wide				Long				Large			
	C-C	U-U	N-C	N-U	C-C	U-U	N-C	N-U	C-C	U-U	N-C	N-U	C-C	U-U	N-C	N-U
Deactivated	52.9	53.6	55.0	55.0	50.6	50.5	52.7	52.7	43.3	46.0	47.8	47.8	42.6	44.8	46.3	46.3
Activated	52.9	53.5	55.5	55.9	50.5	50.4	53.1	53.5	42.8	46.3	49.0	49.7	42.2	45.1	47.6	48.2
R1P3A0	52.6	51.7	55.6	55.6	50.3	48.9	53.3	53.3	43.3	43.4	48.2	48.2	42.4	42.4	46.6	46.6
R1P2A1	40.9	40.7	44.6	43.8	39.7	38.1	43.3	42.3	31.7	33.8	39.1	38.1	31.5	32.8	38.8	37.8
R1P3A1	51.9	51.5	55.3	54.5	50.3	48.5	53.7	53.0	41.6	44.0	49.3	48.1	41.2	42.6	48.9	47.6

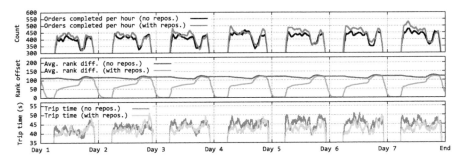

Fig. 2 Time-wise comparison of layout long and mechanisms N-U with (colored lines) and without (gray lines) active repositioning at night

4 Conclusion

The results suggest that active repositioning may boost throughput performance of RMFS. If the system faces regular down periods, costs for repositioning (energy costs, robot wear) are reasonable and charging times allow it, active repositioning can make a reasonable contribution to a system's overall performance. Since the introduced mechanisms greedily search for repositioning moves, more moves are conducted than necessary to obtain a desired inventory well-sortedness. For future research we suggest to predetermine moves before starting repositioning operations, e.g. by using a MIP formulation matching pods with storage locations and selecting the best moves. The source-code of this publication is available at https://github.com/merschformann/RAWSim-O.

References

1. Gu, J., Goetschalckx, M., & McGinnis, L. (2007). Research on warehouse operation: A comprehensive review. *European Journal of Operational Research, 177*(1), 1–21.
2. Hazard, C., Wurman, P., & D'Andrea, R. (2006). Alphabet soup: A testbed for studying resource allocation in multi-vehicle systems. In *AAAI Workshop on Auction Mechanisms for Robot Coordination* (pp. 23–30).
3. Lamballais, T., Roy, D., & de Koster, M. B. M. (2016). Estimating performance in a robotic mobile fulfillment system. *European Journal of Operational Research, 256*(3), 976–990.
4. Merschformann, M., Xie, L., & Erdmann, D. (2017). Path planning for robotic mobile fulfillment systems. arXiv:1706.09347.
5. Merschformann, M., Xie, L., & Rawsim-o, H. Li. (2017). A simulation framework for robotic mobile fulfillment systems: Working paper. arXiv:1710.04726.
6. Tompkins, J. (2010). *Facilities planning* (4th ed.). New York: Wiley.
7. Wurman, P., D'Andrea, R., & Mountz, M. (2008). Coordinating hundreds of cooperative, autonomous vehicles in warehouses. *AI Magazine, 29*(1), 9.

How to Control a Reverse Logistics System When Used Items Return with Diminishing Quality

Imre Dobos, Grigory Pishchulov and Ralf Gössinger

1 Introduction

We study an integrated production–inventory system that manufactures new items of a particular product and receives some of the used items back after a period of use. These can be either remanufactured on the same production line or disposed of. Used items awaiting remanufacturing need to be held in stock. Both manufacturing and remanufacturing operations require setting up accordingly the production equipment. Remanufactured items are considered to be as good as new and can serve the product demand on a par with the new ones. New and as-good-as-new items are kept in stock, from which the product demand is satisfied.

Controlling such a system involves decisions with regard to disposal of used items, succession of manufacturing and remanufacturing operations, and the choice of respective lot sizes. Existing research has studied control policies for such production–inventory systems in a variety of different settings. Specifically, beginning with the work of Schrady [5] and Richter [4], significant attention has been devoted to settings assuming deterministic constant demand and return rates. A more recent work has referred to settings assuming variation in quality of returned items [1] as

---

I. Dobos (✉)
Institute of Economic Sciences, Budapest University of Technology and Economics, Budapest, Hungary
e-mail: dobos@kgt.bme.hu

G. Pishchulov
Alliance Manchester Business School, The University of Manchester, Manchester, UK

G. Pishchulov
St. Petersburg State University, 7/9 Universitetskaya nab., 199034 St. Petersburg, Russia

R. Gössinger
Department of Production Management and Logistics, University of Dortmund, Dortmund, Germany

well as a limited number of remanufacturing cycles that an item can undergo due to wear and tear [3]. We extend this line of research by studying a setting in which used items return in a condition that depends on the number of remanufacturing cycles an item underwent and determines the inventory holding costs of that item. We determine optimal lot sizing for such a system and derive sufficient conditions for an optimal policy to forego remanufacturing.

The paper is organized as follows. Section 2 presents the model and the optimal total lot size. Section 3 gives sufficient optimality conditions. Section 4 concludes.

2 The Model and the Optimal Solution for the Total Lot Size

Consider a production–inventory system with two shops. The first shop is capable of producing new items of a certain product and also remanufacturing used items, which are then considered as good as new. Both kinds of items are used to satisfy market demand. They return after a period of use and are either disposed of or accumulated in the second shop during the collection interval [0, T], at the end of which these items are shipped to the first shop. The shipment time is constant and normalized to zero. The collection interval thus determines the length of a manufacturing–remanufacturing cycle in the first shop. A cycle begins with remanufacturing used items, if any. The latter are distinct w.r.t. the number of times i they have been remanufactured already before and are respectively called *type-i* items (Fig. 1). We assume that remanufacturing proceeds in the order of increasing type index i. We seek to minimize time-average system costs over an infinite horizon.

Model parameters:

d deterministic demand and product return rate,
L maximum possible number of times an item can be remanufactured,
β_i fraction of items that return and are of type $i - 1$ ($i = 1, \ldots, L+1$),
h holding cost of new and as-good-as-new items, per unit per time unit,
u_i holding cost of returned type items of type $i - 1$ ($i = 1, \ldots, L$), per unit per time unit, non-increasing in i,
s_i setup cost of manufacturing ($i = 0$) or remanufacturing a batch of used items of type $i - 1$ ($i = 1, \ldots, L$).

Decision variables:

l how many times to remanufacture used times, $l \leq L$,
T length of the manufacturing-remanufacturing cycle,
$q_0(l)$ manufacturing lot size, positive,
q_i remanufacturing lot size of used items of type $i - 1$ ($i = 1, \ldots, l$),
$n_0(l)$ number of manufacturing lots, positive integer,
n_i number of remanufacturing lots of type $i - 1$ ($i = 1, \ldots, l$), positive,
x total lot size of the new and as-good-as-new products, $x = d \cdot T$, positive,

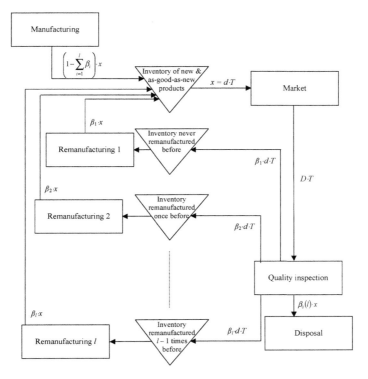

Fig. 1 Material flow when returned items of types above l are disposed of

$\beta_0(l)$ raction of returned items that are disposed of

It holds for the respective fractions of returned items that $\beta_0(l) + \sum_{i=1}^{l} \beta_i = 1$. Note that type-$L$ items are disposed of due to wear and tear, while any other items are disposed of for economic reasons. Manufacturing and remanufacturing volumes within a cycle are then, respectively:

$$n_0(l) \cdot q_0(l) = \left(1 - \sum_{i=1}^{l} \beta_i\right) \cdot x \quad \text{and} \quad n_i \cdot q_i = \beta_i \cdot x \quad (i = 1, \ldots, l). \quad (1)$$

Average total costs per time unit can be expressed as (Dobos and Richter [2]):

$$AC\left(l, \{n_i\}_{i=0}^{l}, x\right) = \left[\frac{d}{x} s_0 n_0(l) + \frac{x}{2} h \frac{\beta_0(l)^2}{n_0(l)}\right]$$
$$+ \sum_{i=1}^{l} \left\{\frac{d}{x} s_i n_i + \frac{x}{2}\left[(h - u_i)\frac{\beta_i^2}{n_i} + u_i\left(\beta_i + \beta_i^2\right)\right]\right\} \quad (2)$$

where the first bracketed expression represents the setup and holding costs associated with manufacturing new items, and the expression in the curly braces represents the respective costs associated with the items of type $i-1$.

Assume for now that $\beta_i > 0$ holds for all $i = 1, \ldots, l$. By expressing the lot sizes $\{q_i\}_{i=0}^{l}$ from (1) and substituting these into (2), we get:

$$AC\left(l, \{q_i\}_{i=0}^{l}, x\right) = \beta_0(l)\left(s_0\frac{d}{q_0} + h\frac{q_0}{2}\right) \\ + \sum_{i=1}^{l}\beta_i\left[s_i\frac{d}{q_i} + (h - u_i)\frac{q_i}{2} + \frac{x}{2}u_i(1+\beta_i)\right] \quad (3)$$

where the numbers of lots have to satisfy: $n_0(l) = x\beta_0(l)/q_0 \geq 1$, $n_i = x\beta_i/q_i \geq 1$ $(i = 1, \ldots, l)$. Given a total lot size x, the optimal lot sizes can be derived from (3):

$$q_0(l, x) = \min\left\{x\beta_0(l), \sqrt{\frac{2s_0 d}{h}}\right\}, \quad q_i(x) = \min\left\{x\beta_i, \sqrt{\frac{2s_i d}{h - u_i}}\right\} \quad (i = 1, \ldots, l).$$

It can be shown that $q_0(l, x) \geq q_0(l+1, x)$, i.e. the manufacturing lot sizes are declining in the remanufacturing limit l. Substituting the lot sizes into (3) gives:

$$AC'(l, x) = \beta_0(l)\left[s_0\frac{d}{q_0(l, x)} + \frac{h}{2}q_0(l, x)\right] \\ + \sum_{i=1}^{l}\left\{\beta_i\left[s_i\frac{d}{q_i(x)} + \frac{h - u_i}{2}q_i(x)\right] + x\frac{u_i}{2}\left(\beta_i + \beta_i^2\right)\right\}$$

Define $u_0 = 0$, and let

$$x_1 = \frac{1}{1 - \sum_{i=1}^{l}\beta_i}\sqrt{\frac{2s_0 d}{h - u_0}} = \frac{1}{\beta_0(l)}\sqrt{\frac{2s_0 d}{h}}, \quad x_{i+1} = \frac{1}{\beta_i}\sqrt{\frac{2s_i d}{h - u_i}} \quad (i = 1, \ldots, l).$$

The value x_{i-1} represents the total lot size that optimizes the type-i lot size. Let x_i $(i = 1, \ldots, l+1)$ be arranged in a non-decreasing order: $0 < x_{i_1} \leq \ldots \leq x_{i_{l+1}}$. Then

$$AC'(l, x) = \begin{cases} \sum_{j=1}^{l+1}\left(\frac{s_{i_j} d}{x} + \frac{h\beta_{i_j}^2 + u_{i_j}\beta_{i_j}}{2}x\right) & 0 < x < x_{i_1} \\ \ldots & \ldots \\ \sum_{j=1}^{k-1}\left[\beta_{i_j}\sqrt{2ds_{i_j}(h - u_{i_j})} + x\frac{u_{i_j}}{2}\left(\beta_{i_j} + \beta_{i_j}^2\right)\right] + \sum_{j=k}^{l+1}\left(\frac{s_{i_j} d}{x} + \frac{h\beta_{i_j}^2 + u_{i_j}\beta_{i_j}}{2}x\right) & x_{i_{k-1}} \leq x < x_{i_k} \\ \ldots & \ldots \\ \sum_{j=1}^{l+1}\left[\beta_{i_j}\sqrt{2ds_{i_j}(h - u_{i_j})} + x\frac{u_{i_j}}{2}\left(\beta_{i_j} + \beta_{i_j}^2\right)\right] & x_{i_{l+1}} \leq x \end{cases}$$

$AC'(l, x)$ is convex and continuously differentiable in x. Thus for a given l, the optimal total lot size x^l satisfies $dAC'(l, x^l)/dx = 0$. Define $x_{i_0} = 0$. Then it holds that

$$x^l = \sqrt{\frac{2d \sum_{j=k^\circ}^{l+1} s_{i_j}}{\sum_{j=1}^{k^\circ-1} u_{i_j}\left(\beta_{i_j} + \beta_{i_j}^2\right) + \sum_{j=k^\circ}^{l+1}\left(h\beta_{i_j}^2 + u_{i_j}\beta_{i_j}\right)}} \quad \text{and} \quad x_{i_{k^\circ-1}} \leq x^l < x_{i_{k^\circ}}$$

for some $k^\circ \in \{1, \ldots, l+1\}$ that can be easily determined by evaluating the sign of the derivative $dAC'(l, x)/dx$ at the boundary points x_{i_k} for $k = 1, \ldots, l+1$. The minimal average total costs for the given l are accordingly found to be

$$AC'\left(l, x^l\right) = \sum_{j=1}^{k^\circ-1} \beta_{i_j} \sqrt{2ds_{i_j}(h - u_{i_j})}$$
$$+ \sqrt{2d \left(\sum_{j=k^\circ}^{l+1} s_{i_j}\right) \left[\sum_{j=1}^{k^\circ-1} u_{i_j}\left(\beta_{i_j} + \beta_{i_j}^2\right) + \sum_{j=k^\circ}^{l+1}\left(h\beta_{i_j}^2 + u_{i_j}\beta_{i_j}\right)\right]}.$$

The above results can now be easily extended to the case of any $\beta_i = 0$ by excluding the respective type from (3) and all subsequent expressions. Below we address the problem of determining an optimal $l^o = \arg\min_{0 \leq l \leq L} AC'(l, x^l)$.

3 A Type of Optimal Solutions: No Remanufacturing

In this section we are looking for sufficient optimality conditions of a policy that excludes any remanufacturing. It is easy to see that the following trivially holds:

$$AC'(l+1, x^{l+1}) - AC'(l, x^l) \geq AC'(l+1, x^{l+1}) - AC'(l, x^{l+1}) \quad (l = 0, \ldots, L-1). \tag{4}$$

If the right-hand side in (4) is nonnegative then $AC'(l, x^l) \leq AC'(l+1, x^{l+1})$ holds for all $l = 0, \ldots, L-1$, and hence an optimal solution excludes remanufacturing.

It can be shown that the right-hand side in (4) can be rewritten as $AC_1(l+1) - AC_0(l)$, where $AC_0(l) = \beta_0(l)\left(s_0 d/q_0(l, x^{l+1}) + hq_0(l, x^{l+1})/2\right)$,

$$AC_1(l+1) = \begin{cases} \sqrt{2d(s_0 + s_{l+1})\left\{h\left[\beta_0(l) - \beta_{l+1}\right]^2 + h\beta_{l+1}^2 + u_{l+1}\beta_{l+1}\right\}} & x < \min(x_0; x_{l+1}) \\ [\beta_0(l) - \beta_{l+1}]\sqrt{2s_0 dh} + \sqrt{2ds_{l+1}\left(h\beta_{l+1}^2 + u_{l+1}\beta_{l+1}\right)} & x_0 \leq x < x_{l+1} \\ \beta_{l+1}\sqrt{2s_{l+1}d(h - u_{l+1})} + \sqrt{2ds_0\left\{h[\beta_0(l) - \beta_{l+1}]^2 + u_{l+1}\left(\beta_{l+1}^2 + \beta_{l+1}\right)\right\}} & x_{l+1} \leq x < x_0 \end{cases}$$

Fig. 2 Convex envelope $\widetilde{K}^l(\beta_{l+1})$ of function $K^l(\beta_{l+1})$

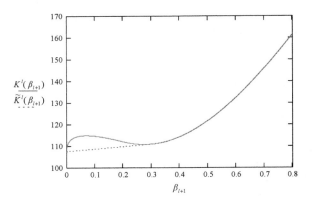

and x, x_0, x_{l+1} are suitably chosen lot sizes whose closed-form expressions are omitted for reasons of space and available from the authors on request.

It is easy to see that $AC_1(l+1)$ converges to $AC_1(l)$ as β_{l+1} approaches zero. Now defining $K^l(\beta_{l+1}) := AC_1(l+1)$, we can immediately use Lemma 1 by Dobos and Richter [2], which yields the following result.

Lemma 1 $K^l(\beta_{l+1})$ can be computed as follows:

(i) $s_{l+1}h\,[\beta_0(l) - \beta_{l+1}]^2 \geq s_0\left(h\beta_{l+1}^2 + u_{l+1}\beta_{l+1}\right)$
$\Rightarrow K^l(\beta_{l+1}) = \beta_{l+1}\sqrt{2s_{l+1}d\,(h-u_{l+1})}$
$+ \sqrt{2ds_0\left\{h\,[\beta_0(l)-\beta_{l+1}]^2 + u_{l+1}\left(\beta_{l+1}^2 + \beta_{l+1}\right)\right\}}$

(ii) $s_{l+1}h\,[\beta_0(l)-\beta_{l+1}]^2 - s_0 u_{l+1}\beta_{l+1}(1+\beta_{l+1}) \leq s_0(h-u_{l+1})\beta_{l+1}^2$
$\leq s_{l+1}\left\{h\,[\beta_0(l)-\beta_{l+1}]^2 + u_{l+1}\beta_{l+1}(1+\beta_{l+1})\right\}$,
$\Rightarrow K^l(\beta_{l+1}) = \sqrt{2d\,(s_0+s_{l+1})\left\{h\,[\beta_0(l)-\beta_{l+1}]^2 + h\beta_{l+1}^2 + u_{l+1}\beta_{l+1}\right\}}$

(iii) $s_{l+1}(h-u_{l+1})\beta_{l+1}^2 \geq s_{l+1}\left\{h\,[\beta_0(l)-\beta_{l+1}]^2 + u_{l+1}\beta_{l+1}(1+\beta_{l+1})\right\}$
$\Rightarrow K^l(\beta_{l+1}) = \beta_{l+1}\sqrt{2s_{l+1}d\,(h-u_{l+1})}$
$+ \sqrt{2ds_0\left\{h\,[\beta_0(l)-\beta_{l+1}]^2 + u_{l+1}\left(\beta_{l+1}^2+\beta_{l+1}\right)\right\}}$.

Using Lemma 1, we can evaluate $AC_1(l+1)$ and thus the right-hand side in (4). If the latter is nonnegative for all l then it is optimal to forego remanufacturing.

Further observe that disposing of all returns of type l is equivalent to enforcing $\beta_{l+1} = 0$ in our model. Below we obtain a sufficient optimality condition for abandoning remanufacturing regardless of the specific return fractions. To this end, we let β_{l+1} vary within the range $[0, 1]$, and refer to the concept of a convex envelope function $\widetilde{K}^l(\beta_{l+1})$ for $K^l(\beta_{l+1})$ (see Fig. 2 for an illustration).

Lemma 2 *If the convex envelope functions $\widetilde{K}^l(\beta_{l+1})$ are monotone increasing for $l = 0, \ldots, L-1$ then it is optimal not to remanufacture given any return fractions β_{l+1} ($l = 0, \ldots, L$).*

The proof is obvious because $AC'(l+1, x^{l+1}) - AC'(l, x^{l+1}) \geq 0$ proves to hold true for all β_{l+1} and $l = 0, \ldots, L-1$, which implies optimality of disposing of type-l items, while type-L items are disposed of due to wear and tear. Of course, if the convex envelopes have minimum points then it can be optimal to remanufacture.

4 Conclusion

We have investigated a manufacturing–remanufacturing model with quality considerations. The goal of the decision maker is to minimize the relevant EOQ related costs, and to choose an optimal number of remanufacturing cycles. We have examined a special case when remanufacturing is not optimal and obtained sufficient conditions for that.

A further study could address settings where the decision maker acquires used items depending on the number of remanufacturing cycles they have undergone. Another extension could be the search for an optimal policy in case of variable return rates.

Acknowledgements The first author gratefully acknowledges financial support by the Gambrinus Fellowship Programme during his research stay at the University of Dortmund, Germany in 2016–2017. The second author gratefully acknowledges financial support by the same university.

References

1. Dobos, I., & Richter, K. (2006). A production/recycling model with quality considerations. *International Journal of Production Economics, 104,* 571–579.
2. Dobos, I., & Richter, K. (2000). The Integer EOQ repair and waste disposal model: A further analysis. *Central European Journal of Operations Research, 8,* 174–193.
3. El Saadany, A. M. A., Jaber, M. Y., & Bonney, M. (2013). How many times to remanufacture? *International Journal of Production Economics, 143,* 598–604.
4. Richter, K. (1996). The EOQ repair and waste disposal model with variable setup numbers. *European Journal of Operational Research, 96,* 313–324.
5. Schrady, D. A. (1967). A deterministic inventory model for repairable items. *Naval Research Logistics Quarterly, 14,* 391–398.

Consistent Inventory Routing with Split Deliveries

Emilio Jose Alarcon Ortega, Michael Schilde, Karl F. Doerner and Sebastian Malicki

1 Introduction

In this paper we face an important problem, referred to in the literature as Inventory Routing Problem (IRP). We extend it by adding the consistency aspect, where we consider that a customer has consistent deliveries if the arrival times of all deliveries are the same or similar. We allow split deliveries in order to satisfy all customers demands and to deal with deliveries that exceed the capacity of a single vehicle. Furthermore, we include time windows in the problem because bars, restaurants, and other retailers have different opening hours and days. It is remarkable that, in the literature, there are still very few solution methods that are able to solve IRPs with time windows or other real-world problem characteristics. We denote our problem as Consistent Inventory Routing Problem with Time-Windows and Split Deliveries (CIRPTWSD). We solve the problem by applying an order up to level policy in order to minimize the total number of deliveries made to the customers.

Previous work about the IRP and variants was summarized in [1]. A similar problem was presented in [2] where the authors assumed that the inventory levels are monitored by the company. In [3], the authors dealt with the arrival-time consistency aspect of the problem by proposing template-based routes. Consistency and its variants were also reported in [4]. Furthermore, in [6], authors propose an algorithm to solve the split delivery vehicle routing problem.

E. J. Alarcon Ortega (✉) · M. Schilde · K. F. Doerner
Faculty of Business Administration, University of Vienna, 1090 Vienna, Austria
e-mail: emilio.jose.alarcon.ortega@univie.ac.at

S. Malicki
TUM School of Management, Technical University of Munich, 80333 Munich, Germany

2 Problem Description

The CIRPTWSD can be formulated as a mixed integer program. Given a directed graph $G = (A, V)$, where $V = 0, \ldots, n$ is the set of nodes with 0 being the depot and A is the set of arcs given by all pairs of nodes i and j. The planning horizon covers P periods, and at the same time these periods are divided into R subperiods. Subperiods are introduced to state a maximum driving time and also to deal with the different working shifts that beer companies have. Due to the complexity of this problem we consider only a single product. We assume that demands are known for each period and subperiod and consumption rates are continuous.

The decision variables in our model are:

- x_{ij}^{kpr}: binary variables that indicate if vehicle k drives from customer i to customer j in period p and subperiod r.
- y_i^{kpr}: binary variables that indicate if customer i is visited by vehicle k in each period and subperiod.
- q_i^{kpr} continuous variables that show the amounts delivered.
- t_i^{kpr}: continuous variables that show the arrival times.
- I_i^{pr}: continuous variables that indicate the final inventory levels at the end of each subperiod.
- t_i^{max} and t_i^{min}: earliest and latest arrival times at each customer i.
- o_i^{pr}: amounts of demand lost at every customer due to stock-out situations.

The objective of the model is to minimize the costs related to the routing, inventory holding, consistency and stock-outs. Routing costs are measured as the total travel distance. The inventory cost of each customer is the difference between the initial and the final inventory levels, if the final inventory level is lower than the initial. Thus, the inventory cost helps us to create a solution which can be repeated as a rolling horizon solution. Otherwise the model would tend to avoid late deliveries in order to save extra routing costs. Both, stock-out cost and inventory holding cost, are multiplied by a penalty factor L in order to integrate them in the objective function. Consistency costs are measured as the difference between the latest and the earliest arrival time to each customer, weighted with a parameter α. In our objective function, in order to unify all costs and balance them, we use a parameter $L = 3$ that will represent a penalty of 3€ for each liter of beer (assuming that the beer price on the market is around 1 €/liter). To measure the impact of consistency in our objective, we set $\alpha = 1$.

Apart from the different costs in this problem, we have different groups of constraints. Some of these constraints are well known and widely commented in the literature. Routing, time, and inventory flow constrains are considered in the mathematical model as well as constraints related to the use of a maximum number of capacitated vehicles. Furthermore, apart from these constraints, we include some others to calculate out of stock amounts and to forbid overstock situations at the retailers by taking into account the possibility of split deliveries and the time continuous consumption of commodity at every customer location. We also consider a group of constraints used to satisfy the order up to level policy, while taking into account the split delivery characteristic of the problem. This way we ensure that at the moment the last delivery to a customer in a subperiod is performed, the amount delivered by the vehicle satisfies the order up to level policy.

3 Solution Approach

To solve the CIRPTWSD, we develop a matheuristic solution approach based on the concept of Adaptive Large Neighborhood Search (ALNS). The initial solution is generated using an adaptation of the cheapest insertion heuristic combined with a local search. After applying the ALNS we solve a reduced problem based on the problem formulation to repair the obtained solution by improving possible inconsistent deliveries, stock-outs, and excessive inventory holding costs.

3.1 Constructive Heuristic

To construct an initial solution for each period and sub-period, we create a list of customers which require service in this subperiod. These customers either run out of stock in the current subperiod, or they run out of stock before the end of the next possible delivery time window. We then calculate an upper and lower bound for the delivery amount for each customer. The upper bound is the difference between the current inventory level and the order up to level. The lower bound is the amount necessary to avoid a stock-out before the end of the next delivery time window. Then, we use cheapest insertion to insert the customers with the largest possible amount. The customers that remain are inserted with the lowest possible amount. If even this is not possible, we split the remaining customers' deliveries to two vehicles. After finishing this procedure, we apply two improve operators to the obtained solution. An operator to destroy single-customer routes and a 2-opt algorithm.

3.2 Adaptive Large Neighborhood Search

After the run of the constructive heuristic, we have an initial feasible solution. Then, we apply the ALNS procedure in order to improve the quality of the obtained solution. In the proposed ALNS we use two sets of operators to destroy and repair the current solution, and a procedure to update the inventory after applying an operator.

We propose five destroy operators and three repair operators that are selected using a roulette-wheel selection. There are five destroy operators are: 1. "Remove worst" operator: deletes the p worst customers (with respect to the detour a customer causes between its preceding and succeeding customer visit along the route).

2. "Remove random" operator: deletes p random customers. 3. "Remove vehicle" operator: removes all customers in all routes of a randomly selected vehicle. 4. "Remove subperiod" operator removes all routes in a randomly selected subperiod. 5. "Remove least consistent" operator deletes p customers whose arrival times are very inconsistent. After we apply a random destroy operator we update the inventory levels of the removed customers and we create a list of customers that must be repaired because of stock-out situations. We then apply a randomly selected repair operator. The three operators are: 1. "Repair best before stock-out day" operator: creates a list of possible insertions in the day a stock-out occurs and the preceding days for customers that must be repaired, the possible insertions are sorted by the total distance of the detour this insertion causes, where the amount delivered is at least equal to a minimum amount calculated in the solution evaluation. We then select randomly one of the three best possible insertion positions and insert the customer into the route. If, after evaluating the new inventory flow of the customer, it still presents stock-out situations in the succeeding days, we repeat the process between the new stock-out day and the last insertion. 2. "Repair random before stock-out day" operator: this operator create a list of possible insertions as in the previous operator and then randomly selects one of them from the list. Also here we repeat the process as long as a feasible solution is reached. 3. "Repair consistency" operator: inserts customer such that the difference between earliest and the latest arrival times to this customer is minimized.

For each iteration of the ALNS we evaluate the new solution obtained and update the "best solution" as well as the weights of the destroy and repair operators that have been used in this iteration, if required. The stopping criterion for the ALNS is given as an overall time limit or as a maximum duration since the last solution improvement was found.

3.3 Postprocessing

After terminating the ALNS, we use the best solution obtained to solve a reduced variant of the mathematical problem formulation to optimality. This problem is solved in order to minimize consistency costs, that can be avoided by introducing waiting times, as well as to improve the performance of the algorithm with respect to the amounts delivered and final inventory levels. In this model, the visited customers and the route sequences for each vehicle, day, and subperiod are given by the solution obtained during the ALNS. We solve a linear program where we do not have binary or integer variables which allows a relatively short solving time.

Consistent Inventory Routing with Split Deliveries 399

Table 1 Results 4 days 1 sub-period

Cust	Inv	CPLEX			ALNS					
		Best	Time(s)	Best	Routing	Consistency	Inventory	Stockout	Gap
5	25	798.14*	0.10	829.68	812.89	–	–	16.80	0.04
	50	915.88*	0.09	962.73	962.73	–	–	–	0.05
	75	855.27*	0.09	855.27	715.07	–	117.09	23.10	0
	100	1606.55*	0.10	1606.55	680.06	–	926.49	–	0
10	25	864.10*	95.56	864.10	697.40	–	94.95	71.75	0
	50	1006.96*	2953.25	1006.96	946.96	–	–	60.00	0
	75	1066.76*	24.60	1066.76	678.56	–	349.12	39.08	0
	100	2805.18*	77.51	2805.18	678.56	–	2096.62	30.00	0
15	25	1141.95*	2741.97	1196.36	1090.95	–	7.12	98.28	0.05
	50	1199.52*	20181.65	1324.88	1207.38	–	65.04	52.45	0.10
	75	1417.54*	965.26	1493.26	811.29	–	567.85	114.11	0.05
	100	3576.99*	415.11	3882.97	985.08	–	2702.21	195.67	0.08
20	25	1720.731	36000	2585.49	2014.04	120.07	93.63	357.74	0.5
	50	1720.73	36000	2324.3	1595.03	–	301.32	427.94	0.35
	75	2019.99	36000	3060.07	1403.71	–	1488.10	159.05	0.50
	100	4105.08	36000	5263.95	1337.07	–	3767.84	159.05	0.28
48	25	4663.22	36000	4597.91	2757.82	162.18	637.37	1040.54	−0.01
	50	4270.64	36000	5349.59	3124.15	–	1563.63	661.81	0.25
	75	–	36000	9110.29	2843.03	–	5003.32	1263.94	–
	100	8782.94	36000	13020.7	1934.78	–	10425.10	660.90	0.48

Table 2 Results 4 days 2 subperiod

Cust	CPLEX			ALNS					
	Inv	Best	Time(s)	Best	Routing	Consistency	Inventory	Stockout	Gap
5	25	1358.71*	0.30	1443.52	1433.27	–	–	10.24	0.06
	50	1476.46*	0.38	1598.17	1598.17	–	–	–	0.08
	75	1463.81*	0.31	1470.58	1335.18	–	117.09	18.30	0.01
	100	2226.91*	0.32	2361.47	1434.98	–	926.49	–	0.06
10	25	1640.53	36000	1915.93	1825.93	–	–	90.00	0.16
	50	1783.38	36000	1804.20	1684.20	–	–	120.00	0.01
	75	–	36000	1903.95	1425.74	–	353.65	124.55	–
	100	3581.61	36000	4833.83	2091.67	68.48	2553.68	120.00	0.35
15	25	2127.12	36000	3272.07	3028.71	126.24	66.26	50.86	0.5
	50	–	36000	2531.28	2321.96	–	98.58	110.74	–
	75	2481.63	36000	2698.73	1927.16	–	607.42	164.15	0.08
	100	4620.07	36000	5938.68	2698.64	230.68	2843.55	165.81	0.28
20	25	–	36000	4632.85	3921.51	125.52	258.74	569.69	–
	50	3118.24	36000	4892.85	3721.51	242.91	358.74	569.69	0.56
	75	3510.75	36000	4851.75	2990.72	326.75	1067.11	467.17	0.38
	100	5613.05	36000	8213.61	3697.65	843.91	3434.13	237.91	0.46

4 Computational Experiments and Conclusion

Preliminary computational tests using the described algorithm implemented in C++ have been performed using a benchmark set for the periodic vehicle routing problem with time windows [5]. These instances were adapted to include the inventory holding information needed to solve our problem. In Table 1 we summarize the results obtained by considering a time horizon of 4 periods and 1 subperiod. The results are compared to the best solution obtained by solving the mathematical problem using CPLEX with a time limit of 10 h. The time limit for ALNS was set to 10 min. The algorithm has been tested by solving instances with 5, 10, 15, 20, and 48 customers with an initial inventory level of 25, 50, 75, and 100 percent of the inventory capacity. In Table 2 we present results for a time horizon of 4 periods and 2 subperiods on instances of 5, 10, 15 and 20 customers, where for instances of 48 customers CPLEX cannot find any feasible solution within 10 h (an asterisk indicates an optimal solution in both tables). In both tables we can see that the consistency cost are reduced to 0 in almost all instances while inventory holding costs are still significant with a time limit of 10 min. Further computational experiments including more realistic solution techniques to deal with stochastic demands and multi-product scenarios are planned for the future.

Acknowledgements Financial support from the Austrian and German Science Fund (FWF and DFG, D-A-CH) under Grant I 2248-N32 is gratefully acknowledged.

References

1. Coelho, L. C., Cordeau, J. F., & Laporte, G. (2014). Thirty years of inventory routing. *Transportation Science, 48*(1), 1–19.
2. Hemmelmayr, V., Doerner, K. F., Hartl, R. F., & Savelsbergh, M. W. P. (2009). Delivery strategies for blood products supplies. *OR Spectrum: Quantitative Approaches in Management, 31*(4), 707–725.
3. Kovacs, A., Parragh, S. N., & Hartl, R. F. (2014). A template-based adaptive large neighborhood search for the consistent vehicle routing problem. *Networks, 63*(1), 60–81.
4. Coelho, L. C., Cordeau, J. F., & Laporte, G. (2014). Consistency in multi-vehicle inventory-routing. *Transportation Science, 24*, 270–287.
5. Cordeau, J. F., Laporte, G., & Mercier, A. (2001). A unified tabu search heuristic for vehicle routing problems with time windows. *Journal of the Operational Research Society, 52*(8), 928–936.
6. Archetti, C., Speranza, M. G., & Hertz, A. (2006). A tabu search algorithm for the split delivery vehicle routing problem. *Transportation Science, 40*(1), 64–73.

Order Picking with Heterogeneous Technologies: An Integrated Article-to-Device Assignment and Manpower Allocation Problem

Ralf Gössinger, Grigory Pishchulov and Imre Dobos

1 Problem Description

In a warehouse multiple devices (non-automated or automated equipment) are used to pick articles according to customer orders. At each *device* the activities of order picking and storage slot replenishment are performed. Devices vary in the number of workplaces and storage slots, slot dimension, time per pick and per slot replenishment. Their capacity can only be utilized, if manpower is allocated to it. There are a number of specialized *operators* qualified to work at a certain device only and a pool of generalists able to work at all devices, yet with a lower efficiency. *Article* data includes information on demand, dimension and eligibility for being picked at specific devices. Regular fluctuations of *workload* induce a sequence of slack and peak periods per day. In slack periods all slots of devices are replenished so as to reduce the number of replenishments during the peak periods.

Two basic *decisions* are relevant for the article-to-device assignment and manpower allocation (ADAMA) problem: Which articles have to be picked at which devices? How much manpower of which kind has to be allocated to each device? Further two decisions have to be made for each device: How many slots have to

---

R. Gössinger (✉)
Department of Production Management and Logistics, University of Dortmund, Dortmund, Germany
e-mail: ralf.goessinger@udo.edu

G. Pishchulov
Alliance Manchester Business School, The University of Manchester, M1 3BU Manchester, UK

G. Pishchulov
St. Petersburg State University, 7/9 Universitetskaya nab., 199034 St. Petersburg, Russia

I. Dobos
Institute of Economic Sciences, Budapest University of Technology and Economics, Budapest, Hungary

be occupied by each article? How many replenishments have to be performed for each article? All decisions have to be made subject to the operational *objective* of minimizing the makespan.

Despite of its relevance, the ADAMA problem has not been discussed yet in the literature, but there is research on structurally similar problems: *Forward-reserve assignment and allocation problems* (FRAAP) occur in warehouses with two types of storage areas: *Reserve areas* hold the bulk storage and replenish forward areas. Order picking at these areas is possible, but time-consuming. *Forward areas* allow for fast order picking, but have very limited storage space. This induces replenishments, which are the more frequent, the more articles share the forward area [3]. Accordingly, there is a *trade-off* between picking and replenishment time when the article assignment changes [8]. The question is, how much of the limited forward area space has to be allocated to each article in order to minimize the total demand fulfillment time [5]. To solve this ADA problem a heuristic based on a ranking index is developed in [5]. A branch-and-bound procedure to find optimal solutions is developed in [3]. Further FRAAP approaches consider objects with limited divisibility [4, 8, 9]. Such discrete problems are found to be generalized knapsack problems that require heuristics for solving real-world instances in reasonable time. In comparison to the ADAMA problem, existing FRAAP approaches do not consider the following aspects: manpower needed for picking and replenishment activities; more than two picking devices (except for [4]); articles can be picked simultaneously at several devices.

More general analyses are performed under the topic *dual resource constrained systems* (DRC). DRC are production systems with capacity restricted by both, machine and labor [10]. From this point of view the interdependent sub-problems of machine loading (ML), job dispatching (JD) and manpower allocation (MA) are to be solved. Two approaches are similar to the ADAMA problem. Integrative ML-MA decisions in a cellular manufacturing system (CMS) with multiple work zones and a pool of differently skilled workers are analyzed in [1]. A simultaneous and a sequential ML-MA optimization approach as well as a heuristic approach are developed and compared. In the context of CMS a sequential approach for the MA-ML problem is developed in [7] and remarkably generalized in [2]. The ADAMA problem substantially differs from the situations analyzed in [1, 2, 7] in two regards: Instead of loading the whole system by releasing orders, articles are assigned to multiple types of devices and occupy one or multiple storage slots there; replenishment activities have to be considered in addition to picking activities.

With regard to the problem discussed in this paper FRAAP and DRC are complementary approaches. This paper aims at combining both in order to allow for a more efficient manpower and device utilization. The extent of efficiency improvement is dependent from the ability to coordinate interdependent deployment decisions. Planning approaches that make both decisions sequentially cannot bring about a better coordination than a simultaneous one, but will reduce computational effort. Hence, the question is, how a sequential approach balances the trade-off between coordination deficit and solution time.

The remainder of the paper is organized as follows: Sect. 2 presents the simultaneous and the sequential decision models for the ADAMA problem. Both approaches are compared with respect to solution quality and solution time under different operating conditions (Sect. 3). Finally conclusions on the applicability of the approaches are drawn in Sect. 4.

2 Decision Models

Based on the problem description a simultaneous ADAMA model can be formulated as follows (cf. Table 1 for notations and co-domains of variables):
ADAMA

$$\min m \quad (1)$$

s.t.

$$m \geq d_i \quad \forall i \quad (2)$$

$$d_i \leq \bar{d} \quad \forall i \quad (3)$$

$$\sum_i e_{ij} \cdot a_{ij} = 1 \quad \forall j \quad (4)$$

$$\sum_i s_i^f \leq 1 \quad (5)$$

$$w_i \cdot s_i^s + w^f \cdot s_i^f \leq p_i \quad \forall i \quad (6)$$

$$y_j \cdot a_{ij} \leq \rho_{ij} \cdot \lfloor h_i / g_j \rfloor \quad \forall i, j \quad (7)$$

$$\sum_j o_{ij} \leq l_i \quad \forall i \quad (8)$$

$$o_{ij} \leq \rho_{ij} \quad \forall i, j \quad (9)$$

$$o_{ij} \geq a_{ij} \quad \forall i, j \quad (10)$$

$$\sum_j \left(y_j \cdot a_{ij} \cdot t_{ij}^p + (\rho_{ij} - o_{ij}) \cdot t_{ij}^r \right) \leq \left(w_i \cdot s_i^s + \lambda \cdot w^f \cdot s_i^f \right) \cdot d_i \quad \forall i \quad (11)$$

The model aims at minimizing the makespan (1), which is the longest time one device needs for fulfilling demand of assigned articles (2). Constraints (3, 4) prevent tardy *demand fulfillment*. The time needed must not exceed the peak period's duration (3). All suitable devices can be used to completely fulfill article's demand (4). Constraints (5, 6) avoid infeasible *MA*. Manpower of flexible operators can be allocated to each device up to its maximum extent (5). The number of specialists and generalists deployed at one device must not exceed its number of workplaces (6). Constraints (7–10) prohibit unrealizable *ADA*. Storage slot requirements of an ADA have to be

Table 1 Notations

Indices	
i	Device $i = 1, \ldots, I$
j	Article $j = 1, \ldots, J$
Parameters	
$\bar{d}$	Duration of peak period
e_{ij}	Eligibility of j to be picked at i
g_j	Size of j
h_i	Length of one storage slot at i
λ	Output ratio between generalists and specialists $0 < \lambda < 1$
l_i	Number of storage slots available at i
t_{ij}^p	Time per piece to pick j at i
t_{ij}^r	Time to replenish one slot at i with j
p_i	Number of workplaces at i
w_i	Number of specialists available for i, with $w_i < p_i$
w^f	Number of available generalists
y_j	Demand of j
Variables	
a_{ij}	Share of jth demand assigned to i, $a_{ij} \in [0, 1]$
d_i	Total time to fulfill article demand assigned to i, $d_i \in \mathcal{R}_0^+$
m	Makespan $m \in \mathcal{R}_0^+$
o_{ij}	Number of storage slots occupied by j at i, $o_{ij} \in \mathcal{N}_0$
r_{ij}	Number of storage slot replenishments for j at i, $r_{ij} \in \mathcal{R}_0^+$
ρ_{ij}	Total storage slot usage for j at i, with $\rho_{ij} = r_{ij} \cdot o_{ij}$, $\rho_{ij} \in \mathcal{R}_0^+$
s_i^f	Share of generalists allocated to i, $s_i^f \in [0, 1]$
s_i^s	Share of specialists allocated to i, $s_i^s \in [0, 1]$
Indicators	
$\tilde{s}_i^f$	Estimated share of generalists allocated to i
$\tilde{s}_i^s$	Estimated share of generalists allocated to i

fulfilled by occupying and replenishing storage slots (7). At a device the number of occupied storage slots cannot be greater than the number of available storage slots (8). Constraint (9) requires using all occupied slots, while constraint (10) requires an article to occupy at least one slot at a device if any fraction of the article's demand is assigned to that device. Constraint (11) reflects the *ADA-MA interdependency*: The workload induced by ADA has to be met by MA within device's utilization time. ADAMA represents a mixed-integer quadratically constrained program (MIQCP).

In order to avoid non-linearity the described planning problem can be decomposed to a sequential approach. At its *top level* the ADA problem is solved assuming that at each device an estimated number of workplaces is manned. Therefore, in the *ADA model* constraints (5) and (6) are not relevant and (11) becomes linear:

$$\sum_j \left(y_j \cdot a_{ij} \cdot t_{ij}^p + (\rho_{ij} - o_{ij}) \cdot t_{ij}^r \right) \leq \left(w_i \cdot \tilde{s}_i^s + \lambda \cdot w^f \cdot \tilde{s}_i^f \right) \cdot d_i \quad \forall i \quad (12)$$

The $\tilde{s}$-values are estimated based on *anticipated decision behavior of the base level* [6] which is assumed to be in line with preferring (a) manpower of more efficient operators and (b) manpower allocation to more productive devices. From top level's objective and preference (a) follows $\tilde{s}_i^s = 1$. For setting $\tilde{s}_i^f$ the allocation rule AR represents preference (b).

AR

1. Initialize: $\tilde{s}_i^f := 0$, $s^e := 1$.
2. Determine: $U = \left\{ i \mid i = 1, \ldots, I \wedge p_i - \tilde{s}_i^s \cdot w_i - \tilde{s}_i^f \cdot w^f > 0 \right\}$.
3. If $U = \emptyset$, go to 7, else go to 4.
4. Calculate:

$$q_i = \frac{q_i^u}{\sum_{i \in U} q_i^u} \quad \text{with} \quad q_i^u = \frac{\max_i \left(\sum_j y_j \cdot e_{ij} \cdot t_{ij}^p / \sum_j y_j \cdot e_{ij} \right)}{\sum_j y_j \cdot e_{ij} \cdot t_{ij}^p / \sum_j y_j \cdot e_{ij}} \quad \forall i \in U$$

$$\Delta \tilde{s}_i^f = \min(q_i \cdot s^e; (p_i - \tilde{s}_i^s \cdot w_i - \tilde{s}_i^f \cdot w^f)/w^f) \quad \forall i \in U$$

5. Update:

$$s^e := \sum_{i \in U} \max(0; q_i \cdot s^e - (p_i - \tilde{s}_i^s \cdot w_i - \tilde{s}_i^f \cdot w^f)/w^f)$$

$$\tilde{s}_i^f := \tilde{s}_i^f + \Delta \tilde{s}_i^f \quad \forall i \in U$$

6. If $s^e > 0$, go to 2, else go to 7.
7. Stop.

The solution to ADA provides the fixed values $\bar{m}$, $\bar{a}_{ij}$, $\bar{\rho}_{ij}$, $\bar{o}_{ij}$ and the instruction for the *base level* to fulfill workload (induced by $\bar{a}_{ij}$, $\bar{\rho}_{ij}$, $\bar{o}_{ij}$) within $\bar{m}$ with minimum manpower. Hence, the objective of the *MA problem* is

$$\min \sum_i (w_i \cdot s_i^s + w^f \cdot s_i^f) \quad (13)$$

Furthermore constraints (5) and (6) are relevant and (11) becomes linear:

$$\sum_j \left(y_j \cdot \bar{a}_{ij} \cdot t_{ij}^p + (\bar{\rho}_{ij} - \bar{o}_{ij}) \cdot t_{ij}^r \right) \leq \left(w_i \cdot s_i^s + \lambda \cdot w^f \cdot s_i^f \right) \cdot \bar{m} \quad \forall i \quad (14)$$

Table 2 Results of regression analyses regarding makespan

Subject	m	α	β_I	β_{II}	β_{III}	r^2
Sim	Linear	821	13	−307	−1542	0.990
Seq	Linear	228	14	−175	−404	0.991
seq/sim	Linear	1.00	0.00	0.00	0.11	0.999

Table 3 Results of regression analyses regarding solution time

Subject	st	α	β_I	β_{II}	β_{III}	r^2
Sim	expon.	36.12	1.00	0.53	3.94	0.893
Seq	expon.	0.25	1.00	0.05	1.31	0.870
Seq/sim	expon.	0.01	1.00	0.18	0.20	0.499

Thus, the MA problem is a linear program and its solution provides information on s_i^s and s_i^f.

3 Numerical Study

Real data of a pharmaceutical wholesaler is used. Per peak period order picking is performed by a workforce of 12 operators ($\lambda = 0.9$), working at 4 automated and 2 manual devices.[1] A representative sample of demand data reveals that orders are fulfilled from an assortment of over 73,000 articles. We restrict attention in this study to the 4% of articles eligible for both, automated and manual order picking. In each problem instance, y_j is sampled from a Poisson distribution with the parameter equal to the average observed demand, g_j is sampled from the empirical distribution of standardized article size and t_{ij}^r is dependent on g_j and h_i. We conduct a 3^k full-factorial study with $k = 3$ factors that characterize the specific *problem instances*: (I) number of articles (500, 1000, 1500), (II) number of storage slots at automated devices (25, 50, 75% of total slot requirements), and (III) the fraction of flexible workforce (25, 50, 75%). For each combination of factor levels, 3 problem instances are randomly generated, which altogether yields $3^3 \cdot 3 = 81$ instances. In contrast to the sequential approach, the simultaneous one failed to solve 5 of the instances.[2]

A *comparison of approaches* reveals that the sequential approach exceeds the minimum makespan (m) on average by 5.3% (CV 2.1%), but reduces solution time (st) on average to 0.3% (CV 62.7%). Correlations between (I), (II), (III) and m, st are quantified by *multiple regression analyses* based on absolute values observed with each approach and their ratios (Tables 2 and 3).

---

[1] Data for instance generation and generated instances can be provided by the authors.
[2] We used a MINLP solver (BARON 15.9) for solving ADAMA, and a MIP solver (Gurobi 7.0.2) for solving ADA-MA on a MacBook Pro computer (2 GHz Intel Core i5 with two cores).

In both approaches the same *tendencies* can be noticed: (I) is by far the strongest factor and positively correlated with m and st. (II) is negatively correlated with both indicators and concerning m weaker than (III). The correlation directions of (III) are negative/positive for m, resp. st. *Ratios* of observed values are not correlated with (I), but positively/negatively correlated with (II) and (III) in case of m, resp. st. Correlations of (III) are much stronger than those of (II) in case of m. That is, the coordination deficit is noticeable positively correlated with (III). Since (III) is considered in the sequential approach at the top level by anticipating the base level, a deficit reduction could be achieved by improving AR.

4 Conclusions

For warehouses with heterogeneous order picking technologies we propose two approaches that assign articles and allocate manpower to devices in an integrative way. The simultaneous approach is a MIQCP. To avoid non-linearity, a hierarchical decomposition leads to a sequential approach composed of a MIP (top level) and a LP (base level). A numerical study reveals that the simultaneous approach cannot handle real-world problems in acceptable time and fails sometimes. In contrast, the sequential approach was able to solve all instances, allows for a strong reduction of solution time, but slightly reduces solution quality. A regression analysis indicates the fraction of flexible workforce as a driver of this coordination deficit. Therefore, continuing research will be directed to the sequential approach, in particular towards a better anticipation of the base level's behavior.

References

1. Davis, D. J., & Mabert, V. A. (2000). Order dispatching and labor assignment in cellular manufacturing systems. *Decision Sciences, 31,* 745–771.
2. Egilmez, G., Erenay, B., & Süer, G. A. (2014). Stochastic skill-based manpower allocation in a cellular manufacturing system. *Journal of Manufacturing Systems, 33,* 578–588.
3. Gu, J., Goetschalckx, M., & McGinnis, L. F. (2010). Solving the forward-reserve allocation problem in warehouse order picking systems. *Journal of the Operational Research Society, 61,* 1013–1021.
4. Hackman, S. T., & Platzman, L. K. (1990). Near-optimal solution of generalized resource allocation problems with large capacities. *Operations Research, 38,* 902–910.
5. Hackman, S. T., & Rosenblatt, M. J. (1990). allocating items to an automated storage and retrieval system. *IIE Transactions, 22,* 7–14.
6. Schneeweiss, C. (1998). Hierarchical planning in organizations: elements of a general theory. *International Journal of Production Economics, 56*(57), 547–556.
7. Süer, G. A., & Sánchez-Bera, I. (1998). Optimal operator assignment and cell loading when lot-splitting is allowed. *Computers and Industrial Engineering, 35,* 431–434.
8. van den Berg, J. P., et al. (1998). Forward-reserve allocation in a warehouse with unit load replenishments. *European Journal of Operational Research, 111,* 98–113.

9. Walter, R., Boysen, N., & Scholl, A. (2013). The discrete forward-reserve problem—allocating space, selecting products, and area sizing in forward order picking. *European Journal of Operational Research, 229,* 585–594.
10. Xu, J., Xu, X., & Xie, S. Q. (2011). Recent developments in dual resource constrained (DRC) system research. *European Journal of Operational Research, 215,* 309–318.

Part XII
Metaheuristics

Metaheuristic for the Vehicle Routing Problem with Backhauls and Time Windows

José Brandão

1 Introduction

The vehicle routing problem with backhauls and time windows (VRPBTW) consists in finding a set of routes, in order to serve a given number of dispersed customers, whose geographical location, demand, and time window for the service are known. Each route is travelled by one vehicle assigned to it, which starts the trip at the depot, visits each customer of the route according to a given schedule, and, in the end returns to the depot. It is well established that the VRPBTW is *NP-hard*.

In this problem there are two distinct sets of customers—those that require the delivery of goods, who are called linehauls, and those that require the collection of goods, named backhauls. In a route, the linehaul customers must be served first, followed by the backhauls (this is called a precedence constraint), but a route may contain only linehauls or only backhauls. In this paper, the objective is to minimise the number of routes and, for the same number of routes, to minimise the total distance travelled by the vehicles. So far as we know, the most relevant papers that used the same objective function are the following: Thangiah et al. [8], Reimann et al. [5], Ropke and Pisinger [6] and Vidal et al. [10]. Real-world applications and surveys of this and similar types of problems can be found at Casco et al. [2], Toth and Vigo [9] and Battarra et al. [1].

J. Brandão (✉)
Departamento de Gestão, Escola de Economia e Gestão, Universidade do Minho, Largo do Paço, 4704–553 Braga, Portugal
e-mail: sbrandao@eeg.uminho.pt

J. Brandão
CEMAPRE, ISEG, University of Lisbon, Lisbon, Portugal

The remainder of this paper is organised as follows. In Sect. 2, we describe our iterated local search algorithm. In Sect. 3, we present the computational experiments and compare the quality of our algorithm with the best algorithms published, and in the final section, we draw the main conclusions.

2 The Iterated Local Search Algorithm

The reader can find a detailed description of the iterated local search in Lourenço et al. [4]. Very succinctly, this metaheuristic consists of applying iteratively and sequentially *local search* and *perturbation*. Given a solution s, the local search finds a local optimum, s^*, by exploring the neighbourhood of s and performing a set of descent moves. The perturbation consists of applying a set of non-descent moves to given local optimum, s^*, generating s_p^*, in order to escape from this optimum. In the following, first we describe the components of our iterated local search algorithm and then we present its general framework.

The initial solution is generated by a sequential insertion method that takes into account the insertion cost of each unrouted customer in the route under construction and its time window wideness.

Our algorithm comprises the following neighbourhood structures for performing *local search*: (i) *cross over*, (ii) *swap*, (iii) *insertion*, (iv) *interchange of chains (2, 0), (2, 1), and (2, 2)*; (v) *intra swap*, (vi) *shift* and (vii) *2-opt*. The first four kinds of structures consist of moves between the routes of s, whilst the other three are performed inside each route of s. All the solutions generated along the search process must be feasible.

In our algorithm, the following five types of perturbation procedures were used: *P1—ejection chain, P2—swap of linehauls with backhauls and interchange of chains, P3—direct swap, P4—insertion, and P5—filling*. Note that some of these procedures are similar to those already defined to perform the local search, but in the perturbation non improving moves are executed. The ejection chain is applied in every iteration before one of the other four, which are applied in this sequence alternately. All the operators used in our algorithm are deterministic. Therefore, in order to reduce the likelihood of cycling a large number of operators have been used and the way of applying some of them depends on the iteration.

During the execution, after finishing the local search cycle, a set of elite solutions is selected. The composition of this set may change at each iteration, depending on the quality of the solution just found. This quality is evaluated according to the following hierarchical criteria: (i) *Phase 1*—number of routes, number of customers of the least route, distance of the least route; (ii) *Phase 2*—number of routes, total distance of the solution. After K iterations without improving the best known solution, the route chosen to apply the perturbation is the best solution of this elite set not yet explored. After some experiments, the cardinality of the elite set, E, and K were defined as 5 and 100, respectively.

With the purpose of minimising the number of routes the following *mechanisms* were used: 1—In any iteration, if the trial move generates a feasible solution with one route less than the best known feasible solution, this move is immediately accepted, independently of its cost. 2—In order to reinforce the natural ability of the insertion move to eliminate a route, the following is done in phase 1: the trial insertion moves always start with the least route, then if a customer is removed from this route the (trial) insertion cost is reduced by a fixed cost (F_c), but if a customer is inserted in this route the insertion cost is increased by F_c, which has been defined as $F_c = RD/N$. Where R is the total number of routes of the initial solution and D is its total distance.

Our algorithm comprises *phase 1* and *phase 2*. These two phases are executed in this sequence twice. The only difference between them is that phase 1 tries directly to minimise the number of routes, while phase 2 tries to minimise the distance. This is why mechanism 2 is not applied in phase 2. The execution of each phase stops after T iterations without improving the best known solution. After repeating this cycle twice, phase 2 is executed again during T iterations without improving the best known solution. After some experiments, T was set equal 350 during the first two cycles (phase 1 followed by phase 2) and 1000 in the last application of phase 2.

Framework of the Algorithm

s_b—best solution found.

t—counter of the number of iterations without improving the best solution. It restarts from zero in the beginning of each phase and when a new best solution is found.

1. Generate the initial solution, s; $s_b = s$.
2. Apply *phase 1* or *phase 2* while $t < T$//phase 1 and 2 are repeated twice sequentially.
3. Repeat while there is an improvement of the current solution, s:

    - Apply cross over;
    - Apply swap;
    - Apply interchange (2, 0) and (2, 1);
    - Repeat while there is an improvement of s:
      - Apply insertion and then intra swap to each route $r \in s$.
    - Apply interchange (2, 2);
    - Apply intra shift to each $r \in s$ and then 2-opt if r contains one type of customers only.

4. Insert s in the elite set if it is better than any of those already in it.
5. Perturb s or the best solution of the elite set not yet explored if $t \mod 100 = 0$, set $t = t + 1$ and go to step 2.
6. Update T, set $s = s_b$ and execute phase 2 again, going to step 2.

3 Computational Experiments

The benchmark problems were created by Gélinas et al. [3] and Thangiah et al. [8] based on Solomon [7] data for the VRPTW. There are three sets of problems, containing 100, 250 and 500 customers, respectively, where the first set contains 15 problems and the other two sets contain 12 problems each.

Our algorithm (JB, for short) was programmed in the C language, and was executed on a desktop computer with an Intel i7-3820 processor at 3.6 GHz, and 32 GB of RAM. In order to evaluate the performance of JB, we compare it, in terms of solution cost and computing time, with the best algorithms found in the literature that assume the same objective function, namely the following: Thangiah et al. [8]—Thangiah, for short, Reimann et al. [5]—Reimann, Ropke and Pisinger [6]—Ropke, and Vidal et al. [10]—Vidal. These algorithms were executed on the following computers, respectively: NeXT, Pentium III at 900 MHz, Pentium IV at 1.5 GHz, and Opteron 250 at 2.4 GHz. Following the literature, their speeds, measured in millions of floating-point operations per second (Mflop/s), can roughly be estimated as presented in Table 1.

JB is deterministic and, therefore, the results presented in the tables were obtained with one execution of the algorithm, using always the same parameter values defined in Sect. 2. Since in these articles there are several versions of the algorithm, we compare with the version that produces better results, except in the case of Thangiah where we use the best overall. The algorithms of Reimann, Ropke and Vidal contain several stochastic parameters. Therefore, the authors executed them 10 times, but they presented the results in different ways, as follows: Reimann—best results of the 10 executions; Ropke—average number of routes (but not the average distance), best solutions (routes and distance) and average computing time; Vidal—average solutions (routes and distance), best solutions and average computing time. The average results are the most relevant for the sake of comparison with our algorithm, but if they are not provided, we present the best solutions and the computing time required by the 10 executions. Note also that Reimann and Vidal only solved the first set of problems. In the tables, R is the total number of routes of the solutions for each set of problems, CPU is the total computing time, in seconds. The computing times presented in Table 3 results from scaling the original times given by the authors, according to the relative speeds of the computers defined by Table 1.

Table 1 Relative speeds of the computers used by the algorithms

Algorithm	Processor	Mflop/s	Speed scaled
JB	Intel i7-3820, 3.6 GHz	1960	1
Thangiah	NeXT	1	0.0005
Reimann	Pentium III, 900 MHz	234	0.1194
Ropke	Pentium IV, 1.5 GHz	326	0.1663
Vidal	Opteron 250, 2.4 GHz	1385	0.7066

Table 2 Global results for the three sets of problems

	Thangiah		Reimann		Ropke		Vidal		JB	
	R	Distance	R	Distance	R	Distance	R	Distance	R	Distance
N = 100	274	24051.9	265[b]	23514.9[b]	259.9[a]	23416.7[b]	258.4[a]	23488.7[a]	259	23672.6
N = 250	517	57688.9	–	–	449[a]	54499[b]	–	–	446	55062.0
N = 500	799	93156.1	–	–	680[a]	82796[b]	–	–	663	84446.7

[a]This is the average of 10 runs
[b]This the best of 10 runs

Table 3 Computing times for each set of problems, in seconds

Problem set	Thangiah[a]	Reimann[b]	Ropke[c]	Vidal[c]	JB
N = 100	0.4	2686	283	2607	206
N = 250	6.0	–	1006	–	1504
N = 500	63.4	–	3492	–	7381

[a]Time to execute all the versions
[b]Time to execute 10 runs
[c]Average time of 10 runs

Looking at the first set, Table 2 shows that JB produces very slightly worse solutions than the algorithm of Vidal, a little bit better than the algorithm of Ropke (note that the distance presented for this algorithm is the best from 10 executions and, consequently, it is not directly comparable with the distance yield by JB), better than Reimann, and substantially better than Thangiah. In terms of computing time, Table 3 shows that Thangiah is much faster than JB, but JB is faster than any of the others, being much faster than the algorithms of Vidal and of Reimann (about ten times).

The results for the other two sets show that the quality of the solutions given by JB becomes increasingly better than the yield by the other two algorithms (Thangiah and Ropke) as long as the number of customers increases. For example, the total number of routes given by Thangiah is 5.8, 15.9 and 20.5% more than the yield by JB, for the problems with 100, 250 and 500 customers, respectively, and a similar trend is observed with Ropke. In what concerns to the computing time, the behaviour in relation to the algorithm of Thangiah is just the opposite, i.e., the percentage difference decreases when the dimension of the problem increases, but it is still much faster than JB. On the contrary, JB becomes increasingly slower than the algorithm of Ropke.

As final conclusion, we can say that, in general, JB produces better solutions than the other algorithms and it is rather fast. Note that the algorithm of Thangiah is much faster, but this is not sufficient to compensate the difference in the quality of the solutions. For example, for the set of problems with 500 customers the initial solution of JB is better (34 routes less) and takes less than one twentieth of computing time.

4 Conclusions

This paper presents an iterated local search algorithm for the VRPBTW that has proven to be very competitive with the existing algorithms, both in terms of quality of the solutions and computing time. This performance owes a lot to the effectiveness of the perturbations applied and also to the use of elite solutions. Furthermore, our algorithm is deterministic, what means that the results are fully reproducible.

References

1. Battarra, M., Cordeau, J-F. & Iori, M. (2014). Pickup-and-delivery problems for goods transportation. In P. Toth & D. Vigo (Eds.), Vehicle Routing: Problems, Methods, and Applications. 2nd ed., (pp. 161–192). MOS/SIAM Series on Optimization.
2. Casco, D., Golden, B., & Wasil, E. (1988). Vehicle routing with backhauls: models, algorithms and case studies. In B. Golden & A. Assad (Eds.), *Vehicle routing: Methods and studies* (pp. 127–147). Elsevier Science Publishers.
3. Gélinas, S., Desrochers, M., Desrosiers, J., & Solomon, M. (1995). A new branching strategy for time constrained routing problems with application to backhauling. *Annals of Operations Research, 61,* 91–109.
4. Lourenço, H., Martin, O., & Stützle, T. (2003). Iterated local search. In F. Glover & G. Kochenberger (Eds.), *Handbook of Metaheuristics* (pp. 321–353). Alphen aan den Rijn, Netherlands: Kluwer.
5. Reimann, M., Doerner, M., & Hartl, R. (2002). Insertion based ants for vehicle routing problems with backhauls and time windows. In Dorigo et al. (Eds.), *Ant algorithms*. Lecture Notes in Computer Science (Vol. 2463, pp. 135–147). Berlin/Heidelberg: Springer.
6. Ropke, S., & Pisinger, D. (2006). A unified heuristic for a large class of vehicle routing problems with backhauls. *European Journal of Operational Research, 171,* 750–775.
7. Solomon, M. (1987). Algorithms for the vehicle routing and scheduling problems with time window constraints. *Operations Research, 35,* 254–265.
8. Thangiah, R., Potvin, J., & Sun, T. (1996). Heuristic approaches to vehicle routing with backhauls and time windows. *Computers & Operations Research, 23,* 1043–1057.
9. Toth, P., & Vigo, D. (2002). VRP with backhauls. In P. Toth & D. Vigo (Eds.), The vehicle routing problem, SIAM Monographs on Discrete Mathematics and Applications (Vol. 9, pp 195–224). Philadelphia, PA: SIAM.
10. Vidal, T., Crainic, T., Gendreau, M., & Prins, C. (2014). A unified solution framework for multi-attribute vehicle routing problems. *European Journal of Operational Research, 234,* 658–673. *Logistics and Transportation Review, 48,* 233–247.

Part XIII
Optimization Under Uncertainty

Convex Approach with Sub-gradient Method to Robust Service System Design

Jaroslav Janáček and Marek Kvet

1 Introduction

Public service systems are established to provide users by necessary service in emergency situations. Since the traversing time between service center and the affected user might be impacted by various random events, the system must be resistant to such critical events [7]. One of common approaches to robust system designing uses a set of scenarios to model various combinations of failures in service deliveries [1, 8]. Whereas the standard service system can be designed by solving min-sum location problem [5, 6], the robust design is usually formulated as min-max problem, where maximum of objective functions associated with the individual scenarios is minimized. The transformation of the original min-sum problem to the new more complex min-max problem brings difficulties into the computational process. The min-max link-up constraints represent an undesirable burden, because the branch-and-bound method converges much slower than the computational process solving a simple service system design problem [3, 4]. Slow convergence may be caused by the link-up constraints, which formalize relation between the individual scenario objective functions and their upper bound. Here, we focus on bad convergence of the branch-and-bound method and try to improve the computational process by replacing the waste relaxed model by a smaller one obtained by a convex combination.

J. Janáček · M. Kvet (✉)
Faculty of Management Science and Informatics,
University of Žilina, Univerzitná 8215/1, 010 26 Žilina, Slovak Republic
e-mail: marek.kvet@fri.uniza.sk

J. Janáček
e-mail: jaroslav.janacek@fri.uniza.sk

2 Robust Service System Design and Lagrangean Relaxation

The robust service system design problem can be described by the following denotations, where symbol J denotes the set of users' locations and symbol I denotes the set of possible service center locations. We denote by b_j the volume of service demand at the user location j. To solve the problem, p locations must be chosen from I so that the maximal scenario objective function value is minimal. The scenario objective function value is defined as a sum of users' distances from the location of the nearest center, where each distance is multiplied by the associated demand b_j. Let symbol U denote the set of possible detrimental scenarios. The distance between locations i and j under a specific scenario $u \in U$ is denoted by d_{iju}. The variable $y_i \in \{0, 1\}$ models the decision on service center location at place $i \in I$ by the value of 1 if a service center is located at i and by the value of 0 otherwise. The variable h represents an upper bound of the objective function issues corresponding to the individual scenarios. The nonlinear model of the problem follows.

$$\text{Minimize} \quad h \tag{1}$$

$$\text{Subject to:} \quad \sum_{i \in I} y_i \leq p \tag{2}$$

$$\sum_{j \in J} b_j \min \{d_{iju} : i \in I, y_i = 1\} \leq h \quad \text{for } u \in U \tag{3}$$

$$y_i \in \{0, 1\} \quad \text{for } i \in I \tag{4}$$

$$h \geq 0 \tag{5}$$

The objective function (1) represented by single variable h gives an upper bound of all objective function values corresponding to the individual scenarios. The constraint (2) limits the number of located service centers by p. The link-up constraints (3) ensure that each scenario objective function value is less than or equal to the upper bound h. As the min-max link-up constraints (3) represent an undesirable burden in any integer programming problem, we applied the Lagrangean relaxation on the constraints (3). Each of these constraints is associated with a nonnegative Lagrangean multiplier λ_u and the following relaxed problem is formulated.

$$\text{Minimize} \quad h + \sum_{u \in U} \lambda_u \left(\sum_{j \in J} b_j \min \{d_{iju} : i \in I, y_i = 1\} - h \right) \tag{6}$$

$$\text{Subject to:} \quad (2), (4) \text{ and } (5)$$

According to [3], we can restrict ourselves on such setting of non-negative multipliers, which sum equals to one. The commonly known propositions claim that the optimal objective function value of (6), (2), (4) and (5) gives lower bound of the optimal solution of the model (1)–(5) and the optimal solution of (6), (2), (4) and (5) is the optimal solution of (1)–(5), if and only if the gap (7) equals to zero for h determined as maximum of the individual objective functions.

$$gap = -\sum_{u \in U} \lambda_u \left(\sum_{j \in J} b_j \min \{d_{iju} : i \in I, y_i = 1\} - h \right) \quad (7)$$

The original method employing Lagrangean relaxation minimizes the gap by iterative adjusting the Lagrangean multipliers. In the convex approach, we use the inequality (8) to surrogate problem (6), (2), (4) (5) by a smaller "convex" problem. Thus, the problem (9), (2) and (4) will be solved instead of (6), (2), (4), (5).
11.3

$$\sum_{u \in U} \lambda_u \left(\min \{d_{iju} : i \in I, y_i = 1\} \right) \leq \min \left\{ \sum_{u \in U} \lambda_u d_{iju} : i \in I, y_i = 1 \right\} \quad \text{for } j \in J \quad (8)$$

$$\text{Minimize} \quad \sum_{j \in J} b_j \min \left\{ \sum_{u \in U} \lambda_u d_{iju} : i \in I, y_i = 1 \right\} \quad (9)$$

3 Radial Formulation of the Min-Sum Location Problems

Using the denotations, variables and constraints introduced in Sect. 2, we will present the radial formulation of the problem (6), (2), (4) and (5) taking into account that the variable h leaves the model due to the sum of considered λ_u equals to one. To formulate the radial model, the integer range $[0, m]$ of all possible distances of the matrices $\{d_{iju}\}$ is partitioned into m unit zones according to [2, 3]. The zone s corresponds to the interval $(s, s + 1]$. Further, auxiliary zero-one variables x_{jus} for $s = 0 \ldots m - 1$ and $u \in U$ are introduced. The variable x_{jus} takes the value of 1, if the distance of the user at $j \in J$ under scenario $u \in U$ from the nearest located center is greater than s and it takes the value of 0 otherwise. Then the expression $x_{ju0} + x_{ju1} + \cdots + x_{ju(m-1)}$ constitutes the distance d_{ju*} from user location j to the nearest located service center under scenario $u \in U$. Let us introduce a zero-one constant a_{iju}^s under scenario $u \in U$ for each $i \in I, j \in J, s \in [0, m-1]$. The constant a_{iju}^s is equal to 1, if the distance d_{iju} between the user location j and the possible center location i is less than or equal to s, otherwise a_{iju}^s is equal to 0. Then the radial model of the problem takes the form of (10)–(12), (2) and (4).

$$\text{Minimize} \quad \sum_{u \in U} \lambda_u \sum_{j \in J} b_j \sum_{s=0}^{m-1} x_{jus} \quad (10)$$

$$\text{Subject to:} \quad x_{jus} + \sum_{i \in I} a_{iju}^s y_i \geq 1 \quad \text{for } j \in J, u \in U, s = 0, \ldots, m-1 \quad (11)$$

$$x_{jus} \geq 0 \quad \text{for } j \in J, u \in U, s = 0, \ldots, m-1 \quad (12)$$

The problem (9), (2) and (4) can be formulated in similar way, where auxiliary zero-one variables x_{js} for $s \in [0, m-1]$ are introduced. The zero-one constant a_{ij}^s for each $i \in I, j \in J, s \in [0, m-1]$ equals 1, if the convex combination of d_{iju} for $u \in U$ with coefficients λ_u is less than or equal to s for the user location j and the possible center location i, otherwise $a_{ij}^s = 0$. Then, the model takes the form of (13)–(15), (2) and (4). Obviously, this model is $|U|$ times smaller than the previous one.

$$\text{Minimize} \quad \sum_{j \in J} b_j \sum_{s=0}^{m-1} x_{js} \quad (13)$$

$$\text{Subject to:} \quad x_{js} + \sum_{i \in I} a_{ij}^s y_i \geq 1 \quad \text{for } j \in J, s = 0, \ldots, m-1 \quad (14)$$

$$x_{js} \geq 0 \quad \text{for } j \in J, s = 0, \ldots, m-1 \quad (15)$$

4 Sub-gradient Iterative Method

We try to minimize the *gap* (7) so that h is set to the value of the best found solution of the problem (1)–(5) (further denoted as upper bound UB) and we aim to reach estimation (13) of (10) as big as possible by suitable adjustment of λ_u. The algorithm performs according to the following steps.

0. Initialize $\lambda_u = 1/|U|$ for $u \in U$ and set the bounds $LB = 0$ and $UB = +\infty$.
1. Solve the problem (13)–(15), (2) and (4) obtaining solution y. Compute associated objective function values $f_u(y)$ and $h = max \{f_u(y) : u \in U\}$. Compute the estimation ULB as the value of (6) for given y.
2. If $UB > h$, update $UB = h$ and $y^{best} = y$. If $LB < ULB$, then update $LB = ULB$ and update components $g_u = f_u(y) - h$ of gradient and determine magnitude α of the step in direction given by the gradient. Otherwise, set $\alpha = \alpha/(1 + \beta)$.
3. If termination rule is met, finish, otherwise update λ_u for $u \in U$ and go to 1.

Table 1 Comparison of three studied approaches to robust service system design problem solved for each self-governing region of Slovakia

Region	$|I|$	p	EXACT		LAGRANGE		L-CONVEX-OLD		L-CONVEX-NEW	
			Time	h	Time	dif	Time	dif	Time	dif
BA	87	9	26.9	25417	36.1	3.07	1.5	2.70	19.3	2.70
BB	515	52	963.8	18549	149.6	1.67	39.5	2.17	69.6	3.38
KE	460	46	1154.7	21286	193.7	1.11	11.3	3.27	25.2	6.15
NR	350	35	1883.7	24193	110.8	1.02	2.7	0.86	2.8	0.86
PO	664	67	1052.2	21298	219.4	0.57	26.9	1.20	62.8	0.97
TN	276	28	241.7	17524	38.5	2.23	4.3	1.72	14.8	0.91
TT	249	25	402.5	20558	138.3	0.01	3.8	2.87	9.7	1.80
ZA	315	32	1143.9	23004	33.0	0.78	4.0	1.53	11.7	1.16
Average			858.7		114.9	1.3	11.8	2.0	27.0	2.2

5 Numerical Experiments

The goal of performed experiments was to compare suggested approaches concerning the computational time and the objective function value of the resulting solution. The studied instances were solved using the optimization software FICO Xpress 8.0 (64-bit, release 2016) and the experiments were run on a PC equipped with the Intel Core i7 5500U processor with the parameters: 2.4 GHz and 16 GB RAM. The experiments were performed with the pool of benchmarks obtained from the road network of Slovakia corresponding to the self-governing regions, i.e. Bratislava - BA, Banská Bystrica - BB, Košice - KE, Nitra - NR, Prešov - PO, Trenčín - TN, Trnava - TT and Žilina - ZA. The set of communities represents both the set J of users' locations and the set I of possible center locations. Due to the lack of common benchmarks, ten detrimental scenarios for each self-governing region were generated randomly [3, 4]. An individual experiment was organized so that the exact solution of the problem (1)–(5) was obtained first. Then, the approaches based on Lagrangean relaxation were applied. The results are summarized in Table 1.

The exact solution is reported in the columns denoted by "EXACT". All columns denoted by "time" contain the computational time in seconds. The optimal objective function value is reported in the column denoted by h. The right part contains the results of suggested approaches based on Lagrangean relaxation. The basic version of the solving method is denoted by "LAGRANGE" and the reduced ones combined with the convex combination of scenarios are denoted by "L-CONVEX-OLD" and "L-CONVEX-NEW" respectively. The difference between "L-CONVEX-OLD" and "L-CONVEX-NEW" approaches consists in the way of parameter α adjustment in λ_u updating. The multipliers λ_u are generally updated according to (16), assuming that the denominator is greater than zero.

$$\lambda_u^{updated} = \frac{max\{0, \lambda_u + \alpha g_u\}}{\sum_{w \in U} max\{0, \lambda_w + \alpha g_w\}} \quad for \ u \in U \qquad (16)$$

The original way (OLD) of α adjustment assures that all positive λ_u stay positive after update [3]. The alternative way (NEW) chooses α so that at least one of positive λ_u stays positive. The parameter β was set to the value of 1 in all presented experiments. Each method using Lagrangean relaxation was evaluated by computational time and *dif* defined as the percentual difference of the obtained objective function value from the objective function of the exact solution taken as the base.

6 Conclusions

The paper deals with reduction of the robust service system design problem with the goal of reaching good design in acceptable time. We suggested an approximate method based on Lagrangean relaxation and sub-gradient adjustment of Lagrangean multipliers. The convex combination of scenarios was used to reduce the problem size. Presented results show success in reducing the computational time, which was approximately ten times lower than computational time of the standard Lagrangean relaxation. The deviations of the results obtained by methods based on Lagrangean relaxation from the exact solution were almost the same. Concerning tested parameter β, no significant impact on the results was found. None of the alternative strategies of α setting proved to dominate the other. Future research may be aimed at some other forms of α setting strategies and possible adapting or learning methods.

Acknowledgements This work was supported by the research grants VEGA 1/0518/15 "Resilient rescue systems with uncertain accessibility of service", VEGA 1/0463/16 "Economically efficient charging infrastructure deployment for electric vehicles in smart cities and communities" and APVV-15-0179 "Reliability of emergency systems on infrastructure with uncertain functionality of critical elements".

References

1. Correia, I., & Saldanha da Gama, F. (2015). In Laporte, G., Nickel, S., & Saldanha da Gama, F. (Eds.), *Facility locations under uncertainty, location science* (pp. 177–203).
2. García, S., Labbé, M., & Marín, A. (2011). Solving large p-median problems with a radius formulation. *INFORMS Journal on Computing, 23*(4), 546–556.
3. Janáček, J., & Kvet, M. (2016). Designing a robust emergency service system by Lagrangean relaxation. *Mathematical Methods in Economics*, 349–353. (Liberec).
4. Janáček, J., & Kvet, M. (2016). Min-max robust emergency service system design. *Communications: Scientific Letters of the University of Žilina, 8*(3), 12–18.
5. Jánošíková, Ľ. (2007). Emergency medical service planning. *Communications: Scientific Letters of the University of Žilina, 9*(2), 64–68.
6. Marianov, V., & Serra, D. (2002). Location problems in the public sector. In Z. Drezner, et al. (Eds.), *Facility location: Applications and theory* (pp. 119–150). Berlin: Springer.
7. Marsh, M., & Schilling, D. (1994). Equity measurement in facility location analysis. *European Journal of Operational Research, 74*, 1–17.
8. Pan, Y., Du, Y., & Wei, Z. (2014). Reliable facility system design subject to edge failures. *American Journal of Operations Research, 4*, 164–172.

Leasing with Uncertainty

Christine Markarian

1 Introduction

Traditionally, companies used to *buy* their resources at start-up and then update them whenever needed (e.g., when new resources were released). Nowadays, due to rapid technological advances, such updates have become necessary quite often. This has led companies to *lease* their resources rather than buying them, thus maintaining up-to-date resources for reasonable costs. Consequently, smart leasing decisions that need to be made *without knowing the future* were constantly needed, i.e., *when* to lease, *which* resource, and for *how long*, while not knowing future demands and without paying much.

Recent Work. The first attempt to answer these questions was by Meyerson in 2005 [6]. Meyerson introduced the first theoretic leasing model with a simple problem in which one resource is leased, the Parking Permit Problem. Each day, depending on the weather, we have to either use the car (if it is rainy) or walk (if it is sunny). In the former case, we must have a valid parking permit, which we choose among K different types of permits (leases), each having a different duration and price. On any day, lease prices respect *economy of scale* such that a longer lease costs less per day. The goal is to buy a set of leases in order to serve all rainy days while minimizing the total cost of purchases and without using weather forecasts. Since the seminal work of Meyerson [6], there have been a number of works that extend the Parking Permit Problem to problems including more resources [1, 2, 7]. All of these works assume that *all* arriving demands must be served, and *immediately* upon arrival. In some circumstances, however, it is possible to *decline* serving some demands at the

---

This work was partially supported by the German Research Foundation (DFG) within the Collaborative Research Centre "On-The-Fly Computing" (SFB 901).

C. Markarian (✉)
Heinz Nixdorf Institute and Computer Science Department, Paderborn University,
Fürstenallee 11, 33102 Paderborn, Germany
e-mail: christine.markarian@upb.de

cost of paying a *penalty* associated with them. Penalties have been studied in the context of many online optimization problems (e.g., [4]) and scheduling problems (e.g., [3]). Moreover, in many scenarios, demands need not be served immediately. Abshoff et al. [5] introduced *deadlines* into the leasing model by Meyerson, such that each demand is associated with a deadline and is allowed to be served anytime before its deadline.

Our Contribution. In this paper we incorporate penalties into existing leasing models by introducing the Lease-or-Decline and Lease-or-Delay leasing models, described as follows. In the Lease-or-Decline model, not all demands need to be served, i.e., the algorithm may *decline* a demand as long as a penalty associated with it is paid. In the Lease-or-Delay model, each demand has a deadline and can be served any day before its deadline as long as a penalty is paid for each delayed day. Note that this is a generalization of the model with deadlines proposed by Abshoff et al. [5] in which no penalty is incurred for delays. The goal is to minimize the total costs of leases and penalties. Should we know the sequence of demands in advance (offline version), the two problems can easily be solved optimally using Dynamic Programming. Nevertheless, the demands are only revealed to us with time and so we seek algorithms that provide provably good solutions without knowing the future (online version). We give deterministic online primal-dual algorithms, evaluated using standard *competitive analysis* in which an online algorithm is compared to the optimal offline algorithm which knows the entire sequence of demands in advance and is optimal. Given an input sequence σ, let $\mathcal{C}_A(\sigma)$ and $\mathcal{C}_{OPT}(\sigma)$ denote the cost incurred by an algorithm A and an optimal offline algorithm OPT, respectively. We say algorithm A is c-competitive if $\mathcal{C}_A(\sigma) \leq c \cdot \mathcal{C}_{OPT}(\sigma)$ for all input sequences σ. We seek algorithms that achieve competitive ratios independent of time.

2 The Lease-or-Decline Model

In this section we start be formally defining the Lease-or-Decline model which we formulate as a primal-dual program. Then we describe an online deterministic primal-dual algorithm for the problem and analyze its competitive ratio.

Problem Description. The Lease-or-Decline model is an online problem defined as follows. Given a set L of different lease types each with a fixed duration and price. The duration of a lease of type $i \in L$ is denoted as l_i and its price as c_i. A lease of type i that starts on day t is represented as a pair (i, t). These pairs form the set Q. There is a set D of demands that need to be covered by the algorithm. Each day the algorithm is either given one of these demands or no demand (it does not know the set D in advance). A demand arriving on day j (we say demand j) is *covered* if the algorithm buys some lease (i, t) such that $j \in [t, t + l_i]$. A lease that can cover demand $j \in D$ is called j's *candidate*. The algorithm has the option not to cover a demand j (*decline* it) by paying a penalty p_j associated with it. The goal is to minimize the total costs of leasing and penalties. The linear programming formulation of the problem is depicted in Fig. 1.

$$\min \sum_{(i,t)\in Q} \mathbf{x}_{(i,t)} \cdot \mathbf{c_i} + \sum_{j\in D} \mathbf{x_j} \cdot \mathbf{p_j}$$

Subject to: $\forall j \in D$: $\sum_{(i,t)\in Q, j\in[t,t+l_i]} \mathbf{x}_{(i,t)} + \mathbf{x_j} \geq 1$

$\forall (i,t) \in Q$: $\mathbf{x}_{(i,t)} \in \{\mathbf{0}, \mathbf{1}\}$; $\forall j \in D$: $\mathbf{x_j} \in \{\mathbf{0}, \mathbf{1}\}$

$$\max \sum_{j\in D} \mathbf{y_j}$$

Subject to: $\forall (i,t) \in Q$: $\sum_{j\in D, j\in[t,t+l_i]} \mathbf{y_j} \leq \mathbf{c_i}$

$\forall j \in D$: $\mathbf{y_j} \leq \mathbf{p_j}$

$\forall j \in D$: $\mathbf{y_j} \geq \mathbf{0}$

Fig. 1 Linear program (Lease-or-Decline)

Lease configuration: we simplify any given instance of the problem by assuming that no two leases of the same type overlap and all lease lengths are powers of two. Meyerson showed that by doing so we only lose a constant factor in the competitive ratio of the problem (Theorem 2.2 in [6]).
Online Primal-dual Scheme. Whenever a demand $j \in D$ arrives, we increase its corresponding dual variable y_j until some dual constraint becomes tight. If such a constraint corresponds to the demand j: $y_j = p_j$ (and not a candidate lease), we set its corresponding primal variable to 1 ($x_j = 1$). Else, we set the primal variables of all leases corresponding to a tight dual constraint to 1. Next we show that the primal-dual algorithm above has an $O(|L|)$-competitive ratio, where $|L|$ is the number of available leases. Note that this is the best ratio any deterministic algorithm for the problem can achieve due to the lower bound of $\Omega(|L|)$ by Meyerson for the Parking Permit (which is a special case of the Lease-or-Decline model in which all penalties are set to infinity). We bound the costs of leasing and the costs of penalties separately, as follows. Let $P \subseteq Q$ denote the set of leases bought by the algorithm. Because the dual constraint is tight for each $(i,t) \in P$, we have $c_i = \sum_{j\in D, j\in[t,t+l_i]} y_j$. Thus, $\sum_{(i,t)\in P} c_i = \sum_{(i,t)\in P} \sum_{j\in D: j\in[t,t+l_i]} y_j = \sum_{j\in D} y_j \sum_{(i,t)\in P: j\in[t,t+l_i]} 1$. Notice that due to the configuration of the leases, there are exactly $|L|$ leases covering any single day. So the algorithm does not buy more than $|L|$ leases for each demand. This means $\sum_{(i,t)\in P: j\in[t,t+l_i]} 1 \leq |L|$. Now we bound the costs of penalties. Since the dual constraint is tight for each demand j declined, we have $p_j = y_j$. Thus, $\sum_{j\in D} x_j \cdot p_j \leq \sum_{j\in D} y_j$.
Hence, since both the primal and dual solutions constructed are feasible, we can apply Weak Duality Theorem - any feasible solution to the primal (minimization)

program is a lower bound for any feasible solution to the corresponding dual program - and conclude that:

Theorem 1 *There is a deterministic $O(|L|)$-competitive primal-dual algorithm for the Lease-or-Decline model.*

3 The Lease-or-Delay Model

In this section we start be formally defining the Lease-or-Delay model which we formulate as a primal-dual program. Then we describe an online deterministic primal-dual algorithm for the problem and analyze its competitive ratio.

Problem Description. The Lease-or-Delay model is an online problem defined as follows. Given a set L of different lease types each with a fixed duration and price. The duration of a lease of type $i \in L$ is denoted as l_i and its price as c_i. A lease of type i that starts on day t is represented as a pair (i, t). These pairs form the set Q. We assume the same lease configuration as in the Lease-or-Decline model. There is a set D of demands that need to be covered by the algorithm. Each day the algorithm is either given one of these demands or no demand (it does not know the set D in advance). Each demand arriving on day j (we say demand j) is associated with a deadline duration d. We denote by d_{max} the longest available deadline duration. A demand j with deadline duration d is *covered* if $\exists$ day $t' \in [j, j+d]$ and lease (i, t) bought by the algorithm such that $t' \in [t, t+l_i]$. A lease that can cover demand j is called j's *candidate*. Each day $t' \in [j, j+d]$ is associated with a penalty $p_{t'}^j$ that needs to be paid if the algorithm covers demand j on day t'. These penalties, associated with each demand j, are increasing with time (the later the demand is covered, the higher the penalty paid) and given to the algorithm as soon as demand j arrives. The goal is to minimize the total costs of leasing and penalties. The linear programming formulation of the problem is depicted in Fig. 2.

Online Primal-dual Scheme. Whenever a demand $j \in D$ arrives, we uniformly increase the dual variables z_t^j corresponding to each $t \in [j, j+d]$ until some dual constraint corresponding to a candidate lease becomes tight. At this point there is at least one such lease (i, t') that covers day j. Moreover, we always maintain $(y_j - z_t^j) \geq 0$ for each $t \in [j, j+d]$. Hence, we increase the dual variable y_j corresponding to the demand until the dual constraint corresponding to some $t \in [j, j+d]$ becomes tight. Note that the dual constraint corresponding to no other day $t \in [j+1, j+d]$ will become tight *strictly before* the dual constraint corresponding to day j. We then set the primal variables $x_{(i,t')}$ and x_j^j to 1. Notice that the algorithm covers a demand on the day of its arrival. Had it been the case that it knew future demands (offline instance), it might have considered delaying a demand (pay more penalties but buy leases shared by more demands). Next we show that the primal-dual algorithm above has an $O(|L| + \frac{d_{max}}{l_{min}})$-competitive ratio, where $|L|$ is the number of available leases, d_{max} is the longest available deadline duration, and l_{min} is the shortest lease length. Since the Parking Permit Prob-

$$\min \sum_{(i,t) \in Q} \mathbf{x_{(i,t)}} \cdot \mathbf{c_i} + \sum_{j \in D} \sum_{t=j}^{j+d} \mathbf{x_t^j} \cdot \mathbf{p_t^j}$$

Subject to: $\forall (j,d) \in D$: $\sum_{t=j}^{j+d} \mathbf{x_t^j} \geq 1$

$\forall (j,d) \in D, t \in [j, j+d]$: $\mathbf{x_t^j} \leq \sum_{(i,t') \in Q, t \in [t'+t'+l_i]} \mathbf{x_{(i,t')}}$

$\forall (i,t) \in Q$: $\mathbf{x_{(i,t)}} \in \{\mathbf{0,1}\}$; $\forall (j,d) \in D, t \in [j, j+d]$: $\mathbf{x_t^j} \in \{\mathbf{0,1}\}$

$$\max \sum_{(j,d) \in D} \mathbf{y_j}$$

Subject to: $\forall (i,t) \in Q, (j,d) \in D, t' \in [j, j+d]$: $\sum_{t' \in [t, t+l_i]} \mathbf{z_{t'}^j} \leq \mathbf{c_i}$

$\forall (j,d) \in D, t \in [j, j+d]$: $\mathbf{y_j} - \mathbf{z_t^j} \leq \mathbf{p_t^j}$

$\forall (j,d) \in D$: $\mathbf{y_j} \geq \mathbf{0}$; $\forall (j,d) \in D, t \in [j, j+d]$: $\mathbf{z_t^j} \geq \mathbf{0}$

Fig. 2 Linear program (Lease-or-Delay)

---

lem is a special case of the Lease-or-Delay model in which $d_{max} = 0$ and all penalties for all demands are set to 0, the lower bound of $\Omega(|L|)$ by Meyerson holds here as well. We bound the costs of leasing and the costs of penalties separately, as follows. Let $P \subseteq Q$ denote the set of leases bought by the algorithm. Because the dual constraint is tight for each $(i,t) \in P$, we have $c_i = \sum_{j \in D, j \in [t, t+l_i]} y_j$. Thus, $\sum_{(i,t) \in P} c_i = \sum_{(i,t) \in P} \sum_{j \in D: j \in [t, t+l_i]} y_j = \sum_{j \in D} y_j \sum_{(i,t) \in P: j \in [t, t+l_i]} 1$. Unlike in the previous model, for each demand j, the algorithm may buy more than $|L|$ leases. Although it buys leases covering only day j to cover demand j, the latter may have more than $|L|$ candidates shared by other demands arriving later. Hence the algorithm may end up buying some of j's candidates on a later day. Abshoff et al. [5] showed an upper bound of $(|L| + \frac{d_{max}}{l_{min}})$ on the total number of these candidates, for the special case in which demands also have deadlines but can be covered any day before their deadline *without* incurring any penalty costs. Thus, $\sum_{(i,t) \in P: j \in [t, t+l_i]} 1 \leq |L| + \frac{d_{max}}{l_{min}}$. Now we bound the costs of penalties. Since the algorithm covers a demand j on only one day t and the dual constraint is tight for that day, we have $p_t^j \leq y_j$. Thus, $\sum_{j \in D} \sum_{t=j}^{j+d} x_t^j \cdot p_t^j \leq \sum_{j \in D} y_j$. Hence, by Weak Duality Theorem (both primal and dual solutions are feasible), we conclude that:

Theorem 2 *There is a deterministic $O(|L| + \frac{d_{max}}{l_{min}})$-competitive primal-dual algorithm for the Lease-or-Delay model.*

4 Conclusion

The algorithms in this paper perform under the uncertainty of the future in terms of both the arrival times of demands as well as the penalties associated with them. In certain scenarios, it might be the case that the algorithm is partially aware of the future (e.g., it may not know the arrival times of demands in advance, but has some information about the penalties (such as penalties are drawn from some probability distribution)). It is interesting to know whether better solutions can be attained under such assumptions. Furthermore, we are curious to know how our algorithms will perform in actual leasing scenarios from real markets and what modifications might be necessary in terms of both models and algorithms, in order to close the gap between theory and practice.

References

1. Abshoff, S., Kling, P., Markarian, C., Meyer auf der Heide, F., & Pietrzyk, P. (2015). Towards the price of leasing online. *Journal of Combinatorial Optimization*, 1–20.
2. Bienkowski, M., Kraska, A., & Schmidt, P. (2017). A deterministic algorithm for online steiner tree leasing. In *Algorithms and Data Structures Symposium (WADS 2017), August 2017, Proceedings*.
3. Epstein, L., & Zebedat-Haider, H. (2011). Online scheduling with rejection and withdrawal. *Theoretical Computer Science*, *412*(48), 6666–6674.
4. Hajiaghayi, M. T., Liaghat, V., & Panigrahi, D. (2014). Near-optimal online algorithms for prize-collecting steiner problems. In *ICALP*.
5. Shouwei, L., Mäcker, A., Markarian, C., auf der Heide, F. M., & Riechers, S. (2015). Towards flexible demands in online leasing problems. In *Computing and Combinatorics - 21st International Conference, COCOON 2015, Beijing, China, August 4-6, 2015, Proceedings*, 277–288.
6. Meyerson, A. (2005). The parking permit problem. In *46th Annual IEEE Symposium on Foundations of Computer Science (FOCS 2005), 23-25 October 2005, Pittsburgh, PA, USA, Proceedings*, 274–284.
7. Nagarajan, C., & Williamson, D. P. (2013). Offline and online facility leasing. *Discrete Optimization*, *10*(4), 361–370.

Risk Averse Scheduling with Scenarios

Mikita Hradovich, Adam Kasperski and Paweł Zieliński

1 Preliminaries

We are given a set J of n jobs, which can be partially ordered by some precedence constraints. Namely, $i \to j$ means that job j cannot start before job i is completed. For each job $j \in J$ a nonnegative processing time p_j, a nonnegative due date d_j and a nonnegative weight w_j can be specified. A schedule π is a feasible permutation of the jobs and Π is the set of all feasible schedules. We will use $C_j(\pi)$ to denote the *completion time* and $T_j(\pi) = [C_j(\pi) - d_j]^+$ the *tardiness* of job j in schedule π, where $[x]^+ = \max\{0, x\}$. Let $f(\pi)$ be a nonnegative cost of $\pi \in \Pi$. The following cost functions are commonly used (see, e.g., [2]): *total weighted completion time* $f(\pi) = \sum_{j \in J} w_j C_j(\pi)$, *total weighted tardiness* $f(\pi) = \sum_{j \in J} w_j T_j(\pi)$, and *maximum weighted tardiness* $f(\pi) = \max_{j \in J} w_j T_j(\pi)$. We will denote scheduling problems $\mathcal{P}$ by using the standard Graham's notation (see, e.g., [2]).

In this paper we assume that job processing times and due dates can be uncertain. The uncertainty is modeled by a discrete *scenario set* $\mathcal{U} = \{\xi_1, \xi_2, \ldots, \xi_K\}$. Each realization of the parameters $\xi \in \mathcal{U}$ is called a *scenario*. For each scenario $\xi \in \mathcal{U}$ a probability $\Pr[\xi] > 0$ of its occurrence is known. We will use $p_j(\xi)$ and $d_j(\xi)$ to denote the processing time and due date of job j under scenario $\xi \in \mathcal{U}$. We will also denote by $C_j(\pi, \xi)$ the completion time of job π under $\xi \in \mathcal{U}$ and by $f(\pi, \xi)$ the cost of schedule π under ξ. Given a feasible schedule $\pi \in \Pi$, we denote by $F(\pi)$ a

---

M. Hradovich · P. Zieliński
Department of Computer Science, Faculty of Fundamental Problems of Technology,
Wrocław University of Science and Technology, Wrocław, Poland
e-mail: Mikita.Hradovich@pwr.edu.pl

P. Zieliński
e-mail: Pawel.Zielinski@pwr.edu.pl

A. Kasperski (✉)
Department of Operations Research, Faculty of Computer Science and Management,
Wrocław University of Science and Technology, Wrocław, Poland
e-mail: Adam.Kasperski@pwr.edu.pl

random cost of π. Notice that $F(\pi)$ is a discrete random variable with the probability distribution induced by the probability distribution in $\mathcal{U}$.

The scenario-based representation allows us to avoid assumptions on distributions of random parameters. It has been applied to capture the randomness of uncertain parameters in several discrete optimization problems (see, e.g., [1, 6, 12]). A frequent goal, in this case, is to minimize the *expected cost* of a solution built. This criterion assumes that a decision maker is risk neutral and leads to a solution that guarantees an optimal long run performance. However, sometimes a solution found may be questionable, especially when it is implemented only once (see, e.g., [7]). Moreover, the expected cost criterion does not take a decision maker's risk aversion into account [3]. In order to compute a risk averse schedule that hedges against the uncertainty, we adopt performance measures of schedule π, called the *value at risk* and the *conditional value at risk*. We recall their definitions, following [10, 11].

Given a random variable Y with a fixed level α, we define the Value at Risk as the α-quantile:

$$\mathbf{VaR}_\alpha[Y] = \inf\{t : \Pr[Y \leq t] \geq \alpha\}, \alpha \in (0, 1]. \tag{1}$$

The Conditional Value at Risk can be defined as follows:

$$\mathbf{CVaR}_\alpha[Y] = \inf\{\gamma + \frac{1}{1-\alpha}\mathbf{E}[Y - \gamma]^+ : \gamma \in \mathbb{R}\}, \alpha \in [0, 1). \tag{2}$$

Let Y be a discrete random variable taking the values $b_1, \ldots, b_K$. Then the values of $\mathbf{VaR}_\alpha[Y]$ and $\mathbf{CVaR}_\alpha[Y]$ can be computed by solving the following programs, respectively (see, e.g., [1, 9, 11]):

$$
\begin{aligned}
&\text{(a)} \quad \min \theta &&\text{(b)} \quad \min \gamma + \frac{1}{1-\alpha} \sum_{i \in [K]} \Pr[Y = b_k] u_k \\
&\text{s.t. } b_k - \theta \leq M\beta_k, \quad k \in [K] &&\text{s.t. } \gamma + u_k \geq b_k, \quad k \in [K] \\
&\sum_{k \in [K]} \Pr[Y = b_k]\beta_k \leq 1 - \alpha &&u_k \geq 0, \quad k \in [K] \\
&\beta_k \in \{0, 1\}, \quad k \in [K] &&
\end{aligned}
\tag{3}
$$

where $M \geq \max\{b_1, \ldots, b_K\}$. The following property will be used in Sect. 2:

Property 1 *Let X and Y be two discrete random variables taking nonnegative values $a_1, \ldots, a_K$, and $b_1, \ldots, b_K$, respectively, with $\Pr[X = a_k] = \Pr[Y = b_k]$ and $a_k \leq \gamma b_k$ for each $k \in [K]$ and some fixed $\gamma \geq 0$. Then $\mathbf{CVaR}_\alpha[X] \leq \gamma\mathbf{CVaR}_\alpha[Y]$ for each $\alpha \in [0, 1)$ and $\mathbf{VaR}_\alpha[X] \leq \gamma\mathbf{VaR}_\alpha[Y]$ for each $\alpha \in (0, 1]$.*

Proof We will show the proof for the value at risk criterion (the proof for the conditional value at risk is straightforward). Let $\theta^*, \beta_k^*, k \in [K]$, be an optimal solution to (3a). Since $\gamma \geq 0$, the constraint $\gamma b_k - \gamma \theta^* \leq \gamma M \beta_k^*$ holds for each $k \in [K]$. By $a_k \leq \gamma \beta_k$ for each $k \in [K]$, we get $a_k - \gamma \theta^* \leq M'\beta_k^*$, where $M' =$

$\gamma M \geq \max\{a_1, \ldots, a_K\}$, $k \in [K]$. We also have $\sum_{k \in [K]} \Pr[X = a_k] \cdot \beta_k^* \leq 1 - \alpha$. Hence $\mathbf{VaR}_\alpha[X] \leq \gamma \theta^* = \gamma \mathbf{VaR}_\alpha[Y]$. □

The value at risk is an estimate of the maximum potential loss with a certain confidence level α. It is known that the following relations among the conditional value at risk, the value at risk, the expected cost and the maximum cost criteria hold (see, e.g., [9]), namely, $\mathbf{CVaR}_0[F(\pi)] = \mathbf{E}[F(\pi)] = \sum_{i \in [K]} \Pr[\xi_i] f(\pi, \xi_i)$, is the expected cost of schedule π and $\mathbf{VaR}_1[F(\pi)] = \lim_{\alpha \to 1} \mathbf{CVaR}_\alpha[F(\pi)] = \mathbf{Max}[F(\pi)] = \max_{i \in [K]} f(\pi, \xi_i)$ is the maximum cost of π under scenario set $\mathcal{U}$, which is a popular criterion used in robust optimization [7]. For a deeper motivation of using the risk criteria in decision making and a description of their various properties, we refer the reader to [10].

In this paper we will discuss the problems MIN- VAR_α $\mathcal{P}$, MIN- CVAR_α $\mathcal{P}$, MIN- EXP $\mathcal{P}$, and MIN- MAX $\mathcal{P}$, in which we minimize the corresponding performance measure for a fixed α and a specific single machine scheduling problem $\mathcal{P}$ under a given scenario set $\mathcal{U}$. Scheduling problems with risk criteria have been recently discussed in [1, 12].

2 Complexity and Approximation Results

We first prove the following result:

Theorem 1 *If* MIN- EXP $\mathcal{P}$ *is approximable within* $\sigma > 1$ *(for* $\sigma = 1$ *it is polynomially solvable), then* MIN- CVAR_α $\mathcal{P}$ *is approximable within* $\sigma \rho$, *where* $\rho = \min\{\frac{1}{\Pr_{\min}}, \frac{1}{1-\alpha}\}$, *for each constant* $\alpha \in [0, 1)$.

Proof We first show that for any π and $\alpha \in [0, 1)$, it holds

$$\mathbf{E}[F(\pi)] \leq \mathbf{CVaR}_\alpha[F(\pi)] \leq \min\left\{\frac{1}{\Pr_{\min}}, \frac{1}{1-\alpha}\right\} \mathbf{E}[F(\pi)], \quad (4)$$

where $\Pr_{\min} = \min_{k \in [K]} \Pr[\xi_k]$. The first inequality follows directly from the definition of the conditional value at risk. We now show the second inequality. The value of $\mathbf{CVaR}_\alpha[F(\pi)]$ can be computed by solving the following linear program. Indeed, it is easily seen that (5) is the dual to (3b).

$$\begin{aligned}
\max \quad & \sum_{k \in [K]} r_k f(\pi, \xi_k) \\
\text{s.t.} \quad & r_1 + \cdots + r_K = 1 \\
& 0 \leq r_k \leq \frac{\Pr[\xi_k]}{(1-\alpha)}, \quad k \in [K]
\end{aligned} \quad (5)$$

Let $r_1^*, \ldots r_k^*$ be an optimal solution to (5). Thus

$$\mathbf{CVaR}_\alpha[F(\pi)] = \sum_{k \in [K]} r_k^* f(\pi, \xi_k) \leq \sum_{k \in [K]} \frac{\Pr[\xi_k]}{(1-\alpha)} f(\pi, \xi_k) = \frac{1}{1-\alpha} \mathbf{E}[F(\pi)].$$

Moreover, $\mathbf{CVaR}_\alpha[F(\pi)] \leq \sum_{k \in [K]} \frac{\Pr[\xi_k]}{\Pr_{\min}} f(\pi, \xi_k) = \frac{1}{\Pr_{\min}} \mathbf{E}[F(\pi)]$ and (4) holds. Let π^* minimize the expected cost and π' minimize the conditional value at risk for a fixed $\alpha \in [0, 1)$. We will denote by $\hat{\pi}$ a σ-approximation schedule for MIN- EXP $\mathcal{P}$. Using (4), we get $\mathbf{CVaR}_\alpha[F(\hat{\pi})] \leq \rho \mathbf{E}[F(\hat{\pi})] \leq \sigma \rho \mathbf{E}[F(\pi^*)] \leq \sigma \rho \mathbf{E}[F(\pi')] \leq \sigma \rho \mathbf{CVaR}_\alpha[F(\pi')]$, and the theorem follows. □

In the following, we will show some applications of Theorem 1. Consider the problems MIN- EXP $1|prec|\sum w_j C_j$ and MIN- EXP $1|p_j = 1|\sum w_j T_j$. The first problem is equivalent to the deterministic counterpart $1|prec|\sum w_j C_j$ with the average processing times $\hat{p}_j = \sum_{k \in [K]} p_j(\xi_k) \Pr[\xi_k]$. Hence it is polynomially solvable for some particular structure of the precedence constraints (see [2]) and approximable within 2 in the general case (see [4]). It is not difficult to verify that the second problem is equivalent to the minimum assignment with costs $c_{ij} = \sum_{k \in [K]} \Pr[\xi_k] w_j [i - d_j(\xi_k)]^+$, $i, j \in [n]$, where c_{ij} is the cost of placing job i at position j. We can thus obtain approximation algorithms for MIN- CVAR$_\alpha$ $1|prec|\sum w_j C_j$ and MIN- CVAR$_\alpha$ $1|p_j = 1|\sum w_j T_j$ by applying Theorem 1. Notice that the former problem is NP-hard due to the results obtained in [7, 8].

We now show a sketch of 2-approximation algorithms for the problems with the weighted total flow time criterion. Our analysis will be similar to that in [8]. First, for each processing time scenario ξ_k, $k \in [K]$, we invert the role of processing times and weights obtaining the weight scenario ξ_k'. Formally, $p_j = w_j$ and $w_j(\xi_k') = p_j(\xi_k)$ for each $k \in [K]$. The new scenario set $\mathcal{U}'$ contains scenario ξ_k' with $\Pr[\xi_k'] = \Pr[\xi_k]$ for each $k \in [K]$. We also invert the precedence constraints, i.e. if $i \to j$ in the original problem, then $j \to i$ in the new one. Given a feasible schedule $\pi = (\pi(1), \ldots, \pi(n))$, let $\pi' = (\pi(n), \ldots, \pi(1))$ be the corresponding inverted schedule. Of course, schedule π' is feasible for the inverted precedence constraints. It is easy to verify that $f(\pi, \xi_k) = f(\pi', \xi_k')$ for each $k \in [K]$. In consequence $\mathbf{CVaR}_\alpha[F(\pi)] = \mathbf{CVaR}_\alpha[F'(\pi')]$ and $\mathbf{VaR}_\alpha[F(\pi)] = \mathbf{VaR}_\alpha[F'(\pi')]$, where $F'(\pi')$ is the random cost of π' for scenario set $\mathcal{U}'$. Since now the processing times are deterministic, we can express the set of feasible job completion times by the following system of constraints:

$$VC: \begin{array}{ll} C_j = p_j + \sum_{i \in J \setminus \{j\}} \delta_{ij} p_i & j \in J \\ \delta_{ij} + \delta_{ji} = 1 & i, j \in J, i \neq j \\ \delta_{ij} + \delta_{jk} + \delta_{ki} \geq 1 & i, j, k \in J \\ \delta_{ij} = 1 & i \to j \\ \delta_{ij} \in \{0, 1\} & i, j \in J, \end{array} \quad (6)$$

where $\delta_{ij} = 1$ if j is processed after i in the schedule constructed. We now relax the constraints $\delta_{ij} \in \{0, 1\}$ with $0 \leq \delta_{ij} \leq 1$ obtaining a system of linear constraints VC' and plug it into (3) with $b_k = \sum_{j \in J} w_j(\xi_k)C_j$. We get a linear programming problem for the case (3b) and a mixed integer problem, with K binary variables, for the case (3a). In the latter case the problem is polynomially solvable when K is constant. Suppose that $(C_1^*, \ldots, C_n^*)$ are the optimal values in the resulting program with the objective value of z^*. Let Y be discrete random variable taking the values of $\sum_{j \in J} w_j(\xi_k)C_j^*$, $k \in [K]$, with probabilities $\Pr[\xi_k]$, $k \in [K]$, respectively. Clearly, $\mathbf{VaR}_\alpha[Y] = z^*$ ($\mathbf{CVaR}_\alpha[Y] = z^*$). We can relabel the jobs so that $C_1^* \leq \cdots \leq C_n^*$. Consider a feasible schedule $\pi = (1, 2, \ldots, n)$. Applying the same reasoning as in [8], we can show that $C_j(\pi) \leq 2C_j^*$, which implies $f(\pi, \xi_k) \leq 2 \sum_{j \in J} w_j(\xi_k)C_j^*$ for each $k \in [K]$. Now, Property 1 implies $\mathbf{VaR}_\alpha[F(\pi)] \leq 2z^*$ ($\mathbf{CVaR}_\alpha[F(\pi)] \leq 2z^*$). Because z^* is a lower bound on the optimal objective value, π is a 2-approximate schedule. The following theorem summarizes the obtained results (we also use Theorem 1):

Theorem 2 MIN-$\mathrm{VAR}_\alpha 1|prec| \sum w_j C_j$ is approximable within 2 when K is constant. MIN-$\mathrm{CVAR}_\alpha 1|prec| \sum w_j C_j$ is approximable within 2 and approximable within $\min\{2, \frac{1}{1-\alpha}\}$ when the deterministic $1|prec| \sum w_j C_j$ problem is polynomially solvable.

We now address a scheduling problem $\mathcal{P}$ with the maximum weighted tardiness criterion. Hence $f(\pi, \xi_k) = \max_{j \in J} w_j[C_j(\pi, \xi_k) - d_j(\xi_k)]^+$, $k \in [K]$. All job processing times, due dates, and weights under scenarios are assumed to be nonnegative integers, and $w_j > 0$ for each $j \in J$. The set of jobs can be partially ordered by arbitrary precedence constraints. Minimizing the expected cost in such a problem is NP-hard [5]. Let $f_{\max}$ be an upper bound on the schedule cost over all scenarios. Let $h : \mathbb{Q}_+^K \to \mathbb{Q}_+$ be a nondecreasing with respect to $\mathbb{Q}_+^K$ function. Suppose that h can be evaluated in $g(K)$ time for a given vector $(t_1, \ldots, t_K) \in \mathbb{Q}_+^K$. Consider the corresponding scheduling problem $\mathcal{PS}$, in which we seek a feasible schedule $\pi \in \Pi$ minimizing $H(\pi) = h(f(\pi, \xi_1), \ldots, f(\pi, \xi_K))$. We can find such a schedule by solving a number of the following auxiliary problems: given a vector $\boldsymbol{t} \in \mathbb{Z}_+^K$, check if $\Pi(\boldsymbol{t}) = \{\pi \in \Pi : f(\pi, \xi_k) \leq t_k, k \in [K]\}$ is nonempty, and if so, return any schedule $\pi_{\boldsymbol{t}} \in \Pi(\boldsymbol{t})$. Given any $\boldsymbol{t} \in \mathbb{Z}_+^K$, we first form scenario set $\mathcal{U}'$ by specifying the following parameters for each $\xi_k \in \mathcal{U}$ and $j \in J$: $p_j(\xi_k') = p_j(\xi_k)$, $w_j' = 1$, $d_j(\xi_k') = \max\{C \geq 0 : w_j(C - d_j(\xi_k)) \leq t_k\} = t_k/w_j + d_j(\xi_k)$. The scenario set $\mathcal{U}'$ can be built in $O(Kn)$ time. We then solve MIN-MAX $\mathcal{P}$ with the scenario set $\mathcal{U}'$, which can be done in $O(Kn^2)$ time by using the algorithm constructed in [5]. If the maximum cost of the schedule π returned over $\mathcal{U}'$ is 0, then $\pi_{\boldsymbol{t}} = \pi$; otherwise $\Pi(\boldsymbol{t})$ is empty. From the monotonicity of the function h, it follows that for each $\pi \in \Pi(\boldsymbol{t})$ the inequality $h(f(\pi, \xi_1), \ldots, f(\pi, \xi_K)) \leq h(\boldsymbol{t})$. Thus, in order to solve the problem $\mathcal{PS}$, it suffices to enumerate all possible vectors $\boldsymbol{t} = (t_1, \ldots, t_K)$, where $t_i \in \{0, \ldots, f_{\max}\}$, $i \in [K]$, and compute $\pi_{\boldsymbol{t}} \in \Pi(\boldsymbol{t})$ if $\Pi(\boldsymbol{t})$ is nonempty. A schedule $\pi_{\boldsymbol{t}}$ with the minimum value of $H(\pi_{\boldsymbol{t}})$ is returned. Clearly, this can be done in $O(f_{\max}^K(Kn^2 + g(K)))$ time. Since all the risk criteria are nondecreasing functions with respect to schedule costs over scenarios and in this case $g(K)$ is negligible in comparison with Kn^2, we conclude that MIN-$\mathrm{VAR}_\alpha \mathcal{P}$ and MIN-$\mathrm{CVAR}_\alpha \mathcal{P}$ are solvable in $O(f_{\max}^K Kn^2)$ time. This running time is pseudopolynomial if K is constant. Notice that the special cases of the problems, when $\mathcal{P}$ is $1|prec, p_j = 1|T_{\max}$ are solvable in $O(Kn^{K+2})$ time, which is polynomial if K is constant (in this case $f_{\max} = n$).

We now show that problem $\mathcal{PS}$ admits an FPTAS if K is a constant and $h(\gamma t) \leq \gamma h(t)$, for any $t \in \mathbb{Q}_+^K$, $\gamma \geq 0$. First we partition the interval $[0, f_{\max}]$ into geometrically increasing subintervals: $[0, 1) \cup \bigcup_{\ell \in [\eta]}[(1+\epsilon)^{\ell-1}, (1+\epsilon)^{\ell})$, where $\eta = \lceil \log_{1+\epsilon} f_{\max} \rceil$ and $\epsilon \in (0, 1)$. Then we enumerate all possible vectors $t = (t_1, \ldots, t_K)$, where $t_i \in \{0, 1\} \cup \bigcup_{\ell \in [\eta]}\{(1+\epsilon)^{\ell}\}$, $i \in [K]$, and find $\pi_t \in \Pi(t)$ if $\Pi(t) \neq \emptyset$. Finally, we output a schedule $\pi_{\hat{t}}$ that minimizes value of $H(\pi_t)$ over the nonempty subsets of schedules. Obviously, the running time is $O((\log_{1+\epsilon} f_{\max})^K (Kn^2 + g(K))) = O((\epsilon^{-1} \log f_{\max})^K (Kn^2 + g(K)))$. Let π^* be an optimal schedule to $\mathcal{PS}$. Fix $\ell_i \in \{0, \ldots, \eta\}$ for each $i \in [K]$, such that $(1+\epsilon)^{\ell_i-1} \leq f(\pi^*, \xi_i) < (1+\epsilon)^{\ell_i}$, where we assume that $(1+\epsilon)^{\ell_i-1} = 0$ for $\ell_i = 0$. This clearly forces $\Pi((1+\epsilon)^{\ell_1}, \ldots, (1+\epsilon)^{\ell_K}) \neq \emptyset$. Moreover, $(1+\epsilon)^{\ell_i} \leq (1+\epsilon)f(\pi^*, \xi_i)$ for $\ell_i, i \in [K]$. By the definition of $\pi_{\hat{t}}$, we get $H(\pi_{\hat{t}}) \leq h((1+\epsilon)^{\ell_1}, \ldots, (1+\epsilon)^{\ell_K})$. Since h is a nondecreasing function and $h(\gamma t) \leq \gamma h(t)$, $h((1+\epsilon)^{\ell_1}, \ldots, (1+\epsilon)^{\ell_K}) \leq (1+\epsilon)h(f(\pi^*, \xi_1), \ldots, f(\pi^*, \xi_K))$. Hence, $H(\pi_{\hat{t}}) \leq (1+\epsilon)H(\pi^*)$. By Observation 1, the risk criteria satisfy the additional assumption on the function $h(t)$. In consequence, MIN- VAR$_\alpha$ $\mathcal{P}$ and MIN- CVAR$_\alpha$ $\mathcal{P}$ admit an FPTAS, when the number of scenarios is constant.

Acknowledgements Mikita Hradovich was supported by Wrocław University of Science and Technology, Grant 0401/0086/16.

References

1. Atakan, S., Bulbul, K., & Noyan, N. (2017). Minimizng value-at-risk in single machine scheduling. *Annals of Operations Research, 248*, 25–73.
2. Brucker, P. (2007). *Scheduling algorithms* (5th ed.). Heidelberg: Springer.
3. Daniels, R. L., & Kouvelis, P. (1995). Robust scheduling to hedge against processing time uncertainty in single-stage production. *Management Science, 41*, 363–376.
4. Hall, L. A., Schulz, A. S., Shmoys, D. B., & Wein, J. (1997). Scheduling to minimize average completion time: Off-line and on-line approximation problems. *Mathematics of Operations Research, 22*, 513–544.
5. Kasperski, A., & Zieliński, P. (2016). Single machine scheduling problems with uncertain parameters and the OWA criterion. *Journal of Scheduling, 19*, 177–190.
6. Katriel, I., Kenyon-Mathieu, C., & Upfal, E. (2008). Commitment under uncertainty: Two-stage matching problems. *Theoretical Computer Science, 408*, 213–223.
7. Kouvelis, P., & Yu, G. (1997). *Robust discrete optimization and its applications*. Dordrecht: Kluwer Academic Publishers.
8. Mastrolilli, M., Mutsanas, N., & Svensson, O. (2013). Single machine scheduling with scenarios. *Theoretical Computer Science, 477*, 57–66.
9. Ogryczak, W. (2012). Robust decisions under risk for imprecise probabilities. In Y. Ermoliev, M. Makowski, & K. Marti (Eds.), *Managing safety of heterogeneous systems* (pp. 51–66). Berlin: Springer.

10. Pflug, G. C. (2000). Some remarks on the value-at-risk and the conditional value-at-risk. In S. P. Uryasev (Ed.), *Probabilistic constrained optimization: Methodology and applications* (pp. 272–281). Dordrecht: Kluwer Academic Publishers.
11. Rockafellar, R. T., & Uryasev, S. P. (2000). Optimization of conditional value-at-risk. *The Journal of Risk*, *2*, 21–41.
12. Sarin, S., Sherali, H., & Liao, L. (2014). Minimizing conditional-value-at-risk for stochastic scheduling problems. *Journal of Scheduling*, *17*, 5–15.

Part XIV
OR in Engineering

A Nonlinear Model for Vertical Free-Flight Trajectory Planning

Liana Amaya Moreno, Armin Fügenschuh, Anton Kaier and Swen Schlobach

1 Introduction

In recent years free-flight trajectory planning came into the focus for the commercial airline industry. It provides a new way to deal with the rapid growth of the air traffic in Europe [2] and the resulting difficulties that this entails for the air traffic management (ATM). Although the priority of the ATM is to ensure the safety of the flight operations, other factors such as CO_2 emissions and, directly related to this, fuel costs, could benefit if all three goals are considered simultaneously by an integrating approach. This translates into computing fuel optimal trajectories that reduce the environmental degradation due to carbon fuel combustion and might further lead to a reduction in costs given the ever growing prices of fuel in the last years. From a computational point of view, the challenge is to find trajectories, composed of adjacent segments connecting two points (on the earth's surface), that avoid head-winds and benefit from tail-winds. Moreover, a time constraint is always enforced in order not to incur extra costs due to early or late arrival. This 4-dimensional problem (3 space dimensions plus time) is computationally difficult, and it is solved in practice typically in two subsequent stages: a horizontal phase, in which the segments of a 2-dimensional trajectory are computed, and then a vertical phase, in which different

---

L. A. Moreno (✉) · A. Fügenschuh
Professorship of Applied Mathematics, Department of Mechanical Engineering,
Helmut Schmidt University/University of the Federal Armed Forces, Hamburg, Germany
e-mail: lamayamo@hsu-hh.de

A. Fügenschuh
e-mail: fuegeschuh@hsu-hh.de

A. Kaier · S. Schlobach
Lufthansa Systems AG, Kelsterbach, Germany
e-mail: anton.kaier@lhsystems.com

S. Schlobach
e-mail: swen.schlobach@lhsystems.com

altitudes are assigned to the segments. Moreover, fuel consumption data is needed to optimally assign speed and altitude in order to minimize the amount of fuel used during the flight. Fuel information is given by the aircrafts manufactures, as a black box function which provides data only for a grid of points depending on speed, altitude and weight levels. In general during the optimization process, fuel consumption data is required for values that do not coincide with the given grid points, hence some techniques must be applied to obtain the required intermediate fuel consumption values. To come up with this continuous formulation of the data, different interpolation and approximation techniques are used. It is important to note that these drastically affect the computation times. In this study we concentrate on the vertical flight planning of commercial aircrafts. We propose an NLP model in which we integrate local and global interpolation and approximation techniques as continuous formulations of the problem's input data. We discuss briefly the characteristics of these formulations. Moreover, we compare different available commercial solvers for nonlinear programming for our test instances.

2 Mathematical Model

Our work is based on a model for vertical flight planning [1, 5], where speed and altitude are assigned to each of the segments that compose the trajectory, and the wind is assumed to be equal in all altitudes over one segment (but can vary from segment to segment). The fuel consumption is a bivariate function that depends on the current weight of the airplane (which is decreasing during the flight, since fuel is consumed) and the selected speed, see Fig. 1. The fuel also depends on the flown altitude, which is determined in a post-processing step, once the optimal speed and weight are computed. Hence we do not need to consider altitude as a variable in our model. The objective of the model is to assign to each of the segments that compose the trajectory a speed and a weight value. Let n be total number of segments, and

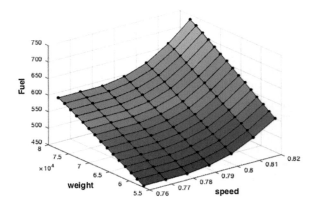

Fig. 1 Unit fuel consumption (kg per nautical mile) for the airbus 320. The horizontal axis is the aircrafts speed (Mach number from optimal speed to maximal speed), and the vertical axis is the weight (kg)

A Nonlinear Model for Vertical Free-Flight Trajectory Planning   447

let $S = \{1, \ldots, n\}$ denote the segment indexes. The nodes that link the segments then have the indexes $N = S \cup \{0\}$. Besides the fuel consumption data, the instance is further specified by the following data: L_i is the length of segment $i \in S$. The minimum and maximum duration of the entire trip are given by $\underline{T}$ and $\overline{T}$, respectively. The dry weight of the loaded airplane including the contingency fuel is W^{dry}.

We introduce the following variables: For each segment $i \in S$ the variable $v_i \in \mathbb{R}_+$ models the velocity of the airplane in this segment (the velocity can only be set once for the entire segment). The weight of the airplane at node $i \in N$ is denoted by $w_i \in \mathbb{R}_+$, and $w_i^{\text{mid}} \in \mathbb{R}_+$ is the "middle weight" of the airplane within segment $i \in S$, which is an auxiliary variable that is used in the computation of the fuel consumption $f_i \in \mathbb{R}_+$. The mathematical model reads as follows:

$$\min \quad w_0 - w_n \tag{1}$$
$$\text{s.t.} \quad t_0 = 0, \quad \underline{T} \leq t_n \leq \overline{T} \tag{2}$$
$$\forall i \in S: \quad \Delta t_i = t_i - t_{i-1} \tag{3}$$
$$\forall i \in S: \quad L_i = v_i \cdot \Delta t_i \tag{4}$$
$$w_n = W^{dry} \tag{5}$$
$$\forall i \in S: \quad w_{i-1} = w_i + f_i \tag{6}$$
$$\forall i \in S: \quad w_{i-1} + w_i = 2 \cdot w_i^{\text{mid}} \tag{7}$$
$$\forall i \in S: \quad f_i = L_i \cdot \widehat{F}(v_i, w_i^{\text{mid}}) \tag{8}$$

The objective function (1) minimizes the fuel consumed during the trip. It is computed as the difference between the start and arrival weight. In Eq. (2) the starting time t_0 is set to zero, and the final time t_n is forced to be within the arrival time window. Equation (3) enforces the time consistency, and the equation of motion is given by (4). With Eq. (5) all the fuel is consumed during the flight. The weight consistency is enforce by Eq. (6). The middle weight is computed in (7) which is required to calculate the fuel consumption in each segment in (8), where $\widehat{F}(v_i, w_i^{\text{mid}})$ is the continuous approximation or interpolation of the discrete fuel consumption data. This function offers intermediate data points within the corresponding ranges. Both interpolation and approximation techniques accomplish this purpose, however, the choice between one or the other depends on user. Nonlinear solvers require information about the first and sometimes the second order partial derivatives of all functions used in the problem formulation. In our model, Eq. (1)–(7) are either linear or quadratic equations, therefore its derivatives are easy to compute and this is done automatically by the solver. For Eq. (8), we need to explicitly compute first and second derivatives so they can be passed on to the solver. In the following we briefly describe the interpolation and approximation techniques used in this work. Each of them yields a polynomial function in two dimensions for the fuel consumption function. The first and second derivatives of the fuel function are then approximated by taking the first and second derivatives of these approximations.

Bilinear: Bilinear interpolation is a local technique, where intermediate values computed based on the four neighboring points. The interpolant is obtained by performing a linear interpolation along each of the dimensions of the table, which leads to a second degree polynomial. Further details can be found in [4].

Bicubic: Bicubic interpolation is a local technique where intermediate values are computed based on the four neighboring points, the first and second derivatives of these points (which are approximated). This leads to a linear system of 16 equations where the variables are the coefficients of a 6th order polynomial in two dimensions. In this work we have approximated the derivatives at the points of the table by two different methods, using finite differences and using cubic splines. For more details we refer the reader to [4].

Cubic Splines: Intermediate values are computed based on the information of the whole table. Therefore we refer to it as a global technique. The idea is to construct one-dimensional cubic splines along all the rows of the table and evaluate them at one of the first coordinate of the intermediate point. With these new values, another one-dimensional spline is constructed and finally evaluated at the second coordinate of the intermediate point. If smoothing is desired, a smoothing parameter is used for the construction of the cubic splines (approximation method). This results in a new set of points that best approximates the surface using cubic splines. For further details we refer the reader to [3, 4].

3 Numerical Results

The models were written using AMPL as modeling language and solved by the NLP solvers SNOPT 7.2-5, CONOPT 3.5C, KNITRO 8.1.1, and MINOS 5.51. We have used similar test instances as in [1], that is, the airplanes Airbus 320, 380, Boeing 737 and 772. For each airplane several travel distances were tested ranging form 800 Nautical Miles (NM) for the B737. to 7500 NM for A380 and B772. Two different time windows were used for each distance, for a total of 42 instances. Each flight is divided into equidistant segments of 100 NM. Table 1 summarizes the features of the test instances. All instances were solved using a 6-core Intel Xeon E5 at 3.5 GHz and 16 GB RAM computing machine. In Table 2, we give the percentage of the instances that were actually solved within a 10% error of the global optimal

Table 1 Maximal speed (in Mach number), dry weight and maximal weight in (kg), maximal distance (in NM) and number of segments $|S|$ for each instance

Type	Max. speed	Dry weight	Max. weight	Max. distance	$|S|$
A320	0.82	56614	76990	3500	15, 20, 30, 35
A380	0.89	349750	569000	7500	30, 40, 50, 60, 70 ,75
B737	0.76	43190	54000	1800	8, 12, 15, 18
B772	0.89	183240	294835	7500	25, 35, 45, 55 ,65,75

Table 2 Percentage of solved instances with each solver and each method

	SNOPT	MINOS	KNITRO	CONOPT
Bilinear	100	100	43	48
Splines1	100	100	67	100
Splines2	17	31	5	5
Bicubic1	0	12	0	0
Bicubic2	0	10	0	0

values reported on [1] by each solver using the different methods. We have used the following abbreviations: *Splines1* refers to cubic splines interpolation (no smoothing of the data); *Splines2* refers to the method of smoothing cubic splines; *Bicubic1* refers to bicubic interpolation using finite differences approximations for the value of the derivatives and finally, *Bicubic2* refers to bicubic interpolation using cubic splines to approximate the value of the derivatives. The results in Table 2 indicate that the most successful methods are *Splines1* followed by *bilinear*, both interpolating techniques. *Splines1* is consistently, among all the solvers, the one that allows to solve the greatest number of instances. In order to compare these two methods, and the solvers as well, we give a graphical evaluation of the solution times in Fig. 2. On the x-axis of these plots, the instances are listed in ascending order according to their size, i.e., according the number of segments used for the trip. The data points, whose solution time are 100 s, represent the instances that were not solved, within a 10% gap from the global optimum. For both methods, the solution times of most instances are below 12 s. For the bilinear method, the solver Snopt outperforms the others. Note that the squared-shaped data points are consistently below all other data points. Most of the instances are solved within one second; the rest, within five. Minos is also very successful using bilinear interpolation, as the solution of all instances requires at most 10 s. The solution times of our instances using cubic splines interpolation are below 25 s. In this case, there is no straightforward outperformance of one solver over the others. On the contrary, the solvers take similar time to compute the (same) optimal solution. Knitro fails sometimes this purpose.

In conclusion, these two methods provide suitable continuous formulations of the input data, that can efficiently be integrated into our NLP models. Bilinear interpolation is very simple to implement and the number of computations needed is very low in comparison to the cubic splines method. The latter one requires the solution of many systems of linear equations (in the order of rows or columns in the input data table). Thus one can expect that cubic splines takes a longer solving time. Nevertheless, the derivatives provided with this last method, are smoother than the ones obtained with the bilinear interpolation, which can explain its faster convergence to a local optimum. What is important to note here, is that the approximations of the derivatives with bilinear technique are still good enough for the solvers to search in good directions for local minima. As for the bicubic methods, it is obvious, they are

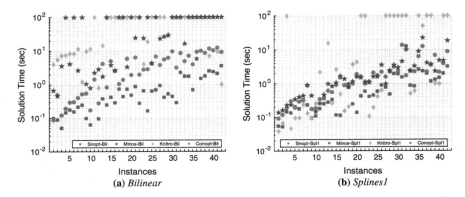

Fig. 2 Solution times of all test instances using **a** bilinear and **b** cubic splines interpolation with each solver

not successful. A reason behind this, might be the inaccurate approximations of the required first and second derivatives at the points of the table.

In our ongoing work we extend our methods to a full 4-dimensional trajectory planning. That is, to include a vertical optimization phase where a more realistic wind field (which can deviate also in altitude) is taken into account. This adds one more dimension to the fuel consumption data. On the other hand, introducing dynamic wind also increases one dimension to the wind data, therefore we need to study if the interpolation and approximation techniques, here presented, extend efficiently to more dimensions.

References

1. Moreno, L. A., Yuan, Z., Fügenschuh, A., Kaier, A., & Schlobach, S. (2017). *Combining NLP and MILP in vertical flight planning* (pp. 273–278). Cham: Springer International Publishing.
2. EUROCONTROL. (2013). Challenges of Growth 2013: Task 4: European Air Traffic in 2035.
3. Knott, D. (2000). *Interpolating cubic splines* (Vol. 18)., Progress in computer science and applied logic New York: Springer-Science+Business Media, LLC.
4. Press, W. H., Teukolsky, S. A., Vetterling, W. T., & Flannery, B. P. (1992). *Numerical recepies in C. The art of scientific computing* (2nd ed.). Cambridge: Cambridge University Press.
5. Yuan, Z., Fügenschuh, A., Kaier, A., & Schlobach, S. (2016). *Variable speed in vertical flight planning* (pp. 635–641). Cham: Springer International Publishing.

Lattice Structure Design with Linear Optimization for Additive Manufacturing as an Initial Design in the Field of Generative Design

Christian Reintjes, Michael Hartisch and Ulf Lorenz

1 Introduction and Motivation

The design process is not as automated as it can be as shown in [1]. Over the last decade technical optimization became increasingly important for the modern, energy-efficient and therefore material-saving construction. The factors essentially responsible for this are increasing computing capacity and constant improvement of construction software. Both factors are at a level that allows us to radically rethink the entire design process [1].

The traditional approach uses construction software combined with human knowledge to produce drawings and to display all technical data. The results can be considered in succeeding analysis, e.g. numerical or nonlinear topology optimization. Irrespective of the quality of those methods, the initial drawing still is a product of the human intellectual capacity.

The profitability of Lattice Structure Design with optimization for additive manufacturing has been shown by recent publications [1, 3, 4]. However, the development of an approach which aims at solving a considerable amount of nodes in reasonable time – and according to that a large construction volume or high granularity of the lattice structure – by the use of appropriate simplifications for the elastic deformation seems to be unconsidered.

Our approach is to use optimization in the first instance to generate the design itself. Therefore, we want to provide optimization tools to compute the design, since it might be possible that the computer-based design outperforms the human-based design. As mixed-integer linear optimization is not able to take aesthetic and manufacturing-friendly design into account some of the resulting structures still can not be build using conventional manufacturing methods. However, due to the rapid development in the field of additive manufacturing there is now the opportunity to

---

C. Reintjes (✉) · M. Hartisch · U. Lorenz
Institute of Technology Management, University of Siegen, Siegen, Germany
e-mail: christian.reintjes@student.uni-siegen.de

combine optimization with this technology due to the high level of geometrical freedom which is needed to manufacture complex structures. An overview on research in this field can be found in [2, 3].

2 Problem Setting

In this contribution we discuss a mixed-integer programming model to describe the static load distribution in a two-dimensional space as a first approach. The required amount of material should be minimized, as the number of truss beams is minimized. It is important to note here, that this model is made to create an initial design and the sizing of a lattice structure instead optimizing a given structure.

A two-dimensional assembly space consists of nodes and edges, whereas nodes represent truss joints and edges represent beams, as specified in Fig. 1.

It is assumed that the truss joints are adequately dimensioned and ensure the point of action of the forces to work in the center of the truss joint profile. The force axis

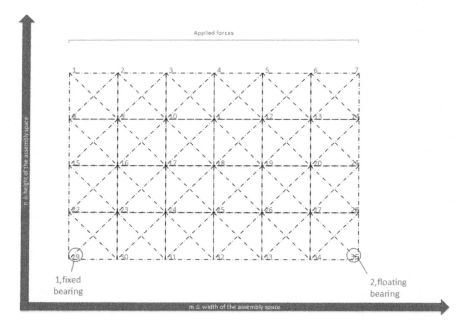

Fig. 1 The two-dimensional assembly space as an exemplary application with a width of 7 length units and a height of 4. The two bearings and the nodes for external forces are marked

and the central axis of the truss girders are identical. Girder non-linearities (yielding, tearing, slippage etc.) are not possible. This is achieved by assigning a permissible compressive and tensile loading at each truss girder. A safety factor for the limited elasticity of the material is considered. The two-dimensional grid is mounted on a locating/non-locating bearing arrangement. Introduced forces are only possible at truss joints to avoid torques in the beams itself. The statical determinacy is considered as an external and internal determinacy, whereby both conditions have to be fulfilled (Tables 1 and 2).

3 Model Formulation

Table 1 Parameters

Symbol	Definition
$m \in \mathbb{N}$	Width of the assembly space
$n \in \mathbb{N}$	Height of the assembly space
$V = \{1, \ldots, nm\}$	Set of connecting nodes (possible truss joints)
$T = \{0, 1, \ldots, s\}$	Set of different beam types
$c_t \in \mathbb{R}_+$	Capacity of beam type t
$M \in \mathbb{R}$	Big M - maximum capacity of the most robust beam type
$Cost_t \in \mathbb{R}$	Costs of beam type t
$Q_i \in \mathbb{R}_+$	Source force at node $i \in V$
$V_A \in \mathbb{R}_+$	Vertical reaction force at fixed bearing $A = mn - m + 1$ (bottom left)
$V_B \in \mathbb{R}_+$	Vertical reaction force at floating bearing $B = mn$ (bottom right)
$A = \{\vert, -\}$	Set of possible lines of action of the force in the horizontal and vertical directions
$O = \{\vert, -, /, \backslash\}$	Set of possible lines of action of the force at a node
$NB(i) \subseteq V$	Set of neighboring nodes of i
$NB_o(i) \subseteq NB(i)$	Set of neighboring nodes of i regarding the orientation o
$r_{i,j,a} \in [0, 1]$	Force component at node i relative to the reference plane $a \in A$, caused by beam structure between $i \in V$ and $j \in V$

Table 2 Variables

Symbol	Definition
$B_{t,i,j} \in \{0,1\}$	Binary variable indicating whether bar of type $t \in T$ is present between $i \in V$ and $j \in V$
$F_{i,j} \in \mathbb{R}$	Flow of forces between nodes $i \in V$ and $j \in V$
$x_{i,j} \in \{0,1\}$	Binary variable indicating whether a bar is present between nodes $i \in V$ and $j \in V$
$y_i \in \{0,1\}$	Indicator whether at least one bar is present at node $i \in V$
$Z_i \in \{0,1\}$	Indicator whether at least one of four possible zero-force member bar combinations are present at node $i \in V$

$$\min \sum_{i \in V} \sum_{j \in V} \sum_{t \in T} B_{t,i,j} \cdot cost_t$$

$$\text{s.t.} \sum_{j \in NB(i)} r_{i,j,a} F_{i,j} - Q_i \mathbb{1}_{(a=|)}$$
$$+ V_A \mathbb{1}_{(i=mn-m+1 \wedge a=|)} + V_B \mathbb{1}_{(i=mn \wedge a=|)} = 0 \qquad \forall i \in V, a \in A \quad (1)$$

$$F_{i,j} \leq M \cdot x_{i,j} \qquad \forall i,j \in V \quad (2)$$

$$F_{i,j} = -F_{j,i} \qquad \forall i,j \in V \quad (3)$$

$$B_{t,i,j} = B_{t,j,i} \qquad \forall i,j \in V, t \in T \quad (4)$$

$$x_{i,j} = x_{j,i} \qquad \forall i,j \in V \quad (5)$$

$$F_{i,j} \leq \sum_{t \in T} c_t \cdot B_{t,i,j} \qquad \forall i,j \in V \quad (6)$$

$$\sum_{t \in T} B_{t,i,j} = x_{i,j} \qquad \forall i,j \in V \quad (7)$$

$$2y_i \leq \sum_{j \in NB(i)} x_{i,j} \leq 8 y_i \qquad \forall i,j \in V \quad (8)$$

$$2\ell_i^o \leq \sum_{j \in NB_o(i)} x_{i,j} \leq \ell_i^o + 1 \qquad \forall \in V, o \in O \quad (9)$$

$$\sum_{o \in O} \ell_i^o \leq 4 Z_i \qquad \forall i \in V \quad (10)$$

$$\sum_{j \in NB(i)} x_{i,j} \geq 3 Z_i \qquad \forall i \in V \quad (11)$$

$$x_{i,j}, y_i, \ell_i^o, z_i \in \{0,1\} \qquad \forall i,j \in V, o \in O \quad (12)$$

The objective function aims at minimizing the costs of the used truss griders which are needed to create a statically determined structure under the influence of external forces exerted perpendiculary.

Restriction (1) implies a force equilibrium point due to external forces and beam forces in the horizontal and vertical directions exactly in the center of each truss joint. The forces are determined by using vector arithmetic and the angle a between the currently observed beam between the nodes i and j relative to the reference plane R_- or $R_|$. The imaginary ray starting at i towards j is denoted $g(i, j)$ and the parameters $r_{i,j,a}$ can be precalculated as follows. Equations (14) and (15) describe the ray of the horizontal and vertical reference planes, respectively.

$$r_{i,j,a} = \cos\left(\angle\left(R_a, g(i, j)\right)\right) \tag{13}$$
$$R_- = g(mn - m + 1, mn) \tag{14}$$
$$R_| = g(1, mn - m + 1) \tag{15}$$

For determining the two bearing reaction forces V_A and V_B the following equations are used:

$$-\sum_{i=1}^{m} Q_i \cdot i + V_B \cdot (m - 1) = 0 \tag{16}$$

$$\sum_{i=1}^{m} Q_i \cdot (m - i) - V_A \cdot (m - 1) = 0 \tag{17}$$

V_A and V_B are essential elements of the moment equilibrium condition for each bearing and thus known from the beginning. Constraint (2) ensures that only used beams can transfer forces. Constraint (3) represents Newton's third law, whereas Constraint (4) and Constraint (5) simply demand a used beam to go both ways. Constraint (6) limits the force in a beam with regard to the permissible force of the used beam type. Constraint (7) guarantees that a specific beam type is selected if a beam is used.

Constraint (8) forbids the construction of a cantilever bearing, since (16) and (17) would not be fulfilled. Besides that a bending moment distribution from the element with the largest lever-arm to the bearing itself (maximum torque) would arise, which is not covered by this purely statically approach of the model.

In order to avoid unstrained members Constraints (9), (10) and (11) are defined. Possible unstrained members are identified by (9). If the direction of action in two truss griders is the same, which implies they are mounted at the same angle in relation to their joints, there is a combination possibility to use a unstrained member. It is used to have a better case of the Euler buckling. Condition (9) and (10) can identify the four combination possibilities, whereas (11) forces the model to add an unstrained member in an arbitrary spatial arrangement if at least one combination possibility to use an unstrained member is present. All these factors result in a lower buckling risk.

4 Computational Results

Using the provided model we investigated an assembly space of 164 ($m = 41, n = 4$) nodes. Three vertical external forces of 250 kN were used: one in the center at node 21 and two symmetrically arranged forces at nodes 11 and 31. We expected a symmetrical design and strength curve profile as a result. The given beam types were four different round bars with 4.0, 20.0, 40.0 and 50.0 mm average diameters with their corresponding tensile/compressive strengths and costs. The 4.0 mm bar represents a zero-force member bar. For space reasons only the left-hand side of the resulting design is displayed in Fig. 2, whereat the second half is arranged symmetrically.

As a result of the symmetric design in the solution each bearing is loaded equally. A horizontal force component does not exist. The torque path spreads symmetrically from the midpoint of the longitudinal axis to the outer bearings. The experiments were executed on a PC with an Intel i7-4790 (3.6 GHz) processor and 32 GB RAM with IBM ILOG CPLEX Optimization Studio Version 12.6.1. Due to a practical calculation time and a planned integration to a CAD-software to receive a first construction proposal the calculation time was limited to 4 h .

In a second example a cantilever bridge with a triple articulated arch has been calculated, which is a quite common construction solution to achieve a wide span and total length of the bridge. A simple cantilever bridge has two cantilever arms extending from opposite sides of the span and they meet at the center.

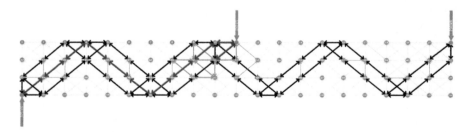

Fig. 2 Optimized structure with symmetrical design. Red arrows at the top level represent source forces. The different beam types are marked blue, green, black and brown. The right-hand side of the design is omitted

5 Conclusion and Outlook

We presented a mixed-integer linear programming model to describe the static load distribution in a two-dimensional space. The results show that it is possible to solve a considerable amount of nodes in feasible time by the use of appropriate simplifications for the elastic deformation. The computer-based design is consistent and we focus on extending our solution approach to design three-dimensional test cases controlled through an integrated software interface to mechanical CAD software (see [5]). This creates the possibility of doing a subsequent FEM-analysis, analyze the elastic behavior and export the 3D document as an STL file to use additive manufacturing methods. It might be possible to outperform the human-based design.

References

1. Li, W. (2016). *Beyond System Modeling - Modeling System Knowledge for Generative Design*, Bd. Global Product Data Interoperability Summit, Autodesk Research.
2. Merkt, S. J. (2015). Qualifizierung von generativ gefertigten Gitterstrukturen für maßgeschneiderte Bauteilfunktionen: Dissertation RWTH Aachen.
3. Zegard, T., & Paulino, G. H. (2016). Bridging topology optimization and additive manufacturing. *Structural and Multidisciplinary Optimization*, *53*(1), 175–192.
4. Smith, C., Todd und, I., & Gilbert, M. (2013). Utilizing additive manufacturing techniques to fabricate weight optimized components designed using structural optimization methods. In *Solid Freeform Fabrication Symposium*, 879–894.
5. Hadi, A., Vignat und, F., & Villeneuve, F. (2015). Design configurations and creation of lattice structures for metallic additive manufacturing. *14ème Colloque National AIP PRIMECA*.

Co-allocation of Communication Messages in an Integrated Modular Avionic System

Elina Rönnberg

1 Introduction

Electronics in an aircraft is called avionics and nowadays the majority of the avionics industry uses an integrated architecture called Integrated Modular Avionics (IMA) where applications share hardware resources on a common avionic platform. In such architectures it is vital to prevent faults from propagating between different aircraft functions and one component used to ensure this is pre-runtime scheduling of the tasks and the communication in the system. For more details about the industrial background of the IMA-system addressed in this paper, see [2], and for further reading about resource allocation in hard real-time avionic systems, see [1].

In [2], a mathematical model and a constraint generation procedure for an industrially relevant IMA-system is presented. The contribution of this paper is to extend that model and solution approach to include the possibility of co-allocation of communication messages to enable sending and receiving more than one message at a time. This possibility is of practical relevance since it induces capacity savings by reducing the execution requirements of certain tasks.

The addressed IMA-system can be considered as a multi-processing system that constitutes of nodes connected by a Communication network (CN). In each node there is a Communication module (CM) which handles both the inter-node and the intra-node communication as well as communication with external systems. Each node also hosts a set of Application modules (AMs) that run applications (software processes).

The system executes periodically with a period referred to as a major frame for which the schedule is cyclically repeated. The mathematical model for the system

---

E. Rönnberg (✉)
Department of Mathematics, Linköping University, 581 83 Linköping, Sweden
e-mail: elina.ronnberg@liu.se

E. Rönnberg
Saab AB, 581 88 Linköping, Sweden

considers only one major frame and is formulated such that an infinite repetition of the schedule for one major frame provides a valid infinite schedule for the system. The schedule is created pre-runtime and each task has been beforehand assigned to its module.

A full model for this problem can be divided into four components: Communication network scheduling, CM task scheduling, AM task scheduling, and Precedence relations. The two latter components are of no interest for the scope of this paper and will therefore not be presented here, instead the labels [AM-scheduling] and [Precedence relations] are used in the models to refer to them, for details see [2].

A short summary of the solution approach developed in [2] is given in Sect. 2. The existing approach exploits known characteristics of the problem and is designed to cope with the main computational challenge of this problem, which is the huge number of tasks to be sequenced on the CMs. The contribution of this paper, which is how to extend the solution approach to facilitate co-allocation of messages, is given in Sect. 3. Section 4 presents some preliminary results and conclusions.

2 Existing Solution Approach

For the industrially relevant instances that the solution approach in [2] is derived for, it is known that the CMs have huge numbers of tasks and that a large portion of these are fixed. In the complete model, tailored for constraint generation, sequencing of CM-tasks is achieved by two collaborating requirements. One is that each non-fixed task is assigned to a section in-between fixed tasks and the other is sequencing requirements for each subset of tasks that, for at least one section, can be assigned to a section together.

In the solution strategy, a relaxed problem is obtained by omitting the sequencing requirements for the subsets of tasks. In a solution to the relaxed problem, there will be a subset of tasks assigned to each section. For such solution, a subproblem is solved with the aim to sequence the tasks within the sections, still obeying all other constraints of the original model. If the subproblem is successfully solved, a solution to the original problem is obtained. If the subproblem fails to sequence all tasks, sequencing constraints for subsets that include failed tasks are permanently added to both models.

2.1 Common Components

The components [AM-scheduling] and [Precedence relations] are used in both of the models together with the components presented in this section.

Denote the set of CMs by $\mathcal{H}^{CM}$ and the set of tasks at CM h by $\mathcal{I}_h^{CM}$, $h \in \mathcal{H}^{CM}$. Task i has an execution requirement e_i and must execute within the interval between its release time t_i^r and deadline t_i^d, $i \in \mathcal{I}_h^{CM}$, $h \in \mathcal{H}^{CM}$. For task i let the variable

x_i = start time of task i, and constrain it by $t_i^r \leq x_i \leq t_i^d - e_i$, $i \in \mathcal{I}_h^{CM}$, $h \in \mathcal{H}^{CM}$. If $t_i^r = x_i = t_i^d - e_i$ hold, task i is referred to as fixed.

Let $\mathcal{M}$ be an ordered set of CN-message indices and let l_m^{msg} be the capacity required to send CN-message m, $m \in \mathcal{M}$. Denote the set of CN-slots by $\mathcal{N}$ and let the capacity of slot n be l_n^{slot}, $n \in \mathcal{N}$. Let the binary variable z_{nm} indicate if CN-message m is assigned to CN-slot n ($= 1$) or not ($= 0$), $n \in \mathcal{N}$, $m \in \mathcal{M}$. Constraint (1) assigns each CN-message to a slot and constraint (2) ensures that the capacity of the slots are respected.

There are four types of tasks involved in communicating a CN-message, these are indexed by the set $\mathcal{K} = \{1, 2, 3, 4\}$ and have to execute in this given order, ensured by using [Precedence relations]. Let the set $\mathcal{I}_k^K$ include all tasks of type k, $k \in \mathcal{K}$, and let $\mathcal{I}_m^M$ include all tasks used to communicate CN-message m, $m \in \mathcal{M}$. Introduce t_{nk}^{M-r} and t_{nk}^{M-d} to respectively denote the release time and deadline of task i, $i \in \mathcal{I}_m^M \cap \mathcal{I}_k^K$, if CN-message m is assigned to slot n, $n \in \mathcal{N}$, $k \in \mathcal{K}$, $m \in \mathcal{M}$. Constraint (3) makes sure that these times are respected.

$$\sum_{n \in \mathcal{N}} z_{nm} = 1, \qquad\qquad\qquad\qquad m \in \mathcal{M} \quad (1)$$

$$\sum_{m \in \mathcal{M}} l_m^{msg} z_{nm} \leq l_n^{slot}, \qquad\qquad\qquad\qquad n \in \mathcal{N} \quad (2)$$

$$\sum_{n \in \mathcal{N}} t_{nk}^{M-r} z_{nm} \leq x_i \leq \sum_{n \in \mathcal{N}} t_{nk}^{M-d} z_{nm} - e_i, \quad i \in \mathcal{I}_m^M \cap \mathcal{I}_k^K, \ k \in \mathcal{K}, \ m \in \mathcal{M} \quad (3)$$

2.2 Relaxed Problem

Let $\mathcal{R}_h^{CM}$ be the disjoint sections that correspond to intervals in-between pairs of adjacent fixed tasks at CM h, and denote their lengths by l_r^{sec}, $r \in \mathcal{R}_h^{CM}$, $h \in \mathcal{H}^{CM}$. Denote the set of tasks that can execute within section r by $\mathcal{I}_r^{sec}$ and for task i let t_{ir}^r and t_{ir}^d respectively be the release time and deadline in section r, $i \in \mathcal{I}_r^{sec}$, $r \in \mathcal{R}_h^{CM}$, $h \in \mathcal{H}^{CM}$. Introduce the binary variable α_{ir} that indicate if task i is assigned to section r ($= 1$) or not ($= 0$), $i \in \mathcal{I}_r^{sec}$, $r \in \mathcal{R}_h^{CM}$, $h \in \mathcal{H}^{CM}$.

Constraint (4) assigns each non-fixed task to a section and constraint (5) makes sure that the capacities of the sections are respected. Constraint (6) makes a task respect the release time and deadline within the section it is assigned to. The objective function of the relaxed problem is given in Sect. 3.

$$\sum_{r \in \mathcal{R}_h} \alpha_{ir} = 1, \qquad\qquad i \in \cup_{r \in \mathcal{R}_h} \mathcal{I}_r^{sec}, \ h \in \mathcal{H}^{CM} \quad (4)$$

$$\sum_{i \in \mathcal{I}_r^{\text{sec}}} e_i \alpha_{ir} \leq l_r^{\text{sec}}, \qquad r \in \mathcal{R}_h, \ h \in \mathcal{H}^{\text{CM}} \qquad (5)$$

$$\sum_{r \in \mathcal{R}_h} t_{ir}^{\text{r}} \alpha_{ir} \leq x_i \leq \sum_{r \in \mathcal{R}_h} t_{ir}^{\text{d}} \alpha_{ir} - e_i, \qquad i \in \cup_{r \in \mathcal{R}_h} \mathcal{I}_r^{\text{sec}}, \ h \in \mathcal{H}^{\text{CM}} \qquad (6)$$

2.3 Subproblem

Introduce a set $\mathcal{S}_h$ that includes an index for each subset of non-fixed tasks that can, for at least one section, be assigned together in the same section, and let the set $\mathcal{I}_s^{\text{sub}}$ include the tasks of subset s, $s \in \mathcal{S}_h$, $h \in \mathcal{H}^{\text{CM}}$. Given a solution to the relaxed problem there is one set $\bar{s}_r$ for each section r and each non-fixed task i belongs to exactly one set $\bar{s}_r$ and must respect $t_{ir}^{\text{r}} \leq x_i \leq t_{ir}^{\text{d}} - e_i$, $i \in \mathcal{I}_{\bar{s}_r}^{\text{sub}}$, $r \in \mathcal{R}_h^{\text{CM}}$, $h \in \mathcal{H}^{\text{CM}}$. For each subset $\bar{s}_r$ denote the set of possible immediate successors and predecessor of task i by $\mathcal{I}_{i\bar{s}_r}^{+}$ and $\mathcal{I}_{i\bar{s}_r}^{-}$ respectively, $r \in \mathcal{R}_h^{\text{CM}}$, $h \in \mathcal{H}^{\text{CM}}$.

Let the binary variable $\beta_{i\bar{s}_r}$ indicate if task i is successfully sequenced within section r ($= 1$) or not ($= 0$), and let the binary variable $y_{ij\bar{s}_r}$ indicate if task i is the immediate predecessor of j in subset $\bar{s}_r$ ($= 1$) or not ($= 0$), $j \in \mathcal{I}_{i\bar{s}_r}^{+}$, $i \in \mathcal{I}_{\bar{s}_r}^{\text{sub}}$, $r \in \mathcal{R}_h^{\text{CM}}$, $h \in \mathcal{H}^{\text{CM}}$. Further, introduce the tasks $\tilde{p}$ and $\tilde{q}$ placed first and last respectively, with $e_{\tilde{p}} = e_{\tilde{q}} = 0$ and $\beta_{\tilde{p}\bar{s}_r} = \beta_{\tilde{q}\bar{s}_r} = 1$.

The objective function of the subproblem, $\max \sum_{h \in \mathcal{H}^{\text{CM}}} \sum_{r \in \mathcal{R}_h^{\text{CM}}} \sum_{i \in \mathcal{I}_{\bar{s}_r}^{\text{sub}}} \beta_{i\bar{s}_r}$, is to maximise the number of tasks that are successfully sequenced, and constraints (7)–(9) creates sequences for these tasks. If a task is not successfully sequenced it means that it overlaps another task within its subset.

$$\sum_{j \in \mathcal{I}_{i\bar{s}_r}^{+}} y_{ij\bar{s}_r} = \beta_{i\bar{s}_r}, \qquad i \in \mathcal{I}_{\bar{s}_r}^{\text{sub}} \setminus \{\tilde{q}\}, \ r \in \mathcal{R}_h^{\text{CM}}, \ h \in \mathcal{H}^{\text{CM}} \qquad (7)$$

$$\sum_{j \in \mathcal{I}_{i\bar{s}_r}^{-}} y_{ji\bar{s}_r} = \beta_{i\bar{s}_r}, \qquad i \in \mathcal{I}_{\bar{s}_r}^{\text{sub}} \setminus \{\tilde{p}\}, \ r \in \mathcal{R}_h^{\text{CM}}, \ h \in \mathcal{H}^{\text{CM}} \qquad (8)$$

$$x_j \geq x_i + e_i - (t_{ir}^{\text{d}} - t_{jr}^{\text{r}})(1 - y_{ij\bar{s}_r}), \quad j \in \mathcal{I}_{i\bar{s}_r}^{+}, \ i \in \mathcal{I}_{\bar{s}_r}^{\text{sub}}, \ r \in \mathcal{R}_h^{\text{CM}}, \ h \in \mathcal{H}^{\text{CM}} \qquad (9)$$

3 Co-allocation of CN-messages

Both the relaxed problem and the subproblem include the requirement that *CN-messages are co-allocated if and only if they are assigned to the same slot*, enforced by constraints (10)–(12). To avoid symmetries in the model, and without loss of generality, co-allocated CN-messages are assumed to be placed in ascending order with respect to CN-message number. To formulate the constraints, introduce the binary variables $w_{mm'}$ that indicate if m is placed immediately before m' in a slot ($= 1$) or not ($= 0$), and $w_{nmm'}^{\text{slot}}$ that indicate if m is placed immediately before m' in slot n ($= 1$) or not ($= 0$), $n \in \mathcal{N}, m \in \mathcal{M} : m < m', m' \in \mathcal{M}$.

$$z_{nm} + z_{nm'} - 1 \leq \sum_{m'' \in \mathcal{M} : m < m'' \leq m'} w_{nmm''}^{\text{slot}} \leq 1,$$

$$n \in \mathcal{N},\ m \in \mathcal{M} : m < m',\ m' \in \mathcal{M} \quad (10)$$

$$w_{nmm'}^{\text{slot}} \leq z_{nm}, \quad w_{nmm'}^{\text{slot}} \leq z_{nm'}, \quad n \in \mathcal{N},\ m \in \mathcal{M} : m < m',\ m' \in \mathcal{M} \quad (11)$$

$$w_{mm'} = \sum_{n \in \mathcal{N}} w_{nmm'}^{\text{slot}}, \quad m \in \mathcal{M} : m < m',\ m' \in \mathcal{M} \quad (12)$$

The subproblem includes the requirement that *a set of CN-messages are co-allocated if and only if their respective tasks that are of the same type and on the same module are merged*. To merge tasks means that they are placed immediately after each other in ascending order with respect to CN-message number (constraint (13)) and that their total execution requirement is reduced as follows. The execution requirement of task i, $i \in \mathcal{I}_m^{\text{M}} \cap \mathcal{I}_k^{\text{K}}, k \in \mathcal{K}, m \in \mathcal{M}$, constitutes of two terms of similar size: initialisation e_i^{init} and a CN-message specific part $e_i - e_i^{\text{init}}$. When a set of tasks is merged, the initialisation time e_i^{init} can be omitted for all tasks but for the one placed first, and this is achieved by replacing constraint (9) by constraint (14) and by adding constraint (15).

Use the auxilliary notation $\mathcal{I}_{mkh}^{\text{aux}} = \mathcal{I}_m^{\text{M}} \cap \mathcal{I}_k^{\text{K}} \cap \mathcal{I}_h^{\text{CM}}, m \in \mathcal{M}, k \in \mathcal{K}, h \in \mathcal{H}^{\text{CM}}$, and for $m' \in \mathcal{M}, m \in \mathcal{M} : m < m'$ introduce the constraints

$$w_{mm'} \leq y_{ii'\bar{s}_r},\ i \in \mathcal{I}_{mkh}^{\text{aux}},\ i' \in \mathcal{I}_{m'kh}^{\text{aux}},\ k \in \mathcal{K},\ r \in \mathcal{R}_h^{\text{CM}},\ h \in \mathcal{H}^{\text{CM}}, \quad (13)$$

$$x_i + e_i - e_{i'}^{\text{init}} w_{mm'} - (t_{ir}^{\text{d}} - t_{i'r}^{\text{r}})(1 - y_{ii'\bar{s}_r}) \leq x_{i'}, \quad (14)$$

$$x_{i'} \leq x_i + e_i - e_{i'}^{\text{init}} + (t_{i'r}^{\text{d}} - t_{ir}^{\text{r}} - e_i + e_{i'}^{\text{init}})(1 - w_{mm'}),$$

$$i \in \mathcal{I}_{mkh}^{\text{aux}},\ i' \in \mathcal{I}_{m'kh}^{\text{aux}},\ k \in \mathcal{K},\ r \in \mathcal{R}_h^{\text{CM}},\ h \in \mathcal{H}^{\text{CM}}. \quad (15)$$

For tasks to be eligible for merging in the subproblem, a necessary condition in the relaxed problem is that they are assigned to the same section. Let the variable $w_{ii'r}^{\text{sec}}$ indicate if tasks $i \in \mathcal{I}_{mkh}^{\text{aux}}$ and $i' \in \mathcal{I}_{m'kh}^{\text{aux}}$ are both placed in section r on CM h ($= 1$) or not ($= 0$), $k \in \mathcal{K}, m \in \mathcal{M} : m < m', m' \in \mathcal{M}, r \in \mathcal{R}_h^{\text{CM}}, h \in \mathcal{H}^{\text{CM}}$. The relaxed problem includes, by constraints (16)–(18), the requirement that *a set of CN-messages are co-allocated if and only if all their respective tasks that are of the same type and on the same module are assigned to sections that facilitate merging*. The objective function of the relaxed problem, max $\sum_{m' \in \mathcal{M}} \sum_{m \in \mathcal{M} : m < m'} w_{mm'}$, is to maximise the number of CN-messages that are co-allocated in this sense.

For $m' \in \mathcal{M}, m \in \mathcal{M} : m < m'$ introduce the constraints

$$\alpha_{ir} + \alpha_{i'r} - 1 \leq w^{\text{sec}}_{ii'r}, \quad w^{\text{sec}}_{ii'r} \leq \alpha_{ir}, \quad w^{\text{sec}}_{ii'r} \leq \alpha_{i'r},$$
$$i \in \mathcal{I}^{\text{aux}}_{mkh}, \ i' \in \mathcal{I}^{\text{aux}}_{m'kh}, k \in \mathcal{K}, \ r \in \mathcal{R}^{\text{CM}}_h, \ h \in \mathcal{H}^{\text{CM}}, \quad (16)$$

$$\sum_{h \in \mathcal{H}^{\text{CM}}} \sum_{k \in \mathcal{K}} \sum_{i' \in \mathcal{I}^{\text{aux}}_{m'kh}} \sum_{i \in \mathcal{I}^{\text{aux}}_{mkh}} \left(\sum_{r \in \mathcal{R}^{\text{CM}}_h} w^{\text{sec}}_{ii'r} - 1 \right) + 1 \leq \sum_{m'' \in \mathcal{M}: m < m'' \leq m'} w_{mm''} \leq 1, \quad (17)$$

$$w_{mm'} \leq \sum_{r \in \mathcal{R}^{\text{CM}}_h} w^{\text{sec}}_{ii'r}, i \in \mathcal{I}^{\text{aux}}_{mkh}, \ i' \in \mathcal{I}^{\text{aux}}_{m'kh}, \ k \in \mathcal{K}, \ h \in \mathcal{H}^{\text{CM}}. \quad (18)$$

Further, constraint (5) is adjusted by subtracting from its left-hand side the term $\sum_{k \in \mathcal{K}} \sum_{m' \in \mathcal{M}} \sum_{m \in \mathcal{M}: m < m'} \sum_{i' \in \mathcal{I}^{\text{aux}}_{m'kh}} \sum_{i \in \mathcal{I}^{\text{aux}}_{mkh}} e^{\text{init}}_{i'} w^{\text{sec}}_{ii'r}$.

4 Preliminary Computational Results and Concluding Comments

For the computational results, the instances that were thoroughly described in [2] have been used. The implementation is made in Python Version 3.6 and the models have been solved by Gurobi Optimizer Version 7.5.1.

As can be seen from Table 1, the co-allocation component of the model was successfully included and achieved co-allocation of CN-messages, without the solution times becoming much higher than before. The current instances have only a moderate number of CN-messages and for future work it is of interest to combine co-allocation of CN-messages with other enhancements of the solution strategy in order to solve much larger instances where co-allocation is expected to be required for finding a solution to the problem.

Table 1 Instance characteristics (for details see [2]) and co-allocation results

Instance	I	II	III
# [nodes, CM-tasks, CN-messages]	[2, 6536, 64]	[5, 14167, 96]	[7, 19894, 96]
[Best, Worst] solution time (s) of [2]	[164, 467]	[1025, 29676]	[2210, 52269]
Solution time (s) with co-allocation	452	2798	18663
Number of co-allocated CN-messages	25	32	13

Acknowledgements This work was supported the Center for Industrial Information Technology (CENIIT) and by the Swedish Armed Forces, the Swedish Defence Materiel Administration and the Swedish Governmental Agency for Innovation Systems under grant number NFFP6-2014-00917.

References

1. Al-Sheikh, A. (2011). Resource allocation in hard real-time avionic systems. Scheduling and routing problems. Ph. D thesis, INSA de Toulouse.
2. Blikstad, M., Karlsson, E., Lööw, T., & Rönnberg, E. (2017). *An optimisation approach for pre-runtime scheduling of tasks and communication in an integrated modular avionic system.* Under review and available as Technical report LiTH-MAT-R–2017/03–SE.

Computing Pareto-Optimal Transit Routes Through Mathematical Algorithms

M. Fawad Zazai and Armin Fügenschuh

1 Aim and Idea

Afghanistan is located geographically in the center of Asia. The country has great potential to transform in South and Central Asia into a "logistical crossroad". The aim of this study is to develop trajectories for optimal transit routes in Afghanistan by mathematical optimization methods. In the present research phase, the focus is to apply algorithms for the shortest path problem, which compute point-to-point connections between two cities. The shortest path problem belongs to the class of graph problem and deals with the issue of how to find an optimal route between two nodes or points (start and end point) within a graph $G = (V, E, w)$ with respect to a cost function that is the sum of non-negative weights $w_{i,j}$ of each edge $\{i, j\} \in E$ that is used in the route. The edge weight can represent (a) its length, (b) its construction cost, or (c) the height variation. In order to estimate the construction cost of an edge, that may become part of a route, several factors are taken into account, in particular the national land use of Afghanistan and the elevation profile of the terrain (topography). These data are taken from publicly available sources. For the design and modeling of the routes, a computer program named "CONTRA" (**C**omputing an **O**ptimal **N**etwork of **T**ransit **R**outes through mathematical **A**lgorithms) was developed. CONTRA transforms the input data (land use, terrain) into a weighted graph and applies Dijkstra's shortest path algorithm [1] to find an optimal routes between any two given nodes. Details of this are given in the next section.

M. F. Zazai (✉) · A. Fügenschuh
Helmut Schmidt University, Hamburg, Germany
e-mail: zazaif@hsu-hh.de

M. F. Zazai · A. Fügenschuh
University of the Federal Armed Forces, Hamburg, Germany
e-mail: fuegenschuh@hsu-hh.de

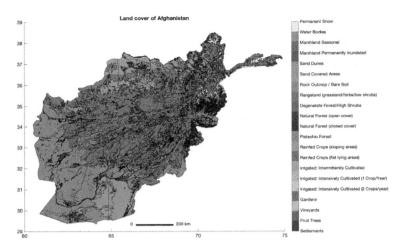

Fig. 1 A map of Afghanistan with different land cover. Each color represents a specific land cover (e.g., deserts or forests)

2 Input Data

To determine a route automatically with the help of mathematical optimization, the following input data has to be provided to the program CONTRA:

1. Coordinates of start and destination points (given as latitude and longitude).
2. National land cover of Afghanistan in shapefiles (currently from date 1997) of the organization AIMS, originally of the Afghan Geodesy and Cartography Head Office [2]. The whole country is separated in polygonal shapes that describe the respective type of land, see Fig. 1.
3. Topographic representation of Afghanistan, the SRTM data of USGS/NASA [3]. The resolution of the SRTM data for Afghanistan is 14000 × 18001 pixels. As an example, in Fig. 2 the area of the Uruzgan province in Afghanistan is depicted.
4. The costs of construction and maintenance of a road. According to reports of the Asian Development Bank the construction of a (two-lane) road in Afghanistan on average is approximately 1 Mio. USD per km [4]. The cost amount can be dependent on the height and the land surface on which it is built [5]. For this research the assumption is an estimated cost ratio among the different land covers, and thus determine the construction costs of the routes. In general, it is possible to modify the estimated cost ratio.

3 Creating a Graph and a Shortest-Path-Problem

A weighted graph $G = (V, E, w)$ is created as follows. A regular mesh grid Γ is spanned over the terrain of a selected geographical area A, so that $\Gamma(A)$ defines a

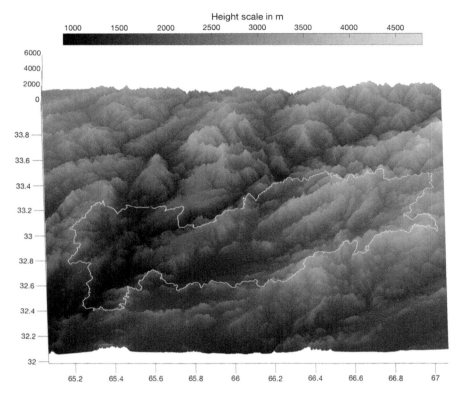

Fig. 2 The topography of the Uruzgan province in Afghanistan. The color spectrum from dark to light indicates the height of the terrain

geographical mesh grid of A. A grid consists of rectangles, which has corner points in $\mathbb{R}^3$. These points are added as nodes to the set V. The four side lines of each grid rectangle is equidistantly subdivided into $m \in \mathbb{N}$ segments. The end points of these segments are also added as nodes to V. Note that each grid rectangle has $4m$ associated nodes (see Fig. 3). All pairs of these nodes (but excluding those on the same side of the rectangle's boundary) are now connected by edges and added to the edge set E of G, which gives $6m^2$ edges for each rectangle of the grid. The problem of the construction of optimal routes in Afghanistan leads to the shortest path problem. A shortest path is a route that is minimal with respect to the sum of all costs of all segments that are used in the entire route. The non-negative cost per segment (edge weight) $w_{i,j}$ are determined from the construction cost, the length or the height variation (depending on the desired goal of the optimization). To solve the shortest path problem Dijkstra's algorithm [1] is used. The number of rectangular subdivisions of the grid Γ as well as the value of m are set by the user. Clearly, the finer the resolution, the more properly the route can follow the topography of

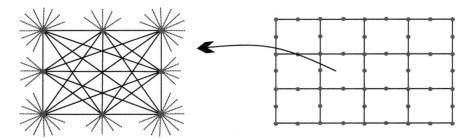

Fig. 3 A single rectangle with m = 2 subdivision, hence 4 m = 8 nodes and 6 m^2 = 24 connecting edges. The edges represent the basic building segments of a route in CONTRA

the area. With modern computers, the solution time is not so much a bottleneck, however, the memory consumption is very high and can easily touch the limits also of modern workstations (64 GByte), even when special programming techniques (such as sparse data structures for storing all edges) were applied. Besides focusing on a single optimization goal, there are several conflicting objectives to consider. This could be, for instance, minimizing the total length of the route as well as the construction cost. These two can be in conflict, because a shorter route may go through more difficult terrain that a slightly longer route would have avoided, and thus turns out to be lest costly. In general, we consider three objectives "route length", "construction cost" and "elevation variation" that are in conflict with each other. That means, there is no route that is simultaneously optimal for all three. Multi-objective optimization (Pareto optimization) is an area of multi-criteria decision making, that is concerned with mathematical optimization problems involving more than one objective function, that have to be optimized simultaneously [6]. Here one seeks for a Pareto optimum of a route, which is a route that cannot be improved with respect to one criterion without worsening at least another. Using this concept, one can analyze the trade-off between them.

4 Results

As an example, we compute routes between the city of *Khas Uruzgan* and the city of *Kabul*. We first calculate optimal solutions for the three different single objectives, e.g., the shortest route (red), the cost-minimal or cheapest route (blue), and the most convenient route w.r.t. the elevation change (black), see Fig. 4. The columns in Table 1 describe a) the construction costs of the routes in million USD, b) the lengths of the routes in km and c) the absolute elevation changes of the routes, i.e., the sum of all height changes from the starting point to the ending point along the route. Each of these three routes has a certain length and elevation profile. Fig. 5 shows the height profile along these three routes and Fig. 6 shows the Pareto front of

Computing Pareto-Optimal Transit Routes Through Mathematical Algorithms

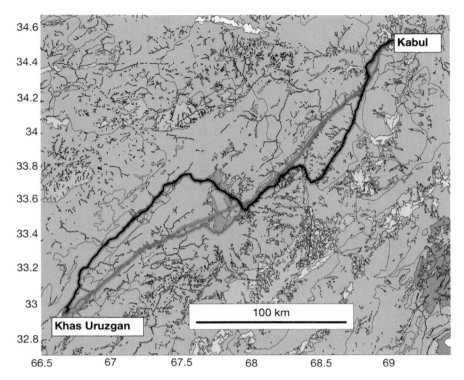

Fig. 4 Three single-objective optimal routes: shortest (red), cheapest (blue), and most convenient (black)

Table 1 The result of the three routes regarding the cost, length and absolute elevation change

Route	Cost in million USD	Length in km	Absolute elevation change in m
red	312	**310**	18,570
blue	**291**	318	20,652
black	426	388	**10,494**

the routes from Khas Uruzgan to Kabul regarding the two objectives "route length" and "construction cost". This chart shows on the horizontal axis the length of the routes in km and on the vertical axis the construction cost of the routes in million USD. The small red circles represent specific routes between these two cities. The leftmost circle represents the shortest route and the rightmost circle represents the cheapest route. The circles that lie between these two extremal circles, are other optimal routes that are a combination of the shortest and cheapest route from Khas Uruzgan to Kabul. As the project acronym CONTRA indicates, our future work is to extend these point-to-point connections to automatically design large *networks* that connect several cities with optimally located transit routes.

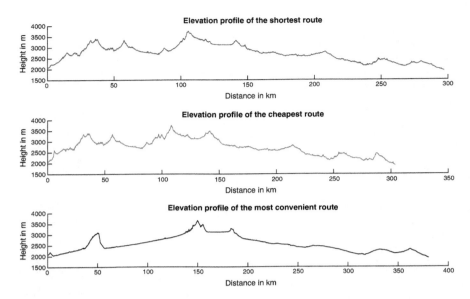

Fig. 5 The height and length diagram of the shortest, the cheapest and the most convenient w.r.t. the elevation route. The line in the middle of each chart is the average elevation

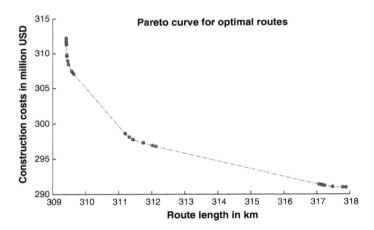

Fig. 6 The Pareto front of the shortest and cheapest routes regarding length and construction cost from Khas Uruzgan to Kabul. The shortest route costs 312.5 Mio. USD and has a length of 309.5 km, whereas the cost for a slightly longer route of 317 km already drops to 291 Mio. USD, and even longer routes do not save much further cost anymore

References

1. Dijkstra, E. (1959). A Note on Two Problems in Connexion with Graphs. *Numerische Mathematik*, *1*(1), 269–271.
2. Afghanistan Information Management Services (AIMS), Land Cover of Afghanistan (1997), 2014, http://www.aims.org.af/. Retrieved July 3, 2017.
3. Consortium for Spatial Information (CGIAR-CSI), SRTM 90m Digital Elevation Database v4.1, https://goo.gl/g2LQYC. Retrieved July 3, 2017.
4. Asian Development Bank (ADB), Proposed Asian Development Fund Grant Islamic Republic of Afghanistan: Road Network Development Project 1, Report and Recommendation of the President to the Board of Directors 40333, 2007, https://goo.gl/U9E9jR. Retrieved July 3, 2017.
5. Minister of Public Works and Government Services Canada, Northern Land Use Guidelines - Access: Roads and Trails, Tech. Rep. 5, Minister of Indian Affairs and Northern Develop. and Federal Interlocutor for Métis and Non-Status Indians, Ottawa, 2010, https://goo.gl/6Xd19b. Retrieved July 3, 2017.
6. Ehrgott, M. (2005). *Multicriteria optimization* (2nd ed.). Berlin: Springer.

Energy-Efficient Design of a Water Supply System for Skyscrapers by Mixed-Integer Nonlinear Programming

Philipp Leise, Lena C. Altherr and Peter F. Pelz

1 Introduction

Especially in high buildings like skyscrapers, the supply pressure of the waterworks is not sufficient to supply the higher floors with water. In this case, booster systems consisting of one or more pumps are used to increase the pressure. While there are references for the placement of booster stations in buildings [3, 4], the determination of the exact number and position of pumps is mostly still done manually for each application setting. Especially for larger buildings, it is not feasible for a human designer to assess all different options, but the decision is made based e.g. on experience or other heuristics. Technical Operations Research (TOR) [2, 5] however ensures the consideration of all possible choices and allows to find the global optimal solution. The optimization problem is formulated as a Mixed-Integer Nonlinear Program (MINLP). The result after solving this model is the cost-optimal combination, placement and control strategy of pumps, given a set of different load cases and their frequencies, cf. Fig. 1. All possible layout options of the water supply network are shown in Fig. 2. In this application example, a skyscraper with 9 floors has to be supplied. Three adjacent floors are combined to one level of the abstract graph on the left. The question marks denote that the optimization algorithm can choose the location, type and number of parallel pumps at each pipe between the main levels.

P. Leise (✉) · L. C. Altherr · P. F. Pelz
Chair of Fluid Systems, Department of Mechanical Engineering,
Technische Universität Darmstadt, Otto-Berndt-Str. 2, 64287 Darmstadt, Germany
e-mail: philipp.leise@fst.tu-darmstadt.de

L. C. Altherr
e-mail: lena.altherr@fst.tu-darmstadt.de

P. F. Pelz
e-mail: peter.pelzg@fst.tu-darmstadt.de

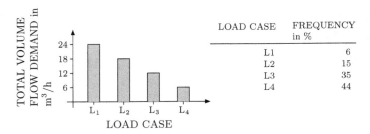

Fig. 1 Load cases in the building: volume flow demands and their respective frequency

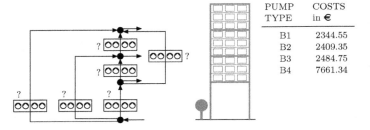

Fig. 2 Layout options and costs of different pump types of the construction kit

Given is a construction kit consisting of different pump types with different costs and characteristic curves (pressure head and power consumption vs. volume flow).

2 Optimization Model

Our MINLP for modeling the water supply system for the skyscraper is shown in Eqs. (1)–(28). In this model, capital letters denote variables and sets, lower-case characters denote parameters, cf. Table 1. We minimize the total costs of the water supply system, consisting of the energy costs of all pumps, and the investment costs for pumps and pipes, cf. Eq. (1).

In the building, it is possible to place a pipe between every two levels, but the rooms in one level can only be supplied by one pipe. This leads to a tree-shaped graph $\mathcal{G}(\mathcal{V}, \mathcal{E})$ in which the pipes are represented by edges $\mathcal{E}$, and the levels are represented by nodes $\mathcal{V}$. A binary variable $K_{i,j}^{\text{pipe}}$ indicates whether levels i and j are connected by a pipe. On each pipe, it is possible to place multiple pumps of the same type. Equations (2)–(10) are used to model these logic properties of the system. Equations (11) and (12) set the normalized speed $N_{i,j,b,l}$ to zero if the respective pump is not used in load case $l \in \mathcal{L}$, i.e. $K_{i,j,b,l}^{\text{load}} = 0$. Equations (13)–(15) add boundary conditions for the pressure head on each level. The continuity equation and general flow constraints are represented by Eqs. (16)–(22).

$$\min \sum_{b \in \mathcal{B}} \sum_{(i,j) \in \mathcal{E}} c_b^{\text{pump}} K_{i,j,b}^{\text{pump}} + \sum_{(i,j) \in \mathcal{E}} c_{i,j}^{\text{pipe}} K_{i,j}^{\text{pipe}} + c^{\text{energy}} \tau \sum_{l \in \mathcal{L}} \sum_{b \in \mathcal{B}} \sum_{(i,j) \in \mathcal{E}} \phi_1 P_{i,j,b,l} \quad \text{subject to} \tag{1}$$

$$\sum_{(i,j) \in \mathcal{E}: j=v} K_{i,j}^{\text{pipe}} \leq 1 \tag{2}$$

$$\sum_{b \in \mathcal{B}} K_{i,j,b}^{\text{pump}} \leq 1 \tag{3}$$

$$K_{i,j,b}^{\text{pump}} \leq K_{i,j}^{\text{pipe}} \tag{4}$$

$$K_{i,j,b,l}^{\text{load}} \leq K_{i,j,b}^{\text{pump}} \tag{5}$$

$$K_{i,j,b}^{\text{pump}} \leq Y_{i,j,b}^{\text{pump}} \tag{6}$$

$$K_{i,j,b}^{\text{pump}} n^{\text{P}} \geq Y_{i,j,b}^{\text{pump}} \tag{7}$$

$$K_{i,j,b,l}^{\text{load}} \leq Y_{i,j,b,l}^{\text{load}} \tag{8}$$

$$K_{i,j,b,l}^{\text{load}} n^{\text{p}} \geq Y_{i,j,b,l}^{\text{load}} \tag{9}$$

$$Y_{i,j,b,l}^{\text{load}} \leq Y_{i,j,b}^{\text{pump}} \tag{10}$$

$$K_{i,j,b,l}^{\text{load}} \geq N_{i,j,b,l} \tag{11}$$

$$n^{\min} K_{i,j,b,l}^{\text{load}} \leq N_{i,j,b,l} \tag{12}$$

$$H_{\text{In},l} = h^0 \tag{13}$$

$$H_{1,l} \geq h_1^{\min} \tag{14}$$

$$\forall v \in \mathcal{V} \setminus 1 : H_{v,l} \geq h^{\min} \tag{15}$$

$$Q_{i,j,l} - q^{\max} K_{i,j}^{\text{pipe}} \leq 0 \tag{16}$$

$$Q_{i,j,l}^{\text{part}} \leq Q_{i,j,l} \tag{17}$$

$$\sum_{(v,j) \in \mathcal{E}} Q_{v,j,l} + q_{v,l}^{\text{load}} - \sum_{(i,v) \in \mathcal{E}} Q_{i,v,l} = 0 \tag{18}$$

$$Y_{i,j,b,l}^{\text{load}} Q_{i,j,l}^{\text{part}} + q^{\max}(1 - K_{i,j,b,l}^{\text{load}}) \geq Q_{i,j,l} \tag{19}$$

$$Y_{i,j,b,l}^{\text{load}} Q_{i,j,l}^{\text{part}} - q^{\max}(1 - K_{i,j,b,l}^{\text{load}}) \leq Q_{i,j,l} \tag{20}$$

$$Q_{i,j,l}^{\text{part}} - Q_{i,j,l} - q^{max} \sum_{b \in \mathcal{B}} K_{i,j,b,l}^{\text{load}} \leq 0 \tag{21}$$

$$Q_{i,j,l}^{\text{part}} - Q_{i,j,l} + q^{max} \sum_{b \in \mathcal{B}} K_{i,j,b,l}^{\text{load}} \geq 0 \tag{22}$$

$$\alpha_b^{Q,1} N_{i,j,b,l} + \beta_b^{Q,1} K_{i,j,b,l}^{\text{load}} + q^{\max}(1 - K_{i,j,b,l}^{\text{load}}) \geq Q_{i,j,l}^{\text{part}} \tag{23}$$

$$\alpha_b^{Q,2} N_{i,j,b,l} + \beta_b^{Q,2} K_{i,j,b,l}^{\text{load}} - q^{\max}(1 - K_{i,j,b,l}^{\text{load}}) \leq Q_{i,j,l}^{\text{part}} \tag{24}$$

$$\alpha_b^{P}(Q_{i,j,l}^{\text{part}})^2 + \beta_b^{P} N_{i,j,b,l}^2 + \gamma_b^{P} Q_{i,j,l}^{\text{part}} N_{i,j,b,l} + \delta_b^{P} Q_{i,j,l}^{\text{part}} + \epsilon_b^{P} N_{i,j,b,l} + \mu_b^{P} \leq P_{i,j,b,l}^{\text{part}} \tag{25}$$

$$Y_{i,j,b,l}^{\text{load}} P_{i,j,b,l}^{\text{part}} = P_{i,j,b,l} \tag{26}$$

$$K_{i,j,b,l}^{\text{load}} \alpha_b^{H}(Q_{i,j,l}^{\text{part}})^2 + \beta_b^{H} Q_{i,j,l}^{\text{part}} N_{i,j,b,l} + \gamma_b^{H} N_{i,j,b,l}^2 = \Delta H_{i,j,b,l} \tag{27}$$

$$(H_{j,l} - \sum_{b \in \mathcal{B}} \Delta H_{i,j,b,l} + \Delta h_{i,j} - H_{i,l}) K_{i,j}^{\text{pipe}} = 0 \tag{28}$$

with $v \in \mathcal{V}, (i,j) \in \mathcal{E}, b \in B, l \in L$ if not stated otherwise.

They ensure on the one hand that the flow through the network meets the demand $q_{v,l}^{\text{load}}$ on vertex $v \in \mathcal{V}$ and in load case $l \in \mathcal{L}$, and on the other hand that the total flow rate $Q_{i,j,l}$ in each pipe is the sum of all partial flow rates $Q_{i,j,l}^{\text{part}}$ through each parallel pump on the specific pipe (i,j) in the network. Equations (23)–(27) are characteristic curves dependent on the pump type b. They define the relationship between volume flow, pressure head, and power consumption for each pump. The head-volume flow characteristic is modeled according to [6] as a quadratic relationship between the pressure head, the volume flow and normalized speed, cf. Eq. (27). The power characteristic is modeled by a fully quadratic relation between the power and the volume flow and normalized speed, cf. Eq. (25). This approach represents a quadratic approximation to the cubic model in [6]. If a booster station consists of multiple parallel pumps of the same type b, the highest energy-efficiency can be achieved by operating all pumps with the same rotational speed, resulting in the same volume flow and power consumption of each pump. Assuming equal rotational speeds, Eq. (26) calculates the total power consumption of a booster station with $Y_{i,j,b,l}^{\text{load}}$ pumps of type b operating in load case l. Equation (28) ensures that the head between two different levels decreases proportional to the height difference between them.

Table 1 Decision variables and parameters

Index set	Description	
$\mathcal{V}$	Set of vertices of system graph $\mathcal{G}$	
$\mathcal{E}$	Set of edges of system graph $\mathcal{G}$	
$\mathcal{L}$	Set of load cases l	
$\mathcal{B}$	Set of pump types b	
Variable	Description	Domain
$Q_{i,j,l}$	Volume flow on edge (i, j) in load case l	$[0, q^{\max}]$
$Q^{\text{part}}_{i,j,l}$	Volume flow in pump on edge (i, j) in load case l	$[0, q^{\max}]$
$H_{v,l}$	Pressure head on level v in load case l	$[0, h^{\max}]$
$N_{i,j,b,l}$	Normalized speed on edge (i, j) for pump b in load case l	$[0, 1]$
$P_{i,j,b,l}$	Total power of pumps of type b on edge (i, j) in load case l	$[0, p^{\max}]$
$P^{\text{part}}_{i,j,b,l}$	Power of each pump of type b on edge (i, j) in load case l	$[0, p^{\max}]$
$K^{\text{pipe}}_{i,j}$	Indicator whether pipe (i, j) is used	$\{0, 1\}$
$K^{\text{pump}}_{i,j,b}$	Indicator whether pumps of type b is used in pipe (i, j)	$\{0, 1\}$
$K^{\text{load}}_{i,j,b,l}$	Indicator if pump b are used in pipe (i, j) in load case l	$\{0, 1\}$
$Y^{\text{pump}}_{i,j,b}$	Number of parallel pumps of type b on edge (i, j)	$\{0, n^{\text{P}}\}$
$Y^{\text{load}}_{i,j,b,l}$	Number of active pumps of type b on edge (i, j) in load case l	$\{0, n^{\text{P}}\}$
Parameter	Description	Domain
$q^{\max}$	Upper bound for volume flow	$\mathbb{R}^+$
$q^{\text{load}}_{v,l}$	Volume flow demand on level v in loadcase l	$\mathbb{R}$
h^0	Pressure head by water supplier	$\mathbb{R}^+$
$h^{\min}$	Minimal pressure head in each level, except the first one	$\mathbb{R}^+$
$h^{\min}_1$	Minimal pressure head in first level	$\mathbb{R}^+$
$h^{\max}$	Maximum height of the building	$\mathbb{R}^+$
$\Delta h_{i,j}$	Pressure loss between level i and level j	$\mathbb{R}^+$
$p^{\max}$	Maximum power consumption	$\mathbb{R}^+$
n^{P}	Maximum number of parallel pumps	$\mathbb{R}^+$
$n^{\min}$	Minimum normalized speed of all pumps	$[0, 1]$
c^{energy}	Energy costs per Watt	$\mathbb{R}^+$
c^{pump}_b	Investment cost for each pump	$\mathbb{R}^+$
$c^{\text{pipe}}_{i,j}$	investment costs per pipe	$\mathbb{R}^+$
ϕ_l	frequency of load case l	$\mathbb{R}^+$
τ	usage period	$\mathbb{R}^+$
Quantity	Fitting parameters for each pump $b \in B$	Domain
Power	$\alpha^{\text{P}}_b, \beta^{\text{P}}_b, \gamma^{\text{P}}_b, \delta^{\text{P}}_b, \epsilon^{\text{P}}_b, \mu^{\text{P}}_b$	$\mathbb{R}$
Head	$\alpha^{\text{H}}_b, \beta^{\text{H}}_b, \gamma^{\text{H}}_b$	$\mathbb{R}$
Flow	$\alpha^{\text{Q},1}_b, \alpha^{\text{Q},2}_b, \beta^{\text{Q},1}_b, \beta^{\text{Q},2}_b$	$\mathbb{R}$

3 Application and Results

We analyze the potential energy savings in an exemplarily chosen hotel building. The maximum daily water consumption is derived by taking into account the number of floors, rooms, and consumers and the available guidelines [3]. The hotel starts 20 m above the input valve of the supply by the waterwork. It consists of 9 levels which are each 3 m high. We combine three adjacent floors into one node of the corresponding graph. The daily water consumption is approximated by four different load cases, cf. Fig. 1. A minimum pressure head of 15.25 m is required at each level. This value consists of the minimum pressure needed at the output of the network (i.e. a faucet) and the additional pressure needed to overcome the pipe resistance from the rising pipe to the rooms. The pressure head provided by the waterworks is 28.63 m. The optimizer can chose between four different booster stations with different costs, cf. Fig. 2. B4 consists of up to four identical pumps and would conventionally be used. B1 to B3 are single pumps, with different characteristics. The parameters for all pump models in Eq. (25) and (27) are derived by a least-square fit based on data provided by the manufacturer. The costs for pipes are estimated with 50 €/m, energy costs with 0.3 €/kWh. We chose Scip (version 4.0.0) [1] for solving since it ensures a global optimal solution, also for nonconvex MINLPs. We set the optimization gap to 5%. The optimized layout is shown in Fig. 3 on the left side, the right side shows the conventional layout for comparison. The conventional layout uses four pumps of booster station B4, while in the optimized layout, only three pumps of this station are used and another pump of type B2 is added. The algorithm increases the number of pressure zones (floors with the same pressure). With the optimized layout, overall energy savings of ≈18.9% can be achieved. The total costs (investment plus energy) decrease by 12% for a usage period of $\tau = 15$ a.

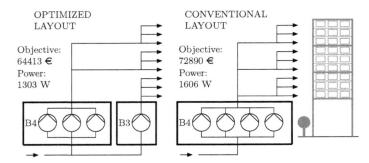

Fig. 3 Comparison of conventional and optimized system layouts. With our approach, energy savings of almost 20 % can be achieved compared to the conventional layout

4 Conclusion

This work shows that the usage of mathematical programming in the engineering domain can lead to significant improvements in the overall energy consumption of technical systems. With the presented model , not only an optimal pump selection and operation can be computed automatically, but also the division of the different building levels in individual pressure zones. The developed model can therefore be seen as a support tool for engineers.

Acknowledgements The authors thank the German Research Foundation for funding this research within the Collaborative Research Center SFB 805 "Control of Uncertainties in Load-Carrying Structures in Mechanical Engineering".

References

1. Achterberg, T. (2009). SCIP: Solving constraint integer programs. *Mathematical Programming Computation, 1*, 1–41.
2. Altherr, L. C., et al. (2016). Experimental validation of an enhanced system synthesis approach. In *Operations Research Proceedings 2014* (pp. 1–7). Berlin: Springer International Publishing.
3. DIN 1988-300. (2012). Codes of practice for drinking water installations - Part 300: Pipe sizing; DVGW code of practice.
4. DIN 1988-500. (2011). Codes of practice for drinking water installations - Part 500: Pressure boosting stations with RPM-regulated pumps; DVGW code of practice.
5. Pöttgen, P., et al. (2016). Examination and optimization of a heating circuit for energy efficient buildings. *Energy Technology, 4*, 136–144.
6. Ulanicki, B., Kahler, J., & Coulbeck, B. (2008). Modeling the efficiency and power characteristics of a pump group. *Journal of Water Resources Planning and Management, 134*, 88–93.

Mixed Integer PDE Constrained Optimization for the Control of a Wildfire Hazard

Fabian Gnegel, Michael Dudzinski, Armin Fügenschuh and Markus Stiemer

1 Introduction

In a forest close to inhabited regions is an ongoing wildfire spread. Leaving it burning uncontrolled might endanger the local population and their properties, hence the firefighters are trying to plan their response in an optimal way, without endangering themselves. A road network is passing through the forest that can now be used for firefighting operations. The forest itself cannot be crossed; all movements are restricted to the said road network. In order to prevent endangering the firefighters, no movement should take place on roads leading through or too close to burning territory. The resources necessary to control the fire (water, equipment, and manpower) are limited, therefore an optimal resource allocation and proper scheduling might make the difference between getting the fire under control or a major disaster.

In this situation an optimal planning has to take two different types of dynamics into account: Firstly, the physics of the fire, which allows to predict the spread direction and velocity, and secondly, the movement of the firefighters and their extinguishing agents (water). Those two systems cannot be considered separately. The ultimate goal of any firefighter mission is to influence the spread of the fire, but during this mission, the fire might temporarily prevent the firefighters from reaching certain areas.

For the modeling of the fire a time dependent PDE is used, and a dynamic network flow is used to model the movements of the firefighters, or more precisely, the water that they use. In order to express the interdependencies, the flow variables of the network are used as control variables for the PDE and additional constraints are imposed on the network flow which include the state of the PDE. The inclusion of these interdependencies make our model unique in comparison to other recent work.

F. Gnegel (✉) · M. Dudzinski · A. Fügenschuh · M. Stiemer
Helmut Schmidt Universität/Universität der Bundeswehr Hamburg,
Holstenhofweg 85, 22043 Hamburg, Germany
e-mail: gnegelf@hsu-hh.de

For example, Göttlich et al. [3] studied an evacuation planning problem in response to a gas hazard, where the latter is modeled by a PDE, which is independent of the network dynamics (i.e., flows of evacuating people). Frank et al. [2] consider the coolest path problem, where an object traverses a network graph while being heated or cooled on the arcs. Here the heat PDE gives rise to objective function coefficients for a shortest path problem, but does not constrain the combinatorial decisions.

2 The Mathematical Model

In order to solve the planning problem of the response to the wildfire, an integrated model for the spread of the fire and for the movement of the firefighters is formulated. We define the sets and variables of this model, and then the constraints and the objective function.

For the dynamic flow of the water used by the firefighters we assume that the road network is given in form of a graph $G := (V, A)$ with capacities $c_{i,j}$ and traversing times $\delta_{i,j}$ for all arcs $(i, j) \in A$. Graph G is embedded in the plane by endowing each vertex $i \in V$ with a coordinate $x_i \in \Omega$, where $\Omega := [0, L]^2$ is a square area of interest. The arcs $(i, j) \in A$ are associated with a straight lines between the coordinates of their respective incident vertices. The flow can start in source nodes, denoted by $S \subset V$, and ends in demand nodes $D \subset V$, which are nodes suitable for extinguishing the fire. We introduce a discretization of the time horizon $[0, T]$ by the set of time $\mathbb{T} := \{0, \Delta t, \ldots, n_t \Delta t = T\}$.

The variables $v_{i,j,t} \in \mathbb{R}_+$ represent the flow (of water) on arc (i, j), starting in i at time $t \in \mathbb{T}$. In nodes $i \in S$ the flow can enter the network, and the intensity at time $t \in \mathbb{T}$ is specified by the variables $w_{i,t} \in \mathbb{R}_-$. Vice versa, the flow leaves the network in nodes $i \in D$ at time $t \in \mathbb{T}$, with an intensity given by $w_{i,t} \in \mathbb{R}_+$. The temperature in the forest at location $x \in \Omega$ at time $t \in \mathbb{T}$ is given by $u(x, t)$. Finally, binary decision variables $z_{i,t} \in \{0, 1\}$ for $i \in V$ and $t \in \mathbb{T}$ are introduced to link the temperature to the flow, with $z_{i,t} = 0$ if and only if the temperature at x_i at time t exceeds a certain threshold U_B, at which further firefighter operations have to be terminated for safety reasons, that is, the area is burning.

The following constraints are now used to ensure the desired behavior of the firefighter operations (i.e., the flow of water), where we use a dynamic maximum flow formulation (see [6] for a survey):

$$v_{i,j,0} = 0 \qquad \forall (i,j) \in A, \quad (1a)$$

$$\sum_{i \in V:(i,k)\in A, \delta_{i,k}\leq t} v_{i,k,t-\delta_{i,k}} = \sum_{j \in V:(k,j)\in A} v_{k,j,t} + w_{k,t} \qquad \forall k \in V, t \in \mathbb{T}, \quad (1b)$$

$$u(x_i, t) - (1 - z_{i,t})M \leq U_B \qquad \forall i \in V, t \in \mathbb{T}, \quad (1c)$$

$$\sum_{s \in \mathbb{T}: s \leq \min(\delta_{i,j}, T-t)} v_{i,j,t+s} \leq c_{i,j} z_{j,t} \qquad \forall (i,j) \in A, t \in \mathbb{T}. \quad (1d)$$

The initial condition (1a) guarantees that no flow is inside the network at $t = 0$. The flow conservation is ensured by (1b). Constraints (1c) and (1d) prevent flow from passing through a burning area, where the first sets the binary switch variable $z_{i,t}$ to zero if the threshold temperature is reached, and the second only allow for flow (w.r.t. to the capacity restriction) as long as $z_{j,t} = 1$.

The dynamics of the fire is modeled by the following PDE system:

$$u_t(x, t) - c \cdot \nabla u(x, t) - d \Delta u(x, t) = y(x, t, w) \quad \forall (x, t) \in \Omega \times (0, T), \quad (2a)$$

$$\frac{\partial}{\partial n} u(x, t) = h_R(u_R - u(x, t)) \quad \forall (x, t) \in \partial\Omega \times (0, T), \quad (2b)$$

$$u(x, 0) = f(x) \quad \forall x \in \Omega, \quad (2c)$$

$$u(x, t) \geq 0, \quad \forall (x, t) \in \Omega \times (0, T). \quad (2d)$$

This is a convection-diffusion equation with Robin type boundary conditions on the spacial domain Ω and the time domain $[0, T]$. The fire model is able to express the effect of the wind and the diffusive behavior of fire, while still being a linear PDE (which we need later for computational reasons[1]). Condition (2b) imposes that the normal derivative at the boundary is directly proportional to the difference of the temperature on the boundary and the temperature u_R. Parameter d is the coefficient of the diffusion term, it determines the speed of the fire spread. Parameter c is the velocity-vector of the wind. Furthermore condition (2d) ensures that, when the fire is extinguished (at temperature zero), the control function cannot push the temperature any lower thereafter.

The term in this PDE that represents an outer influence is $y(x, t, w)$, which depends on the outflow of water $w_{i,t}$ for $i \in D$ of a nearby node ($x \approx x_i$) as follows: The controls at the different vertices and different points in time are independent of each other, hence y is the sum of several individual control functions. It is further assumed that each outflow variable $w_{i,t}$ ($i \in D$) has only a local effect with a peak at the coordinate of its vertex and acts only for a certain duration T_E. We assume that the spatial effect follows a Gaussian distribution with the coordinates of the vertex x_i at its center:

$$y(x, t, w) = \lambda \sum_{\tau \in \mathbb{T}} \sum_{i \in D} -w_{i,\tau} \chi_{[t, t+T_E)}(t) \exp\left(-\frac{\|x - x_i\|_2^2}{\sigma^2}\right), \quad (3)$$

where λ and σ are parameters that represent the spatial influence of the outflow of water on the surrounding fire (more precisely, its temperature), and χ_I is the characteristic function (i.e., $\chi_I(t) = 1$ for $t \in I$ and 0 otherwise) that restricts the duration of the influence to a time interval of size T_E.

---

[1] We remark that there are more complex fire models known, for example [4], where a further nonlinear term expresses the consumption of fuel (here: wooden trees), but on such models our presented computational techniques do not work. Their adaptation is a direction for future research.

The objective is to minimize the damage caused by the fire. We assume that the damage is proportional to a weighted integral of the temperature $u(x, t)$ in Ω over a time horizon $[0, T]$.

$$\min \int_0^T \int_\Omega \omega(x) u(x, t) \, dx \, dt, \qquad (4)$$
$$\text{s.t.} (1a) - (1d), (2a) - (2d).$$

We present two approaches to solve (4). Neither of them solves this model directly. Instead, we derive suitable finite dimensional systems that approximate (4), which turn out to be linear mixed-integer problems (MILP) and thus can be solved using a state-of-the-art MILP solver.

Finite Differences. The first approach uses a one-to-one replacement of the constraints and objective with a discrete counterpart. The PDE is replaced by a linear system obtained from a convergent finite difference method [5] and the integral is replaced by a quadrature formula. The domain is discretized by replacing Ω with an equidistant grid of length $\Delta x = \frac{L}{n_x}$ with $n_x \in \mathbb{N}$. The interval $[0, T]$ is replaced by the discrete time set $\mathbb{T}$, which was already used for setting up to the network flow. Then for each point $(i\Delta x, j\Delta x, t)$ of the grid a variable $u_{i,j,t}$ is added. The function $u(x, t)$ is approximated at each gridpoint, i.e., $u((i\Delta x, j\Delta x), t) \approx u_{i,j,t}$. All constraints that depend on u have to be adjusted for those discrete variables. The PDE and its initial and boundary conditions (2a)–(2c) are replaced by a linear system

$$\begin{pmatrix} A_1 & A_2 \end{pmatrix} \begin{pmatrix} u \\ w \end{pmatrix} = b, \qquad (5)$$

where the coefficients in the matrices A_1, A_2 and the vector b are derived according to a finite difference scheme. Condition (2d) is converted by enforcing it for the discrete variables:

$$u_{i,j,t} \geq 0, \qquad \forall i, j \in \{0, \ldots n_x\}, t \in \mathbb{T}. \qquad (6)$$

From the network conditions only (1c) has to be adjusted as

$$u_{j,k,t} - (1 - z_{i,t})M \leq U_B, \qquad \forall i \in V, x_i = (j\Delta x, k\Delta x), t \in \mathbb{T}. \qquad (7)$$

Note that we assume here for simplicity that the coordinates x_i of the nodes $i \in V$ are aligned to the grid. More generally, then one can take the weighted sum of the neighboring grid points according to their distance to the position of the vertex x_i, which still gives a linear constraint. The objective function can be approximated by the trapezoidal rule applied at the grid points. Then the first linear mixed-integer approximation of (4) is:

$$\min \ (\Delta x)^2 \Delta t \sum_{t \in \mathbb{T}} \sum_{i,j=0}^{n_x} \lambda_t \mu_i \nu_j \omega(i\Delta x, j\Delta x) u_{i,j,t}, \tag{8}$$

s.t. (1a) − (1d), (5), (6), (7),

where: $\lambda_0 = 0.5$, and $\lambda_t = 1$ if $t > 0$; $\mu_0, \mu_{n_x}, \nu_0, \nu_{n_x} = 0.5$, and $\mu_i, \nu_j = 1$ otherwise.

Finite Elements. The second approach is based on the observation that because of the principle of superposition for linear PDEs the continuous state u can be defined as

$$u = u_{inh} + \sum_{t \in \mathbb{T}} \sum_{i \in V} w_{i,t} \hat{u}_{i,t}, \tag{9}$$

where u_{inh} is the solution of (2a) for $w = \mathbf{0}$, and $\hat{u}_{i,t}$ are the solutions of (2a) for each individual summand of u and homogeneous boundary and initial conditions. Since the summands of the control functions for a fixed vertex i can be obtained by shifting $\hat{u}_{i,0}(t)$ to the right it holds for all $\tau \in \mathbb{T}$

$$\hat{u}_{i,\tau}(t) = \begin{cases} 0, & 0 \leq t \leq \tau, \\ \hat{u}_{i,0}(t-\tau), & \tau < t \leq T. \end{cases}$$

Therefore only $|V|+1$ PDEs have to be solved in order to obtain u, and (9) can be used to replace (2a)–(2c) in the continuous model. This also makes it possible to separate the solution of the PDE from the optimization process, which opens up the possibility to use adaptive finite element methods instead of finite differences. Finite element methods in contrast to finite differences define a linear combination of base functions and thus can be used to derive values anywhere in Ω and not only on a grid. So independent on the meshes of the finite element method, it is possible to define the discrete variables as:

$$u_{i,j,t} = u_{inh}(i\Delta x, j\Delta x, t) + \sum_{\tau \in \mathbb{T}} \sum_{k \in V} w_{k,\tau} \hat{u}_{k,\tau}(i\Delta x, j\Delta x, t). \tag{10}$$

With this we define the MILP for the second approach:

$$\min \ (\Delta x)^2 \Delta t \sum_{t \in \mathbb{T}} \sum_{i,j=0}^{n_x} \lambda_t \mu_i \nu_j \omega(i\Delta x, j\Delta x) u_{i,j,t}, \tag{11}$$

s.t. (1a) − (1d), (6), (7), (10).

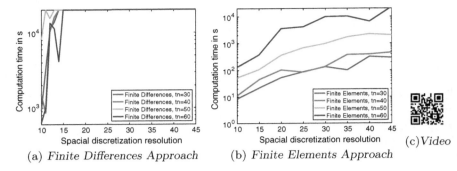

Fig. 1 Computational results

3 Computational Results and Conclusion

Two different MILP were derived that approximate the continuous problem (4). The second model (11) has much less constraints and variables compared to model (8) based on finite differences. Yet it remains to be shown that the second model indeed outperforms the first one. For solving the required PDEs, the object oriented software package oFEM [1] has been employed. The computational results for a problem formulated for the two models are included in the Fig. 1a, b. The different graphs show the runtimes for different degrees of time and space discretizations. The figures illustrate that the finite difference method was only able to solve problems with only a 10×10 spacial grid and up to 60 timesteps within a time limit of 20, 000 s. In contrast, the second method still solves problems with a 45×45 spacial grid and 50 timesteps within the same timeframe, using IBM ILOG CPLEX 12.6.3.0 on a 2014 Mac mini with a 2.6 GHz Intel Core i7 CPU and 16 GB RAM.

References

1. Dudzinski, M., Rozgić, M., & Stiemer, M. (2016). *o*FEM: An object oriented package in MATLAB. *Applied Mathematics and Computation*, (submitted).
2. Frank, M., Fügenschuh, A., Herty, M., & Schewe, L. (2010). The coolest path problem. *Networks and Heterogeneous Media*, 5(1), 143–162.
3. Göttlich, S., Kühn, S., Ohst, J. P., Ruzika, S., & Thiemann, M. (2011). Evacuation dynamics influenced by spreading hazardous material. *Networks and Heterogeneous Media*, 6, 443–464.
4. Mandel, J., Bennethum, L., Beezley, J., Coen, J., Douglas, C., Kim, M., et al. (2008). A wildland fire model with data assimilation. *Mathematics and Computers in Simulation*, 79, 584–606.
5. Noye, B. J., & Tan, H. H. (1989). Finite difference methods for solving the two-dimensional advection-diffusion equation. *International Journal for Numerical Methods in Fluids*, 9, 75–98.
6. Skutella, M. (2008). An introduction to network flows over time. In W. J. Cook, L. Lovász, & J. Vygen (Eds.), *Research trends in combinatorial optimization*. Berlin: Springer.

The Multiple Traveling Salesmen Problem with Moving Targets and Nonlinear Trajectories

Anke Stieber and Armin Fügenschuh

1 Problem Description of the MTSPMT

The MTSPMT is a dynamic variant of the classical TSP and the time plays an important role here. The nodes (targets, objects) are not fixed as in the classical TSP, they move over time on arbitrary trajectories. Each target is associated with a certain speed value and a visibility time window. We consider hard time windows, so that a target can only be intercepted by a salesman within its respective time window. This variant also considers more than one salesman. Each salesman is assigned a certain speed value. All salesmen start their tours from an initial depot, located w.l.o.g. in the middle of the considered area or space. Each target has to be visited once and by exactly one salesman. The aim is to find a tour for each salesman in order to minimize the total traveled distance aggregated by all salesmen. If we restrict the number of salesmen to one, fix each target to a certain local position and extend their time windows to infinity, we obtain the classical TSP, which is NP-hard. Thus, the MTSPMT as a generalization of the TSP is also NP-hard.

Applications can be found in the defense sector, e.g., protection of an airport or a security zone (for details see Stieber et al. [9]) or in the logistic sector, e.g., supplying a fleet of boats or mobile ground units. For many such applications, MTSPMT should be treated as an online optimization problem, that is, the targets are not known before the optimization starts ("offline"), instead they occur afterwards. Still, a fast routine to solve the "offline" variant could serve as the backbone of an online solver with a moving horizon approach. Here new data is integrated into the offline algorithm at

---

A. Stieber (✉) · A. Fügenschuh
Helmut Schmidt University/University of the Federal Armed Forces Hamburg,
Holstenhofweg 85, 22043 Hamburg, Germany
e-mail: anke.stieber@hsu-hh.de

A. Fügenschuh
e-mail: fuegenschuh@hsu-hh.de

run-time, and a fast offline algorithm can be used to get a tentative decision to the "online" problem that is re-optimize anytime new targets emerge.

In the literature the MTSPMT is only addressed considering small instances and with many restrictions to the problem parameters, e.g., movement and speed of the targets or the one-dimensional case is considered, see for example [1, 2, 4, 5, 7]. The MTSPMT modeled with discrete time steps is also very similar to the asymmetrical equality generalized multiple depot TSP (E-GMDTSP), where targets are assigned to clusters and exactly one target from each cluster has to be visited by a salesman, see for example [6, 8, 10]. To the best of our knowledge, there is no exact algorithms to the asymmetrical E-GMDTSP.

2 Model Formulation

We presented a mixed-integer linear programming formulation for the MTSPMT in Stieber et al. [9]. Therefor, the underlying graph is embedded in a time-expanded network and the MTSPMT is formulated as a multi-commodity flow problem. We recall the formulation in a concise way.

The set of salesmen is denoted by $\mathcal{W} = \{1, \ldots, w\}$ and the set of targets by $\mathcal{V} = \{1, \ldots, n\}$. All salesmen start their tour from the same depot location o, hence, $\mathcal{V}_o = \mathcal{V} \cup \{o\}$. Then we have the set of arcs (roads) as $\mathcal{A} \subseteq \mathcal{V}_o \times \mathcal{V}$. We consider a finite time horizon $[0, T]$. The distance for salesman k traveling from target i to target j starting at time s in i and arriving at time t in j is given by the function $c_{i,j,k} : [0, T] \times [0, T] \to \mathbb{R}_+ \cup \{\infty\}$. Since each target $i \in \mathcal{V}$ is assigned a visibility time window $[\underline{t}_i, \overline{t}_i]$, we have

$$c_{i,j,k}(s, t) = \infty \quad \text{if } s \notin [\underline{t}_i, \overline{t}_i] \text{ or } t \notin [\underline{t}_j, \overline{t}_j] \text{ or } (t - s)\overline{v} < \|v_j(t) - v_i(s)\|_2,$$

where $v_i(s)$ and $v_j(t)$ are the respective locations of the targets at the times s and t and $\overline{v}$ is the maximum speed value of all salesmen. The arrival time of any salesman at a target is equal to his departure time at the same target, because waiting times are included in the traveling times. Thus, salesmen do not necessarily use the maximum speed $\overline{v}$.

The time horizon is discretized into $m + 1$ equidistant time steps $\mathcal{T} = \{0, \ldots, m\}$ with step length Δt, hence $T = m\Delta t$. We evaluate c only at these:

$$c_{i,j,k}^{p,q} := c_{i,j,k}(p\Delta t, q\Delta t).$$

In the time-expanded network arcs go from one time layer to a later time layer, hence the time-dependent set of arcs is denoted by $\tilde{\mathcal{A}}$ and an arc is specified by (i, p, j, q) with $i \in \mathcal{V}_o$, $j \in \mathcal{V}$ and $p, q \in \mathcal{T}$. This means the length of an arc (i.e., distance) is dependent on the departure and arrival times. We introduce a family of binary decision variables $x_{i,j,k}^{p,q} \in \{0, 1\}$, where $x_{i,j,k}^{p,q} = 1$ represents the decision of sending salesman k from target i to j, departing in i at time step p and arriving in j at time step q. Then the optimization problem is given by

$$\sum_{k \in \mathcal{W}} \sum_{(i,p,j,q) \in \tilde{\mathcal{A}}} c_{i,j,k}^{p,q} x_{i,j,k}^{p,q} \to \min. \tag{1}$$

$$\text{s.t.} \sum_{k \in \mathcal{W}} \sum_{(i,p,q):(i,p,j,q) \in \tilde{\mathcal{A}}} x_{i,j,k}^{p,q} = 1, \quad \forall j \in \mathcal{V}. \tag{2}$$

$$\sum_{(i,j,q):(i,p,j,q) \in \tilde{\mathcal{A}}} x_{i,j,k}^{p,q} \leq 1, \quad \forall k \in \mathcal{W}, p \in \mathcal{T}. \tag{3}$$

$$\sum_{(i,p):(i,p,j,q) \in \tilde{\mathcal{A}}} x_{i,j,k}^{p,q} \geq \sum_{(i,p):(j,q,i,p) \in \tilde{\mathcal{A}}} x_{j,i,k}^{q,p},$$

$$\forall j \in \mathcal{V}, q \in \mathcal{T}, k \in \mathcal{W}. \tag{4}$$

$$x \in \{0,1\}^{\tilde{\mathcal{A}} \times \mathcal{W}}. \tag{5}$$

The objective function (1) is the sum of all traveled distances of all salesmen. Constraints (2) ensure, that every target is reached once. Inequalities (3) guarantee, that a tour is not split up and (4) are the flow conservation constraints. The presented model is not restricted to particular shapes of the target trajectories. Thus, it can handle linear and non-linear trajectories.

3 Instance Generation

The operating space for our test instances is a square of 500 length units. A test instance is specified by the number of targets, the number of salesmen and the discretization level. The discretization level is a measure of how dense the discretization is done. We used 3 different levels D32, D16 and D8. The first one is based on a discretization every 32 length units (arc length) on each trajectory, the other discretization levels use a step size of 16 and 8 length units respectively. The targets are assigned a constant speed value of 32 length units per time step and the salesmen can travel with at most 200 length units per time step.

For the non-linear trajectories we used polynomial functions and trigonometric functions and a combination by sum and product. We created 16 non-linear trajectories. Then, instances for 6, 8, 10, 12, 14, and 16 targets were created in a way, that we started with 6 trajectories and gradually added two more until we had 16. See the left picture of Fig. 1 for their visualization. In the linear case we used randomly generated trajectories. For reasons of visibility we avoided the straight lines from intercepting each other. We created 5 instances per setting. The trajectories were added the same way as for the nonlinear ones when the number of targets rises. All generated trajectories have a length between 100 and 400 length units and time steps

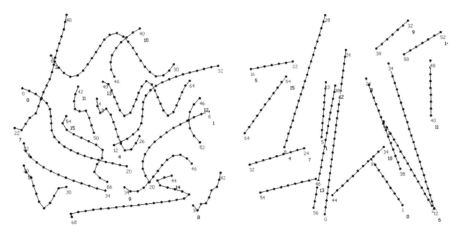

Fig. 1 Generated trajectories. Left picture: 16 nonlinear trajectories. Right picture: 16 linear trajectories (one out of 5 instances). The trajectory number is given in blue and the first and last time step of the trajectory is given in black. All trajectories were generated with the discretization level D16

are distributed in a way that all instances are solvable instances. An instance with 16 linear trajectories is visualized in the right picture of Fig. 1 as an example.

4 Computational Results

All computational experiments were carried out on a 2014 Apple Mac mini computer with an Intel Core i7 CPU running at 2.6 GHz on 4 cores, and 16 GB 1600 MHz RAM. The model was implemented in C++ and instances were solved with the MILP solver IBM ILOG CPLEX 12.7.0 [3]. The computations were performed on a single thread, the CPLEX parameter for the MIP gap was set to 0.0. All other CPLEX parameters were used with their default values.

The computational results are listed in Table 1. Here, the first two columns define the number of targets (nbt) and the number of salesmen (nbs), columns 3 to 5 contain the running times for different discretization levels (dl) and column 6 and 7 the objective function values (ofv) for the nonlinear instances. The running times for the linear instances are given in columns 8 to 10 (for different discretization levels). Since the run-time values for the linear trajectories are averaged values over 5 different instances, we do not provide objective function values. All values in Table 1 are rounded to one digit after point. The results show for both nonlinear and linear trajectories, that the instances become more complex, when the discretization level rises. But there is another effect, that is apparent in the results. For bigger instances (for nonlinear trajectories 10 targets and greater with D16 and D8, for linear trajectories 12 targets and greater with D16 and D8) the running times for a salesman number

Table 1 Running times in seconds. Values for linear instances averaged over 5

Instance		Run-time nonlinear			ofv nonlinear		Run-time linear		
nbt	nbs	D32	D16	D8	D32	D8	D32	D16	D8
6	2	0.1	0.2	0.9	449.2	431.1	0.0	0.1	0.6
6	4	0.1	0.5	1.3	437.6	422.0	0.0	0.1	0.7
6	6	0.2	0.8	2.2	437.6	422.0	0.0	0.3	0.9
8	2	0.1	0.5	3.4	584.3	521.9	0.1	0.7	5.0
8	4	0.1	0.8	3.8	531.6	484.0	0.1	0.4	1.3
8	6	0.1	1.3	4.8	531.6	484.0	0.1	0.6	2.8
10	2	0.4	3.9	44.3	849.3	818.3	0.1	0.9	16.1
10	4	0.4	1.7	4.5	716.1	666.2	0.2	1.2	13.6
10	6	0.6	1.7	9.1	716.1	666.2	0.2	1.9	18.7
12	2	0.5	6.9	34.3	998.2	965.5	0.3	6.2	77.9
12	4	0.6	1.7	6.5	865.1	813.2	0.2	3.0	15.3
12	6	0.9	2.6	10.9	865.1	813.2	0.4	3.2	22.7
14	2	0.8	13.0	80.5	1187.3	1136.9	0.6	16.0	3757.5
14	4	0.8	3.3	10.2	955.1	911.8	0.4	2.9	37.5
14	6	0.4	3.2	14.6	948.6	898.3	0.5	3.9	63.9
16	2	1.0	12.8	1763.4	1321.4	1276.7	0,5	14,3	4939,1
16	4	1.8	5.6	88.3	1071.8	1022.0	0,6	4,2	31,0
16	6	1.7	5.9	31.1	1039.1	982.5	0,6	4,6	41,4

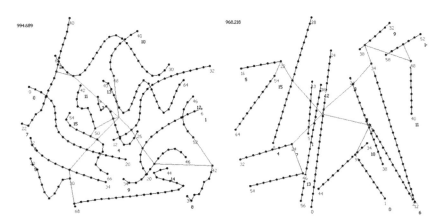

Fig. 2 Solution tours with 16 targets and 6 salesmen with D16 is given by a red line. Left picture: nonlinear trajectories. Right picture: linear trajectories

of 2 is much higher than for 4 and 6 salesmen. The run-time for 2 salesmen is more than twice as much as the run-time for 4 or 6 salesmen in the nonlinear case. Also, in the linear case most of the instances follow this behavior. Considering the smaller instances with 6 and 8 targets, the behavior is completely reversed, the run-times increase when the salesman number rises. In the linear case this behavior is similar but not so apparent as for the nonlinear trajectories, because of the instances with 8 targets. The optimal solutions of the trajectories visualized above are given in Fig. 2.

5 Conclusion

We considered the MTSPMT with a model as a multi-commodity flow problem. In the literature and in [9] only linear trajectories were considered. Here, we solved instances with nonlinear trajectories and compared them with linear ones. From the computational results we can conclude, that the running time is not dependent on the shape of the trajectories and that instances with 2 targets are often more complex than with 4 or 6 targets.

References

1. Helvig, C., Robins, G., & Zelikovsky, A. (1998). Moving-target TSP and related problems. In A. P. G. Bilardi, G . F. Italiano, & G. Pucci (Eds.), *Proceedings of the European symposium on algorithms* (Vol. 1461, pp. 453–464) Lecture notes in computer science. Berlin: Springer.
2. Helvig, C., Robins, G., & Zelikovsky, A. (2003). The moving-target traveling salesman problem. *Journal of Algorithms, 49*(1), 153–174.
3. IBM ILOG CPLEX. User's Manual, (2017). https://goo.gl/TpwkRq; Retrieved 5.7.2017.
4. Jiang, Q., Sarker, R., & Abbass, H. (2005). Tracking moving targets and the non-stationary traveling salesman problem. *Complexity International, 11*, 171–179.
5. Jindal, P., Kumar, A., & Kumar, S. (2011). Multiple target intercepting traveling salesman problem. *International Journal of Computer Science and Technology, 2*(2), 327–331.
6. Laporte, G., Mercure, H., & Nobert, Y. (1987). Generalized travelling salesman problem through n sets of nodes: The asymmetrical case. *DAM, 18*(2), 185–197.
7. Liu, C. H. (2013). The Moving-Target Traveling Salesman Problem with Resupply. Technical report, The National Chung Cheng University Library. http://ccur.lib.ccu.edu.tw/handle/987654321/7877; Retrieved 5.7.2017.
8. Noon, C. E., & Bean, J. C. (1991). A lagrangian based approach for the asymmetric generalized traveling salesman problem. *Operations Research, 39*(4), 623–632.
9. Stieber, A., Fügenschuh, A., Epp, M., Knapp, M., & Rothe, H. (2014). The multiple traveling salesmen problem with moving targets. *Optimization Letters, 9*(8), 1569–1583.
10. Sundar, K., & Rathinam, S. (2016). Generalized multiple depot traveling salesmen problem - polyhedral study and exact algorithm. *COR, 70*, 39–55.

Product Family Design Optimization Using Model-Based Engineering Techniques

David Stenger, Lena C. Altherr, Tankred Müller and Peter F. Pelz

1 Introduction

Product families (PFs) consist of a number of different products which are derived from a common platform and satisfy different customer requirements. When designing a scale-based PF (cf. Fig. 1) engineers need to find the optimal platform configuration, specifying which platform parameters are identical for which products as well as the optimal designs for each product. A product platform is defined here as the set of all different platform parameter values across the PF.

Total PF costs, C^{tot}, need to be minimised while maintaining technical feasibility of single product designs. $\quad C^{\text{tot}} = C^{\text{ind}}(X_{i,j}) + C^{\text{var}}(\Lambda)$, with:

$X_{i,j}$ Matrix defining a specific PF. i: Product index. j: Parameter index
$\Lambda(X_{i,j}) = [\Lambda_1(X_{*,1}), \Lambda_2(X_{*,2}), \ldots, \Lambda_m(X_{*,m})]$ Platform variance vector specifying the number of different values for each platform parameter of a given PF.
V_i Expected production volume for product i.
$C^{\text{ind}} = \sum_{i=1}^{n} c_i^{\text{ind}}(X_{i,*}) V_i$ Cumulated individual product costs.
$C^{\text{var}}(\Lambda)$ Platform variance related costs.

D. Stenger (✉) · L. C. Altherr · P. F. Pelz
Chair of Fluid Systems, Dept. of Mechanical Engineering, Technische Universität Darmstadt,
Otto-Berndt-Str. 2, 64287 Darmstadt, Germany
e-mail: david.stenger@gmx.net

L. C. Altherr
e-mail: lena.altherr@fst.tu-darmstadt.de

P. F. Pelz
e-mail: peter.pelz@fst.tu-darmstadt.de

T. Müller
Robert Bosch GmbH, Robert-Bosch-Straße 1, 77815 Bühl, Germany
e-mail: tankred.mueller@de.bosch.com

	Platform parameters		Individual parameters	
	$X_{*,1}$	$X_{*,2}$	$X_{*,3}$	$X_{*,4}$
Product 1 ($X_{1,*}$)	1.2	200	20	10
Product 2 ($X_{2,*}$)	1.2	200	25	8
Product 3 ($X_{3,*}$)	1.5	200	27	7
Platform variance (Λ)	2	1		

Platform: □ Platform configuration:

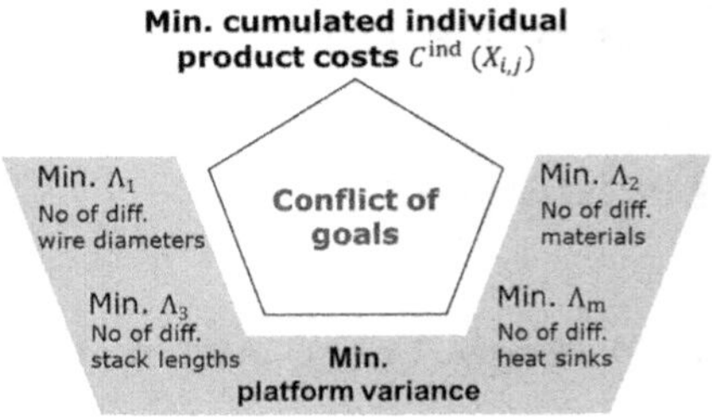

Fig. 1 Generic product family (left). Conflict of goals in PF optimization (right)

$C^{\text{var}}(\Lambda)$, consisting e.g. of tooling, logistic and development costs, decreases with increasing commonality. In contrast, single product individual costs $c_i^{\text{ind}}(X_{i,*})$, which only depend on each product's parameters $X_{i,*}$, typically increase with increasing commonality due to overdesign caused by restrictions from the platform. They consist e.g. of material and value added costs assuming constant quantities. This study focuses on the optimization of a Bosch product family of electric drives. In early design phases, a general cost model of production lines and logistics is not available. Therefore the function $C^{\text{var}}(\Lambda)$ is unknown and the described trade-off cannot be resolved by aggregation of costs. Instead, the conflict of goals between low individual product costs and high commonality, cf. Fig. 1 right, is visualised and used during the product development process as a basis for commonality decisions.

2 State of the Art

Extensive research has been conducted in the field of PF design optimization. For a comprehensive literature review we refer to [1]. Common approaches to simultaneously optimize platform configuration and design parameters of individual products include the usage of meta-heuristics such as genetic algorithms [2] and sensitivity and cluster analysis [3]. Commonality indices are widely used to aggregate the high commonality goal in one objective function and therefore reduce the problem complexity significantly. Little research has been conducted for the optimization of product families with expensive black box simulations. In [4] meta-models are employed to efficiently perform sensitivity analysis.

3 Problem Description

In this work the trade-off between low individual product costs and high commonality in each platform parameter is examined. Commonality indices and sensitivity and cluster analysis are not used here, because they are unable to capture the whole pareto frontier. Instead, the task is formulated as a vector optimization problem:

$$\min_{X_{i,j} \in \mathcal{D}_{i,j}} [C^{\text{ind}}(X_{i,j}), \Lambda_1(X_{*,1}), \ldots, \Lambda_m(X_{*,m})] \ \forall l \in \{m, T, \eta \ldots\} : g_{l,i}(X_{i,*}) \leq g_{l,i}^{\max} \quad (1)$$

Individual product cost $c_i^{\text{ind}}(X_{i,*})$ and other design properties $g_{l,i}(X_{i,*})$, such as mass m, efficiency η and temperature T are restricted by upper bounds $g_{l,i}^{\max}$. They cannot be calculated analytically with acceptable accuracy due to complex non linear product behaviour. Instead individual designs $X_{i,*}$ of each product i are evaluated using multi-domain (mechanical, thermal, electromagnetic...) and multi-component (E-motor, gearbox, ECU...) transient numerical simulations. The simulation of one design for one set of customer requirements takes approximately one minute.

4 Exploration of Single Product Design Spaces and Tree Search

The problem described in Eq. 1 can be solved by exploring the design spaces of the single products and combining feasible designs to pareto-optimal PFs using bounded depth first tree search (DFS). The DFS-algorithm used guarantees global pareto-optimal solutions given a set of evaluated designs. Therefore this approach is used to generate reference solutions for the algorithm described in Sect. 5.

First the single product design spaces are discretised using expert-knowledge and evaluated full-factorial. The resolution of discretisation is limited by the available computational resources needed for the black-box simulations. The evaluated designs are filtered for technical feasibility. Given sets D_i of feasible designs for each product i, the following discrete optimization problem needs to be solved:

$$\min_{d_1, d_2, \ldots, d_n} (\sum_{i=1}^{n} c_i^{\text{ind}}(d_i) V_i, \Lambda_1, \Lambda_2, \ldots \Lambda_m) \quad d_i \in D_i \quad (2)$$

It can be seen as a decision tree with a depth equal to the number of products n and a branching factor at depth i of the number of different evaluated designs for product i. A path $p_i \in P_i = D_1 \times D_2 \times \cdots \times D_i$ from the root to a node at depth i corresponds to a partial product family with a partial platform variance $\Lambda^{\text{part}}(p_i)$ and partial cumulated product costs $C^{\text{ind,part}}(p_i) = \sum_{k=1}^{i}(V_k c_k^{\text{ind}}(d_k))$. A path from the root to a leaf at depth n corresponds to one possible PF solution. To avoid complete enumeration in the DFS-algorithm, an upper bound $B_i^{\text{upper}}(\Lambda^{\text{part}})$ at depth i is introduced as a function of the partial PF variance:

$$B_i^{\text{upper}}(\Lambda^{\text{part}}) = \max_{r \in N^+ | \Lambda^r \geq \Lambda^{\text{part}}} (C_{\min, \Lambda^r}^{\text{ind}}) - \sum_{k=i+1}^{n} V_k \min_{d \in D_k} c_k^{\text{ind}}(d) \quad (3)$$

Λ^r is a list of all possible platform variance vectors. $C^{\text{ind}}_{\min, \Lambda^r}$ are the cumulated individual costs of the cheapest PFs already found for platform variance Λ^r. The subtrahend is the sum over products $i + 1$ to n of the cheapest of all feasible designs for each product k. If for an explored node p_i at any time of the search the cut off rule $C^{\text{ind,part}}_i(p_i) > B^{\text{upper}}_i(\Lambda^{\text{part}}(p_i))$ is true, p_i cannot lead to an improved solution and Its sub tree is not searched. Finally pareto optimal solutions are extracted by sorting all nodes explored at depth n.

5 Meta-model Based Algorithm

In order to avoid the full-factorial exploration of single product design spaces on the simulation model, we propose an optimization algorithm based on the globally searching maximal expected improvement sampling method [5]. The steps outlined in Fig. 2 are as following:

The adaptive *initial sampling* ① is performed with the objective to achieve high global meta-model accuracy. Designs with high model uncertainty are sampled successively in a number of iterations. As a local measure of model uncertainty the kriging variance is used. All design evaluations are stored in an archive. An initial product family solution set is calculated by using the DFS-algorithm described in Sect. 4 on the archived single product designs. The expected mean and estimated prediction error of individual product costs, $\overline{c^{\text{ind}}_i}(X_{i,*})$ and $\sigma_{c^{\text{ind}}_i}(X_{i,*})$, and of all other constrained system responses $\overline{g_{l,i}}(X_{i,*})$ and $\sigma_{g_{l,i}}(X_{i,*})$ are modelled using an anisotropic interpolating *kriging meta-model* ② with a squared exponential kernel. At each iteration of the optimization loop the model is fitted on the current archive of system simulations. The length scales of the kernel are optimised by maximizing the marginal log likelihood. In order to *generate candidate solutions* ③, two statistical lower bounds corresponding to the adaptive confidence intervals $Y_1\sigma$ and $Y_2\sigma$ are minimised, cf. Fig. 3. Y_1 is adapted so that new candidate PFs incorporate at least two not yet evaluated designs. Y_2 is gradually increased at each iteration until one PF candidate with at least one not yet evaluated design is found. Y_1 enforces a global and Y_2 a local search component. Each single product design in the discretised product-specific optimization domain is evaluated on the kriging model. A product design is considered feasible given a confidence interval $Y_k, k \in \{1, 2\}$, if $\forall l \in L : (\overline{g_{l,i}}(X_{i,j}) - Y_k\sigma_{g_{l,i}(X_{i,j})}) \leq g^{\max}_{l,i}$. The DFS-algorithm described in Sect. 4 is used to combine these probably feasible designs to find new PF candidate solutions for evaluation by simulation. Therefore, $c^{\text{ind}}_i(X_{i,*})$ in Eq. 2 is replaced by the statistical lower bound of the individual costs: $\overline{c^{\text{ind}}_i}(X_{i,*}) - Y_k\sigma_{c^{\text{ind}}_i(X_{i,*})}$. All candidate solutions are *sorted w.r.t. maximal expected improvement* ④. The expected improvement with constraints, EIC, for a product family given the current best solution for its platform variance $C^{\text{ind}}_{\min, \Lambda(X_{i,j})}$ can be written as:

Fig. 2 Scheme of the proposed algorithm

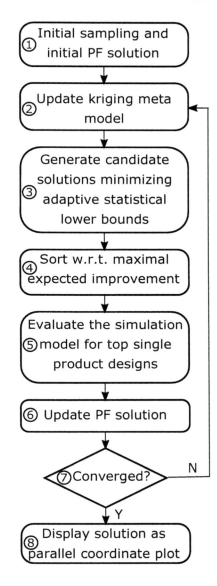

$$\text{EIC}(X_{i,j}) = \prod_{i=1}^{n}\prod_{l=1}^{k} \text{P}(g_{l,i}(X_{i,*}) < g_{l,i}^{\max}) \cdot \text{E}[\max\{0, C_{\min,\Lambda(X_{i,j})}^{\text{ind}} - \tilde{C}^{\text{ind}}(X_{i,j})\}] \quad (4)$$

The first factor is the probability that each design of the candidate PF complies with every constraint. The second factor denotes the expected improvement. The random variable $\tilde{C}^{\text{ind}}(X_{i,j})$ describes the normally distributed total cumulated individual PF

Fig. 3 Kriging Model with candidate solutions belonging to two lower bounds for a discretised 1-D example

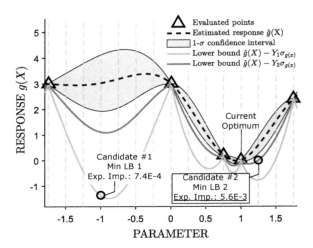

costs. Its mean and standard deviation are calculated from $\overline{c^{\text{ind}}}_i$ and $\sigma_{c_i^{\text{ind}}}$. Single product designs belonging to candidate PFs with maximal expected improvement are *evaluated on the simulation model* ⑤ and added to the archive of evaluated designs. The number of design evaluations per iteration is controlled by the ratio of overhead time (time for fitting the kriging model and for searching the response surface) and the system simulation time. The *solution is updated* ⑥ by combining all evaluated product designs stored in the archive to pareto-optimal product families using the DFS-algorithm described in Sect. 4. The usage of the kriging meta-model in combination with the minimization of the adaptive statistical lower bounds allows the formulation of a probabilistic *convergence criterion* ⑦. If Y_2 is consistently bigger than the maximum cross validation error of the standard deviation predictor, it is likely that the optimum is reached.

6 Results

The proposed meta-model based algorithm is benchmarked on a Bosch product family with 8 different products. Each of them is defined by 7 descritised design parameters, resulting in 2016 different designs per product and $2016^8 \approx 10^{26}$ possible different PFs. Out of the 7 design parameters, one and two are chosen to be platform parameters, respectively. For both cases the reference solutions are calculated using the approach presented in Sect. 4. In this example the reference solution contains a valid PF for every possible platform variance. The pareto-optimal set contains 4 and 11 PFs, respectively. In Fig. 4 the normalised deviations from the reference solution are displayed. Within Ⓐ 4.4% and Ⓑ 7.1% of the function evaluations needed for the reference solution, a PF is found for every possible platform variance Λ. The average cumulated individual product cost of the reference solution is reached

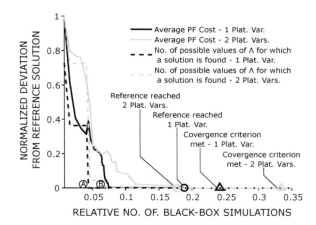

Fig. 4 Convergence plot of the meta-model based algorithm

within 18.9 and 17.7%. Although after 8 and 12% a reasonable approximation of the optimal solution is achieved. The convergence criterion is met at 24.2 and 33.4% of the maximal number of function evaluations.

7 Discussion and Outlook

The DFS-algorithm allows designers to quickly search discretised problem domains for pareto optimal PFs. It was shown that applying the kriging meta-model reduces simulation effort significantly while in our test case also achieving the optimal reference solution. Additionally the introduced statistical convergence criterion enables designers to conservatively estimate when global optimality is reached. Further research will include the usage of meta-heuristics to directly search the response surface for candidate solutions with maximal expected improvement.

Acknowledgements The authors thank the German Research Foundation DFG for funding this research within the Collaborative Research Center SFB 805 "Control of Uncertainties in Load-Carrying Structures in Mechanical Engineering" and the Robert Bosch GmbH.

References

1. Simpson, T. W., Jiao, J., Siddique, Z., & Hölttä-Otto, K. (Eds.). (2014). *Advances in product family and product platform design*. New York: Springer.
2. Khajavirad, A., Michalek, J. J., & Simpson, T. W. (2009). An efficient decomposed multiobjective genetic algorithm for solving the joint product platform selection and product family design problem with generalised commonality. *Structural and Multidisciplinary Optimization*, *39*(2), 187201. https://doi.org/10.1007/s00158-008-0321-9.

3. Dai, Z., & Scott, M. J. (2007). Product platform design through sensitivity analysis and cluster analysis. *Journal of Intelligent Manufacturing, 18*(1), 97113. https://doi.org/10.1007/s10845-007-0011-2.
4. Pirmoradi, Z. (2015). Metamodel-based product family design optimization for plug-in hybrid electric vehicles. Ph.D. thesis. Simon Fraser University.
5. Jones, D. R. (2001). A taxonomy of global optimiation methods based on response surfaces. *Journal of Global Optimization, 21*(4), 345383. https://doi.org/10.1023/A:1012771025575.

The Multistatic Sonar Location Problem and Mixed-Integer Programming

Emily M. Craparo and Armin Fügenschuh

1 Introduction to Sonar

Sonar is a technique to detect objects that are under water or at the surface using sound propagation. In active sonar systems, a sound is emitted from a source and its echoes are detected by a receiver, revealing information about nearby objects. Active sonar has been in use for nearly 100 years and has become a key component of undersea detection. The basic operating principle of active sonar is that acoustic energy is emitted from a source and its echoes are detected by a receiver; these echoes reveal information about surrounding objects. In a monostatic system, the source and the receiver are collocated in the same place. Bistatic sonar uses a source and a receiver pair in different locations. Multistatic sonar uses several sources and receivers simultaneously as a network. For the surveillance of a large area of the ocean, a number of both types of devices must be deployed. This leads to an optimization problem to find the least costly multistatic network that is able to cover all of a desired area. No algorithm currently in the literature provides an optimal placement of an arbitrary number of sources and receivers. In a discretized setting, we describe mathematical models designed to determine the minimum-cost sensor layout that will cover a portion of the ocean (a tile) by sonar surveillance, with adequate detection probability throughout the tile. We model the physical properties of sound traveling between sources, target, and receivers, the ocean (temperature, density, salinity) as well as geometrical considerations (obstacles such as islands or coastlines). Details are given in Sect. 2. We formulate an integer nonlinear program

E. M. Craparo
Naval Postgraduate School, Operations Research Department, 1411 Cunningham Road, Monterey, CA 93943, USA
e-mail: emcrapar@nps.edu

A. Fügenschuh (✉)
Helmut Schmidt University/University of the Federal Armed Forces Hamburg,
Holstenhofweg 85, 22043 Hamburg, Germany
e-mail: fuegenschuh@hsu-hh.de

for the multistatic sonar source-receiver location problem and discuss several linearizations in Sect. 3. We compare these formulations empirically using topological data from coastal areas around the world and a state-of-the-art solver MIP solver and give concluding remarks in Sect. 4.

2 Input Data

We obtain ocean topography data from [10]. At present, we do not use sea level information and only distinguish in a binary fashion between the ocean (negative elevation value) and the dry land (positive elevation value). A desired part of the ocean and shoreline (a tile) is taken from the database. Since the resolution of the data is too fine to let each data pixel become a possible target/source/receiver location, we aggregate the raw input data into larger rectangular areas (also called grid cells). We then average the elevation data from all pixels within a cell and apply the resulting elevation to the entire cell. Denote the set of rectangles with negative elevation (i.e., those that are underwater) by G (for grid) and the number of elements in G by $n := |G|$.

The sonar signal is characterized by the range of the day ϱ_0, which indicates how quickly the signal diminishes as the target, source, and receiver become farther apart. In a definite range ("cookie-cutter") sensor model, a target in a cell $k \in G$ is detected by a source placed in cell $i \in G$ and a receiver placed in cell $j \in G$ with probability $p_{i,j,k} \in \{0, 1\}$. Denote by $d_{i,j}$ the Euclidean distance between (the centers of) cell i and j. Necessary for detection ($p_{i,j,k} = 1$) is that the target k is inside the Cassini oval defined by the equation $d_{i,k} \cdot d_{k,j} \leq \varrho_0^2$, c.f. [7]. If the target is too close to the line from source to receiver, then the original signal and its reflection at the target become indistinguishable at the receiver. This phenomenon is known as the direct blast effect. The pulse length κ_b determines the severity of this effect, since longer pulses are more prone to overlapping with the reflected signal. The direct blast zone is defined by the ellipsoid $d_{i,k} + d_{k,j} \leq d_{i,j} + 2\kappa_b$, c.f. [6]. To account for the direct blast effect, we say that $p_{i,j,k} = 0$, if the target lies within the direct blast zone. Additionally, if an obstacle lies on either straight-line path of source to target, target to receiver, or source to receiver, then $p_{i,j,k} = 0$.

The cost for each source is c_s, and the cost for each receiver is c_r. Typically, $c_s \gg c_r$, i.e., a source is much more costly than a receiver, usually by a factor of 5.

3 Model Formulations

All model formulations below have in common the binary decision variables $s_i, r_i \in \{0, 1\}$ for each $i \in G$, which model the decision whether to place a source ($s_i = 1$) or a receiver ($r_i = 1$) in cell i. The objective (in all formulations) is to minimize the total deployment cost, which we calculate as follows:

$$c_s \sum_{i \in G} s_i + c_r \sum_{j \in G} r_j. \tag{1}$$

An Integer Nonlinear Model. In the first nonlinear formulation the binary variables s_i and r_j are multiplied in order to represent the joint decision of placing a source at i **and** a receiver in j:

$$\sum_{i \in G} \sum_{j \in G} p_{i,j,k} s_i r_j \geq 1, \quad \forall k \in G. \tag{2}$$

Each constraint of (2) is a quadratic knapsack constraint. In general, for any given $k \in G$ the non-negative matrix $(p_{i,j,k})_{i,j}$ is indefinite. The solver CPLEX is able to process constraints of this form since version 12.6 [2]. Thus, the base run for comparison with the other reformulation approaches is to solve the model:

$$\min\{(1)|(2); s, r \in \{0, 1\}^G\}. \tag{3}$$

The Oldest Linearization Technique. The first documented linearization of a product of binaries $s_i r_j$ by [1, 3] (and independently by others later on) introduces a new binary variable $h_{i,j} \in \{0, 1\}$ with $h_{i,j} = 1$ if and only if $s_i = 1$ and $r_j = 1$. In this method the constraints $2h_{i,j} \leq s_i + r_j$ and $s_i + r_j \leq 1 + h_{i,j}$ (for all $i, j \in G$) are a linear description of this relationship. In our case, because of the non-negativity of all $p_{i,j,k}$, only the first constraint is necessary. Thus the first linear version of (3) is

$$\min\ (1), \text{ s.t. } \sum_{i \in G} \sum_{j \in G} p_{i,j,k} h_{i,j} \geq 1, \quad \forall k \in G, \tag{4a}$$

$$2h_{i,j} \leq s_i + r_j, \quad \forall i, j \in G, \tag{4b}$$

$$s \in \{0, 1\}^G, r \in \{0, 1\}^G, h \in \{0, 1\}^{G \times G}. \tag{4c}$$

Compared to the nonlinear integer formulation (3), this binary linear model has an additional n^2 binary variables and n^2 constraints.

Standard Linearization of the Model. A linearization for $s_i r_j$ similar to the previous one from [5] introduces continuous auxiliary variables $h_{i,j} \in [0, 1]$ together with the constraints $h_{i,j} \leq s_i$, $h_{i,j} \leq r_j$ and $s_i + r_j \leq 1 + h_{i,j}$. This is perhaps the first and most natural formulation to come to mind (and for good reason: Padberg [9] showed that the constraints are facet defining), and is hence called "standard linearization." As before, the third constraint is not required in our case. Then, the second linear version of (3) is

$$\min\ (1), \text{ s.t. } \sum_{i \in G} \sum_{j \in G} p_{i,j,k} h_{i,j} \geq 1, \quad \forall k \in G, \tag{5a}$$

$$\left. \begin{array}{l} h_{i,j} \leq s_i \\ h_{i,j} \leq r_j \end{array} \right\}, \quad \forall i, j \in G, \tag{5b}$$

$$s \in \{0, 1\}^G, r \in \{0, 1\}^G, h \in [0, 1]^{G \times G}. \tag{5c}$$

Compared to the nonlinear binary formulation (3), this mixed-integer linear model has an additional n^2 continuous variables and $2n^2$ constraints.

Glover's Linearization. To adapt a linearization technique from Glover [4], we set $L_{j,k} := \sum_{i \in G} p_{i,j,k}$ for all $j, k \in G$, and the model reads:

$$\min \ (1), \ \text{s.t.} \ \sum_{j \in G} z_{j,k} \geq 1, \quad \forall k \in G, \tag{6a}$$

$$\left. \begin{array}{r} \sum_{i \in G} p_{i,j,k} s_i \geq z_{j,k} \\ L_{j,k} r_j \geq z_{j,k}, \end{array} \right\}, \ \forall j, k \in G, \tag{6b}$$

$$s \in \{0,1\}^G, r \in \{0,1\}^G, z \in \mathbb{R}_+^{G \times G}. \tag{6c}$$

This model introduces n^2 additional continuous variables and $2n^2$ additional constraints (compared to (3)).

Oral–Kettani's Linearization. Oral and Kettani [8] proposed two formulations that come with n^2 additional continuous variables, but fewer constraints compared to Glover's formulation; namely, only n^2 (not counting the trivial bound on $z_{j,k}$ as constraint). The first of the two formulations is:

$$\min \ (1), \ \text{s.t.} \ \sum_{j \in G} (L_{j,k} r_j - z_{j,k}) \geq 1, \quad \forall k \in G, \tag{7a}$$

$$\left. \begin{array}{r} z_{j,k} \geq L_{j,k} r_j - \sum_{i \in G} p_{i,j,k} s_i \\ L_{j,k} \geq z_{j,k}, \end{array} \right\}, \ \forall j, k \in G, \tag{7b}$$

$$s \in \{0,1\}^G, r \in \{0,1\}^G, z \in \mathbb{R}_+^{G \times G}. \tag{7c}$$

The second Oral–Kettani linearization is:

$$\min \ (1), \ \text{s.t.} \ \sum_{j \in G} \left(\sum_{i \in G} p_{i,j,k} s_i - z_{j,k} \right) \geq 1, \quad \forall k \in G, \tag{8a}$$

$$\left. \begin{array}{r} z_{j,k} \geq \sum_{i \in G} p_{i,j,k} s_i - L_{j,k} r_j \\ L_{j,k} \geq z_{j,k}, \end{array} \right\}, \ \forall j, k \in G, \tag{8b}$$

$$s \in \{0,1\}^G, r \in \{0,1\}^G, z \in \mathbb{R}_+^{G \times G}. \tag{8c}$$

4 Computational Results and Conclusions

We compare the above six formulations on a test set of 22 instances. The ocean topography data from various regions all over the world were extracted from a global map, collected by Ryan et al. [10]. The computations were carried out on a 2014 MacBookPro with 16 GB RAM and a 2.8 GHz Intel Core i7 processor. We set a time limit of 1,000 s and default settings of the solver IBM ILOG CPLEX 12.7.1 otherwise.

Table 1 Computational results

Instance	n	(3)	(4)	(5)	(6)	(7)	(8)
BabAlMandabStrait	29	1000.01	2.03	1.02	0.85	1.76	0.53
ChoctawhatcheeBay	31	1000.01	2.5	0.8	0.53	1.9	0.49
Dardanelles	19	3.04	0.24	0.07	0.1	0.07	0.04
EnglishChannel	48	1000.35	11.32	9.44	4.1	13.24	7.5
Falklandsund	57	1005.49	66.23	296.46	65.5	36.6	59.9
GulfOfAkaba	22	39.74	0.15	0.1	0.09	0.16	0.07
GulfOfFinland	37	1000.09	582.37	4.65	1.91	12.58	2.29
GulfOfSirte	45	1002.06	295.68	70.69	16.04	48.42	9.71
KarkinytskaGulf	34	1000.01	3.5	0.33	0.69	1.95	0.38
KerchStrait	36	1000.01	1.05	0.25	0.33	0.55	0.29
LagoDeMaracaibo	48	1000.02	4.45	1.46	1.44	2.35	1.6
Lesbos	30	1000.02	1.88	0.43	0.4	1.09	0.69
MontereyPeninsular	45	1000.26	12.19	5.06	4.29	12.74	2.66
NewYork	38	1000.52	6.59	1.25	1.57	8.27	2.83
OpenSea-Biscaya	54	1002.77	22.52	151.63	24.59	22.34	30.56
Oresund	71	1000.84	33.69	20.45	40.03	37.98	16.08
Ruegen	37	1000.02	34	7.3	2.94	45.32	1.19
Smalandsfarvandet	58	1000.73	229.88	31.74	26.09	32.02	7.53
Storebaelt	40	1000.25	57.58	10.63	2.26	12.66	2.57
StraitOfGibraltar	52	1000.49	28.66	43.62	7.72	70.45	15.63
StraitOfHormuz	41	1000.02	0.97	0.58	0.99	2.64	0.5
TaedongGang	39	1000.02	6.76	2.8	1.77	4.88	3.13
SUM		20056.77	1404.24	660.76	204.23	369.97	166.17
RANK		6	5	4	2	3	1

The results can be found in Table 1, with the second Oral–Kettani formulation slightly ahead that of Glover, and CPLEX failing to solve most instances within the time limit. An example result appears in Fig. 1.

When facing a bilinear constraint of the type $x^T A y \leq b$ with binary variable vectors x, y and an indefinite matrix A, several techniques for their linearization were developed by researchers over the last five decades. Today, classical MILP solvers (such as CPLEX) offer features to automatically deal with such nonlinear constraints, lifting the burden of going to the library from the user. As our results demonstrate, it is still worthwhile to consider the knowledge of the past, and not to blindly rely on the solver. Since it is unclear to determine a priori which of the method outperforms the others, it is necessary to implement and test all of them.

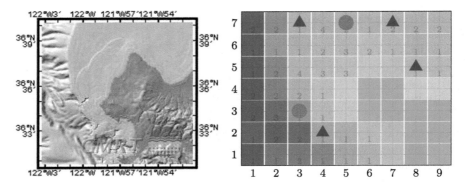

Fig. 1 Left: The Monterey Peninsula area tile [10] as raw input data (365 cols, 285 rows). Right: Optimal placement of 2 sources (red circles) and 4 receivers (blue triangles) on a 9 × 7 grid. Numbers ≥ 1 at each coordinate show multiplicity of coverage

References

1. Balas, E. (1964). Extension de l'algorithme additif à la programmation en nombres entiers et à la programmation non linéaire. Technical report. Comptes rendus de l'Académie des Sciences, Paris.
2. Bliek, C., Bonami, P., & Lodi, A. (2014). Solving mixed-integer quadratic programming problems with IBM-CPLEX: A progress report. In *Proceedings of the Twenty-Sixth RAMP Symposium Hosei University, Tokyo, October 16–17, 2014*.
3. Fortet, R. (1959). L'algèbre de Boole et ses applications en recherche opérationelle. *Cahiers du Centre d'Études de Recherche Opérationelle, 4*, 5–36.
4. Glover, F. (1975). Improved linear integer programming formulations of nonlinear integer problems. *Management Science, 22*(4), 455–460.
5. Glover, F., & Woolsey, E. (1974). Converting the 0–1 polynomial programming problem to a 0–1 linear program. *Operations Research, 22*(1), 180–182.
6. Karatas, M., & Craparo, E. M. (2015). Evaluating the direct blast effect in multistatic sonar networks using monte carlo simulation. In L. Yilmaz et al. (Eds.), *Proceedings of the 2015 Winter Simulation Conference*. Piscataway, NJ: IEEE Press.
7. Karatas, M., Craparo, E. M., & Washburn, A. (2014). A cost effectiveness analysis of randomly placed multistatic sonobuoy fields. In C. Bruzzone et al. (Eds.), *The International Workshop on Applied Modeling and Simulation*.
8. Oral, M., & Kettani, O. (1992). A linearization procedure for quadratic and cubic mixed-integer problems. *Operations Research, 40*(1), 109–116.
9. Padberg, M. (1989). The Boolean quadric polytope: Some characteristics, facets and relatives. *Mathematical Programming, 45*, 139–172.
10. Ryan, W. B. F., Carbotte, S. M., Coplan, J. O., O'Hara, S., Melkonian, A., Arko, R., et al. (2009). Global multi-resolution topography synthesis. *Geochemistry, Geophysics, Geosystems, 10*(3), Q03014.

Using Mixed-Integer Programming for the Optimal Design of Water Supply Networks for Slums

Lea Rausch, John Friesen, Lena C. Altherr and Peter F. Pelz

1 Introduction

Currently, the UN estimates that 663 million people are still without sufficient water supply and at least 1.8 billion people globally use a source of drinking water that is fecally contaminated [1]. Especially, slums, which in many countries are a defining part of urban areas, are often characterized by the lack of an appropriate water supply [2]. We developed a multidisciplinary approach to design an optimal water supply system for slums within a city. For this purpose, the required information on the slum location as well as its size is taken from remote sensing and used as input data for the decision problem. Out of different central and decentral approaches with combined water supply by motorized vehicles as well as installed pipe systems, we find the solution yielding minimal total costs. The technical application is detailed further in Sect. 2 and the modeling as a mixed-integer linear problem (MILP) is given in Sect. 3. Finally, we show optimization results for a slum cluster in Dhaka in Sect. 4.

L. Rausch (✉) · J. Friesen · L. C. Altherr · P. F. Pelz
Department of Mechanical Engineering, Technische Universität Darmstadt,
Otto-Berndt-Str. 2, 64287 Darmstadt, Germany
e-mail: lea.rausch@fst.tu-darmstadt.de

J. Friesen
e-mail: john.friesen@fst.tu-darmstadt.de

L. C. Altherr
e-mail: lena.altherr@fst.tu-darmstadt.de

P. F. Pelz
e-mail: peter.pelz@fst.tu-darmstadt.de

© Springer International Publishing AG, part of Springer Nature 2018
N. Kliewer et al. (eds.), *Operations Research Proceedings 2017*, Operations
Research Proceedings, https://doi.org/10.1007/978-3-319-89920-6_68

2 Technical Application

The presented approach leverages the location data of slums as input, which is provided by applying algorithms to identify slum areas on remote sensing data [3]. In addition, the size of the slums derived from this data is used to calculate the daily water need for each slum by multiplying it with an estimated population density and the daily water need per person. The aim is to find a network describing the water supply infrastructure for all slums within one large city [4]. Between any two slums as well as between the waterworks and any slum, one or more different connections can be chosen: The water can be transported via a selection of pipes with different diameters or via a variety of motorized vehicles combined with water tanks of different sizes. The objective is to reduce the total costs over a specified period of time, including investment and operating costs. Additional requirements ensure the full functionality of the water supply system, such as flow conditions and capacity restrictions.

3 Water Supply Design via MILP

The previously described water supply network design is modeled as a MILP which is introduced in the following.

Objective function The objective of the optimization model is to minimize the total costs, including investment and operating costs, within a fixed term:

$$minimize \quad \sum_{i \in N_w} \sum_{j \in N} \sum_{k \in K_{truck}} x_{\text{truck}}(i, j, k) \cdot Cost_{\text{truck}}(i, j, k) \\ + \sum_{k \in K_P} x_{\text{pipe}}(i, j, k) \cdot Cost_{\text{P,fix}}(i, j, k) \\ + Q_{\text{cubic}}(i, j) \cdot Cost_{\text{P,var}}(i, j, k) \\ + \sum_{i \in N} \sum_{k \in K_{tank}} x_{\text{tank}}(i, k) \cdot Cost_{\text{tank}}(k)$$

Parameters and variables The parameters of this model are given in Table 1 and the decision variables in Table 2. All costs are given in Euro. Since the water requirements are examined on a day to day basis, volume flows as well as capacities are given in liters per day. Connections from slum i to slum j are modeled as directed edge (i, j) in the complete graph.

Constraints The volume flow is the sum of the flow via pipes and trucks:

$$Q_P(i, j) + Q_T(i, j) = Q(i, j) \quad \forall i \in N_w, j \in N \tag{1}$$

The capacity of a connection needs to exceed the volume flow:

Table 1 Parameters of the mixed-integer linear problem

Parameter	Description
N	Set of all slums, $N \subset \mathbb{N}$
N_w	Set of slums and waterworks w, $N_w = N \cup \{w\}$
K_{pipe}	Set of available pipe types with different diameters
$Capa_{\text{pipe}}(k)$	Volume flow capacity of pipes of type $k \in K_{\text{pipe}}$
$Cost_{\text{pipe,fix}}(i, j, k)$	Fixed pipe costs of type $k \in K_{\text{pipe}}$, with length equal to distance from slum i to slum j
$Cost_{\text{pipe,var}}(i, j, k)$	Variable pipe cost factor of type $k \in K_{\text{pipe}}$
K_{truck}	Set of available truck types
$Capa_{\text{truck}}(k)$	Volume flow capacity of truck of type $k \in K_{\text{truck}}$
$Cost_{\text{truck}}(i, j, k)$	Truck cost of type $k \in K_{\text{truck}}$ for distance from i to j
$N_{\text{truck}}^{\max}$	Maximal number of trucks allowed between two slums
K_{tank}	Set of available tank types
$Capa_{\text{tank}}(k)$	Capacity of tank of type $k \in K_{\text{tank}}$
$Cost_{\text{tank}}(k)$	Tank cost for type $k \in K_{\text{tank}}$
$Q_{\text{daily}}(i)$	Daily water need in slum i
$\mathbb{M}$	Sum of daily water needs over all slums, $\mathbb{M} = \sum_{i \in N} Q_{\text{daily}}(i)$
N_{Q^3}	Grid point set for linearization of cubic volume flow
$Q_{\text{lin}}(m)$	Volume flow at linearization grid point $m \in N_{Q^3}$
$Q_{\text{lin}}^3(m)$	Cubic volume flow at linearization grid point $m \in N_{Q^3}$
$B_{\text{pipe}}(i, j, k)$	Binary indicator if connection on (i, j) is forbidden for pipe of type k due to geographic barriers
$B_{\text{truck}}(i, j, k)$	Binary indicator if connection on (i, j) is forbidden for truck of type k due to geographic barriers

$$\sum_{k \in K_{\text{pipe}}} Capa_{\text{pipe}}(k) \cdot x_{\text{pipe}}(i, j, k) \geq Q_P(i, j) \qquad \forall i \in N_w, j \in N \quad (2)$$

$$\sum_{k \in K_{\text{truck}}} Capa_{\text{truck}}(k) \cdot x_{\text{truck}}(i, j, k) \geq Q_T(i, j) \qquad \forall i \in N_w, j \in N \quad (3)$$

Pipes and trucks can only be chosen if connection between the slums i and j is used. Number of trucks per edge is limited by $N_{\text{truck}}^{\max}$ and only one pipe is allowed:

$$\sum_{k \in K_T} x_{\text{truck}}(i, j, k) + \sum_{k \in K_P} x_{\text{pipe}}(i, j, k) \geq x_{\text{use}}(i, j) \qquad \forall i \in N_w, j \in N \quad (4)$$

$$\sum_{k \in K_P} x_{\text{pipe}}(i, j, k) \leq x_{\text{use}}(i, j) \qquad \forall i \in N_w, j \in N \quad (5)$$

$$\sum_{k \in K_T} x_{\text{truck}}(i, j, k) \leq x_{\text{use}}(i, j) \cdot N_{\text{truck}}^{\max} \qquad \forall i \in N_w, j \in N \quad (6)$$

Table 2 Decision variables of the mixed-integer linear problem

Variable	Description
$x_{\text{use}}(i, j)$	Binary indicator if any connection is chosen on (i, j)
$x_{\text{pipe}}(i, j, k)$	Binary indicator if pipe of type k is chosen on (i, j)
$x_{\text{truck}}(i, j, k)$	Number of trucks of type k chosen on (i, j)
$x_{\text{tank}}(i, k)$	Number of tanks of type k chosen in i
$Q(i, j)$	Total volume flow on (i, j)
$Q_P(i, j)$	Volume flow in pipes on (i, j)
$Q_T(i, j)$	Volume flow carried by trucks on (i, j)
$Q_{\text{cubic}}(i, j)$	Cubic volume flow approximated by piecewise linearization on (i, j) with $Q(i, j)^3 \approx Q_{\text{cubic}}(i, j)$
$Q_{T_{\text{out}}}(i)$	Volume flow carried out of i by trucks
$Q_{T_{\text{in}}P_{\text{out}}}(i)$	Volume flow carried into i by trucks and out of i via pipes
$Q_{T_{\text{in}}\text{daily}}(i)$	Volume flow carried into i by trucks and used in slum i
$Q_{P_{\text{in}}!P_{\text{out}}}(i)$	Volume flow carried into i via pipe but not out of i via pipe
$\lambda_{P_{\text{in}}!P_{\text{out}}}(i)$	Binary auxiliary variable for modeling maximum relation
$\lambda_{Q_{\text{lin}}}(i, j, k, m)$	Auxiliary variable $\in [0, 1]$ of grid point $m \in N_{Q^3}\setminus\{1\}$ to linearize cubic volume flow on (i, j) for pipe type k
$z_{Q_{\text{lin}}}(i, j, k, m)$	Binary auxiliary variable for grid point $m \in N_{Q^3}$ to linearize cubic volume flow on (i, j) for pipe type k

The costs in the objective function scale cubical with the volume flow and are hence modeled with constraints of an incremental linearization method:
$\forall i \in N_w, j \in N, k \in K_P$:

$$Q_P(i, j) = Q_{\text{lin}}(1) + \sum_{m=2}^{|N_{Q^3}|} (Q_{\text{lin}}(m) - Q_{\text{lin}}(m-1))\lambda_{Q_{\text{lin}}}(i, j, k, m) \quad (7)$$

$$Q_{\text{cubic}}(i, j) = Q_{\text{lin}}^3(1) + \sum_{m=2}^{|N_{Q^3}|} (Q_{\text{lin}}^3(m) - Q_{\text{lin}}^3(m-1))\lambda_{Q_{\text{lin}}}(i, j, k, m) \quad (8)$$

$\forall i \in N_w, j \in N, k \in K_P, m \in N_{Q^3}\setminus\{1\}$:

$$\lambda_{Q_{\text{lin}}}(i, j, k, m) \leq x_{\text{pipe}}(i, j, k) \quad (9)$$
$$\lambda_{Q_{\text{lin}}}(i, j, k, m) \geq z_{Q_{\text{lin}}}(i, j, k, m) \quad (10)$$
$$\lambda_{Q_{\text{lin}}}(i, j, k, m) \leq z_{Q_{\text{lin}}}(i, j, k, m-1) \quad (11)$$

A slum cannot supply itself, therefore loops are not allowed:

$$x_{\text{use}}(i, i) = 0 \quad \forall i \in N_w \tag{12}$$
$$x_{\text{pipe}}(i, i, k) = 0 \quad \forall i \in N_w, k \in K_P \tag{13}$$
$$x_{\text{truck}}(i, i, k) = 0 \quad \forall i \in N_w, k \in K_T \tag{14}$$

The flow condition requires that the incoming volume flow equals the sum of outgoing flow and daily need for each slum:

$$\sum_{j \in N_w} Q(j, i) = \sum_{j \in N} Q(i, j) + Q_{\text{daily}}(i) \quad \forall i \in N_w \tag{15}$$

Geographic barriers, like rivers, prevent specific connections. These information are inserted into the model as manual input for the individual connections. The compliance with these requirements is modeled in the following equations:

$$x_{\text{pipe}}(i, j, k) \leq 1 - B_{\text{pipe}}(i, j, k) \quad \forall i, j \in N_w, k \in K_P \tag{16}$$
$$x_{\text{truck}}(i, j, k) \leq 1 - B_{\text{truck}}(i, j, k) \quad \forall i, j \in N_w, k \in K_T \tag{17}$$

A tank is required in a slum if water is carried by trucks into the slum or out of the slum, i.e for the following four volume flows: (I) Water delivered by a truck, (I-a) for the daily need of the slum itself $Q_{T_{in}\text{daily}}$, (I-b) to be carried onward by truck $Q_{T_{in}T_{out}}$ or (I-c) to be carried onward by pipe $Q_{T_{in}P_{out}}$, and (II) water coming into the slum by pipe and then being carried onward by a truck ($Q_{P_{in}T_{out}}$). To reduce the overall tank capacity, the following prioritization logic is applied for water coming in by pipe $Q_{P_{in}}$: First continue by pipe $Q_{P_{in}P_{out}}$, the remaining water $Q_{P_{in}!P_{out}}$ is used for the slum itself $Q_{P_{in}\text{daily}}$. Still remaining water is further transported by truck $Q_{P_{in}T_{out}}$. This logic is represented by the following three relationships. Firstly, $Q_{P_{in}!P_{out}} = Q_{P_{in}} - Q_{P_{in}P_{out}} = \max\{0, Q_{P_{in}} - Q_{P_{out}}\}$ which is modeled by applying a BigM-Method [5]. $\forall i \in N$:

$$Q_{P_{in}!P_{out}}(i) \geq \sum_{j_1 \in N} Q_P(j_1, i) - \sum_{j_2 \in N} Q_P(i, j_2) \tag{18}$$

$$Q_{P_{in}!P_{out}}(i) \leq \sum_{j_1 \in N} Q_P(j_1, i) - \sum_{j_2 \in N} Q_P(i, j_2) + \mathbb{M} \cdot \lambda_{P_{in}!P_{out}}(i) \tag{19}$$

$$Q_{P_{in}!P_{out}}(i) \leq \mathbb{M} \cdot (1 - \lambda_{P_{in}!P_{out}}(i)) \tag{20}$$

Secondly, $Q_{T_{in}\text{daily}} = \max\{0, Q_{\text{daily}} - Q_{P_{in}!P_{out}}\}$ represented by:

$$Q_{T_{in}\text{daily}}(i) \geq Q_{\text{daily}}(i) - Q_{P_{in}!P_{out}}(i) \quad \forall i \in N \tag{21}$$

Thirdly, $Q_{T_{in}P_{out}} = Q_{P_{out}} - Q_{P_{in}P_{out}} = \max\{0, Q_{P_{out}} - Q_{P_{in}}\}$ which is modeled in the following equation:

$$Q_{T_{in}P_{out}}(i) \geq \sum_{j_1 \in N} Q_P(i, j_1) - \sum_{j_2 \in N} Q_P(j_2, i) \quad \forall i \in N \tag{22}$$

For the second and third equation, only the lower bounds of the maximum relation need to be modeled since the variables are automatically pushed down by the objective function. Finally, $Q_{T_{out}}$ is calculated by:

$$Q_{T_{out}}(i) = \sum_{j \in N_w} Q_T(i, j) \quad \forall i \in N \tag{23}$$

Based on these calculations, the tank capacity requirement in each slum, which can be provided by a number of tanks, is given by $\forall i \in N$:

$$\sum_{k=1}^{K_{tank}} x_{\text{tank}}(i, k) \cdot Capa_{tank}(k) \geq Q_{T_{out}}(i) + Q_{T_{in}P_{out}}(i) + Q_{T_{in}daily}(i) \tag{24}$$

Additional constraints set all volume flow variables to zero if the corresponding edge is not used:

$$Q_{T_{in}daily}(i) \leq \sum_{j \in N} \sum_{k \in \{1,\ldots,K_{\text{truck}}\}} x_{\text{truck}}(j, i, k) \cdot \mathbb{M} \quad \forall i \in N \tag{25}$$

$$Q_{P_{in}!P_{out}}(i) \leq \sum_{j \in N} \sum_{k \in \{1,\ldots,K_{\text{pipe}}\}} x_{\text{pipe}}(j, i, k) \cdot \mathbb{M} \quad \forall i \in N \tag{26}$$

$$Q_{T_{in}P_{out}}(i) \leq \sum_{j \in N} \sum_{k \in \{1,\ldots,K_{\text{pipe}}\}} x_{\text{pipe}}(i, j, k) \cdot \mathbb{M} \quad \forall i \in N \tag{27}$$

4 Results and Conclusion

For illustration, we applied our approach to a slum cluster in Dhaka, the capital of Bangladesh and one of the world's most populated cities.

The classification of slums within the city area is based on remote sensing via quickbird satellite images with a resolution of 0.6 m from the year 2010 [6]. The optimization problem was modeled in GMPL and solved with the IBM optimization software CPLEX. The results show the optimal supply systems for two different instances. In the original problem without barriers in Fig. 1, a network of only truck connections was chosen for the supply of 14 slums.

Fig. 1 Optimization result for a slum cluster in Dhaka, Bangladesh, with 14 slums

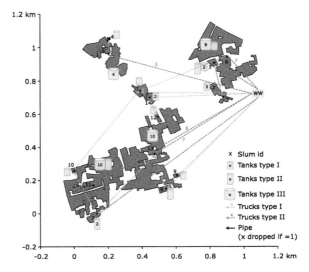

Fig. 2 Optimization result with additional consideration of geographic barriers

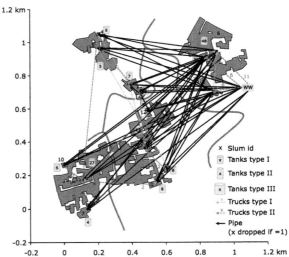

In Fig. 2, additional constraints blocking specific truck connections were introduced to simulate natural barriers, such as rivers. The result is a very entangled network, which however adheres to the barrier restriction, but would be reconsidered from an engineer perspective due to its complexity. Currently, the run time is too long to solve the model for a whole city. For instance, this optimization with 14 slums took a day whereas a large city, like Sao Paulo, can have up to 2000 slums.

Therefore, two adaptions of the approach employing primal heuristics are in development. Firstly, clustering the slums to split the master problem into various sub-problems, which then can individually be solved with the current optimization model. This adaption compromises between loosing global optimality and being able to solve larger instances. Secondly, using a minimal spanning tree for the network as start solution for the optimization and employing specialized graph algorithms, e.g. known from the field of large-scale logistic networks [7].

Acknowledgements The authors thank the German Research Foundation (DFG) for funding this research within the Collaborative Research Center SFB 805 "Control of Uncertainties in Load-Carrying Structures in Mechanical Engineering" as well as the KSB Foundation.

References

1. UN-HABITAT: World Cities Report 2016 - Urbanization and Development. http://unhabitat.org/books/world-cities-report/.
2. Wissenschaftlicher Beirat der Bundesregierung - Globale Umweltveränderungen (WBGU): Der Umzug der Menschheit - Die transformative Kraft der Städte (2016). http://www.wbgu.de/hauptgutachten/
3. Taubenböck, H., Esch, T., et al. (2012). Monitoring urbanization in mega cities from space. *Remote Sensing of Environment, 117*, 162–176.
4. Friesen, J., Rausch, L., & Pelz, P. F. (2017). Providing water for the poor - towards optimal water supply infrastructures for informal settlements by using remote sensing data. In *Joint Urban Remote Sensing Event (JURSE), Conference Proceedings*.
5. Suhl, L., & Mellouli, T. (2013). *Optimierungssysteme - Modelle, Verfahren, Software, Anwendungen*. Berlin: Springer.
6. Gruebner, O., Sachs, J., et al. (2014). Mapping the slums of Dhaka from 2006 to 2010. *Dataset Papers in Science*. https://www.hindawi.com/journals/dpis/2014/172182/
7. Matuschke, J. (2014). Network flows and network design in theory and practice. TU Berlin. https://depositonce.tu-berlin.de/bitstream/11303/4265/1/matuschke_jannik.pdf.

Polyhedral 3D Models for Compressors in Gas Networks

Tom Walther, Benjamin Hiller and René Saitenmacher

In gas networks, compressors are used to increase the pressure of the incoming gas to a higher outflow pressure, thus counteracting the pressure loss caused by friction in pipes. This allows for the gas to be transported over long distances. A *compressor machine* (or compressor, for short) is powered by an associated *compressor drive*. The technical models for compressors and drives are highly nonlinear [1, 5, 7, 8]. As optimization models for gas networks usually involve switching compressors, this leads to hard-to-solve MINLPs. It is thus desirable to use simpler (i.e. polyhedral) yet accurate models for a compressor.

In this paper, we construct a polyhedral model for the operating range of a compressor machine in the three-dimensional space (q, p^{in}, p^{out}) of mass flow rate, inlet and outlet pressure. We will closely follow the steps as in [2] and analyse some of the assumptions that are made therein. In contrast to the construction in [2], we are considering technical restrictions from the drive and a non-constant compressibility factor. Moreover, we suggest a complexity-reducing postprocessing algorithm and provide computational results based on publicly available compressor data from the GASLIB [3].

For an overview on gas network optimization problems and the modeling of compressors we refer to [4, 6] and the references therein.

The authors thank the DFG for their support within project A04 in CRC TRR154 and the BMBF Research Campus Modal (fund number 05M14ZAM).

T. Walther · B. Hiller (✉) · R. Saitenmacher
Zuse Institute Berlin, 14195 Berlin, Germany
e-mail: hiller@zib.de

T. Walther
e-mail: walther@zib.de

R. Saitenmacher
e-mail: saitenmacher@zib.de

1 Physical Compressor Model

For every compressor machine $m \in \mathcal{M}$, our starting point of interest is a flow and pressure tuple $(q_m, p_m^{\text{in}}, p_m^{\text{out}})$. An accurate modelling of a compressor machine involves many nonlinear and nonconvex constraints as well as several physical variables and quantities [1, 5, 8]. An overview of the quantities that we consider constant or a variable together with their units is given in Table 1.

The physical and technical capabilities of m are given in a so-called characteristic diagram $\mathcal{D}_m$ in the space of (Q, H_{ad}). The physical model of a turbo compressor machine that we are using as a reference is given as follows [1, 7]:

$$z = z(p_m^{\text{in}}; T) \tag{1a}$$

$$Q = q_m R_s T z (p_m^{\text{in}})^{-1} \tag{1b}$$

$$H_{\text{ad}} = R_s T z \frac{\kappa}{\kappa - 1} \left[\left(\frac{p_m^{\text{out}}}{p_m^{\text{in}}} \right)^{\frac{\kappa-1}{\kappa}} - 1 \right] \tag{1c}$$

$$(Q, H_{\text{ad}}) \in \mathcal{D}_m \tag{1d}$$

$$n \in [n^{\min}, n^{\max}] \tag{1e}$$

$$H_{\text{ad}} = \chi(Q, n; A^{\text{speed}}) \tag{1f}$$

$$\eta_{\text{ad}} = \chi(Q, n; A^{\text{eff}}) \tag{1g}$$

$$P = q H_{\text{ad}} \eta_{\text{ad}}^{-1} \tag{1h}$$

$$P \leq \chi(n, T_{\text{amb}}; A^{\text{power}}) \tag{1i}$$

Table 1 General physical quantities and constants

Physical and model constants		
Temperature	T	[K]
Ambient temperature	T_{amb}	[K]
Isentropic exponent	κ	[−]
Specific gas constant	R_s	[kJ/(kg · K)]
Gas- and compressor-specific physical variables		
Pressure	p	[bar]
Mass flow rate	q	[kg/s]
Compressibility factor	z	[−]
Volumetric flow rate	Q	[m³/s]
Adiabatic head	H_{ad}	[kJ/kg]
Compressor speed	n	[rot./min]
Adiabatic efficiency	η_{ad}	[−]
Power	P	[kW]

In constraint (1a), the compressibility factor z can be computed according to different formulas, see Sect. 2.2. The constraints (1a), (1b) and (1c) relate the problem variables $(q_m, p_m^{\text{in}}, p_m^{\text{out}})$ to the characteristic diagram variables (Q, H_{ad}). The constraints (1d) typically comprise a set of quadratic and possibly nonconvex inequalities. A restriction on the available power for compression as induced by the compressor drive is given by (1h) and (1i). In (1f), (1g) and (1i), $\chi(\cdot, \cdot; A)$ denotes a biquadratic function with some coefficient matrix $A \in \mathbb{R}^{3 \times 3}$.

2 Reformulated Compressor Model in $(q_m, p_m^{\text{in}}, p_m^{\text{out}})$

The physical model is highly nonlinear, making it hard to find globally optimal solutions for large-scale gas network optimization problems. Moreover, in most cases, we are not directly interested in the values of most compressor-specific quantities of the physical model, which motivates the construction of a less complex compressor model as in Sect. 7.3.4 in [2]. Starting with a compressor machine $m \in \mathcal{M}$, every point (Q, H_{ad}) of its characteristic diagram can be transformed into a (curved) ray in the space of $(q_m, p_m^{\text{in}}, p_m^{\text{out}})$ by inverting the Eq. (1b) and (1c). There is one degree of freedom in this transformation, denoted by p:

$$\begin{pmatrix} q \\ p^{\text{in}} \\ p^{\text{out}} \end{pmatrix} = g(Q, H_{\text{ad}}; p) = \begin{pmatrix} \frac{Qp}{R_s T z} \\ p \\ \left(\frac{H_{\text{ad}}}{R_s T \frac{\kappa}{\kappa-1}} + 1 \right)^{\frac{\kappa}{\kappa-1}} p \end{pmatrix} \in \mathbb{R}^3 \quad \text{for } p \geq 0. \quad (2)$$

Practically, in order to obtain a polyhedral approximation of the operating range in $(q_m, p_m^{\text{in}}, p_m^{\text{out}})$, we apply (2) to a set of sample points within the characteristic diagram and on its boundary for a set of different values of p. This yields a set of points $\{q_m^k, p_m^{\text{in},k}, p_m^{\text{out},k}\}$. From this set, we remove all points that violate some of the technical bounds $p_m^{\text{in}} \geq p^{\text{in,min}}$, $p_m^{\text{out}} \leq p^{\text{out,max}}$, or $q_m^{\text{min}} \leq q_m \leq q_m^{\text{max}}$.

2.1 Impact of Restricted Compressor Power

As mentioned before, the power for the compression process is provided by a compressor drive. This power ist limited by an upper bound on the power that depends on the compressor speed n and the ambient air temperature T_{amb} [1]. The lower the ambient temperatures, the more power can be provided. In order to account for this power bound, we compute the required power P for every point of our set $\{q_m^k, p_m^{\text{in},k}, p_m^{\text{out},k}\}$ according to (1h) and discard all points that violate (1i). The convex hull $\mathcal{P}_m$ of the remaining points yields the desired representation.

In Table 2, we show the impact of the power bound on our operating ranges in the space of $(q_m, p_m^{\text{in}}, p_m^{\text{out}})$, based on a set of 5000 sampled instances on compressor

Table 2 Percentage of instances feasible for the compressor machine but infeasible due to the power bound of the compressor drive for different air temperatures

Compressor	Cold (%)	Warm (%)	Hot (%)
m_1	2.71	5.66	9.77
m_2	1.85	3.97	5.83
m_3	2.50	5.55	9.43
m_4	1.86	3.98	5.90
m_5	2.43	5.37	9.33
m_6	1.85	3.97	5.83

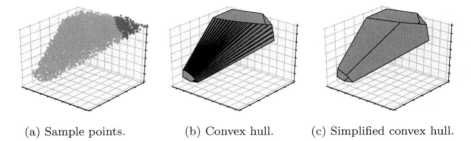

(a) Sample points.   (b) Convex hull.   (c) Simplified convex hull.

Fig. 1 Construction of the operating range of a compressor machine

data taken from the GASLIB [3]. It can be seen that roughly 2–10% of the feasible instances for a compressor machine are rendered infeasible by the compressor drive restrictions. Typically, these instances are characterized by high throughput q_m and high inlet pressure p_m^{in}. Figure 1a shows a set of feasible and infeasible sample points.

2.2 Impact of Variable Compressibility Factor

There exist several formulas for computing the compressibility factor z. Due to its accuracy for the pressure range that we are considering, we take the formula of PAPAY [1] as our reference:

$$z(p, T) = 1 - 3.52 \frac{p}{p_c} e^{-2.26 \frac{T}{T_c}} + 0.247 \left(\frac{p}{p_c} \right)^2 e^{-1.878 \frac{T}{T_c}}. \tag{3}$$

p_c and T_c denote the pseudocritical pressure and temperature of the gas, respectively, that are constant in our case. Since we also assume T to be constant, (3) reduces to a quadratic equation in the pressure p.

Table 3 Percentage of false-positive/-negative instances for different z formulas

Compressor	PAPAY		AGA		const $z = 0.9$	
	False-pos (%)	False-neg (%)	False-pos (%)	False-neg (%)	False-pos (%)	False-neg (%)
m_1	8.62	0.00	12.22	0.03	5.66	5.52
m_2	10.30	0.00	11.56	0.00	6.30	3.29
m_3	9.02	0.00	11.52	0.00	5.64	6.64
m_4	9.24	0.00	11.46	0.00	5.92	3.42
m_5	8.72	0.00	12.62	0.03	5.52	5.51
m_6	9.18	0.00	11.62	0.02	6.26	3.13

Another formula, which is linear in p and suitable for pressure values up to 70 bar, has been proposed by the AMERICAN GAS ASSOCIATION (AGA) [1]:

$$z(p, T) = 1 + 0.257 \frac{p}{p_c} - 0.533 \frac{p}{p_c} \frac{T_c}{T}. \qquad (4)$$

As a third and most simple alternative, used also in [2], the compressibility factor can also be considered constant with a value around $z = 0.9$.

Using the different z-factor formulas for the construction of our operating ranges leads to different polytopes in $(q_m, p_m^{in}, p_m^{out})$. There are errors in two directions: the polytopes may contain points that are infeasible for the physical model (false-positive; due to working with convex hulls), but they may also exclude feasible points (false-negative; due to using (4) or a constant value for z as compared to our reference formula (3)). Again, we have sampled 5000 instances and determined the percentage of false-positive and false-negative feasibility outcomes for the different z-factor computation variants. The results are shown in Table 3. It can be seen that the polytope obtained from using a constant compressibility factor contains the least percentage of instances that are technically infeasible. On the other hand, it also excludes some amount of technically feasible instances, which is not the case for AGA and PAPAY. Moreover, it turns out that the polyhedral approximation of the more exact PAPAY formula is better than the one for the AGA formula.

2.3 Reduction of Polytope Facets

The resulting polytope becomes more precise for a large set of sampling points in (Q, H_{ad}) and many values of p. As a downside, the convex hull representation gets more complex, i.e., the number of its vertices and facets sharply increases, blowing up the model formulation. Therefore, we propose Algorithm 1 to reduce the number of polytope facets until a given volume error tolerance is reached.

Table 4 Computational results of facet reduction algorithm

Compressor	(Original)	$\tau = 0.1$		$\tau = 0.01$		$\tau = 0.001$	
	Facets	Facets	Time[s]	Facets	Time[s]	Facets	Time[s]
m_1	256	7	5.20	9	9.71	23	128.56
m_2	286	6	2.75	8	6.51	23	102.48
m_3	259	6	3.37	10	12.71	16	50.96
m_4	284	6	4.39	9	9.67	23	109.44
m_5	261	6	3.80	9	9.60	18	74.32
m_6	284	6	3.40	9	10.59	21	81.96

Tentatively trying different halfspaces is the most computationally expensive part of the algorithm. Hence, if the number of facets of the input polytope $\mathcal{P}$ is large, it may be advantageous to only consider a (random) subset of all facets in every iteration. Our experience has shown that usually very few facets suffice to approximate any given polytope $\mathcal{P}$ with a volume error of only 1% in reasonable time. Some results and computation times for our GASLIB compressors are shown in the following Table 4.

Algorithm 1: Polytope Facet Reduction$(\mathcal{P}, \tau)$

Input: A polytope $\mathcal{P}$ and a volume error tolerance $\tau \geq 0$.
Output: A polytope $\mathcal{Q} \supseteq \mathcal{P}$ with $\frac{\text{vol}(\mathcal{Q})}{\text{vol}(\mathcal{P})} - 1 \leq \tau$ and less facets than $\mathcal{P}$.

Compute $\text{vol}(\mathcal{P})$.
Let $\mathcal{F}$ be the set of facets of $\mathcal{P}$ (given as halfspace inequalities).
Initialize $\mathcal{Q}$ to be some box around $\mathcal{P}$.
while $\frac{\text{vol}(\mathcal{Q})}{\text{vol}(\mathcal{P})} - 1 > \tau$ **do**
$\quad f^* := \arg\min_{f \in \mathcal{F}} \text{vol}(\mathcal{Q} \cap f)$
$\quad \mathcal{Q} := \mathcal{Q} \cap f^*$
end
Return $\mathcal{Q}$.

---

3 Conclusion

We have presented a method to construct a small yet quite accurate 3D polyhedral model for a compressor machine that approximates the true operating range as given by a nonlinear physical reference model. We computationally quantified the impact of neglecting the drives power restrictions as well as the impact of the choice of the formula for computing the compressibility factor. It turns out that the polyhedral model approximates the true nonlinear operating range within an error of roughly 10%, which is acceptable for many applications.

References

1. Fügenschuh, A., Geißler, B., Gollmer, R., Morsi, A., Pfetsch, M. E., Rövekamp, J., et al. Physical and technical fundamentals of gas networks. In T. Koch, et al. [4].
2. Humpola, J., Fügenschuh, A., Hiller, B., Koch, T., Lehmann, T., Lenz, R., et al. The specialized MINLP approach. In T. Koch, et al. [4].
3. Humpola, J., Joormann, I., Kanelakis, N., Oucherif, D., Pfetsch, M.E., Schewe, L., et al. (2017). GasLib – a library of gas network instances. Report, optimization online. http://www.optimization-online.org/DB_HTML/2015/11/5216.html.
4. Koch, T., Hiller, B., Pfetsch, M., & Schewe, L. (Eds.). (2001). *Evaluating gas network capacities*. MOS-SIAM series on optimization. Philadelphia: SIAM.
5. Králik, J. (1993). Compressor stations in SIMONE. In *Proceedings of 2nd International Workshop SIMONE on Innovative Approaches to Modeling and Optimal Control of Large Scale Pipeline Networks* (pp. 93–117). Prague.
6. Ríos-Mercado, R. Z., & Borraz-Sánchez, C. (2015). Optimization problems in natural gas transportation systems: A state-of-the-art review. *Applied Energy, 147*, 536–555.
7. Rose, D., Schmidt, M., Steinbach, M. C., & Willert, B. M. (2016). Computational optimization of gas compressor stations: MINLP models versus continuous reformulations. *Mathematical Methods of Operations Research, 83*(3), 409–444.
8. Traupel, W. (2001). *Thermische Turbomaschinen*. Berlin: Springer.

Part XV
Pricing and Revenue Management

Stochastic Dynamic Multi-product Pricing Under Competition

Rainer Schlosser

1 Introduction

In lots of markets, sellers have to deal with competition. Typically, there is limited demand information and market dynamics are unknown. To maximize their expected profits, sellers are required to constantly decide on prices for multiple products. Although, in e-commerce it has become easy to observe competitors' prices and to adjust prices, it is challenging to estimate demand and to compute intelligent pricing strategies. Decision-making needs to take into account competitors' prices as well as substitution effects between a firm's own products. Applications can be found in a variety of contexts that involve perishable (e.g., fashion goods, seasonal products, event tickets) as well as durable goods (e.g., technical devices, licenses, natural resources).

In this paper, we study competitive multi-product pricing models in a stochastic dynamic framework. We focus on durable goods. Our aim is to deal with the following assumptions: (i) limited demand information, (ii) unknown competitors' strategies, and (iii) substitution effects in demand. To compute robust pricing strategies in competitive settings, we use data-driven demand estimations and a dynamic programming model that circumvents the curse of dimensionality.

The best way to sell products is a classical application of revenue management theory. The problem is closely related to the field of dynamic pricing, see, e.g., Talluri and van Ryzin [9]. The literature on stochastic dynamic pricing strategies that incorporate multi-product settings is limited. Such models are characterized by mutual dependent demand intensities and hence, are much more complex than single-product models.

The survey by Chen [3] provides an excellent overview of recent multi-product models under competition. For the finite horizon case, monopolistic multi-product

---

R. Schlosser (✉)
Hasso Plattner Institute, Potsdam, Germany
e-mail: rainer.schlosser@hpi.de

models with fairly general demand functions have been analyzed in studies by Gallego and van Ryzin [5] and Maglaras and Meissner [7]. In order to take the substitutability of (horizontally or vertically differentiated) products into account, in recent literature customer choice models are used. In case of horizontally differentiated products commonly multinomial logit (e.g., [1]) or nested logit models (e.g., [6]) are applied.

In most existing models, the demand intensity is assumed to be known. Dynamic pricing competition models with limited demand information are analyzed by, e.g., Adida and Perakis [2] or Chung et al. [4] using robust optimization and learning approaches. For a more comprehensive review, we refer to Chen [3].

In contrast to the assumptions of most papers, in real-life applications, specific information is not observable: sales dynamics or price reactions are typically unknown, and customers as well as sellers might not act rational. Moreover, when dealing with dynamic pricing competition models, the most critical problem is their high complexity and the size of the state space (curse of dimensionality). Hence, most solution approaches are just applicable if highly stylized assumptions can be verified and the number of competitors is small. To our knowledge, there is a lack of publications providing applicable solutions for real-life multi-product models under dynamic pricing competition.

The main contribution of this paper is twofold. We (i) present a data-driven approach to measure substitution effects and to predict sales intensities, and (ii) we derive effective pricing strategies that are even applicable when the number of competitors is large and their strategies are unknown.

This paper is organized as follows. In Sect. 2, we describe the stochastic dynamic multi-product model for a fairly general setting. We allow sales probabilities to depend on our prices, on competitors' prices as well as on time (e.g., due to seasonal effects). In Sect. 3, we show how observable market data can be analyzed in order to estimate sales probabilities for various market situations. In Sect. 4, we derive our feedback pricing heuristic to be applied in competitive markets. Based on estimated sales probabilities, we set up a dynamic model and demonstrate how to compute powerful feedback pricing heuristics. As computation times are small, our heuristic is able to quickly react to changing market environments. Final conclusions are summarized in Sect. 5.

2 Model Description

We consider the situation in which a firm seeks to sell different types of goods (e.g., books) on a digital market platform (e.g., Amazon or eBay). We assume several competitors for our products. In our model, we include substitution effects in demand as customers might compare prices. The number of different types of products is denoted by J, $J < \infty$. We assume that there are no inventory restrictions, i.e., items can be reproduced or reordered. If a sale takes place shipping costs c have to be paid, $c \geq 0$. A sale of one item at price $a^{(j)}$ leads to profit of $a^{(j)} - c$, $j = 1, \ldots, J$. Since in real-life applications prices cannot be adjusted arbitrarily often, we use a discrete time model. For the length of one period, we use the discount factor δ, $0 < \delta < 1$. The time horizon is assumed to be infinite.

The sales intensity of product j is denoted by $\lambda^{(j)}$, $j = 1, \ldots, J$. Due to customer choice, the sales intensity particularly will depend on our offer prices $\mathbf{a} = (a^{(1)}, \ldots, a^{(J)})$ and the competitors' prices $\mathbf{p} = (p^{(1)}, \ldots, p^{(I)})$ of I potential substitutes. Moreover, we allow the sales intensity also to depend on time, e.g., the time of the day or the weekday. We assume that the time-dependence is periodic and has a finite cycle length of L periods. In our model, the sales intensity is a general function of our offer prices $\mathbf{a}$ and the current market situation characterized by the current competitors' prices $\mathbf{p}$. Given such a market situation $\mathbf{p}$ in period t, we consider the sales intensity of product j, $a^{(j)} \geq 0$, $j = 1, \ldots, J$, $t = 0, 1, 2, \ldots$,

$$\lambda_t^{(j)}(\mathbf{a}, \mathbf{p}) = \lambda_{t \bmod L}^{(j)}(\mathbf{a}, \mathbf{p}). \tag{1}$$

With loss of generality, we assume sales probabilities (for one period) to be Poisson distributed. I.e., the probability to sell exactly k items of product j in a specific market situation $\mathbf{p}$ is given by $P_t^{(j)}(k, \mathbf{a}, \mathbf{p}) := \frac{\lambda_t^{(j)}(\mathbf{a},\mathbf{p})^k}{k!} \cdot e^{-\lambda_t^{(j)}(\mathbf{a},\mathbf{p})}$, $a^{(j)} \geq 0$, $j = 1, \ldots, J$, $p^{(i)} \geq 0$, $i = 1, \ldots, I$, $k = 0, 1, 2, \ldots$. For each period t, a price $a_t^{(j)}$ has to be chosen, $j = 1, \ldots, J$. We call strategies $(\mathbf{a}_t)_t$ admissible if they belong to the class of Markovian feedback policies; i.e., pricing decisions $a_t^{(j)} \geq 0$ may depend on time and the current market situation $\mathbf{p}$, which contains the prices of the competitors. The set of admissible prices is denoted by A.

By $X_t^{(j)}$ we denote the random number of sales of product j in period t, $j = 1, \ldots, J$. Depending on the chosen pricing strategy $(\mathbf{a}_t)_t$, the random accumulated profit from time t on (discounted on time t) amounts to, $t = 0, 1, 2, \ldots$,

$$G_t := \sum_{s=t}^{\infty} \delta^{s-t} \cdot \sum_{j=1}^{J} (a_s^{(j)} - c) \cdot X_s^{(j)}. \tag{2}$$

The objective is to determine a non-anticipating (Markovian) feedback pricing policy that maximizes the expected total profit $E\left(G_0 \mid \mathbf{p}_0\right)$, where $\mathbf{p}_0$ denotes the initial market situation in $t = 0$. In Sect. 4, we will solve dynamic pricing problems that are related to (1)–(2). In the next section, we show how sales probabilities can be estimated in competitive markets with incomplete demand information.

3 Estimation of Substitution Effects and Sales Probabilities

The goal of this section is to estimate sales probabilities from market data. In competitive online markets, competitors' prices are typically observable. The sales data, however, is a firm's private knowledge. The idea is to ascribe the number of realized sales (of our firm) within different time intervals to the relation of our offer prices and the competitors' prices (observed at the beginning of the corresponding intervals). We assume that there is data for B time intervals and J products of our firm as well as I different products offered by (one or more) competitors, $j = 1, \ldots, J$, $i = 1, \ldots, I, t = 0, 1, \ldots, B - 1$. In our framework, the availability of a product is indicated by a positive offer price. Data is supposed to consist of competitor prices $p_t^{(i)}$, $i = 1, \ldots, I$, our prices $a_t^{(j)}$, $j = 1, \ldots, J$, and the realized sales $y_t^{(j)}$, i.e., the number of products sold of type j in period t, i.e., $(t, t + 1)$.

In the following, we show how to estimate sales intensities (for specific market situations) that can be applied in our dynamic model, cf. Sect. 2. To explain the dependent variable $y_t^{(j)}$, $t = 0, \ldots, B - 1$, $j = 1, \ldots, J$, for product j, we use a simple least squares regression model (OLS model). Following the OLS model, we aim to specify the sales intensities, $j = 1, \ldots, J$,

$$\lambda_t^{(j)}(\mathbf{a}, \mathbf{p}; \boldsymbol{\beta}^{(j)}) := \mathbf{x}_t^{(j)}(\mathbf{a}, \mathbf{p})' \boldsymbol{\beta}^{(j)}, \tag{3}$$

where $\boldsymbol{\beta}^{(j)} = (\beta_1^{(j)}, \ldots, \beta_M^{(j)})$ is the unknown parameter vector that is associated to the vector $\mathbf{x}_t^{(j)} = (x_{t,1}^{(j)}, \ldots, x_{t,M}^{(j)})$ of M explanatory variables. The regressors $\mathbf{x}_t^{(j)}(\mathbf{a}, \mathbf{p})$ can be a function of time t, the prices $\mathbf{a}$, and the market situation $\mathbf{p}$. For each product j the optimal coefficients $\boldsymbol{\beta}^{(j)*} = (\beta_1^{(j)*}, \ldots, \beta_M^{(j)*})$, $j = 1, \ldots, J$, can be easily obtained using standard methods. Finally, the resulting intensities $\lambda_t^{(j)*}$, cf. (3), are used to estimate sales probabilities: For one period, we let $P_t^{(j)}(\cdot, \mathbf{a}, \mathbf{p})$ be Poisson distributed with rate $\lambda_t^{(j)}(\mathbf{a}, \mathbf{p}; \boldsymbol{\beta}^{(j)*})$. Note, the time dependence of $\lambda^{(j)}$ can be captured by time-dependent explanatory variables. To illustrate the approach, in the following definition, we give simple examples of explanatory variables.

Definition 1 We define the following regressors (besides the intercept $x_{t,1}^{(j)} = 1$):

$x_{t,2}^{(j)}(\mathbf{a}, \mathbf{p}) := a_t^{(j)} - \min_{k=1,\ldots,J, k \neq j} \{a_t^{(k)}\}$ price gap between $a_t^{(j)}$ and best own price

$x_{t,3}^{(j)}(\mathbf{a}, \mathbf{p}) := a_t^{(j)} - \min_{i=1,\ldots,I} \{p_t^{(i)}\}$ price gap between $a_t^{(j)}$ and best competitor

$x_{t,3+k}^{(j)}(\mathbf{a}, \mathbf{p}) := 1_{\{a_t^{(k)} > 0\}}$ availability of our product k, $k = 1, \ldots, J$

$x_{t,3+J+l}^{(j)}(\mathbf{a}, \mathbf{p}) := 1_{\{t \bmod L = l\}}$ time-effect/seasonality l, $l = 0, \ldots, L - 1$

Our framework allows to measure substitution effects that may originate from competitors' products and our own products. Besides cross price elasticity effects also out-of-stock substitution effects can be taken into account. The goal of the next section is to find the best allocation of prices $a_t^{(j)}$ by taking all (mutual) substitution effects of our products as well as the competitors' products into account.

4 Dynamic Model and Heuristic Strategy

In this section, we derive heuristic pricing strategies. We circumvent the curse of dimensionality by using a separation approach that is based on sticky prices. Following the Bellman approach, the best expected future profits $E(G_t|\mathbf{p})$, cf. (2), are described by the value function $V_t(\mathbf{p})$ of the stochastic control problem. The time dependence in our model is assumed to be seasonal (daily/weekly effects) with a given cycle length of L periods. Hence, for all t, where $t \bmod L = k$, we have $P_t^{(j)}(i, \mathbf{a}, \mathbf{p}) = P_k^{(j)}(i, \mathbf{a}, \mathbf{p}), k = 0, 1, \ldots, L-1, i = 0, 1, \ldots,$ cf. Sect. 3. Since, we can assume $V_t(\mathbf{p}) = V_{t \bmod L}(\mathbf{p})$ for all t, we just have to determine the values $V_t(\mathbf{p})$, $t = 0, 1, \ldots, L-1$, which are characterized by the associated Bellman equation,

$$V_t(\mathbf{p}) = \max_{\mathbf{a} \in A^J} \left\{ \sum_{j=1}^{J} \sum_{i=0}^{\infty} P_t^{(j)}(i, \mathbf{a}, \mathbf{p}) \cdot (a^{(j)} - c) \cdot i + \delta \cdot V_{(t+1) \bmod L}(\mathbf{p}) \right\}. \quad (4)$$

The system (4) can be written as

$$V_t(\mathbf{p}) = \sum_{k=0}^{L-1} \delta^k \cdot \max_{\mathbf{a} \in A^J} \left\{ \sum_{j=1}^{J} \sum_{i=0}^{\infty} P_{(t+k) \bmod L}^{(j)}(i, \mathbf{a}, \mathbf{p}) \cdot (a^{(j)} - c) \cdot i \right\} + \delta^L \cdot V_t(\mathbf{p}) \quad (5)$$

and from (5) we finally obtain $V_t(\mathbf{p})$ in *explicit* form, $t = 0, 1, \ldots, L-1$,

$$V_t(\mathbf{p}) = \left(1 - \delta^L\right)^{-1} \cdot \sum_{k=0}^{L-1} \delta^k \cdot \max_{\mathbf{a} \in A^J} \left\{ \sum_{j=1}^{J} \sum_{i=0}^{\infty} P_{(t+k) \bmod L}^{(j)}(i, \mathbf{a}, \mathbf{p}) \cdot (a^{(j)} - c) \cdot i \right\}. \quad (6)$$

The associated pricing strategy $\mathbf{a}_t^*(\mathbf{p})$, $t = 0, 1, \ldots, L-1$, is determined by (6) and the arg max of (4). If prices are not uniquely determined, we choose the largest one. When the number of competitors is large, i.e., the market situation has many dimensions the state space can grow exponentially (cf. curse of dimensionality)! The advantage of our approach is that the value function does not need to be computed for all states $\mathbf{p}$ in advance. The value function and the associated pricing policy can be computed separately for specific market situations $\mathbf{p}$ (i.e., just when they occur).

As competitors' strategies are not known (which is usually the case), it is not possible to anticipate potential price reactions. However, by regularly adjusting our prices we are able to react immediately if market situations change as prices can be easily recomputed for new states. Thus, our approach makes it possible to derive applicable pricing strategies in competitive markets with a large number of competitors. Moreover, our approach can be extended to problems with (i) finite horizon or (ii) limited supply.

The performance of our strategy is promising, cf. Schlosser et al. [8], especially when short-term profits are maximized and competitors do not use high adjustment frequencies. Note, in some businesses it can be observed that the prices of many competitors' are kept constant over large time spans. Some suppliers do not adjust prices at all.

5 Conclusion

We analyzed stochastic dynamic infinite horizon multi-product oligopoly models characterized by realistic assumptions: (i) demand intensities are mutually dependent and not explicitly known, (ii) the competitors' offer prices can be observed, and (iii) the competitors' pricing strategies are unknown.

We combine private sales data with observable data of the competitors' offers to efficiently predict sales probabilities of multiple products in competitive markets. Based on such market data, various explanatory variables that capture substitution effects of our products as well as the competitors' products can be defined. Furthermore, our model allows inclusion of time-dependent effects in the demand.

Using estimated sales probabilities, we have set up a time-dependent dynamic model including discounting and shipping costs. We have shown how to compute powerful feedback pricing strategies, which are even applicable if the number of competitors' products is large. Our solution approach is characterized by a decomposition approach, in which only current market situation have to be considered.

The big advantage of our model is that the pricing strategy can depend on a large number of competitors' offers with multiple dimensions (price, quality, rating, shipping time, etc.). Our technique to estimate substitution effects in demand and to compute prices remains simple and is easy to implement, since the relevant state space solely consists of time. The approach is successfully applied on a large online marketplace, cf. Schlosser et al. [8].

References

1. Ackay, Y., Natarajan, H. P., & Xu, S. H. (2010). Joint dynamic pricing of multiple perishable products under consumer choice. *Management Science*, 56(8), 1345–1361.
2. Adida, E., & Perakis, G. (2010). Dynamic pricing and inventory control: Uncertainty and competition. *Operations Research*, 58(2), 289–302.
3. Chen, M., & Chen, Z.-L. (2015). Recent developments in dynamic pricing research: Multiple products, competition, and limited demand information. *Production and Operations Management*, 24(5), 704–731.
4. Chung, B. D., Li, J., Yao, T., Kwon, C., & Friesz, T. L. (2012). Demand learning and dynamic pricing under competition in a state-space framework. *IEEE Transactions on Engineering Management*, 59(2), 240–249.
5. Gallego, G., & van Ryzin, G. (1997). A multi-product dynamic pricing problem and its application to network yield management. *Operations Research*, 45(1), 24–41.
6. Gallego, G., & Wang, R. (2014). Multi-product optimization and competition under the nested logit model with product-differentiated price sensitivities. *Operations Research*, 62(2), 450–461.

7. Maglaras, C., & Meissner, J. (2006). Dynamic pricing strategies for multi-product revenue management problems. *Manufacturing and Service Operations Management, 8*, 136–148.
8. Schlosser, R., Boissier, M., Schober, A., Uflacker, M. (2016). How to survive dynamic pricing competition in e-commerce. In *Poster Proceedings of the 10th ACM Conference on Recommender Systems, Boston, MA, USA*.
9. Talluri, K. T., & van Ryzin, G. (2004). *The Theory and Practice of Revenue Management*. Boston: Kluver Academic Publishers.

Part XVI
Production and Operations Management

In-Line Sequencing in Automotive Production Plants—A Simulation Study

Marcel Lehmann and Heinrich Kuhn

1 Introduction

The continuous development of the automotive industry towards higher numbers of models, variants and equipment options cause a tremendous raise in the complexity from a manufacturing point of view [2, 6]. Additionally, this trend is enhanced by the recent transformation process in the automotive industry including electrification, autonomous driving and digitalization. Many car producers deal with the arising problem by using a stabilized production, which is present in different specifications and known under several names such as In-Line Vehicle Sequencing (ILVS) [5] or Pearl Necklace Concept (PNC) [3]. Core of this approach is a unaltered production sequence, which is the baseline for the entire manufacturing process, especially for the just in sequence (JIS) material flow. Even though the underlying principle seems to be simple at the first glance, the planning and realization of such facilities features a high complexity [4]. The transformation of an existing plant towards the Pearl Necklace Concept however seems to be even more demanding since many original equipment manufactures (OEM) are still struggling to adapt their high volume facilities to this concept [7]. From the lack of possibilities to measure and evaluate the achievable stability level of an existing production configuration arises the necessity to simulate a stable production within the framework of the current production setting. This article provides an introduction of the key aspects of the ILVS/PNC in the first section followed by the description of the case study, the modeling approach and a preview of the results. The last section presents the conclusion and further prospects.

M. Lehmann · H. Kuhn (✉)
Department of Operations, Catholic University of Eichstätt-Ingolstadt,
Auf der Schanz 49, 85049 Ingolstadt, Germany
e-mail: heinrich.kuhn@ku.de
URL: http://www.ku.de/wwf/pw/

M. Lehmann
e-mail: marcel.lehmann@ku.de

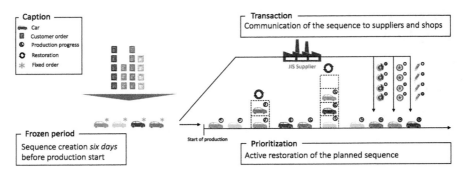

Fig. 1 Concept of a stable production (see also [6])

2 Stabilized Order Production System

Key point of a stabilized production is a number of customer orders, which is brought into a fixed production sequence several days before the beginning of the manufacturing process. The time span between the determination of the sequence and start of the production is called frozen period [7]. Once the sequence is frozen it is communicated towards the JIS suppliers, who are expected to provide the parts in exactly the same order [7]. The concept is visualized in Fig. 1.

Benefits. The consistent application of the ILVS system reduces the handling expenses since the material can be delivered directly to the assembly line from distant suppliers [7]. Otherwise the parts would have to be stored close to the assembly and brought into sequence by their employees based on the realized production sequence, which is available only a few hours before the start of the final assembly. Alternatively, suppliers would have to produce right next to the OEM. Both of these options result in higher logistics or production costs.

Challenges. Highly individualized cars, prototypes, quality samples, rework, down times and unsynchronized production steps cause deterministic and stochastic lead time extensions [7]. Therefore the original sequence changes during the production process since some cars overtake or fall behind. This phenomenon is called scrambling [5].

External factors cause additional sequence scrambling. Missing parts for example lead to detents, which means a car cannot be built properly and has to be locked down in the sequencer in front of the assembly line. Since other cars from the sequencer are used instead in order to maintain the production volume, the sequence is scrambled.

Arising research question. In order to pursue a stable manufacturing strategy a high and constant level of stability is required. The achievable level of stability can only be measured if the plant is currently operating under these premises. Since this is not the case in several existing plants the stability level has to be predicted. The model described in the next section is designed to answer the two arising research questions:

- What level of stability can be achieved assuming the current production control process?
- Which parameters influence the achievable stability level?

3 Modeling Approach

In order to understand the underlying data set the real world facility is described in the first section, followed by a short introduction to the data set. Afterwards the modeling approach is displayed.

3.1 Real World Production System

The observed plant is a fully appointed production side containing all four shop types, i.e., sheet-metal shop, body shop, paint shop and final assembly line. In front of the assembly line is a facultative buffer a so called AS/RS. For the current production system, the buffer fulfills four different functions. The first is to decouple volumes loses from previous production stages caused by machine down times. The second is to hold back bodies, that cannot be assembled properly, due to missing parts. Another aspect of this buffer is to compensate unmatched production volumes caused by divergent shifts and production rates. The last function and most influential in terms of sequence instability is to ensure a valid sequence. This means a sequence without any violations of car sequencing rules. Since a completely new sequence is created based on the available cars in the sequencer it differs heavily from the originally planned sequence.

3.2 Data Set

In order to get representative results a bipartite data set over half a year has been collected. The first part contains car body related production data. Since order and body are irrevocably matched from the start of production every body equates one customer order. Amendatory to the order data set exists another record with all detent information.

Each entry of the car data set represents one order consisting of four values. The first one is a unique serial number to distinguish all orders. The second value is the sequence number reflecting the position in the originally planned sequence. The last two values are so called time stamps, consisting of the exact date and time of a certain event during the production process. One value determines the entrance in the AS/RS, the other represents the exit time.

The second data record contains all relevant detent information. This information can be matched to a car body via the car serial number. One value defines the category of the detent e.g. logistic detent and the remaining two values define the lock down period. During this time a car, effected by this detent, is not allowed to leave the sequencer towards the final assembly line.

3.3 Modeling Methodology, Features and Assumptions

The underlying modeling technique is a discrete event based simulation, since this method fits well to the specifications of a car manufacturing plant and the derived real world data set [1]. Every event in this simulation is triggered by a real world event taken from the data sets. The stochastic of the production process is not modeled explicitly, but is enclosed in the input data. Therefore the sample needs a certain size to ensure explanatory power.

Since the purpose of this model is to reveal the achievable sequence stability under the current production and storage configurations the main difference to the real world setting is the sequencing logic. The conceptual model as shown in Fig. 2 consists of the three production sides and an AS/RS. The first area containing the body and paint shop is designed as black box, which means they are not explicitly modeled but their effects are recognizable in the used data set. In this case the information lies in the entrance time of the sequencer for each order. The entrance time stamp determines the order of cars and therefore contains all the scrambling of the previous sides. This is caused by varying lead times, down times and steering effects.

The exit times of the sequencer correspond to real world events as well. Thus, the number of bodies, which are stored in the sequencer should be the same as the real world system. The main difference to the existing production system is the prioritizing logic for the AS/RS release. Instead of an assembly line oriented approach

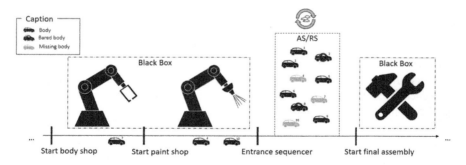

Fig. 2 Modeling approach

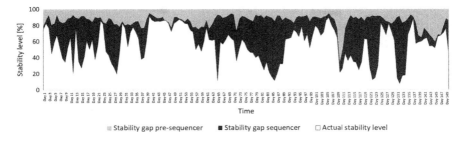

Fig. 3 Possible stability levels

the sequencer focuses on the sequence number. Every time a car body enters or leaves the sequencer the whole content of the sequencer is sorted in an ascending order by the sequence number. Before a body leaves the sequencer it is checked whether or not the car with the highest priority is affected by one or more detents. Therefore, the sequencing logic compares the current simulation time with the start and ending time of all attached detents of the car. If the simulation time lies within a detent interval the car is locked down. If a car is not allowed to exit the sequencer the car with the next highest priority is checked.

4 Results

To evaluate the sequence stability accurately a measurement logic is needed. Therefore, the difference of the sequence numbers of every car and its direct successor were calculated. If the difference does not equal one, the car is is out of sequence. Afterwards the whole sequence was separated into equal batches each representing the production volume of one day. The number of cars in sequence during one production day is divided by the everyday production volume in order to get a daily, percentaged representation of the stability.

The results for half a year displayed in Fig. 3 show that the stability level of the two facilities body shop and paint shop are stable except for some days (Stability gap pre-sequencer). The drop of the stability due to detents on the other hand is significantly stronger and also more volatile (Stability gap sequencer).

Figure 4 shows, that a significant loss of stability is caused by logistic detents. It can be seen that the logistic detents have a seasonal character with heavy impact from time to time.

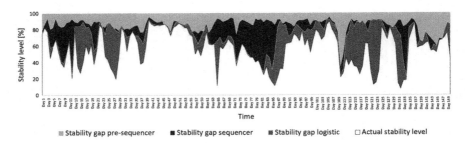

Fig. 4 Possible stability levels without logistic detents

5 Conclusions and Prospects

This article shows that the potential stability levels of a plant can be revealed with a discrete event based simulation and an appropriate data set. The simulation results demonstrate that the scrambling caused by the production process and the effect of the detents can be measured separately. The overall stability level is below the recommended 98% [4] and volatile, which is manly caused by the detents in this case. To generate more insights two additional data sets will be observed with different car types, life cycle states and additional detents. In order to gain further knowledge a sensitivity analysis concerning the sequencer volume will be done to calculate the stability gains through additional capacity.

References

1. Bayer, J., Collisi, T., & Wenzel, S. (2012). *Simulation in der Automobilproduktion*. Berlin: Springer.
2. Boysen, N., Golle, U., & Rothlauf, F. (2011). The car resequencing problem with pull-off tables. *Business Research*, *4*(2), 276–292.
3. Boysen, N., Scholl, A., & Wopperer, N. (2012). Resequencing of mixed-model assembly lines: Survey and research agenda. *European Journal of Operational Research*, *216*(3), 594–604.
4. Gusikhin, O., Caprihan, R., & Stecke, K. (2008). Least in-sequence probability heuristic for mixed-volume production lines. *International Journal of Production Research*, *46*(3), 647–673.
5. Inman, R. (2003). ASRS sizing for recreating automotive assembly sequences. *International Journal of Production Research*, *41*(5), 847–863.
6. Klug, F. (2010). *Logistikmanagement in der Automobilindustrie: Grundlagen der Logistik im Automobilbau*. Berlin: Springer.
7. Meissner, S. (2010). Controlling just-in-sequence flow-production. *Logistics Research*, *2*(1), 45–53.

Maintenance Planning Using Condition Monitoring Data

Daniel Olivotti, Jens Passlick, Sonja Dreyer, Benedikt Lebek and Michael H. Breitner

1 Introduction

A main goal of the manufacturing industry is to keep machines at a high availability level and to ensure the planned output. This is achieved through an intelligent maintenance policy. Maintenance is also an important cost factor in manufacturing companies [1]. Infrequent maintenance activities can increase the probability of a machine breakdown and results in very high costs. Too frequent maintenance activities can lead to unnecessary high expenditures for maintenance. Optimal maintenance activities in the manufacturing industry are dependent from various influencing factors. A common approach is to group several machines, this helps reducing set-up costs and fix costs during maintenance [6]. It is not enough to know the optimal maintenance plan for a single machine but to group maintenance activities of several machines on the same maintenance time [2]. A meaningful grouping can be supported by the information received from sensor data. This sensor values are necessary for condition-based maintenance [5]. In practice, a challenge for condition-based maintenance is to combine actual sensor data with prediction methods in order to forecast optimal maintenance activities [3, 4].

The presented decision support system, including an optimization model, determines the optimal maintenance policy for several machines. The actual condition of the machine is determined by sensor data and a breakdown probability is calculated for several periods. The trade-off between grouping of machines to save set-up and fixed costs and moving away from the individual optimal time for maintenance is considered.

D. Olivotti (✉) · J. Passlick · S. Dreyer · M. H. Breitner
Information Systems Institute, Leibniz Universität Hannover,
Königsworther Platz 1, 30167 Hannover, Germany
e-mail: olivotti@iwi.uni-hannover.de

B. Lebek
BHN Dienstleistungs GmbH & Co. KG, Hans–Lenze–Straße 1,
31855 Aerzen, Germany

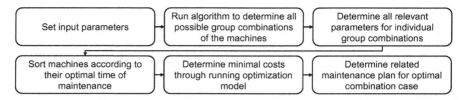

Fig. 1 General procedure to determine the optimal maintenance activities

2 Optimizing Maintenance Activities

The main goal is to determine optimally grouped maintenance activities considering the costs of maintenance and possible breakdowns. The general procedure is shown in Fig. 1 and explained in detail in the following. First all input parameters are set. The number of machines is important to build possible groups and to do further calculations. We do not consider machines which do not need to be maintained in the set time horizon. General costs which do not depend on the considered machine have to be defined and are called maintenance costs for groups. This could be for example general set-up costs for maintenance activities. For all machines the individual breakdown costs and the estimated probability of failure for future periods serve as input. The probabilities of failure can either be retrieved through a function or set by values for each period. A developed algorithm to determine all possible group combination cases is executed then. We assume that machines with similar optimal time for maintenance are grouped together. The machines are sorted according to their individual optimal maintenance time before executing the optimization model. This results in a number of 2^{M-1} different possible combination cases for groups according to the number of machines. With the help of the groups all parameters that are further used in the optimization model can be determined. The optimization model minimizes the costs for all possible combination cases and provides the optimal group combination case. The decision support system shows the optimal grouping combination and the associated maintenance periods for all groups and consequently for all machines. Based on this, a maintenance plan for all considered machines can be created.

$$\min \left\{ \sum_{j_i=1}^{g_i} C_{j_i} + g_i \times F \,\middle|\, i = 1, \ldots, 2^{M-1} \right\} \quad (1)$$

$$C_{j_i} = \min \left\{ \sum_{m=x_{j_i}}^{n_{j_i}+x_{j_i}-1} \left| D_m \times p_{mt} - \frac{F}{n_{j_i}} - V_m \right| \,\middle|\, t = 0, \ldots, T \right\} \quad (2)$$

$$0 \leq p_{mt} \leq 1 \quad \forall\, m \text{ and } t \quad (3)$$

$$p_{m(t-1)} \leq p_{mt} \quad \forall\, m \text{ and } t \quad (4)$$

Sets:
$i = (1, \ldots, 2^{M-1})$: Considered combination case
$j_i = (1, \ldots, g_i)$: Considered group in combination case i
$m = (1, \ldots, M)$: Considered machine with M: total number of machines
$t = (0, \ldots, T)$: Considered period with T: number of periods in the future

Parameters:
C_{j_i}: Total costs of a group j in a combination case i
D_m: Downtime costs of the machine m
F: Maintenance costs per group
g_i: Number of groups in a combination case i
n_{j_i}: Number of machines in a group j in a combination case i
p_{mt}: Probability of failure for machine m in period t
V_m: Maintenance costs for machine m
x_{j_i}: Smallest machine number in a group j in a combination case i

$$\frac{F}{n_{j_i}} + V_m \leq D_m \quad \forall \; j_i \text{ and } m \tag{5}$$

$$V_m, F \geq 0 \quad \forall \; m \tag{6}$$

$$1 \leq j_i \leq M \quad \forall \; i \tag{7}$$

$$1 \leq x_{j_i} \leq M \quad \forall \; j_i \tag{8}$$

$$g_i, \; n_{j_i}, \; x_{j_i} \in \mathbb{N} \setminus \{0\} \tag{9}$$

$$M, \; T \in \mathbb{N} \setminus \{0\} \tag{10}$$

The objective function (1) minimizes the costs over all possible combination cases. The costs for each group in a combination case are summed up and the fix costs per group are added. As a result the combination case with lowest costs is determined. The costs for each group in a combination case is defined in (2). It is described as the absolute value of the difference between expected downtime costs and costs for maintenance for each machine. The maintenance costs consist of the maintenance costs per machine and the maintenance costs per group. The maintenance costs per group are divided by the number of machines in that group and assigned partially to the machines. The parameters n_{j_i} and x_{j_i} are received considering the possible group combinations. The probability of failure is important for the model to determine the expected downtime costs. It is zero when there is a new machine and one for a broken machine (3). It is assumed that the probabilities of failure are already predicted through sensor data. We assume that a machine can only degrade and the probability of failure can only be reduced while performing maintenance activities (4). The probability of failure has to be a function over time which is monotonically increasing. Downtime costs of a machine are always equal or greater than maintenance costs of the machine (5). Maintenance costs per machine and group have to be positive, ensured by (6). Constraints (7) and (8) prohibit that j_i and x_{j_i} are greater than the number of machines. The number of machines and periods as well as the number of groups in a combination, the number of machines in a group and the smallest machine number in a group are defined as a positive integer (9) and (10).

3 Experimental Results

The presented decision support system including the optimization model is used in a demonstration case to optimize the maintenance activities. The general input parameters which are not period dependent are stated in Table 1. The used probabilities of failure for the future periods are shown in Table 2.

All input parameters are fixed except the maintenance costs per group to figure out the influence on the number of groups. Figure 2 shows the results of the test case. Herein, the optimal number of groups in dependence of the maintenance costs per group are shown. In the case of higher maintenance costs per group the number of the groups formed is usually reduced. The results of the system demonstrate that for certain values the number of groups can decrease and afterwards rise again. This is caused by two effects. First, if there are machines with similar optimal times for maintenance the grouping has stronger effects. This can be seen in Fig. 2 where the maintenance costs per group are low. The number of groups did not become four but jumps from five to three. Second, the higher the number of machines in a group the lower are the proportionate maintenance costs per group for each machine.

In the presented model it is not considered how many periods the time for maintenance for a group is away from the individually optimum. Only the costs difference between expected breakdown costs and maintenance costs are considered for each period. The difference in the costs depend strongly on the progress of degradation of

Table 1 Maintenance costs and downtime costs for each machine

Number of machines	Periods	Maintenance costs for a group
8	10	Varied
Machine	Maintenance costs	Downtime costs
1	600	1,000
2	1,500	5,000
3	1,200	2,000
4	800	1,600
5	450	1,150
6	1,050	1,200
7	950	2,400
8	650	8,000

Maintenance Planning Using Condition Monitoring Data 547

Table 2 Probability of failure for machines and periods

Period	Machine							
	1	2	3	4	5	6	7	8
0	0.01	0.03	0.01	0.01	0.28	0.15	0.02	0.02
1	0.01	0.04	0.02	0.05	0.31	0.23	0.05	0.08
2	0.02	0.07	0.52	0.07	0.41	0.29	0.21	0.16
3	0.03	0.10	0.61	0.09	0.52	0.32	0.36	0.28
4	0.15	0.20	0.63	0.15	0.56	0.38	0.39	0.34
5	0.26	0.33	0.67	0.23	0.60	0.41	0.43	0.39
6	0.39	0.50	0.69	0.25	0.63	0.45	0.48	0.41
7	0.65	0.60	0.70	0.25	0.69	0.56	0.51	0.49
8	0.74	0.75	0.75	0.48	0.72	0.59	0.56	0.57
9	0.81	0.78	0.78	0.59	0.80	0.82	0.66	0.62
10	0.92	0.83	0.80	0.72	0.92	0.96	0.70	0.75

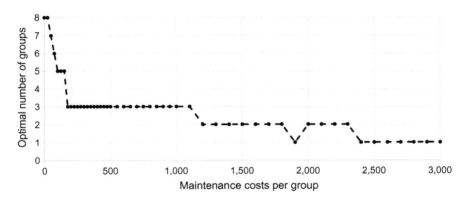

Fig. 2 Different costs per group in relation to the number of groups

the machines. The fact that the absolute value is used can shift the maintenance time of a machine forward or backward. Due to this, a wide range of group combinations is possible.

4 Conclusions and Further Research

This paper presents a decision support system to determine the optimal maintenance policy based on the actual condition of the machine. An optimization model is the main part of the presented system. A trade-off between potential breakdown costs and costs for too early or too late maintenance activities is addressed. The grouping of machines is a key aspect of the presented model. A demonstration case of the developed optimization model is performed. The decision support system helps to plan maintenance activities based on sensor data. It has to be taken into account that sensor data can be used for the probability of failure. Reliable data is essential for meaningful results. Influencing factors of the model are possible breakdown costs. In practice it could be difficult to estimate these costs exactly. The experimental results show that the grouping of machines seems reasonable. Especially when machines with similar probabilities of failure and maintenance costs are in the production site scale effects through grouping can occur. In a future version of the decision support system additional influencing factors such as the production plan should be considered to improve maintenance scheduling further. Designing an interface for maintenance planners in a production site would contribute to an easier application in practice. In particular, chained production lines could be represented in more detail.

References

1. Bousdekis, A., Magoutas, B., Apostolou, D., & Mentzas, G. (2015). Review, analysis and synthesis of prognostic-based decision support methods for condition based maintenance. *Journal of Intelligent Manufacturing, 26*, 1–14.
2. Bouvard, K., Artus, S., Berenguer, C., & Cocquempot, V. (2011). Condition-based dynamic maintenance operations planning and grouping. Application to commercial heavy vehicles. *Reliability Engineering and System Safety, 96*, 601–610.
3. Kaiser, K., & Gabraeel, N. (2009). Predictive maintenance management using sensor-based degradation models. *IEEE Transactions on Systems, Man, and Cybernetics - Part A: Systems and Humans, 39*(4), 840–849.
4. Kothamasu, R., Huang, S. H., & VerDuin, W. H. (2006). System health monitoring and prognostics - a review of current paradigms and practices. *The International Journal of Advanced Manufacturing Technology, 28*, 1012–1024.
5. Peng, Y., Dong, M., & Zuo, M. (2010). Current status of machine prognostics in condition-based maintenance: A review. *The International Journal of Advanced Manufacturing Technology, 50*, 297–313.
6. Wildeman, R., Dekker, R., & Smit, A. (1997). A dynamic policy for grouping maintenance activities. *European Journal of Operational Research, 99*(3), 530–551.

Tactical Planning of Modular Production Networks with Reconfigurable Plants

Tristan Becker, Stefan Lier and Brigitte Werners

1 Introduction

Process industries are facing strong global competition in recent years. In order to stay competitive, it is important to quickly react to new market developments and to serve specialized product demands. To meet these requirements, flexible production capacity is required. Recently, the use of modular plants has received much attention as an answer to these new challenges. Modular plants consist of standardized process modules, which allow for quick assembly, disassembly and relocation of production plants. Using modular production concepts, the flexibility of the production network is greatly increased [6]. The technical feasibility of modular plant concepts has been proven by several projects, like the EU funded "F^3-Factory" project and the "CoPIRIDE" project [3]. Modular plants offer an increased manufacturing flexibility in type, volume and location of production capacities.

The remainder of this paper is structured as follows. Section 2 gives an overview of the implications of modular production concepts on the production network. In Sect. 3, we give a problem description and a brief overview of key aspects of our mathematical formulation for production network planning. In Sect. 4, we illustrate the framework of our case study and discuss results and implications dependent on different degrees of freedom. Section 5 concludes with some remarks and a perspective for future research on modular production plants.

T. Becker (✉) · B. Werners
Chair of Operations Research and Accounting, Faculty of Management and Economics, Ruhr University Bochum, Bochum, Germany
e-mail: tristan.becker@rub.de

S. Lier
Department of Engineering and Economics, University of Applied Sciences Südwestfalen, Meschede, Germany
e-mail: Lier.Stefan@fh-swf.de

2 Tactical Production Network Flexibility Using Modular Plants

In a production network with modular production plants, the entire network structure can be changed in the short-term. This is due to transformable plant designs, which are significantly more flexible compared to conventional large-scale plants [7]. A single production module, which can be installed in an ISO transportation container, consists of several process modules. A production module is associated with specific production capabilities regarding type and volume of capacity. To assemble a production module, several process modules are combined. When a production module is disassembled, the released process modules can be reused in different production modules. This process is called *reconfiguration*. It can be carried out short term at a workshop. Module reconfigurations allow for a greater flexibility of the production network regarding the type of commodities produced. The set of process modules available to a company is defined in a central database, as well as the production processes, which can be created with different combinations of process modules. Another important degree of freedom associated with modular production plants is their geographical flexibility. Since modular plants may be installed in ISO transportation containers, they can be transported with standard transportation. Before a modular plant can start operating at a production location, it has to be setup with peripherals, which are made available at the decentralized production locations. A capacity shift, which removes a plant from one location and transfers it to another, can be carried out short-term.

Using the flexibility options associated with modular production concepts, the structure of the entire production network can be altered short-term. Since both reconfigurations and capacity shifts are associated with cost, it is important to carefully plan the future layout of the production network and anticipate transitions over time on the basis of customer demands.

In the literature, few works consider the described types of capacity flexibility. In a recent review on tactical manufacturing flexibility, plant relocation and reconfiguration decisions are not recognized as tactical manufacturing flexibility options treated by the existing literature [2].

Reference [5] presented a modification to their strategic facility location framework, which allows for shifts of modules between production locations. In a recent work by [4], modular production capacities have been considered, which are modular only with respect to the level of production capacity at each facility. The model does not allow for an exchange of capacity between facilities or modular plant reconfigurations. The modular production network configuration problem has been considered with a strategic scope and limited flexibility options by [8]. We are not aware of any model, which considers the degrees of freedom associated with modular plants for tactical planning of the production network.

3 Problem Description and Model Formulation

The modular production network configuration problem [1] is modeled as an extension of the dynamic multi-commodity capacitated facility location problem. The set of locations is denoted by I, while the set of customers is denoted by J. Each customer demands a subset of the set of products P in every period. The customer demands have to be satisfied by selecting the subset of locations to operate, the amount of production modules to install at each location as well as the modules to operate in each period. The goal is to minimize all network and production costs. Each production location can hold a number of up to L_i production modules of arbitrary configuration. It is assumed that the process modules necessary for assembly of production modules can be acquired for the duration of the planning horizon for a fee. Production modules can be assembled, disassembled and reconfigured only at the modular hub. The cost for acquiring and transporting raw materials, production and transportation of product to a customer is represented by c_{ijpt}. In order to model the possibilities associated with plant reconfiguration accurately, we consider individual process modules. All demand has to be met, since demands represent customer orders on the tactical level and must be fulfilled.

Since the amount of production modules is flexible, we introduce integer variables for the available capacity at each location. The integer variable y_{ist} captures the number of production modules of configuration s available at location i in time period t. To make a production module available at a production location, a complex multi-step process is involved. First, the necessary process modules for the desired configuration have to be acquired. The amount of process modules of type m acquired in period t is represented by a_{mt}. Process modules are ordered to arrive at the modular hub, where they can be assembled to the desired configuration. Module configuration, which includes assembly, disassembly and reconfiguration, is represented by the integer variable $\gamma_{ss't}$. It indicates how many production modules change from configuration s to s' in period t. After a module is built, the process modules cannot be used for other production modules. The number of process modules required for a production module of configuration s is defined by W_{mt}. The following constraint ensures that the amount of production modules built is covered by the amount of process modules:

$$\sum_{i \in I \cup \{0\}} \sum_{s \in S} W_{ms} y_{ist} \leq a_{mt} \quad \forall m \in M, t \in T \qquad (1)$$

The configuration $s = 0$ is used to represent assembly and disassembly, i. e., production modules changing from configuration 0 to a configuration $s \neq 0$ are assembled. Modules changing from a configuration $s \neq 0$ to configuration 0 are disassembled. After assembly of a production module using the necessary process modules, it has to be deployed to a production location to serve customers. Such a capacity shift from location i to i' of production modules in configuration s in period t is represented by $\eta_{ii'st}$, whereas the modular hub is represented by index $i = 0$. The number

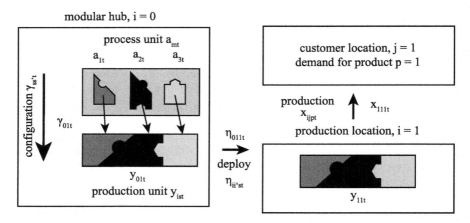

Fig. 1 The relation of the different decision variables from process module to production unit

of production modules available at the modular hub is derived by the following constraints:

$$y_{0st} - y_{0s(t-1)} = \sum_{s' \in S \cup \{0\}} (\gamma_{s'st} - \gamma_{ss't}) \sum_{i' \in I \cup \{0\}} (\eta_{i'0st} - \eta_{0i'st}) \quad s \in S, t \in T \quad (2)$$

Finally, production modules can commence production at the target location. Figure 1 depicts the relation of the decision variables from module assembly to production.

4 Case Study

To evaluate the cost advantage associated with the flexibility of modular production concepts, we have conducted several computational experiments. We compare the amount of network cost using modular production concepts with those of a less flexible model. The base model is obtained by removing the possibility of capacity shifts and plant reconfiguration from the previously presented model. Production plants of a certain configuration can be obtained for the price of process module acquisition, assembly and transportation to production locations at each location. The base model allows only for capacity expansions and reductions. The modular model additionally considers capacity shifts and reconfigurations.

Based on a real-world data set from the chemical industry, we constructed a large amount of different test instances by varying the number of customers and the demand development. Demand data were varied systematically with regard to location, type and amount of demands. Our demand data consider customers all over Europe, while a large fraction of customers is located in Germany, as is the company which provided the data. To construct data sets with more geographical variation

Fig. 2 Cost reduction compared in test scenarios with static and dynamic customer locations

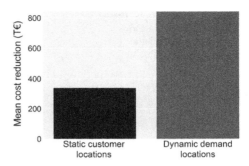

of customer locations, density-based clustering was applied. Transportation costs for our models were estimated for truck transportation between all production and customer locations using road distances from Google DistanceMatrix API. All test instances were solved using *Gurobi 7.0* on a *Windows 10* Computer with an *Intel i7 I7-6700K* processor and 16 GB of RAM. It turns out that the additional flexibility considered in our model formulation strongly increases the computational difficulty. The solution characteristics of both model formulations are listed in Fig. 5. Linking capacity variables across periods and the high number of integer variables render the problem hard to solve for large instances. Strong differences in total network and demand fulfillment costs can be identified between the base model and the modular production formulation. Further, the individual demand patterns strongly influence the utilization of flexibility options and consequently the costs.

Both, capacity shifts and reconfigurations, are frequently used in the modular production model to react on dynamic changes in demand. As a result, the modular production formulation obtains lower total network cost in all test instances. In fact, any feasible solution for the base model can be transformed into a feasible solution for the modular production formulation with the same cost. For our examples, an average cost reduction of 589.51 T€ can be realized by adding modular flexibility options. The high standard deviation of 811.33 T€ associated with the cost reduction, implies that the advantageousness of modular flexibility options highly depends on the respective demand patterns and development. The cost difference between the models is strongest in test cases with highly dynamic customer locations and type of product demand (Figs. 2, 3). Without the possibility of reconfiguration, new modular plants are acquired in reaction to dynamic demands. With modular production concepts, unused modules are transported back to the modular hub and can be reconfigured for the production of different products. In regions with declining overall customer demands, modular plants are removed and transported to more favorable production locations regarding customer proximity (Fig. 4).

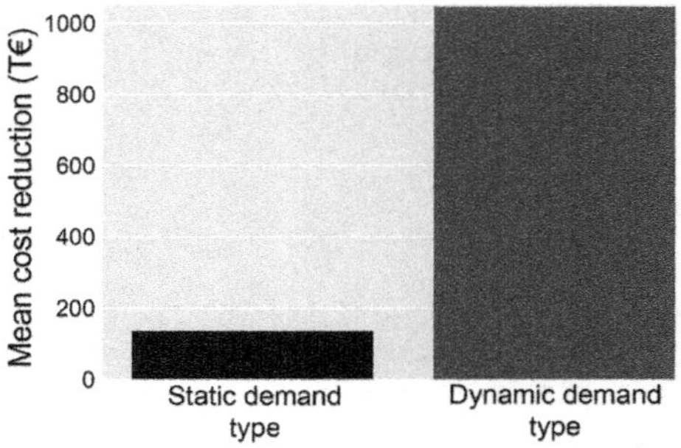

Fig. 3 Cost reduction compared in test scenarios with static and dynamic product demands

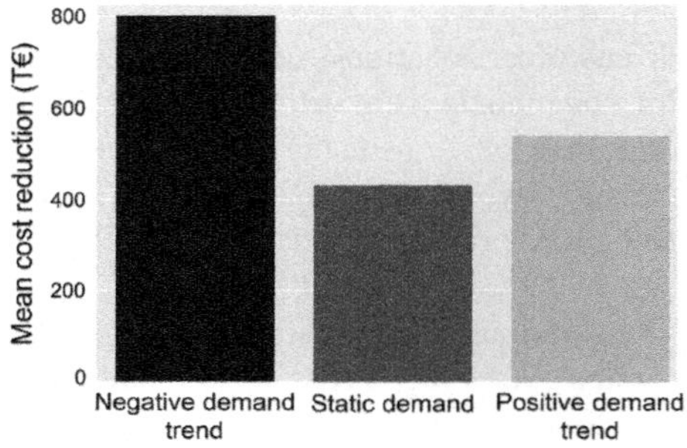

Fig. 4 Cost reduction compared with regard to different demand trends

Fig. 5 Computational characteristics of the base model and modular formulation

	Base case		Modular flexibility	
Cus.	Time (s)	Gap (%)	Time (s)	Gap (%)
30	19.41	0.06	65.46	0.07
40	2.72	0.04	45.02	0.07
50	3.43	0.06	10.88	0.06
60	4.84	0.05	23.65	0.08
70	14.65	0.08	415.48	0.08
80	3.91	0.05	28.71	0.07
90	6.06	0.05	17.13	0.04
100	12.05	0.05	207.9	0.08

5 Conclusion and Outlook

Prior work on tactical production network configuration has rarely considered flexibility regarding location or type of production plants. This paper confirms the importance of adequate consideration of the flexibility options that go along with modular production in tactical planning. The flexibility provided by both capacity shifts and reconfigurations is very useful as a short-term reaction to dynamic demand changes. In our case study with realistic data from the chemical industry, high cost reductions were realized by utilization of modular production concepts. The implementation of fast mounting and dismounting options for modular plants, as well as streamlined reconfiguration processes may be critical for the further technical development of modular plants from a managerial perspective. In order to solve large-scale problems in reasonable time, future research should focus on solution methodologies for the tactical planning of modular production networks (Fig. 5).

References

1. Becker, T., Lutter, P., Lier, S., & Werners, B. (2018). *Optimization of modular production networks considering demand uncertainties* (pp. 413–418). Cham: Springer International Publishing.
2. Esmaeilikia, M., Fahimnia, B., Sarkis, J., Govindan, K., Kumar, A., & Mo, J. (2016). Tactical supply chain planning models with inherent flexibility: Definition and review. *Annals of Operations Research*, *244*(2), 407–427.
3. European Commission: Final report for flexible, fast and future production processes. Technical report (2014)
4. Jena, S. D., Cordeau, J. F., & Gendron, B. (2015). Dynamic facility location with generalized modular capacities. *Transportation Science*, *49*(3), 484–499.
5. Melo, M. T., Nickel, S., & da Gama, F. S. (2006). Dynamic multi-commodity capacitated facility location: A mathematical modeling framework for strategic supply chain planning. *Computers and Operations Research*, *33*(1), 181–208.
6. Seifert, T., Lesniak, A. K., Sievers, S., Schembecker, G., & Bramsiepe, C. (2014). Capacity flexibility of chemical plants. *Chemical Engineering and Technology*, *37*(2), 332–342.
7. Sievers, S., Seifert, T., Schembecker, G., & Bramsiepe, C. (2016). Methodology for evaluating modular production concepts. *Chemical Engineering Science*, *155*, 153–166.
8. Wörsdörfer, D., Lutter, P., Lier, S., & Werners, B. (2017). *Optimized modular production networks in the process industry* (pp. 429–435). Cham: Springer International Publishing.

Part XVII
Project Management and Scheduling

A CTMDP-Based Exact Method for RCPSP with Uncertain Activity Durations and Rework

Xiaoming Wang, Roel Leus, Stefan Creemers, Qingxin Chen and Ning Mao

1 Introduction

Product design and many other practical projects incorporate random rework, which leads to a stochastic project network structure. Although a lot of research on stochastic RCPSP has been done over the past few decades [1], only a few of them considered activity rework or uncertain project network structure. The few existing methods for RCPSP with random rework including priority rules [2], genetic algorithms [3], stochastic dynamic programming (SDP) [4], etc. However, these studies not only require some particular assumptions such as deterministic rework time [4] and maximum number of reworks [3], but can also only obtain approximate solutions. To the best of our knowledge, until now there is no study that has proposed on exact method for RCPSP with both random activity durations and rework.

In recent years, MDP-based methods have been widely applied to stochastic RCPSP due to their inherent advantage of modeling sequential decision problems under uncertainty [5–7]. However, the biggest challenge in applying this approach is the curse of dimensionality that comes into play upon solving large-scale instances. In this study, we will explore how to model the RCPSP with random activity durations and rework, as well as how to efficiently solve it.

X. Wang (✉) · Q. Chen · N. Mao
Key Laboratory of Computer Integrated Manufacturing, Guangdong University of Technology, Guangzhou 510006, People's Republic of China
e-mail: simonwang@gdut.edu.cn

X. Wang · R. Leus
ORSTAT, KU Leuven, 3000 Leuven, Belgium

S. Creemers
IESEG School of Management, 59000 Lille, France

2 Problem Definition

The basic RCPSP consists of a project and a renewable resource set $R = \{1, 2, \ldots, K\}$. The project can be represented by an activity-on-node (AON) network $G(N, E)$, where $N = \{0, 1, 2, \ldots, n\}$ denotes the activity set and $E = \{(j, i) \mid j, i \in N\}$ denotes the zero-lag precedence constraints between the activities. The capacity of resource type $k \in R$ is M_k. In this study, the duration of activity $j \in N$ is an exponentially distributed variable d_j. During the execution of activity $j \in N$, it will occupy r_{jk} units of resource $k \in R$.

We further consider two types of random rework, where the first type of rework is caused by the failed quality inspection itself and the second type of rework is results from the discovery of error in another activity. When the second type of rework activity was finished, it will return to the original activity to be redone. A rework activity has the same resource requirements as the corresponding regular activity, a constant rework probability and an exponentially distributed duration.

The decision objective of the above RCPSP is to minimize the expected project makespan.

3 CTMDP-Based Decision Model

3.1 Modeling

We build the decision model based on CTMDP owing to the memoryless property of exponential distribution. The CTMDP is the four-tuple given by $\{X, A, q(x' \mid x, a), c(x' \mid x, a)\}$, each element is defined below.

The state space X. Since the rework activity may have a different mean duration and it may not optimal to start it first, we need to transform the project network by adding the rework activities and virtual precedence constraints. On the basis of the transformed network, we can give a definition of the state. In previous work [5], the state of the decision process is composed of an idle and a processing activity set. Although this definition can also be applied to this study, we introduce a new definition of the state $x = \{W, P\}$ for RCPSP with rework, where W and P is the waiting and processing activity set, respectively. The so-called waiting activity is an activity that not started yet but all of its predecessor activities are finished.

The action space A. Each state $x \in X$ has an action set $A_x \subset A$, where action $a \in A_x$ is a feasible combination of starting activities. $A_x = \emptyset$ is a special action when none of waiting activity satisfies the resource constraints in state x.

The transition rates $q(x' \mid x, a)$. $0 \leq q(x' \mid x, a) < \infty$ is the transition rate from state x to x' (for $x' \neq x, x, x' \in X$) under action $a \in A_x$. The state transition will occur once a processing activity is finished or a waiting activity is started. For the latter case, the system will immediately transfer to an intermediate state from current state with probability one, and afterwards transfer to next state with corresponding rate. In order to reduce the state space size to improve computational efficiency, we will aggregate the current state and intermediate state. For a given state $x \in X$ and an action $a \in A_x$, the rate of next state transition is $\lambda_x = \sum_{j \in P_x \cup a} 1/E(d_j)$. Furthermore, $q(x' \mid x, a) = p(x' \mid x, a)\lambda_x$ for $x' \neq x$ and $\sum_{x' \in X} q(x' \mid x, a) = 0$, where $p(x' \mid x, a)$ is the state transition probability. Suppose the state transition from x to x' after action a is caused by activity $j \in P_x \cup a$ be finished first with inspection result α_j. Then $p(x' \mid x, a) = p(\alpha_j)/(E(d_j)\lambda_x)$ due to the independence of these two events.

The cost rates $c(x' \mid x, a)$. The cost rate of transition from state x to x' is defined as the corresponding transition rate according to the decision objective, namely $c(x' \mid x, a) = q(x' \mid x, a)$.

3.2 Equivalent DTMDP and Optimal Policy

The decision time points in the above CTMDP are random variables in $[0, \infty]$. From another perspective, an action can only be taken when a processing activity is finished except the initial time. Hence, we convert the above CTMDP into an equivalent discrete-time MDP (DTMDP) which is easier to handle.

The decision period of the equivalent DTMDP is one unit of time, the state space, action space and transition probability is the same as it in the above CTMDP. Since the expected project makespan is equal to the sum of mean sojourn times of all visited transient states, we define the cost function of taking action $a \in A_x$ as the mean sojourn time of state x, namely $\tilde{c}(x, a) = 1/\sum_{x' \in X, x' \neq x} q(x' \mid x, a)$. It is noted that all of the self-transitions due to rework are removed under this cost function definition. The advantage of this operation lies in reducing the number of iterations when obtaining an optimal policy by solving the following Bellman equations.

$$v^{i+1}(x) = \min_{a \in A_x} E\{\tilde{c}(x, a) + v^i(x' \mid x, a)\} \quad (1)$$

$$a^* = \arg\min_{a \in A_x} E\{\tilde{c}(x, a) + v^*(x' \mid x, a)\} \quad (2)$$

3.3 Decomposition and Parallel Method

When compared to the modeling of the studied problem, the computation of an optimal policy is much more difficult due to the state and action space explosion. Specifically, memory consumption rather than CPU time is usually the bottleneck of computation [5].

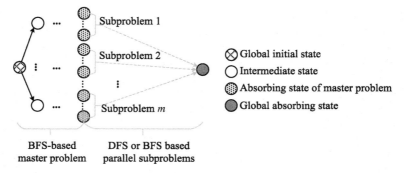

Fig. 1 Diagram of decomposition and parallel method

According to the optimality principle of dynamic programming, the optimal actions in a subproblem are also optimal in the original problem [8]. This indicates that we can solve the original problem in a divide and conquer style. The advantage of doing so lies in decomposing the large intractable state space into multiple tractable subspaces. In addition, since the subproblems can be solved independently, we can also exploit the advantages of parallel computing. The proposed decomposition and parallel method started with a breadth-first-search (BFS) based state transition network traversal, and trigger the parallel computing of subproblems when the real memory consumption exceeds the predefined threshold ϑ, as illustrated in Fig. 1.

The reason why we use a BFS-based approach for state transition network traversal in the master problem is to balance the computational load between subproblems. At the time that the parallel computing is triggered, there would be a great number of absorbing states of the master problem. The task of subproblems is to continue exploring the decision processes from these states to the global absorbing state. Obviously, it is unreasonable to treat each of these states as the initial state of a subproblem because it will result in too many redundant states and waste of computation time. A rational way is to determine the number of subproblems by considering the state space size of the original problem and the number of available computing resources.

4 Computational Experiment

The computational experiment is used to analyze the impact of activity rework on project makespan and optimal actions, as well as compare the traditional SDP with the proposed decomposition and parallel method. The methods have been implemented in Visual C♯ and performed on a workstation with Intel Xeon CPU E5-2695 v4 and 128G RAM.

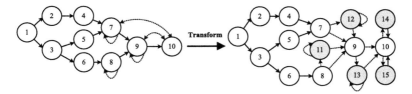

Fig. 2 The considered project network

4.1 Experimental Data

We consider a project consisting of three concurrent projects, where each of them has the same network structure as shown in Fig. 2.

The mean durations $E(d) = (5, 3, 5, 4, 4, 3, 6, 4, 10, 3)$. There are two types of renewable resources ($M_1 = M_2 = 3$). The resource requirements are randomly generated where each activity require one unit of one or two types of resource.

Let τ_d and τ_p denote the coefficient of rework duration and probability, respectively. The duration of rework activity $j \in N$ is $E(d_j)' = \tau_d E(d_j)$ and all rework probabilities are τ_p. Here, we set $\tau_d = \{0.25, 0.5\}$ and $\tau_p = \{0.1, 0.2\}$.

4.2 Results and Analysis

We first use the traditional SDP to solve the instance and collect the expected makespan E(MS), the number of all states N(all), states with waiting rework activities N(Re), states that is optimal to start rework activities first N(Re*), and the computation time CT in seconds, the results are shown in Table 1.

We see that both of the rework probability and rework duration have an impact on project makespan and rework activity decisions. In contrast, rework probability has a greater impact on project makespan, while rework duration has a greater impact on rework activity decisions. In addition, the state space size and computation time is significantly increased when consider random rework.

Table 1 Computational results under different rework parameter settings

τ_d, τ_p	E(MS)	N(all)	N(Re)	N(Re*)	CT(s)
0, 0	51.77	667,809	–	–	54.87
0.25, 0.1	53.85	2,509,373	1,311,402	125,904	322.54
0.25, 0.2	57.18	2,509,373	1,311,402	126,555	354.97
0.5, 0.1	55.27	2,509,373	1,311,402	131,136	325.26
0.5, 0.2	60.62	2,509,373	1,311,402	132,439	351.87

Afterwards, we use the proposed decomposition and parallel method to solve the instance under the following parameter settings: $\theta = \{0.8 \times 10^6, 1.0 \times 10^6, 1.5 \times 10^6\}$, $m = \{2, 4\}$, $\tau_d = 0.25$, $\tau_p = 0.1$. We find that although the total number of visited states is larger than it in traditional SDP, the proposed method achieves a shorter computation time and a smaller number of states (memory consumption) on each computing node. For example, the CPU time of this method under the parameter setting $\theta = 1.0 \times 10^6$, $m = 4$ is 258.88 s, the total number of visited states is 5,570,305 and the average state space size in subproblems is 1,210,923. In addition, we also find that evenly dividing the states will result in a computational load imbalance between subproblems.

5 Conclusions

We studied the RCPSP with uncertain activity durations and rework which rarely studied before although it is a common practical problem. There are two main contributions of this study. The first one is the representation and simplification of complex rework process by converting a CTMDP into an equivalent DTMDP. The second one is a decomposition and parallel method which improve the ability to handle large instances. However, there are still many potential works to be done in future. Next, we will study the theoretical analysis of optimal parameter setting, as well as effective action elimination procedures for the proposed method.

Acknowledgements This research is jointly supported by the National Natural Science Foundation of China under Contract No. 51505090 and 61573109.

References

1. Herroelen, W., & Leus, R. (2005). Project scheduling under uncertainty: Survey and research potentials. *European Journal of Operational Research, 165*(2), 289–306. https://doi.org/10.1016/j.ejor.2004.04.002.
2. Browning, T. R., & Yassine, A. A. (2016). Managing a portfolio of product development projects under resource constraints. *Decision Sciences, 47*(2), 333–372. https://doi.org/10.1111/deci.12172.
3. Yassine, A. A., Mostafa, O., & Browning, T. R. (2017). Scheduling multiple, resource-constrained, iterative, product development projects with genetic algorithms. *Computers and Industrial Engineering, 107*, 39–56. https://doi.org/10.1016/j.cie.2017.03.001.
4. Luh, P. B., Liu, F., & Moser, B. (1999). Scheduling of design projects with uncertain number of iterations. *European Journal of Operational Research, 113*(3), 575–592. https://doi.org/10.1016/S0377-2217(98)00027-7.
5. Creemers, S. (2015). Minimizing the expected makespan of a project with stochastic activity durations under resource constraints. *Journal of Scheduling, 18*(3), 263–273. https://doi.org/10.1007/s10951-015-0421-5.

6. Creemers, S., Leus, R., & Lambrecht, M. (2010). Scheduling Markovian PERT networks to maximize the net present value. *Operations Research Letters*, *38*(1), 51–56. https://doi.org/10.1016/j.orl.2009.10.006.
7. Wang, X., Chen, Q., Mao, N., Chen, X., & Li, Z. (2015). Proactive approach for stochastic RCMPSP based on multi-priority rule combinations. *International Journal of Production Research*, *53*(4), 1098–1110. https://doi.org/10.1080/00207543.2014.946570.
8. Puterman, M. L. (2005). *Markov decision processes: Discrete stochastic dynamic programming*. New York: Wiley.

Explicit Modelling of Multiple Intervals in a Constraint Generation Procedure for Multiprocessor Scheduling

Emil Karlsson and Elina Rönnberg

1 Introduction

A constraint generation approach for multiprocessor scheduling of tasks with multiple intervals is presented in this paper. The characteristic of multiple intervals has been studied in different settings, for example the travelling salesman problem in [1, 4] and ship scheduling in [3]. The problem studied in this paper is a relaxation of an industrially relevant multiprocessor scheduling problem of interest for the development of future avionic systems, presented in [2].

The objective of this paper is to improve a model used in a constraint generation procedure presented in [2] by an explicit modelling of multiple intervals. This enables us to improve the relaxed problem used in the constraint generation procedure, with the aim of reducing the number of constraints that need to be generated. The preliminary results that we present indicate that this explicit modelling yields better computational performance.

2 Preliminaries

Since the focus of this paper is the explicit modelling of multiple intervals, we here simplify the setting by addressing a relaxation of the industrial problem presented in [2]. The relaxation is made with respect to the details of the communication network scheduling and the resulting problem is as follows.

E. Karlsson (✉) · E. Rönnberg
Department of Mathematics, Linköping University, 581 83 Linköping, Sweden
e-mail: emil.karlsson@liu.se

E. Karlsson · E. Rönnberg
Saab AB, 581 88 Linköping, Sweden

The system consists of a set of nodes and executes periodically with period P, called a major frame. For each node there is a module called communication module (CM) and the set of all such modules is denoted by $\mathcal{H}^{\text{CM}}$. The set of tasks assigned to CM h is denoted by $\mathcal{I}_h$, $h \in \mathcal{H}^{\text{CM}}$, and all tasks on a CM has the period of a major frame. Task i needs to be granted non-preemptive execution time for the duration of its execution requirement e_i on its module h between its release time t_i^{r} and deadline t_i^{d}, $i \in \mathcal{I}_h$, $h \in \mathcal{H}^{\text{CM}}$. Further, for each node there is a set of modules, called application modules (AM). Each task on an AM has a period of $P/64$ and when scheduling AMs there is a minimum idle time between each pair of tasks that need to be respected.

The CMs in the system communicate through a single communication network (CN) where each CN-message is assigned to a discrete CN-slot in which it is transmitted through the CN. Let the set of CN-messages be denoted by $\mathcal{M}$ and the set of CN-slots be denoted by $\mathcal{N}$. The set of tasks required to transmit and receive CN-message m is denoted by $\mathcal{I}_m^{\text{msg}}$. If CN-message m is assigned to CN-slot n, then task i has to obey the release time t_{in}^{r} and deadline t_{in}^{d} for CN-slot n, $i \in \mathcal{I}_m^{\text{msg}}$, $n \in \mathcal{N}$, $m \in \mathcal{M}$. There are two types of precedence relations between tasks. A dependency restricts the duration from the start of a task to the next start of another task. A chain specifies that certain tasks, linked by dependencies, have to execute in a given order.

A known characteristic of the instances in [2] is that the CMs have a huge number of tasks and that a large portion of these are fixed. The constraint generation procedure in [2] is designed to be efficient for problems with these characteristics and the mathematical model exploits the existence of fixed tasks on the CMs. In the complete mathematical model for the CMs, the major frame is divided into sections in-between fixed tasks and each non-fixed task is to be assigned to a section. To ensure that no tasks overlap within a section, a requirement is added stating that the tasks of each subset that can be assigned together in the same section do not overlap. The constraints of each such subset is referred to as a sequencing formulation. Since the number of sequencing formulations required in a complete formulation is huge and not all of them are expected to be needed for solving the problem, this formulation lends itself to constraint generation as follows.

A relaxed model, called the α-model, is obtained by removing the sequencing formulations from the complete model. In a solution to the α-model, each non-fixed task is assigned to a section. Given such a solution, a subproblem called the β-model is obtained by restricting each non-fixed task to its assigned section and by introducing a sequencing formulation for the subset of tasks in each section. A solution to the β-model is either a feasible schedule or it generates at least one new sequencing formulation to be added to the α- and the β-model. The added sequencing formulations are called generated sequences. For details, see [2].

The α-model used in this paper consists of the following components.

min/max	Objective function
s.t.	AM-scheduling
	Precedence relations
	Generated sequences
	CM-assignment (Constraints (1)–(3))
	Relaxed CN-scheduling (Constraints (4)–(5))

The CM-assignment and the relaxed CN-scheduling components are of interest in this paper. For details about the other components, see [2].

In order to describe the CM-assignment component, the following notation is introduced. Let the set of sections of CM h be denoted by $\mathcal{R}_h$ and let the set of fixed task on CM h be denoted by $\mathcal{I}_h^{\text{fix}}$, $h \in \mathcal{H}^{\text{CM}}$. Let l_r^{sec} denote the duration of section r and let $\mathcal{I}_r^{\text{sec}}$ denote the set of tasks that can be assigned to section r, $r \in \mathcal{R}_h, h \in \mathcal{H}^{\text{CM}}$. The set of sections that task i can be assigned to is denoted by $\mathcal{R}_i^{\text{task}}$ and the release time and deadline of task i in section r is denoted by t_{ir}^r and t_{ir}^d respectively, $i \in \mathcal{I}_r^{\text{sec}}$, $r \in \mathcal{R}_h, h \in \mathcal{H}^{\text{CM}}$.

For task i, $i \in \mathcal{I}_h, h \in \mathcal{H}^{\text{CM}}$, introduce a variable

$$x_i = \text{start time of task } i \text{ offset its period start.}$$

For a fixed task i, it is required that $x_i = t_i^r, i \in \mathcal{I}_h^{\text{fix}}, h \in \mathcal{H}^{\text{CM}}$. Introduce, for $i \in \mathcal{I}_r^{\text{sec}}$, $r \in \mathcal{R}_h, h \in \mathcal{H}^{\text{CM}}$, a binary variable

$$\alpha_{ir} = \begin{cases} 1 & \text{if task } i \text{ assigned to section } r, \\ 0 & \text{otherwise.} \end{cases}$$

The following constraints ensure that each task is assigned to one section, that the capacity of each section is respected, and that each task obeys its release time and deadline within its section.

$$\sum_{r \in \mathcal{R}_i^{\text{task}}} \alpha_{ir} = 1, \ i \in \mathcal{I}_h \setminus \mathcal{I}_h^{\text{fix}}, \ h \in \mathcal{H}^{\text{CM}} \quad (1)$$

$$\sum_{i \in \mathcal{I}_r^{\text{sec}}} e_i \alpha_{ir} \leq l_r^{\text{sec}}, \ r \in \mathcal{R}_h, \ h \in \mathcal{H}^{\text{CM}} \quad (2)$$

$$\sum_{r \in \mathcal{R}_i^{\text{task}}} t_{ir}^r \alpha_{ir} \leq x_i \leq \sum_{r \in \mathcal{R}_i^{\text{task}}} t_{ir}^d \alpha_{ir} - e_i, \ i \in \mathcal{I}_h \setminus \mathcal{I}_h^{\text{fix}}, \ h \in \mathcal{H}^{\text{CM}} \quad (3)$$

The relaxed CN-scheduling component is modelled as follows. For each pair of CN-message m and CN-slot n, $m \in \mathcal{M}, n \in \mathcal{N}$, introduce a binary variable

$$z_{nm} = \begin{cases} 1 & \text{if CN-message } m \text{ is assigned to CN-slot } n, \\ 0 & \text{otherwise.} \end{cases}$$

The following constraints ensure that each CN-message will be assigned to a CN-slot and that the tasks involved in the sending and receiving of the CN-message obey their release times and deadlines with respect to their assigned slot.

$$\sum_{n \in \mathcal{N}} z_{nm} = 1, \ m \in \mathcal{M} \tag{4}$$

$$\sum_{n \in \mathcal{N}} t_{in}^{r} z_{nm} \leq x_i \leq \sum_{n \in \mathcal{N}} t_{in}^{d} z_{nm} - e_i, \ i \in \mathcal{I}_m^{\text{msg}}, \ m \in \mathcal{M} \tag{5}$$

3 Explicit Modelling of Tasks with Multiple Intervals

In the pre-processed instance data it can be observed that a task on a CM can have multiple disjoint intervals where it can execute, as illustrated in Fig. 1. By explicitly considering these intervals, the α-model can be improved as follows. Instead of assigning a task to a section, it is assigned to one of its intervals. With this formulation, Constraints (4)–(5) can be incorporated into the interval assignment instead of being explicitly modelled. This also enables us to strengthen the relaxed problem by introducing capacity constraints for intervals of the major frame that are more fine-grained than the sections. These fine-grained intervals are referred to as segments.

The improved formulation of the α-model is as follows.

  min/max Objective function
  s.t.   AM-scheduling
      Precedence relations
      Generated sequences
      CM-interval assignment (Constraints (6)–(8))

In order to describe the CM-interval assignment component, the following notation is introduced. Task i has a set of multiple intervals, denoted by $\mathcal{Q}_i$, and task i must execute within one of these intervals, $i \in \mathcal{I}_h$, $h \in \mathcal{H}^{\text{CM}}$. Each interval q has a release time $t_{iq}^{\text{r-int}}$ and a deadline $t_{iq}^{\text{d-int}}$ that task i must obey if it is assigned to interval q, $q \in \mathcal{Q}_i$, $i \in \mathcal{I}_h$.

For each pair of a release time $t_{iq}^{\text{r-int}}$, $q \in \mathcal{Q}_i$, $i \in \mathcal{I}_h$ and a deadline $t_{jq'}^{\text{d-int}}$, $q' \in \mathcal{Q}_j$, $j \in \mathcal{I}_h$, create a segment, and let $\mathcal{R}_h$ denote the set of all such segments on CM h, $h \in \mathcal{H}^{\text{CM}}$. Denote the duration of segment r by l_r^{seg}, $r \in \mathcal{R}_h$, $h \in \mathcal{H}^{\text{CM}}$. Introduce the

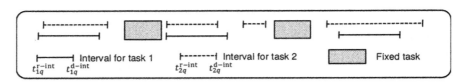

Fig. 1 An illustration of tasks with multiple intervals

set $\mathcal{I}_r^{\text{seg}}$ that includes all tasks that have at least one interval that is a subset of segment r, $r \in \mathcal{R}_h$, $h \in \mathcal{H}^{\text{CM}}$. Further, let $\mathcal{Q}_{ir}$ denote the set of intervals that belong to task i and is a subset of segment r, $r \in \mathcal{R}_h$, $i \in \mathcal{I}_h$, $h \in \mathcal{H}^{\text{CM}}$. Introduce, for interval q and task i, $q \in \mathcal{Q}_i$, $i \in \mathcal{I}_h$, $h \in \mathcal{H}^{\text{CM}}$, a binary variable

$$\alpha_{iq} = \begin{cases} 1 & \text{if task } i \text{ is assigned to interval } q, \\ 0 & \text{otherwise.} \end{cases}$$

The following constraints ensure that each task is assigned to one of its intervals, that each task respect the release time and deadline of its assigned interval and that the capacity of segments are respected.

$$\sum_{q \in \mathcal{Q}_i} \alpha_{iq} = 1, \ i \in \mathcal{I}_h, \ h \in \mathcal{H}^{\text{CM}} \tag{6}$$

$$\sum_{q \in \mathcal{Q}_i} t_{iq}^{\text{r-int}} \alpha_{iq} \leq x_i \leq \sum_{q \in \mathcal{Q}_i} t_{iq}^{\text{d-int}} \alpha_{iq} - e_i, \ i \in \mathcal{I}_h, \ h \in \mathcal{H}^{\text{CM}} \tag{7}$$

$$\sum_{i \in \mathcal{I}_r^{\text{seg}}} \sum_{q \in \mathcal{Q}_{ir}} e_i \alpha_{iq} \leq l_r^{\text{seg}}, \ r \in \mathcal{R}_h, \ h \in \mathcal{H}^{\text{CM}} \tag{8}$$

4 Results

In this section we present performance comparisons between the original model presented in Sect. 2 and the new model introduced in Sect. 3. To further analyse the impact of segment capacity constraints we also present results for the model in Sect. 3 with Constraint (8) removed. The results are presented for the largest of the instances, Instance III, introduced and thoroughly described in [2]. This instance has 7 CMs, 8 AMs, and 19919 tasks.

The tests were conducted using the scheduling tool developed in [2] with Gurobi Optimizer Version 7.0.2 and Python 3.6.0. The objective function for the first iteration was the Center-task objective with different choices of Δ and for later iterations the stabilise objective, for details see [2]. The time limit in the β-model is initially 2 h and whenever an improved integer solution is found, it is reset to 4 hours. The time-out was set to 8 hours in the α-model, with a MIP-gap of 0.1 and the scheduling tool was terminated after 48 hours if no valid schedule was found.

The results in Table 1 show that the model in Sect. 3 finds a valid schedule within the time limit for all choices of the objective function while the model in Sect. 2 only finds a valid schedule when $\Delta = 0.00$. Comparing the model in Sect. 3 with and without Constraint (8) shows that by including the constraint, the number of iterations decreases and so does the total running time.

The results indicate that the performance of the formulation in Sect. 3 is better than the formulation in Sect. 2, suggesting that explicit modelling of multiple intervals is

Table 1 Comparison of models. Time limit exceeded is denoted by TLE

Measurements	Instance III					
	$\Delta = 0.00$	$\Delta = 0.05$	$\Delta = 0.10$	$\Delta = 0.15$	$\Delta = 0.20$	$\Delta = 0.25$
Model in Sect. 2						
Total time (m)	757	TLE	TLE	TLE	TLE	TLE
Iterations	2	TLE	TLE	TLE	TLE	TLE
Time α-model (m)	483	TLE	TLE	TLE	TLE	TLE
Time β-model (m)	260	TLE	TLE	TLE	TLE	TLE
Model in Sect. 3 without Constraint (8)						
Total time (m)	725	1421	1128	TLE	850	194
Iterations	2	4	4	TLE	4	3
Time α-model (m)	449	482	482	TLE	485	136
Time β-model (m)	262	915	622	TLE	343	39
Model in Sect. 3						
Total time (m)	515	1391	769	614	511	77
Iterations	1	4	2	1	1	1
Time α-model (m)	480	482	481	480	480	31
Time β-model (m)	22	866	266	121	18	33

preferred. Future research includes to strengthen the α-model further by deriving valid inequalities based on the explicit modelling of multiple intervals.

Acknowledgements The work is part of a project funded by the Center for Industrial Information Technology (CENIIT) and the work of Emil Karlsson is supported by the Research School in Interdisciplinary Mathematics at Linköping University.

References

1. Baltz, A., El Ouali, M., Jäger, G., Sauerland, V., & Srivastav, A. (2015). Exact and heuristic algorithms for the travelling salesman problem with multiple time windows and hotel selection. *Journal of the Operational Research Society, 66*(4), 615–626.
2. Blikstad, M., Karlsson, E., Lööw, T., & Rönnberg, E. (2017). An optimisation approach for pre-runtime scheduling of tasks and communication in an integrated modular avionic system. Technical report LiTH-MAT-R–2017/03–SE, Linköping University.
3. Christiansen, M., & Fagerholt, K. (2002). Robust ship scheduling with multiple time windows. *Naval Research Logistics, 49*(6), 611–625.
4. Pesant, G., Gendreau, M., Potvin, J.-Y., & Rousseau, J.-M. (1999). On the flexibility of constraint programming models: From single to multiple time windows for the traveling salesman problem. *European Journal of Operational Research, 117*(2), 253–263.

Multi-objective Large-Scale Staff Allocation

Roberto Anzaldua, Christina Burt, Harry Edmonds, Karsten Lehmann and Guangyan Song

1 Introduction: A Staff Allocation Problem

Our client has a large-scale staff allocation problem. As a multinational, the company aims to work on jobs for 2,000 projects. Each year, our client must perform the basic resourcing task of allocating its existing workforce of 1,000 employees to these projects, which comprise of 10,000 jobs. Projects and employees may each have several special requests, totalling 50,000 requests, and the problem is subject to employee work rules and quality requirements.

There are many stakeholders within the client, each with a different perspective on what the priorities should be. While one might favour maximising project demand because this relates to bottom-line revenue, another might also be concerned with impact on employees, such as fair work allocation and less travel, or with meeting policy (we will go on to show its possible to satisfy them all simultaneously). Each of these preferences can be expressed as a soft constraint which can pull the search direction along contradicting axes. Effectively capturing these constraints was a big challenge. In fact, the majority of our client meetings were dedicated to fully understanding and translating business rules such that they could be cast into mathematical expressions. The problem is challenging, most obviously because of its scale, such that no optimisation software can solve it. Computationally, the soft constraints (at least 16 of them) contribute to numerical instability which challenges even state-of-the-art solvers.

For this task we created a pilot tool for our client. This tool extracts the relevant data with encryption, preprocesses the data, runs our proposed heuristic, delivers an optimised schedule and then runs metrics on the output solution. The main contributions of this paper include (a) a description of a real case study, including lessons

R. Anzaldua · C. Burt · H. Edmonds (✉) · K. Lehmann · G. Song
Satalia, London, UK
e-mail: harry@satalia.com
URL: http://www.satalia.com

learned; and, (b) an approach for solving large-scale allocation problems with a large number of soft constraints.

In Sect. 2, we describe the problem in more detail, including problem definition and base models. In Sect. 3, we broadly describe our approach, including an outline of our algorithm. In Sect. 4, we describe practical challenges of the project, before presenting some results in Sect. 5, followed by a discussion.

2 Problem Description and Model

The basis, or academic, version of the problem we consider is as follows.

Definition 1 *(Feasible region of the Staff Allocation Problem)* Given a set of jobs, J; a set of tuples specifying hours required per week, and preferred employee skill level, per job, $(h, w, s) \in W$; a set of employees, E; a set of employee skills, E_s; the staff allocation problem is to find a matching of employees to jobs such that a job can only be satisfied if all hours are satisfied; employee work conditions are satisfied as much as possible; and employee-job required skills are matched as much as possible.

The employee conditions cover both basic and complex rules about how many hours each employee should work, such as maximum weekly allotment, maximum 13 weeks allotment, average work-hours, public holidays, leave requests, training periods and the quantity of work that can be allocated during travel periods. The job preferences cover the number of employees allocated per job, the preferred skills of the employee, fixed requests for specific employees and job conflicts.

In its simplest form the optimisation version of the problem is to satisfy the hard constraints, such that the maximum number of jobs can be satisfied. For simplicity of prose and to better understand the structure of the problem, we disregard many of the complex hard constraints to obtain the following academic version of the problem.

We define the main decisions as follows:

$t_{j,e,w}$ the proportion of work hours allocated to job j, for employee e in week w;
$y_{j,e}$ a binary variable indicating if employee e works on job j;
q_j a binary variable indicating if the job j will be scheduled.

Other parameters include:

$H_{j,w}$ the hours required by job j in week w;
$H_{e,w}^{\max}$ the maximum weekly working hours for employee e.

If a maximum of two employees may work on a job, then the staff allocation problem can be represented by:

$$\max \sum_{e,w,j} t_{j,e,w}$$

$$H_{j,w} q_j = \sum_e t_{j,e,w} \qquad \forall\, j, w, \qquad (1)$$

$$\sum_j t_{j,e,w} \le H_{e,w}^{max} \qquad \forall\, e, w, \qquad (2)$$

$$\sum_e y_{j,e} \le 2 \qquad \forall\, j, \qquad (3)$$

$$\sum_{j,w} t_{j,e,w} \le y_{j,e} \sum_w H_{j,w} \qquad \forall\, j, e,$$

$$t_{j,e,w} \ge 0, \quad y_{j,e}, q_j \in \{0,1\}. \qquad (4)$$

The objective maximises satisfied demand. The first constraint ensures that all the hours for each job of that project are allocated to a set of employees, per week. The second constraint places a limit on the employees weekly work hours. The third constraint limits the number of employees per job to two. The last two constraints ensure that the binary indicator of an employee-job pair is true if and only if there is at least one hour of work allocated to that job. The remaining, hidden constraints place restrictions on the sets of jobs that each employee can work, and can in many cases be preprocessed into these sets rather than explicitly represented as constraints.

This problem has more than 16 soft constraints, ranging from preferences such as continuity of employees in ongoing jobs, the use of alternate skill levels, target working hours, minimising travel time and fair allocation of hours. With this many soft constraints, there arise two challenges. The first of these is how to prioritise, or apply weight, to each of them. To resolve this, we relied on our client to provide their insights, and allowed them to iteratively adjust the weightings after viewing over 20 rounds of solutions. The second challenge arises from interacting and competing constraints. One such example could be fair allocation of work and minimising long-distance work. To minimise long-distance work, it could be preferential to allocate two jobs to the employee that lives in the town where the project is based. However, this can lead to an unfair allocation of work with an employee not based in that town being allocated zero hours. These interactions are complex, and required us to sit down with our client and prioritise each constraint in interacting sets.

3 Method: LNS Inspired Heuristics

Even though the mixed-integer programming model is simple, the scale of the problem renders a model size too large for desktop computers and is so large that even solving the linear programming relaxation at root took longer than 2 h and consumed up to 16GB RAM. To overcome this hurdle, we developed a decomposition strategy as follows.

Input: $i \leftarrow 0$; Initial solution, $sol_0 \leftarrow 0$; create a set of jobs, l_0, ordered by demand in decreasing order; $p_0 \leftarrow 0$; $d_0 = \sum_{0,n_c} D_c$ with $n_c = 5$.

while *time limit has not been reached* **do**
    Choose a subset of jobs in l starting from a specific job index, p_i, until we have enough jobs to fill demand d_i;
    All jobs not in subset, fix to solution sol_i;
    Solve $\rightarrow sol_{i+1}$. Get new random p_{i+1}. Add sol_{i+1} to solution pool;
    If $p_i > |l_i|$, set $p_i = 0$ and random shuffle the list;
    Increase the iterator, i;
end

This approach is guaranteed to be feasible, since we maintain a solution pool that at least contains the initial feasible solution. However, there is no termination criterion, so the solve time must be manually pre-set. Also, the approach is not guaranteed to improve, since it is not enumerative but random. Thus, it is not guaranteed to converge on the optimal solution. In the remaining paragraphs, we will outline solutions for each of these issues. This approach is dependent on having a good initial solution. Since there are no hard constraints on minimum demand satisfied and most of the constraints are soft, not allocating any jobs at all is actually feasible. A very basic and efficient greedy start solution orders the jobs by demand in decreasing order. From the available employee set, we allocate employees to the jobs in turn until no more jobs can be fulfilled.

In order to obtain convergence criteria, we must be able to guarantee that, in the worst case, we completely enumerate all possible solutions, and in the best case, we have a well-defined search direction that improves the solution at each iteration. We influence the search direction by considering the residual capacity in the employee set, as well as the set of unsatisfied jobs. We first find an upper bound on the procedure by solving this partial problem. We then look for employee candidates to swap out of satisfied jobs, to work for hitherto unsatisfied jobs, such that our objective function improves. There are simple ways to identify these candidates, such as looking at travel times and hours worked. This search direction improvement, however, does not guarantee convergence.

In each iteration of our LNS procedure, we have an opportunity to choose the size of the next subproblem. This is useful if the last iteration has not found an improvement. Thus, at each iteration, it is possible that we increase the size of the subproblem (specifically, increase d_i), until we cover the entire problem. In the latter case, we solve the full problem and convergence is guaranteed.

4 Overcoming the Hurdles of Real-World Problems

A Startup working with a multinational corporation can at times face particular obstacles to success. It was essential to ensure our client was engaged right from the beginning and explaining how optimisation works was critical since none of the clients team had experience in this field. Before we could focus on the maths, we faced multiple hurdles such as: obtaining funding approval; vendor onboarding, procurement and passing data security requirements. The multinational client is accustomed to receiving detailed documentation from large software enterprises; Satalia on the other hand is used to being lean on documentation. Although it was labour intensive, we bought into the process, understanding that our client has a level of compliance we must meet.

In the world of consulting, we do not have the luxury of stripping down a problem to an academic version; we must solve the entirety of the real problem for our client. We knew from the outset that the problem was mathematically hard, but it took some time to arrive at the formal description. Since the previous scheduling method was conducted manually, the problem was only defined tacitly and thus had no unified definition among different stakeholders. Equally challenging was understanding all the stakeholders needs and then calibrating the model to produce a solution that satisfied them all. A large part of our initial work was in consulting to extract a formal problem definition. Neither party knew exactly what logic was needed to produce the desired quality, but we had our best estimate. Once we began to produce results that could be analysed, the real learning began. With each new solution, it would become apparent to our client that a constraint was missing from the problem definition, or needed to be reworded. Sharp pivots in approach cost precious development time, but meant our model was moulded closer and closer to reality with each iteration.

We quickly learnt that change could be minimised by ensuring we understood the root purpose for each constraint, and delved beyond surface explanations. On several occasions we re-developed a constraint after discovering there were unintended consequences or better approaches to achieve the true intention. Since Mixed Integer Programming technology is very rigid, once a model is formulated it is hard to change basic decisions, such as what variables mean. This forced us to model constraints generically and independently so we could easily respond to change. Agility in mindset and a mutual understanding that disruption is to be expected and should be accommodated flexibly enabled our team to withstand this level of turbulence.

With such a large problem and so many soft constraints, just understanding whether an output schedule was of good quality or not was hard — the scale of the problem prevented us from obtaining a bound using linear programming. Therefore we built multiple metrics for each constraint that analysed performance, which enabled us to clearly communicate the solution to our client and to receive informed feedback. These metrics were also vital for spotting data, mathematical and implementation bugs.

We overcame the hurdles outlined here, and many more, by augmenting our team and the clients team (using Slack as a communication tool) and by ensuring our mutual ambition of solving the problem as best we could took priority over anything else.

5 Results: Improvements and Impact

We implemented the project in Java8, using the Fico Xpress libraries v.8.2 to solve subproblems. We built the model and ran the heuristic on Google Cloud, using a 4 core machine with 26GB RAM. Our client provided two case studies for the pilot tool phase. The focus of our experiments was to refine the weighting of the soft constraints, in order to find good quality solutions from the perspective of our client. We ran the heuristic for 24 h. Unfortunately, owing to the size of the problem we cannot run the root relaxation to obtain a bound on the quality of our solution. However, we can calculate the amount of demand satisfied in terms of the amount requested, and evaluate several metrics along the axis that the soft constraints try to capture.

The biggest improvement on the manual approach is clearly to save time. Instead of using 10 people to create a schedule in 4 months (i.e. 6,400 work-hours), our heuristic automatically creates better schedules in just 24 h. When looking at comparable datasets, our heuristic satisfies 92% of demand, while our clients solution for the same period only satisfied 80%. Our solutions also obtained higher average employee hours than theirs, showing that our solutions make better use of employee availability in order to fulfill demand. Even though we achieved better demand satisfied, we also reduced total travel time by 15%. Their solution was less compliant with policy, violating 3.94% of heavily weighted rules on average, while ours violated 0%. This result is significant as it flags policy violations for review.

Our solution provides great impact, and therefore the greatest value, through the increase in efficiency of the schedule, while also increasing all aspects covered by our metrics. The drastic improvement in solve time also permits more experimentation in the generation of each schedule. Rather than running the heuristic once, the client will be able to tweak their weightings and explore alternative schedules that better suit their needs for the period in question, allowing easy response to the turbulent external environment.

6 Discussion

In this work, simply arriving at an accurate problem description that adequately captures the problem from the perspective of all stakeholders has been a monumental achievement. Beyond this, our achievements include:

- Building a stable data parser that validates data as it comes in;
- Designing a mathematically sound heuristic that results in high satisfied demand for our client;
- Surpassing performance of the existing approach across all key metrics;
- Enabling our client to incorporate new and interesting soft constraints that they could not have otherwise considered.

The main impact of our work is recognisable as increased efficiency of the schedule. The employees who were previously assigned to this task are now free to work on more important and satisfying activities. By enabling our client to consider new soft constraints, they chose to consider constraints that have social impact on their employees, such as fair allocation of work and the minimising of travel time. This new ability to include arbitrary soft constraints will open up yet more possibilities for our client.

A Permutation-Based Neighborhood for the Blocking Job-Shop Problem with Total Tardiness Minimization

Julia Lange and Frank Werner

1 Introduction

Motivated by applications in production and logistics, the job-shop problem is one of the well-studied models in scheduling research. The increasing complexity of real-world production systems leads to an interest in additional constraints to classical scheduling problems. This is why researches start to regard application-inspired restrictions like setup times, limited buffer capacities and machine flexibility during the last decades. Furthermore, practically relevant optimization criteria based on the earliness and tardiness of jobs or the costs of production are taken into account.

A job-shop scheduling problem with blocking constraints describes production systems with a lack of storage capacity. Jobs have to move directly from one machine to another. In case that the succeeding machine is not idle, the job will block the machine until its processing can be continued. To increase customer satisfaction, the minimization of the total tardiness of all jobs with regard to given due dates is the objective. Since this corresponds to a regular optimization criterion a solution to the problem is a schedule defined by the operation sequences on the machines.

Such a job-shop model is tackled by researchers following a variety of different approaches. In [1], the authors present a generalized graph formulation for the blocking job-shop problem. Based on this, a branch-and-bound-approach is applied to a train scheduling problem with a tardiness-based objective in [2]. Following a similar real-world application, a constructive approach based on a shifting bottleneck procedure for the blocking job-shop with a makespan objective is presented in [3]. Different mixed-integer programming formulations are compared with regard to model size and computation time in [4]. These results give evidence to the necessity

J. Lange (✉) · F. Werner
Otto-von-Guericke-Universität Magdeburg, Magdeburg, Germany
e-mail: julia.lange@ovgu.de

F. Werner
e-mail: frank.werner@ovgu.de

of efficient heuristic methods to obtain good solutions in reasonable run time. In line with this idea, several researchers present tabu search approaches for the blocking job-shop (see [5, 6]), an iterative improvement algorithm is shown in [7] and the problem is solved by an iterated greedy metaheuristic in [8].

Two main difficulties arise during the application of heuristics using a permutation-based solution representation. While any given permutation satisfying the technological routes corresponds to a feasible schedule for the classical job-shop scheduling problem, such a permutation is not necessarily feasible with regard to blocking constraints. In the following, a procedure is presented, which constructs a feasible schedule from any given list of operations. This repair may result in necessary changes of the operation sequences on the machines.

A well-known strategy in neighborhood construction is the adjacent pairwise interchange (API) of two operations on one machine. It is also applied here to the blocking job-shop problem, but may lead to infeasible solutions. The repairing procedure mentioned above is not directly applicable, since regaining feasibility equals reverting the API and reconstructing the initial solution in many cases. The challenging problem is to repair the permutation while preserving the given API, which corresponds to the problem of completing a partial solution to a feasible schedule. Many researchers, e.g. in [5, 6] and [8], ascertain that it is not always possible to set up a feasible schedule from a partial solution without doing any changes in the given part. The corresponding decision problem is shown to be NP-complete in [1].

In the following, a permutation-based neighborhood for the blocking job-shop problem with swaps regarding a total tardiness objective is presented. A procedure to regain the feasibility of a neighbor is shown and with this partially defined solutions are completed to feasible schedules. The neighborhood is embedded in a simulated annealing. Computational experiments are done on randomly generated instances based on a single-track train scheduling problem as well as on benchmark instances. The results are compared to those obtain by solving the corresponding MIP formulations given in [4].

2 Problem Description and Representation of Schedules

The problem involves a set of jobs $\mathcal{J} = \{J_i \mid i = 1, ..., n\}$ having to be processed on a set of machines $\mathcal{M} = \{M_k \mid k = 1, ..., m\}$. Each job J_i consists of a set of operations, where $O_{i,j}$ describes the j-th operation of job J_i. A machine $M_k \in \mathcal{M}$ is assigned to every operation $O_{i,j}$ defining the technological route of every job $J_i \in \mathcal{J}$. Additionally, release dates r_i and due dates d_i are given for each job and recirculation is allowed. Following the three field notation, the blocking job-shop problem tackled here can be denoted by $J \mid r_i, d_i, block, recr \mid \sum T_i$.

A solution S is a schedule, which is uniquely defined by a permutation of the operations. This permutation is expressed by list indexes $lidx(O_{i,j}) \in \{0, 1, \ldots, n_{op} - 1\}$ of all n_{op} operations $O_{i,j}$. With it, the operation sequences on the machines are defined, where $midx(O_{i,j}) \in \{0, 1, \ldots, R_k - 1\}$ denote the machine indexes of the operations with R_k indicating the number of operations on machine M_k.

With regard to the solution encoding applied in the neighborhood construction, the permutation or list of operations refers to the operation-based representation and the corresponding operation sequences constitute the machine-based representation. For each operation $O_{i,j}$ a predecessor $pred(O_{i,j}) = O_{i,j-1}$ and a successor $succ(O_{i,j}) = O_{i,j+1}$ are defined by the technological route of job J_i and a machine predecessor $\alpha(O_{i,j})$ and a machine successor $\beta(O_{i,j})$ are derived from the operation sequences on the machines. Both encodings are shown below for a small instance of four jobs and three machines with the technological routes $J_1 : M_1 \rightarrow M_3$, $J_2 : M_3 \rightarrow M_2$, $J_3 : M_1 \rightarrow M_2 \rightarrow M_3$, $J_4 : M_2 \rightarrow M_1 \rightarrow M_2$ and $O = 10$, $R_1 = 3$, $R_2 = 4$, $R_3 = 3$, respectively.

operation-based representation

$S =$	$O_{1,1}$	$O_{3,1}$	$O_{4,1}$	$O_{3,2}$	$O_{4,2}$	$O_{2,1}$	$O_{1,2}$	$O_{2,2}$	$O_{3,3}$	$O_{4,3}$
$lidx(O_{i,j})$	0	1	2	3	4	5	6	7	8	9

machine-based representation

M_3	$O_{2,1}$	$O_{1,2}$	$O_{3,3}$	
M_2	$O_{4,1}$	$O_{3,2}$	$O_{2,2}$	$O_{4,3}$
M_1	$O_{1,1}$	$O_{3,1}$	$O_{4,2}$	

$midx(O_{1,1}) = midx(O_{4,1}) = midx(O_{2,1}) = 0$
$midx(O_{3,1}) = midx(O_{3,2}) = midx(O_{1,2}) = 1$
$midx(O_{4,2}) = midx(O_{2,2}) = midx(O_{3,3}) = 2$
$midx(O_{4,3}) = 3$

The solution given by the operation-based representation fulfills the technological routes but it is infeasible with regard to blocking constraints. The second operation in the list $O_{3,1}$ cannot be scheduled, since the first operation $O_{1,1}$ is blocking machine M_3 until the processing of $O_{1,2}$ begins on M_1. Solution S is infeasible for the blocking job-shop problem. For every pair of operations $O_{i,j}$ and $O_{i',j'}$ on machine M_k with $midx(O_{i,j}) < midx(O_{i',j'})$ the following blocking-related index constraint has to be fulfilled.

$$lidx(succ(O_{i,j})) \leq lidx(\beta(O_{i,j})) \Leftrightarrow lidx(O_{i,j+1}) \leq lidx(O_{i',j'})$$

The successor of an operation has to be scheduled at the same time or earlier than its machine successor. This constraint is defined by an inequality here, since swaps are allowed to be part of a feasible solution. A simultaneous movement of an operation from machine M_k to M_{k+1} and another operation from machine M_{k+1} to M_k is regarded to be possible. Considering the generalized graph representation of the problem (see [1]), this refers to a special type of cycles in the solution, which are defined to be feasible.

The procedure to regain feasibility is applied to all operations one by one with increasing list indexes to ensure the blocking-related index constraints. For every operation $O_{i,j}$ in the permutation, the index constraints of the form $lidx(succ(\alpha(O_{i,j}))) \leq lidx(O_{i,j})$ are set up and checked. Violated constraints are fulfilled by left-shifts of operations required at a lower list index. Since the procedure is applied to the permutation from the left to the right, it is guaranteed that the list is feasible for all list indexes less than or equal to the current index and this feasibility cannot be destructed by the end of the repair.

In the example, the first violated blocking-related constraint is $lidx(O_{1,2}) \leq lidx(O_{3,1})$ at list index 1. $O_{1,2}$ is shifted to the left and inserted at list index 1 in the permutation. With this, the operation sequence on machine M_1 changes to $O_{1,2} \to O_{2,1} \to O_{3,3}$. This indicates that adapting a solution to be feasible might cause significant changes in the permutation.

In the example, swaps are appearing with operations $O_{3,2}$ and $O_{4,2}$ on machines M_1 and M_2 and operations $O_{2,2}$ and $O_{3,3}$ on machines M_2 and M_3. In the repairing procedure, these operations are defined to form swap groups $(O_{3,2}, O_{4,2})$ and $(O_{2,2}, O_{3,3})$ with $lidx(O_{3,2}, O_{4,2}) = 4$ and $lidx(O_{2,2}, O_{3,3}) = 6$, so that the index constraints for both pairs of operations are fulfilled with equality. The solution $\bar{S} = [O_{1,1}, O_{1,2}, O_{3,1}, O_{4,1}, (O_{3,2}, O_{4,2}), O_{2,1}, (O_{2,2}, O_{3,3}), O_{4,3}]$ is a feasible solution to the blocking job-shop problem. The corresponding schedule with $r_1 = r_2 = r_3 = r_4 = 0$ is given in the following Gantt-chart.

M_3		$O_{1,2}$	$O_{2,1}$	$O_{3,3}$	
M_2	$O_{4,1}$	╳	$O_{3,2}$	$O_{2,2}$	$O_{4,3}$
M_1	$O_{1,1}$	$O_{3,1}$	$O_{4,2}$	╳	

3 Neighborhood Construction

Neighbors are defined by reverting the order of two operations on one machine. This approach is applied regarding the additional conditions that the operation shifted to the lower machine index belongs to a tardy job and that there is no idle time on the machine between the operations chosen for the API. The neighborhood is called 'Tardy Adjacent Pairwise Interchange' (TAPI). As an example, the operations $O_{3,2}$ and $O_{2,2}$ are interchanged on machine M_2. This API changes the machine indexes to $midx(O_{2,2}) = 1$ and $midx(O_{3,2}) = 2$. It can be transferred to the operation-based representation by a left shift (shifting $O_{3,2}$ to a list index higher than $lidx(O_{2,2})$) or a right shift (shifting $O_{2,2}$ to a list index lower than $lidx(O_{3,2})$). These transfer options may lead to different neighbors, since the following repair procedure strongly depends on particular list index relations.

Breaking up the swap groups and transferring the API to the operation-based representation by left shift results in the neighbor $N = [O_{1,1}, O_{1,2}, O_{3,1}, O_{4,1}, O_{2,1}, O_{2,2}, O_{3,2}, O_{4,2}, O_{3,3}, O_{4,3}]$. This permutation is infeasible with regard to the blocking constraints, since $lidx(O_{4,2}) \leq lidx(O_{2,2})$ is not fulfilled. The application of the repair procedure shifts $O_{4,2}$ and $O_{3,2}$ to the left as a swap group, reverts the API and constructs the same schedule $\bar{S}$. Due to this, it is necessary to set $O_{2,2} \to O_{3,2}$ as a non-reversible precedence constraint for the new schedule. The technological routes together with the given API constitute the partial solution, which has to be completed to a feasible schedule, while changing the operation sequences on the machines as little as possible.

The strategy is to perform additional APIs between the operation shifted to the left in the initial API (here $O_{2,2}$) and its new machine predecessors (here $O_{4,1}$). Once the repair procedure is going to revert the given precedence relation, the operation is interchanged with its current machine predecessor and the repair procedure is restarted. If the operation is already at machine index 0, additional APIs are applied to its job predecessors (here $O_{2,1}$) as well. Repair and restart are repeated until the procedure constructs a feasible solution involving the given API. In the best case, there is no additional API necessary, while in the worst case, all operations of one job are shifted to machine index 0 on their machines. This approach is similar to the job insertion technique presented in [5, 6], but it is less restricted and swaps of operations are allowed.

In the example, operation $O_{2,2}$ is shifted to the left before $O_{4,1}$ on M_2 and after restarting the repair procedure $\bar{N} = [O_{1,1}, O_{1,2}, O_{2,1}, O_{2,2}, O_{3,1}, O_{4,1}, (O_{3,2}, O_{4,2}), O_{3,3}, O_{4,3}]$ defines the feasible neighbor of $\bar{S}$.

4 Computational Experiments and Results

The neighborhood is embedded in a simulated annealing (SA) metaheuristic. The SA algorithm is applied to solve train-scheduling-inspired instances (TS instances), that are randomly generated following a given network structure with the number of machines $m = 11$ and the number of jobs $n \in \{10, 15\}$. As a benchmark, the Lawrence instances (see [9]) are solved with additional release dates and due dates determined by the rules given in [4]. In total, there are 5 distinct instances to solve for 10 different instance sizes (m, n).

The initial solution is determined as the best solution found by several priority rules. A geometric cooling scheme is applied, where the starting temperature and multiplier are chosen in accordance to the absolute measure of the mean objective function value of the instances. The length of the Markov chain set up per temperature level varies dependent on the total number of operations of the instance. Thus, the total number of iterations ranges dependent on the instance size for the TS instances between 30000 and 60000 and for the Lawrence instances between 11000 and 64000 iterations.

In the following table, the performance of SA is compared to the MIP results presented in [4]. The number of instances, for which the optimal (opt) or a feasible solution is obtained by MIP, is given for each instance size. Similarly, the number of instances, for which SA found the optimal solution or improved the MIP result (opt/im), is stated together with the number of instances, for which SA obtained a feasible solution with a gap less than 10% compared to the MIP result.

The SA algorithm performs well with regard to feasibility and optimality while applying the presented neighborhood. The comparison shows that SA found optimal

	TS inst.		Lawrence instances							
(m, n)	(11, 10)	(11, 15)	(5, 10)	(5, 15)	(5, 20)	(10, 10)	(10, 15)	(10, 20)	(10, 30)	(15, 15)
total	5	5	5	5	5	5	5	5	5	5
MIP										
opt	5	3	5	1	–	5	1	–	–	2
feasible	–	2	–	4	5	–	4	5	1	3
SA										
opt/im	4	1	4	2	2	3	1	3	5	–
< 10%	1	3	1	1	3	1	1	–	–	–

and near-optimal solutions for the small instances and outperforms the MIP solver for larger instance sizes, e.g. (5, 20), (10, 20) and (10, 30). The application of a pure SA algorithm and a hybrid approach combining a heuristic method and a MIP solver seem to be promising to increase the solution quality and the size of solvable problems.

References

1. Mascis, A., & Pacciarelli, D. (2002). Job-shop scheduling with blocking and no-wait constraints. *European Journal of Operational Research*, *143*(3), 498–517.
2. D'Ariano, A., Pacciarelli, D., & Pranzo, M. (2007). A branch and bound algorithm for scheduling trains in a railway network. *European Journal of Operational Research*, *183*(2), 643–657.
3. Liu, S. Q., & Kozan, E. (2009). Scheduling trains as a blocking parallel-machine job shop scheduling problem. *Computers and Operations Research*, *36*(10), 2840–2852.
4. Lange, J., & Werner, F. (2017). Approaches to modeling train scheduling problems as job-shop problems with blocking constraints. *Journal of Scheduling*. https://doi.org/10.1007/s10951-017-0526-0.
5. Bürgy, R. (2017). A neighborhood for complex job shop scheduling problems with regular objectives. *Journal of Scheduling*, *20*(4), 391–422.
6. Groeflin, H., & Klinkert, A. (2009). A new neighborhood and tabu search for the blocking job shop. *Discrete Applied Mathematics*, *157*(17), 3643–3655.
7. Oddi, A., Rasconi, R., Cesta, A., & Smith, S. F. (2012). Iterative improvement algorithms for the blocking job shop. In *ICAPS*.
8. Pranzo, M., & Pacciarelli, D. (2013). An iterated greedy metaheuristic for the blocking job shop scheduling problem. *Journal of Heuristics*, 1–25.
9. Lawrence, S. (1984). *Supplement to resource constrained project scheduling: an experimental investigation of heuristic scheduling techniques*. Pittsburgh: GSIA, Carnegie Mellon University.

Preemptive Scheduling of Jobs with a Learning Effect on Two Parallel Machines

Marcin Żurowski

1 Introduction

An important part of scheduling theory concerns preemptive scheduling algorithms, while the schedule length is the most popular schedule optimality criterion. Job preemption usually decreases the length of a schedule. In literature one can find many examples of scheduling problems with job preemption. The most known example is McNaughton's algorithm for finding an optimal preemptive schedule for jobs with fixed processing times [1]. There is assumed the most frequently encountered definition of job preemption, saying that any job can be preempted at any time and resumed later without any cost. Although it is the most popular definition, there are also known other ones such as the restricted job preemption [2] or job preemption at integer time moments [3].

Definitions mentioned above are applied only to scheduling problems with fixed job processing times. Recently, there is a growing interest to scheduling problems with variable processing times, since such problems appear in many applications [4, 5]. The variable job processing times may depend on the number of already executed jobs [6], the starting times of jobs [7] or the positions of jobs in a schedule [5]. Among the models of position-dependent job processing times the most common is the one with a learning effect. In this case, the processing time of each job is a product of a fixed basic processing time of the job and the value of a non-increasing function of the job position in a schedule.

Literature on scheduling with variable job processing times is focused on non-preemptive cases only. To the best of our knowledge, the problems of preemptive scheduling of jobs with variable processing times, in particular those with a learning effect, have not been considered earlier. Therefore, in this paper we consider a scheduling problem, where a learning effect and job preemption exist together. We

M. Żurowski (✉)
Faculty of Mathematics and Computer Science, Adam Mickiewicz
University in Poznań, Umultowska 87, 61-614 Poznań, Poland
e-mail: marcin.zurowski@amu.edu.pl

will call it *problem PSLE* (*Preemptive Scheduling with a Learning Effect*). Similarly to [8, 9], we assume that any job can be preempted at any time without cost, but some additional conditions must be satisfied. For this problem, we present a few of its properties and an exact algorithm.

The remaining sections of the paper are organized as follows. In Sect. 2, we formulate our problem. In Sect. 3, we introduce a new definition of job preemption. In Sect. 4, we present basic properties of the problem. In Sect. 5, we present an exact algorithm for this problem.

2 Problem PSLE Formulation

The problem under consideration can be formulated as follows. We are given $n \geq 3$ jobs $J_1, J_2, \ldots, J_n$ and two parallel identical machines M_1, M_2 available from time 0. All jobs are independent, and the processing time of each job depends on the position of the job in a schedule. More precisely, the processing time of job J_j scheduled without preemption on the rth position in schedule equals $p_{j,r} = p_j r^a$, where p_j is the basic processing time of job J_j, r is the position of this job in the schedule, and $a < 0$ is the learning index. We schedule the jobs on the both machines without idle times. The basic processing time of job J_j after preemption, defined similarly, is the sum of basic processing times of two parts of the job: the one completed before preemption and the one after preemption. We allow the preemption of jobs but, in view of reasons stated in the next section, only one job can be preempted in a given schedule. Both parts of the preempted job are treated like two new independent jobs, which cannot be executed on both machines at the same time or on the same machine in different time intervals. The criterion of schedule optimality is the maximum completion time, $C_{\max} = \max\{C_j\}$, where C_j denotes the completion time of job J_j.

3 Job Preemption in Problem PSLE

In this section, we introduce a definition of job preemption in problem PSLE. We begin with some remarks on job preemption in scheduling problems with fixed job processing times.

In classical case [1], when job processing times are fixed, the sum of processing times of both parts of a preempted job is equal to the processing time of the job without preemption. However, when job processing times are variable and depend on their positions in a schedule, the sum of the processing times of both parts of a preempted job may not be equal to the processing time of the job without preemption. This is caused by the fact that the processing times of position-dependent jobs are non-increasing functions of their positions. Hence, a sequence of preemptions of the same job iteratively decreases the processing time of the job. Therefore, we will apply the following definition of job preemption.

Definition 1 A job with a variable position-dependent processing time is said to be preemptable if one can interrupt its execution at most once at any time before its completion and resume it later in such a way that both parts of this job cannot be executed on both machines at the same time or at the same machine in different time intervals, the basic processing time of this job equals the sum of the basic processing times of both parts of the job and the processing times of parts of the preempted job are described by functions of the same form as the processing time of the job without preemption.

Let us notice that Definition 1 does not specify on which machine the preemption of a job occurs or on which machine the second part of the preempted job is executed. In the paper, we assume that preemption concerns only the job scheduled in the first position on machine M_2 and the second part of a preempted job is executed only in the last position on machine M_1.

4 Properties of the PSLE Problem

In this section, we present a few properties of problem PSLE.

We will use the following notation. By J we will denote the set of all jobs, by J^i we will denote the sequence of non-preempted jobs scheduled on the ith machine, $i = 1, 2$, in the SPT order, by $J^i_{[j]}$ we will denote the jth job in this sequence. Finally, by $L^i_{[q]} = \sum_{j=q}^{|J^i|} p^i_{[j]} j^a$ we will denote the sum of processing times of all the jobs from set J^i except of jobs $J^i_{[1]}, \ldots J^i_{[q-1]}$.

The first property follows from a single machine problem result [9].

Property 1 *In any optimal schedule for problem PSLE all non-preempted jobs assigned to machine M_i $i = 1, 2$, are scheduled in the SPT order with respect to the basic processing times of the jobs.*

The next property is a consequence of Definition 1 and describes the *division factor* of a job, which is the ratio of the basic processing time of the first part of a preempted job and the basic processing time of the whole job. This property relates the division factor with total machine loads. Let us recall that the total load of a machine is the sum of the actual processing times of all jobs assigned to the machine.

Property 2 *Let job J_k be scheduled in a feasible schedule for problem PSLE as the last one on machine M_1 and let s be the position of J_k in the schedule. If total loads of machines M_1 and M_2 are the same, then for the division factor x_k of J_k we have* $0 \leq x_k = \frac{L^2_{[2]} - L^1_{[1]} + p_k}{p_k(s^a + 1)} < 1.$

The next property describes how much the total loads of machines differ.

Property 3 *Let job J_k be scheduled in a feasible schedule for problem PSLE as the last one, let s be the position of J_k in the schedule and let $D = L^2_{[2]} - L^1_{[1]}$. Then $D + p_k \geq 0$ and $D < p_k s^a$.*

The last property shows that problem PSLE is not a direct generalization of the $P2|pmtn|C_{\max}$ problem of two-machine preemptive scheduling with fixed job processing times and the $C_{\max}$ criterion. Namely, in the latter case [1], the order of jobs assigned to a particular machine is not important. Indeed, in the case of two machines for a given optimal schedule σ we have

$$C_{\max}(\sigma) = \max \left\{ \max_{1 \leq j \leq n} \{p_j\}, \frac{1}{2} \sum_{j=1}^{n} p_j \right\}. \tag{1}$$

Moreover, any rearrangement of non-preempted jobs assigned in such a schedule to the same machine leads also to an optimal schedule, since the rearrangement does not change the value of $C_{\max}$. Hence, there are many optimal schedules, and all of them have the same length satisfying (1). However, if jobs have variable position-dependent processing times, then schedules obtained by rearrangements of jobs may have different schedule lengths.

Property 4 *Consider a feasible schedule for problem PSLE. If total loads of machines M_1 and M_2 are equal, then length of the schedule may be affected by the sequences of jobs assigned to the machines.*

We illustrate Property 4 by an example. Let $n = 3$, $a = -1$ and the basic processing times of jobs be equal to $p_1 = 1$, $p_2 = 2$, $p_3 = 3$. Figure 1a shows a schedule σ_1 in which machines M_1 and M_2 complete the processing of jobs at the same time, and the division factor of J_3 is $x_3 = \frac{2}{3}$. Then $C_{\max}(\sigma_1) = 2$. However, as it is shown in Fig. 1b, if we will change the sequence of jobs on machines M_1 and M_2, and preempt J_2 with the division factor $x_2 = \frac{5}{6}$, then in the new schedule, σ_2, both the machines complete the processing of jobs at the same time again but the schedule is shorter, $C_{\max}(\sigma_2) = 1\frac{5}{6}$.

The time complexity of problem PSLE is unknown, though we conjecture that the problem is at least NP-hard in the ordinary sense.

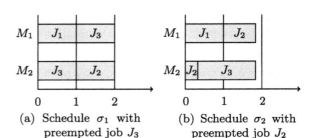

Fig. 1 Two different schedules with a single preempted job

(a) Schedule σ_1 with preempted job J_3

(b) Schedule σ_2 with preempted job J_2

5 Exact Algorithm for Problem PSLE

In this section, we present an exact algorithm for problem PSLE. Given on input the basic processing times of jobs, this algorithm generates on output two sequences of jobs and two parts of a preempted job.

Below we present a pseudocode of this algorithm. In the pseudocode we use the following functions and procedures. Function allSubsequences generates all subsequences of the set given as argument. Procedure setSubsequences returns the sequence of jobs scheduled on machine M_2 and the division factor of a preempted job. Function sortSPT sorts a corresponding sequence of jobs in the SPT order. Finally, function getLength returns the length of the schedule depending on the sequences of jobs assigned on the both machines, the basic processing time of preempted job and the value of the division factor: if $x_k < 0$ then getLength returns $L^1_{[1]}$, if $x_k \geq 1$ then it returns $\max\{L^1_{[1]} + p_k j^a, L^2_{[1]}\}$, otherwise it returns $L^1_{[1]} + x_k p_k j^a$, where $j = |J^1| + 1$.

Algorithm SolvePSLE($n, J_1, J_2, \ldots, J_n$)
1 $\bar{J} = \text{sortSPT}(J)$;
2 $J^1 = \bar{J}; J^2 = \bar{J} \setminus J^1$
3 $C_{\max} = L^1_{[1]}; k = 1$
4 **for** $i = 1$ **to** n **do**
5    $\underline{J} = \bar{J} \setminus \{J_i\}$
6   **forall** $R \in \text{allSubsequences}(\underline{J})$ **do**
7   setSubsequences(R, i, J^2, x_i)
8      $c = \text{getLength}(R, J^2, p_i, x_i)$
9   **if** $c < C_{\max}$ **then**
10        $C_{\max} = c; J^1 = R; J^2 = \underline{J} \setminus J^1; k = i$
11 setSubsequences(J^1, k, J^2, x_k)
12 $p'_k = x_k p_k$; add to $\bar{J}$ job J'_k with processing time p'_k
13 $p''_k = (1 - x_k) p_k$; add to $\bar{J}$ job J''_k with processing time p''_k
14 **return** J^1, J^2, J'_k, J''_k

Algorithm 1: Exact algorithm for problem PSLE

Algorithm 1 works as follows. In line 1 function sortSPT sorts all jobs. In line 3 initial assignments are made. In line 4 the loop for iterates through all jobs. The job J_i is preempted. In line 5 Algorithm 1 determines the sequence $\underline{J}$ of non-preemptive jobs. In line 6 the loop forall iterates through the all subsequences of the sequence $\underline{J}$. For each such a subsequence scheduled on M_1 in line 7, setSubsequences generates subsequence J^2 scheduled on M_2 and the division factor x_i of job J_i. Based on the sequence of jobs scheduled on the both machines, the basic processing time of division job p_i and the division factor x_i, in line 8 getLength generates the length of the schedule c. If c is less than $C_{\max}$, then in line 10 we update $C_{\max}$, J^1 and the index k of the preempted job. When both loops are completed in line 11, we generate the sequence of jobs for M_2 and the division factor x_k of the job J_k. Finally, algorithm generates the both parts, J'_k and J''_k, of the preempted job in lines 12 and 13. Results of Algorithm 1, i.e. J^1, J^2, J'_k and J''_k, are returned in line 14.

Theorem 1 *Algorithm 1 solves problem PSLE in time $O(n^2 2^n)$.*

Theorem 1 follows from the fact that Algorithm 1 generates all feasible schedules for problem PSLE. Functions `setSubSequences` and `getLength` have linear complexity, provided that the values of powers r^a have been calculated in advance. The loop `for` in line 4 is executed n times and the loop `forall` in line 6 is executed $O(2^n)$ times. Hence, the total running time of Algorithm 1 is $O(n^2 2^n)$.

Acknowledgements I thank to Stanisław Gawiejnowicz (Adam Mickiewicz University in Poznań) for introducing me to problem PSLE and for help in improving the presentation in the paper. I also thank to Bartłomiej Przybylski (Adam Mickiewicz University in Poznań) for remarks, which resulted in a more strict formulation of some results.

References

1. McNaughton, R. (1959). Scheduling with deadlines and loss functions. *Management Science*, 6, 1–12. https://doi.org/10.1287/mnsc.6.1.1.
2. Ecker, K., & Hirschberg, R. (1993). Task scheduling with restricted preemptions. *Lecture Notes in Coputer Science*, 694, 464–475. https://doi.org/10.1007/3-540-56891-3_37.
3. Baptiste, Ph., Carlier, J., Kononov, A., Queyranne, M., Sevastyanov, S., & Sviridenko, M. (2012). Integer preemptive scheduling on parallel machines. *Operations Reasearch Letters*, 40, 440–444. https://doi.org/10.1016/j.orl.2012.06.011.
4. Agnetis, A., Billaut, J.C., Gawiejnowicz, S., Pacciarelli, D., & Soukhal, A. (2014). *Multiagent scheduling: models and algorithms*. Berlin: Springer. https://doi.org/10.1007/978-3-642-41880-8.
5. Strusevich, V., Rustogi, K. (2017). *Scheduling with time-changing effects and rate-modifying activities*. Berlin: Springer. https://doi.org/10.1007/978-3-319-39574-6.
6. Gawiejnowicz, S. (1996). A note on scheduling on a single processor with speed dependent on a number of executed jobs. *Information Processing Letters*, 57, 297–300. https://doi.org/10.1016/0020-0190(96)00021-X.
7. Gawiejnowicz, S. (2008). *Time-dependent scheduling*. Berlin: Springer. https://doi.org/10.1007/978-3-540-69446-5.
8. Coffman jr, E.G. (Ed.). (1976). *Computer and job-shop scheduling theory*. New York: Wiley.
9. Mosheiov, G. (2001). Scheduling problems with a learning effect. *European Journal of Operational Research*, 132, 687–693. https://doi.org/10.1016/S0377-2217(00)00175-2.

Part XVIII
Simulation and Statistical Modelling

An Agent-Based Simulation Using Conjoint Data: The Case of Electric Vehicles in Germany

Markus Günther, Marvin Klein and Lars Lüpke

Agent-based models are currently in wide use in innovation and technology diffusion research, as they are able to capture the inherent complexity arising from adoption processes and they allow the consideration of various influences of the underlying social systems. While they are sometimes criticized as "toy models", agent-based models often do not reach their full potential if they lack an empirical foundation. Therefore, we present an agent-based simulation that addresses consumers' adoption behavior of electric and plug-in hybrid electric vehicles in Germany using various empirical data sources for parametrization and validation. In particular, we conducted a focus group and a choice-based conjoint study. Additionally, our model is to our knowledge the first that takes into account explicitly and comprehensively the supply of home charging options.

1 Introduction

Agent-based simulation (ABS) has become increasingly popular in innovation and technology diffusion research; it enriches traditional approaches (like those based on differential equations or system dynamics approaches) by explicitly modelling the diffusion process at a micro-level (for a review of agent-based models of innovation diffusion, see, e.g., [9], and for some recent applications see [13, 16, 17, 19]). Such

M. Günther (✉) · M. Klein · L. Lüpke
Faculty of Business Administration and Economics, Bielefeld University, Universitaetsstr. 25, 33615 Bielefeld, Germany
e-mail: markus.guenther@uni-bielefeld.de

M. Klein
e-mail: marvin.klein@uni-bielefeld.de

L. Lüpke
e-mail: lars.luepke@uni-bielefeld.de

models allow in particular the consideration of heterogeneity of consumers who differ in their preferences, are geographically distributed across regions, are connected to each other in various ways within a social system, and act as well as react based on their limited available information. At the same time, agent-based approaches are sometimes criticized as "toy models" [4], especially if they lack an empirical foundation and therefore do not adequately capture actual behavior in real markets. Even when they are empirically grounded, parameters are often derived from (aggregated) sociodemographic datasets, and individual choice-behavior is therefore not taken into account. Thus, profound parametrization as well as validation of agent-based models often receives less attention.

We present an ABS that builds on empirical data derived primarily from a choice-based conjoint study (CBC). CBC is a (realistic) purchase-decision simulation in which participants repeatedly choose between products with different attribute levels. Our application case focuses on generation Y's (future) adoption behavior on electric (EV), plug-in hybrid electric (PHEV), and conventional vehicles (CV) in Germany. Although several studies on EVs using agent-based simulation have been published (e.g., [13, 16, 18]), they generally lack the abovementioned empirical foundation and thus might not fully represent actual market behavior. To the best of our knowledge, [18] is the only work using conjoint data to calibrate an ABS for EVs. However, as they have a different focus than our study, they do not consider aspects such as density of charging infrastructure. Our model, in contrast, explicitly considers station density, fast charging, and the capability of home charging as well as technological progress.

The remainder of this paper is structured as follows. In Sect. 2 we illustrate the CBC study, which we use to parametrize our agent-based model (Sect. 3). The results of our study are discussed in Sect. 4 and the paper concludes with a summary and outlook toward future research.

2 Using a Choice-Based Conjoint Study for Parametrizing an Agent-Based Simulation

In the process of empirical parametrization of our ABS, we first identified the relevant product attributes through a literature review and by conducting a focus group. This resulted in seven attributes that differ across the three vehicles (CV, EV, PHEV): (i) engine type, (ii) price, (iii) consumption costs, (iv) station density, (v) charging time, (vi) range, and (vii) home charging option. Interestingly, although several studies on EVs have addressed the market potential of EVs by using discrete-choice experiments (e.g., [5, 7, 18]), effects arising from the possibility of home charging have been, so far, largely unconsidered, even though this option is a prominent beneficial differentiation between EVs/PHEVs and CVs (e.g., [3, 10, 12]).

Based on these findings, we conducted a CBC study to address consumers' main adoption barriers and discover their preference structure for parameterizing our simulation. As the importance of several attributes and their levels differ massively among vehicle types, we focus on new cars of the compact class, which is the most popular class in terms of sales in Germany [11]. Note, as we used an alternative-specific design, range and charging time were only displayed for EVs and the attribute home charging option was exclusively shown for EVs and PHEVs.

Respondents were asked to choose their preferred car concept in twelve different choice tasks. Every choice task consisted of a randomly constituted set of one EV, one PHEV, and one CV. Additionally, we collected data on driving behavior, current vehicle, anticipated next vehicle purchase, home charging options, and communication behavior (e.g., trust in advertising and personal communications concerning vehicles).

The CBC study was conducted in July 2016 with the target sample group of young German potential vehicle buyers. Following [14], we used Hierarchical Bayes estimation to determine the individual parameters for the 552 participants. The root likelihood (geometric mean of the predicted probabilities) of 0.72 and a percent certainty of 0.70 indicate solid goodness of fit of the data. In-depth analysis of the CBC data indicates high standard deviations concerning preferences and importance of attributes, especially for engine type. This finding, inter alia, calls for a simulation approach that is capable of capturing heterogeneity, such as an agent-based approach.

3 Agent-Based Model

Products—in our case, CV, PHEV, and EV—are characterized by various attributes that differ not only in their performance levels (e.g., range) but also in their availability (e.g., home charging is not available for CVs). Attributes may change over time due to technological advances between product generations [6] or a change in station density or price. Excepting price, the true performance of a given attribute may not be instantly observable. However, once consumers have adopted a product, they learn about the attribute levels through first-hand experience.

As **Consumers** are initially only aware of the product they own and its corresponding attributes (in our case, CVs only). They learn about the other products and form attitudes over time (i) by being exposed to marketing activities, (ii) by receiving information on the attitudes of their peers via word of mouth, and, finally, (iii) by first-hand experience when using the product after adoption. Note, consumers do not trust all sources of information equally, and—as observed in our study—the biggest impact on the perception of products (certainly) originates from first-hand experience, followed by word of mouth, and, finally, marketing activities. To exchange information, consumers must be embedded into a social network. As we use geographic positioning of agents, we used the extended Barabási-Albert [1] network algorithm described in [17], which anticipates a higher probability of interconnectedness based on geographic proximity.

Based on individual buying behavior, consumers enter the buying process and evaluate all available products of which they are aware. Evaluation of a product is done using an additive utility function that takes into account the attitudes of a consumer agent concerning each known attribute of each available product (using individual part-worth utilities from the conjoint study) and (the obviously objective) available price. Note, unknown product attributes are neglected in the utility estimation and the benefit from home charging is only considered if consumers have the physical ability for it. The product with the highest total utility is purchased. We allow for repurchases but exclude a non-buying option. After adoption, consumers start using the product and thus lean more about the "true" performance of the attributes. Note that our model distinguishes all five phases of the adoption of an innovation as described by [9].

Finally, **marketing's** main purpose (mainly mass media advertising) is to inform consumers about new products, their attributes, and their performance.

4 Simulation Results and Discussion

The model was implemented using AnyLogic 7.0.3. Parametrization was done using data from our CBC study (for the part-worth utilities; see Sect. 2), an online survey (for, e.g., parametrizing personal communication on the topic vehicles), census data (for the geographical distribution of the agents), and various sources of technical data about the vehicles. We simulate a time horizon of 15 years.

Different approaches for model validation were conducted, for instance, a cross-model validation using the choice simulator of [15], which results in almost identical market shares.

The first of four different scenarios on the diffusion and the development of the market shares of EVs is the baseline scenario (Fig. 1, upper left) and shows for both EVs and PHEVs the typical s-shaped curve of innovation diffusion. In scenario 2, the product attributes of EVs improve due to technological progress and increasing charging station density. Additionally, a subsidy on price is granted for the first three years, which reflects the current situation in Germany, and marketing efforts are increased. This subsidy initially pushes diffusion of PHEVs, but after three years, the market share is slowly overtaken by the technologically improving EVs (Fig. 1, upper right). In scenario 3, we substitute the price subsidy with a measure that allows all consumer agents to have the option of home charging. This additional benefit for EVs and PHEVs compared to CVs leads to a significant increase in market shares for both alternative vehicles (Fig. 1, lower left). Finally, in the last scenario we also assume technological progress without a subsidy on the initial price, but we lower the charging time at every station to 30 min. This measure strongly accelerates the diffusion of EVs at the expense of PHEVs, especially as the performance of EVs is relatively close to that of CVs (Fig. 1, lower right).

Besides technological progress (e.g., in range), charging time is a critical factor hindering the diffusion of EVs; charging station density is of less importance, as long as a sufficient number of charging points exist. In contrast, our findings reveal that home charging would be a substantial benefit for both PHEVs and EVs. In the

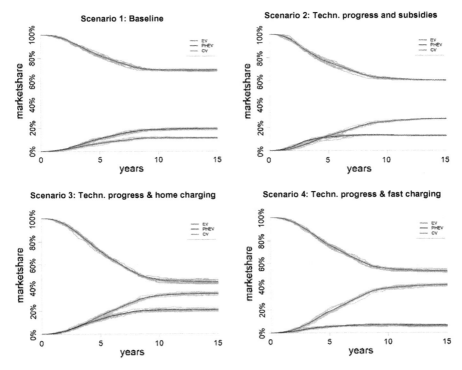

Fig. 1 Market share of CVs, PHEVs, and EVs

near term, technological progresses in EV range, charging time, and charging station density are likely to cannibalize market shares of PHEVs but not of CVs. However, while the technological capabilities of EVs are still low, a purchase price subsidy (like the current German one) will initially promote PHEVs—similar to what has been observed in the Netherlands [8]. However, EVs profit later from high PHEV market shares due to that promotional effect [5].

5 Conclusions

In this paper, we introduced an agent-based simulation approach calibrated with empirical data derived from a CBC study in order to investigate the future market potential of EVs in Germany. Based on our data, technological progress and the possibility of charging at home fosters the acceptance of EVs as well as PHEVs, whereas fast charging promotes EVs only.

One limitation is that our study focuses primarily on young potential car buyers, which might lead to an effect similar to pro-innovation bias [2]. Currently, we do not consider station density as a spatial parameter but as an attribute of vehicle type, which could be relaxed with real-world data and extended with consumers' actual driving, parking, and charging/refueling behavior. Moreover, we only simulate either a fast or a regular charging network. Optimizing the charging infrastructure in terms of locations and development of an economically efficient ratio of fast and regular charging stations seem promising future directions for research, in which ABS based on empirical data can be a very helpful approach.

References

1. Barabási, A.-L., Albert, R., & Jeong, H. (1999). Mean-field theory for scale-free random networks. *Physica A, 272*(1–2), 173–187.
2. Ellen, P. S., Bearden, W. O., & Sharma, S. (1991). Resistance to technological innovations: An examination of the role of self-efficacy and performance satisfaction. *Journal of the Academy of Marketing Science, 19*(4), 297–307.
3. Frenzel, I., Jarass, J., Trommer, S., & Lenz, B. (2015). *Erstnutzer von Elektrofahrzeugen in Deutschland: Nutzerprofile, Anschaffung, Fahrzeugnutzung (First-time users of electric vehicles in Germany: user profiles, acquisition, vehicle use)*. Berlin: Deutsches Zentrum für Luft- und Raumfahrt e. V.
4. Garcia, R., & Jager, W. (2011). From the special issue editors: Agent-based modeling of innovation diffusion. *Journal of Product* Innovation Management, *28*(2), 148–151.
5. Götz, K., Sunderer, G., Birzle-Harder, B., & Deffner, J. (2012). Attraktivität und Akzeptanz von Elektroautos. Ergebnisse aus dem Projekt OPTUM - Optimierung der Umweltentlastungspotenziale von Elektrofahrzeugen (Attractiveness and acceptance of electric cars. Results from the OPTUM project - Optimizing the environmental impact potential of electric vehicles). ISOE-Studientexte, vol 18. ISOE - Institut für sozial-ökologische Forschung, Frankfurt am Main.
6. Günther, M., & Stummer, C. (in press). Simulating the diffusion of competing multi-generation technologies: An agent-based model and its application to the consumer computer market in Germany. In A. Fink, A. Fügenschuh & M.J. Geiger (Eds.), *Operations Research Proceedings* 2016.
7. Hidruea, M. K., Parsons, G. R., Kempton, W., & Gardner, M. P. (2011). Willingness to pay for electric vehicles and their attributes. *Resource and Energy Economics, 33*(3), 686–705.
8. Kaiser, A. (2016). Warum Holland grün angemalten Spritschluckern 7000 Euro schenkt (Why Holland pays greenly sprinkled 7,000 euros), manager magazin online. Retrieved July 14, 2017, from http://www.manager-magazin.de/politik/europa/elektromobilitaet-so-setzt-der-elektroauto-boom-hollands-fiskus-zu-a-1072200.html.
9. Kiesling, E., Günther, M., Stummer, C., & Wakolbinger, L. M. (2012). Agent-based simulation of innovation diffusion: a review. *Central European Journal of Operations Research, 20*(2), 183–230.
10. Krupa, J. S., Rizzo, D. M., Eppstein, M. J., Lanute, B. D., Galeema, D. E., Lakkaraju, K., et al. (2014). Analysis of a consumer survey on plug-in hybrid electric vehicles. *Transportation Research Part A: Policy and Practice, 64*(14), 31.
11. Kraftfahrt Bundesamt. (2016). Fahrzeugzulassungen im Juni 2016 (Vehicle registrations in June 2016), 21/2016.
12. Morrissey, P., Weldon, P., & O'Mahony, M. (2016). Future standard and fast charging infrastructure planning: An analysis of electric vehicle charging behaviour. *Energy Policy, 89*, 257–270.
13. Noori, M., & Tatari, O. (2016). Development of an agent-based model for regional market penetration projections of electric vehicles in the United States. *Energy, 96*, 215–230.

14. Orme, B. (2000). Hierarchical Bayes: Why All the Attention? Sawtooth Software, Research Paper.
15. Sawtooth. (2016) Lighthouse Studio v9.0. Sawtooth Software, Manual.
16. Silvia, C., & Krause, R. M. (2016). Assessing the impact of policy interventions on the adoption of plug-in electric vehicles: An agent-based model. *Energy Policy, 96,* 105–118.
17. Stummer, C., Kiesling, E., Günther, M., & Vetschera, R. (2015). Innovation diffusion of repeat purchase products in a competitive market: An agent-based simulation approach. *European Journal of Operational Research, 245*(1), 157–167.
18. Zhang, T., Gensler, S., & Garcia, R. (2011). A study of the diffusion of alternative fuel vehicles: An agent-based modeling approach. *Journal of Product Innovation Management, 28*(2), 152–168.
19. Zsifkovits, M., & Günther, M. (2015). Simulating resistances in innovation diffusion over multiple generations: an agent-based approach for fuel-cell vehicles. *Central European Journal of Operations Research, 23*(2), 501–522.

Hybrid Agent-Based Modeling (HABM)—A Framework for Combining Agent-Based Modeling and Simulation, Discrete Event Simulation, and System Dynamics

Joachim Block

1 Hybrid Methods in OR

Decision and policy makers in our modern world are facing complex and unstructured problems in a volatile and differentiated environment. The decision process for these hard problems is still dominated by intuition and judgement. However, replacing judgement and intuition with algorithms would result in much better solutions as Kahneman et al. state [1].

Standard optimization algorithms such as linear programming fail in solving hard problems when a certain size is exceeded. Despite impressive progress in computational power, computational space and time complexity often do not permit identifying an optimal solution in an acceptable amount of time. On the other hand, heuristics have a much lower complexity but cannot guarantee to find an optimal solution.

In order to keep a balance between efficiency and quality of the solution, hybrid algorithms gain more and more interest within the OR community [2]. This kind of algorithms blends different OR techniques into new algorithms. For instance, by mixing exact mathematical methods with heuristics the conflict between accuracy and reliability on the one hand and computational time needed on the other hand can be resolved [3].

In addition to hybrid algorithms, hybridization spreads to simulation as well. Simulation is still one of the most widely used quantitative approaches for decision making [4]. Similar to hybrid algorithms, hybrid simulation integrates different simulation paradigms to exploit the individual strengths by overcoming the inherent limitations at the same time. For instance, system dynamics (SD) handles time continuously [5] while discrete event simulation (DES) is restricted to discrete time steps or events. On the other hand, DES, due to its micro perspective, is able to consider

---

J. Block (✉)
Department of Computer Science, Universität der Bundeswehr München, COMTESSA, Werner-Heisenberg-Weg 39, 85579 Neubiberg, Germany
e-mail: archibald@ieee.org

more details than highly aggregated SD. In contrast to these two paradigms, agent-based modeling and simulation (ABMS) can cope with flexible structures. Overall aim of hybrid simulation is to offer more realistic simulation models and, therefore, to contribute to better decisions [6].

The scientific literature offers a growing number of publications where hybrid simulation is applied to OR problems. However and in contrast to hybrid algorithms, this field of research seems still to be in its infancy. Significantly, a recent review of 192 simulation models for sustainability reveals only eight hybrid models between 2000 and 2015 [7].

It is not due to a lack of programming tools why hybrid simulation is not very commonly used. Indeed, tools such as AnyLogic® do support the implementation of hybrid simulation models. Rather, it is the absence of an established formalism to specify models of this kind. While the different simulation paradigms are well elaborated the formalisms to connect and integrate two or more paradigms into holistic models are not.

We aim to foster the use of hybrid simulation by introducing a formalism to specify models based on ABMS, DES, and SD. Our hybrid agent-based modeling (HABM) framework deliberately focuses on these three paradigms. Besides being capable of explicitly handling time, they are the dominant simulation paradigms in many OR fields such as supply chain management [8] or healthcare [9].

2 Building a Dynamic Agent World

Different ways to combine ABMS, DES, and SD into a hybrid simulation model do exist. However, we propose to embed DES and SD into ABMS. More precisely, the internals of an agent are modeled by DES and SD with agents forming a network of interacting components. This approach is in line with widely found so called low-level teamwork hybrid algorithms where teamwork exploration is extended by a low-level exploitation algorithm. In HABM, ABMS builds the frame and both DES and SD the guiding rules.

An agent is an entity that is embedded into and interacts with an environment. It is capable of flexible reactive, proactive, and social behavior to satisfy its intentions or design objectives (derived from [10]). Hence, an agent-based model consists of a set of agents A, possibly a set of objects O, and relations R between them. All these elements together form a world W. It is important to note that a world is not static but changes as time evolves.

2.1 Agents and Objects

Agents and objects have observable as well as non-observable attributes and exhibit behavior. Attributes and behavior can be both discrete and continuous. Let us think about reading a book. Reading a page is a continuous behavior while turning the page can be classified as a discrete action.

The theory of hybrid systems [11] offers an established approach to specify continuous state systems that are disturbed by transitions in discrete states. This in mind, we model an agent as a hybrid automaton with a discrete subsystem (DES) and a continuous one (SD). Modeling agents and objects by hybrid automata enables us among others to take use of verification tools for instance to identify critical Zeno executions where the simulation would get locked [12].

Definition 1 An **agent** $a \in A$ is defined as a hybrid automaton in the form

$$a = (id, X^{disc}, Q, E, \Sigma_{cd}, Trans, Out, Y^{disc},$$
$$Guard, Reset, X^{cont}, U, f, h, Y^{cont}, s_0, q_\omega)$$

where

- id is the **unique identifier** of the agent
- X^{disc} is the **set of input events and ports**
- Q is the finite **set of discrete states**
- $E \subseteq Q \times Q$ is the **set of directed edges**
- Σ_{cd} is the **set of events caused by the internal continuous dynamics**
- $Trans : E \to 2^{(X^{disc} \cup \Sigma_{cd})}$ is the **discrete state transition condition**
- $Out : E \to Y^{disc}$ is the **discrete output function**. An output is sent when a discrete state transition takes place.
- Y^{disc} is the **set of output events and ports**
- $Guard : Q \times U \to \Sigma_{cd} \cup \emptyset$ is the **guard condition**. It describes the events that result from the internal continuous dynamics.
- $Reset : E \times U \to U$ is a **reset map**. This function describes the value to which the continuous state is set in case of a discrete state transition.
- X^{cont} is the **set of continuous input values and ports**
- $U = \mathbb{R}^n$ is the **set of continuous states**
- $f : U \times X^{cont} \to U$ which describes, through a differential equation, the **continuous evolution** of the continuous state vector
- $h : U \times X^{cont} \to Y^{cont}$ is the **continuous output function**
- Y^{cont} is the **set of continuous output values and ports**
- $s_0 \in Q \times U$ is the **initial state** and $q_\omega \in Q$ is the **exit state**

The discrete subsystem is defined by the structure $DES = (X^{disc}, Q, E, Trans, Out, Y^{disc})$ and the continuous one by $SD = (X^{cont}, U, f, h, Y^{cont})$. The integration is as follows. When the continuous state variable exceeds a given threshold an internal event is triggered ($Guard$). On the other hand, a discrete state transition ($Trans$) can force the continuous state variable to perform a jump ($Reset$).

The definition of an object matches that of an agent except that a dedicated discrete exit state q_ω - indicating that the agent will leave the world - does not exist. In contrast to an agent, an object does not exhibit rational behavior and, hence, cannot deliberately leave the world.

2.2 Dynamic Networks of Agents and Objects

Emergent behavior of an agent-based system is mainly determined by interactions between agents and objects and, moreover, within the agent population. However, the relations between the different elements of an agent world are often not static. In fact, the underlying interaction network can be highly dynamic as agents leave the world and new ones enter and existing links vanish while new relations are established as time passes by.

For HABM, we take a similar approach as used for describing dynamic discrete event system specification (DEVS) networks [13]. So, a network executive χ, which controls the structure of the network, is introduced (see Fig. 1).

Definition 2 A **dynamic network of hybrid automata** is a structure

$$N_{dyn} = (X_N^{disc}, X_N^{cont}, Y_N^{disc}, Y_N^{cont}, \chi, M_\chi)$$

where

- X_N^{disc} is the network's **set of input events and ports**
- X_N^{cont} is the network's **set of continuous input values and ports**
- Y_N^{disc} is the network's **set of output events and ports**
- Y_N^{cont} is the network's **set of continuous output values and ports**
- χ is the **identifier of the network executive**
- M_χ is the **model of the network executive**

Again, the model of the network executive M_χ consists - similar to our hybrid automaton definition - of a discrete and a continuous state system for internal decisions.

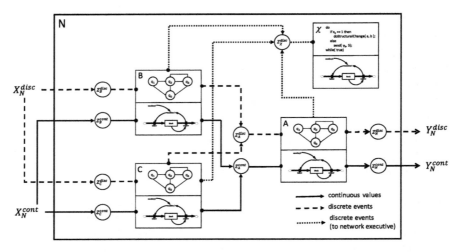

Fig. 1 Example of a dynamic agent-based network with the executive χ

In addition, this component has encoded the state of the whole network of hybrid automata N at time t.

Definition 3 A **network of hybrid automata** is a structure

$$N = (D, \{M_d\}, \{I_d\}, \{Z_d^{disc}\}, \{Z_d^{cont}\})$$

where

- D is the set of **component identifiers**
- $\{M_d\}$ is the set of **hybrid automata definitions** for all $d \in D$
- $\{I_d\}$ is the set of **influencers** for all $d \in D$
- $\{Z_d^{disc}\}$ is the set of **discrete input interface maps** for all $d \in D$
- $\{Z_d^{cont}\}$ is the set of **continuous input interface maps** for all $d \in D$

As the network is an internal state of the executive it can be reconfigured. The reconfiguration can for instance be an adjustment of the interface maps or the addition and removal of network components. The transformation of the network is implemented in the internal decision system. However, only discrete state transitions can force the network specification to change from N^t to N^{t+1}.

Three mechanisms can initiate a structural change of the network. First of all, a network component sends a request to the executive. Second, the internal decision process of the network executive forces it. Finally, an external event to the network, e.g. by the simulation user, triggers a structural change.

3 Discussion and Conclusion

We have successfully applied the HABM framework to the OR field of strategic workforce planning in a public sector context [14]. People in such organizations pass different grades during their professional life while performing at work. The challenge for human resource (HR) managers is to implement sustainable HR practices and policies in order to optimize overall performance of the workforce.

In our simulation model the workforce is modeled as a set of agents, each representing an employee. The actual grade a person is in and the promotion into a higher grade is implemented in DES while continuous performance is realized by SD. The underlying SD model [15] is founded on the well established AMO theory where performance is a function of ability, motivation and opportunity. New employees are hired while older ones retire. Hence, the workforce changes constantly. Besides removing and inserting agents, the network executive applies HR policies to the workforce by among others deciding which employee to promote in case of an existing vacancy. The simulation model enables HR managers to simulate different policies, e.g. promotion by seniority or by performance or even an increase in training, and to study the effects on overall performance. Simulation runs show for instance that an increase in training does not necessarily lead to a higher overall performance. Rather, if not flanked with other actions - for instance a higher salary or better career opportunities - particularly well trained high potentials will quit the organization. In the end, the opposite results: a drop in organizational performance.

HABM provides a framework for hybrid ABMS, DES, and SD models. The graphical representation of DES and SD facilitates the critical discussion with and the understanding by model stakeholders with a weak mathematical background. This is important as managers should not trust a model that they do not understand. Furthermore, HABM fosters the reuse of well validated models offered by the SD community. However, HABM is not limited to simulation models for fields such as strategic workforce planning or supply chain networks. Moreover, it can be used for the specification of agent-based metaheuristics.

References

1. Kahneman, D., Rosenfield, A.M., Gandhi, L., & Blaser, T. (2016, October). Noise: how to overcome the high, hidden cost of inconsistent decision making. *Harvard Business Review*, 36–43.
2. Blum, Ch., Puchinger, J., Raidl, G. R., & Roli, A. (2011). Hybrid metaheuristics in combinatorial optimization: a survey. *Applied Soft Computing, 11*, 4135–4151.
3. Sörensen, K., & Glover, F. W. (2013). Metaheuristics. In S. I. Gass & M. C. Fu (Eds.), *Encyclopedia of Operations Research and Management Science* (3rd ed., pp. 960–970). New York: Springer.
4. Anderson, D. R., Sweeney, D. J., Williams, Th A, Camm, J. D., Cochran, J. J., Fry, M. J., et al. (2016). *An Introduction to Management Science - Quantitative Approaches to Decision Making* (14th ed.). Mason: South-Western.
5. Sterman, J. D. (2000). *Business Dynamics - Systems Thinking and Modeling for a Complex World*. Boston: McGraw-Hill.
6. Lättilä, L., Hilletofth, P., & Lin, B. (2010). Hybrid simulation models - when, why, how? *Expert Systems with Applications, 37*(12), 7969–7975.
7. Moon, Y. B. (2017). Simulation modeling for sustainability: a review of the literature. *International Journal of Sustainable Engineering, 10*(1), 2–19.
8. Tako, A. A., & Robinson, S. (2012). The application of discrete event simulation and system dynamics in the logistic and supply chain context. *Decision Support Systems, 52*(4), 802–815.
9. Brailsford, S. C., Harper, P. R., & Pitt, M. (2009). An analysis of the academic literature on simulation and modelling in healthcare. *Journal of Simulation, 3*, 130–140.
10. Wooldridge, M., & Jennings, N. R. (1995). Intelligent agents: theory and practice. *The Knowledge Engineering Review, 10*(2), 115–152.
11. Witsenhausen, H. S. (1966). A class of hybrid-state continuous-time dynamic systems. *IEEE Transactions on Automatic Control, 11*(2), 161–167.
12. Johansson, K. H., Egerstedt, M., Lygeros, J., & Sastry, S. (1999). On the regularization of Zeno hybrid automata. *System and Control Letters, 38*, 141–150.
13. Zeigler, B. P., Praehofer, H., & Kim, T. G. (2000). *Theory of Modeling and Simulation - Integrating Discrete Event and Continuous Complex Dynamic Systems* (2nd ed.). San Diego: Academic Press.
14. Block, J. (2016). A hybrid modeling approach for incorporating behavioral issues into workforce planning. *In: IEEE International Conference on Systems, Man, and Cybernetics* (pp. 326–331).
15. Block, J., & Pickl, S. (2014). The mystery of job performance: a system dynamics model of human behavior. *In: 32nd International Conference of the System Dynamics Society*.

The European Air Transport System: A Methodological Perspective on System Dynamics Modeling

Gonzalo Barbeito, Ulrike Kluge, Marcia Urban, Maximilian Moll, Martin Zsifkovits, Kay Plötner and Stefan Pickl

1 Introduction

Due to an ever-increasing complexity in aviation related operations, modeling and simulation became widely used tools for analyzing the underlying systems and their associated behavior. System Dynamics (SD) is a well-known technique capable of qualitatively assessing the system as a whole by studying the interrelations of different internal operations and exogenous effects, and their contribution to the systems general behavior. The semantic simplicity and clear cause-effect structures in SD, as well as its potential for decision support, made this technique a popular choice among modelers to represent the non-linear dynamics arising from complex transport systems [1]. This paper describes challenges and insights from the development process of the model for the MATS project (Modeling the Air Transport System), a SD based approach to the aviation industry. The objective of this model is to understand how each major stakeholder in the air transport system is affected by the complex dynamics involved in this industry. This paper considers the aspects required for building a meaningful model, based not only on expert knowledge and modelers intuition, but also aided by data science techniques to extract information from exhaustive datasets describing the system.

G. Barbeito (✉) · M. Moll · M. Zsifkovits · S. Pickl
Universität der Bundeswehr München, 85577 Neubiberg, Germany
e-mail: gonzalo.barbeito@unibw.de

U. Kluge · M. Urban · K. Plötner
Bauhaus Luftfahrt, 82024 Taufkirchen, Germany

2 Literature Review

Previous SD models in the commercial aviation industry take a broad range of aspects into consideration, such as the effect of policies regarding carbon footprint [2] and the identification of business cycles [3–5] among others. In [6–8] highly comprehensive SD models are introduced. These models are aimed to forecast air travel demand and terminal capacity expansion analyzing the relation between passenger demand and airport capacity and considering GDP and population growth as principal demand drivers. The aggregated logic behind these works was used as the starting point of the MATS model. However, it stands out in its holistic approach to the air transport system.

3 Methodology

This work describes the devised solution to cope with three main challenges of the project. The first one involved defining a clear strategy for interdisciplinary team management, because of the complexity of the subject and the background diversity of each member of the involved workgroups. The second task, information gathering, accounted for a significant share of data collection and analysis, as well as understanding the studied system, while becoming familiar with the dynamics governing its behavior. The third and final task was actually building the model, i.e. both developing the structure and determining the right parameters throughout the model.

3.1 Modeling Approach: System Dynamics

Developed by Jay W. Forrester in the 1950s, this approach allows a high-level study of complex systems over time through modeling and simulation of sufficiently aggregated structures [9]. Using a combination of simple modeling elements (i.e. Variables, Stocks or Levels, and Material and Information Flows), the results of this technique are highly approachable models, ideal for the introduction of stakeholders to the modeling process.

3.2 Team Management: The SCRUM Methodolgy

The SCRUM methodology has been established as an agile method in software development. It enables a faster and more flexible development process of new software or products [10]. This methodology accepts the existence of uncertainty in the development process (e.g. during aircraft conceptual design [11]) and a certain degree of unpredictability in the steps required to reach a final product or software. The concept of roles is also of importance, with three main roles required for the workflow: Product Owners or Stakeholders are responsible for defining the customer-centric aspects that will add value to the product; Developers are in charge of carrying the operative tasks that will result in the actual product; and a Scrum Master acts as resource facilitator for the whole team [12]. This technique allows to consider not only initial but also additional requirements, which might arise during later steps of the process [13]. For the concrete development of the SD model and according to the SCRUM roles, the team was divided into two sub-teams: the modelers team acting as developers, with core competencies in SD modelling, and the input team as a combination of aviation experts with strong data handling competences and developers.

3.3 Data Analysis and System Understanding

Two types of information sources were required for developing the model: expert knowledge and system data. The first one, also including modelers insight and intuition, was mostly used for the structure and logic development of the model. Expertise on the air transport system was provided by the Bauhaus Luftfahrt workgroup. The downside of expert-driven modeling is the potential introduction of a certain degree of bias, given the particular mental models that each person possesses [9]. To overcome this, data characterizing the system was used. As a compromise between quantity and quality, data was collected by the input team between the years 2000 and 2015, setting the monetary base on the first year collected. This particular scope was selected for its high availability of data, and was capped to fifteen years in order to avoid the inclusion of past, non-related dynamics.

For the European MATS model, 43 European countries were included. The country set was retrieved from the UN Report World Population Prospects: The 2015 Revision [14]. From the original list by the UN, the countries Holy See, Channel Islands, Faroe Islands, Gibraltar, and Isle of Man were excluded for the purpose of the MATS model, as these are countries with proportionally small GDP and gathering reliable data for them proved a rather challenging task.

3.4 Model Parameterization: Analysis and Sources

This section presents a short description of a few parameters requiring a more detailed explanation on the consulted sources and their inclusion process.

GDP Growth Rate: The gdpGrowthRate for the year 2000–2015 was calculated using data from the World Bank. The sum of the weighted gdpGrowthRate was implemented into the model for each year accordingly. The gdpGrowthRate for 2016 and 2017 was retrieved from a forecast by the European Commission [15]. Years 2018 to 2050 were covered by an analysis by PWC [16]. No further information was provided on which countries have been included in the PWC study. Due to a lack of data, gdpGrowthRate covering all 43 countries in scope could not directly be implemented in the model for all years 2000–2050. For this reason, especially for forecasted data, in some cases the EU28 was taken as a proxy for all 43 countries in scope.

Traveler Conversion Rate: For the travelerConversionRate, an OLS - regression was conducted to test the statistical influence of GDP per capita (independent variable) on number of trips (depended variable, taken here as a proxy for general trips). Data on the number of trips for all 43 countries was retrieved from Eurostat [17] and GDP per capita data for all 43 countries from the World Bank. Population data for all 43 countries was used to calculate a new variable, trips per capita with at least one overnight stay, as detailed in [18]. Both variables, GDP per capita and trips per capita have been logged. Hence, the coefficient can be used as an elasticity in the model [19].

Mode Share Aviation: The modeShareAviation variable disaggregates from the passenger demand for all transportation modes the number of people travelling using air transport only. This variable is calculated by taking a European Mode Share percentage for the year 2000, 2010 and 2014 from a report by the European Commission on EU transport in figures (euModeShareAviation). It is worth noting that these numbers are for intra-EU flights on EU27 only and taken as a proxy for general modeShareAviation [20].

4 Model Description

The model relates a variable demand to several dimensions or stakeholders within the air transport industry, such as aircraft manufacturers, airports and airlines, in order to qualitatively assess the impact on each one of these dimensions. All units are in the metric system, and the currency is United States Dollars. As an example, the Passenger Subsystem will be described in more detail and shown in Fig. 1.

Passenger Subsystem: The stronger dynamics of the model, those driving the temporal evolution of the system, are handled by the Passenger subsystem, converting the effects of internal changes into variations in the air travel demand. Passenger demand is driven mainly by GDP growth, and calculated as a percentage of the sum of all travelers. While there are several other factors driving passengers air transport demand [21], GDP is the most well-known and studied of them. Moreover, testing further demand drivers with an OLS-regression, Population and Population Growth were also found to strongly influence the commercial air industry demand and were included as secondary demand driver. The results are then passed to a single variable forecasting the number of air passengers for a certain year. Once the passenger

Passenger Model

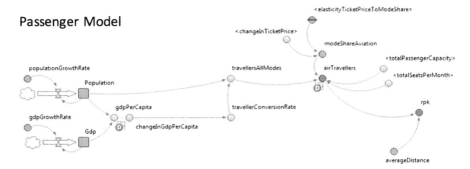

Fig. 1 Passenger Subsystem

demand is disaggregated in each transportation mode, changes in the flight ticket price were also found to affect the demand of this particular mode. This behavior is generated in the airline model, and integrated as a factor in the passengers subsystem, externally influencing the number of air travelers over time. The final output of this subsystem is the weighted combination of all passengers through the number of kilometers expected to travel. This results in a new variable, measured in kilometers, commonly known in the industry as Revenue Passenger Kilometer (RPK).

Airline Subsystem: The resulting number of air passengers as well as the resulting RPK are linked with the airline model. An increasing number of air passengers will have a positive impact on the number of aircraft an airline will order in a subsequent simulation step. Threshold values for a desired load factor are included in order to influence the decision to order a new aircraft. Furthermore, the RPK serves as an input to calculate the airline seat load factor.
The model considers furthermore factors like the number of flights, operating costs and ticket prices.

Airport Subsystem: The airport subsystem is in charge of keeping track of the structural capacity variation in all airports throughout the region. This includes also a logic for runway expansion and new airport construction. The main objective is to provide the required terminal capacities for the number of air travelers and the required runways for the number of requested flights.

Aircraft Manufacturer Subsystem: This subsystem was created to model the expansion logic of the aircraft supply side considering mainly economic variables from the manufacturer's perspective. The manufacturer model is currently narrowed down to the production of additional aircraft when ordered from an airline, with a lead time of 6 months.

5 Conclusion and Future Work

This paper presents the methodological aspects involved in the development process of a new System Dynamics (SD) model for the air transport industry. While the integration of a team with significant background diversity presented several challenges, the SCRUM methodology, coupled with a well-thought management strategy provided several benefits that significantly outweighed the challenges. The next step on the project is to conduct a validation of the model, and to adapt parameters and structure for any behavior not matching the data. The disaggregation of average elements is also a priority on the project. One example being the introduction of a second aircraft manufacturer, which is expected to create a market with competition. Moreover, through a combination of variants of the model with different data and a network system connecting them, the model will be extended to the global level following an Operations Research based approach.

References

1. Shepherd, S. (2014). A review of system dynamics models applied in transportation. *Transportmetrica B: Transport Dynamics*, *2*(2), 83–105.
2. Sgouridis, S., Bonnefoy, P. A., & Hansman, R. J. (2011). Air transportation in a carbon constrained world: Long-term dynamics of policies and strategies for mitigating the carbon footprint of commercial aviation. *Transportation Research Part A: Policy and Practice*, *45*(10), 1077–1091.
3. Lyneis, J. M. (2000). System dynamics for market forecasting and structural analysis. *System Dynamics Review*, *16*(1), 3–25.
4. Liehr, M., Größler, A., & Klein, M. (1999). Understanding Business Cycles in the Airline Market. In: *17th International Conference of the System Dynamics Society*.
5. Pierson, K., & Sterman, J. D. (2013). Cyclical dynamics of airline industry earnings. *System Dynamics Review*, *29*(3), 129–156.
6. Bießlich, P., Schröder, M., Lütjens, K., & Gollnick, V. (2014). Systemdynamische Abbildung von Langfristszenarien des Flughafens Hamburg. In: *63. Deutscher Luft- und Raumfahrtkongress (DLRK)*.
7. Biesslich, P., Schröder, M., Gollnick, V., & Lütjens, K. (2014). A system dynamics approach to airport modeling. In: *14th AIAA Aviation Technology, Integration, and Operations Conference*. Reston, Virginia: American Institute of Aeronautics and Astronautics.
8. Suryani, E., Chou, S. Y., & Chen, C. H. (2010). Air passenger demand forecasting and passenger terminal capacity expansion: a system dynamics framework. *Expert Systems with Applications*, *37*(3), 2324–2339.
9. Sterman, J. (2000). Business Dynamics: systems thinking and modeling for a complex world. Irwin/McGraw-Hill.
10. Takeuchi, H., & Nonaka, I. (1986). The New New Product Development Game.
11. Glas, M., & Seitz, A. (2012). Application of agile methods in conceptul aircraft design. dglr.de (pp. 1–11). http://www.dglr.de/publikationen/2012/281384.pdf.
12. Deemer, P., Benefield, G., Larman, C., & Vodde, B. (2012). A lightweight guide to the theory and practice of scrum version 2.0.
13. Schwaber, K. (1995, April). Scrum development process. Business Object Design and Implementation (1987), 10–19.

14. United Nations: World population prospects: the 2015 revision (2015). https://esa.un.org/unpd/wpp/
15. European Comission: winter 2017 economic forecast (2017). https://ec.europa.eu/info/business-economy-euro/economic-performance-and-forecasts/economic-forecasts/winter-2017-economic-forecast_en
16. PWC: The World in 2050. Will the shift in global economic power continue? Technical report (2015). http://www.pwc.com/gx/en/issues/the-economy/assets/world-in-2050-february-2015.pdf
17. Eurostat: data explorer-number of trips by Country/destination (2017). http://appsso.eurostat.ec.europa.eu/nui/show.do?dataset=tour_dem_ttw&lang=en
18. World bank: World development indicators — DataBank (2017). http://databank.worldbank.org/data/reports.aspx?source=2&country=EUU
19. Wooldridge, J. M. (2009). *Introductory econometrics : a modern approach*. South Western: Cengage Learning.
20. European Comission: EU transport in figures - statistical pocketbook. Technical report (2012).
21. Kluge, U., Paul, A., Cook, A., & Cristobal, S. (2017). Factors influencing European passenger demand for air transport. *In: Air Transport Research Society Conference.*

A Galerkin Method for the Dynamic Nash Equilibrium Problem with Shared Constraint

Zhengyu Wang and Stefan Pickl

1 Problem Formulation

The dynamic Nash equilibrium problem with shared constraint (NEPSC) is ubiquitous in engineering and economics [5]. For instance in real traffic situation, the guidance and control of several autonomous vehicles can be perfectly modelled by the dynamic NEPSC. Such a problem involves N agents, each of which (the ν-th agent, $\nu = 1, \ldots, N$) solves the optimal control problem of the form

$$\begin{aligned}\min_{(y_\nu, u_\nu)} \;& \psi_\nu(y_\nu(T), y_{-\nu}(T)) + \int_0^T \varphi_\nu(t, y_\nu(t), y_{-\nu}(t), u_\nu(t), u_{-\nu}(t)) dt \\ \text{s.t.} \;& \dot{y}_\nu(t) = F_\nu(t, y_\nu, u_\nu) \; \text{with} \; y_\nu(0) = y_\nu^0 \in R^{n_\nu} \\ & u_\nu(t) \in U_\nu(u_{-\nu}(t)) \; \text{a.e. in } [0,T],\end{aligned} \quad (1)$$

where $(y_\nu, u_\nu) \in R^{n_\nu} \times R^{m_\nu}$ denotes the state-control pair of the ν-th agent. The (1) is parameterized by $y_{-\nu} = (y_{\nu'})_{\nu' \neq \nu}$ and $u_{-\nu} = (u_{\nu'})_{\nu' \neq \nu}$, which denote the state and control variables of all the rivals, respectively. We write $y = (y_\nu, y_{-\nu})$ and $y = (y_\nu, y_{-\nu})$ when emphasizing the ν-th agent's state/control variables. In this article we focus on the linear-quadratic case: the dynamic $F_\nu(t, y_\nu, u_\nu) = A^\nu y_\nu + B^\nu u_\nu + f^\nu(t)$ is linear and the two cost functionals are quadratic:

This work was supported in part by the Fundamental Research Funds for the Central Universities (Grant No.14380011), by National Natural Science Foundation of China (Grant No.11571166), and by the Priority Academic Program Development of Jiangsu Higher Education Institutions.

Z. Wang (✉)
Department of Mathematics, Nanjing University, Nanjing, China
e-mail: zywang@nju.edu.cn

S. Pickl
Department of Computer Science, Core Competence Center for Operations Research, Universität der Bundeswehr München, Munich, Germany
e-mail: stefan.pickl@unibw.de

© Springer International Publishing AG, part of Springer Nature 2018
N. Kliewer et al. (eds.), *Operations Research Proceedings 2017*, Operations Research Proceedings, https://doi.org/10.1007/978-3-319-89920-6_82

$$\psi_\nu(y) = \tfrac{1}{2} y^T E^\nu y + y^T c^\nu$$
$$\varphi_\nu(t, y, u) = \tfrac{1}{2} u^T M^\nu u + \tfrac{1}{2} y^T K^\nu y + u^T [Q^\nu y + q^\nu(t)],$$

where $A^\nu \in R^{n_\nu \times n_\nu}$, $B^\nu \in R^{n_\nu \times m_\nu}$, $Q^\nu \in R^{m \times n}$, $f^\nu : [0, T] \to R^{n_\nu}$, $q^\nu : [0, T] \to R^m$, $c^\nu \in R^n$, the matrices $E^\nu, M^\nu \in R^{n \times n}$ and $K^\nu \in R^{m \times m}$ are symmetric. Due to the shared constraint, the ν-th agent's control set is a set-valued mapping

$$U_\nu(u_{-\nu}) = \{u_\nu \in R^{m_\nu} \mid L^\nu u_\nu + l^\nu \le 0,\ G^\nu u + g^\nu \le 0\}, \tag{2}$$

where $L^\nu \in R^{k_\nu \times m_\nu}$, $l^\nu \in R^{k_\nu}$, $G^\nu \in R^{r_\nu \times m}$, $g^\nu \in R^{r_\nu}$.

The dynamic NEPSC is a hard problem, the standard optimization techniques can not be directly applied to it because of the coupled cost functionals. Well known is that the Pontryagin's minimum principle can reduce each (1) into a system consisting of ordinary differential equations (ODE) and a minimization. In [3], by formulating the optimality of the minimization as variational inequality (VI), and by the all-in-one method of collecting the reduced systems into a larger one, we reformulate the dynamic NEPSC into the following system

$$\begin{cases} \dot{y}(t) = Ay(t) + Bu(t) + f(t) \quad \text{with } y(0) = y^0 \\ \dot{v}(t) = -Ky(t) - A^T v(t) - Q_1^T u(t) \quad \text{with } v(T) = Ey(T) + c \\ u(t) \in \text{SOL}\,(U(u), q(t) + Q_2 y(t) + B^T v(t) + M(\cdot)) \quad \text{with } t \in [0, T], \end{cases}$$

where $A, E, K \in R^{n \times n}$, $B \in R^{n \times m}$, $Q_1, Q_2 \in R^{m \times n}$, $M \in R^{m \times m}$, $y^0, c \in R^n$, $f : [0, T] \to R^n$, $q : [0, T] \to R^m$, these data are reconstructed from those of the (1). Here $\text{SOL}\,(U, q(t) + Q_2 y(t) + B^T v(t) + M(\cdot))$ denotes the set of the solutions of the quasi VI: Find $u \in U(u)$ such that for fixed (t, y, v), it holds

$$(z - u)^T [q(t) + Q_2 y + B^T v + Mu] \ge 0. \quad (\forall z \in U(u) = \prod_{\nu=1}^N U_\nu(u_{-\nu}))$$

Solving the quasi VI requires the fixed point condition $u \in U(u)$, which yields

$$u \in U = \{z \in R^m : Lz + l \le 0,\ Gz + g \le 0\}.$$

Replacing the set-valued mapping $U(u)$ by the set U, we obtain a standard VI, which is more readily to be treated than the quasi one. Somehow, the coupled nature of the control set is removed by using U enlarged from the $U(u)$. Now the dynamic NEPSC is reduced into the *differential variational inequality* (DVI) [9]

$$\begin{cases} \dot{y}(t) = Ay(t) + Bu(t) + f(t) \quad \text{with } y(0) = y^0 \\ \dot{v}(t) = -Ky(t) - A^T v(t) - Q_1^T u(t) \quad \text{with } v(T) = Ey(T) + c \\ u(t) \in \text{SOL}\,(U(u), q(t) + Q_2 y(t) + B^T v(t) + M(\cdot)) \quad \text{with } t \in [0, T]. \end{cases} \tag{3}$$

The DVI reformulation offers a very powerful access to the dynamic NEPSC [8], however, solving the DVI is still very challenging. Actually, among the few existing methods, the time stepping method is the most popular for the DVIs [1, 9], which applies the Euler scheme to the ODEs and solves the VI at the grid points. Such a method was shown convergent of 1-order in the best case [2].

For contributing a remedy of such an inadequacy, we propose a VI-based Galerkin method for solving the equilibrium solution of the dynamic NEPSC, which can have high order convergence. This is done by reformulating the DVI as a VI posed in an Hilbert space. Such a VI reformulation is sensible since the well-posed property of the ODE is utilized, and it is advantageous because abundant numerical methods, like the Newton-type methods, are available for VIs [7, 10]. The numerical performance of the proposed method is illustrated by a two-agent zero-sum Nash equilibrium problem.

2 Galerkin Approximation

2.1 Variational Inequality Reformulation

Denote by $X = L^2(0, T; R^m)$ the Hilbert space of the m-dimensional vector-valued square integrable functions equipped with the inner product $\langle \cdot, \cdot \rangle_{L^2}$. Let (3) have a weak solution (y^*, v^*, u^*), namely, y^* and v^* are absolutely continuous and u^* integrable such that the ODEs in (3) are fulfilled in the weak sense of integral equation and the VI condition is fulfilled almost everywhere in $[0, T]$. Then by applying the constant variation formula to the state equation, we obtain:

$$y^*(t) = e^{tA} y^0 + \int_0^t e^{(t-s)A} Bu^*(s) ds. \tag{4}$$

Again by applying this formula to the adjoint equation with the terminal value

$$v^*(T) = c + Ey^*(T) = c + E\left(e^{TA} y^0 + \int_0^T e^{(T-s)A} Bu^*(s) ds\right), \tag{5}$$

we obtain

$$v^*(t) = e^{-(t-T)A^T} v^*(T) + \int_t^T e^{(s-t)A^T} Qu^*(s) ds. \tag{6}$$

By plugging the form of $y^*(t)$ and $v^*(t)$ into the VI in (3), we can see that u^* is a solution of the VI(Ω, Φ) posed in X: find $u \in \Omega \subseteq X$ such that for any $w \in \Omega$

$$\langle w - u, \Phi(u) \rangle_{L^2} = \int_0^T (w(t) - u(t))^T \Phi(u(t)) dt \geq 0, \tag{7}$$

where $\Omega = \{u \in X : u(t) \in U \text{ a.e. in } [0, T]\}$, $\Phi(u(t)) = Mu(t) + Lu(t) + \hat{g}(t)$, and where

$$\hat{g}(t) = B^T e^{-(t-T)A^T}\left(c + Ee^{TA}y^0\right) + Ke^{tA}y^0,$$

$$Lu(t) = \int_0^T B^T e^{-(t-T)A^T} E e^{(T-s)A} Bu(s)ds + \int_t^T B^T e^{(s-t)A^T} Qu(s)ds$$

$$+ \int_0^t K e^{(t-s)A} Bu(s)ds.$$

Theorem 1 *If (y^*, v^*, u^*) is a weak solution of the DVI (3), then u^* is a solution of the VI(Ω, Φ). Conversely, if u^* is a solution of the VI(Ω, Φ), then (y^*, v^*, u^*) is a weak solution of (3), and (y^*, u^*) is an equilibrium solution of the dynamic NEPSC (1) in the weak sense, where y^* and v^* are given by (4)–(6).*

Proof As indicated above, (y^*, v^*, u^*) is a weak solution of the DVI (3) if and only if u^* is a solution of the VI(Ω, Φ); and if (y^*, v^*, u^*) is a weak solution of the DVI (3), then (y^*, u^*) is an equilibrium solution of the dynamic NEPSC (1) in the weak sense. For the details we refer to Theorem 3.1 in [3]. □

2.2 Approximation Scheme and Convergence Analysis

Let X^h be a finite-dimensional subspace of X dependent of the stepsize of the subdivision of the interval $[0, T]$. The orthogonal projection onto the subspace X^h is denoted by $P_h : X \to X^h$, with $\|I - P_h\|_{L^2} = C(h) \to 0$ as $h \downarrow 0$. Take $\Omega^h = P_h \Omega$ as the orthogonal projection of Ω onto X^h. Note that Ω^h is convex and closed because so is Ω and P_h is linear and bounded. Denote by P_{Ω^h} the metric projection onto Ω^h. It is easy to see that $\lim_{h \downarrow 0} \|P_{\Omega^h} u - u\|_{L^2} = 0$ for any $u \in \Omega$, since $P_h u \in \Omega^h$ and

$$\|P_{\Omega^h} u - u\|_{L^2} \leq \|P_h u - u\|_{L^2} \leq C(h)\|u\|_{L^2} \to 0. \quad (h \to 0)$$

Denote by $\Phi^h = P_h \Phi$. Now we are in the position to apply the Galerkin approximation to the VI(Ω, Φ), which yields a finite-dimensional VI(Ω^h, Φ^h): find $u^h \in \Omega^h$ such that

$$\langle v^h - u^h, \Phi^h(u^h) \rangle_{L^2} \geq 0. \quad (\forall v^h \in \Omega^h)$$

Theorem 2 *The VI(Ω^h, Φ^h) has a unique solution if Φ is strongly monotone, and has a solution if Φ is pseudo-monotone and if there is a $u_0 \in \Omega$ such that*

$$\frac{\langle Mu + \mathfrak{L}u, u - u_0 \rangle_{L^2}}{\|u\|_{L^2}} \to +\infty \text{ as } \|u\|_{L^2} \to \infty, u \in \Omega. \tag{8}$$

The VI(Ω, Φ) and its Galerkin approximation VI(Ω^h, Φ^h) have their unique solutions if the operator Φ is strongly monotone, which is fulfilled if, for example, the matrix M is positive definite and T is small enough. The VI(Ω, Φ) has a solution if Φ is pseudo-monotone and (8) holds. We mention that Φ is pseudo-monotone when M is positive semi-definite, since L is a compact operator. We then can establish the following convergence results, for which the proof can be found in the supplement material [11].

Theorem 3 *Let u^h be a solution of the VI(Ω^h, Φ^h).*
(1) If Φ is strongly monotone, then the VI(Ω, Φ) has a solution u and there is a constant $C \geq 0$ such that

$$\|u^h - u\|_{L^2} \leq C\sqrt{\|P_{\Omega^h}u - u\|_{L^2}}.$$

(2) If Φ is monotone and $\{u^h\}$ is uniformly bounded for h small enough, then $\{u^h\}$ has a subsequence, which weakly converges to a solution u of the VI(Ω, Φ).

3 Numerical Experience

In this section we apply the Galerkin method and the time stepping method for the example arising from a two-agent zero-sum dynamic NEPSC [3], which generate respectively the numerical solutions (x_g^h, u_g^h) and (x_e^h, u_e^h), where the subscript "e" stands for "Euler" since the time stepping method actually makes the use of implicit Euler method to dicretize the involved ODEs. The exact solution of the problem is always denoted by (x, u).

For realizing the Galerkin method, we take X^h as the subspace of piecewise linear functions in $X = L^2(0, T; R^m)$. The numerical solution u^h of VI(Ω^h, Φ^h) is obtained by using the PATH solver [4], the approximate state y^h and the approximate costate v^h are given by

$$y^h(t) = e^{tA}y^0 + \int_0^t e^{(t-s)A}Bu^h(s)ds,$$

and

$$v^h(t) = e^{-(t-T)A^T}v^h(T) + \int_t^T e^{(s-t)A^T}Qu^h(s)ds,$$

where $v^h(T) = c + Ey^h(T)$. Given the computed u^h, the two approximate states $x^h = (y^h, v^h)$ are computed by utilizing the Krylov subspace approximation to evaluate the matrix exponential, for attaining a high precision [6]. The both two numerical methods are coded and performed in the setting of Octave 4.0.

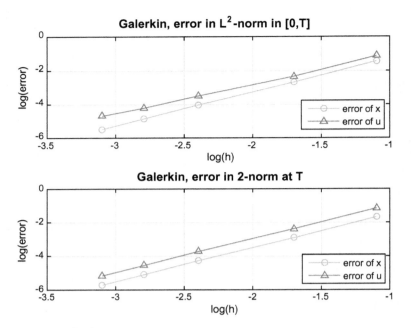

Fig. 1 Errors of (x_g^h, u_g^h) in $\|\cdot\|_{L^2}$ and $\|\cdot\|_2$ at $T = 3\pi$

Here we are interested in the errors of the state and the control, in $\|\cdot\|_{L^2}$ and in $\|\cdot\|_2$, namely we compute the values of

$$\|x_g^h - x\|_{L^2}, \quad \|u_g^h - u\|_{L^2}, \quad \|x_g^h(T) - x(T)\|_2, \quad \|u_g^h(T) - u(T)\|_2,$$
$$\|x_e^h - x\|_{L^2}, \quad \|u_e^h - u\|_{L^2}, \quad \|x_e^h(T) - x(T)\|_2, \quad \|u_e^h(T) - u(T)\|_2.$$

If, for example, $\log(\|x_g^h - x\|_{L^2})$ is affine w.r.t. $\log(h)$, then the slope gives the order of the state convergence in $\|\cdot\|_{L^2}$. Therefore we report the logarithms of the errors in different h. The numerical results are plotted in Fig. 1, which approximately suggest a 2-order convergence of the Galerkin method, and show that the method numerically outperforms the time stepping scheme with much better precision in different magnitude of scale.

4 Conclusions and Outlook on Future Research

This paper proposes a Galerkin method for the dynamic NEPSC, offering a powerful access to this problem. It is necessary and promising to extend our method to the general nonlinear case where the constraints are also dependent of the state and could be non-convex. The non-convexity of the constraints happens, i.e., in the control of autonomous vehicles where we have to avoid their collisions.

References

1. Acary, V., & Brogliato, B. (2008). *Numerical methods for nonsmooth dynamical systems*. Berlin: Springer.
2. Chen, X., & Wang, Z. (2011). Error bounds for a differential linear variational inequality. *IMA Journal Numerical Analysis, 32*, 957–982.
3. Chen, X., & Wang, Z. (2014). Differential variational inequality approach to dynamic games with shared constraints. *Mathematical Programming, 146*, 379–408.
4. Dirkse, S., Ferris, M., & Munson, T. (2017). The PATH solver. Retrieved July 20, 2017, from http://pages.cs.wisc.edu/~ferris/path.html.
5. Dreves, A., & Gerdts, M. (2017). A generalized nash equilibrium approach for optimal control problems of autonomous cars. *Optimal Control Applications and Methods*. https://doi.org/10.1002/oca.2348.
6. Higham, N. J., & Al-Mohy, A. H. (2010). Computing matrix functions. *Acta Numerica, 19*, 159–208.
7. Hintermueller, M., Ito, K., & Kunisch, K. (2003). The primal-dual active set strategy as a semismooth Newton method. *SIAM Journal of Optimization, 13*(2002), 865–888.
8. Leitmann, G., Pickl, S., & Wang, Z. (2014). Multi-agent optimal control problems and variational inequality based reformulations. *Dynamic games in economics* (pp. 205–217). Berlin: Springer.
9. Pang, J.-S., & Stewart, D. E. (2008). Differential variational inequalities. *Mathematical Programming, 113*, 345–424.
10. Ulbrich, M. (2003). Semismooth Newton methods for operator equations in function spaces. *SIAM Journal of Optimization, 13*, 805–842.
11. Wang, Z., & Pickl, S. (2017). Proof of convergence of Galerkin method for dynamics nash equilibrium problem with shared constraint, supplement material.
12. Zeidler, E. (1990). *Nonlinear functional analysis and its applications*. Leipzig: Teubner Verlag.

Part XIX
Software Applications and Modelling Systems

Generic Construction and Efficient Evaluation of Flow Network DAEs and Their Derivatives in the Context of Gas Networks

Tom Streubel, Christian Strohm, Philipp Trunschke and Caren Tischendorf

1 Introduction

The dynamic behavior of flow networks is often modeled by differential-algebraic equations, cf. [1]. The network is considered as an oriented graph $\mathcal{G} = (\mathcal{N}, \mathcal{E})$ with a node set $\mathcal{N}$ and a hyper edge set $\mathcal{E}$. A hyper edge $E \in \mathcal{E}$ is a non-empty ordered tuple of nodes from $\mathcal{N}$. Each hyper edge $E_i \in \mathcal{E}$ represents a network element such as a junction, pipe, valve or compressor station. The element model is then given by an *element function* $\tilde{f}_i : \mathbb{R}^{m_i} \times \mathbb{R}^{m_i} \times \mathbb{R} \to \mathbb{R}^{n_i}$ imposing

$$\tilde{f}_i(\dot{x}_i(t), x_i(t), t) = 0. \tag{1}$$

When simulating gas networks, t refers to the time and x usually contains pressures and flows. Depending on the topology some element functions may share some of their variables with others. If, for example, two pipes represented by $E_i, E_j \in \mathcal{E}$ are sharing the same junction, then their pressures associated to that junction are equal.

Further, it is important to mention that $\tilde{f}_i$ may not depend on all components of $\dot{x}_i$. And in the case of static elements $\dot{x}_i$ has even no influence. By taking the union x of all needed x_i, i. e. ignoring redundant variables, and incerting hyperedge functions that describe certain x_i explicitly, we obtain the whole flow network model as

$$f(\dot{x}(t), x(t), t) = \begin{pmatrix} f_1(\dot{x}(t), x(t), t) \\ \vdots \\ f_m(\dot{x}(t), x(t), t) \end{pmatrix} = 0 \tag{2}$$

T. Streubel (✉) · P. Trunschke
Department of Optimization, Zuse Institute Berlin, Berlin, Germany
e-mail: streubel@zib.de

T. Streubel · C. Strohm · P. Trunschke · C. Tischendorf
Department of Mathematics, Humboldt University of Berlin, Berlin, Germany

© Springer International Publishing AG, part of Springer Nature 2018
N. Kliewer et al. (eds.), *Operations Research Proceedings 2017*, Operations Research Proceedings, https://doi.org/10.1007/978-3-319-89920-6_83

with $f_j(\dot{x}(t), x(t), t)$ for $j = 1, ..., m \leq |\mathcal{E}|$ being deduced from $\tilde{f}_i(\dot{x}_i(t), x_i(t), t)$ for $i = 1, ..., |\mathcal{E}|$. Notice that the functions f_j may be not smooth at certain points. This is particularly the case when valves and limiting bounds are described (usually by min- and max-evaluations).

Several solvers have been developed to solve DAEs of the form (2), e. g. DASPK from L. Petzold, ode15i (in Matlab) from L.F. Shampine and IDAS from SUNDIALS. Such solvers often run more efficiently and more stable if the user provides not only evaluations of the residual function $f(y, x, t)$ but also evaluations of the partial derivatives $f_y(y, x, t)$ and $f_x(y, x, t)$.

In this paper we present a concept that automatically provides functions f, f_y and f_x. The user has to provide only the network graph $\mathcal{G}$, the element functions $\tilde{f}_i$ and their sparsity patterns. Thereby, the sparsity patterns of f_y and f_x are determined prior to their evaluation. Previously determined values of f, f_y and f_x can be exploited. The presented approach focusses on the use of automatic differentiation [2] but could also use other variants of differentiation. For treating non-smooth functions as min() and max() we use an approach via their abs-normal-form representation, see Sect. 3.

2 Jacobian Representation

We consider the structure of the nonlinear functions $\mathfrak{f}$ to be differentiated for the determination of f_y and f_x. Fixing $x = x_*$, $t = t_*$ and $y = y_*$, $t = t_*$, respectively, we have to differentiate the functions

$$\mathfrak{f}(y) := \begin{bmatrix} f_1(y, x_*, t_*) \\ \vdots \\ f_m(y, x_*, t_*) \end{bmatrix}, \quad \mathfrak{f}(x) := \begin{bmatrix} f_1(y_*, x, t_*) \\ \vdots \\ f_m(y_*, x, t_*) \end{bmatrix} \quad (3)$$

in order to provide f_y and f_x. Since we are interested in an element-wise computation of $\mathfrak{f}'(y)$ and $\mathfrak{f}'(x)$ the CSR format (compressed row format [3]) is a suitable choice to represent f_y and f_x.

3 Treatment of Switching Elements Using the abs-Normal-Form

In the case of switching elements, we need min/max-evaluations. Consequently, the element functions $\tilde{f}_i$ are only *piecewise differentiable* (PD). In order to treat them, we introduce the following representation of functions.

A function $\mathfrak{f}$ is called in abs-normal-form (ANF, [4]) if there exist twice differentiable functions F and G such that the function value $\mathfrak{f}(x)$ can be computed via

$$z = G(x, |z|), \quad \mathfrak{f}(x) = F(x, |z|)$$

where $G_w(x, w) \equiv \frac{\partial}{\partial w} G(x, w)$ is of strictly lower triangular form. The vector z represents *switching variables* and is uniquely determined. Moreover it can be evaluated component-wise in a forward and explicit fashion, because of the special nilpotent form of G_w. So $z = G(x, |z|)$ may be understood as an explicit evaluation of z for given input x. If a function $\mathfrak{f}$ has an abs-normal-form representation we note $\mathfrak{f} \in \mathcal{C}^2_{abs}$. Notice that all piecewise linear functions have an ANF representation [5].

A first order Taylor expansion of F and G at $(\mathring{x}, \mathring{w}) \in \mathbb{R}^{n+s}$ followed by a subsequent substitution $\mathring{w} = |\mathring{z}|$, $\Delta w \equiv |\mathring{z} + \Delta z| - |z(\mathring{x})|$, where $\mathring{z} \equiv z(\mathring{x}) = G(\mathring{x}, |\mathring{z}|)$ leads to a piecewise linear operator in ANF mapping $\Delta x \equiv x - \mathring{x}$ to $\mathfrak{f}(\mathring{x}) + \Delta \mathfrak{f}$:

$$\begin{pmatrix} \mathring{z} + \Delta z \\ \mathfrak{f}(\mathring{x}) + \Delta \mathfrak{f} \end{pmatrix} = \begin{pmatrix} G(\mathring{x}, |\mathring{z}|) \\ F(\mathring{x}, |\mathring{z}|) \end{pmatrix} + \begin{bmatrix} G_x(\mathring{x}, |\mathring{z}|) & G_w(\mathring{x}, |\mathring{z}|) \\ F_x(\mathring{x}, |\mathring{z}|) & F_w(\mathring{x}, |\mathring{z}|) \end{bmatrix} \cdot \begin{pmatrix} x - \mathring{x} \\ |\mathring{z} + \Delta z| - |\mathring{z}| \end{pmatrix}, \quad (4)$$

that satisfies the approximation property $\mathfrak{f}(x) = \mathfrak{f}(\mathring{x}) + \Delta \mathfrak{f} + \mathcal{O}(\|x - \mathring{x}\|^2)$. The block matrix of the piecewise linear operator (4) can be stored in a CSR fashion as well as the Jacobians in the differentiable case.

For standard DAE solvers we have to provide one suitable representative $\mathfrak{f}'(x)$ for the Bouligand subdifferential $\partial_B \mathfrak{f}(x)$. This can be derived from equation (4)

$$\mathfrak{f}'(x) := J + Y \Sigma (I - L \Sigma)^{-1} Z, \quad \begin{bmatrix} Z & L \\ J & Y \end{bmatrix} := \begin{bmatrix} G_x(\mathring{x}, |\mathring{z}|) & G_w(\mathring{x}, |\mathring{z}|) \\ F_x(\mathring{x}, |\mathring{z}|) & F_w(\mathring{x}, |\mathring{z}|) \end{bmatrix}$$

using a suitable signature Σ, see [5].

A better way would be to exploit (4) directly in the numerical integration scheme for the differential-algebraic equation. It is demonstrated in [6] for the implicit Trapezoidal method for the integration of ordinary differential equations.

Since we pursue an approach by treating flow networks generically via its elements, the ANF operators propagated from network structures appear in a more complex form compared to (4):

$$\begin{pmatrix} z_1 \\ y_1 \\ \hline z_2 \\ y_2 \\ \hline \vdots \\ \hline z_m \\ y_m \end{pmatrix} = \begin{pmatrix} c_1 \\ b_1 \\ \hline c_2 \\ b_2 \\ \hline \vdots \\ \hline c_m \\ b_m \end{pmatrix} + \begin{bmatrix} \bullet & & G_w^1 & * & & * \\ \bullet & & F_w^1 & * & & * \\ \hline * & & \bullet & G_w^2 & & * \\ * & & \bullet & F_w^2 & & * \\ \hline \vdots & & \vdots & \vdots & & \vdots \\ \hline * & & * & & \bullet & G_w^m \\ * & & * & & \bullet & F_w^m \end{bmatrix} \cdot \begin{pmatrix} x_1 \\ |z_1| \\ \hline x_2 \\ |z_2| \\ \hline \vdots \\ \hline x_m \\ |z_m| \end{pmatrix}. \quad (5)$$

Here $*$ (typically empty or sparse) and $\bullet$ (typically sparse or dense) are sub-matrices of G_x^j and F_x^j, respectively. The horizontal lines indicate *element blocks*.

4 Network Structure Preserving Representation and Implemention

First, for each f_i, we realize the Jacobian evaluations or, if necessary, their ANF representations by a new class, which we call *partial CSR*. Contrary, each ANF representation of f is stored in a so-called *complete CSR* class, obtained by merging all the corresponding partial CSRs.

These CSR classes implement slightly modified versions of the CSR format, each comprising a data-, indices- and indptr-array as well as a shape-attribute. Further, there is implemented a new attribute nabs containing the number of switching variables. In contrast to the classical CSR format, the indices-array shall be initialized as a signed array to mark all indices of nonzero entries from G_w and F_w by signs. In doing so we can distinguish coefficients for x from those of the absolute value of the switching vector $|z|$.

The relationship between both classes and their individual attributes are illustrated in Fig. 1. Here it becomes clear that the corresponding partial CSRs are collected in a list partialCSRs and parsed, as the only argument, to the constructor of complete CSR. On the other hand partial CSR objects are created with the arguments nnzPerRow, ncols and nabs. It is nnzPerRow a list containing the numbers of variable dependencies per component of the element function $\tilde{f}_i$. Further, ncols is the amount of variables contributed to the whole DAE system (2).

The partial CSR object proceeds as follows: A local indptr is created as the cumulative sum of nnzPerRow. Additional informations are derived, such as nnz the number on non zero entries of the local CSR and nrows the number of rows of

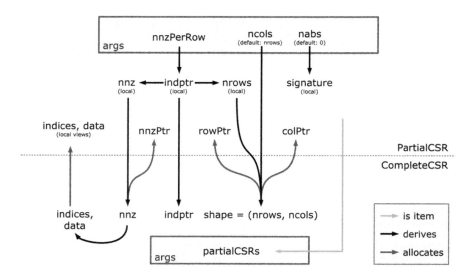

Fig. 1 Schema to manage Jacobians and ANFs of flow networks in CSR format

the CSR. The `signature` = sign(z) stores the sign-vector of switching variables and is needed for certain evaluation routines.

The complete CSR proceeds in a different manner: Its `indptr`-array gets aggregated from the `indptr`-arrays of the elemental partial CSR instances. Thereafter, the lengths of the `indices`- and `data`-array is determined, the arrays can be allocated and local views are provided to the partial CSRs. In this way an arbitrary number of complete CSR instances for any purpose, e. g. all arguments, can be created dynamically.

5 Jacobians for a Gas Network Example

We tested the GasLib40 instance from the open gas network library [7]. Figure 2 shows the topology of the network and fingerprints of the Jacobians f_y and f_x. The Jacobian f_y is constant. The Jacobians f_x is in ANF representation, due to check valve functionality of two (modified) resistors. Their partial CSRs are displayed as enlarged section on top of Fig. 2. The first and third row of the partial CSRs represent the data of G_x and G_w for the determination of the two switching variables.

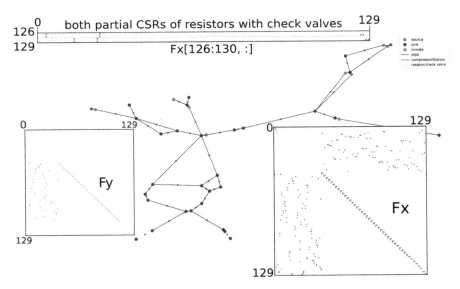

Fig. 2 Modified version of GasLib-40 [7], Fy and Fx are fingerprints of f_y and f_x, respectively. On the top is a zoom of the fingerprints of the two partial CSRs belonging to the switching elements of the nework (two check valve resistors)

Acknowledgements This work was supported by the German Federal Ministry of Education and Research (BMBF) within the Research Campus MODAL (fund number 05M14ZAM) and by the Deutsche Forschungsgemeinschaft through the Collaborative Research Centre TRR154 Mathematical Modelling, Simulation and Optimization Using the Example of Gas Networks.

References

1. Jansen, L., & Tischendorf, C. (2014). A unified (P)DAE modeling approach for flow networks. In S. Schöps, A. Bartel, M. Günther, E. J. W. ter Maten, & P. C. Müller (Eds.), *Progress in Differential-Algebraic Equations* (pp. 127–151)., Differential-Algebraic Equations Forum Berlin: Springer.
2. Griewank, A., & Walther, A. (2008). *Evaluating derivatives* [Second edition]. Society for industrial and applied mathematics
3. Golub, H. G., & Van Loan, C. F. (2012). *Matrix computations*. Wiley: JHU Press.
4. Griewank, A., & Walther A. (2016). First and second order optimality conditions for piecewise smooth objective functions. *Optimization Methods and Software*
5. Griewank, A., Bernt, J. -U., Radons, M., & Streubel, T. (2015). Solving piecewise linear systems in abs-normal form. *Linear Algebra and its Applications*
6. Griewank, A., Hasenfelder, R., Radons, M., & Streubel, T. (2017). Integrating lipschitzian dynamical systems using piecewise algorithmic differentiation
7. Humpola, J., Joormann, I., Oucherif, D., Pfetsch, M. E., Schewe L., Schmidt, M. & Schwarz R. GasLib – A library of gas network instances.

On the Performance of NLP Solvers Within Global MINLP Solvers

Benjamin Müller, Renke Kuhlmann and Stefan Vigerske

1 Introduction

We consider nonconvex mixed-integer nonlinear programs (MINLPs) of the form

$$\min_{x \in [\ell,u] \subseteq \mathbb{R}^n} \left\{ c^\top x \mid g_j(x) \leq 0 \; \forall j \in \mathcal{M}, \, x_i \in \mathbb{Z} \; \forall i \in \mathcal{I} \right\}, \quad (1)$$

where $c \in \mathbb{R}^n$, $\mathcal{M} := \{1, \ldots, m\}$, $\mathcal{N} := \{1, \ldots, n\}$, $\mathcal{I} \subseteq \mathcal{N}$, $\ell_i, u_i \in \mathbb{R} \cup \{\pm\infty\}$, $i \in \mathcal{N}$, and $g_j : [\ell, u] \to \mathbb{R}$, $j \in \mathcal{M}$, differentiable. MINLPs have applications in many areas, we refer to [1] for an overview. The state-of-the-art algorithm for solving MINLPs to global ϵ-optimality is spatial branch-and-bound, see, e.g., [2–4]. Solvers that implement this method typically need to compute local optimal solutions of nonlinear programs (NLPs). For example, primal heuristics [5] may require the solution of an NLP sub-problem of (1) and bounding methods may require the solution of a convex NLP relaxation [6, 7]. Two important solution methods for NLPs are the Inter-Point Method (IPM), which has been shown to be very efficient, and Sequential Quadratic Programming (SQP), which is said to be more robust and has better warm-starting properties than IPM.

The goal of this paper is to investigate the impact of different NLP solvers on the performance of an MINLP solver. For that, we consider the use of a portfolio of NLP solvers to solve a sequence of – sometimes very similar – NLPs as they arise in various

B. Müller (✉)
Zuse Institute Berlin, Takustr. 7, 14195 Berlin, Germany
e-mail: benjamin.mueller@zib.de

R. Kuhlmann
University Bremen, Bibliothekstr. 5, 28195 Bremen, Germany
e-mail: renke.kuhlmann@math.uni-bremen.de

S. Vigerske
GAMS Software GmbH, c/o Zuse Institute Berlin, Takustr. 7, 14195 Berlin, Germany
e-mail: svigerske@gams.com

components of an MINLP solver. With *dual components* we refer to algorithms that aim to strengthen a relaxation of the problem in each node of the spatial branch-and-bound tree, while with *primal components* we refer to algorithms that aim on finding an improving feasible solution. Naturally, for dual components, finding dual feasible solutions for convex NLPs is important, while for primal components finding a primal feasible solution of an NLP is sufficient, though the NLP might be nonconvex.

1.1 Dual Components

Often, a convex NLP relaxation of (1) is obtained by replacing constraints where $g_j(x)$ is nonconvex over $[\ell, u]$ by a convex relaxation, e.g., by using convex underestimators of $g_j(x)$ [6, 8, 9]. As the tightness of these underestimators depends on the variable bounds, branching decisions in the branch-and-bound tree search can allow to update the convex underestimators and thus improve the bound that the relaxation provides for the corresponding node in the tree.

The dual components that we consider in this paper are, first, the solution of convex nonlinear relaxations to bound the objective function $c^\top x$ in a node in the branch-and-bound tree. Thus, this component solves an NLP in potentially all nodes of the branch-and-bound tree, each being a convex relaxation of (1) when restricted to the variable bounds that are defining the node. Second, we consider Optimization-Based Bounds Tightening (OBBT), where possibly tighter bounds on selected variables are computed by minimizing and maximizing each of them over a convex relaxation of (1). Improved variable bounds can help to tighten the relaxation that is used to bound the optimal value of (1).

1.2 Primal Components

The NLPs that are solved by primal components are often obtained after fixing some or all of the integer variables x_i, $i \in \mathcal{I}$, in (1) to a given value and relaxing the integrality requirement on all non-fixed integer variables. A locally optimal (or at least feasible) solution to this NLP can be used to update the incumbent for the original MINLP, if all non-fixed integer variables take an integral value.

Many different strategies to find a good variable fixing have been developed. For our experiments, we consider an algorithm that uses the solution point computed by any primal heuristic applied to a MILP relaxation of (1) to fix all integer variables in (1) and to provide a starting point for the NLP solver. Additionally, we consider an NLP-diving heuristic, where first all integrality requirements are relaxed and then iteratively the NLP is solved and, based on its solution, additional integer variables are fixed, until either the NLP solution is feasible for the MINLP or the NLP solver fails to find a feasible solution to the NLP. In the latter case, a backtrack strategy may be applied to investigate an alternative for the latest variable fixing decision.

2 Computational Results

We used a development version of SCIP[1] [9] as MINLP solver. This version is based on SCIP 4.0, but next to the already existing interface to the IPM solver Ipopt [10], it includes new interfaces to the IPM and SQP solvers of WORHP [11] and the SQP solver FilterSQP [12]. For our experiments, we have used SCIP with the NLP solvers FilterSQP 20010817, Ipopt 3.12.7, WORHP-IP (IPM algorithm of WORHP 1.10.3), and WORHP-SQP (SQP algorithm of WORHP 1.10.3). Except for FilterSQP, all solvers use MA97 from HSL[2] to solve systems of linear equations. For all solvers, we disabled scaled termination tolerances and used a feasibility tolerance of 10^{-6}, an optimality tolerance of 10^{-7}, and equal limits on the number of NLP iterations. We used a time limit of one hour for SCIP.

As test set we consider all instances of MINLPLib2[3] (as of 2017/7/10) which can be handled by SCIP. When comparing the NLP solvers on dual components, we additionally discarded instances where SCIP does not detect any convex nonlinear constraint, since the convex relaxations would otherwise be linear (SCIP uses only polyhedral relaxations for nonconvex constraints). This leaves 327 instances. Further, when comparing on primal components, we disregard instances with only continuous variables, i.e., $\mathcal{I} = \emptyset$, since the primal components would not be applied otherwise. This leaves 938 instances.

The experiments were conducted on a cluster of 64bit Intel Xeon X5672 CPUs at 3.2 GHz with 12 MB cache and 48 GB main memory.

2.1 Dual and Primal Components

To ensure that all NLP solvers solve the same sequence of NLPs, we solve each NLP that occurs in the considered components of SCIP independently by all solvers, but pass only the result from the first solver back to SCIP. Table 1 contains aggregated results for the consumed time and the success of each NLP solver. Per instance, we collect the overall time spent in each solver and the number of NLPs where a feasible or locally optimal solution has been found. Next, we compute the shifted geometric mean over all instances with a shift value of one second. To reduce the impact of trivial instances, we disregarded all instances where the sum of NLP solving times was at most one second for the virtual worst, which is the theoretical worst performing solver on each NLP– this leaves 607 instances for the primal and 201 instances for the dual components.

Comparing the running time of the NLP solvers, FilterSQP is on both, the dual and primal components, the fastest solver. Further, it is more than 3.1 times faster than the second fastest solver, WORHP-IP, on the dual components. Interestingly,

---

[1] Solving Constraint Integer Programs, http://scip.zib.de.
[2] Harwell Subroutine Library, http://www.hsl.rl.ac.uk.
[3] MINLP Library 2, http://www.gamsworld.org/minlp/minlplib2.html.

Table 1 Aggregated results for dual and primal components

Solver	Dual components			Primal components		
	Time	nfeas	nopt	Time	nfeas	nopt
Ipopt	37.0	663.4	613.4	4.4	65.5	58.5
FilterSQP	7.9	634.3	616.0	2.7	58.4	51.3
WORHP-IP	24.1	610.9	494.6	4.7	52.3	47.3
WORHP-SQP	179.2	568.3	235.9	21.7	57.7	31.7
Virtual best	3.9	814.4	766.5	1.0	78.7	73.8

WORHP-IP performs 34.9% faster than Ipopt on the dual components and 6.4% slower on the primal components. Furthermore, the variability in performance of the NLP solvers is quite large. Choosing the best performing solver for each NLP yields a speed-up of at least a factor of 2.0 compared to FilterSQP.

Regarding the solution quality, Ipopt found more often than all other solvers feasible and local optimal solutions. On dual components, Ipopt found between 4.4% and 14.3% more feasible and up to 61.5% more local optimal solutions than the other solvers. Even though WORHP-SQP finds many feasible points, it frequently fails to converge to a local optimum. Again, the variability with respect to the solution quality is large. Choosing the best NLP solver increases the success rate of finding a feasible solution by 17.7% on average, and finding a local optimal point by around 20.0% compared to Ipopt. This indicates that a dynamic and smart choice between a portfolio of NLP solvers could allow for a considerably better performance than deciding for a single NLP solver in advance.

Figure 1 shows different performance profiles comparing the sum and the shifted geometric mean of NLP solving times per instance. As already observed above, FilterSQP outperforms all solvers when considering the sum of NLP solving times on all NLPs. On the primal components, we see that Ipopt performs more robust than the other solvers because its worst case ratio to the virtual best is bounded by a factor of 100. A considerable part of the good performance of FilterSQP seems to come from NLPs that might be infeasible. The speed-up on instances for which at least one solver provided a certificate of infeasibility is much higher than on all NLPs. This phenomena is more distinct on the dual components than on the primal components.

Due to fixing integer variables heuristically, many NLPs that appear in the primal components turn out to be infeasible. Quickly detecting their infeasibility is important. FilterSQP seems to be the fastest solver on these NLPs, too, but WORHP-IP performs significantly better than Ipopt. This is due to the Penalty-IP approach of WORHP-IP, which is able to converge to infeasible stationary points quickly without the necessity of a separate feasibility restoration phase [13].

In contradiction to the previous results, Ipopt and WORHP-IP perform better than FilterSQP when considering the shifted geometric mean of solving times. The performance profiles in the second column of Fig. 1 show that the difference between

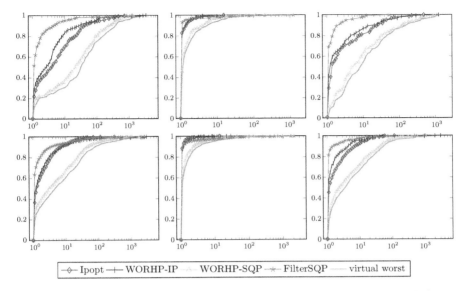

Fig. 1 Performance profiles for dual (first row) and primal components (second row). Left: sum of NLP solving times for all NLPs per MINLP instance. Middle: shifted geometric mean of same solving times. Right: sum of solving times for NLPs where at least one solver returned a certificate of (local) infeasibility

the solvers is less distinct as when considering the sum of solving times. This can be explained by the reduced impact of outliers in the shifted geometric mean. Thus, the superior performance of FilterSQP could be caused by the absence of sometimes expensive fallback strategies, which are implemented by the other solvers. Within a MINLP solver, where not every NLP needs to be solved to optimality, such a "fast fail" strategy seems to be advantageous. This presumption is reinforced by our observation that tuning a solver to find more local optimal points decreased its average performance considerably.

Finally, we want to emphasize that the NLPs that arise within our experiments are typically small. This might be a disadvantage for solvers like Ipopt and WORHP, which are designed to solve large-scale NLPs.

2.2 Overall Performance

The impact of using different NLP solvers in SCIP is summarized in Table 2. For this comparison, we used the selection of 115 instances from MINLPLib2 that is also used in a publicly available MINLP benchmark[4] and set a gap limit of 10^{-3}.

---

[4]H. Mittelmann MINLP Benchmark, http://plato.asu.edu/ftp/minlp.html.

Table 2 Aggregated results for SCIP using different NLP solvers. The entries of the *time* and *nodes* columns are relative to the first row

Setting	All		All optimal		
	# solved	Time	# solved	Nodes	Time
SCIP + Ipopt	55	984.0s	41	108312	227.7 s
SCIP + FilterSQP	53	0.92%	41	0.93%	0.90%
SCIP + WORHP-IP	50	1.06%	41	1.06%	0.98%
SCIP + WORHP-SQP	49	1.17%	41	0.95%	1.09%

Choosing a different NLP solver has a large impact on the performance and solvability of MINLPs. Table 2 shows that SCIP with Ipopt could solve the largest number of instances. However, SCIP performed fastest when using FilterSQP. On all instances the speed-up is 8% compared to using Ipopt, and on all instances that could be solved by all settings, the speed-up is even larger, namely 10%.

Acknowledgements This work has been supported by the Research Campus MODAL *Mathematical Optimization and Data Analysis Laboratories* funded by the Federal Ministry of Education and Research (BMBF Grant 05M14ZAM).

References

1. Grossmann, I. E., & Sahinidis, N. V. (2002). Special issue on mixed integer programming and its application to engineering, part I. *Optimization and engineering*, 3(4).
2. Horst, R., & Tuy, H. (1996). *Global optimization: deterministic approaches*. Berlin: Springer.
3. Quesada, I., & Grossmann, I. E. (1995). A global optimization algorithm for linear fractional and bilinear programs. *Journal of Global Optimization*, 6, 39–76.
4. Ryoo, H. S., & Sahinidis, N. V. (1995). Global optimization of nonconvex NLPs and MINLPs with applications in process design. *Computers and Chemical Engineering*, 19(5), 551–566.
5. Berthold, T. (2014). Heuristic algorithms in global MINLP solvers. Ph.D. thesis, Technische Universität Berlin
6. Ryoo, H. S., & Sahinidis, N. V. (1996). A branch-and-reduce approach to global optimization. *Journal of Global Optimization*, 8(2), 107–138.
7. Zamora, J. M., & Grossmann, I. E. (1999). A branch and contract algorithm for problems with concave univariate, bilinear and linear fractional terms. *Journal of Global Optimization*, 14, 217–249.
8. McCormick, G. P. (1976). Computability of global solutions to factorable nonconvex programs: part I - Convex underestimating problems. *Mathematical Programming B*, 10(1), 147–175.
9. Vigerske, S., & Gleixner, A. SCIP: Global optimization of mixed-integer nonlinear programs in a branch-and-cut framework. *Optimization Methods and Software* (to appear)
10. Wächter, A., & Biegler, L. T. (2006). On the implementation of an interior-point filter line-search algorithm for large-scale nonlinear programming. *Mathematical Programming*, 106(1), 25–57.
11. Büskens, C., & Wassel, D. (2013). The ESA NLP solver WORHP. In G. Fasano & J. D. Pintér (Eds.), *Modeling and optimization in space engineering* (Vol. 73, pp. 85–110). New York: Springer.

12. Fletcher, R., & Leyffer, S. (1998). User manual for filterSQP. Numerical analysis report NA/181, Department of Mathematics, University of Dundee, Scotland.
13. Kuhlmann, R., & Büskens, C. (2017). *A primal-dual augmented Lagrangian penalty-interior-point filter line search algorithm*. Technical report, Universität Bremen.

Optimizing Large-Scale Linear Energy System Problems with Block Diagonal Structure by Using Parallel Interior-Point Methods

Thomas Breuer, Michael Bussieck, Karl-Kiên Cao, Felix Cebulla, Frederik Fiand, Hans Christian Gils, Ambros Gleixner, Dmitry Khabi, Thorsten Koch, Daniel Rehfeldt and Manuel Wetzel

1 Introduction

Energy system models (ESMs) have versatile fields of application. For example they can be utilized to gain insights into the design of future energy supply systems. Increasing decentralization and the need for more flexibility caused by the temporal fluctuations of solar and wind power lead to increasing spatial and temporal granularity of ESMs. In consequence, state-of-the-art solvers meet their limits for certain model instances.

A distinctive characteristic of many linear programs (LPs) arising from ESMs is their block-diagonal structure with both linking variables and linking constraints. This article sketches extensions of the parallel interior-point solver PIPS-IPM [6] to handle LPs with this characteristic. The extended solver is designed to make use of the massive parallel power of high performance computing (HPC) platforms.

Furthermore, this article introduces an interface between PIPS-IPM (including its new extension) and energy system models implemented in GAMS. In particular, it will be described how users can communicate the model's problem structure to

---

T. Breuer
Jülich Supercomputing Centre (JSC), Forschungszentrum
Jülich GmbH, Jülich, Germany

M. Bussieck · F. Fiand
GAMS Software GmbH, Frechen, Germany

K.-K. Cao · F. Cebulla · H. C. Gils · M. Wetzel (✉)
German Aerospace Center (DLR), Cologne, Germany
e-mail: Manuel.Wetzel@dlr.de

A. Gleixner · T. Koch · D. Rehfeldt
Zuse Institute Berlin/Technical University Berlin, Berlin, Germany
e-mail: rehfeldt@zib.de

D. Khabi
High Performance Computing Center Stuttgart (HLRS), Stuttgart, Germany

PIPS-IPM. Since finding a proper block structure annotation for a complex ESM is not trivial, we will exemplify the annotation process for the ESM REMix [4]. With many ESMs implemented in GAMS, the new interface between GAMS and PIPS-IPM makes the solver available to the energy modeling community.

2 A Specialized Parallel Interior Point Solver

When it comes to solving linear programs (LPs), the two predominant algorithmic approaches to choose from are Simplex and interior-point, see e.g. [7]. Since interior-point methods are often more successful for large problems, in particular for ESM [1], this method was chosen for the LPs at hand. Mathematically, a salient characteristic of these LPs is their block-diagonal structure with both linking constraints and linking variables, as depicted below

$$
\begin{aligned}
\min \quad & c^T x \\
\text{s.t.} \quad & T_0 x_0 && = h_0 && (eq_0) \\
& T_1 x_0 + W_1 x_1 && = h_1 && (eq_1) \\
& T_2 x_0 + W_2 x_2 && = h_2 && (eq_2) \\
& \quad \vdots && \quad \vdots \\
& T_N x_0 + W_N x_N && = h_N && (eq_N) \\
& F_0 x_0 + F_1 x_1 + F_2 x_2 + \cdots + F_N x_N && = h_{N+1}, && (eq_{N+1})
\end{aligned}
$$

with $x = (x_0, x_1, ..., x_N)$. The linking variables are represented by the vector x_0, whereas the linking constraints are described by the matrices $F_0, ..., F_N$ and the vector h_{N+1}. The approach to solve this LP is based on the parallel interior-point solver PIPS-IPM [6] that was originally developed for solving stochastic linear programs. Such problems also exhibit a block-diagonal structures, although only with linking variables and without linking constraints. In this way, PIPS-IPM in its original form cannot handle problems with linking constraints. In the last months, the authors of this paper have extended PIPS-IPM in order to handle LPs with both linking constraints and linking variables.

PIPS-IPM and also its new extension make use of the Message Passing Interface (MPI) for communication between their (parallel) *MPI-processes*. An important feature of PIPS-IPM is the distribution of the LP among the MPI-processes with no process needing to store the entire problem. This allows to tackle problems that are too large to even be stored in the main memory of a single desktop machine. The main principle is that for each index $i \in \{0, 1, ..., N\}$ all x_i, h_i, T_i, and W_i (for $i > 0$) need to be available in the same MPI-process—h_{N+1} needs to be assigned to the MPI-process handling $i = 0$. Moreover, each MPI-process needs access to the current value of x_0. The distribution is in the following exemplified for the case of the information to both $i = 0$ and $i = 1$ being assigned to the same MPI-process (in gray). The vectors and matrices that need to be processed together are marked in gray, black, and bold, respectively.

$$\begin{aligned}
\min \quad & c_0^T x_0 + c_1^T x_1 + c_2^T x_2 + \cdots \quad \mathbf{c_N^T x_N} \\
\text{s.t.} \quad & T_0 x_0 & = h_0 \\
& T_1 x_0 + W_1 x_1 & = h_1 \\
& T_2 x_0 + W_2 x_2 & = h_2 \\
& \quad\vdots \qquad\qquad \ddots & \vdots \\
& \mathbf{T_N x_0} + \mathbf{W_N x_N} & = \mathbf{h_N} \\
& F_0 x_0 + F_1 x_1 + F_2 x_2 + \cdots \quad \mathbf{F_N x_N} & = h_{N+1}
\end{aligned}$$

The maximum of MPI processes that can be used is N; in the opposite border case the whole LP is assigned to a single MPI-process

The extension of PIPS-IPM has already been successfully tested on medium-scale ESM problems with up to a million constraints and variables and up to 90 blocks. Since the number of MPI-processes is bounded by the number of blocks, the maximum number of MPI-processes we have used so far is also 90.

3 Communicating Block Structured GAMS Models to PIPS-IPM

A recently implemented GAMS/PIPS-IPM interface that considers the special HPC platform characteristics makes the solver available to a broader audience. This section is twofold. It outlines how users can annotate their GAMS models to provide a processable representation of the model block structure and provides insights in some technical aspects of the GAMS/PIPS-IPM-Link.

3.1 Annotating GAMS Models to Communicate Block Structures

Automatic detection of block structures in models is challenging [3], hence, a processable block structure information based on the user's deep understanding of the model is often preferable. It is important to note that there is no unique block structure in a model but there are many of them, depending on how rows and columns of the corresponding matrix are permuted. For ESMs blocks may for example be formed by regions or time steps as elaborated in Sect. 4.

GAMS provides facilities that allow complex processable model annotations [2]. The modeler can assign stages to variables via an attribute `<variable name>`

.stage. That functionality originates from multistage stochastic programming and can also be used to annotate the block structure of a model to be solved with PIPS-IPM. Once the block membership for all variables is annotated, the block membership of the constraints can in principle be derived from that annotation. However, manual annotation of constraints in a similar fashion is also possible and allows to run consistency checks on the annotation to detect potential mistakes. The annotation assignment can be demonstrated with a simple example based on the block structure introduced in Sect. 2. The following pseudo-annotation would assign stages to all variables x_i to indicate their block membership.

$$x_i.stage = i \qquad \forall i \in \{0, 1, ..., N\}$$

Linking variables are those assigned to stage 0. Similarly, constraints could also be annotated where stage 0 constraints are those containing only linking variables. Constraints assigned to stages 1,...,N are those incorporating only variables from the corresponding block plus linking variables and finally constraints assigned to stage N + 1 are the linking ones. Note that the exemplary pseudo-annotation may seem obvious and simple but finding a good block structure annotation for a complex model is not trivial. The challenge is not mainly to find an annotation that is correct in the mathematical sense but to find one where the power of PIPS-IPM is exploited best. A desirable annotation would reveal a block structure with many independent blocks of similar size while the set of linking variables and linking constraints is small.

3.2 The GAMS/PIPS-IPM-Link

Currently, the GAMS/PIPS-IPM-Link implements the connection between modeling language and the solver in a two-phase process. Phase 1, the model generation, is followed by phase 2 where PIPS-IPM pulls the previously generated model via its callback interface and solves the problem.

So far, model generation used to be a sequential process where GAMS generates one constraint after another. For the majority of applications this is fine as model generation is usually fast and the time consumption is negligible compared to the time consumed to solve the actual problem. However, some ESMs may result in sizeable LPs where model generation time becomes relevant. Hence, it is worthwhile to mention that the previously introduced annotation can also serve as a basis to generate the model in a distributed fashion. Instead of generating one large monolithic model, many small model blocks can be generated in parallel to exploit the power of HPC architectures already during model generation.

4 Structuring Energy System Models for PIPS-IPM

In order to distribute all blocks of the full-scale ESM to the computing nodes of a HPC architecture a problem-specific model annotation has to be provided. Based on the modeler's knowledge about the problem at hand the number of blocks and block structure has to be decided upon corresponding directly to the assignment of variables to blocks.

The concurrency of supply and demand of electrical energy necessitates a balancing for every region and time step. While in theory these balancing constraints can be solved independently, transport of energy between regions and storage of energy require a integrated optimization of all regions and time steps. The number of variables and constraints linked by the annotation depends strongly on these spatial and temporal interconnections. Transport of energy between two regions is typically represented by dispatch variables leading to linking variables if their respective regions have been assigned to different blocks. State of charge variables for energy storages consider the state of charge in the previous time step and therefore lead to a large number of linking constraints if each time step is represented by a single block. Typically, ESM also comprise boundary conditions that link both regions and time steps, e.g. by the consideration of global and annual emission limits. The high number of linking variables and constraints lead to a trade-off between speed-up and parallelism that needs to be studied systematically in future numerical experiments.

Figure 1 shows the non-zero entries matrix of the ESM REMix [4] on the left side and the revealed underlying block structure after permutation of the matrix on the right side. Linking variables and constraints are marked in dark gray while PIPS-IPM blocks are marked in light gray. The ESM represents the electricity sector for Germany with 21 spatial regions, 17 technologies per region and 168 time steps respectively 7 blocks of 24 time steps in the annotated case.

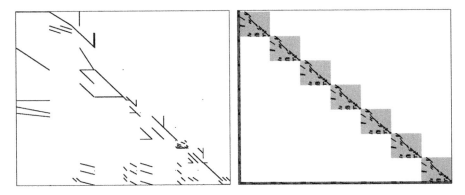

Fig. 1 Non-zero entries of the ESM and permuted matrix with block structure

5 Summary and Outlook

Large-scale LPs emerging from ESMs that are computationally intractable for today's state-of-the-art LP solvers motivate the need for new solution approaches. To serve those needs, extensions to the parallel interior point solver PIPS-IPM that exploits the parallel power of high performance computers have been implemented. In the future, the solver will be made available to the ESM community by a GAMS/PIPS-IPM interface.

The integration of HPC specialists in the development process ensures consideration of peculiarities of several targeted HPC platforms at an early stage of development. PIPS-IPM is developed and tested on several target platforms like the petaflops systems Hazel Hen at HLRS and JURECA at JSC as well as on many-core platforms like JUQUEEN and modern Intel Xeon Phi Processors. Workflow automation tools explicitly designed for HPC applications like JUBE [5] support the development and execution by simplifying the usage of workflow managers like PBS and Slurm.

Initial computational experiments already show the capability of the extended PIPS-IPM version to solve the ESM problems at hand, although so far only on a small scale. However, the good scaling behavior and the results of the original PIPS-IPM in solving large-scale problems [6] suggest that the approach described in this article might ultimately lead to a solver that can tackle currently unsolvable large-scale ESMs. Extensions to the GAMS/PIPS-IPM-Link will finally integrate the current multi-phase workflow (see Sect. 3.2) into one seamless process to give energy system modelers a similar workflow compared to the use of conventional LP solvers.

Acknowledgements The described research activities are funded by the Federal Ministry for Economic Affairs and Energy within the BEAM-ME project (ID: 03ET4023A-F). Ambros Gleixner was supported by the Research Campus MODAL *Mathematical Optimization and Data Analysis Laboratories* funded by the Federal Ministry of Education and Research (BMBF Grant 05M14ZAM).

References

1. Cao, K., Gleixner, A., & Miltenberger, M. (2016). Methoden zur Reduktion der Rechenzeit linearer Optimierungsmodelle in der Energiewirtschaft - Eine Performance-Analyse. In *EnInnov 2016 Symposium Energieinnovation* (p. 14).
2. Ferris, M. C., Dirkse, S. P., Jagla, J., & Meeraus, A. (2009). An extended mathematical programming framework. *Computers & Chemical Engineering, 33*, 1973–1982. https://doi.org/10.1016/j.compchemeng.2009.06.013.
3. Ferris, M. C., & Horn, J. D. (1998). Partitioning mathematical programs for parallel solution. In *Mathematical programming* (Vol. 80, pp. 35–61). https://doi.org/10.1007/BF01582130.
4. Gils, H. C., et al. (2017). Integrated modelling of variable renewable energy-based power supply in Europe. *Energy, 123*, 173–188 (2017). https://doi.org/10.1016/j.energy.2017.01.115.
5. Luehrs, S., et al. *Flexible and generic workflow management.* https://doi.org/10.3233/978-1-61499-621-7-431.

6. Petra, C. G., Schenk, O., & Anitescu, M. (2014). Real-time stochastic optimization of complex energy systems on high performance computers. *Computing in Science & Engineering (CiSE), 16*(5), 32–42.
7. Vanderbei, R. J. (2014). *Linear programming: Foundations and extensions*. Springer.
8. Schenk, O., & Gartner, K. (2004). *On Fast Factorization Pivoting Methods for Sparse Symmetric Indefinite Systems*. Technical Report, Department of Computer Science, University of Basel (2004)

WORHP Zen: Parametric Sensitivity Analysis for the Nonlinear Programming Solver WORHP

Renke Kuhlmann, Sören Geffken and Christof Büskens

1 Introduction

Nonlinear optimization problems that arise in real-world applications usually depend on parameter data. Parametric sensitivity analysis is concerned with the effects on the optimal solution caused by changes of these. The calculated sensitivities are of high interest because they improve the understanding of the optimal solution and allow the formulation of real-time capable update algorithms. Examples of applications are therefore real-time optimal control [9] or limited computational environments like in space [10] or automotive industry [1].

In this paper, we present WORHP Zen, a sensitivity analysis module for the nonlinear programming solver WORHP [3] that is capable of calculating sensitivity derivatives with exploitation of special structured parameters, performing real-time approximations of parameter perturbed optimization problems and estimating the allowed parameter space. We consider the parameter dependent nonlinear programming problem

$$\min_{x \in \mathbb{R}^n} \quad f(x; p) := \tilde{f}(x; t) - r^\top x$$
$$\text{s.t.} \quad c(x; p) := \tilde{c}(x; t) - q \leq 0 \tag{1}$$

with parameters $p := (t, r, q) \in \mathbb{R}^{k+n+m}$ and twice continuously differentiable functions $\tilde{f} : \mathbb{R}^n \times \mathbb{R}^k \to \mathbb{R}$ and $\tilde{c} : \mathbb{R}^n \times \mathbb{R}^k \to \mathbb{R}^m$. For a fixed $p_0 = (t_0, r_0, q_0)$

R. Kuhlmann (✉) · S. Geffken · C. Büskens
University Bremen, Bibliothekstr. 5, 28195 Bremen, Germany
e-mail: renke.kuhlmann@math.uni-bremen.de

S. Geffken
e-mail: sgeffken@math.uni-bremen.de

C. Büskens
e-mail: bueskens@math.uni-bremen.de

problem (1) is called unperturbed and can be solved by standard nonlinear programming techniques.

In Sect. 2 we will present the main properties of the sensitivity analysis and the special features of WORHP ZEN. Section 3 provides an example application in the field of parameter identification.

2 Sensitivity Analysis with WORHP ZEN

2.1 Sensitivity Properties

The Lagrangian function of (1) is $L(x, y; p) = f(x; p) + y^\top c(x; p)$ with multipliers $y \in \mathbb{R}^m$ and the first-order optimality conditions of (1) are:

$$\nabla_x \widetilde{f}(x; t) + \nabla_x \widetilde{c}(x; p) - r = 0$$
$$0 \leq y \perp \widetilde{c}(x; t) - q \leq 0 \qquad (2)$$

The sensitivity analysis is valid only locally and requires the linear independence constraint qualification (LICQ) and the second order sufficient condition (SOSC). We define the active set as $\mathcal{I}(x, p) := \{j \mid c_j(x; p) = 0\}$ and the special case $\mathcal{B}(x, p) := \{i \in \{1, \ldots, n\} \mid \exists j : c_j(x; p) = x_i - q_j = 0\}$.

Definition 1 The LICQ holds for solution of (1) x^*, if the gradients $\nabla_x c_i(x^*; p_0)$ and $i \in \mathcal{I}(x^*, p_0)$ are linearly independent.

Definition 2 The SOSC holds for p_0 and (x^*, y^*) satisfying the first-order optimality conditions (2), if $d^\top \nabla^2_{xx} L(x^*, y^*; p_0) d > 0$ for all $d \neq 0$ with $\nabla_x c_i(x^*; p_0)^\top d = 0$ for $i \in \mathcal{I}(x^*, p_0)$.

With these definitions we can present the main theorem of the sensitivity analysis, which shows the local existence of parameter dependent functions that solve (1) for a perturbation p. It has been introduced and proven by Fiacco [4].

Theorem 1 *Let (x^*, y^*) satisfy the first-order optimality conditions (2) of the unperturbed problem (1), which satisfies the LICQ, the SOSC and strict complementarity, i.e. $y^* - c(x^*; p_0) > 0$. Suppose, that f and c are twice continuously differentiable w.r.t. x and that $\nabla_x f$, $\nabla_x c$ and c are continuously differentiable w.r.t. p in a neighborhood of p_0. Then a neighborhood of p_0, $P \subset \mathbb{R}^{k+n+m}$, and continuously differentiable functions $x : P \to \mathbb{R}^n$, $y : P \to \mathbb{R}^m$ exist that satisfy:*

1. $x(p_0) = x^*$ and $y(p_0) = y^*$.
2. *The active set does not change, i.e. $\mathcal{I}(x(p), p) \equiv \mathcal{I}(x^*, p_0)$ for $p \in P$.*
3. $x(p)$ *satisfies the LICQ for all $p \in P$.*
4. $x(p)$ *satisfies the SOSC together with $y(p)$ with $y_i(p) > 0$, $i \in \mathcal{I}(x(p), p)$ and, thus, $x(p)$ is a solution of (1).*

2.2 Calculation of Sensitivity Derivatives

Sensitivity derivatives of the optimal solution can be obtained by the application of the implicit function theorem, which yields

$$\begin{bmatrix} \nabla^2_{xx} L(x^*, y^*; p_0) & \nabla_x c(x^*; p_0) \\ Y^* \nabla_x c(x^*; p_0)^\top & \Gamma \end{bmatrix} \begin{bmatrix} \frac{dx}{dp} \\ \frac{dy}{dp} \end{bmatrix} = - \begin{bmatrix} \nabla_{xp} L(x^*, y^*; p_0) \\ Y^* \nabla_p c(x^*; p_0) \end{bmatrix} \quad (3)$$

with $Y^* := \mathrm{diag}(y^*)$ and $\Gamma := \mathrm{diag}(c(x^*; p_0))$. The matrix in (3) is invertible under the conditions of Theorem 1 (see Büskens [2]) and already exists in factored form for both optimization algorithms in WORHP, the sequential quadratic programming [3] and the penalty-interior-point algorithm [7]. In case of the latter, the factored matrix actually belongs to a barrier problem with a small barrier parameter $\mu > 0$. The calculated sensitivity derivatives therefore have an error of $\mathcal{O}(\mu)$ which is negligible in practice. For a detailed study see Fiacco [5] or for an overview Pirnay et al. [9].

Further sensitivity derivatives can be provided for the objective function and the constraints. For the constraints these are $\frac{dc_i}{dp}(p_0) = 0$ for $i \in \mathcal{I}(x^*, p_0)$ and $\frac{dc_i}{dp}(p_0) = \nabla_x c_i(x^*; p_0)^\top \frac{dx}{dp}(p_0) + \nabla_p c_i(x^*; p_0)$ otherwise. For the objective function we have $\frac{df}{dp}(p_0) = \nabla_p L(x^*, y^*; p_0)$ and, thus, $\frac{df}{dr}(r_0) = -x^*$ and $\frac{df}{dq}(q_0) = -y^*$. For a complete overview and also second-order sensitivity derivatives of the objective function see Büskens [2]. It is also possible to exploit quadratic perturbations in the objective function or linear ones in the constraints (see Geffken [6]), but this is not considered in WORHP ZEN.

2.3 Storage of Sensitivity Derivatives

WORHP ZEN not just provides sensitivity derivatives, but also uses them for further calculations (see below). Therefore, the sensitivity derivatives are once calculated and then stored, which can be a memory costly operation for large-scale problems. However, from Theorem 1 (2) we know that the sensitivity matrices are sparse. Furthermore, symmetry and equivalence of sensitivity derivatives can be exploited. Table 1 gives an overview.

Table 1 Sparsity, symmetry and equivalence of sensitivity derivatives for special structured perturbations. Symmetry is considered for the appropriate sensitivity matrices (all i and all j). $\mathcal{I}$ and $\mathcal{B}$ are abbreviations for $\mathcal{I}(x^*, p_0)$ and $\mathcal{B}(x^*, p_0)$, respectively

Sensitivity	$\frac{dx_i}{dt_j}$	$\frac{dx_i}{dr_j}$	$\frac{dx_i}{dq_j} = \frac{dy_j}{dr_i}$	$\frac{dy_i}{dq_j}$	$\frac{dy_i}{dt_j}$
Symmetric	–	yes	–	yes	–
Zero if	$i \in \mathcal{B}$	$i \in \mathcal{B}$ or $j \in \mathcal{B}$	$i \in \mathcal{B}$ or $j \notin \mathcal{I}$	$i \notin \mathcal{I}$ or $j \notin \mathcal{I}$	$i \notin \mathcal{I}$

2.4 Real-Time Approximations

Using the sensitivity derivatives and Taylor's theorem, WORHP ZEN can provide a real-time approximation of (2) with perturbation p by

$$x(p) = x^* + \frac{\mathrm{d}x}{\mathrm{d}p}(p - p_0) + \mathcal{O}(\|p - p_0\|^2)$$

and analogously for $y(p)$, $z(p)$, $f(x(p); p)$ and $c(x(p); p)$. Due to the error and the nonlinearity of c, the approximation is likely to violate the constraints, i.e. for $i \in \mathcal{I}(x^*, p_0)$ we have $c_i(x(p); t) - q_i = \varepsilon \neq 0$. Büskens [2] proposes to interpret this as a perturbation $(0, 0, \varepsilon)$ and iteratively refine the approximation, again by using a first order Taylor approximation. Since $\frac{\mathrm{d}x}{\mathrm{d}q}$ is available in WORHP ZEN, this strategy can directly be applied.

2.5 Estimation of the Neighborhood P

The main restriction of the sensitivity analysis is the limitation to a local and unknown neighborhood P, see Theorem 1. Due to the equivalence of the active set for all $p \in P$, however, WORHP ZEN can estimate P by calculating the maximal perturbation without active set changes using real-time approximations (see Büskens [2]), i.e. for a perturbation p_j

$$P \approx \left[\max\{p_j \in \bar{P}_j \mid p_j < (p_0)_j\}, \min\{p_j \in \bar{P}_j \mid p_j > (p_0)_j\}\right],$$
$$\bar{P}_j := \{p_j^i \mid i = 1, \ldots, m\} \cup \{-\infty, \infty\},$$
$$p_j^i := (p_0)_j - \frac{c_i(x^*, p_0)}{\frac{\mathrm{d}c_i}{\mathrm{d}p_j}(p_0)}, \text{ if } i \notin \mathcal{I}(x^*), \text{ or } p_j^i := (p_0)_j - \frac{y_i}{\frac{\mathrm{d}y_i}{\mathrm{d}p_j}(p_0)}, \text{ if } i \in \mathcal{I}(x^*).$$

3 Example Application

In this example we show the application of post-optimality analysis to the parameter identification of characteristic maps. These maps are often used by engineers to simplify complex physical relationships within their simulations. The main goal of this identification is to fit the parameters such that some measurement can be reconstructed using the model (see Nelles [8]).

A *characteristic map* represents a function $f : \mathbb{R}^2 \times \mathbb{R}^{N_v} \to \mathbb{R}$, where the output is computed using piecewise interpolation in two-dimensional space. It consists of N_x nodes in x-direction and N_y nodes in y-direction with associated values $v \in \mathbb{R}^{N_v} =$

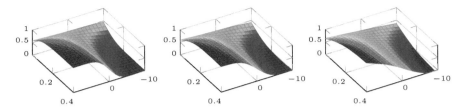

Fig. 1 Optimal solutions of the characteristic map for $p \in \{0.1, 1, 10\}$ with increasing p from left to right

$\mathbb{R}^{N_x \cdot N_y}$. Let a measurement consisting of data points $(\bar{x}, \bar{y}, \bar{z}) \in \mathbb{R}^{K \times 3}$ be given. The maps can be determined using optimization techniques with the objective function

$$\min_{v \in \mathbb{R}^v} \frac{1}{K} \sum_{i=1}^{K} (f(\bar{x}_i, \bar{y}_i; v) - \bar{z}_i)^2 + \frac{p}{2|N_x||N_y|} \sum_{x,y} \kappa_x(x, y; v)^2 + \kappa_y(x, y; v)^2,$$

with second order difference quotients κ_x and κ_y with an equidistant grid, i.e. $\kappa_x(x, y; v) := \Delta x^{-2}(f(x + \Delta x, y; v) - 2f(x, y; v) + f(x - \Delta x, y; v))$ and analogously in y-direction, each with a suitable scaling. The two contrary goals of the optimization are error of the fit (first summand) and smoothness of the map (second summand) controlled by a parameter p. The latter is of special interest for the integration of these maps into extensive models. Another option to control the smoothness is by using constraints. However, this has the major drawback of the requirement to formulate absolute bounds on κ_x and κ_y, usually to be determined by trial and error (cf. [8]). In this example, let $x \in [-10, 10]$ be discretized as $\bar{x} = -10, -9, \ldots, 10$ and $y \in [0.01, 0.4]$ as $\bar{y} = 0.01, 0.0533, \ldots, 0.3567, 0.4$ and let $z(x, y) = \frac{1}{2}(1 + \tanh(xy))$ with some additional random noise. Figure 1 shows the optimal solutions for varying parameters $p \in \{0.1, 1, 10\}$.

The strong shape of the $\tanh(xy)$ is flattened by the smoothing term if $p = 10$, showing the conflict of the two objective parts. The key questions for an user are: How should the weighting parameter p be chosen? How much can p be perturbed for real-time updates?

In Fig. 2 the sensitivity fields $\frac{dx}{dp}$ reveal the areas which are affected most by the applied smoothing term and, thus, help to understand the influence of the parameter p. Red regions will heighten, while blue regions will lower if p is increased. Using this knowledge helps to choose p accordingly.

To show the estimation of the neighborhood P we use the stricter approach for controlling the smoothness (here just in x-direction) by using the constraint $c(x, y) := \Delta x^{-2} (f(x + \Delta x, y; v) - 2f(x, y; v) + f(x - \Delta x, y; v))$ for all interior points x of the map. The special perturbation q is used to examine two different cases of interest: The constraints are (1) $-1 \leq c(x, y) \leq 1$ (inactive constraints, may become active) and (2) $-0.35 \leq c(x, y) \leq 0.35$ (active constraints, may become inactive). The allowed perturbation is shown in Fig. 3 and reveals that the constraint is mostly influencing the strong hyperbolic shape ($y \approx 0.4$).

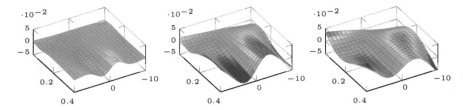

Fig. 2 Sensitivity fields of the characteristic map for $p \in \{0.1, 1, 10\}$ with increasing p from left to right

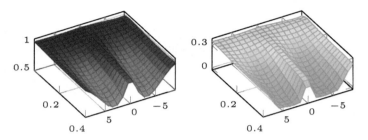

Fig. 3 Estimation of maximum allowed perturbation with respect to second order smoothness condition (left: constraint (1); right: constraint (2))

References

1. Böhme, T. J., & Frank, B. (2017). *Optimal design of hybrid powertrain configurations* (pp. 481–518). Cham: Springer International Publishing.
2. Böskens, C. (2002). Real-time optimization and real-time optimal control of parameter-perturbed problems. Habilitation thesis, Universität Bayreuth.
3. Büskens, C., & Wassel, D. (2013). The ESA NLP solver WORPH. In G. Fasano & J. D. Pintér (Eds.), *Modeling and optimization in space engineering* (Vol. 73, pp. 85–110). Springer optimization and its applications. New York: Springer.
4. Fiacco, A. V. (1983). Introduction to sensitivity and stability analysis in nonlinear programming. *Mathematics in science and engineering* (Vol. 165). New York: Academic Press.
5. Fiacco, A. V., & Ishizuka, Y. (1990). Sensitivity and stability analysis for nonlinear programming. *Annals of Operations Research*, 27(1), 215–235.
6. Geffken, S.: Effizienzsteigerung numerischer Verfahren der nichtlinearen Optimierung. Ph.D. thesis, Universität Bremen, Bremen (to appear)
7. Kuhlmann, R., & Büskens, C. (2017). Primal-dual augmented Lagrangian penalty-interior-point algorithm. Technical report, Universität Bremen.
8. Nelles, O. (2001). *Nonlinear system identification*. Berlin: Springer.
9. Pirnay, H., López-Negrete, R., & Biegler, L. T. (2012). Optimal sensitivity based on ipopt. *Mathematical Programming Computation*, 4(4), 307–331.
10. Seelbinder, D., & Büskens, C. (2016). Real-time atmospheric entry trajectory computation using parametric sensitivities. In *International Conference on Astrodynamics Tools and Techniques*.

Part XX
Supply Chain Management

Joint Optimization of Reorder Points in n-Level Distribution Networks Using (R, Q)-Order Policies

Christopher Grob, Andreas Bley and Konrad Schade

1 The Distribution Network

A central question in distribution networks is how to allocate safety stock along the different echelons and warehouses in order to fulfill the required service level targets at minimal investment in capital. We present an optimization algorithm which efficiently computes optimal reorder points taking into consideration the wait time approximation of Kiesmüller et al. [2]. This algorithm builds on insights obtained during the development of an algorithm for the simpler 2-level case [1].

We consider a n-level distribution network where all warehouses use an (R,Q)-order policy. Reorder points are denoted as $R_{i,j}$, where i is the echelon, $i = 1, \ldots, n$ and j is the index of the warehouses in the respective echelon. Accordingly, the order quantity is denoted as $Q_{i,j}$ and $\mu_{i,j}$ and $\sigma_{i,j}$ are the mean and standard deviation of demand at a warehouse (i, j) per time unit. The lead time from warehouse j at echelon i from its predecessor is $L_{i,j} = T_{i,j} + W_{i,j}$, where $T_{i,j}$ is the transportation time and $W_{i,j}$ is the wait time due to stock-outs at the predecessor. For $i = 1$ we assume $L_{11} = T_{11}$.

C denotes the set of all local warehouses, i.e., warehouses without successors. Only local warehouses fulfill external customer demand and have fill rate targets $\bar{\beta}_{i,j}$. The fill rate is the fraction of orders that is satisfied immediately from stock on hand. The set of preceding warehouses of a warehouse is referred to as $P_{i,j}$.

In industrial applications it is common practice to determine reorder points such that the investment in capital is minimized for given fill rate targets at the local warehouses. We therefore introduce $p_{i,j}$ as the price of the considered part in warehouse (i, j). An alternative interpretation could be the cost of stocking a unit of this item

---

C. Grob (✉) · K. Schade
Volkswagen AG, Wolfsburgx, Germany
e-mail: christopher.grob@volkswagen.de

A. Bley
Universität Kassel, Heinrich-Plett-Straße 40, 34123 Kassel, Germany

in the location. Let $\beta_{i,j}(R_{i,j}, R_{l,m})$ be the fill rate, given the reorder point $R_{i,j}$ and the reorder points of all predecessors (l, m). Then we have to solve the following optimization model.

$$\min \sum_i \sum_j p_{i,j} \cdot R_{i,j} \tag{1}$$

$$\text{s.t. } \beta_{i,j}(R_{i,j}, R_{l,m}) \geq \bar{\beta}_{i,j} \quad \forall (i, j) \in C, \ (l, m) \in P_{i,j} \tag{2}$$

To approximate the random wait times W_{ij}, we use the method by Kiesmüller et al. [2]. Note that we assume $Q_{i,j} >> E[D_{i,j}]$ and, therefore, that the average replenishment order size is $Q_{i,j}$. This simplifies notation but can easily be relaxed as in [2] and does not influence our subsequent analysis.

2 Optimization Algorithm

We construct an underestimating ILP for Model defined by Eqs. (1) and (2) by creating an approximate linear relationship between the reorder point of the local warehouse and each of its predecessor. For each of the local warehouses, we introduce an underestimating piecewise linear function which assumes that an increase of a predecessors reorder point has maximal impact on the wait time.

The first two moments of the wait time are functions of the reorder point of the preceding warehouse. First, we analyze how big the maximum decrease in the two moments of the wait time is, when we increase the respective reorder point by one. Second, we consider the effect on the local warehouses that do not have successors and fulfill external customer demand. For $(i, j) \in C$ and $(l, m) \in P_{i,j}$, we are able to show that

$$\max \delta(E[W_{i,j}](R_{l,m})) = \frac{E[L_{l,m}]}{\sum_{(n,o)\in P_{l,m}} Q_{n,o}} \text{ and}$$

$$\max \delta(E[W_{i,j}^2](R_{l,m})) = \frac{E[(L_{l,m})^2]}{\sum_{(n,o)\in P_{l,m}} Q_{n,o}}.$$

From this we can derive the largest possible decrease in mean and variance of the effective lead time demand of a local warehouse if we increase a reorder point $R_{l,m}$ of any preceding warehouse $(l, m) \in P_{i,j}$ as

$$\max_{\Delta R_{l,m}} \delta(E[D_{i,j}(L_{i,j})]) = \mu_{i,j} \frac{E[L_{l,m}]}{\sum_{(n,o)\in P_{l,m}} Q_{n,o}} \text{ and} \tag{3}$$

$$\max_{\Delta R_{l,m}} \delta(Var[D_{i,j}(L_{i,j})]) = \sigma_{i,j}^2 E[L_{l,m}] / \sum_{(n,o)\in P_{l,m}} Q_{n,o}$$

$$+ \mu_{i,j}^2 \left[E[(L_{l,m})^2]/\sum_{(n,o)\in P_{l,m}} Q_{n,o} - (E[L_{l,m}]/\sum_{(n,o)\in P_{l,m}} Q_{n,o})^2 \right]. \quad (4)$$

To determine $L_{l,m}$ in Eqs. (3) and (4), we need in principle to set all reorder points of warehouses in $P_{l,m}$ and calculate the respective wait times. To avoid this, we assume, that we always have to wait at every preceding stage, i.e., we incorporate the full transportation time of all predecessors in $L_{l,m}$. By doing so, we overestimate the possible impact.

For each (i, j) and $(l, m) \in P_{i,j}$ the range of $R_{l,m}$ in which the wait time of (i, j) is influenced, assuming maximum impact, is limited and smaller than the actual range. We calculate for all predecessors $(l, m) \in P_{i,j}$ the upper bounds, up to which the reorder point of the respective predecessor can influence the mean of the wait time and therefore, implicitly, the variance. Using Eqs. (3) and (4) we obtain:

$$\bar{u}_{l,m}^{i,j} := E[L_{l,m}] \sum_{(n,o)\in P_{l,m}} Q_{n,o}/E[L_{l,m}] = \sum_{(n,o)\in P_{l,m}} Q_{n,o}. \quad (5)$$

We also assume that the lower bound for each $R_{l,m}$ is 0. $R_{l,m}$ then is the range $[0, \bar{u}_{l,m}^{i,j}]$.

Let $R_{i,j}|(L_{i,j} = L)$ be the reorder point needed at a local warehouse (i, j) assuming $L_{i,j} = L$, such that the fill rate constraint is fulfilled. We calculate a general lower bound for all local warehouses as $l_{i,j} := R_{i,j}|(L_{i,j} = T_{i,j})$. This assumes that the wait time caused by all preceding warehouses $(l, m) \in P_{i,j}$ is 0. We also calculate an upper bound by assuming the order has to wait at every stage, i.e., $\bar{R}_{i,j} := R_{i,j}|(L_{i,j} = \sum_{(l,m)\in P_{i,j}} T_{l,m})$.

For each predecessor (l, m) of a local warehouse (i, j), we calculate the following.

1. Set the lead time $L_{l,m}$ of the predecessors (l, m) to its upper bound, i.e. as the sum of all transportation times to this predecessor, and assume all intermediate warehouses between (l, m) and (i, j) act as cross-docks only, i.e., full transportation time on this path applies.
2. Calculate $R_{i,j}$ assuming $R_{l,m} = 0$. Call this reorder point $\hat{R}_{i,j}^{l,m}$
3. Calculate the slope $b_{l,m}^{i,j}$ of the linear relationship as $b_{l,m}^{i,j} = (\hat{R}_{i,j}^{l,m} - l_{i,j})/\bar{u}_{l,m}^{i,j}$.
4. With the above assumptions, the following function is an underestimator for $R_{i,j}$, given a $R_{l,m}$, $(l, m) \in P_{i,j}$ and all $R_{n,o} = -1$, $(n, o) \in P_{i,j} \setminus \{(l, m)\}$:

$$e_{l,m}^{i,j}(R_{l,m}) = \begin{cases} (\hat{R}_{i,j}^{l,m} - 1) - b_{l,m} R_{l,m}, & \text{if } \bar{u}_{l,m}^{i,j} \geq R_{l,m} \geq 0 \\ (\hat{R}_{i,j}^{l,m} - 1) - b_{l,m} \bar{u}_{l,m}^{i,j}, & \text{if } R_{l,m} > \bar{u}_{l,m}^{i,j} \end{cases} \quad (6)$$

Step 1 ensures that we capture the highest possible slope $b_{l,m}^{i,j}$ and exclude all effects of the intermediate warehouses except the transportation time. The function, which supplies $R_{i,j}$ under the above assumptions for a given $R_{l,m}$, approximated by $e_{l,m}^{i,j}(R_{l,m})$ is a step function. If a local reorder point is decreased by 1 right at the start when $R_{l,m}$ is increased from 0 to 1, $e_{l,m}^{i,j}(R_{l,m})$ can intersect this function. By subtracting 1 in Eq. (6) from the y-intercept, we prevent this from happening.

Combining Eq. (6) for all $(l, m) \in P_{i,j}$, we get an underestimator for $R_{i,j}$ given the reorder points of all predecessors.

$$R_{i,j} \geq \bar{R}_{i,j} - \sum_{P_{i,j}} (g_{l,m}^{i,j}(R_{l,m})) \ \forall (i, j) \in C, \tag{7}$$

where $g_{l,m}^{i,j}(R_{l,m}) := -(e_{l,m}^{i,j}(R_{l,m}) - \hat{R}_{i,j}^{l,m})$.

In the construction of the linear relationship between (i, j) and (l, m), we ensured that the slope $b_{l,m}^{i,j}$ is overestimating the real decrease of $R_{i,j}$ if $R_{l,m}$ is increased. We limited the range in which $R_{l,m}$ influences $R_{i,j}$ by assuming the wait time is always decreased by the maximum possible value (Eq. (5)) and in a second step, calculated an upper bound for the decrease of $R_{i,j}$ caused by $R_{l,m}$ and applied this upper bound to the limited range.

Additionally, we need to impose an upper bound on the sum of the reductions of a local reorder point caused by each preceding warehouse and all its predecessors. This prevents the local reorder point to be reduced to an extent that implies a negative overall wait time.

$$g_{l,m}^{i,j}(R_{l,m}) + \sum_{(n,o) \in P_{l,m}} (g_{n,o}^{i,j}(R_{n,o})) \leq \bar{b}_{l,m}^{i,j} \ \forall (i, j) \in C, \ (l, m) \notin C \cup \{(1, 1)\} \tag{8}$$

Here $\bar{b}_{l,m}^{i,j}$ is the maximum reduction of a local reorder point that can be achieved by the warehouses (l, m) and $(n, o) \in P_{l,m}$. Defining $W_{l,m}^{i,j}$ as the set of all warehouses that connect (i, j) and (l, m), we can calculate the maximum reduction as

$$\bar{b}_{l,m}^{i,j} = R_{i,j} | (L_{i,j} = \sum_{(n,o) \in P_{i,j}} T_{n,o}) - R_{i,j} | (L_{i,j} = \sum_{(n,o) \in W_{l,m}^{i,j}} T_{n,o}). \tag{9}$$

Now we can construct an ILP, that is an underestimator of the Model defined by Eqs. (1) and (2):

$$\min \quad \sum_i \sum_j p_{i,j} \cdot R_{i,j} \tag{10}$$

$$\text{s.t.} \quad R_{i,j} \geq \bar{R}_{i,j} - \sum_{(l,m) \in P_{i,j}} (g_{l,m}^{i,j}(R_{l,m})) \quad \text{for all } (i, j) \in C \tag{11}$$

$$R_{i,j} \geq l_{i,j} \quad \text{for all } (i, j) \in C \tag{12}$$

$$g_{l,m}^{i,j}(R_{l,m}) \leq \bar{b}_{l,m}^{i,j} - \sum_{(n,o) \in P_{l,m}} (g_{n,o}^{i,j}(R_{n,o})) \quad \text{for all } (i, j) \in C$$

$$\text{for all } (l, m) \notin C. \tag{13}$$

We construct an algorithm to determine optimal reorder points by iteratively refining the linear functions and introducing new constraints.

Algorithm 1 (Optimization)

1. *Construct initial ILP (Eqs. (10)–(13) as described in this section.*
2. *Solve ILP and obtain the solution $R_{i,j}^*, \forall (i,j)$.*
3. *Based on all non-local reorder points $R_{i,j}^*$, $(i,j) \notin C$, calculate the actual local reorder points $\tilde{R}_{i,j}$, $(i,j) \in C$ needed to fulfill the fill rate targets of Eq. (2).*
4. *Compare the objective function for the solution of the ILP and the solution incorporating $\tilde{R}_{i,j}$, $(i,j) \in C$. If the gap is sufficiently small or if the solution did not change compared to the previous iteration, terminate the algorithm and return the solution $\{R_{i,j}^*, (i,j) \notin C, \tilde{R}_{i,j}, (i,j) \in C\}$.*
5. *Otherwise, refine the model as described in the following and continue with Step 2.*

For each local warehouse $(i,j) \in C$, we refine the function $g_{l,m}^{i,j}(R_{l,m})$ for all $(l,m) \in P_{i,j}$ based on the optimal solution determined in Step 2. Recall that the function $g_{l,m}^{i,j}(R_{l,m})$ gives an upper bound for the reduction of the local reorder point $R_{i,j}$ by setting a reorder point $R_{l,m}$ of a predecessor to a certain value. With the solution $R_{l,m}^*$ of the ILP obtained in Step 2, we can update the function $g_{l,m}^{i,j}(R_{l,m})$.

We have to charge the reduction to the different predecessors. We assume that $R_{n,o} = 0$, $(n,o) \in P_{l,m}$ and that all intermediate warehouses between (i,j) and (l,m) act as cross-docks only. Furthermore, we set the variance of the wait time caused by $R_{l,m}^*$ to 0. With these settings, we make sure that we still strictly overestimate the reduction of $R_{i,j}$ by an increase of $R_{l,m}$ and our overall model is still underestimating the original problem. First, we recalculate the local reorder point $R_{i,j}$ given $R_{l,m}^*$ for each $(l,m) \in P_{i,j}$. Then we calculate the reduction as $\bar{r}_{l,m}^{i,j} = \hat{R}_{i,j}^{l,m} - R_{i,j} + 1$ and refine $g_{l,m}^{i,j}(R_{l,m})$ by inserting the new point $(R_{l,m}^*, \bar{r}_{l,m}^{i,j})$ using the following two properties: If $(R_{l,m})$ increases by 1, the maximum possible decrease of $R_{i,j}$ is $b_{l,m}^{i,j}$. If $(R_{l,m})$ decreases, a decrease of $R_{i,j}$ is not possible. Finally, we have to update the upper bound $\bar{u}_{l,m}^{i,j} = \max(\bar{u}_{l,m}^{i,j}, (\bar{R}_{i,j} - \bar{r}_{l,m}^{i,j})/b_{l,m}^{i,j} + R_{l,m}^*)$.

Additionally, we can calculate the actual local reorder points $R_{i,j}^{act}$, $(i,j) \in C$ needed to fulfill the fill rate targets given the non-local reorder points from the optimal solution. If we reduce the reorder points $R_{l,m}$, $(l,m) \in P_{i,j}$, the required local reorder point (i,j) can not be smaller:

$$R_{i,j} \geq R_{i,j}^{act}, \text{ if } R_{l,m} < R_{l,m}^* \text{ f. a. } (l,m) \in P_{i,j}. \tag{14}$$

Furthermore, we can introduce constraints for all other cases, i.e., if we increase some or many $R_{l,m}$, $(l,m) \in P_{i,j}$:

$$R_{i,j} \geq R_{i,j}^{act} - \sum_{(l,m) \mid R_{l,m} \geq R_{l,m}^*} R_{l,m}. \tag{15}$$

Equations (14) and (15) can be modeled in the ILP as indicator constraints or with the help of auxiliary binary variables. Equation (15) is based on the property, that a local reorder point can be reduced by at most by 1, if a non-local reorder point is increased by 1. Instead of this simple relationship, we could also model a constraint using functions $g_{l,m}^{i,j}(R_{l,m})$ re-based on $R_{i,j}^{act}$. While this would imply a better approximation, the generated ILP is more difficult to solve as we require many more auxiliary variables and special order set constraints. By introducing Eqs. (14) and (15), our ILP is now exact at $\{R_{l,m}^*\}$, much tighter in the area around this spot and we can guarantee optimality of our algorithm. However, those constraints are expensive and we advise on only introducing them, when refining the functions $g_{l,m}^{i,j}(R_{l,m})$ does not change the solution anymore.

3 Summary and Conclusion

In this paper we have developed an efficient optimization algorithm that is able to determine reorder points in n-level divergent distribution networks. This algorithm is, to our knowledge, the first one, that is able to determine optimal reorder points for general n-level distribution networks using the wait time approximation by Kiesmüller et al.

Experiences from our work with 2-level distribution networks show that fill rates at non-local warehouses should be much lower than what is common in practice and stock can be drastically reduced compared to prescribed fill rates. We plan to validate this for the n-level case using real world data. Additionally, the performance of the algorithm can be further improved.

References

1. Grob, C., & Bley, A. (2017). On the optimality of reorder points - a solution procedure for joint optimization in 2-level distribution networks using (R, Q) order policies. Working paper
2. Kiesmüller, G. P., de Kok, T. G., Smits, S. R., & van Laarhoven, P. J. (2004). Evaluation of divergent n-echelon (s, nq)-policies under compound renewal demand. *OR-Spektrum*, 26(4), 547–577.

An Integrated Loss-Based Optimization Model for Apple Supply Chain

P. Paam, R. Berretta and M. Heydar

1 Introduction

Food supply chain (FSC) is more complex compared to other kinds of supply chain (SC) because of the perishable nature of foodstuff. Therefore, any drawback within the FSC may lead to food loss. Food loss refers to "the decrease in edible food mass throughout the part of the SC that specifically leads to edible food for human consumption" [1]. It is a worldwide phenomenon, which affects both developed and developing countries.

Agricultural fruit supply chain (AFSC) "constitutes the processes from production to delivery of the fruit products from farm to market" [2]. Fruit loss in AFSC is dependent on how the products are dealt with throughout the SC processes and can happen due to problems in different stages. This may cause the production of unavoidable second and third-grade fruits, fruit damage across the chain or perished fruits. A remedy for this problem is to develop planning mathematical models in AFSC.

Few studies have previously focused on developing mixed-integer linear programming (MILP) models for AFSC, which consider food loss. For instance, Ferrer et al. [3] developed a model for grape harvest scheduling with the objective of total cost minimization, including a penalty cost for grapes, harvested before or after the optimal harvest period. Ahumada and Villalobos [4] developed a tomato SC model,

---

P. Paam (✉) · R. Berretta · M. Heydar
School of Electrical Engineering and Computing, The University of Newcastle, Callaghan, Australia
e-mail: Parichehr.Paam@uon.edu.au

R. Berretta
e-mail: regina.berretta@newcastle.edu.au

M. Heydar
e-mail: mojtaba.heydar@newcastle.edu.au

applying a penalty cost for having tomatoes with unacceptable color. Rong et al. [5] investigated food quality deterioration in a distribution problem for a bell pepper SC.

Only few papers considered inventory management of fruit products. For example, Herbon et al. [6] managed inventory by tracking the age and quality of perishable products using RFID. Muriana [7] investigated the impact of time-temperature indicators on food quality changes in inventory. Recently, Paam et al. [8] conducted a comprehensive survey on planning and optimization models in the agricultural fresh food SC considering the concept of food loss.

The contribution of this research is developing a mathematical model for postharvest handling and storage of an apple SC with two different time periods of harvesting and planning and quantifying apple losses based on the time gap between their harvest and delivery.

2 Problem Statement and Mathematical Formulation

In this study, we consider a typical apple SC company in Australia, where harvested apples are transported to the plant and assigned to three different types of storage rooms, including conventional cold (CC), smart fresh (SF), and controlled atmosphere (CA). The energy cost and shelf life of apples vary in different room types. After being stored, apples are processed through a processing line to meet the demand in each time period. The line consists of various stages, with one designated to classify apples into different quality grades based on their size and color. Among these grades, grade-1 has the best quality and gains the most profit, grade-2 has lower quality and price, and grade-3 has almost the lowest quality, which is used for producing apple juices. Therefore, the grades of apples are not known in storage rooms and the quantity of each grade is specified at the end of the processing line, where apples are packed and directly transported to the market.

In the proposed model, we define two separate time periods, which are harvesting and planning time periods. Harvesting periods are a subset of planning periods. Apples are assigned to the storage rooms during harvesting periods and depleted from rooms and sold to the market during planning periods. In our problem, the time interval is defined as one week. We assume that the planning horizon is 46 weeks, in which the first 16 weeks are harvesting periods.

In our model, apple losses happen when apples are kept more than their shelf life in the storage rooms. Each storage room must be only of one type and its optimal type is decided by the model. We also assume that all parameters such as the quantity of shipped apples to the plant, the demand of each apple grade, percentages of different grades of apples at the grading stage and costs are known. Finally, it is assumed that apple losses are kept inside the storage rooms until the end of the planning period and processed apples with no demand at the end of the line are discarded without incurring any cost.

The indices and sets, parameters, and decision variables of the developed MILP model are as follows:

2.1 Indices and Sets

$t \in T$: planning periods (week), where $T = \{1, \ldots, 46\}$; $h \in H \subseteq T$: harvesting periods (week), where $H = \{1, \ldots, 16\}$; $s \in S$: type of storage rooms, where $S = \{1, 2, 3\}$; $n \in N$: number of storage rooms, where $N = \{1, \ldots, 39\}$; and $g \in G$: quality grades, where $G = \{1, 2, 3\}$.

2.2 Parameters

qop_h: quantity of input apples to the plant in harvesting period h (kg/week); mxp: maximum capacity of the processing line (kg/week); mxs_n: maximum capacity of storage room number n (kg); sl_s: shelf life of apples in storage room type s (week); dm_{gt}: demand of apple grade g in period t (kg/week); pg_{gt}: percentage of apple grade g at grading stage in period t (%); pct: processing cost of one unit of apple ($/kg); ict: inventory cost of one unit of apple ($/kg); ect_s: energy cost of storage room type s ($/week); kct_g: packing cost of one unit of apple grade g ($/kg); and lct_s: penalty cost for one unit of apple loss in storage room type s ($/kg).

2.3 Decision Variables

qin_{nsh}: quantity of input apples to storage room number n of type s in harvesting period h; qou_{nsht}: quantity of output apples from storage room number n of type s to the processing line harvested in period h in period t; qpm_{ght}: quantity of apple grade g at the end of the processing line harvested in period h in period t; $qloss_{ns}$: quantity of apple losses in storage room number n of type s; inv_{nsht}: inventory level of apples in storage room number n of type s harvested in period h at the end of period t; x_{nst}: equal to 1, if storage room number n of type s is on in period t, 0 otherwise; and y_{ns}: equal to 1, if storage number n is of type s, 0 otherwise.

2.4 Mixed-Integer Linear Programming Model

$$OBJ = Min \begin{bmatrix} \sum_n \sum_s \sum_h \sum_t pct \times qou_{nsht} + \sum_n \sum_s \sum_h \sum_t ict \times inv_{nsht} \\ + \sum_n \sum_s \sum_t ect_s \times x_{nst} + \sum_g \sum_s \sum_t kct_g \times qpm_{ght} + \sum_n \sum_s lct_s \times qloss_{ns} \end{bmatrix} \quad (1)$$

Subject to:

$$qop_h = \sum_n \sum_s qin_{nsh} \quad \forall h \in H \quad (2)$$

$$inv_{nsht} = qin_{nsh} - qou_{nsht} \quad \forall n \in N, s \in S, h \in H, t \in T \text{ where } h = t \quad (3)$$

$$inv_{nsht} = inv_{nsh,t-1} - qou_{nsht} \quad \forall n \in N, s \in S, h \in H, t \in T \text{ where } h < t \quad (4)$$

$$\sum_n \sum_s \sum_h qou_{nsht} \geq \sum_h qpm_{ght}/pg_{gt} \quad \forall t \in T, g = 1 \text{ where } h \leq t \quad (5)$$

$$\sum_n \sum_s \sum_h qou_{nsht} \geq \sum_h \sum_g qpm_{ght} \quad \forall t \in T \text{ where } h \leq t \quad (6)$$

$$\sum_n \sum_s \sum_h qou_{nsht} \leq mxp \quad \forall t \in T \text{ where } h \leq t \quad (7)$$

$$qin_{ns,h=t} + \sum_{h=1}^{t-1} inv_{nsh,t-1} \leq mxs_n \times x_{nst} \quad \forall n \in N, s \in S, t \in H, \quad inv_{nsh,t=0} = 0 \quad (8)$$

$$\sum_h inv_{nsh,t-1} \leq mxs_n \times x_{nst} \quad \forall n \in N, s \in S, t \in T - H \text{ where } h \leq t \quad (9)$$

$$\sum_h qpm_{ght} \geq dm_{gt} \quad \forall t \in T, g \in G \text{ where } h \leq t \quad (10)$$

$$\sum_s y_{ns} \leq 1 \quad \forall n \in N \quad (11)$$

$$\sum_t x_{nst} \leq T \times y_{ns} \quad \forall n \in N, s \in S \quad (12)$$

$$qloss_{ns} = \sum_h \sum_t inv_{nsht} \quad \forall n \in N, s \in S \text{ where } t - h = sl_s + 1 \quad (13)$$

$$\sum_h \sum_t qou_{nsht} = 0 \quad \forall n \in N, s \in S \text{ where } h + sl_s + 1 \leq t \quad (14)$$

$$qin_{nsh}, qou_{nsht}, qpm_{ght}, qloss_{ns}, inv_{nsht} \geq 0 \quad \forall n \in N, s \in S, g \in G, h \in H, t \in T \quad (15)$$

$$x_{nst}, y_{ns} \in \{0, 1\} \quad \forall n \in N, s \in S, t \in T \quad (16)$$

In the above formulation, the objective function (1) minimizes the total costs, including processing costs, inventory costs, the energy cost of different types of storage rooms, packing costs of different grades of apples, and penalty cost for apple losses.

Constraint (2) states that all apples shipped from orchards to the plant should be assigned to storage rooms. Constraints (3) and (4) are inventory balance where $h = t$ and $h < t$, respectively. In the former, there is an input to the storage rooms without any inventory from the previous period, while in the latter, it is the other way around. Constraints (5) and (6) state that in each planning period, the amount of apple coming out of the storage rooms should be the maximum amount between two values: the quantity of grade-1 apples at the end of the processing line divided by its percentage and the sum of all apple grades combined at the end of the line. So, these two constraints guarantee that grade-1 (the most profitable grade) and total demand are satisfied. Constraint (7) is the capacity constraint of the processing line. Constraints (8) and (9) are the capacity constraints of storage rooms during and after harvesting periods, respectively. These two constraints also ensure that if a storage room is off in a period, there should not be any assignment or inventory from the previous period to that room. Constraint (10) guarantees that, in each period, the demand for each grade is satisfied from the processing line. Constraint (11) determines the type of storage rooms. Constraints (12) guarantees that if a storage room is of one type and is on in period t, its type cannot change anymore. Constraint (13) states that in each period those apples staying more than their shelf life in a storage room are considered as loss. Constraint (14) ensures that if apples are kept more than their shelf life in a storage room, they are not used anymore to satisfy the demand. Constraints (15) and (16) are the non-negativity and binary constraints on the decision variables, respectively.

3 Computational Implementation and Results

To solve the model, we used python programming language and Gurobi optimization solver 7.0 [9], and to evaluate the performance of it, we used a real-world instance of the Australian apple company. We assessed the results using four key performance measures: (1) total output of the storage rooms, (2) total loss, (3) fresh end stock (= end stock—total losses), and (4) objective function value (OFV). Besides, to analyze the behavior of the model, we defined four different scenarios (modifying the real data). Scenario (1): increasing the processing cost, Scenario (2): increasing the penalty cost of apple losses, Scenario (3): the shelf life in all types of storage room increased by 2 weeks, and Scenario (4): the shelf life in all types of storage room decreased by 2 weeks.

Table 1 shows the results of mentioned key performance measures for the real-world instance and the scenarios with the optimality gap of 0.04% in 900 s. In scenario 1, the processing cost is increased in a way to have the least amount of output from storage rooms to the line (9716988.57 kg), so total loss increments and consequently, total cost increases. In scenario 2, the total output increases by raising the penalty cost of apple losses so that total loss reaches to zero. However, we do not witness any improvement in the OFV. In scenario 3, there is no apple loss, when the shelf life of all three storage types is increased by 2 weeks. So, there are more fresh apples in storage rooms at the end of the planning horizon (2521883.36 kg). Accordingly, unlike scenario 2, this time the OFV decreases as a result of loss reduction. Finally, in scenario 4, both total loss and total cost increment, as shelf life falls by 2 weeks. Moreover, optimal types of storage rooms are 14,15 and 10 for CC, SF and CA, respectively. It is interesting to note that when apple loss exists, it only happens in CA rooms. So, the total loss value is so dependent on the apples' shelf life in this room. We can conclude that the OFV decreases by increasing the shelf life (decreasing apple loss) in CA storage room.

4 Conclusion

In this paper, we developed a mixed-integer linear programming (MILP) model for inventory management of apple SC in Australia with the aim of minimizing total

Table 1 The results of the key performance measures

Instance	Total rooms' output (kg)	Total loss (kg)	Fresh end stock (kg)	OFV ($)
Real instance	11580757.5	446409	1030033.5	18385339.4
Scenario 1	9716988.57	2310177.92	1030033.5	20752686.42
Scenario 2	12027166.5	0	1030033.5	18400032.06
Scenario 3	10535316.63	0	2521883.36	18252587.67
Scenario 4	11578060.57	1332905.92	146233.48	18506269.64

cost, while satisfying the demand of different apple grades. The model determines the type of storage rooms and quantifies apple losses of each type according to the time gap between their harvest and delivery with an associated penalty cost in the objective function. We solved the model using Gurobi optimization solver 7.0, evaluated the performance of the model using a real-world instance, and tested its behavior by analyzing different scenarios. The results indicated that the total cost decreases by incrementing the shelf life of apples (decreasing apple loss) in CA storage rooms. This outcome may aid policy makers on investment decisions for regional development.

As future study, developing stochastic mathematical models or integrated ones for AFSC such as, harvest-inventory or inventory-transportation can be considered.

Acknowledgements This research was supported under Australian Research Council's Industrial Transformation Training Centre funding scheme (project number IC140100032). We thank Dr. Rodolfo García-Flores and Dr. Pablo Juliano for their cooperation.

References

1. Gustavsson, J., Cederberg, C., Sonesson, U., Van Otterdijk, R., & Meybeck, A. (2011). *Global food losses and food waste*. Gothenburg, Sweden: Food and Agriculture Organization of the United Nations.
2. Shukla, M., & Jharkharia, S. (2013). Agri-fresh produce supply chain management: A state-of-the-art literature review. *International Journal of Operations & Production Management, 33*(2), 114–158.
3. Ferrer, J.-C., Mac Cawley, A., Maturana, S., Toloza, S., & Vera, J. (2008). An optimization approach for scheduling wine grape harvest operations. *International Journal of Production Economics, 112*(2), 985–999.
4. Ahumada, O., & Villalobos, J. R. (2011). Operational model for planning the harvest and distribution of perishable agricultural products. *International Journal of Production Economics, 133*(2), 677–687.
5. Rong, A., Akkerman, R., & Grunow, M. (2011). An optimization approach for managing fresh food quality throughout the supply chain. *International Journal of Production Economics, 131*(1), 421–429.
6. Herbon, A., Levner, E., & Cheng, T. C. E. (2014). Perishable inventory management with dynamic pricing using time–temperature indicators linked to automatic detecting devices. International Journal of Production Economics, vol. 147, Part C, pp 605-613. (2014).

7. Muriana, C. (2015). Effectiveness of the food recovery at the retailing stage under shelf life uncertainty, An application to Italian food chains. *Waste Management, 41,* 159–168.
8. Paam, P., Berretta, R., Heydar, M., Middleton, R. H., García-Flores, R., & Juliano, P. (2016). Planning models to optimize the agri-fresh food supply chain for loss minimization: A review. *Reference Module in Food Sciences,* 1–16. Elsevier.
9. http://www.gurobi.com

Simulating Fresh Food Supply Chains by Integrating Product Quality

Magdalena Leithner and Christian Fikar

1 Introduction

Supply chain management has gained importance to strengthen competitiveness in the fresh fruit sector [21]. Efficiently planned inventory management, transportation and distribution are indispensable for the profitability of food supply chains [18]. Nevertheless, food logistics is challenged by thin margins, various uncertainties as well as changing qualities and quantities influenced by harvest-times and unpredictable weather conditions [3, 6, 9, 18].

In Europe, nearly one third of produced fresh fruits and vegetables (FFVs) gets lost along postharvest handling [10]. Furthermore, the quality of FFVs decreases over time, resulting in limited shelf lives [13]. Temperature has the greatest impact on postharvest life due to its effect on the biological reaction rates and microbial growth. These specific requirements and process characteristics complicate the integration of decision support systems [16]. Operational research methods present powerful tools to handle the complexity of food logistics [4]. Consequently, when modelling FFV supply chains, approaches which integrate the specific characteristics of fresh produces are required to reduce food losses and maintain high product qualities [6, 18].

Most modelling approaches applied to improve food supply chains, however, do not consider changes in product quality and interdependencies between quality and chain design [21]. Simulation techniques are able to incorporate such uncertainties and allow the integration of food quality models to enable holistic approaches [11, 21]. In this work, commonly applied stock rotation schemes and distribution strategies are jointly investigated within a regional FFV supply chain. Strawberries, due to their high perishability, act as a sample setting. Consequently, the contribution of this work is twofold: (i) it embeds a generic keeping quality model in a discrete

---

M. Leithner (✉) · C. Fikar
Institute of Production and Logistics, University of Natural Resources and Life Sciences, Vienna, Feistmantelstraße 4, 1180 Vienna, Austria
e-mail: magdalena.leithner@boku.ac.at

event simulation to provide guidance on integrating product quality within food supply chain simulations and (ii) analyzes the impacts of stock rotation and distribution strategies on operations.

2 Related Literature

In a literature review of food aspects in logistics research, [6] note that food products have unique characteristics creating special demands which challenge food supply chain actors. [18] analyzed literature dealing with operational research models applied to the fresh fruit supply chain. The authors note that various works use simulation methods to improve fresh fruit logistics, however, linear programming is the predominant modeling technique. [4] focus on reviewing operations research methods which are able to handle uncertainties in food supply chains. Three main operations research techniques are discussed, namely, stochastic, robust, and simulation-based programming. Furthermore, they categorize simulation-based works into three main functional areas (i) in-field, irrigation, harvest and transportation activities, (ii) prices and profits, and (iii) food supply chain coordination and (re)design. Two types of uncertainties are predominately taken into account, supply (e.g., weather, resource availability) and market uncertainties (e.g., prices, costs, demand fluctuation).

In [21], food quality models and sustainability indicators are embedded in a discrete event simulation environment, investigating a global pineapple supply chain. To calculate the effects of temperature on the quality of FFVs, [20] developed the generic keeping quality model. Therefore, the keeping quality describes the time until a commodity becomes unacceptable and is inversely proportional to the sum of reaction rates which decrease product quality. The limit of acceptance, i.e. the end of keeping quality, depends on the product quality, its intrinsic characteristics like visual appearance, firmness and fungal decay as well as on consumers' perceptions. At constant environmental conditions, known initial quality and a defined quality limit, the same attribute always hits the acceptance limit first. The impacts of temperature on product quality are described by applying the Arrhenius law to calculate the reaction rate at specified temperatures.

3 Method

In this work, a generic food quality model is integrated in a discrete event simulation modeling a regional FFV supply chain consisting of local producers, a central warehouse and retailers. Various entities (e.g., perishable products, producers, trucks, warehouses, retailers) are modeled in detail, which follow and build various process steps. Along the supply chain varying temperatures are assumed, i.e. at harvest, storage, transport and in the warehouse. Figure 1 provides an overview of the implemented process flow. After harvest, a perishable good is shipped either directly or

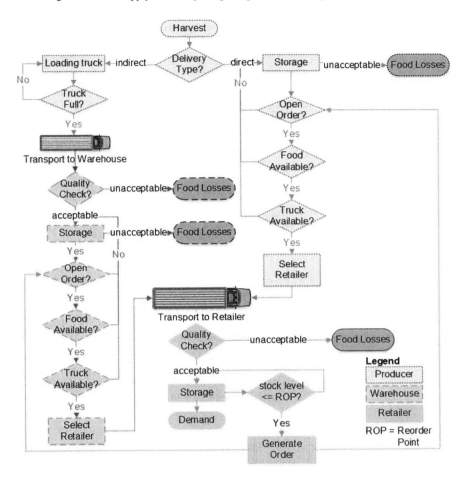

Fig. 1 Fresh fruit supply storage and distribution process

indirectly (via a warehouse) from the place of harvest to several retail stores where goods are in demand. The quality of these products is calculated individually and updated after each process step according to storage temperatures and durations. At different stages of the supply chain, FFVs, that are unacceptable, are removed from the process and counted as food losses. At random times during the simulation, based on a Poisson-distributed arrival rate, customers request products. If goods are available, the inventory is adjusted. Otherwise, the customer request is counted as lost sales. At reorder points, derived from a desired service level and expected lead times based on [5], an replenishment order is generated. This order is, depending on the simulation setting, either fulfilled directly by a producer or via the central warehouse.

Three different strategies are compared on how to fulfill incoming replenishment orders, (i) serving orders in accordance to the arrival time, i.e. the order placed first is assigned first, (ii) by distance to the retailer's location, i.e. the order is served by the location resulting in the lowest travel distance, or (iii) randomly. Additionally, three stock rotation schemes are implemented to model product selection, 'first in, first out' (FIFO), 'last in, first out' (LIFO) and 'least shelf life, first out' (LSFO). Generated food losses along the various stages of the supply chain as well as achieved service levels and travel durations are calculated and reported to the user after each simulation run.

The model is implemented with the simulation software AnyLogic 8.1.0 [2]. Data from the OpenStreetMap [15] are used to generate transportation routes and travel durations. The simulation models two weeks of harvest, with demand equal to supply and an additional day of warm-up phase to initiate the simulation. Average results are reported, based on 100 replications for each setting.

4 Computational Experiments: Strawberry Supply Chain in Lower Austria

Lower Austria is an important strawberry production area in Austria, responsible for nearly half of the nationwide strawberry crops [1, 19]. Strawberries (Fragaria ananassa) are one of the most widely consumed fruits [12]. Even under perfect storage conditions, the strawberry fruit can only be stored and maintain its desired quality for five to seven days [14]. The wrong postharvest temperatures rapidly induce quality changes. Since Botrytis infection (spoilage) is one of the first visible attributes the consumer can assess, it mainly limits the keeping quality of strawberries [8]. As the presence or absence of spoilage is the main criterion concerning spoilage, the keeping quality of strawberries can be described as a batch keeping quality. It gives the percentage of strawberries affected by spoilage within one batch, hence, the quality limit can be defined as the time until the first strawberry in one batch is visibly infected [8]. Therefore, according to Arrhenius law, the spoilage rate depending on temperature is calculated. Considering different temperature conditions and assuming zero order kinetics, the quality losses as well as the remaining shelf lives are calculated. The initial quality of the strawberry batches, the quality limit as well as the spoilage rate constant refer to the used parameters in [8, 17].

To simulate the strawberry supply chain in Lower Austria, data from the GLOBALG.A.P. database [7] are used to refer to major strawberry producers. The retail stores as well as the warehouse refer to the supply chain of a major Austrian grocery chain. A focus on the largest cities of Lower Austria is set. This results in a problem size consisting of ten producers, a central warehouse and 23 retail locations. The production rate of strawberries is calculated according to the production output of 2013 [19]. At each producer one chilled truck is available for shipments, whereas the warehouse operates four chilled trucks.

Table 1 Impact of distribution strategy on service level, travel duration and food losses (indirect deliveries)

Output	Service level (%)	Travel duration (h)	Food losses (items)
Delivery			
Firstorder	86	919	2164
Nearestretailer	92	894	1013
RANDOM	85	921	2292

5 Results and Discussion

Regarding indirect deliveries, the results, shown in Table 1, indicate that regional deliveries, i.e., the nearest retailers get delivered first, positively influence travel duration, the amount of food losses and customer service levels of the stores. Nevertheless, distant retailers are poorly served in such settings. Four warehouse trucks reduce food losses under LSFO and FIFO to zero whereas high amounts of food losses occur under LIFO. If less trucks are available, the LSFO approach produces significantly less food losses than the FIFO approach. Direct deliveries enable high service levels and zero food losses irrespective of the chosen delivery strategy. The travel durations are higher using the first order and the random strategy both with direct and indirect deliveries. In general, indirect deliveries have longer transport routes than direct deliveries, resulting in longer travel durations. The results support the findings of [10] that longer transport routes negatively influence product quality. Therefore, the assignment of low quality products to shorter routes to, further, reduce food losses is recommended.

6 Conclusions

The work shows that the integration of food quality and losses in food supply chain simulations provides the opportunity to improve operations. Applying the LSFO approach can reduce food losses. In addition, regional supply chains reduce travel distances as well as food losses and improve product availability. Future work focuses on the integration of optimization methods within the simulation framework and on expending the product range to consider interactions among various FFVs.

Acknowledgements This work was financially supported by the program Mobility of the Future (grant number 859 148), an initiative of the Austrian Ministry for Transport, Innovation and Technology. The authors are grateful for this support.

References

1. AMA: Erdbeere (2017). Retrieved June 27, 2017, from http://amainfo.at/ama-themen/produktvielfalt/obst/erdbeere/
2. AnyLogic: Anylogic - multimethod simulation software (2017). Retrieved June 27, 2017, from http://www.anylogic.com/
3. Blackburn, J., & Scudder, G. (2009). Supply chain strategies for perishable products: The case of fresh produce. *Production and Operations Managemant, 18*, 129–137.
4. Borodin, V., Bourtembourg, J., Hnaien, F., & Labadie, N. (2016). Handling uncertainty in agricultural supply chain management: A state of the art. *European Journal of Operational Research, 254*(2), 348–359.
5. Chopra, S., & Meindl, P. (2013). *Supply chain management: Strategy, planning, and operation* (5th ed.). Boston: Pearson Education.
6. Fredriksson, A., & Liljestrand, K. (2015). Capturing food logistics: A literature review and research agenda. *International Journal of Logistics, 18*(1), 16–34.
7. GLOBALG.A.P: GLOBALG.A.P. database (2017). Retrieved June 27, 2017, from https://database.globalgap.org/globalgap/search/SearchMain.faces?init=1
8. Hertog, M., Boerrigter, H., van den Boogaard, G., Tijskens, L., & van Schaik, A. (1999). Predicting keeping quality of strawberries (cv. 'Elsanta') packed under modified atmospheres: An integrated model approach. *Postharvest Biology and Technology, 15*(1), 1–12.
9. Hertog, M. L., Lammertyn, J., De Ketelaere, B., Scheerlinck, N., & Nicola, B. M. (2007). Managing quality variance in the postharvest food chain. *Trends in Food Science and Technology, 18*(6), 320–332.
10. Jedermann, R., Nicometo, M., Uysal, I., & Lang, W. (2014). Reducing food losses by intelligent food logistics. *Philosophical Transactions of the Royal Society A, 372*(2017), 20130302.
11. Long, Q., & Zhang, W. (2014). An integrated framework for agent based inventory-production-transportation modeling and distributed simulation of supply chains. *Information Sciences, 277*, 567–581.
12. Nunes, M., Brecht, J., Sargent, S., & Morais, A. (1995). Effects of delays to cooling and wrapping on strawberry quality (cv. Sweet Charlie). *Food Control, 6*(6), 323–328.
13. Nunes, M. C., Nicometo, M., Emond, J. P., Melis, R. B., & Uysal, I. (2014). Improvement in fresh fruit and vegetable logistics quality: Berry logistics field studies. *Philosophical Transactions- Royal Society of London Series A, 372*(2017), 20130307.
14. Nunes, M. C. N. (2008). *Color atlas of postharvest quality of fruits and vegetables* (1st ed.). Ames, Iowa: Blackwell Publishing.
15. OpenStreetMap: Openstreetmap (2017). Retrieved June 27, 2017, from www.openstreetmap.org/
16. Rong, A., Akkerman, R., & Grunow, M. (2011). An optimization approach for managing fresh food quality throughout the supply chain. *International Journal of Production Economics, 131*(1), 421–429.
17. Schouten, R. E., Tijskens, L., & van Kooten, O. (2002). Predicting keeping quality of batches of cucumber fruit based on a physiological mechanism. *Postharvest Biology and Technology, 26*(2), 209–220.
18. Soto-Silva, W. E., Nadal-Roig, E., Gonzlez-Araya, M. C., & Pla-Aragones, L. M. (2016). Operational research models applied to the fresh fruit supply chain. *European Journal of Operational Research, 251*(2), 345–355.
19. Statistik Austria: Obstproduktion aus Erwerbsanlagen, endgltiges Ergebnis 2015 (2015)
20. Tijskens, L. M. M., & Polderdijk, J. J. (1996). A generic model for keeping quality of vegetable produce during storage and distribution. *Agricultural Systems, 51*(4), 431–452.
21. van der Vorst, J., Tromp, S. O., & Zee, D Jvd. (2009). Simulation modelling for food supply chain redesign; integrated decision making on product quality, sustainability and logistics. *International Journal of Production Research, 47*(23), 6611–6631.

Window Fill Rate with Compound Arrival and Assembly Time

Michael Dreyfuss and Yahel Giat

1 Introduction

Exchangeable-item repair systems are systems to which customers bring a failed item and exchange it for a serviceable item.

In this paper, the service measure that we consider is a generalization of the fill rate. The fill rate is defined as the fraction of customers who are served upon arrival. In many cases, however, the service contract allows for a certain period of time until service is rendered. Moreover, even absent contractual agreement, customers will tolerate a certain wait and therefore there is no loss to the firm if a customer waits less than the tolerable wait. The window fill rate incorporates this tolerable wait and is defined as the probability that the customer is served *within* the tolerable wait.

Formulas for the window fill rate are developed in [1] where it is assumed that item assembly and disassembly (i.e., removal and installation) is instantaneous. The goal of this paper is to extend these results and develop the window fill rate formula for the case of nonzero assembly and disassembly times. The main result of the analysis is that a Δ *increase* in the assembly and disassembly time is equivalent to a Δ *decrease* in the tolerable wait.

Our paper contributes to the research of exchangeable-item repair systems originated by Sherbrook's METRIC model [2] that develops an approximate evaluation of the number of backorders in a multi-echelon system and describes a greedy algorithm to solve the spares allocation problem. This body of research is presented in books such as [3, 4] and recently reviewed in [5]. Except for the fact that we limit the system to a single location, we assume the standard METRIC assumptions, which include compound arrival, ample repair servers, that components fail according to

M. Dreyfuss (✉) · Y. Giat
Jerusalem College of Technology, Jerusalem, Israel
e-mail: dreyfuss@jct.ac.il

Y. Giat
e-mail: yahel@jct.ac.il

a Poisson process with a constant arrival rate and a continuous $(S-1, S)$ review policy.

Many METRIC-based papers focus on the number of back-orders performance measure (e.g., [6–8]) or the fill rate performance measure (e.g., [9, 10]). Our use of the *window* fill rate, i.e., the probability of a random customer to be served within a certain time window, is premised on customers tolerating a certain wait. This assumption lies at the intersection of inventory and customer service models. While the concept of a tolerable wait is rarely considered in inventory models, it is recurrent in the service industry and is associate with terms such as "expectation" [11], "reasonable duration" [12], "maximal tolerable wait" [13] and "wait acceptability" [14]. From a service-oriented approach, the customer's attitude to wait is mainly subjective and has cognitive and affective aspects [14]. From a logistics point of view this wait is more objective and usually stated in the service contract. Indeed, researchers have observed that most inventory models fail "to capture the time-based aspects of service agreements as they are actually written" [15, p. 744]. Our paper fills this void by incorporating the tolerable wait into the optimization criterion.

2 The Model

Customers arrive with failed items. Upon arrival, the items are removed and sent to repair. Once an item is repaired it is added to the station's stock. To reduce customer waiting time, the network keeps a number of spare items. Customers are served according to a first-come, first-serve policy (FCFS). Let B_i, $(i = 1, 2, \ldots)$ denote the number of items brought by customer i. We assume that customer arrival follows a Poisson process with parameter λ and that B_i are i.i.d. with a common probability distribution $P_B(\cdot)$ and that the sequence B_i is independent of the customer arrivals and item repair processes. Partial service is not allowed and therefore, the customers leave the system only once they received B_i serviceable items. We assume that there are ample repair servers and that the servers' repair time is i.i.d. The combination of these assumptions is that repair commences once the item is removed and that repair times are independent. Let $R(t)$ denote the cumulative probability for repair to be completed by time t, and let item removal and installation times be t_1 and t_2, respectively. The total assembly and disassembly time, $t_1 + t_2$, is assumed to be no more than the tolerable wait.

In what follows, we develop the window fill rate, i.e., the probability that a customer's demand is satisfied within time t. We first formulate the nonstationary window fill rate and then we take the limits to obtain the stationary window fill rate.

Consider Jane, a customer arriving at time s and let $F^{NS}(s, n, t)$ denote the non stationary probability for her to leave the system by time $s + t$. The FCFS assumption dictates that all the demand of the customers that arrived before Jane must be supplied before her. Since, in addition, it takes t_2 to install Jane's item, $F^{NS}(s, n, t)$ is equal to the probability that by time $s+t-t_2$ she must be at the head of the queue with an operable item ready for her. We distinguish between items that were brought by customers who arrived during [0, s) ("pre-Jane") and items brought by customers who arrived during $(s, s+t-t_2)$ ("post-Jane") and define the following variables:

- W_i ($i = 1, 2, \ldots, N_W(s)$) is the number of items brought by the i'th pre-Jane customer that *were* repaired by date $s+t-t_2$.
- X_i ($i = 1, 2, \ldots, N_X(s)$) is the number of items brought by the i'th pre-Jane customer that *were not* repaired by date $s+t-t_2$.
- $Z(s, t)$ is the number of items brought by Jane that *were not* repaired by date $s+t-t_2$.
- Y_i ($i = 1, 2, \ldots, N_Y(t)$) is the number of items brought by the i'th post-Jane customer and that *were* repaired by date $s+t-t_2$.

$F^{NS}(s, n, t)$ is the probability that the supply of operable items is greater than the demand as follows:

$$F^{NS}(s, n, t) = Pr\left[\sum_{i=1}^{N_W(s)} W_i + \sum_{i=1}^{N_Y(s)} Y_i + B_{Jane} - Z(s,t) + n \geq \sum_{i=1}^{N_W(s)} W_i + \sum_{i=1}^{N_X(s)} X_i + B_{Jane}\right]. \tag{1}$$

In (1), the supply (left side of the inequality) comprises the n spares plus the items that were repaired by date $s+t-t_2$. The demand (right side of the inequality) comprises the items needed to serve the customers that arrived before Jane plus the B_{Jane} items needed by Jane. Cancelling terms and reversing the inequality, we have:

$$F^{NS}(s, n, t) = Pr\left[\sum_{i=1}^{N_X(s)} X_i - \sum_{i=1}^{N_Y(t)} Y_i + Z(s, t) \leq n\right]. \tag{2}$$

The ample servers and the assumptions about the item arrivals distribution guarantee that $X := \sum_{i=1}^{N_X(s)} X_i$, $Y := \sum_{i=1}^{N_Y(t)} Y_i$, $Z(s, t)$ are independent. Further, Y and X are Compound Poisson. To find the probability $Pr[Z(s, t) = j]$ we condition on the number of items that Jane brought, b, and compute the probability that j of these have not been repaired by time $s+t-t_2$. Since it takes t_1 to remove the item, the repair begins at $s + t_1$. Thus, the probability that j items have not been repaired is given by $Bin(j, b, 1 - R(s+t-t_2 - (s+t_1)))$ where $Bin(i, j, p) := \binom{j}{i} p^i (1 - p)^{j-i}$ is the binomial probability mass function. Thus,

$$Pr[Z(s, t) = j] = \sum_{b=j}^{\infty} P_B(b) Bin(j, b, \bar{R}(\hat{t})). \tag{3}$$

where $\hat{t} := t - t_1 - t_2$ and $\bar{R}(\cdot) = 1 - R(\cdot)$.

Conditioning on $Z(s, t)$ and using (3), the probability (2) is given by

$$Pr\left[X - Y + Z(s,t) \leq n\right]$$
$$= \sum_{j=0}^{\infty}\left(\sum_{b=\max\{j,1\}}^{\infty} P_B(b)Bin(j,b,\bar{R}(\hat{t})) \cdot Pr[X - Y + j \leq n]\right)$$
$$= \sum_{b=1}^{\infty}\left(P_B(b) \cdot \sum_{j=0}^{b} Bin(j,b,\bar{R}(\hat{t})) \cdot Pr[X - Y + j) \leq n]\right)$$
$$= \sum_{b=1}^{\infty}\left(P_B(b) \cdot \sum_{j=0}^{b}\left(Bin(j,b,\bar{R}(\hat{t})) \cdot \sum_{y=0}^{\infty}\left(Pr[Y = y] \sum_{x=0}^{n-j+y} Pr[X = x]\right)\right)\right), \quad (4)$$

where the third row is obtained by changing the order of the summations and the fourth row is given by conditioning on Y.

At this point we investigate the distribution of the components of X and Y. To find the probability $Pr[X_i = j]$ we first consider a customer that arrived at the time interval du in $[0, s)$. Notice that for a Poisson arrival the probability for this is du/s. Next, we condition on the number of items brought by the customer, b, and this is multiplied by the probability that exactly j of the b items are not repaired by $s + t - t_2$. The item begins repair at $u + t_1$ and the probability of the item not to be repaired between $[u + t_1, s + t - t_2)$ is $1 - R(s+\hat{t}-u)$ and therefore the probability that j of the b items are not repaired $Bin(j, b, \bar{R}(s+\hat{t}-u))$. Thus,

$$Pr[X_i = j] = \int_{u=0}^{s} \sum_{b=1}^{\infty}\left(P_B(b)Bin(j,b,\bar{R}(s+\hat{t}-u))\right)\frac{du}{s}$$
$$= \int_{v=\hat{t}}^{s+\hat{t}} \sum_{b=1}^{\infty}\left(P_B(b)Bin(j,b,\bar{R}(v))\right)\frac{dv}{s}. \quad (5)$$

For the probability $Pr[Y_i = j]$ we consider a customer that arrived at the time interval du in $[s + t_1, s + t - t_2)$, which happens with probability $du/\hat{t}$. Next, we condition on the number of items brought by the customer, b, and this is multiplied by the probability that exactly j of the b items are repaired by $s + t - t_2$, which is given by $Bin(j, b, R(s+\hat{t}-u))$ since the item begins repair at $u + t_1$. Thus,

$$Pr[Y_i = j] = \int_{u=s}^{s+\hat{t}} \sum_{b=1}^{\infty}\left(P_B(b)Bin(j,b,R(s+\hat{t}-u))\right)\frac{du}{\hat{t}}$$
$$= \int_{v=0}^{\hat{t}} \sum_{b=1}^{\infty}\left(P_B(b)Bin(j,b,R(v))\right)\frac{dv}{\hat{t}}. \quad (6)$$

Let $A(t) := \sum_{i=1}^{N(t)} A_i$ be a Compound Poisson with customer arrival rate λ. By [16, pp 36–7, Eq. 1.48] the distribution of $A(t)$ is given by

$$Pr[A(t) = 0] = \exp^{-\lambda t Pr[A_i > 0]}, \quad Pr[A(t) = j] = \frac{\lambda t}{j}\sum_{k=0}^{j-1}(j-k)Pr[A_i = j-k]Pr[A(t) = k]. \tag{7}$$

By (5) and (7), we have that

$$Pr[X = 0] = \exp^{-\lambda s Pr[X_i > 0]} = \exp^{-\lambda s \sum_{j=1}^{\infty} \int_{v=\hat{t}}^{s+\hat{t}} \sum_{b=1}^{\infty} P_B(b)Bin(j,b,\bar{R}(v))\frac{dv}{s}}$$
$$= \exp^{-\lambda \int_{v=\hat{t}}^{s+\hat{t}} \sum_{b=1}^{\infty} P_B(b)(1-Bin(0,b,\bar{R}(v)))dv},$$

$$Pr[X = j] = \frac{\lambda s}{j} \sum_{k=0}^{j-1}(j-k)Pr[X_i = j-k]Pr[X = k]$$
$$= \frac{\lambda}{j} \sum_{k=0}^{j-1}(j-k)\int_{v=\hat{t}}^{s+\hat{t}}\sum_{b=1}^{\infty} P_B(b)Bin(j-k,b,\bar{R}(v))dv Pr[X=k]. \tag{8}$$

Similarly, using (6) and (7) we have that

$$Pr[Y = 0] = \exp^{-\lambda t Pr[Y_i > 0]} = \exp^{-\lambda \sum_{j=1}^{\infty}\int_{v=0}^{\hat{t}}\sum_{b=1}^{\infty}\left(P_B(b)Bin(j,b,R(v))\right)dv}$$
$$= \exp^{-\lambda \int_{v=0}^{\hat{t}}\sum_{b=1}^{\infty}\left(P_B(b)\left(1-Bin(0,b,R(v))\right)\right)dv},$$

$$Pr[Y = j] = \frac{\lambda \hat{t}}{j}\sum_{k=0}^{j-1}(j-k)Pr[Y_i = j-k]Pr[Y = k]$$
$$= \frac{\lambda}{j}\sum_{k=0}^{j-1}(j-k)\int_{v=0}^{\hat{t}}\sum_{b=1}^{\infty}\left(P_B(b)Bin(j-k,b,R(v))\right)dv Pr[Y = k]. \tag{9}$$

To summarize, the nonstationary window fill rate $F(s, t, n)$ is given by (4), (8) and (9). The (stationary) window fill rate, $F(t, n)$, is obtained by letting $s \to \infty$ in the integral limits of (8).

3 Discussion and Conclusions

The consequence of (4), (8) and (9) is that instead of considering positive assembly times, we can adjust the tolerable wait by subtracting from it the assembly times. Therefore, all the results of [1] for the Compound Poisson case are valid and the algorithm they present for optimizing a system of multiple locations applies also when assembly times are nonzero.

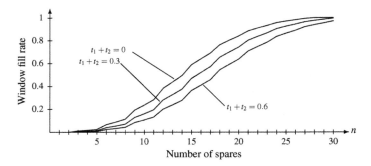

Fig. 1 The window fill rate versus for different values of assembly times

We illustrate the functional form of the window fill rate for the following case. Customer arrival rate is $\lambda = 2$. Each customer arrives with exactly three items ($Pr[B_i = 3] = 1$) and repair time is $R \sim N(3, 0.3^2)$. The tolerable wait is $t = 1$ and we examine three values for the assembly times, $t_1 + t_2 = 0, 0.3, 0.6$, which results with three cases of adjusted waiting times $\hat{t} = 1, 0.7, 0.4$. In Fig. 1, we plot the window fill rate for each case as a function of the number of spares in the station. We can see that when there are too few spares or sufficiently many spares the window fill rate is insensitive to the tolerable wait. For intermediate values, however, the window fill rate is sensitive to the assembly times. For example, when there are 15 spares in the station *increasing* the assembly times from 0 to 0.3 to 0.6, *decreases* the window fill rate from 59 to 47 to 36%, respectively.

In our model, we assume that the assembly times are deterministic. This assumption is reasonable when the item-installment and item-removal procedures are relatively simple and done as a matter of routine. With certain complex systems, this assumption may not be true and one would have to assume that assembly times are stochastic. We leave this for future research.

References

1. Dreyfuss, M., & Giat, Y. (2017). Optimal spares allocation to an exchangeable-item repair system with tolerable wait. *European Journal of Operational Research, 261*(2), 584–594.
2. Sherbrooke, C. C. (1968). METRIC: A multi-echelon technique for recoverable item control. *Operations Research, 16*(1), 122–141.
3. Sherbrooke, C. C. (2004). *Optimal inventory modeling of systems: multi-echelon techniques* (Vol. 72). Springer Science & Business Media.
4. Muckstadt, J. A. (2005). *Analysis and algorithms for service parts supply chains*. Springer Science & Business Media.
5. Basten, R. J. I., & van Houtum, G. J. (2014). System-oriented inventory models for spare parts. *Surveys in Operations Research and Management Science, 19*(1), 34–55.
6. Basten, R. J. I., & van Houtum, G. J. (2013). Near-optimal heuristics to set base stock levels in a two-echelon distribution network. *International Journal of Production Economics, 143*(2), 546–552.

7. Dreyfuss, M., & Giat, Y. (2017). Multi-echelon exchangeable-item repair system optimization. *Military Operations Research*, forthcoming.
8. Ghaddar, B., Sakr, N., & Asiedu, Y. (2016). Spare parts stocking analysis using genetic programming. *European Journal of Operational Research, 252*(1), 136–144.
9. Caggiano, K. E., Jackson, P. L., Muckstadt, J. A., & Rappold, J. A. (2007). Optimizing service parts inventory in a multiechelon, multi-item supply chain with time-based customer service-level agreements. *Operations Research, 55*(2), 303–318.
10. Lien, R. W., Iravani, S. M., & Smilowitz, K. R. (2014). Sequential resource allocation for nonprofit operations. *Operations Research, 62*(2), 301–317.
11. Durrande-Moreau, A. (1999). Waiting for service: Ten years of empirical research. *International Journal of Service Industry Management, 10*(2), 171–194.
12. Katz, K. L., Larson, B. M., & Larson, R. C. (1991). Prescription for the waiting-in-line blues: Entertain, enlighten, and engage. *MIT Sloan Management Review, 32*(2), 44.
13. Smidts, A., & Pruyn, A. (1994). How waiting affects customer satisfaction with service: the role of subjective variables. *Proceedings of the 3rd International Research Seminar in Service Management* (pp. 678–696). xx: Universite d'Aix-Marseille.
14. Demoulin, N. T., & Djelassi, S. (2013). Customer responses to waits for online banking service delivery. *International Journal of Retail & Distribution Management, 41*(6), 442–460.
15. Caggiano, K. E., Jackson, L., Muckstadt, J. A., & Rappold, J. A. (2009). Efficient computation of time-based customer service levels in a multi-item, multi-echelon supply chain: A practical approach for inventory optimization. *European Journal of Operational Research, 199*(3), 744–749.
16. Tijms, H. (1986). *Stochastic modelling and analyis: a computational approach*. NY: Wiley.

Part XXI
Traffic, Mobility and Passenger Transportation

Demand-Driven Line Planning with Selfish Routing

Malte Renken, Amin Ahmadi, Ralf Borndörfer, Güvenç Şahin and Thomas Schlechte

1 Introduction

A common characteristic of most urban transportation systems is the variation in demand during different times of the day and different days of the week. This characteristic is even more notable in bus rapid transit (BRT) systems as such systems typically serve the bulk of the demand, exerting a trunk function for the whole transportation system of a region/area. The demand varies heavily during the day, with morning and evening peaks if it is a weekday while timings of morning and evening peaks may change for the weekends. In addition to fluctuations with respect to time, the demand is typically highly asymmetric with respect to its distribution on the line. Such systems, again typically, cover a central line of a highly populated city. In particularly densely populated cities of such countries, (the population is young and the workforce moves from usually outer regions of the cities to the central parts, i.e.,) the movement of population is typically in reverse directions for morning peaks vs. evening peaks. As a result, the system-wide transportation demand exerts significant changes in both time and space.

M. Renken · R. Borndörfer
Zuse Institute Berlin, Takustr. 7, 14195 Berlin, Germany
e-mail: Renken@zib.de

R. Borndörfer
e-mail: Borndorfer@zib.de

A. Ahmadi · G. Şahin
Sabanci University, Industrial Engineering, 34956 Orhanli, Tuzla/Istanbul, Turkey
e-mail: aminahmadi@sabanciuniv.edu

G. Şahin
e-mail: guvencs@sabanciuniv.edu

T. Schlechte (✉)
LBW Optimization GmbH, Berlin, Germany
e-mail: schlechte@lbw-optimization.de

Traditional line planning is mostly concerned with a static demand and addresses the case of fluctuating demand by constructing a base service, which is augmented in peak hours, or vice versa, i.e., constructing a peak service which is decremented during non-peak hours. In BRT systems with demand sensitivity in both space and time, a demand-driven approach in line planning would be more viable. As a result, line plans are more susceptible to infrastructure capacity and fleet capacity. In this respect, an analytical demand-driven approach should consider the effects of such limitations on the level of demand satisfaction in the system.

In traditional mathematical programming formulations of line planning problems, the relation between capacities and the demand are roughly reflected/considered. There is usually a predetermined frequency requirement rather a demand amount to be satisfied ([1]). We use mathematical formulations where origin-destination (OD) demand between pairs of stations to be satisfied in a finite length planning horizon is explicitly represented as in [2]. In order to investigate the accuracy of demand-driven line planning approaches, we consider a simplistic underlying network structure: a line network. On a line network, each pair of stations is connected via a single path. The alternative mathematical formulations differ from each other with respect to the way demand is represented: an arc-based model where the OD demand is transformed into arc demands without considering the passenger routes explicitly and an OD-based model where the demand is represented in its original form. First, we aim to test the accuracy of the solutions provided by alternative formulations. To this end, we simulate the optimal line plan solutions to observe the differences in basic matrices. In the simulation, we consider passengers with a selfish route choice behavior that is expected to result in a Braess-like paradox.

2 The Braess-Like Paradox with Selfish Routing

The passenger load on lines may significantly change based on the route choice behavior of passengers. The route choice behavior is indeed a very complex problem when there are uncertainties involved; yet, it can be simplified in favor of selecting a certain criterion such as minimizing the number of connections or travel time. When the transport system is sensitive to capacities, the route choice behavior of passengers may play an important role in service levels. We observe a Braess-like paradox with respect to the simple criterion of minimizing connections when the capacity of a system is increased by adding lines.

In order to illustrate our observation, we consider a network consisting of stations 1–5 along a single path with three lines denoted A, B and C as shown in Fig. 1a and OD demands as in Fig. 1b. Each line has a passenger capacity of N.

We compare two different line plans:

- Line plan $\{A, B\}$, consists of line A and line B. Naturally, N passengers from 1 to 3 take line A. Therefore, all passengers from 2 to 4 must use line B. Finally, N passengers from 3 to 5 take line A.

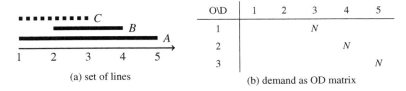

Fig. 1 Alternative line plans and corresponding OD demand matrix

Table 1 Comparison of line plan $\{A, B\}$ and line plan $\{A, B, C\}$

Line plan	Fixed cost	Operational cost	Direct travelers	Travelers with transfer at station 4
$\{A, B\}$	$2 \cdot f$	X	$3 \cdot N$	0
$\{A, B, C\}$	$3 \cdot f$	$X + c_{1,2} + c_{2,3}$	$2 \cdot N + \frac{N}{2}$	$\frac{N}{2}$

- Line plan $\{A, B, C\}$, is likely to change the situation. Passengers from 1 to 3 now have the opportunity to choose between line A and line C. Under a selfish behavior assumption, each passenger takes the first non-full line to come. Supposing that both lines work with the same frequency, this leads to a split of N passengers between A and C: on average $N/2$ passengers on each line. Hence, line A is not full at station 2. Again, passengers taking the first non-full line causes roughly $N/2$ of the demand from 2 to 4 to take line A instead of B. All vehicles on line A are, then, full while traveling from station 2 to station 3 where $N/2$ passengers hop off. In consequence, only about $N/2$ seats are empty on line A at station 3, so only $N/2$ of the passengers waiting at station 3 to go to station 5 can take line A. The rest $N/2$ is forced to take line B from 3 to 4, and connect to line A at station 4 when the passengers traveling from 2 to 4 empty their seats.

We observe that introducing a new line to increase the passenger capacity may lead to a worse situation with respect to minimal number of connections due to selfish behavior of passengers. All passengers could be handled via direct connections with line plan $\{A, B\}$. On the contrary, line plan $\{A, B, C\}$ leads to passengers transfers in station 4. Table 1 summarizes the comparison between the two line plans.

3 Alternative Formulations for Demand-Driven Line Planning

We consider an underlying traffic network represented by a directed graph $G = (N, A)$ where N denotes the stations and A denotes the set of traffic links between the stations. For a finite length planning horizon, OD demand is specified as d_{sq} for each (s, q) pair of stations with $s, q \in N$. When the paths along which passengers travel are known in advance (as is the case for a network with tree structure), OD demand can be transformed into arc demand as denoted by $d_a, a \in A$.

To formulate the line planning problem, we consider a given line set L; c_l^f and c_l^o denote the fixed costs of line l and the operational costs of assigning a vehicle to line l, respectively. In a line plan, at most V vehicles can be used while at most M vehicles can be assigned to any line. A passenger route is defined not only by the physical path on the traffic network but also by each line used to traverse the traffic link along that path. For an OD-pair (s, q), R_{sq} denotes the set of routes from s to q. The set of all routes is denoted by $R = \bigcup_{s,q} R_{sq}$.

In an arc-based demand coverage model (DCM_A) as described in Borndörfer et al. [3], passenger routes are not considered explicitly. Therefore, two decision variables are sufficient in the integer programming formulation of the problem: $x_l \in \{0, 1\}$ taking the value of 1 if line $l \in L$ is selected, and an integer variable $v_l \in \mathbb{N}$ denoting the number of vehicles assigned to line l. However, in an OD-based model (DCM_{OD}), an additional decision variable $z_r \in \mathbb{N}$ is used to denote the number of passengers taking route r. We obtain the following integer programming problem formulation where g_{la}^r is an indicator denoting the traversal of arc a by line l on route r and $\rho(r)$ is some penalty function that characterizes how attractive the route is for passengers.

$$\min \sum_{l \in L} c_l^f x_l + \sum_{l \in L} c_l^o v_l + \sum_{r \in R} z_r \rho(r) \qquad (DCM_{OD})$$

$$\text{subject to} \quad \sum_{r \in R_{sq}} z_r \geq d_{sq} \qquad \forall (s, q) \in D$$

$$\sum_{r \in R} z_r g_{la}^r \leq \kappa v_l \quad \forall l \in L, a \in A$$

$$M x_l - v_l \geq 0 \qquad \forall l \in L$$

$$\sum_{l \in L} v_v \leq V$$

$$x_l \in \{0, 1\}, v_l \in \mathbb{N} \quad \forall l \in L$$

$$z_r \in \mathbb{N} \qquad \forall r \in R$$

It is clear that DCM_{OD} is larger in terms of number of decision variables when compared to DCM_A. On the other hand, DCM_{OD}, when solved to optimality, provides a more accurate solution with respect to demand satisfaction and with respect to minimal number of connections/transfers. To illustrate how DCM_{OD} can be advantageous in comparison to DCM_A, we consider the example in Fig. 2.

In the example, the arc demands are the same for demand set 1 and demand set 2; the arc capacities provided by line plan $\{A_1, B_1\}$ and line plan $\{A_2, B_2\}$ are equivalent. As a matter of fact, according to DCM_A either line plan is optimal for both demand sets. In contrast, only line plan $\{A_1, B_1\}$ is optimal for demand set 1

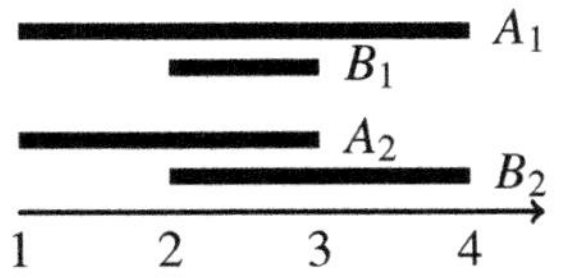

(a) alternative line plans

O\D	1	2	3	4
1				N
2			N	

(b) demand set 1

O\D	1	2	3	4
1			N	
2				N

(c) demand set 2

Fig. 2 Optimal solution examples for two demand sets

with the OD-based demand model, DCM_{OD} and $\{A_2, B_2\}$ for demand set 2 because they allow all passengers to arrive their destination by a direct connection. Using line plan $\{A_1, B_1\}$ for demand set 2 forces the passenger of OD pair (2, 4) to take line B_1 first (as A_1 is full) and then to transfer to A_1 at station 3. Using line plan $\{A_2, B_2\}$ for demand set 1 does not provide a direct connection for the OD pair $(1, 4)$.

4 Computational Results

When the simultaneous effect of ignoring passenger routes on demand satisfaction level and the selfish route choice behavior resulting in a Braess-like paradox is considered, the impact on the outcomes of seemingly optimal line plans might be complicated. In order to test both effects, we run simulations for a real-life system, the Istanbul Metrobüs, which is a BRT system with high fluctuations in demand during the day. We compare the performance of the optimal solutions from DCM_A model against two versions of the DCM_{OD} model. In version I, ρ is set to zero while in version II routes are penalized according to the number of necessary transfers. We calculate both passenger travel times and number of transfers under the assumption that the schedule can be implemented. We select three different time periods of one hour length that are significantly different from each other in terms of the demand characteristics. In the simulation, each one-hour solution is tested for 10 consecutive periods to approximate a continuously run schedule. Most importantly, we assume a selfish route choice behavior and lines arriving in a random order, thus inducing unwanted transfers. The results are shown in Table 2.

Overall, the three models show similar results although the set of optimal lines may differ from each other. As expected, non-zero penalty affects the ratio of transfers to passengers with negligible increase in operational costs. In terms of average travel times, OD-based model results in small savings which are lost when transfer penalty is introduced. To conclude, however, the impact of model choice can only be observed in the number of transfers and when using an appropriate penalty function. Simulation results show that optimal line plan solutions may fall short in response to time-sensitive demand. This observation shall stimulate new ideas and approaches to improve the accuracy of mathematical models in demand-sensitive line planning.

Table 2 Results of computational experiments

Instance	Model	Operational costs (km)	Average travel time (min)	Transfers per passenger
08h–09h	DCM_A	5597.3	37.1	0.19
	DCM_{OD} I	5657.9	35.4	0.19
	DCM_{OD} II	5732.0	36.4	0.08
14h–15h	DCM_A	2263.9	37.5	0.18
	DCM_{OD} I	2284.6	37.3	0.19
	DCM_{OD} II	2309.3	37.9	0.04
18h–19h	DCM_A	6868.2	39.5	0.24
	DCM_{OD} I	6886.8	38.3	0.28
	DCM_{OD} II	6983.7	40.6	0.08

References

1. Borndörfer, R., Hoppmann, H., & Karbstein, M. (2013). A configuration model for the line planning problem. In: D. Frigioni & S. Stiller (Eds.), In: *ATMOS 2013 - 13th Workshop on Algorithmic Approaches for Transportation Modeling, Optimization, and Systems* (Vol. 33, pp. 68–79). https://doi.org/10.4230/OASIcs.ATMOS.2013.68.
2. Karbstein, M. (2016). Integrated line planning and passenger routing: connectivity and transfers. *Operations Research Proceedings, 2014*, 263–269. https://doi.org/10.1007/978-3-319-28697-6_37.
3. Borndörfer, R., Arslan, O., Elijazyfer, Z., Güler, H., Renken, M., Sahin, G., & Schlechte, T. (2016). Line planning on path networks with application to the istanbul metrobüs. In: *Operations Research Proceedings*.

Scheduling of Electric Vehicles in the Police Fleet

Kerstin Schmidt, Felix Saucke and Thomas S. Spengler

1 Introduction

As a pioneer and role model in society, the police integrate electric vehicles into their fleets. These vehicles reduce environmental pollution and have lower energy costs compared to conventional vehicles. Moreover, the electric vehicles are nearly noiseless, leading to advantages for some operation strategies at the police. However, the challenges related to the use of electric vehicles are small distance ranges, long charging times, as well as limited availability of the charging infrastructure, resulting in a limited availability of the vehicles.

The operation tasks of the police vehicles vary from schedulable and time-flexible fiscal runs (e.g. courier services) and schedulable but nearly time-fixed operations in the criminal investigation service (e.g. execution of observations, crime scene work) to non-schedulable and time-fixed operations in service and patrol duty (e.g. road traffic accidents, robbery). Until now, the use of electric vehicles in the police fleet is mainly limited to fiscal runs. To further increase the number of electric vehicles in their fleet, the police aim for using electric vehicles within the criminal investigation service (CIS) as well. Operations of CIS are usually characterized by known starting and residence times of each operation as well as the associated travel time. Since operations are nearly time-fix, delays are to be avoided. Thus, the scheduled operations need to be assigned to the fleet of electric vehicles, which are available as pool vehicles at the police station.

K. Schmidt (✉) · F. Saucke · T. S. Spengler
Technische Universität Braunschweig, 38106 Braunschweig, Germany
e-mail: kerstin.schmidt@tu-braunschweig.de
URL: https://www.tu-braunschweig.de/aip/prodlog

F. Saucke
e-mail: f.saucke@tu-braunschweig.de

T. S. Spengler
e-mail: t.spengler@tu-braunschweig.de

Due to the mentioned challenges of battery driven vehicles, it is not trivial to substitute conventional vehicles by electric vehicles in operation tasks with high demands on the availability of the vehicles such as CIS. We address this issue by developing decision support for fleet operation with electric vehicles at the police. Our contribution is organized as follows: After reviewing the literature in Sect. 2, an extension of the Electric Vehicle Scheduling Problem (E-VSP) for the CIS at the police is presented in Sect. 3. In Sect. 4, an illustrative example is used to gain further insights. Our contribution ends with a conclusion and an outlook in Sect. 5.

2 Literature Review

The scheduling of electric vehicles in fleets is a fairly new but growing field of research. Areas of application can be found in passenger transportation services such as taxi and bus, as well as in postal and nursing services [4, 10, 11]. In the E-VSP a set of timetabled trips is assigned to a set of electric vehicles with limited driving ranges based at different depots under various objective functions (e.g. minimize total costs with regard to the travel distance/time or number of vehicles, maximize utilization of the vehicles) [1, 5, 8, 9, 11]. The E-VSP is based on the well-known Vehicle Scheduling Problem (VSP). Classical VSP variants consider conventional vehicles without allowing recharging [3]. Extensions of the E-VSP can be distinguished with regard to the consideration of time windows (none, fixed, or flexible) and the assumed charging mode (full vs. partial charging). Classic E-VSP do not consider time windows and allow only full charging during a tour [8]. The E-VSP presented by Schneider et al. considers fixed time windows and full charging [6]. Wen et al. are the first to consider partial charging in an E-VSP. Here, the charging time depends on the amount of energy to be charged [11]. Furthermore, they assume fixed time windows for depots and charging stations. To the best of our knowledge, flexible time windows and thus the consideration of delays in the objective function, as presented in Tas et al. [7] for conventional vehicles, have not been taken into account in the literature on E-VSP yet. Since operations of CIS at the police are nearly time-fixed and delays are to be avoided, a new E-VSP with full charging and flexible time windows is presented in the following section (referred to as E-FlexVSP).

3 Problem Statement and Optimization Model

In this section, an extension of the E-VSP for the police, taking into account the special requirements of the CIS, is developed. We consider one police station D and a set of R charging stations. A set of homogenous electric vehicles V is assigned to the police station to conduct a set of police operations T in the planning horizon of one shift. The E-FlexVSP is defined on a directed, closed graph $G = (S, A)$, where

$S = (D, R, T)$ represents the set of nodes and A the set of edges. Each operation $i \in T$ has a scheduled starting time z_i and residence time l_i. We assume that each operation i is visited exactly once by exactly one vehicle $k \in V$. If an operation i starts with a delay, each minute of delay w_i^k is penalized with penalty cost rate p. Each vehicle k has a maximum state of charge (SoC) of $y_D^k = Q$ [in driving minutes] and starts and ends in D. We assume that the vehicles are fully charged when they leave D. The travel time [in driving minutes] from node i to j [$i, j \in S$] is defined as t_{ij} and the cost rate for one minute of driving is given by c. We assume that the cost rate c corresponds to the energy costs. Due to the short planning horizon, no fixed costs are considered. The charging time [in driving minutes] at charging station $j \in R$ of a vehicle k coming from node $i \in \{T, D\}$ is defined as $g_{j,i}^k$ and is assumed to be a linear function of the travel time. Furthermore, we assume that any number of vehicles can be charged in parallel at the charging station, but each vehicle may be recharged at most once during the planning horizon.

The objective function of our E-FlexVSP is given in (1). The objective is to minimize the total cost Z, which consists of the cost of travel and the cost of delay. The decisions to be taken are which vehicle k is assigned to which operation i. To this end, the binary decision variable x_{ij}^k is set to 1, if vehicle k drives from node i to j.

$$\min Z = \sum_{k \in V} \sum_{i,j \in S} c \cdot t_{ij} \cdot x_{ij}^k + \sum_{k \in V} \sum_{i \in T} p \cdot w_i^k \quad (1)$$

The constraints of our E-FlexVSP can be divided into three categories: Route constraints, which involve typical vehicle scheduling constraints, time constraints for all nodes, and charging constraints, which include constraints regarding the SoC of the electric vehicles. The constraints are described in the following:

The **route constraints** include the starting condition for all nodes $i \in \{T, D\}$, the flow conservation constraint for all nodes $i \in S$, and, to eliminate subtours, the Miller–Tucker–Zemlin condition for all nodes $i \in \{T, R\}$ [2].

The **time constraints** include the arrival times at all nodes $i \in S$. Starting and residence time at the police station are set to zero. Furthermore, we assume that the arrival time a_i^k of vehicle k at each operation node $i \in T$ must be greater than or equal to the scheduled starting time z_i. The arrival time $a_{j,i}^k$ of vehicle k at node $j \in \{T, R\}$, coming from node $i \in \{D, T\}$, results from adding the travel time t_{ij} and the residence time l_i to the arrival time a_i^k at the previous node i (2), with M sufficiently big. Respectively, the arrival time a_h^k of vehicle k at node $h \in T$, coming from the charging station $j \in R$, is given in (3). Here, in contrast to (2) the charging time $g_{j,i}^k$ instead of the residence time l_i has to be taken into account. A minute of delay w_i^k is defined as the difference between arrival time a_i^k and scheduled starting time z_i.

$$a_{j,i}^k \geq a_i^k + t_{ij} + l_i - M(1 - x_{ij}^k) \quad \forall j \in \{T, R\}; i \in \{T, D\}; k \in V; i \neq j \quad (2)$$
$$a_h^k \geq a_{j,i}^k + t_{jh} + g_{j,i}^k - M(1 - x_{jh}^k) \quad \forall h \in T; j \in R; i \in \{T, D\}; k \in V; h \neq i \quad (3)$$

The **charging constraints** include the constraints regarding the SoC y_j^k of the electric vehicles. The SoC y_j^k of vehicle k in node $j \in T$ results from the SoC y_i^k at the previous node $i \in S$ minus the travel time t_{ij} (4). Furthermore, the SoC y_i^k of vehicle k in node $i \in \{T, D\}$, needs to be higher than or equal to the travel time t_{ij} to the following node $j \in \{T, D\}$ plus the minimum travel time b_j from all following nodes j to the nearest charging station (5). The charging time $g_{j,i}^k$ of vehicle k at charging station $j \in R$ depends on the maximum SoC Q, the available SoC y_i^k of vehicle k at the previous node i, and the travel time t_{ij} (6). Note that the charging time $g_{j,i}^k$ cannot exceed the maximum SoC Q. In addition, constraints 4–6 only hold if $x_{ij}^k = 1$.

$$y_j^k = y_i^k - t_{ij} \quad \forall j \in T; i \in S; k \in V; i \neq j \qquad (4)$$

$$y_i^k \geq b_j + t_{ij} \quad \forall i, j \in \{T, D\}; k \in V; i \neq j \qquad (5)$$

$$g_{j,i}^k = Q - (y_i^k - t_{ij}) \quad \forall i \in \{T, D\}; j \in R; k \in V \qquad (6)$$

All constraints of the presented E-FlexVSP with full charging and flexible time windows can be formulated as linear constraints. Therefore, the resulting model can be categorized as a mixed-integer linear problem. However, the model is difficult to solve since it is an extension of the NP-hard VSP.

4 Illustrative Example

In this section, an illustrative example is used to gain further insights into the E-FlexVSP. For this purpose, the optimization model is implemented in AIMMS and solved with CPLEX 12.7 on a 2.5 GHz CPU with 8 GB RAM.

We consider one police station D and one charging station R. Two homogenous electric vehicles are assigned to the police station. The vehicles have an average distance range of 200 km (e.g. Volkswagen e-Golf). Since police vehicles are equipped with technical equipment such as siren or emergency lights, the distance range is reduced to approximately 75%. We assume that the vehicles drive with an average speed of 60 km/h. Thus, the maximum SoC of a vehicle equates to $y_D^k = Q = 150$ driving minutes. Seven operations need to be assigned to the vehicles during the planning horizon of one shift. The scheduled starting time z_i and residence time l_i of each operation $i \in T$ as well as the corresponding travel time t_{ij} between all nodes $\{i,j\} \in S$ are given in Table 1.

The analysis has two objectives. First, we take a closer look on how the criticality of the operations, i.e. the importance of starting the operations in time, influences the results. Second, we analyse the influence of the maximum SoC on the travel cost and delay time.

Criticality of operations. Within the first analysis we set the number of vehicles to 2, the cost rate c for one driving minute to 1 and vary the penalty cost rate p for

Table 1 Travel times between the nodes and scheduled starting, residence time [min]

	D	1	2	3	4	5	6	7	R	z_i	l_i
D	–	30	35	50	20	30	60	35	5	–	–
1	30	–	30	50	25	35	40	50	35	9:00	40
2	35	30	–	45	50	60	25	55	40	9:40	30
3	50	50	45	–	25	65	20	30	50	10:40	20
4	20	25	50	25	–	20	60	45	25	11:20	45
5	30	35	60	65	20	–	75	40	25	12:50	10
6	60	40	25	20	60	75	–	45	60	13:50	60
7	35	50	55	30	45	40	45	–	35	14:40	30
R	5	35	40	50	25	25	60	35	–	–	–

Table 2 First analysis: Vehicle $|V| = 2$, $c = 1$, $Q = 150$

p	Z	Total travel time	Total delay time
0	250	250	1055
1	475	295	180
10	1405	405	100

Table 3 Second analysis: $c = 1$, $p = 10$

Q	Z	Total travel time	Total delay time
120	4540	340	420
175	460	310	15
250	340	340	0

one minute of delay. The values for p as well as the results are given in Table 2. The analysis shows that the total cost as well as total travel time increase with an increasing cost rate p. The opposite holds for the total time of delay. Thus, the more critical the operations, i.e. the higher the penalty cost rate for one minute of delay, the higher the acceptance of taking into account longer travel times. Therefore, it is important to consider the penalty costs for delays in the objective function.

Maximum state of charge. Within the second analysis we set the cost rate c for one driving minute to 1, the penalty cost rate p for one minute of delay to 10 and vary the maximum SoC Q. The results are given in Table 3. In comparison to $Q = 150$ min, the second analysis shows that the total cost decreases with an increasing maximum SoC. The opposite holds for the total delay time. Thus, the maximum SoC has a significant impact on operational scheduling of the police vehicles, i.e. the compliance with the scheduled starting times of the operations. It is therefore of great importance to determine the maximum SoC in the police fleet.

5 Conclusion and Outlook

In this contribution, we present an E-FlexVSP for the CIS at the police. The characteristics of the underlying decision situation are described and transferred into a new extension of the E-VSP with full charging and flexible time windows. Furthermore, the application of the model is illustrated in an example. We show that it is important to include penalty costs for delays in the objective function and that the maximum SoC has a crucial influence on the reliability of the police fleet.

Future research will address the following directions. First, the presented E-FlexVSP will be extended to be able to depict more realistic settings. For this purpose, more than one police station as well as constraints for partial charging, and uncertainties with regard to the operations will be included into our model. Second, since our model is difficult to solve, a suitable solution procedure will be developed. Third, the approach will be applied to a case study in order to verify and outline its potential for the CIS at the police.

Acknowledgements This work is part of the research project *lautlos and einsatz-bereit* which is funded by the Federal Ministry for the Environment, Nature Conservation, Building and Nuclear Safety (BMUB). The authors would like to thank for the support.

References

1. Adler, J. D., & Mirchandani, P. B. (2016). The vehicle scheduling problem for fleets with alternative-fuel vehicles. *Transportation Science*, *51*(2), 441–456.
2. Bektas, T. (2006). The multiple traveling salesman problem: an overview of formulations and solution procedures. *Omega*, *34*(3), 209–219.
3. Bunte, S., & Kliewer, N. (2009). An overview on vehicle scheduling models. *Public Transport*, *1*(4), 299–317.
4. Gnann, T., Plötz, P., Funke, S., & Wietschel, M. (2015). What is the market potential of plug-in electric vehicles as commercial passenger cars? A case study from Germany. *Transportation Research Part D: Transport and Environment*, *37*, 171–187.
5. Sassi, O., & Oulamara, A. (2017). Electric vehicle scheduling and optimal charging problem: complexity, exact and heuristic approaches. *International Journal of Production Research*, *55*(2), 519–535.
6. Schneider, M., Stenger, A., & Goeke, D. (2014). The electric vehicle-routing problem with time windows and recharging stations. *Transportation Science*, *48*(4), 500–520.
7. Tas, D., Jabali, O., & Van Woensel, T. (2014). A vehicle routing problem with flexible time windows. *Computers and Operations Research*, *52*, 39–54.
8. van Kooten Niekerk, M. E., van den Akker, J. M., & Hoogeveen, J. A. (2017). Scheduling electric vehicles. Public. *Transport*, *9*(1–2), 155–176.
9. Wang, H., & Shen, J. (2007). Heuristic approaches for solving transit vehicle scheduling problem with route and fueling time constraints. *Applied Mathematics and Computation*, *190*(2), 1237–1249.

10. Wang, Y., Huang, Y., Xu, J., & Barclay, N. (2017). Optimal recharging scheduling for urban electric buses: A case study in Davis. *Transportation Research Part E: Logistics and Transportation Review, 100*, 115–132.
11. Wen, M., Linde, E., Ropke, S., Mirchandani, P., & Larsen, A. (2016). An adaptive large neighborhood search heuristic for the electric vehicle scheduling problem. *Computers and Operations Research, 76*, 73–83.

Location Planning of Charging Stations for Electric City Buses Considering Battery Ageing Effects

Brita Rohrbeck, Kilian Berthold and Felix Hettich

1 Introduction

There are different technical approaches electric buses can be configured and charged by [1]: Buses could be charged over-night-in the bus depot exclusively, their batteries may be swapped [2], or buses can operate based on opportunity charging during the daily service at defined charging points. Opportunity charging comes with the advantage that smaller batteries can be used which have lower vehicle costs, less weight and less technical ageing-effects can be used. On the other hand, a charging infrastructure along the circuit has to be introduced. This circuit structure however makes bus traffic easier to plan than for individual electric vehicles. The main challenge here is to connect logically physically identical stops. This makes the problem more complex than other location problems, and few literature exists. [3, 4] suggest different approaches, the latter with a more technical background. [5] propose a model based on car park sites in order to determine optimal charging locations for individual electric vehicles.

In our paper, we developed a multi-period mixed integer model for one bus line with several buses. We also consider the ageing of batteries with time and the possibility to exchange batteries after some periods.

In the next section we explain the *Charging Stations Location Problem with Battery Ageing* (*CSLP-BA*) and our model in detail. Section 3 focusses on computational tests and evaluation. Finally, we give an outlook on future research.

B. Rohrbeck (✉) · F. Hettich
Karlsruhe Institute of Technology, Institute of Operations Research,
Kaiserstr 89, 76133 Karlsruhe, Germany
e-mail: brita.rohrbeck@kit.edu
URL: http://www.dol.ior.kit.edu/Team_BritaRohrbeck.php

K. Berthold · F. Hettich
Karlsruhe Institute of Technology, Institute of Vehicle System Technology,
Rintheimer Querallee 2, 76131 Karlsruhe, Germany

2 Problem Formulation

The formulation of the *CSLP-BA* is an enhancement by battery ageing of the formulation in [6]. The two main challenges lie in depicting the battery ageing and in connecting logically bus stops that are physically identical. This connection requires that if a charging station is built at location i in a certain circuit, then there also exists a charging station in every other circuit. By contrast, the state of charge of a bus may vary. Therefore, we introduce the set $\mathcal{C} = \{0, 1, \ldots, C, C+1\}$ of circuits. Here, $c = 1, \ldots, C$ stand for the actual circuits a bus drives during one day. $c = 0$ stands for the outbound trip from the depot to the first bus stop and $c = C + 1$ for the inbound path from the terminus back to the depot after the last circuit. $\mathcal{T} = \{1, \ldots, T\}$ are the considered time periods. Let $\mathcal{N}$ subsume all potential charging stations, comprising the respective C replications of the bus stops of the circuit route ($\mathcal{N}^{circ}$) as well as the nodes $\mathcal{N}^{out}$ and $\mathcal{N}^{in}$ from and back to the depot. Hence, $\mathcal{N} = \mathcal{N}^{out} \cup \mathcal{N}^{circ} \cup \mathcal{N}^{in}$, see Fig. 1.

To model the *CSLP-BA* different decisions have to be taken. The decision that is later implemented in practice is the location decision, i.e. whether and if so when charging station is built. Hence, we introduce the decision variables y_{it},

$$y_{it} = \begin{cases} 1 & \text{a charging station in node } i \in \mathcal{N} \text{ in period } t \in \mathcal{T} \text{ is opened,} \\ 0 & \text{otherwise.} \end{cases}$$

Also for the buses resp. their batteries decisions have to be taken: When shall an aged battery be substituted. To depict this, we introduce z_{bt}^{intro}, z_{bt}^{use} and z_{bt}^{out},

$$z_{bt}^{intro} = \begin{cases} 1 & \text{battery } b \in \mathcal{B} \text{ is put into operation in period } t \in \mathcal{T}, \\ 0 & \text{otherwise,} \end{cases}$$

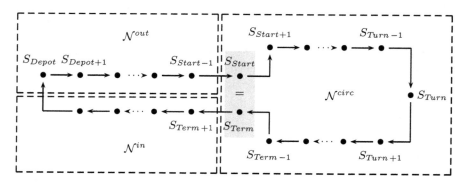

Fig. 1 The route of a bus as graph

$$z^{use}_{bt} = \begin{cases} 1 & \text{battery } b \in \mathcal{B} \text{ is in operation in period } t \in \mathcal{T}, \\ 0 & \text{otherwise,} \end{cases}$$

$$z^{out}_{bt} = \begin{cases} 1 & \text{battery } b \in \mathcal{B} \text{ is taken out of operation in period } t \in \mathcal{T}, \\ 0 & \text{otherwise.} \end{cases}$$

On the buses' side we also have to keep track of their batteries' content. The decision variables e_{icbt} shall designate the amount of energy stored in battery b when its bus leaves node $i \in \mathcal{N}$ in circuit $c \in \mathcal{C}$ during period $t \in \mathcal{T}$. Additionally, to adapt the batteries' capacities with their ageing process, we need the two decision variables $\overline{E}^{soh}_{bt}$ and $\overline{E}^{lim}_{bt}$ for battery $b \in \mathcal{B}$ and period $t \in \mathcal{T}$.

The aim is to find a cost minimal configuration for the location of charging stations and the exchange of batteries. If a charging station is built in node i in period t, fixed costs f^{char}_{it} arise. In case one node refers to a_i physically identical locations, the fixed costs for the station still occur once and have to be discounted: Hence, the installation costs amount to $\sum_{t \in \mathcal{T}} \sum_{i \in \mathcal{N}} \frac{1}{a_i} \cdot f^{char}_{it} \cdot y_{it}$. In addition, fixed costs f^{bat}_t emerge, when a new battery b is taken into operation: $\sum_{t \in \mathcal{T}} \sum_{b \in \mathcal{B}} f^{bat}_t \cdot z^{intro}_{bt}$. Therefore, the objective is

$$\min \sum_{t \in \mathcal{T}} \sum_{i \in \mathcal{N}} \frac{1}{a_i} \cdot f^{char}_{it} \cdot y_{it} + \sum_{t \in \mathcal{T}} \sum_{b \in \mathcal{B}} f^{bat}_t \cdot z^{intro}_{bt}$$

To assure the buses' service several constraints have to be considered: As each bus starts in the depot, we assume a charging station is installed in $t = 1$.

$$y_{s_{Depot},1} = 1 \qquad (1)$$

Accordingly, a bus could leave the depot in the morning with a full battery. The energy level is only bounded by its capacity $\overline{E}^{lim}_{bt}$:

$$e_{icbt} \leq \overline{E}^{lim}_{bt} \quad \forall (i,c) \in \mathcal{N}_1, b \in \mathcal{B}, t \in \mathcal{T} \qquad (2)$$

with $\mathcal{N}_1 = (\mathcal{N}^{out} \times \{0\}) \cup (\mathcal{N}^{circ} \times \{1, ..., C\}) \cup (\mathcal{N}^{in} \times \{C+1\})$. This capacity however depends on two factors: The state of health (SOH) reflects the ageing of the battery. With every period that a battery is in use, its initial capacity E^{intro}_b is reduced by a factor α:

$$\overline{E}^{soh}_{bt} = E^{intro}_b - \alpha \cdot \sum_{\tau=1}^{t} (t-\tau) \cdot z^{intro}_{b\tau} \quad \forall b \in \mathcal{B}, t \in \mathcal{T} \qquad (3)$$

Accordingly, the bound $\overline{E}^{lim}_{bt}$ for the stored energy can be at most $\overline{E}^{soh}_{bt}$:

$$\overline{E}^{lim}_{bt} \leq \overline{E}^{soh}_{bt} \quad \forall b \in \mathcal{B}, t \in \mathcal{T} \qquad (4)$$

However, to avert accelerated battery ageing, batteries are not used to full capacity. Instead, the energy level is to fluctuate between $\underline{E}$ and $\overline{E}$, in practice between 20 and 80% [7]. Hence, $\overline{E}_{bt}^{lim}$ is also bounded by $\overline{E}$ for a battery in use:

$$\overline{E}_{bt}^{lim} \leq \overline{E} \cdot z_{bt}^{use} \quad \forall\, b \in \mathcal{B},\, t \in \mathcal{T} \tag{5}$$

Note that this results in a piecewise linear battery ageing function which is what is effectively applied in practice by our reference instance [8].

Like stated before, the energy level of a battery must not fall below a value $\underline{E}$, neither. The value e_{icbt} of stored energy of a battery b has here also to cover the energy u_{ic} needed to reach the succeeding node to i in circuit c in period t:

$$e_{icbt} - u_{ic} \cdot z_{bt}^{use} \geq \underline{E} \cdot z_{bt}^{use} \quad \forall\, (i,c) \in \mathcal{N}_2,\, b \in \mathcal{B},\, t \in \mathcal{T} \tag{6}$$

with $\mathcal{N}_2 = \left(\mathcal{N}^{out} \times \{0\}\right) \cup \left((\mathcal{N}^{circ} \setminus \{S_{Term}\}) \times \{1, ..., C\}\right) \cup \left(\mathcal{N}^{in} \cup \{S_{Term}\} \times \{C+1\}\right)$. The amount stored in a battery in node i depends on if and how much a bus can charge electricity, but also on the state of charge at a previous point. At each node i in circuit c in period t the stored energy e_{icbt} of battery b is hence determined by the energy level in node $i-1$, $e_{i-1,c,b,t}$, reduced by the energy needed to get to node i, $u_{i-1,c}$, and augmented by the energy x_{icbt} charged in i. x_{icbt} however does not have to be defined, but shall only help here for understanding. We have to assure that the charged energy x_{icbt} is ≥ 0, hence we reformulate the explained energy balance equation towards:

$$e_{icbt} - e_{i-1,c,b,t} + u_{i-1,c} \cdot z_{bt}^{use} \geq 0 \quad \forall\, (i,c) \in \mathcal{N}_3,\, b \in \mathcal{B},\, t \in \mathcal{T} \tag{7}$$

$$e_{S_{Start},1,b,t} - e_{S_{Start}-1,0,b,t} + u_{S_{Start}-1,0} \cdot z_{bt}^{use} \geq 0 \quad \forall\, b \in \mathcal{B},\, t \in \mathcal{T} \tag{8}$$

with $\mathcal{N}_3 = \left((\mathcal{N}^{out} \setminus \{S_{Depot}\}) \times \{0\}\right) \cup \left((\mathcal{N}^{circ} \setminus \{S_{Start}\}) \times \{1, ..., C\}\right) \cup (\mathcal{N}^{in} \times \{C+1\})$. The chargeable energy in node i during circuit c is however also delimited by s_{ic}—or 0 if no charging station is opened in i before period t, i.e. $\sum_{\tau=1}^{t} y_{i\tau} = 0$:

$$e_{icbt} - e_{i-1,c,b,t} + u_{i-1,c} \cdot z_{bt}^{use} \leq s_{ic} \cdot \sum_{\tau=1}^{t} y_{i\tau} \quad \forall\, (i,c) \in \mathcal{N}_3,\, b \in \mathcal{B},\, t \in \mathcal{T} \tag{9}$$

$$e_{S_{Start},1,b,t} - e_{S_{Start}-1,0,b,t} + u_{S_{Start}-1,0} \cdot z_{b,t}^{use} \leq s_{S_{Start},1} \cdot \sum_{\tau=1}^{t} y_{S_{Start},\tau} \tag{10}$$

$$\forall\, b \in \mathcal{B},\, t \in \mathcal{T}$$

The next two constraints assure the right energy level after the transition to a new circuit and the inbound trip, respectively.

$$e_{S_{Start},c+1,b,t} = e_{S_{Term},c,b,t} \quad \forall b \in \mathcal{B}, \ c \in \{1, ..., C\}, \ t \in \mathcal{T} \tag{11}$$

$$e_{S_{Term},C+1,b,t} = e_{S_{Term},C,b,t} \quad \forall b \in \mathcal{B}, \ t \in \mathcal{T} \tag{12}$$

To ensure that the location decisions for two physically identical nodes are equal, their decision variables y_{it} must be identical. The set $\mathcal{I}$ subsumes all these pairs of nodes. State of two location variables must be identical if they belong to physically identical locations:

$$y_{i_1,t} = y_{i_2,t} \quad \forall \ (i_1|i_2) \in \mathcal{I}, \ t \in \mathcal{T} \tag{13}$$

Furthermore, a charging station can be opened at most once:

$$\sum_{t \in \mathcal{T}} y_{it} \leq 1 \quad \forall \ i \in \mathcal{N} \tag{14}$$

For the proper configuration of the batteries' decision variables, their interdependency must be considered:

$$z_{bt}^{use} = \sum_{\tau=1}^{t} z_{b\tau}^{intro} - \sum_{\tau=1}^{t} z_{b\tau}^{out} \quad \forall b \in \mathcal{B}, \ t \in \mathcal{T} \tag{15}$$

To assure that there are always a number of β buses circulating if the bus line requires this, in every period (at least) β batteries must be operating:

$$\sum_{b \in \mathcal{B}} z_{bt}^{use} \geq \beta \quad \forall \ t \in \mathcal{T} \tag{16}$$

Finally, the following domain constraints are needed:

$$z_{bt}^{intro}, \ z_{bt}^{use}, \ z_{bt}^{out} \in \{0, 1\} \qquad \forall b \in \mathcal{B}, t \in \mathcal{T} \tag{17}$$

$$y_{it} \in \{0, 1\} \qquad \forall \ i \in \mathcal{N}, \ t \in \mathcal{T} \tag{18}$$

$$e_{icbt}, \ x_{icbt} \geq 0 \qquad \forall \ (i, c) \in \mathcal{N}_1, \ b \in \mathcal{B}, \ t \in \mathcal{T} \tag{19}$$

$$\overline{E}_{bt}^{soh}, \ \overline{E}_{bt}^{lim} \geq 0 \qquad \forall b \in \mathcal{B}, \ t \in \mathcal{T} \tag{20}$$

3 Computational Results

We tested our model using the data of bus line 63 in the city of Mannheim, Germany. The circuit of line 63 is 9 km long and takes 40 min for its 23 bus stops (way and return). Two buses do service at 20-min intervals. We measured waiting times, passenger numbers and deduced realistic values for energy consumption and chargeable energy. Characteristics of the batteries and the charging stations are taken from the public project description of the PRIMOVE Mannheim Project [9]. For the battery ageing we assumed a factor α of 1.71 kWh/year. This results from the usual practice to replace batteries when their state of health falls below 80%, which is on average after seven years [7].

We ran our tests on a 64-bit Windows 7 Enterprise PC with a 2.6 GHz Intel(R) Xeon(R) processor and 48 GB RAM. To solve our models we used IBM ILOG CPLEX optimization studio 12.6.1. We tested different instances varying in traffic, external factors like temperature and driving behaviour. Solving the basic instance and average values for traffic and auxiliary consumers took 10:14h. Eight charging stations shall be installed, whereas in Mannheim only six are built. Furthermore, over a horizon of twenty years, still the same locations are optimal, and it is cost-efficient to exchange batteries once, i.e. after ten years. This result first shows that it is not necessarily economically reasonable to replace batteries as soon as the producer recommends, i.e. when their SOH falls below 80%. Hence, as long as the usable capacity shrinks just somewhat, this does not make the configuration infeasible. On the other hand, the two extra charging stations do not result from the longer usage of the batteries. Indeed, we get the same configuration if we neglect battery ageing [6]. The smaller number of charging stations in practice with 22% smaller cost does not seem to prove our model right. However, the solution implemented in Mannheim turned out to be insufficient. By changes in timetables, reducing the usage of the heating unit, accepting delays and exploiting more the batteries' capacities, that solution was made feasible, but for the future also more expensive.

Since we intend to extend our model to a whole network, we further improved it with regard to run times. Just by introducing an extra decision variable $\hat{y}_{it} = \sum_{\tau=1}^{t} y_{i\tau}$, adding this constraint to the model and substituting it in the model in Constraints (9) and (14) we could reduce the calculation time tremendously. By adding further valid inequalities run times went down to 1:08h.

4 Conclusion and Outlook

Our model depicts the path and energy course over multiple periods of a single bus line with battery ageing. It results in an optimal solution within reasonable time. We currently test different, very promising reformulations with regard to significantly reduced run times. Additionally, we analyse additional technological configurations for batteries and stations that are realistic for near future. The next step is then to incorporate our one bus line within a bus network. For further research, the timetables of the bus lines or their routes may questioned. Even more synergy effects could be achieve if other municipal electric vehicle could be incorporated into the network [10].

References

1. Müller-Hellmann, A. (2014). Überlegungen zu Ladeverfahren für Batteriebusse im ÖPNV. *Der Nahverkehr, 2014*(7-8), 40–43.
2. Mak, A.-H., Rong, Y., & Shen, Z.-J. (2013). Infrastructure planning for electric vehicles with battery swapping. *Management Science, 59*(7), 1557–1575.
3. Kley, F. (2011). Ladeinfrastrukturen für Elektrofahrzeuge: Entwicklung und Bewertung einer Ausbaustrategie auf Basis des Fahrverhaltens, Fraunhofer.
4. Kunith, A., Mendelevitch, R., & Goehlich, D. (2016). Electrification of a city bus network: an optimization model for cost-effective placing of charging infrastructure and battery sizing of fast charging electric bus systems. *International Journal of Sustainable Transportation*.
5. Chen, D., Kockelman, K., & Khan, M. (2013). The electric vehicle charging station location problem: a parking-based assignment method for seattle. *Transportation Research Board 92nd Annual Meeting, 340*, 13–1254.
6. Berthold, K., Förster, P., & Rohrbeck, B. (2017). Location planning of charging stations. In K. Doerner, I. Ljubic, G. Pflug, & G. Tragler (Eds.), *Operations Research Proceedings 2015*.
7. Rogge, M., Wollny, S., & Sauer, D. U. (2015). Fast charging battery buses for the electrification of urban public transport a feasibility study focusing on charging infrastructure and energy storage requirements. *Energies, 8*(5), 4587–4606.
8. Herb, F. (2010). Alterungsmechanismen in Lithium-Ionen-Batterien und PEM-Brennstoffzellen und deren Einfluss auf die Eigenschaften von daraus bestehenden Hybrid-Systemen, Doctoral dissertation, Universität Ulm.
9. Nationale Organisation Wasserstoff- und Brennstoffzellentechnologie (2012). http://www.now-gmbh.de
10. Pelletier, S., Jabali, O., & Laporte, G. (2017). Charge Scheduling for Electric Freight Vehicle CIRRELT, CIRRELT-2017-37.

On the Benefit of Preprocessing and Heuristics for Periodic Timetabling

Christian Liebchen

1 Introduction

> The timetable is the essence of the service offered by any provider of public transport (Jonathan Tyler, [12]).

Indeed, the timetable marks the interface between service design (including network and line planning) and operations planning (vehicle and crew scheduling). In particular in the context of (European) railways, there are several companies involved in the process of timetable design: many railway undertakings (or, train operating companies) and at least one infrastructure manager. Accordingly, there are more than just one optimization models that fit the different tasks of the various companies, see [1].

The task of a railway undertaking which operates a dense network with an essentially regular line-based service is to define the departure and arrival times of the lines at any of their stations. Since 1989 [11] this is often modeled as Periodic Event Scheduling Problem (PESP). This model covers many relevant requirements on a timetable: safety distances (in the same direction, or in opposite directions on single tracks), vehicle waiting times (correlating to the number of vehicles required to operate the timetable), passenger waiting times (within trains, or at stations during transfers), and many more, see [8].

In particular the latter immediately affect the quality that the (potential) passengers perceive. This is a major motivation to further improve optimization methods aiming to minimize slack times. Already one decade ago, for several networks their timetables had been designed using mathematical optimization techniques [4, 7].

C. Liebchen (✉)
Technische Hochschule Wildau, Ingenieur- und Naturwissenschaften,
Verkehrsbetriebsführung, Hochschulring 1, 15745 Wildau, Germany
e-mail: liebchen@th-wildau.de

2 The Periodic Event Scheduling Problem

The Periodic Event Scheduling Problem (PESP) is defined for a directed graph $D = (V, A)$. The nodes $i \in V$ represent the events (i.e., arrival or departure of a directed traffic line at some station), which are recurring periodically every T time unites, e.g. every 60 min. The constraints are given along the arcs using lower and upper time bounds ℓ_a and u_a, respectively:

$$(\pi_j - \pi_i - \ell_a) \bmod T \leq u_a - \ell_a, \quad \forall a = (i, j) \in A. \qquad (1)$$

In addition, we assume a weight w_a for each arc to be given, which reflects the penalty that is applied to any time unit of slack. In the case of a transfer arc a, w_a might represent the expected number of passengers who desire to use this activity.

In the remainder, for some spanning tree F in D, we are considering the following integer linear optimization problem:

$$\begin{aligned}
\min \; & \sum_{a \in A} w_a x_a \\
\text{s.t.} \; & x_a = \pi_j - \pi_i + T p_a && \forall a = (i, j) \in A \\
& \ell_a \leq x_a \leq u_a && \forall a \in A \\
& p_a = 0 && \text{for all } a \in F \\
& p_a \in \mathbb{Z}
\end{aligned}$$

In the notation of [6, Chap. 9], this particular problem formulation might have been called PESP-IP-π-x-tree.

3 The PESPlib Collection of Instances

We are aware of only one public collection of PESP-instances: the PESPlib [3].[1] As of Apr 30th, 2017, there has been one instance, for which three international research groups[2] provided solutions: the so-called R1L1-instance. Let us collect some properties of R1L1:

- It belongs to the railway instances of PESPlib and according to [3] the larger instance R4L4 is "approximately the size of the German long-distance railway network"
- In the first 3,554 arcs, every second arc a has $[\ell_a, u_a]_T = [1, 5]_{60}$, presumably modeling dwell activities alternating with drive activities
- This interpretation is supported by symmetric trip times for the two opposite directions (e.g. events 409–452 and 453–496, or 1141–1166 and 1167–1192)

---

[1] See http://num.math.uni-goettingen.de/~m.goerigk/pesplib/.
[2] Uni Göttingen, TU Dresden, ETH Zürich.

Fig. 1 Distribution of the weights of the free arcs (transfers)

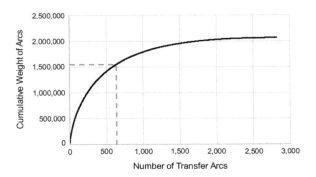

- The remaining arcs are all free arcs a with $[\ell_a, u_a]_T = [3, 62]_{60}$, presumably modeling transfer activities (except for the last three fixed arcs), because none of these is connecting the endpoints of the two directions of the two above-mentioned lines and thus unlikely to model turnarounds.
- When eliminating all the free arcs from R1L1, then the instance does no longer contain any cycles.

Altogether this suggests that for this particular instance the modeling focus had been put on drive, dwell, and transfer activities, while rather neglecting turnaround activities of the trains, headway or single track requirements as they are present in the German long-distance railway network.

Periodic timetabling is easy when applied to cycle-free constraint graphs. On the one hand, the R1L1 instance is too complex for standard MIP solvers to end with an optimum solution. On the other hand, each cycle in R1L1 contains at least one free arc (transfers). Hence, to make this instance better accessible to MIP solvers, we will have to simplify it moderately and thus take a closer look at the distribution of the weights of the free arcs (transfers), presumably being the only linkages between different lines.

Figure 1 illustrates that the distribution of the weights roughly follows the typical 80-20-Pareto rule. In particular, when omitting 77.5% of the free arcs (directed transfers), hereby we are ignoring just 25% of the total weight (number of transferring passengers).

To summarize this investigation, it seems to be promising to simplify PESPlib's R1L1-instance using some simple standard preprocessing and heuristics and then provide it to some standard MIP solver.

4 Preprocessing and Heuristics for PESP Instances

As a heuristic, we are thus proposing to temporarily ignore the free arcs with smallest weight until their weights sum up to some given percentage of the sum W of the weights w_a of all the *free arcs* a, i.e., where $u_a - \ell_a \geq T - 1$. This had been proposed earlier in [5, Chap. 4.2.2] and [10, Chap. 7.4.1].

In addition, we propose to apply graph contractions to the constraint graph as they have been proposed for example in [6, Chap. 14]. In particular, we are contracting nodes with degree one as well as *fixed arcs a* (i.e. where $u_a = \ell_a$). Observe that doing so, there is still a bijection between equivalence classes of optimum solutions of the initial network and of the reduced network.

Moreover, we are contracting nodes with exactly one incoming arc a and one outgoing arc b. There, we are able to preserve the set of feasible solutions, too. Yet, with respect to the objective function, there is a (slight) imprecision, because along the modified arc, for the first units of slack there should apply $\min\{w_a, w_b\}$, whereas $\max\{w_a, w_b\}$ had to apply to the last units of slack. Obviously, this could not be expressed in any linear objective function on the modified arc in the reduced network. In our experiments, we heuristically select $\min\{w_a, w_b\}$ as the weight of the modified arc. For an example illustrating how these contraction steps apply to R1L1, please refer to Fig. 2 in [2].

5 Computational Results

As a general setting, we start by simplifying the PESP constraint graph with the techniques sketched in Sect. 4, i.e., heuristically ignoring light free arcs, and contractions. The MIP formulation for this simplified instance in then stated as given in Sect. 2. We solve this MIP using CPLEX 12.7.0.0 on an Intel Core i5 2.2GHz 8GB RAM (3503 Passmark CPU Mark, Q4/2014), setting the tree memory limit to 2GB.

In our main series of computations, we add the well-established valid inequalities due to Odijk [9] as valid inequalities on the integer variables p_a that correspond to non-tree arcs $a \in A \setminus F$. Then we vary the ignore ratio: The more free arcs that we ignore, the smaller the resulting constrains graphs get (cf. columns nodes and arcs in Table 1), and the better the solution behavior of CPLEX on the simplified network. Yet, when reinterpreting the solution that had been computed for the simplified network back in the initial network, the loss in information translates to worse objective values (cf. column R1L1 objective).

Let us shortly discuss two points: First, notice that the achieved quality is not just due to general improvements that were obtained within the latest versions of CPLEX. This can be seen by solving the lp-file of the 30%-row with the 2012 version of CPLEX (12.3). There, after 900s[3] an objective value of only 35,903,663 is obtained – compared to the 2016 PESPlib benchmark for R1L1 of 37,338,904.

Second, one could ask whether contractions are useful at all in a preprocessing step, because MIP solvers are known for powerful general problem reduction techniques. To this end, observe that the actual R1L1-instance has 3,664 nodes and 6,385 arcs, among which 646 are fixed and 2,827 are free. By only ignoring free arcs such that their weights sum up to 25% of the total free weight W and without

---

[3]Unfortunately, this computation had only been possible on a different machine: Intel Xeon 3.7GHz 16GB RAM (9492 Passmark CPU Mark, Q2/2013).

Table 1 Objective values found by applying different ignore ratios for the free arcs

Ignore (%)	Nodes	Arcs	Odijk	Time	CPLEX gap (%)	R1L1 objective
10	772	1,828	Yes	900	6.89	37,918,546
20	572	1,230	Yes	900	5.69	35,433,189
30	438	862	Yes	900	4.73	36,213,298
40	346	610	Yes	900	3.36	36,720,735
50	257	406	Yes	900	1.77	40,814,013
60	189	251	Yes	900	0.75	41,843,259
70	129	136	Yes	900	0.00	46,010,226
25	501	1,029	No	3,600	4.22	33,711,523

contracting any arc, CPLEX reports a reduced MIP size of 718 rows, 2,844 columns, 5,240 nonzeros, and 535 general integers. Comparing these values to the ones that are obtained, when contractions have been applied, too (708 rows, 1,737 column, 3,923 nonzeros, and 535 general integers), could not seem to make any big difference. Yet, and most important, the larger MIP sizes induce a (much) worse R1L1-solution after one hour of computation time: 38,531,957 versus 33,711,523.

We close by mentioning that on PESPlib's largest railway instance (R4L4, 8,384 nodes and 17,754 arcs), with the very same combination of ignoring (here: 40%) and contracting we were able to come up with another benchmark solution after one hour of computation time,[4] improving the previous benchmark (47,283,768) down to only 43,234,156.

Notice that during the refereeing process of this volume, in a collaboration with Marc Goerigk [2], we achieved further improvements by iterating the method proposed in this paper with the modulo network simplex method.

6 Conclusions and Acknowledgement

Simple preprocessing and heuristics enabled a standard MIP solver to find new benchmark solutions for the smallest and largest railway instances of the PESPlib, R1L1 and R4L4. Hence, these techniques should always be considered when practically solving PESP instances.

Yet, we are aware of the very specific structure of these particular instances. In the case of instances whose cycles do not always contain free arcs, and thus arbitrary variable vectors are (much) more likely to be infeasible, we are convinced, that a much deeper insight will be required to come up with excellent solutions for instances of the size of R1L1 or even R4L4.

---

[4]For this instance, we set the MIP emphasis to "finding hidden feasible solutions".

The author thanks Michel Le (IBM) and Ralf Borndörfer for recently providing him with the CPLEX version of the year 2012 (12.3).

References

1. Caimi, G., Kroon, L., & Liebchen, C. (2017). Models for railway timetable optimization: Applicability and applications in practice. *Journal of Rail Transport Planning and Management, 6,* 285–312.
2. Goerigk, M., & Liebchen, C. (2017). An improved algorithm for the periodic timetabling problem. Submitted to ATMOS 2017.
3. Goerigk, M., & Schöbel, A. (2013). Improving the modulo simplex algorithm for large-scale periodic timetabling. *Computers and Operations Research, 40*(5), 1363–1370.
4. Kroon, L., Huisman, D., Abbink, E., Fioole, P.-J., Fischetti, M., Maróti, G., et al. (2009). The new Dutch timetable: The OR revolution. *Interfaces, 39*(1), 6–17.
5. Liebchen, C. (1998). Optimierungsverfahren zur Erstellung von Taktfahrplänen. Master's thesis, Technical University Berlin, Germany, 1998. In German.
6. Liebchen, C. (2006). Periodic timetable optimization in public transport. Dissertation.de–Verlag im Internet.
7. Liebchen, C. (2008). The first optimized railway timetable in practice. *Transportation Science, 42*(4), 420–435.
8. Liebchen, C., & Möhring, R. H. (2007). The modeling power of the periodic event scheduling problem: Railway timetables-and beyond. *Algorithmic methods for railway optimization* (pp. 3–40). Berlin: Springer.
9. Odijk, M. A. (1996). A constraint generation algorithm for the construction of periodic railway timetables. *Transportation Research Part B: Methodological, 30*(6), 455–464.
10. Opitz, J. (2009). Automatische Erzeugung und Optimierung von Taktfahrplänen in Schienenverkehrsnetzen. PhD thesis, TU Dresden, 2009. In German.
11. Serafini, P., & Ukovich, W. (1989). A mathematical model for periodic scheduling problems. *SIAM Journal on Discrete Mathematics, 2*(4), 550–581.
12. Tyler, J. (2006). Philosophies of timetabling, definitions of bottlenecks and the usefulness of spreadsheets: The experience of a practical strategic timetable planner. In: CASPT 2006, CD-ROM.

Structure-Based Decomposition for Pattern-Detection for Railway Timetables

Stanley Schade, Thomas Schlechte and Jakob Witzig

1 Introduction

The timetable is the starting point of a rotation planner. The objective is to assign job sequences to the available railway vehicles, such that every trip of the timetable is covered. In [4] the timetable is split up into distinct parts that each consist of repeating patterns during the planning process. This leads to the pattern detection problem, which was modeled using a mixed integer program. In this paper we investigate alternative solution strategies for this model in comparison to solving the model using a generic MIP solver.

In Sect. 2 we give an outline what the pattern detection problem is and how it arises. Subsequently, we present two greedy heuristics and a dual reduction that divides the problem into components that can be enumerated. In Sect. 4 we evaluate the run-time and accuracy of the presented algorithms.

The work for this article has been conducted within the Research Campus Modal funded by the German Federal Ministry of Education and Research (fund number 05M14ZAM).

S. Schade (✉)
Zuse Institute Berlin, Mathematics of Transportation and Logistics,
Takustr. 7, 14195 Berlin, Germany
e-mail: schade@zib.de

T. Schlechte
LBW Optimization GmbH, Obwaldener Zeile 19, 12205 Berlin, Germany
e-mail: schlechte@lbw-optimization.de

J. Witzig
Zuse Institute Berlin, Mathematical Optimization Methods,
Takustr. 7, 14195 Berlin, Germany
e-mail: witzig@zib.de

© Springer International Publishing AG, part of Springer Nature 2018
N. Kliewer et al. (eds.), *Operations Research Proceedings 2017*, Operations Research Proceedings, https://doi.org/10.1007/978-3-319-89920-6_95

2 Pattern-Detection

With regard to timetable patterns, it is only relevant whether two days of the timetable are equal with respect to the trips that are operated on these days. Hence, a number can be assigned to each day, such that these numbers for two days are equal if and only if the trips to be operated agree. Note that the timetable has a weekly structure. Thus, generally two days are only compared if they correspond to the same weekday, e.g. two consecutive Mondays. As a result, we define a timetable to be a finite sequence of integers. The length of this finite sequence has to be a multiple of seven. Any subsequence of seven consecutive days of a timetable is a *pattern*. Because of the cyclic structure of patterns, we agree to start with the day that corresponds to Sunday, then Monday, Tuesday and so on if we write them down. An example of a timetable is $T = (1, 1, 1, 1, 1, 2, 2, 1, 1, 1, 1, 1, 2, 2, 2, 1, 1, 1, 1, 1, 2)$. Let us assume that the first day of T is a Sunday. Then, it contains the patterns $A = (1, 1, 1, 1, 1, 2, 2)$, $B = (2, 1, 1, 1, 1, 2, 2)$ and $C = (2, 1, 1, 1, 1, 1, 2)$. The pattern A matches the first 14 days of T, B matches days 8–20 and C matches the last seven days. If a pattern matches at least 8 consecutive days of a timetable, we say that it *covers* these days. Hence, A also covers the first 14 days of T and B also covers the days 8 to 20, but C does not cover any part of T. A more formal definition of the cover relation can be found in [4], but is left out here due to space constraints. More information on cyclic rotation planning with a period of one week can be found in [1]. During the planning process, the timetable for a year is developed gradually to include more and more details. A part of the timetable that is covered by a pattern has a weekly periodic structure. Therefore, one can also use a rotation plan with such a structure for this part. We aim to identify a few patterns that cover as much of the timetable as possible. For each of these patterns a rotation plan needs to be developed. The pattern detection, thus, is a useful tool in early planning stages. Determining patterns that cover parts of a timetable can be done by simple linear preprocessing. In practice, one aims to select only a few relevant patterns that cover as much of the timetable as possible, since each pattern corresponds to a rotation plan that needs to be developed. Parts of the timetable usually also lack a periodic structure and are not covered by patterns, e.g., extended holiday periods like Christmas. The following mixed integer program to identify relevant patterns was presented in [4].

$$\min -\sum_{i=1}^{n} x_i + 8 \sum_{j}^{m} y_i \quad (1)$$

$$s.t. \quad x_i - \sum_{j \,:\, j \text{ covers } i} y_j \leq 0 \quad \forall i = 1, \ldots, n \quad (2)$$

$$x_i \in [0, 1] \quad \forall i = 1, \ldots, n$$

$$y_j \in \{0, 1\} \quad \forall j = 1, \ldots, m$$

Let n be the number of days and m the number of patterns. Setting the binary variable y_j to 1 means that pattern j is selected. In an optimal solution a day i is covered if and only if $x_i = 1$. In this case the constraint (2) ensures the existence of a pattern j that covers i.

3 Structure-Based Propagation

In this section, we present two structure-based procedures for solving the pattern detection problem presented in the previous section. One procedure aims to decompose the search space into independent components, such that the resulting components are (hopefully) easier to solve. The other procedure is a greedy heuristic.

In the following, we denote set of patterns j *covering a day* i by $\mathcal{C}(i)$. Analogous, the set of days i that are *covered by a pattern* j is denoted by $\mathcal{C}^{-1}(j)$. Moreover, the set of days i that are covered by a *unique* pattern j is denoted by

$$\mathcal{U}(j) := \{i = 1, \ldots, n : j \in \mathcal{C}(i) \text{ and } |\mathcal{C}(i)| = 1\}.$$

We call a pattern j a *long pattern* if and only if $|\mathcal{U}(j)| \geq 9$.

Due to the fact that choosing a pattern that covers at most seven days will lead to a deterioration of the objective value, every y_j with $|\mathcal{C}^{-1}(j)| \leq 7$ can be fixed to 0. In fact, such patterns are ruled out by the preprocessing.

Decomposition by Days The special structure of the objective function of (1) allows us to determine patterns that will be part of at least one optimal solution. Selecting a long pattern always leads to an improvement of the objective function value by at least 1, because they cover at least 9 days that are covered by no other pattern. Usually, reductions guaranteeing that at least one optimal solution is preserved are called *dual reductions*, e.g., propagation with the objective function. On the other hand, a reduction that preserves all optimal solutions is called *primal*. In our case, we use a dual argument, i.e., the objective function, but we can guarantee that all optimal solution will be preserved.

Two patterns are called *overlapping* if and only if they mutually cover at least one day of a timetable. Consider a pattern j that has no overlap with any other pattern. Clearly, whether we set y_j to 0 or 1 does not influence the other patterns. If j overlaps with a pattern k, the objective function value may be improved by setting y_j or y_k to 1. But it can be possible that the objective value does not improve if both variables are set to 1 at the same time. Let us call two patterns j and k *connected* if they overlap or they are both connected to a third pattern l. In our decomposition approach we aim to split the search space into smaller pieces that are (hopefully) easier to solve, e.g., by complete enumeration or a MIP solver. To decompose an instance, we first remove all long patterns as depicted in Fig. 1. In the computational experiments, we used an enumeration approach and never had to enumerate more than 400 solutions with this strategy. But potentially, an instance could contain patterns that cover several different parts of the year and foil the decomposition. In this case using a MIP solver would be superior with regard to performance.

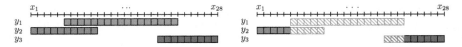

Fig. 1 Structure-based decomposition. Long pattern 1 covers 9 days exclusively (left). This pattern will be part of at least one optimal solution. The search space decomposes into two independent parts (right) after removing pattern 1, all days in $\mathcal{C}^{-1}(1)$, and all days covered by 1 that are in $\mathcal{C}^{-1}(2)$ and $\mathcal{C}^{-1}(3)$

Heuristic A simpler approach is to score patterns and to greedily choose the patterns with the highest score one after another. We use the length of a pattern or the number of days that are uniquely covered as its score. To generalize this, we can use a scoring function $\phi_{\mathcal{Y}}(\alpha, \beta) = (\alpha \cdot |\mathcal{C}^{-1}(j)| + \beta \cdot |\mathcal{U}(j)|)_{j \in \mathcal{Y}}$. The heuristic looks for a pattern j with highest score, such that setting y_j to 1 leads to an improvement of the objective value. If no such pattern can be found, the heuristic terminates. Otherwise, the days covered by j are removed from the timetable. Thus, the scores of the patterns that overlap with j need to be recalculated. To avoid unnecessary update steps, it is reasonable to select all long patterns beforehand. The full heuristic is given as pseudocode in Algorithm 1.

Algorithm 1 Structure-Based Propagation Procedure

1: $(x, y) \leftarrow (0, 0)$ ▷ initialize zero solution
2: $\mathcal{X} \leftarrow \{1, \ldots, n\}, \mathcal{Y} \leftarrow \{1, \ldots, m\}$ ▷ initialize index set of days and patterns
3: **for all** $j \in \{1, \ldots, m\}$ with $|\mathcal{U}(j)| \geq 9$ **do**   ▷ apply trivial fixings
4:    $y_j \leftarrow 1; \mathcal{Y} \leftarrow \mathcal{Y} \setminus j$
5:    **for all** $i \in \mathcal{C}^{-1}(j)$ **do**
6:      $x_i \leftarrow 1; \mathcal{X} \leftarrow \mathcal{X} \setminus i$
7:      **for all** $k \in \mathcal{Y}$ with $i \in \mathcal{C}^{-1}(k)$ **do**
8:        $\mathcal{C}^{-1}(k) \leftarrow \mathcal{C}^{-1}(k) \setminus i$
9: **while** $\mathcal{X} \neq \emptyset$ **do**
10:    $s \leftarrow \phi_{\mathcal{Y}}(\alpha, \beta)$ ▷ get current scores
11:    Get a permutation π such that $s_{\pi_j} \geq s_{\pi_{j+1}}$ for all $j \in \mathcal{Y}$
12:    $success \leftarrow false$
13:    **for** $j = 1, \ldots, |\mathcal{Y}|$ **do**   ▷ find pattern to fix with highest scores
14:      **if** $|\mathcal{C}^{-1}(\pi_j^{-1})| \geq 9$ **then**
15:        $y_{\pi_j^{-1}} \leftarrow 1; \mathcal{Y} \leftarrow \mathcal{Y} \setminus \pi_j^{-1}; success \leftarrow true$
16:        **for all** $i \in \mathcal{C}^{-1}(\pi_j^{-1})$ **do**
17:          $x_i \leftarrow 1; \mathcal{X} \leftarrow \mathcal{X} \setminus i$
18:          **for all** $k \in \mathcal{Y}$ with $i \in \mathcal{C}^{-1}(k)$ **do**
19:            $\mathcal{C}^{-1}(k) \leftarrow \mathcal{C}^{-1}(k) \setminus i$
20:        **break**
21:    **if** $!success$ **then**   ▷ stop if no pattern was chosen
22:      **break**
23: **return** (x, y)

Note that the heuristic may lead to suboptimal solutions. Say, we use the length of patterns as score, i.e., $\alpha = 1, \beta = 0$. We have three patterns 1, 2 and 3 of lengths 10, 11 and 10, respectively. They are arranged in a similar way as the patterns in Fig. 1 with 1 and 2 having an overlap of 3 days and 2 and 3 having an overlap of 2 days. In this case the heuristic would first select 2, because it is the longest pattern, and set y_2 to 1. However, it can easily be checked that we have $y_1 = y_3 = 1$ and $y_2 = 0$ for the optimal solution. A similar example can also be constructed for the case $\alpha = 0, \beta = 1$.

4 Computational Results

The decomposition procedure presented in Sect. 3 is used to decompose the problems into smaller pieces, which are solved by a complete enumeration afterwards. In the following we will refer to this by Enumerate. The heuristic (cf. Algorithm 1) runs with the scoring function ϕ_y and parameters $(1, 0)$ and $(0, 1)$.

A test set of 22 real-world instances provided by DB Fernverkehr AG is used. The number of patterns is shown in Table 1. All instances cover a time horizon of 364 days. All procedures were implemented in Python. The experiments were performed on a Dell Precision Tower 3620 with 3.50 GHz and 32 GB main memory.

In Table 1 we use Enumerate as a base line, for which we give the optimal objective value and running times in *ms*. For the heuristics, we instead give the optimality gap[1] and factors w.r.t. the base line. In [4] the arising MIP (1) was solved using the academic non-commercial mixed integer programming solver SCIP [3] and the according python interface PySCIPOpt [2]. However, all pattern detection problems as described in this article have a time horizon of one year and even for the largest instance the number of arising patterns cannot exceed 52. Such instances are not challenging for a sophisticated MIP solver and, therefore, we omit the SCIP running times in Table 1. Surprisingly, the heuristics determine the optimal solution in all cases, but one. However, as demonstrated in Sect. 3, examples where they do not find an optimal solution are easy to construct. In contrast to that Enumerate guarantees optimality and is still competitive with respect to the running time. Enumerate could even be further improved by solving the independent subproblems in parallel.

5 Conclusion

In this paper we presented an enumerative decomposition method and a heuristic for solving the pattern detection problem that arises in the context of railway rotation planning. It is a pity that the real-world instances are not challenging for a generic

---

[1] Gap to optimality: |primalbound − dualbound/ min{|primalbound|, |dualbound|}| if both bounds have same sign, or infinity, if they have opposite sign.

Table 1 Detailed computational results on 22 real-world instances

Instance		Enumerate		$\phi_y(1,0)$		$\phi_y(0,1)$	
Name	m	ObjVal	Time	Gap (%)	TimeQ	Gap (%)	TimeQ
DB1	23	−201	0.57	0.00	0.60	0.00	0.59
DB2	20	−230	0.53	0.00	0.49	0.00	0.49
DB3	29	−187	0.69	0.00	0.81	0.00	0.81
DB4	26	−205	0.61	0.00	0.66	0.00	0.66
DB5	21	−311	1.27	0.00	0.18	0.00	0.17
DB6	25	−256	1.54	0.00	0.19	0.00	0.19
DB7	18	−295	0.94	0.00	0.20	0.00	0.19
DB8	9	−322	0.44	0.00	0.36	0.00	0.35
DB9	28	−116	0.72	0.00	0.91	0.00	0.91
DB10	18	−281	0.59	0.00	0.35	0.00	0.35
DB11	15	−301	0.64	0.00	0.24	0.00	0.24
DB12	11	−323	0.80	0.00	0.15	0.00	0.19
DB13	16	−312	0.52	0.00	0.27	0.00	0.26
DB14	23	−225	4.84	0.00	0.07	*1.24*	0.09
DB15	24	−117	0.86	0.00	0.46	0.00	0.47
DB16	11	−318	1.34	0.00	0.16	0.00	0.16
DB17	10	−330	0.48	0.00	0.28	0.00	0.28
DB18	21	−215	0.57	0.00	0.45	0.00	0.45
DB19	12	−329	0.77	0.00	0.17	0.00	0.17
DB20	10	−339	2.36	0.00	0.07	0.00	0.07
DB21	16	−250	0.49	0.00	0.42	0.00	0.41
DB22	20	−282	1.01	0.00	0.23	0.00	0.23

MIP solver. However, all presented methods perform quite good and it is funny that there is only one "bad" instance where the heuristic did not find an optimal solution.

Acknowledgements The work for this article has been conducted within the Research Campus Modal funded by the German Federal Ministry of Education and Research (fund number 05M14ZAM).

References

1. Borndörfer, R., Reuther, M., Schlechte, T., Waas, K., & Weider, S. (2016). Integrated optimization of rolling stock rotations for intercity railways. *Transportation Science*, *50*(3), 863–877.
2. Maher, S., Miltenberger, M., Pedroso, J. P., Rehfeldt, D., Schwarz, R., Serrano, F. (2016). PySCIPOpt: Mathematical programming in python with the SCIP optimization suite. In: Mathematical Software ICMS 2016 (Vol. 9725, pp. 301–307)

3. Maher, S.J., Fischer, T., Gally, T., Gamrath, G., Gleixner, A., Gottwald, R.L., & et al. (2017). The SCIP Optimization Suite 4.0. Tech. Rep. 17–12, ZIB, Takustr. 7, 14195 Berlin
4. Schade, S., Borndörfer, R., Breuer, M., Grimm, B., Reuther, M., Schlechte, T., et al. (2017). Pattern detection for large-scale railway timetables. In *Proceedings of the IAROR conference RailLille*

Timetable Sparsification by Rolling Stock Rotation Optimization

Ralf Borndörfer, Matthias Breuer, Boris Grimm, Markus Reuther, Stanley Schade and Thomas Schlechte

1 Facing Capacity Limitations

Planning rolling stock rotations in industrial railway applications is a long-term process that starts with a coarse plan and gains accuracy the closer the day of operation comes. This process is affected by all kinds of unusual events such as natural disasters (floods or snow), technical problems (track or fleet breakdowns), or man-made impediments (strikes). For example, during autumn 2014 and spring 2015, Germany's largest union of train drivers called for not less than nine strikes of varying intensities. In Germany it is possible that different unions for the same class of employees exist such that only a subset of such a class is actually on strike where the other part is still working. Consequently, a strike of a single union is a heavy decrease of capacity than a complete lock down of the railway system. Such events have widespread repercussions on the operation of a railway system: The timetable, the rolling stock

R. Borndörfer · B. Grimm (✉) · S. Schade
Zuse Institute Berlin, Takustr. 7, 14195 Berlin, Germany
e-mail: Grimm@zib.de

R. Borndörfer
e-mail: Borndorfer@zib.de
URL: http://www.zib.de

S. Schade
e-mail: Schade@zib.de

M. Reuther · T. Schlechte
LBW Optimization GmbH, Obwaldener Zeile 19, 12205 Berlin, Germany
e-mail: Reuther@lbw-berlin.de
URL: http://www.lbw-berlin.de

T. Schlechte
e-mail: Schlechte@lbw-berlin.de

M. Breuer
DB Fernverkehr AG, Im Galluspark 15-19, 60326 Frankfurt, Germany

rotations, the maintenance plans, and the crew schedules for the personnel in trains and maintenance facilities all have to be changed.

Finding new or revised rolling stock rotations, i.e., cyclic tours of rolling stock vehicles covering parts of the timetable, after disruptions is a well studied topic in the literature on railway optimization, see [1] for an overview.

In this paper we consider a different, more integrated approach which, to the best of our knowledge, has not been described in the literature before. The idea is to compute revised rolling stock rotations in order to "sparsify" a given undisturbed timetable. The goal is to construct rolling stock rotations that have minimum operational costs while using the limited capacities, in case of a strike the train drivers, as efficient as possible. The balance between these two objectives is controlled by an Analytic Hierarchy Process (AHP) that was developed in cooperation with our industrial partner DB Fernverkehr AG. The AHP can be seen as a key performance indicator (KPI) of the trips in the railway network, which is widely used in economy and operations research. References [2, 3] are examples for applications of KPIs in airline tail assignment. Using the train drivers as efficiently as possible directly leads to a decrease of deadhead trips and deadhead kilometers, since drivers for these kind of movements could not be used for passenger trips.

The paper is organized as follows. The next section deals with the evaluation process of the trips via the Analytic Hierarchy Process (AHP). The main contribution of this paper, the concept to sparsify the timetable according to ensure optimal rotations via mixed integer programming is part of Sect. 3. In Sect. 4 the performance of the algorithm is demonstrated via a case study for the strike period in May 2015 in Germany. Finally, we summarize the results in Sect. 5.

2 Defining Priorities by an Analytic Hierarchy Process

Before tackling the problem how to construct optimized rolling stock rotations, we deal with a subproblem of our optimization procedure. Recall that we want to choose the subset of trips to be operated from all trips of the timetable. Hence, some kind of criterion or evaluation of the trips is necessary to choose the right ones. The idea is to guide the sparsification of the timetable by a prioritization of each trip in terms of certain criteria. Afterwards optimal rolling stock rotations are constructed that cover („i.e., collect) as many trips as possible taking the trip priorities into account.

We use the *Analytic Hierarchy Process* (AHP) by [4] in order to compute trip priorities as described in [5]. The AHP involves several steps. First, criteria that describe different aspects of a trip are identified. Then weights for the importance of one criterion over every other are defined. This information is used to construct a weighting of the criteria that is used to prioritize the trips. In [5] a set of such criteria including weights for their pairwise comparison was defined as well as a sequential approach. The results were reviewed by our industrial partner DB Fernverkehr AG.

The input criteria for the AHP are defined as follows: The *passenger capacity* of the planned railway vehicle for the operation of the trip; the *line coverage* ratio of stops of the trips and stops of the line the trip belongs to; the median of the number of lines that pass each stop of the trip called *network importance*; and the median of the number of *transfer opportunities* at each stop in an time interval after the stop.

These four criteria have the big advantage that they are completely independent from other data sources. Furthermore, it is possible to deduce them directly from existing timetable and network data. The final priority of the trip is then given by $p_t \in \mathbb{Q}^+$ for all trips $t \in T$.

3 Trip Collecting Rolling Stock Rotation Optimization

In this section we consider the *Rolling Stock Rotation Problem* (RSRP) and extend a hypergraph-based integer programming formulation to our setting. We focus on the main modeling ideas and refer the reader to the paper [6] for technical details including the treatment of maintenance and capacity constraints.

We consider a cyclic planning horizon of one *standard week*. The set of timetabled passenger trips is denoted by T. Let V be a set of *nodes* representing timetabled departures and arrivals of vehicles operating passenger trips of T. Trips that could be operated with two or more vehicles have the appropriate number of arrival and departure nodes. Let further $A \subseteq V \times V$ be a set of directed standard arcs, and $H \subseteq 2^A$ a set of *hyperarcs*. Thus, a hyperarc $h \in H$ is a set of standard arcs and includes always an equal number of tail and head nodes, i.e., arrival and departure nodes. A hyperarc $h \in H$ covers $t \in T$ if each standard arc $a \in h$ represents an arc between the departure and arrival of t. Each of the standard arcs a represents a vehicle that is required to operate t. We define the set of all hyperarcs that cover $t \in T$ by $H(t) \subseteq H$. By defining hyperarcs appropriately, vehicle composition rules and regularity aspects can be directly handled by the model. Hyperarcs that contain arrival and departure nodes of different trips are used to model deadhead trips between the operation of two (or more if couplings are involved) trips. The RSRP *hypergraph* is denoted by $G = (V, A, H)$. We define sets of hyperarcs coming into and going out of $v \in V$ in the RSRP hypergraph G as $H(v)^{in} := \{h \in H \mid \exists a \in h : a = (u, v)\}$ and $H(v)^{out} := \{h \in H \mid \exists a \in h : a = (v, w)\}$, respectively. Let finally $k \in \mathbb{N}$ denote a capacity and δ_t the respective capacity consumption of a trip $t \in T$, e.g., a maximum number of trips allowed to be included in the sparsified timetable, a maximum number of aggregated kilometers, or hours of length of the included trips. This number results from the estimate how many employees might be not on strike and thus could drive a train. The *Trip Collecting Rolling Stock Rotation Problem* (TCRSRP) is to find a cost minimal set of hyperarcs $H_0 \subseteq H$ such that the capacity k is not exceeded by the trips $t \in T$ covered by a hyperarc $h \in H_0$ and $\bigcup_{h \in H_0} h \subseteq A$ is a set of *rotations*, i.e., a packing of cycles (each node is covered at most once).

Using a binary decision variable for each hyperarc and a slack variable for each trip, the TCRSRP can be stated as an integer program as follows:

$$\min \sum_{h \in H} c_h x_h + \sum_{t \in T} p_t s_t, \qquad (1)$$

$$\sum_{t \in T} \sum_{h \in H(t)} \delta_t x_h \le k, \qquad (2)$$

$$\sum_{h \in H(t)} x_h = 1 - s_t \quad \forall \ t \in H, \qquad (3)$$

$$\sum_{h \in H(v)^{\text{in}}} x_h - \sum_{h \in H(v)^{\text{out}}} x_h = 0 \quad \forall \ v \in V, \qquad (4)$$

$$x_h \in \{0,1\} \quad \forall \ h \in H, \qquad (5)$$

$$s_t \in \mathbb{Q}_+ \quad \forall \ t \in T. \qquad (6)$$

The objective function of model (1) minimizes a sum consisting of the total cost of the chosen hyperarcs and the priorities of the uncovered trips. For each trip $t \in T$ the covering constraints (3) assign one hyperarc of $H(t)$ or a slack variable to t. Inequality (2) stipulate the capacity consumption of operated trips. (4) are flow conservation constraints for each node $v \in V$ that induce a set of cycles of arcs of A. Finally, (5) and (6) state the domains of the decision variables.

The RSRP, and therefore also the TCRSRP, is $\mathcal{NP}$-hard, even if constraints (3) are trivially fulfilled, i.e., $|H(t)| = 1$ for all trips $t \in T$, see [7].

4 A Case Study at DBF: Strike Period 2015

The proposed model was implemented in our algorithmic framework ROTOR (see [6]) that is integrated in the IT environment of DB Fernverkehr AG. The implementation makes use of the commercial mixed integer programming solver Gurobi 6.5 as an internal LP solver to support a customized column generation and branch and bound procedure. The computations are stopped a after optimality is proved, a fixed number of branching nodes is reached or the LP-IP gap is below 1%.

Our implementation is tested on real-world instances provided by our industrial partner. There are four instances related to the 2014–2015 strike each representing a different fleet of ICE trains, i.e., $ice1$, $ice2$, $ice3$, and $iceT$. Each fleet has different sizes, vehicle characteristics, and different underlying networks which cover wide parts of Germany. To compare our solution approach we run ROTOR without the trip cancelling approach on instances that contain a limited number of trips of the normal DBF timetable. This list of trips was created by planners of DBF with a rough guess which drivers are on strike to offer a maximum customer friendly timetable as possible. Although this list is the result of the planning at DBF there were some changes made before really operating the trips during that period. Reasons for that are a larger number of employees on strike than expected or fine tuning of the rotations by adding additional passenger trips to reduce deadhead kilometres. Nevertheless, these rotations are very close to the operated ones and therefore a most appropriate

Table 1 ice^{DB}: Instance with ≈50% manually cancelled trips by planners of DBF

Name	$|T|$	$|H|$ (×10⁶)	$\sum \delta_t$ (km)	Dh (km)	Cost (×10ˣ)	Gap (%)	CPU (s)	$\sum p_t$
$ice1^{DB}$	379	0.9	296094	8777	1.74	0.14	70	1.82
$ice2^{DB}$	456	4.8	165906	13506	1.00	0.04	622	2.42
$ice3^{DB}$	335	1.6	186653	6279	1.42	0.11	489	2.41
$iceT^{DB}$	232	1.9	131899	9370	0.69	0.47	441	1.16

Table 2 Instances with AHP priorities and integrated trip cancelling no vehicle cost

Name	$|T|$	$|H|$ (×10⁶)	$\sum \delta_t$ (km)	Dh (km)	Cost (×10ˣ)	Gap (%)	CPU (s)	$\sum p_t$
$ice1$	700	1.4	299154	2314	1.71	0.21	519	2.15
$ice2$	973	5.2	155219	6470	0.95	1.21	5381	2.46
$ice3$	922	3.4	166250	3676	1.03	1.00	494	2.22
$iceT$	915	3.1	132798	4413	0.64	1.00	2116	1.32

candidate to compare to. Table 1 shows the main characteristics of the solution process and its outcome. The first three columns show the instance name, respectively fleet, the number of trips, and hyperarcs that were required to model all possible train movements, couplings, and deadhead trips in the hypergraph model. Columns four and five give the sum of the trip and deadhead trip distance of all used vehicles of the solution. Since the costs are confidential column *Cost* shows only a factor of the operational cost of the computed solution. The next two columns *Gap* and *CPU* present the LP-IP gap and the run time of the optimization process. The last column gives the sum of the p_t values for all trips included in the solution.

Table 2 shows the results of the optimization runs with integrated timetable sparsification. We applied a capacity limit for each instance, respectively fleet, equal to the aggregated trip length of all trips included in the corresponding instance with manually canceled trips. Hence, optimized rotations of both approaches have an amount of comparable working hours of the train drivers. Again, columns four and five give the sum of the operated trips and deadhead trips kilometres of all used vehicle of the solution. The aggregated deadhead trip length of the optimized solutions save between ≈41 and ≈74% of the aggregated deadhead km. Also the operational costs of the optimized solutions decrease which is a consequence of the decreased number of deadhead kilometres. Comparing the last columns of the two tables shows that the approach with the included trip cancelling leads to better values for the sum of the p_t values over the trips contained in the solution. Note that the solutions found in the ice· case are most likely not in the solution space of the ice·^{DB} instances, whereas solutions of the ice·^{DB} instances are potential solutions for the ice· instances. The reason for that is the preselection of trips in the ice·^{DB} case. In [5] it was shown that a preselection via the ordering computed with the AHP but without integration into the MIP approach is not sufficient.

5 Conclusion

We presented the integration of a timetable sparsification method into a mixed integer programming approach to solve the TCRSRP. The timetable sparsification is guided by a fast and from external data independent evaluation of the trips. The proposed approach leads to promising results for situations with an heavily decreased offer of passenger railway trips, like strike periods.

Acknowledgements This work has been developed within the Research Campus MODAL [8] funded by the German Ministry of Education and Research (BMBF).

References

1. Cacchiani, V., Huisman, D., Kidd, M., Kroon, L. G., Toth, P., Veelenturf, L., et al. (2014). An overview of recovery models and algorithms for real-time railway rescheduling. *Transportation Research Part B: Methodological, 63*, 15–37.
2. Burke, E. K., De Causmaecker, P., De Maere, G., Mulder, J., Paelinck, M., & Vanden Berghe, G. (2010). A multi-objective approach for robust airline scheduling. *Computers & Operations Research, 37*, 822–832.
3. Rosenberger, J. M., Johnson, E. L., & Nemhauser, G. L. (2004). A robust fleet-assignment model with hub isolation and short cycles. *Transportation Science, 38*(3), 357–368.
4. Saaty, T. L. (1990). How to make a decision: The analytic hierarchy process. *European Journal of Operational Research, 48*, 9–26.
5. Ahmadi, S., Gritzbach, S. F., Lund-Nguyen, K., McCullough-Amal, D. Rolling stock rotation optimization in days of strike: An automated approach for creating an alternative timetable. https://opus4.kobv.de/opus4-zib/files/5642/ZR_15-52.pdf.
6. Reuther, M., Borndörfer, R., Schlechte, T., Weider, S. (2013). Integrated optimization of rolling stock rotations for intercity railways. In *Proceedings of the 5th International Seminar on Railway Operations Modelling and Analysis, Copenhagen, Denmark*.
7. Heismann, O. (2014). The hypergraph assignment problem. PhD thesis, Technische Universität Berlin
8. Research Campus MODAL (Mathematical Optimization and Data Analysis Laboratories) http://www.forschungscampus-modal.de/

Traffic Management Heuristics for Bidirectional Segments on Double-Track Railway Lines

Norman Weik, Stephan Zieger and Nils Nießen

1 Introduction

Temporary closures of one track on double track railway lines pose a severe, yet relatively frequent event on Europe's dense and heavily loaded railway networks. They may arise from train or infrastructure malfunctions or maintenance requirements. Traffic management on the remaining bidirectional track, including determining the admissible traffic load and the train sequence through the bottleneck is a demanding task for dispatchers.

The present work aims to investigate heuristic traffic management strategies for this situation based on a polling system perspective. A polling system corresponds to a queuing system where a single server serves multiple queues. The order in which the server visits (polls) the different queues is given by the *polling table*, whereas the time spent at a given queue is determined by the *polling policy*. When changing from one queue to another a switchover time, where the server does not perform service may be inserted. In addition, the server may need to set up upon arrival at a queue.

In transportation, polling systems have been widely used to analyze waiting times at traffic signals [3], where setup and switchover times are small. The railway case, however, is more closely related to construction zones or underground transport systems, for which approximations of the mean waiting times have been derived in [6] assuming fix service times and switching intervals.

We subsequently take a more general perspective based on polling models with arbitrary switchover and service times and k_i-limited service policy. This class of models does not satisfy the branching property, rendering the solution difficult. Our

---

N. Weik (✉) · S. Zieger · N. Nießen
Institute of Transport Science, RWTH Aachen University,
52062 Aachen, Germany
e-mail: weik@via.rwth-aachen.de
URL: http://www.via.rwth-aachen.de

work is based on previous work by Borst et al. [1], where the authors discuss approximation approaches for finding the set of control parameters $k_1, \ldots, k_N$ that minimizes the expected waiting time. More recently, van Vuuren et al. [7] have presented an iterative approach to determine the queue length distributions for k_i-limited policy. The N-queue polling system is modeled by single queues with server vacations, where vacation times are matched to conditional interarrival times. Vacation, service and arrival intervals are approximated by phase-type distributions and the resulting QBD model is solved using psa [7].

The present work bridges the gap between theory and railway applications. While queueing models have found widespread applications in railway capacity analysis, the use of polling models is new in this context. The contribution of the present paper includes three aspects: First, we compare analytical results based on the Fuhrmann-Wang approximation [4] with simulation results for realistic train programs. In a second step we discuss how priorities, which are typical for railways, affect the results obtained in the previous case. Finally, we analyze the performance of the heuristic, where line orientation is changed according to the optimal control parameters in the polling model.

2 Model

2.1 Polling Model

We subsequently model the bidirectional line segment as a polling system with a single server and two queues. Service times correspond to headway times between successive trains. The two queues are served according to a k_i-limited policy, i.e. the server serves at most k_i trains at a queue and switches to the other queue if the current queue runs empty. The vastly different train separation times between trains of same and opposing directions are modeled using switchover times. Setup times upon arrival at a new queue are not considered. Arrivals are assumed Markovian – modeling the fact that information is scarce in case of heavily perturbed operations. For service and switchover times general independent distributions are admitted.

Notation: For queue i, we denote traffic load by ρ_i, the means of service and switchover times by β_i and s_i and arrival rates by λ_i. The second moments of service and switchover times are denoted by $\beta_i^{(2)}$ and $s_i^{(2)}$, respectively. The total switchover time and traffic load in the system is given by $s = \sum s_i$ and $\rho = \sum \rho_i$.

Stability: For k_i-limited polling models a necessary and sufficient stability criterion reads $\rho + s \cdot \max_i \frac{\rho_i}{\beta_i k_i} < 1$ [3]. This already enforces a lower bound on the control parameter k_i: $k_i \geq s \cdot \frac{\rho_i}{\beta_i(1-\rho)}$ $\forall i$.

2.2 Waiting Times

In [1], Borst et al. compare various approximation formulae for the expected waiting times in k_i-limited polling models. As k-limited systems do not satisfy the branching theorem, no exact formulae for the waiting times are known. We subsequently adopt Approximation (4) in [1], which has been shown to perform best [1]. It has been derived by Fuhrmann and Wang [4] from the pseudo-conservation law for the work as an upper bound on waiting times in k-limited polling systems. While it does not yield a closed formula for the optimal set k, the latter can easily be obtained numerically for a small number of queues.

3 Problem Definition

The traffic management goals are to minimize overall delays while limiting the maximal waiting time for passengers. The quality of service for passengers is ensured by limiting the interarrival time to a given queue, s.t. the time interval between two succeeding trains in the same direction is bounded. In the following we discuss two problems arising in this context.

3.1 Problem 1

Determine the set of control parameters k_i that minimizes the expectation of the overall waiting times under the constraint that the interarrival time in case of overload is bounded by some $C > 0$ on average:

$$k_i \cdot \beta_i + s \leq C. \quad (1)$$

This problem is a variant of the constrained optimization problem in [1].

3.2 Problem 2

Determine the fairest set of control parameters k_i, i.e. choose k such that the maximum expected interarrival time $\max_i E[I_i]$ is minimized, subject to waiting time constraints $E[W_i] < K$ for all queues i and some $K > 0$.

3.3 Input Data

The input data is derived from typical train operating concepts on mixed service railway lines in Europe. For the bidirectional segment different lengths between 3

Table 1 Excerpt of results for asymmetric load in Scenario 2

	$\rho = 0.58, s = 1.19,$ $\beta = (2.5, 3.69),$ $s^{(2)} = 1.54,$ $\beta^{(2)} = (6.25, 13.78)$	$\rho = 0.68, s = 3.78,$ $\beta = (3.01, 6.79),$ $s^{(2)} = 16.29,$ $\beta^{(2)} = (10.16, 48.20)$	$\rho = 0.73, s = 5.99,$ $\beta = (3.46, 9.45),$ $s^{(2)} = 40.41,$ $\beta^{(2)} = (14.90, 94.93)$
k_{simu}	(11,7)	(8,3)	(6,2)
k_{approx}	(11,7)	(8,3)	(6,2)

and 15 km and trains with velocities between 80 and 230 km/h are considered, which gives rise to different service and switchover times in the model.

For the service times of trains traveling in the same direction two profoundly different scenarios can be distinguished. In Scenario 1, the remaining track is equipped with signaling, and hence multiple block segments, in both directions. In this case, the polling system can be assumed to be symmetric as minimum headways in both direction are approximately equivalent. Scenario 2 corresponds to vastly different service times at queue 1 and 2, which occurs if the remaining track is not equipped with signaling in the irregular direction, such that it can only be used one train at a time in this direction.

4 Results

We subsequently present numerical results for the symmetric case (Scenario 1) and the asymmetric case (Scenario 2). In Sect. 4.1 the results based on the Fuhrmann-Wang Approximation are compared to simulation results of polling-based train operations. If both queues run empty in the simulation the server continues at the queue with the next possible service.

In Sect. 4.2 the effects of train priorities are discussed and in Sect. 4.3 the performance of the polling heuristic with optimal k in Problem 1 is compared to the solution for the train scheduling problem.

4.1 Fuhrmann-Wang Approximation for Waiting Times

Problem 1: The solution of Problem 1 in the symmetric scenario can be given analytically. As both the simulated and the approximated waiting time are decreasing in the control parameter k [1], the solution corresponds to $k = \left\lfloor \frac{C-s}{\beta} \right\rfloor$.

For Scenario 2 we consider input parameters resulting from different lengths of the single-track segment as well as different train velocities. An excerpt of the results with $C = 30$ min is presented in Table 1.

Table 2 Optimal set of parameters in symmetric and asymmetric scenario. Comparison of simulation and approximate results. In Scenario 2, the scheduled traffic load ρ increases to ρ_{act} due to the disruption-caused change of service times

		ρ	0.30	0.35	0.40	0.45	0.50	0.55	0.60	0.65	0.70	0.75
Scenario 1	Small var.	k_{simu}	1	1	2	2	2	3	3	4	5	7
		k_{approx}	1	2	2	2	3	3	4	5	6	9
	High var.	k_{simu}	1	2	2	2	3	3	4	5	6	9
		k_{approx}	2	2	2	3	3	4	5	7	9	15
Scenario 2	Small var.	ρ_{act}	0.45	0.52	0.58	0.64	0.70	0.76	0.81	Infeasible		
		k_{simu}	(1,1)	(1,1)	(1,1)	(1,2)	(2,2)	(3,3)	(4,5)			
		k_{approx}	(1,1)	(1,1)	(1,1)	(2,2)	(2,2)	(3,4)	(5,6)			
	High var.	ρ_{act}	0.48	0.55	0.61	0.68	0.74	0.79				
		k_{simu}	(1,1)	(1,1)	(1,2)	(2,2)	(3,4)	(4,6)				
		k_{approx}	(1,1)	(1,2)	(2,2)	(2,3)	(4,4)	(6,8)				

Our results suggest that for the asymmetric scenario with different β_i the optimal strategy corresponds to taking the highest admissible k_i that satisfies the constraints. In [1], it was shown for the unconstrained problem that $\sum_j \rho_j E[W_j]$ is non-increasing in k_i. The experiments we conducted seem to indicate that this strategy is also close to minimizing the constrained waiting times.

Problem 2: For Problem 2, the results for $K = 30$ min, which is a typical train frequency in railway operations, are depicted in Table 2. For both scenarios results with a homogeneous and a heterogeneous traffic mix are shown. The first one is based on a segment of 8 km and train speed differences of max. 60 km/h, the second one to a 15 km segment and very distinct speed differences of up to 130 km/h. The variation coefficients of service and switchover times are ($v_B = 0.2$, $v_S = 0.14$) and ($v_B = 0.53$, $v_S = 0.25$).

It seems that while the results of the Fuhrmann-Wang Approximation [4] are reasonably good for intermediate and low traffic load the approximated results deviate for large ρ. This suggests that the bound obtained from the approximation might not be as tight in the regime of high load and small switchover and service times variation. For this regime, only limited results have been presented in [1] as the analysis is numerically costly due to the large state space.

4.2 On the Effects of Train Priorities on the Control Strategy

In practice, railway dispatchers often employ a priority strategy, where faster passenger trains are given precedence over slower trains. It is unclear, how this affects the waiting times of trains. It is assumed that

- amongst waiting trains high-priority train are given preference over low-priority trains of the same direction.
- high-priority trains waiting at a queue which currently does not receive service do not enforce early switchovers from the other queue.

Table 3 Comparison avg. waiting times polling-heuristic and optimization approach

ρ	0.53	0.56	0.58	0.61	0.64
$T_{W,sched}$	7.81	8.59	9.40	10.13	10.84
$T_{W,heuristic}$	10.22	9.24	11.07	13.38	12.86
$T_{W,heuristic\ (lt)}$	11.99	12.51	13.06	13.78	15.13

Our simulation results for this type of system show that waiting times tend to decrease if priorities are considered. This can probably be explained by the grouping of trains with similar velocities leading to globally smaller service times. The effect is more pronounced the higher the system load and variation of service times. For typical mixed-service operations with 20% priority trains, $\rho \leq 0.6$ and $E[W_i] \leq 30$ min the size of the effect is found to be in the range of $2 - 9\%$.

4.3 Comparison Polling Heuristic and Scheduling Solution

The performance of the polling heuristic with optimal k in Problem 1 is analyzed in a case study by comparing the results to the optimal scheduling solution for a line segment. The time frame includes 17 trains in each direction and arrival intervals are rescaled to account for different ρ. Average service and switchover times are 3 min and 8 min, respectively, and k_i is taken to be 5 for all queues. For the optimization Castillo et al. [2] has been adopted and solved with Gurobi [5]. The objective is to minimize the sum of differences of actual starting time and release time of trains. Train priorities are modeled by increasing the weight in the objective function by 10%. In the polling simulation both the performance for the schedule time frame as well as the long turn average (lt) obtained by periodically repeating the schedule are calculated.

For the same schedule structure and different ρ, Table 3 shows that the polling heuristic performs roughly 10–30 % worse than the optimal solution. However, for the time frame of ca. 3 h the heuristic results exhibit huge variation as they depend on switching decisions. Monotonicity of waiting times is restored in longer time frames as $T_{W,heuristic\ (lt)}$ shows. Still, even for the relatively short schedule, the optimization for $\rho > 0.6$ takes more than 12 h, whereas the polling simulation including long turn average can be calculated within seconds.

5 Conclusion

We have adapted k-limited polling models for analysis and control of single-track line segments of railway lines. The polling-based heuristic provides a fast and easy way to determine maximum admissible traffic load and efficient traffic management.

For the case study considered, the results imply that the quality of the solution is roughly 10–30% above the optimal solution, hence it can only be considered a rough estimate for the sequencing problem if information is scarce.

Acknowledgements This work was supported by DFG grant 283085490.

References

1. Borst, S. C., Boxma, O. J., & Levy, H. (1995). The use of service limits for efficient operation of multistation single-medium communication systems. *IEEE/ACM Transactions on Networking*, *3*(5), 602–612.
2. Castillo, E., et al. (2009). Timetabling optimization of a single railway track line with sensitivity analysis. *TOP*, *17*(2), 256–287.
3. Frigui, I. (1997). Analysis of a time-limited polling system with Markovian Arrival Process and phase type service. Ph.D. thesis, University of Manitoba, Winnipeg.
4. Fuhrmann, S. W., & Wang, Y. T. (1988). Analysis of cyclic service systems with limited service: bounds and approximations. *Performance Evaluation*, *9*(1), 35–54.
5. Optimization, Gurobi. Inc. (2014). Gurobi optimizer reference manual.
6. Van der Heijden, M., van Harten, A., & Ebben, M. (2001). Waiting times at periodically switched one-way traffic lanes. *Probability in the Engineering and Informational Sciences*, *15*(4), 495–517.
7. Van Vuuren, M., & Winands, E. M. M. (2007). Iterative approximation of k-limited polling systems. *Queueing Systems*, *55*(3), 161–178.

Traffic Speed Prediction with Neural Networks

Umut Can Çakmak, Mehmet Serkan Apaydın and Bülent Çatay

1 Introduction

Road transport has various hazardous and threatening impacts on the environment and human life such as resource consumption, pollution, emission, congestion, and noise. Growing concerns in modern societies about these issues and the quality of life in cities call attention to new methods and approaches in traffic management, transportation planning, and route optimization for both commercial and individual drivers. Many of these methods depend on the estimation of travel time, traffic speed and volume. Recent advancements in Global Positioning Systems (GPS), Geographical Information Systems (GIS), image processing, and sensor technologies enable the real-time collection of these massive data, which can be effectively used to improve the accuracy of the prediction methods.

Early studies mainly collected their data from highway sensors, and GPS data was not common until 2011 [5]. The acquired data usually consists of speed, congestion classification, journey time, and volume. The research on the analysis of the collected data can be categorized as discrete and continuous. Discrete analyses include binary or multiclass classification methods while continuous analyses mainly employ function approximation and time series analysis.

U. C. Çakmak (✉) · B. Çatay
Faculty of Engineering and Natural Sciences, Sabanci University, Istanbul, Turkey
e-mail: ucakmak@sabanciuniv.edu

M. S. Apaydın
Department of Computer Science, Istanbul Şehir University, Istanbul, Turkey

U. C. Çakmak · B. Çatay
Smart Mobility and Logistics Lab, Sabanci University, Istanbul, Turkey

© Springer International Publishing AG, part of Springer Nature 2018
N. Kliewer et al. (eds.), *Operations Research Proceedings 2017*, Operations Research Proceedings, https://doi.org/10.1007/978-3-319-89920-6_98

It has been observed that the accuracy of the predictions improve as the number of segments in a route increases [6]. Reference [2] reports around 90% classification accuracy for short-term predictions (up to 5 min) on the highway; however, the results deteriorate in the urban setting. Reference [7] achieves an average of mean absolute deviations (MAD) value of 6.60 km/h for 1-step ahead and 12.47 km/h for 5-step ahead prediction over 20 segments.

In this study, we employ a feedforward neural network (FFNN) to perform a continuous prediction. We are mainly motivated by the work of [7] on irregular data. Our aim is to perform accurate predictions over a relatively longer horizon instead of a fixed point in the future. The remainder of the paper is organized as follows: Sect. 2 introduces the methodology including data collection and cleaning, prediction methods and machine learning concepts. Section 3 presents the experimental setup while Sect. 4 reports and discusses the results. Finally, Sect. 5 concludes with suggestions for future research.

2 Methodology

2.1 Data Collection and Cleaning

The historical speed data is obtained from Başarsoft Information Technologies Inc. It includes floating car speeds collected on Istanbul road network with 1-min time intervals over a 5-month horizon from Oct. 2016 to Feb. 2017.

Since the raw data needed cleaning, we firstly linearly interpolated the missing data and reduced the high speed values to the legal speed limit. Secondly, we used a systematic interpolation technique to smooth out the erratic jumps in observations. For instance, the speed on a particular road segment may change by up to 80 km/h from one minute to the next, which is unrealistic and may be due to data collected from different vehicles en-route or from different road segments nearby. Briefly, our method smoothes the erratic observations by removing the speeds that vary by more than z standard deviations in a given segment, where z is gradually reduced until speed variations are realistic.

2.2 Prediction Methods

In this section, we briefly describe different time-series forecasting methods, where s_t represents the observed speed at time t while f_{t+k} represents the prediction of the speed at time $t + k$.

Naïve Naïve method is the simplest forecasting technique where the prediction is equal to the recently observed speed. This method may perform well for short-term predictions.

$$f_{t+1} = s_t \tag{1}$$

Weighted Moving Average (WMA) The method makes a prediction by taking the weighted moving average of the last n observations as follows:

$$f_{t+1} = w_t s_t + w_{t-1} s_{t-1} + w_{t-2} s_{t-2} + \cdots + w_{t-(n-1)} s_{t-(n-1)} \tag{2}$$

where w_i is the weight associated with the observation at time i with $\sum_{t-(n-1)}^{t} w_i = 1$ and $0 \le w_i \le 1$. The benefit of weighted moving average is that it can be tuned to give the most relevant past data more importance [4].

Simple Exponential Smoothing (SES) This method is similar to the weighted moving average where a weight is associated with the most recent observation and another weight is given to the last forecast. This recursive relationship makes the process take into account the whole set of past observations. The formulation is as follows:

$$f_{t+1} = \alpha s_t + (1-\alpha) f_t \tag{3}$$

where α is the *smoothing constant* and $0 \le \alpha \le 1$ [3].

Triple (Winters) Exponential Smoothing This method is developed to handle trend and seasonality simultaneously and it can also be used when the data shows seasonality but no trend. We use this technique because we observe *microseasons* over the course of five months such as the rush hours of weekdays.

$$L_t = \alpha \frac{s_t}{S_{t-M}} + (1-\alpha)(L_{t-1} + T_{t-1}) \tag{4}$$

$$T_t = \beta(L_t - L_{t-1}) + (1-\beta) T_{t-1} \tag{5}$$

$$S_t = \gamma \frac{s_t}{L_t} + (1-\gamma) S_{t-M} \tag{6}$$

$$f_{t+k} = (L_t + kT_t) S_{t+k-M} \tag{7}$$

where L_i is known as *Level* or *Smoothed Observation* at time i and T_i is known as the *Trend* or *Trend Factor* at time i and β is the *trend smoothing constant* which is similar to α and $0 \le \beta \le 1$ [3]. Here, S_i is the *Seasonal Index* at time i and γ is the *seasonality smoothing constant* and $0 \le \gamma \le 1$ [3]. M is the number of seasons. In our case, the seasons consist of 1-min. time intervals and we have 1440 seasons in a day throughout the entire horizon.

2.3 Machine Learning

Feedforward Neural Networks (FFNN)/Multilayer Perceptrons (MLP) A simple, single layer perceptron has an output unit y_i and input units x_i along with an extra bias unit, a set of weights that connect the inputs and the bias unit to the output [1]. A bias unit is an input unit of $x_0 = 1$. It acts, as can be seen from (8), as the constant in a linear equation.

$$y = \sum_{j=1}^{d} w_j x_j + w_0 x_0 \tag{8}$$

where d is the number of input neurons excluding the bias unit. A multilayer perceptron has the advantage of handling nonlinear functions [1]. The multilayer perceptrons have at least one hidden layer in addition to input and output layers. To train these networks, input and target data are required. In this work, target data is *Cleaned* data, and input data is the forecasts obtained by the methods in Sect. 2.2. Training starts with an initial set of weights and progresses forward over the system to yield an output value. Our network is trained with forward and backpropagation.

3 Experimental Setup

3.1 Route Selection

We performed our analysis on two different routes in Istanbul (see Fig. 1). The first is an urban route with many intersections that covers 324 segments over a distance of 21.49 km, with mean and median segment lengths of 0.07 and 0.05 km, respectively. The second route is a freeway starting from the European side of the city and crossing the Bosphorus Strait through the FSM Bridge. It covers 63 segments over a distance of 22.75 km, with mean and median segment lengths of 0.36 and 0.25 km, respectively.

3.2 Single Segment Approach (Multi-step Ahead Forecast) (SS-M Network)

Our network (see Fig. 2) has 30 input neurons for each prediction method with 50 hidden neurons in the single hidden layer and 30 output neurons. Each neuron in the input and output layers inputs and outputs a k-step ahead prediction, respectively.

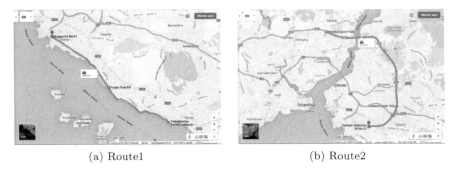

(a) Route1 (b) Route2

Fig. 1 Routes examined

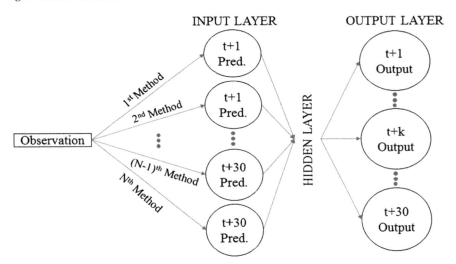

Fig. 2 Single segment approach (multi-step ahead forecast) with N predictive methods (for this work: Naïve, weighted moving average, simple exp. smoothing and winters) (SS-M network)

4 Computational Results

The experiments were carried on a workstation with a 64-bit Windows 7 Professional operating system, a memory of 128 GB, and a 40-core Intel Xeon CPU E5-2640 v4 @ 2.40 GHz processor. We have implemented the FFNN using Keras with Theano and Python 2.7.

We tested NMS (Naïve-Weighted Moving Average-Simple Exponential Smoothing) and NMSW (NMS-Winters) combinations through 30 epochs and a batch size of 1000 with adaptive moment estimation (Adam) optimizer. To prevent overfitting, we also employed a 10% Dropout. We used the following parameters for our prediction methods that are input to the FFNN: Weighted moving average method takes a

Table 1 30-min Test Results by NMS and NMSW predictive methods on SS-M (Proposed Single Segment Multi-step Ahead Forecast Network) and individual predictive methods (Naïve, Weighted Moving Average (WMA), Simple Exponential Smoothing (SES), Winters)

Route	Avg. Seg. Len. (km)	MAD (km/h)						Avg. Train Time (s)	
		Naive	WMA	SES	Winters	NMS	NMSW	NMS	NMSW
1	0.059	0.583	0.595	0.594	0.596	0.470	0.480	738	669
2	0.285	7.867	7.901	7.784	7.749	6.467	6.431	698	647

3-step horizon with three weights: 0.25, 0.50, and 0.25, simple exponential smoothing method takes $\alpha = 0.50$, and Winters method takes $\alpha = 0.45$, $\gamma = 0.20$; thus, only considers seasonality without any trend. In the literature, it is common to assign the parameters intuitively. The first 4.5 months of the dataset were allocated to training while the remaining 15 days were used for testing.

The experimental test results for 30-min prediction horizon are reported in Table 1. Route 1 results are coming from 16 segments spanning 0.95 km while Route 2 results are of 16 segments spanning 4.55 km. In line with [6], we observe that the accuracy of the predictions enhance when the route is split into more segments. This is evident in the fact that the segments of Route 1 return lower error values than those of Route 2. It seems surprising that there is not a significant advantage of employing NMSW over NMS; however, it is worth noting that 30-min-ahead is a relatively short horizon to observe the real effect of seasonality in the prediction.

5 Conclusion

Here we employed FFNN to predict the traffic speed over a 30-min horizon using historical speed data collected in 1-min time intervals. Even though our method requires significant computation effort, its performance is comparable to that of [7], overperforming it on longer term predictions. While their results achieve 12.47 km/h MAD for 5-step ahead prediction over 20 segments, our results for the 16 segments return an average of 0.47–6.43 km/h MAD. To improve our current methods, employing Winters prediction over a longer horizon that reflects seasonal characteristics better than 30-min horizons also seems promising. As further future work, we plan to use recurrent neural networks and also take seasonality into consideration to further improve the prediction accuracy. Random forest regression is also a simple method we can use to combine the individual prediction methods.

Acknowledgements We would like to thank Başarsoft Information Technologies Inc. for providing historical floating car data.

References

1. Alpaydın, E. (2010). *Introduction to Machine Learning*. Cambridge, MA: MIT Press.
2. Khan, R., Landfeldt, B., & Dhamdhere, A. (2012). Predicting travel times in dense and highly varying road traffic networks using STARIMA models. Technical Report.
3. NIST/SEMATECH e-Handbook of Statistical Methods. https://www.itl.nist.gov/div898/handbook/.
4. Stevenson, W. J. (2012). *Operations Management: Theory and Practice*. New York, NY: McGraw-Hill/Irwin.
5. Vlahogianni, E. I., Karlaftis, M. G., & Golias, J. C. (2014). Short-term traffic forecasting: Where we are and where we're going. *Transportation Research Part C: Emerging Technologies, 43*, 3–19. https://doi.org/10.1016/j.trc.2014.01.005.
6. Wang, J., Mao, Y., Li, J., Xiong, Z., & Wang, W. (2015). Predictability of road traffic and congestion in urban areas. *Plos One, 10*(4). https://doi.org/10.1371/journal.pone.0121825
7. Ye, Q., Szeto, W. Y., & Wong, S. C. (2012). Short-term traffic speed forecasting based on data recorded at irregular intervals. *IEEE Transactions on Intelligent Transportation Systems, 13*(4), 1727–1737. https://doi.org/10.1109/tits.2012.2203122.

Part XXII
Business Track

Delivering on Delivery: Optimisation and the Future of Vehicle Routing

Christina Burt, Paul Hart, Desislava Petrova and Adam West

1 Introduction

The growth of online shopping has inevitably led to an increase in the number of products being delivered directly to customers homes or workplaces, presenting retailers with both a challenge to meet customers expectations, but also an opportunity to differentiate their service from their competitors. However, delivering items to individual addresses compared to bulk deliveries to a store is very expensive, and while customers generally accept that free delivery is increasingly unlikely, their view of a reasonable charge rarely covers the retailers costs. Added to this customers expect to be able to choose when their delivery will arrive, and also to be kept informed on its progress. Retailers are increasingly looking at ways to offer this level of service to its customers, but also to minimise their cost in the last mile element of delivery.

Satalia has developed Satalia Delivery - a SaaS solution that uses optimisation and machine learning to calculate the optimal schedule, and routes of any given fleet of vehicles. The solution allows organisations to offer a delivery slot of any duration to their customer at the online checkout, and deploys the latest algorithmic technology to calculate the most efficient schedule, and routes (i.e. which orders should be delivered by which vehicles and in what sequence).

Implementation

The first implementation of this solution was for a large UK furniture retailer who deliver their products directly to customers using their own fleet of 300+ vehicles based at multiple depots. Previously, the retailer could only inform customers of the day their delivery was expected, but could not give customers a specific time of day, or

---

C. Burt · P. Hart · D. Petrova · A. West (✉)
Satalia, London, UK
e-mail: adam@satalia.com

P. Hart
e-mail: paul@satalia.com
URL: http://www.satalia.com

even which part of the day their delivery may arrive. Customers were not happy with this offering as it would require them to wait at home all day for their delivery (often having to take time off from work) with no indication of when it would arrive. Now, Satalia delivery allows retailers to offer a choice of 3-h time windows and while introducing this type of constraint would normally mean a decrease in efficiency, the retailer was actually able to increase the number of orders delivered without increasing their fleet size. In addition, because the fleets capacity and commitment is now known at the time of offering a delivery slot, there is no danger of overcommitting and promising deliveries that cannot be fulfilled.

This is an interesting problem to solve as the heart of the challenge is the infamous 'travelling salesman problem (TSP), an NP-Hard problem that has long challenged academics. As with so many other areas, while advances are made in academia, the commercial world often continues to use ageing systems with algorithms significantly less efficient than have been created recently. As a University College London (UCL) spin-out, Satalia are a company born out of academia and are passionate about bridging the gap between the academic and corporate worlds. It was this approach that made Satalia Delivery an interesting and ultimately successful project.

Why the Problem is Hard

Vehicle routing is a combinatorially layered problem, where each layer may be a cause for a potential inefficiency. Multiple vans have to deliver to multiple destinations. Each van has limited capacity, and may only be suitable for certain routes. Each route has its own restrictions, and may be limited by live events such as unforeseen traffic or accidents. Drivers have certain shifts, and certain preferences as to when they take breaks. Expectations are rising, and customers are increasingly demanding delivery time windows — meaning the accuracy of a schedule has to be far greater than ever before. A vehicle routing system must account for all of these constraints, and produce a schedule that selects the right van, on the right route, for the right delivery, at the right time. To maximise capacity, and to offer customers a reliable, flexible customer experience, schedules must be optimised, and re-optimised in real time.

Approach

Satalia are keen to work closely with clients to fully understand the intricacies of their problem and this was evident when implementing the solution for the furniture retailer. This ensures a solution can be crafted that not only solves the problem, but has a positive impact to all involved. This meant spending time with admin teams, drivers and management to fully understand the current processes and their frustrations with their existing systems. It quickly became apparent that there were many tedious jobs required every day such as re-keying information from one ageing disconnected system to another. Operators were also expected to create and update drivers schedules without the aid of information technology. Unsurprisingly, given the complexity of the task, results were far from optimal.

In this paper we will share details on how this solution was developed, and further discuss the positive impact of our solution on the client. We will also describe the democratisation of efficient vehicle routing, and provide insights into how this will reduce environmental impact, and improve the customer experience, without increasing the costs of those who adopt it.

2 Case Study

One of the UKs leading furniture manufacturer and retailers reviewed their systems to ensure they can continue to lead the market in offering excellent service for all interactions with customers. The retailer had grown rapidly and moved from a warehouse and branch delivery approach to an efficient CDC fulfilment model matching their new multi-channel operation.

The company had invested in the process of customer ordering and seen increases in customer satisfaction, but they recognised the need to match new expectations for home delivery. While customers were very complimentary about the delivery drivers, they were frustrated at not being given a time slot when they booked their delivery; after all who wants to wait in all day for a delivery? But for the retailer this wasnt simply a delivery but a premium installation white gloves service to the exact room required. This meant the time needed at each delivery was going to differ greatly depending on many factors, particularly the number of staircases to navigate, the proximity of parking spaces and the number of items to deliver. This made guaranteeing timeslots, particularly days in advance, impossible using their existing systems.

While satisfying the customer was paramount, it couldn't be achieved by compromising efficiency or increasing costs. Efficiency meant using as few vans as possible to reduce their need for expensive contractor vehicles. It also meant having routes that minimised fuel spend, as well as conforming to other constraints such as the vehicle weight and volumetric limits, shift patterns, loading times and driver breaks.

The retailer quickly recognised Satalia had the expertise to provide a cutting edge solution that offered timeslots without compromising on efficiency. In fact it became clear that this solution could provide improvements to many other processes, such as inter-branch transfers and service visits to maximise the efficiency of their fleet. Satalia and the retailer worked closely together to craft a system that was simple to use for operators, yet solved an incredibly complex problem by utilising the latest algorithmic technology.

It is important to Satalia that they improve the experience of the people that use their products and this reflects in how they work with clients. Satalia spent time with the retailer to customise the UI — presenting only relevant information to users — ensuring the best possible user experience (see Fig. 1 for an illustration).

The retailer now has a system that is intuitive to use, maximises their fleets efficiency and can handle all their transport scheduling. Best of all, customers get a delivery time window that is reliable and available at the time of booking. We have feedback that the system is a pleasure to use, empowers operators and is simple to maintain given Satalia are hosting solution in the Cloud.

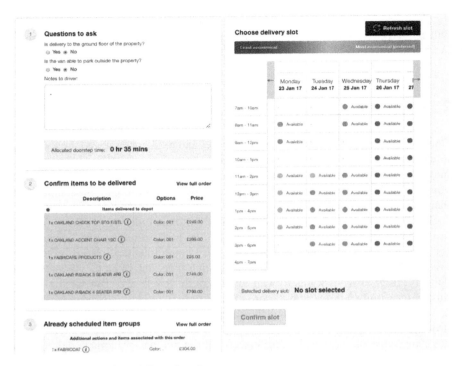

Fig. 1 A screenshot of our delivery interface

3 Our Approach

While the core solution for Satalia Delivery is to provide a discreet solution hosted as a Software as a Service, we recognise that often clients need more support to achieve the results they want. Satalia also provides consultancy, particularly in optimisation and data science, as well as creating products. Therefore clients have the option of a light-touch engagement whereby they submit data to our apis, and our solution will return information, or they can partner with us, allowing us to understand their business processes and thus recommend improvements. We ensure that we account for any limitations, either with budget, regulations or resources so that we never recommend actions that are impossible for the client to achieve.

An example of how we engaged with the furniture retailer was to establish their measures of performance. The leading measure was the average number of orders delivered per trip. Whilst we can use industry data to establish travel times, it was less clear how long a delivery would actually take: we call this the service time though it is sometimes called the dwell time or time at door, depending on the industry. Since furniture can be heavy and cumbersome, the service time for each order could be considerable, in fact the total service time was similar to the total driving time, therefore it is critical for the schedule to be as accurate as possible with this measure.

Delivering on Delivery: Optimisation and the Future of Vehicle Routing

Our Operations Research approach to solving the core problem involved leveraging exact, deterministic approaches (for which there exist efficient algorithms) into a dynamic, online optimisation and repair heuristic. Our heuristic was initialised with a full solution to the routing problem as it was known at a certain date. From this time onwards, new customers can be inserted into the solution and the order of customers is refined. Both of these steps utilise mixed-integer programming technology. The full details of this approach our outlined in [1, 2]. Our computational results show we can compute both the insertion and repair steps in milliseconds, making the approach suitable for online optimisation.

A Satalia data scientist used the data collected from vehicle telematics devices along with data on product details for each order. Those were matched in order to engineer an additional feature containing information about the number of items delivered per order. The final data set was split into training and testing set to which three different machine learning models were applied. The models were then trained and their accuracy was compared using the remaining data for testing. The most accurate and time efficient model was then chosen for the time at door predictions. The results were then implemented into the overall system to enable better optimisation of the delivery schedules. This logic could then be built into the system and represents an example of the customisations available for Satalia Delivery. Additionally custom business intelligence reporting dashboards have been crafted for the client to empower employees to make informed operational decisions, such as that illustrated in Fig. 2.

Companies, understandably need to be convinced that a solution will provide them with benefits before they are willing to commit to an investment, and for that reason, they often want to see a demonstration of the solution. This typically involves significant consultation and data collection between the client and vendor, which is both costly and time consuming for both parties. To prove the capability of Satalia Delivery to future clients, Satalia has built a free to use, single web page

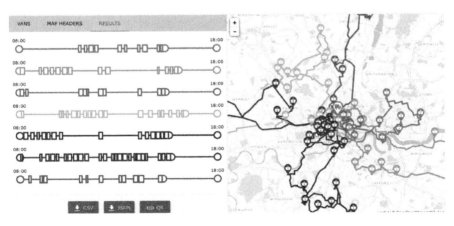

Fig. 2 We provide automated Gantt charts for our client to analyse

demonstration, so that anyone can create schedules using their own data or example data. It walks the users through setting up vehicles, loading orders and seeing the optimisation happening live, before producing a usable schedule. This solution is usable without any licence or fees, and is suitable for smaller clients (as we put a limit of 10 vehicles) to use without further investment but is primarily intended to show the capability of the full solution. A mobile app is also available that displays the optimised schedules and corresponding routes to the drivers of the vehicles.

References

1. Hungerländer, P., Maier, K., Pöcher, J., Rendl, A., & Truden, C. Solving an online capacitated vehicle routing problem with structured time windows. http://www.optimization-online.org/DB_FILE/2016/12/5764.pdf
2. Hungerländer, P., Rendl, A., & Truden, C. On the slot optimization problem in on-line vehicle routing. http://www.optimization-online.org/DB_FILE/2017/04/5962.pdf

Improving on Time Performance at Deutsche Bahn

Christoph Klingenberg

1 Plan Actual Comparisons

For every scheduled operation, be it airline, railway or bus network one can analyze the distribution of actual driving times for a given segment (from A to B) and a given period—say one year—and compare this distribution with the scheduled driving time. All events to the right of the scheduled time are delays. The question arises for the next planning period: where should the new scheduled time from A to B be located in the distribution?

For each segment and train number the distribution of actual driving times has 2 characteristics:

- Difference between average of actual driving times (M) and the planned driving time (P): here the plan must be adjusted to $P = M + x$, where x will be related to the cost of delay, see below.
- Standard deviation σ of the distribution of actual driving times: this can be reduced through improved process stability and higher reliability of the asset base. This improvement constitutes "one half" of the punctuality improvement, but we will focus in this presentation only on the planning aspects.

Usually operators apply static rules of how the planned times are calculated (based on physical parameters of the rolling stock and the infrastructure [7], but these rules have 3 major shortcomings:

- They are not linked to the on-time performance goals (since they only represent one half of the improvement lever besides operational measures like improved technical stability of the rolling stock)

C. Klingenberg (✉)
Deutsche Bahn AG, Gallusanlage 8, 60329 Frankfurt, Germany
e-mail: christoph.klingenberg@deutschebahn.com

- The static supplements (margins) for infrastructure bottlenecks due to construction work and general delays are not dynamic and not verified through simulation
- They are not related to the economics of running into delays.

The central proposition of this presentation covers the third point and reads: the ratio of the cost of one minute delay over the cost of one minute scheduled operation determines the new scheduled time. If for example the cost of running a train per minute is 25€ and the cost of one minute delay is 100€ (including compensation for delayed passengers and revenue loss due to bad reputation), then the new schedule should leave one quarter of the train rides from A to B delayed and three quarters should arrive early.

So once this cost ratio is determined, one can start a mechanistic approach to planning driving times for a segment (between A and B) and the same for stops. Since planners only have limited capacity to make schedule adjustments one must focus on those segments and stops where the effect is most promising.

In the example of long distance trains at Deutsche Bahn, there are over 1,000 segments connecting more than 300 stations with 240,000 train rides—at any given time 150 trains run simultaneously. This yields more than 2 mill plan-actual comparisons for segments and for stops.

The most promising segments are found through sorting by the ratio M/σ (Average of the difference of scheduled time and actual times/standard deviation of the actual times). If this quotient is high, say greater than one, a schedule adjustment is mandatory and will yield a significant on time performance improvement. If this ratio is small, especially because the standard deviation is high, the cause for bad on time performance lies in the erratic nature of delays. This should be dealt with through improvements of the reliability of the operations and not so much by adjusting the plan.

In the case of Deutsche Bahn with an on-time performance of 76% in 2014 (in the 5.59-minute threshold) a theoretical potential from adjusting the plan of 16 PP can be derived. This potential is realized by adjusting driving times between stops to (Average + standard deviation) for all segments and thus improving punctuality by 10 PP. This adjustment would add 10% to the overall driving time. Applying the same reasoning to stop times, that is, adjusting stop times to (Average + standard deviation) yields another 6 PP punctuality improvement, thereby extending stop times by 45%.

This increase in driving time and stop time may seem prohibitive at first glance, but travel time differences in the 10% range are irrelevant for customer choice. The improvement in quality and in connection stability is much more important. The increase in stop times looks even more prohibitive, but bear in mind that usually stop times are 2 min, so the proposal is to extend most of these to 3 min.

There are three important aspect of linking robustness to cost figures (including opportunity costs for lower revenue due to disgruntled passengers over continuing delays):

- The often-quoted antithesis of nominal and robust scheduling options is solved, so there no longer exists the option of constructing either a "cost-optimal, but

delay-prone schedule" or a "robust, but cost-intensive schedule, see Cacchiani and Toth [3].
- The distribution of actual driving or stopping times becomes the primary tool to work with making static calculations of shortest possible physical driving times plus supplements obsolete. In fact, the author believes that thinking in categories like supplements and buffers leads planners to stay in an artificial framework ("the planner's world") instead of confronting themselves with the real world and the sometimes devastating effect of a non-robust schedule (which still may be perfectly planned according to the rules).
- The question of delay propagation comes down to the folding of distributions and needs no extra optimization step. So, if the segments and stops are planned according to the cost optimum, delay propagation is sufficiently dampened, see Chen and Schonfeld [4] and also Goverde [5], p. 236.

2 Practical Aspects of Implementing Plan Changes

In our case there are 11,000 combinations of train numbers and segments and roughly the same number of combinations of train numbers and stops. Changes to driving times and stop times are usually only made in case of traction or infrastructure change, which make up less than 5% of all segments. For every plan adjustment, a new path (time-distance diagram) must be constructed that avoids any conflicts with other paths, especially with regional passenger traffic and network cargo traffic. But even if this would be constructible, the sheer work volume to make some 10,000 changes is not feasible (absent automatic path planning systems). So instead of adjusting the segments and stops in a stand-alone fashion, one must look at train lines. Not only does one recognize certain delay patterns much easier, but one can also differentiate between the first segment of a long-distance train, which should better be planned correctly, because it will influence all the others downstream, and the last segment, where (almost) no ripple-on effects take place.

In addition, one can also make trade-offs between segments and adjacent stops to stabilize the train line.

By applying this to lines we could achieve for some of the worst performing train lines improvements of 16 PP in on time performance. However, some adjustments could not completely be implemented due to infrastructure overload at critical points, compare [1].

To go beyond line optimization, you must look at the network or at least at the connection points of the major long distance lines. The connection patterns with other long distance trains and especially regional trains is the biggest constraint to planning flexibility. So, in many cases the adjustment of a segment or a stop could not be realized due to connections. The fact that those connections had an insufficient degree of realization does not help in arguing for this case—see below.

3 Transfers—Passenger Connections

Comparison of plan to actuals is not limited to segments and stops, but can also be done for passenger connections. In an open system like railways you cannot automatically track individual passengers (this will change to some extent when seat self-check-in is rolled out), so you must use the actual arrival time of the inbound train and actual departure time of the outbound train (and the distance between the actual tracks) to calculate connection success rates. The first and simplest measure for connections with low success rates is simply not to offer them anymore as connections on through tickets. Low success rates happen particularly often on connections where the inbound train is an international train. Connections with medium realization can be improved by schedule changes or by switching tracks with shorter walking distance between trains, preferably from the same platform. Since connections are integral for the passenger journey, it is essential to switch views at some point from train punctuality to passenger punctuality.

Connection success rate is not only determined by system on time performance, but feeds itself back into the system, since waiting for connecting passenger is among the top 10 delay reasons. So, passenger connections constitute a positively reinforcing feedback loop.

4 Turnarounds and Maintenance Stops

These events are special cases of stops and can be treated similarly. The only challenge is that they are planned completely independent of the timetable and thus need special attention. In contrast to stops along a train line the cost for a planned minute only consists of capital costs for the rolling stock and infrastructure cost (or, alternatively the opportunity costs of not offering a revenue train service, which amounts to roughly the same figure), which is below 5€ per minute, whereas a minute delay is 100€. Thus, the planned length for a turnaround should be at least in the 95% quantile of actual turnaround times.

Maintenance stops are calculated on the duration of a planned maintenance program allowing for some unplanned work. In this case the stability of the operations is not only determined by the amount of time allotted to maintenance work, but also—and much more importantly—by the size of the rolling stock reserves. This is because often trains and wagons are pulled from maintenance to cover for trains that cannot operate due to failure. The total number of reserve trains should be sized so that at least 95% of all failures can be covered without delay.

5 Network Optimization

So far, the plan adjustments have been made on an incremental basis. Even if you do this plan adjustment for every major line in the long-distance network, you always fall victim to the restrictions of passenger connections within the long-distance system or to the regional traffic system. So, if you would like to extend driving time from A to B you must sacrifice connections either at A or at B. Even between stops there are many obstacles due to heavy loads on the infrastructure, especially with trains running on different speeds making the construction of paths even more cumbersome.

Let us for the current purpose ignore the infrastructure utilization and focus on passenger connections. There are some indications that the long-distance network optimization can be achieved by applying metaheuristic methods. The objective function should be minimal travel times including the connections. Boundary conditions is that driving and stop times should be at least (median + standard deviation).

As a starting point for the metaheuristic algorithm one could try and take the actual driving times of some arbitrarily chosen day (preferably without heavy weather) and develop this further. This initial schedule would have the advantage over the real schedule that at least on one day in the whole year it was 100% on time, which in real life never happened with any real schedule.

In addition to the cost figures for a planned minute of train operations (around 25€) and for a minute delay (around 100€) we need a cost figure for one minute of passenger delay. Though this (opportunity) cost varies a lot from price sensitive leisure traveler to time sensitive business traveler I would use 10 cts per passenger minute as a proxy for optimization. This cost figure is necessary for calculation the trade-off between establishing a passenger connection through delaying a train or not.

This is still work in progress and it is too early to predict the feasibility and the outcome of a metaheuristic optimization [6].

6 Implications for the Infrastructure Operator

Since from a customer perspective an hourly or half-hourly service seems to be the most use friendly timetable design (despite ubiquitous smartphone use with real-time schedules) it is fundamental to first construct an "ideal timetable" based on an "ideal infrastructure". Since an hourly timetable is very "digital", the construction is rather simple [2]: stations are either multiples of hours apart or driving times must be adjusted to fit this hourly schedule. Infrastructure expansion and enhancement projects are then evaluated and timed to enable hourly (or half-hourly) connections between the major cities. Not only infrastructure expansion projects, but also every infrastructure maintenance project (for example to maintain track quality) must be planned to fit this hourly schedule. This could mean for example, that major track improvements can only be done at night. So again, we are left with trade-off decisions that can only be answered through operations research methods. Whereas the "ideal timetable" can be constructed on paper without IT support, all the intermediate steps and especially the various stages of the infrastructure during maintenance and construction call for trade-off decisions that need IT support and sophisticated OR methods.

7 Future Work and the Role of Operations Research in Improving on Time Performance

On time performance is the result of a huge number of conscious and unconscious trade-off decisions. Most disturbance factors initiate a downward spiral in on time performance. It falls mostly to the planners to achieve robustness and thus to prevent this downward spiral. Since the extra cost of robustness can be calculated it is necessary to use OR to find the optimal point on the trade-off curve. This is not to say that planning is largely automatic or robotic, but there are many inputs needed from management to steer efforts and to reflect their expectations of market developments, customer expectations and competitor reactions. In the case of Deutsche Bahn this initiated a fundamental shift in the mindset of the planners.

References

1. Andersson, E., Peterson, A. & Törnquist Krasemann, J. (2013). Introducing a new quantitative measure of railway timetable robustness based on critical points. In *5th International Seminar on Railway Operations Modelling and Analysis*, Copenhagen, May 13–15, 2013.
2. Breuer, S. (2016). Integrierter Taktfahrplan Deutschland 2030. *Eisenbahn Revue International, 7*(2016), 364–365.
3. Cacchiani, V., & Toth, P. (2012). Nominal and robust train timetabling problems. *European Journal of Operational Research, 219*(3), 727–737; Ferrer, J.-C., Mac Cawley, A., Maturana, S., Toloza, S., & Vera, J. (2008). An optimization approach for scheduling wine grape harvest operations. *International Journal of Production Economics, 112*(2), 985–999.
4. Chen, C.-C., & Schonfeld, P. (2016). A dispatching decision support system for countering delay propagation in intermodal logistics networks. *Transportation Planning and Technology, 39*(3), 254–268.
5. Goverde, R. (2005). *Punctuality of Railway Operations and Timetable Stability Analysis*. Ph.D. Thesis, Technical University of Delft.
6. Hauck, F. (2017ff). Fahrplan-Tuning zur Erhöhung der erwarteten Pünktlichkeit im Schienenverkehr. Ph.D. Thesis in Progress, FU Berlin.
7. Pachl, J. (2013). *Systemtechnik des Schienenverkehrs* (Vol. 7). Auflage. Springer.

Printed by Printforce, the Netherlands